Mechanics of the Cell

Second edition

Exploring the mechanical features of biological cells, including their architecture and stability, this textbook is a pedagogical introduction to the interdisciplinary fields of cell mechanics and soft matter physics from both experimental and theoretical perspectives.

This second edition has been greatly updated and expanded, with new chapters on complex filaments, the cell division cycle, the mechanisms of control and organization in the cell, and fluctuation phenomena. The textbook is now in full color which enhances the diagrams and allows the inclusion of new microscopy images.

With more than 300 end-of-chapter exercises exploring further applications, this textbook is ideal for advanced undergraduate and graduate students in physics and biomedical engineering. A website hosted by the author contains extra support material, diagrams and lecture notes, and is available at www.cambridge.org/Boal.

David Boal is Professor Emeritus in the Department of Physics at Simon Fraser University, Canada. He has been working in the field of biophysics for the past 20 years, and he now studies the mechanical issues in the origin of life.

Mechanics of the Cell

Second edition

DAVID BOAL
Simon Fraser University
Canada

CAMBRIDGE
UNIVERSITY PRESS

University Printing House, Cambridge CB2 8BS, United Kingdom

One Liberty Plaza, 20th Floor, New York, NY 10006, USA

477 Williamstown Road, Port Melbourne, VIC 3207, Australia

314-321, 3rd Floor, Plot 3, Splendor Forum, Jasola District Centre, New Delhi - 110025, India

79 Anson Road, #06-04/06, Singapore 079906

Cambridge University Press is part of the University of Cambridge.

It furthers the University's mission by disseminating knowledge in the pursuit of education, learning and research at the highest international levels of excellence.

www.cambridge.org
Information on this title: www.cambridge.org/9780521113762

First published by Cambridge University Press 2002
Second edition 2012

A catalogue record for this publication is available from the British Library

Library of Congress Cataloging in Publication data
Boal, David H.
Mechanics of the cell / David H. Boal. – 2nd ed.
p. cm.
Includes bibliographical references and index.
ISBN 978-0-521-11376-2 – ISBN 978-0-521-13069-1 (pbk.)
1. Cells – Mechanical properties. 2. Physical biochemistry. I. Title.
QH645.5B63 2012
573.1′536–dc23 2011038238

ISBN 978-0-521-11376-2 Hardback
ISBN 978-0-521-13069-1 Paperback

Contents

Preface

The cells of our bodies represent a very large class of systems whose structural components often are both complex and soft. A system may be *complex* in the sense that it may comprise several components having quite different mechanical characteristics, with the result that the behavior of the system as a whole reflects an interplay between the characteristics of the components in isolation. The mechanical components themselves tend to be *soft*: for example, the compression resistance of a protein network may be more than an order of magnitude lower than that of the air we breathe. Cells have fluid interiors and often exist in a fluid environment, with the result that the motion of the cell and its components is strongly damped and very unlike ideal projectile motion as described in introductory physics courses. While some of the physics relevant to such soft biomaterials has been established for more than a century, there are other aspects, for example the thermal undulations of fluid and polymerized sheets, that have been investigated only in the past few decades.

The general strategy of this text is first to identify common structural features of the cell, then to investigate these mechanical components in isolation, and lastly to assemble the components into simple cells. The initial two chapters introduce metaphors for the cell, describe its architecture and develop some concepts needed for describing the properties of soft materials. The remaining chapters are grouped into three sections. Parts I and II are devoted to biopolymers and membranes, respectively, treating each system in isolation. Part III combines these soft systems into complete, albeit mechanically simple, cells; this section of the text covers the cell cycle, various aspects of cell dynamics, and some molecular-level biophysics important for understanding cell function.

Each chapter begins with an experimental view of the phenomena to be addressed, and a statement of the principal concepts to be developed. Those chapters with a fair amount of mathematical formalism are written such that a reader may skip to the last section of the chapter, without going through traditional physics-style derivations, to see the results applied in a biological context. The approach is to keep the mathematical detail manageable without losing rigor. Care has been taken to choose topics that can be approached within a common theoretical framework; this has forced the omission of some phenomena where a disproportionate effort would be required to assemble the related theoretical machinery. Some of the longer proofs and other support material can be found through the Cambridge University Press website for this book.

Reflecting the multidisciplinary nature of biophysics, four appendices provide quick reviews of introductory material related to the text. For the physicist, Appendices A and B focus on animal cell types and the molecular composition of the cell's mechanical components, respectively. For the biologist, Appendices C and D introduce some core results from statistical mechanics and elasticity theory. The end-of-chapter problems are grouped according to (i) applications to biological systems, and (ii) formal extensions and supplementary derivations. The text is suitable for a one-semester course delivered to senior undergraduate and beginning graduate students with an interest in biophysics. Some formal results from the problem sets may provide additional lecture material for a more mathematical course aimed at graduate students in physics.

Compared to the first edition of this text, the second edition is 50% longer and the number of homework problems per chapter has been doubled. All of the original chapters have been reviewed, updated and reworked as needed. Four new chapters have been added, expanding the treatment of fluids and dynamical processes such as cell growth and division, as well as placing more emphasis on single-molecule aspects of cell mechanics. The mathematical description of soft, fluctuating systems is now part of the text and presented in a dedicated introductory chapter.

Financial support for my research comes from the Natural Sciences and Engineering Research Council of Canada. I am happy to acknowledge extensive discussions with my colleagues John Bechhoefer, Eldon Emberly, Nancy Forde, Michael Plischke, Jenifer Thewalt and Michael Wortis at Simon Fraser University and Myer Bloom and Evan Evans at the University of British Columbia. Gerald Lim deserves credit for reading the first edition of the text and checking its problem sets. This project was started while I enjoyed the hospitality of Rashmi Desai and the Physics Department at the University of Toronto in 1997–98. The late Terry Beveridge at the University of Guelph provided a disproportionate number of the electron microscopy images for the first edition. I thank my wife Heather, along with our children Adrie and Alex, for their support and understanding when the task of writing this text inevitably spilled over into evenings and weekends.

The staff at Cambridge Press, particularly Rufus Neal, Simon Capelin and John Fowler, have helped immensely in bringing focus to the book and placing its content in the larger context. However, the lingering errors, omissions and obfuscations are my responsibility, and I am always appreciative of suggestions for ways to improve the quality of the text.

Dave Boal

List of symbols

A, A_v	area, area per vertex of a network
$\mathbf{a}$	area element of a surface
a, a_o	interface area per molecule (a_o at equilibrium)
$\mathbf{b}_i$, b	monomer bond vector and length
B_{eff}	effective bond length
$b_{\alpha\beta}$	second fundamental form
$\mathcal{C}$	capacitance per unit area
C, C_o	curvature; spontaneous curvature
C_1, C_2	principal curvatures to a surface
C_m, C_p	curvature along, or perpendicular, to a meridian
C_{ijkl}	elastic moduli
C_V	specific heat at constant volume
C_{vdw}	van der Waals interaction parameter
$C(i,j)$	binomial coefficient
c	concentration, units of [*moles/volume*] or [*mass/volume*]
D	diffusion coefficient
D_f	filament diameter for rods or chains
D_s	distance between two membranes or plates
d	dimensionality of a system
d_{bl}, d_p, d_{sh}	thickness of a bilayer, plate or shell
E	energy
$\mathcal{E}$	energy density
$\mathbf{E}$	electric field
E_{bind}	energy to separate an amphiphile from an aggregate
E_{sphere}, E_{disk}	energy of a spherical shell or flat disk
e	elementary unit of charge
$\mathbf{F}$, F	force
F_{buckle}	buckling force of a rod
F, $\mathcal{F}$	free energy; free energy density
$\mathcal{F}_{sol}$	free energy of a solution phase
G	Gibb's free energy; electrical conductance
$G_{channel}$	conductance of an open channel
$G(t)$	time-dependent relaxation modulus

$G'(\omega), G''(\omega)$	shear storage and loss moduli
g, $g_{\alpha\beta}$	metric; metric tensor
H, H_v	enthalpy; enthalpy per vertex
h	Planck's constant
$h(x, y)$	height of a surface in Monge representation
h_x, h_{xx}	first and second derivatives of $h(x, y)$ in the x-direction
$\mathcal{I}$	moment of inertia of cross section
j	flux
K_A, K_V	compression modulus for area, volume
k_B	Boltzmann's constant
k_{on}, k_{off}	capture and release rates in polymerization
k_{sp}	spring constant
L	length of a rod
$\mathcal{L}(y)$	Langevin function
L_c	contour length of a filament
$\mathcal{L}_K$	Kuhn length of a polymer
Lk	linking number
ℓ_B	Bjerrum length
ℓ_{cc}	C–C distance projected on the axis of hydrocarbon chain
ℓ_D	Debye length
ℓ_{hc}	length of hydrocarbon chain along its axis
$\mathcal{M}$	bending moment
$[M]$	monomer concentration
$[M]_c$, $[M]_{ss}$	minimum $[M]$ for filament growth; $[M]$ at treadmilling
m	molecular mass
m_j	mRNA concentration
N, n	number of monomers in a polymer chain
N_A	Avogadro's number
N_K	number of Kuhn lengths in a polymer
n_c	number of carbon atoms in a hydrocarbon chain
$\mathbf{n}$	unit normal vector to a curve or surface
$\mathbf{n}_x$, $\mathbf{n}_y$	derivative of $\mathbf{n}$ in the x or y direction
P	pressure
$\mathcal{P}$, $\mathcal{P}(x)$, $\mathcal{P}_r$	probabilities and probability densities
P_L, P_R, P_{net}	probabilities of motion for thermal ratchets
$\mathbf{p}$	momentum
p	bond or site occupation probability on a lattice; pitch of a helix
p_{bound}	probability of receptor–ligand binding
p_C, p_R	connectivity and rigidity percolation thresholds

p_j	repressor protein concentration
Q, q	electric charge
R, R_p	radius; electrical resistance; pipette radius
$\mathbf{R}$	Reynolds number
R_{hc}	effective radius of a hydrocarbon chain
R_v, R_v^*	vesicle radius, minimum vesicle radius
R_1, R_2	principal radii of curvature
$\mathbf{r}$	position vector with Cartesian components (x, y, z)
S	entropy
S	structure factor
S_{gas}	entropy of an ideal gas
S_{ijkl}	elastic constants
s	arc length
s, s_o	spring length (s_o at equilibrium)
T	temperature
$\mathcal{T}$	torque
$\mathbf{t}$	unit tangent vector to a curve
Tw	twist
U^α	eigenvalues of strain tensor
$\mathbf{u}$	displacement vector in a deformation
u, u_{ij}	strain tensor
V	volume, electrical potential difference
$V(x)$	potential energy function
V_{mol}, V_{slab}	van der Waals potentials
V_o	volume of undeformed object
V_{qss}	quasi-steady-state potential across a membrane
v_{ex}	excluded volume parameter of a polymer
v_{hc}	volume of a hydrocarbon chain
v_{red}	reduced volume parameter
$\mathcal{W}_{ad}$	adhesion energy per unit area
w	width parameter of triangular probability distribution
Wr	writhe
Y	Young's modulus
Z	partition function
α	twist angle per unit length
β	inverse temperature $(k_B T)^{-1}$
γ	surface tension
γ_K, γ_{Na}	membrane conductivities (2D) for Na and K ions
ΔG_{conc}	change in free energy per ion from concentration gradient
ΔG_{coul}	change in free energy per ion from electrical potential

ε, ε_o	permittivity; permittivity of vacuum
ε_h	hydrophobic contact energy
η	viscosity
κ_{ax}	conductivity (3D) of an axon
κ_f	flexural rigidity of filament
κ_b, κ_G	bending rigidity or Gaussian bending rigidity of a membrane
κ_{nl}	non-local bending resistance
κ_{tor}	torsional rigidity of a filament
λ	edge tension of bilayer; protein degradation rate
λ^*	minimum edge tension for membrane stability
λ_p	mass per unit length of polymer
Λ, Λ_x, Λ_y, Λ_z	deformation scaling parameter
μ	mean value of a Gaussian distribution
μ_p, μ_s	pure, simple shear moduli
ν, ν_{Fl}	scaling exponent; Flory exponent for self-avoiding polymers
ξ_p	persistence length
ρ	number density (for chains or molecules)
ρ^*	transition density between dilute and semi-dilute solutions
ρ^{**}	transition density between semi-dilute and concentrated solutions
ρ_{agg}	critical aggregation density
ρ_{ch}	charge per unit volume
ρ_m	mass per unit volume
ρ_L	contour length of polymer per unit volume
ρ_N	transition density between isotropic and nematic phases
ρ_+, ρ_-, ρ_s	number densities of ions in solution
σ	width parameter of a Gaussian distribution
σ_{ij}, σ_θ, σ_z	stress tensor; hoop and axial stress of a cylinder
σ_p	Poisson ratio
σ_s	charge per unit area
τ	two-dimensional tension; decay time in RC circuit
τ_θ, τ_z	hoop and axial tension of a cylinder
Ψ, ψ	electric potential
χ	rotational drag parameter
ζ	roughness exponent for membranes

1 Introduction to the cell

The number of cells in the human body is literally astronomical, about three orders of magnitude more than the number of stars in the Milky Way. Yet, for their immense number, the variety of cells is much smaller: only about 200 different cell types are represented in the collection of about 10^{14} cells that make up our bodies. These cells have diverse capabilities and, superficially, have remarkably different shapes, as illustrated in Fig. 1.1. Some cells, like certain varieties of bacteria, are not much more than inflated bags, shaped like the hot-air or gas balloons invented more than two centuries ago. Others, such as nerve cells, may have branched structures at each end connected by an arm that is more than a thousand times long as it is wide. The basic structural elements of most cells, however, are the same: fluid sheets, sometimes augmented by shear-resistant walls, enclose the cell and its compartments, while networks of filaments maintain the cell's shape and help organize its contents. Further, the chemical composition of these structural elements bears a strong family resemblance from one cell to another, perhaps reflecting the evolution of cells from a common ancestor; for example, the protein actin, which forms one of the cell's principal filaments, is found in organisms ranging from yeasts to humans.

The many chemical and structural similarities of cells tempt us to search for systematics in their architecture and components. We find that the structural elements of the cell are *soft*, in contrast to the hard concrete and steel of buildings and bridges. This is not a trivial observation: the mechanical properties of soft materials may be quite different from their hard, conventional counterparts and may reflect different microscopic origins. For instance, the fact that soft rubber becomes more resistant to stretching when heated, compared with the tendency of most materials to become more compliant, reflects the genesis of rubber elasticity in the variety of a polymer's molecular configurations. The theoretical framework for understanding soft materials, particularly flexible networks and membranes, has been assembled only in the last few decades, even though our experimental knowledge of soft materials goes back two centuries to the investigation of natural rubber by John Gough in 1805.

The functions performed by a cell can be looked upon from a variety of perspectives. Some functions are chemical, such as the manufacture of proteins, while others could be regarded as information processing, such

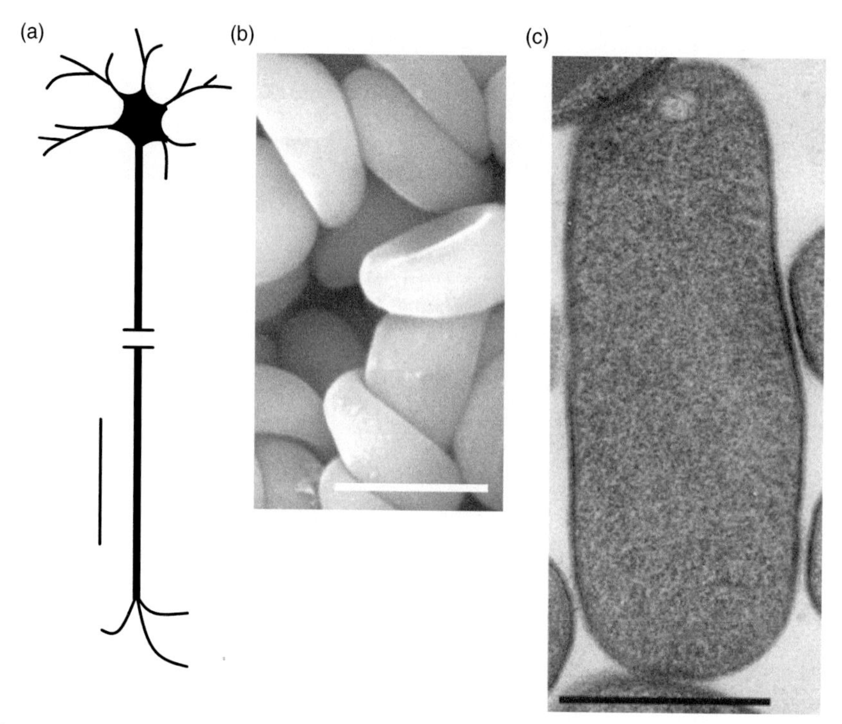

Fig. 1.1 Examples of cell shapes. (a) A neuron is a highly elongated cell usually with extensive branching where it receives sensory input or dispatches signals; indicated by the scale bar, its length may be hundreds of millimeters. (b) Mammalian red blood cells adopt a biconcave shape once they lose their nucleus and enter the circulatory system (bar is 4 μm; courtesy of Dr. Elaine Humphrey, University of British Columbia). (c) Cylindrically shaped, the bacterium *Escherichia coli* has a complex boundary but little internal structure (bar is 0.9 μm; courtesy of Dr. Terry Beveridge, University of Guelph). The image scale changes by two orders of magnitude from (a) to (c).

as how a cell recognizes another cell as friend or foe. In this text, we concentrate on the *physical* attributes of cells, addressing such questions as the following.

- How does a cell maintain or change its shape? Some cells, such as the red blood cell, must be flexible enough to permit very large deformations, while others, such as plant cells, act cooperatively to produce a mildly stiff multicellular structure. What are the properties of the cell's components that are responsible for its strength and elasticity?
- How do cells move? Most cells are more than just inert bags, and some can actively change shape, permitting them to jostle past other cells in a tissue or locomote on their own. What internal structures of a cell are responsible for its movement?
- How do cells transport material internally? For most cells, especially meter-long nerve cells, diffusion is a slow and inefficient means of

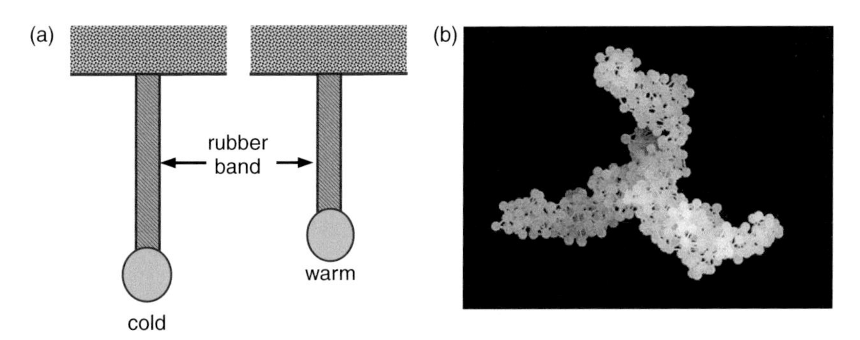

Fig. 1.2 (a) When natural rubber is heated, it becomes more resistant to stretching, an effect which can be demonstrated by hanging an object from an elastic band. When heated, the elastic band contracts, causing the object to rise. (b) A fluid membrane extends arms and fingers at large length scales, although it is smooth at short length scales (simulation from Boal and Rao, 1992a).

transporting proteins from their production site to their working site. What mechanisms, generating what forces, does a cell use for efficient transportation?

- How do cells stick together, as a multicellular organism such as ourselves, or how do they avoid adhering when it is unwanted? Do the thermal fluctuations of the cell's flexible membranes affect adhesion, or is it strictly a chemical process?
- What are the stability limits of the cell's components? A biological filament may buckle, or a membrane may tear, if subjected to strong enough forces. Are there upper and lower limits to the sizes of functioning cells?

To appreciate the mechanical operation of a whole cell, we must understand how its components behave both in isolation and as a composite structure. In the first two sections of this text, we treat the components individually, describing filaments and networks in Part I, followed by fluid and polymerized membranes in Part II. These two sections provide an experimental picture of the cell's structural elements and develop theoretical techniques to interpret and predict their mechanical characteristics. We demonstrate that many properties of soft materials are novel to the point of being counterintuitive; as illustrated in Fig. 1.2, some networks may shrink or stiffen when heated, while fluid sheets form erratic arms and fingers over long distances. Both of these effects are driven by entropy, as we will establish below.

Although it is important to understand the individual behavior of the cell's components, it is equally important to assemble the components and observe how the cell functions as a whole. Often, a given structural element plays more than one role in a cell, and may act cooperatively with other elements to produce a desired result. In Part III, we examine several aspects of multicomponent systems, including cell mobility, adhesion and deformation. The growth and division processes of the cell are of particular importance to its propagation, and these are the topics of the final two chapters of the text.

1.1 Designs for a cell

Although his own plans for Chicago skyscrapers were not devoid of decoration, architect Louis Sullivan (1856–1924) argued that functionally unnecessary embellishments detracted from a building's appeal. His celebrated dictum, "Form follows function", has found application in many areas beyond architecture and engineering, and is particularly obvious in the designs that evolution has selected for the cell.

Let's begin our discussion of the cell, then, by reviewing some strategies used in our own architectural endeavors, particularly in situations where functionality is demanded of minimal materials. Viewing the cell as a self-contained system, we look to the construction of boats, balloons and old cities for common design themes, although we recognize that none of these products of human engineering mimics a complete cell.

1.1.1 Thin membranes for isolating a cell's contents

Sailing ships, particularly older ships built when materials were scarce and landfalls for provisioning infrequent, face many of the same design challenges as cells. For example, both require that the internal workings of the system, including the crew and cargo in the case of a boat, be isolated in a controlled way from the system's environment. As illustrated by the merchant ship of Fig. 1.3(a), the naval architects of the fifteenth century opted for complex, multicomponent structures in their designs. The boundary of the boat is provided by a wooden "membrane" which need not be especially thick to be largely impermeable. However, thin hulls have little structural strength to maintain the boat's shape or integrity. Rather than make the planks uniformly thicker to increase the strength of the hull, naval architects developed a more efficient solution by reinforcing the hull at regular intervals. The reinforcing elements in Fig. 1.3(a) are linear, linked together to form a tension-resistant scaffolding around the hull. In the design of all but the simplest cells, evolution has similarly selected a cytoskeleton or cell wall composed of molecular filaments to reinforce the thin plasma membrane of the cell boundary.

1.1.2 Networks for tensile strength

The rigging of the sailing ship of Fig. 1.3(a) illustrates another design adopted by the cell. Rather than use stout poles placed on either side of the

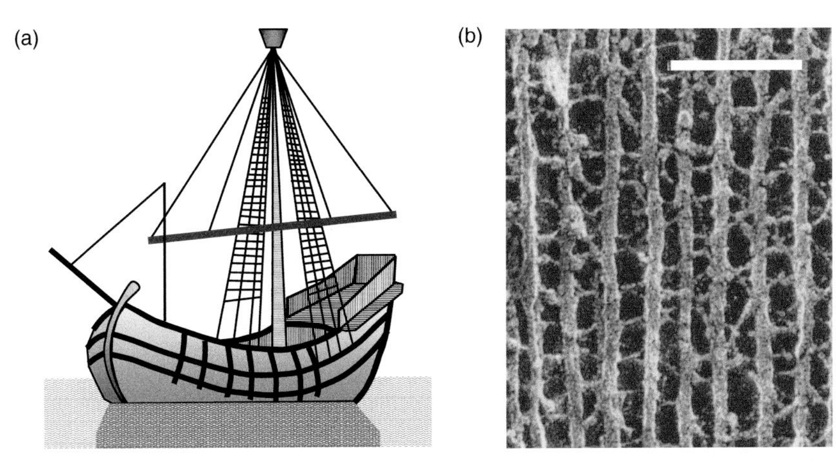

Fig. 1.3

(a) An early fifteenth-century merchant ship displays the efficient use of materials in the design of the reinforced hull and rigging (original illustration by fifteenth-century engraver Israel von Mekenem; redrawn by Gordon Grant in Culver, 1992; ©1994 by Dover Publications). (b) Cross-linked filaments inside a nerve cell from a frog. Neurofilaments (running vertically) are 11 nm in diameter, compared with the cross-links with diameters of 4–6 nm (bar is 0.1 μm; reprinted with permission from Hirokawa, 1982; ©1982 by the Rockefeller University Press).

mast to *push* it into position, a boat uses ropes on either side of the mast to *pull* it into position. By pulling, rather than pushing, the structural elements need only have good tension resistance, rather than the more demanding compression or buckling resistance of poles. Employing strings and ropes with good tensile strength but little resistance to buckling, rigging provides the required functionality with minimal materials. Because the tensile strength of a rope is needed just along one direction, between the top of the mast and the attachment point on the hull, rigging uses only weak lateral links between the strong ropes connected to the mast. This is a design observed in the cross-linking of filaments in nerve cells, as seen in Fig. 1.3(b), and in the cell walls of cylindrical bacteria, which is composed of stiff filaments oriented in the direction bearing the largest stress, linked together transversely by floppy molecular chains.

The relationship between the mast and rigging of a boat exhibits an intriguing balance of tension and compression: the mast has a strong resistance to compression and bending but is held in place by rigging with little resistance to bending. The cytoskeleton of the cell also contains a mix of filaments with strong and weak bending resistance, although these filaments span a more modest range of stiffness than ropes and masts. From Newton's Third Law of mechanics, tension/compression couplets may exist throughout the cell, and there are many examples of thin biofilaments bearing tension while thick ones carry compression without buckling.

(a)

(b)

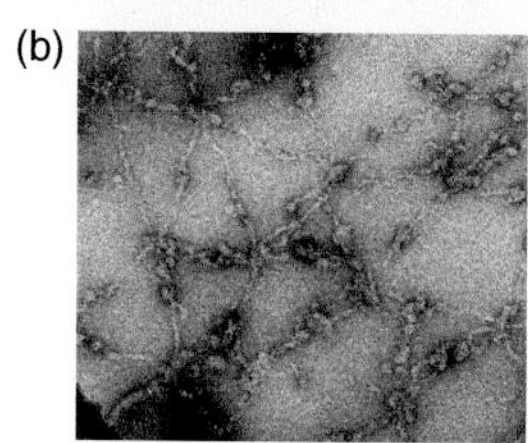

Fig. 1.4

(a) In a hot-air balloon or gas balloon, a thin membrane confines the gas within the balloon, and an external network provides mechanical attachment points and may aid in maintaining the balloon's shape. (b) A two-dimensional network of the protein spectrin is attached to the inside of the red blood cell membrane to provide shear resistance. Shown partially expanded in this image, the separation between the six-fold junctions of the network reaches 200 nm when fully stretched (courtesy of A. McGough and R. Josephs, University of Chicago; see McGough and Josephs, 1990).

1.1.3 Composite structures for materials efficiency

The forces on a boat are not quite the same as the forces experienced by a cell. An important difference is that the external pressure on the hull from the surrounding water is greater than the interior pressure, so that the internal structure of the boat must contain bracing with good compression resistance to prevent the hull from collapsing. In contrast, the interior pressure of some cells, such as many varieties of bacteria, may be much higher than their surroundings. Thus, the engineering problem facing a bacterium is one of explosion rather than collapse, and such cells have a mechanical structure which more closely resembles the hot-air balloon illustrated in Fig. 1.4(a). Balloons have a thin, impermeable membrane to confine the low-density gas that gives the balloon its buoyancy. Outside of the balloon is a network to provide extra mechanical strength to the membrane and to provide attachment sites for structures such as the passenger gondola. By placing the network on the outside of the membrane and allowing the interior pressure to force physical contact between the network and the membrane, the attachment points between the two structural components need not be reinforced to prevent tearing. Again, the network is under tension, so its mechanical strength can be obtained from light-weight ropes rather than heavy poles. Plant cells and most bacteria make use of external walls to reinforce their boundary membrane and balance the pressure difference across it. In a red blood cell, the two-dimensional network illustrated in Fig. 1.4(b) is attached to the membrane's *interior* surface to help the cell recover its rest shape after deformation in the circulatory system.

1.1.4 Internal organization for efficient operation

Advanced cells have a complex internal structure wherein specialized tasks, such as energy production or protein synthesis and sorting, are carried out by specific compartments collectively referred to as organelles. An equivalent system of human design might be a city, in which conflicting activities tend to be geographically isolated. Residential areas might be localized in one part of the city, food distribution in another, manufacturing in yet a third. How can these activities best be organized for the efficient transport of people and material within the city? Consider the plan of the walled city illustrated in Fig. 1.5(a). At this stage in its development, this city still enjoyed fields and open space (green) within its walls, separated from its residential and commercial buildings (magenta). The boundary is defined by the town wall, designed less to confine the inhabitants of the city than to keep hostile forces from entering it. Like the proteins of the cell's plasma membrane, strong gates (pink ovals in the diagram) control much of the

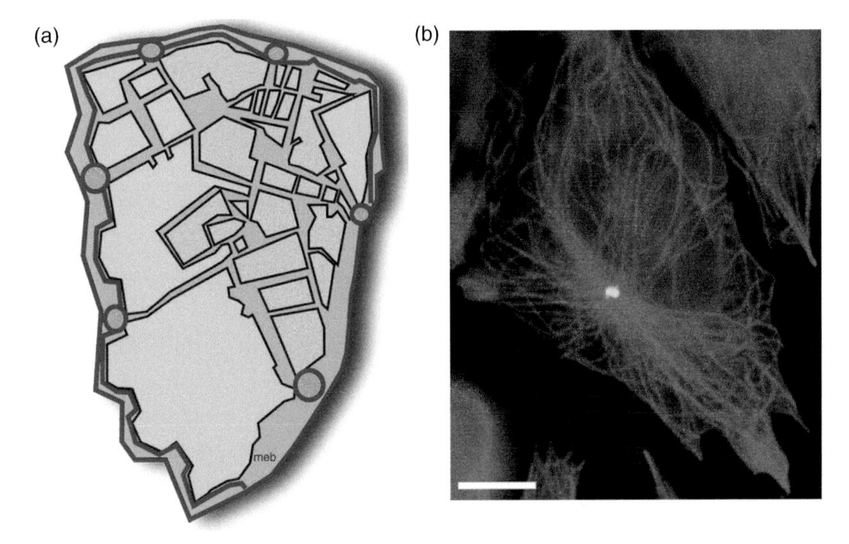

Fig. 1.5 (a) Plan of Quebec in the eighteenth century. Streets within the city walls (red) form an irregular web, with entry points indicated by red disks. (b) The array of microtubules in a cultured fibroblast helps organize the cell's organelles and provides transportation corridors (bar is 10 μm; reprinted with permission from Rodionov *et al.*, 1999; ©1999 by the National Academy of Sciences (USA)).

access to the town's interior. The walls of old cities also reflect the optimal deployment of limited resources, such as the stones used in their construction and the skilled labor needed to assemble them. The minimal town wall needed to enclose a given land area is a circle, just as the minimal cell boundary to enclose a given protein-rich volume is a spherical shell. Of course, other factors, such as their function or mechanisms for growth and division, also influence the design and shape of towns and cells alike. Thus, the design in Fig. 1.5(a) takes advantage of the cliffs and river along the eastern flank of the city so as to concentrate its fortifications along the western side.

An effective transportation system to direct the flow of people and materials is mandatory in an urban setting; for instance, it would be chaotic if visitors arriving at the gates to the city were forced to randomly diffuse through a jumble of houses in search of their destination. Such diffusive processes are very slow: the displacement from the start of a path, as the proverbial crow flies, increases only as the square root of the total path length walked by the visitor. To overcome this problem, cities use dedicated rights-of-way, including roads and railways, to guide traffic between specific locations. Depending on its layout, the most efficient street pattern may be an irregular web, rather than a grid, although the latter has become commonplace in modern times because it simplifies the layout of lots for building construction. In the design of cells, stiff filaments may provide pathways along which specialized molecules can carry their cargo; for instance, microtubules crowd the transportation corridor of the long section of the nerve cell of Fig. 1.1(a) and their layout also can be seen in the fibroblast of Fig. 1.5(b).

1.1.5 Materials to match the expected usage

Lastly, what about the choice of construction materials? Buildings and bridges are subject to a variety of forces that degrade a structure over time and may ultimately cause it to fail. Thus, the engineering specifications for structural materials will depend not only on their cost and availability, but also upon the building's environment and the nature of the forces to which it is subjected. As far as mechanical failure is concerned, each material has its own Achilles' heel, which may limit its applicability to certain structures. For instance, steel provides the flexibility needed to accommodate vibrations from the traffic on a suspension bridge, but has a lifetime imposed by its resistance to corrosion and fatigue. Further, the longevity we expect for our buildings and bridges is influenced by anticipated usage, public taste, and technological change, to name a few criteria. It is senseless, therefore, to overdesign a building that is likely to be torn down long before the mechanical strength of its components is in doubt. Similarly, the choice of materials for the construction of a cell is influenced by many competing requirements or limitations. For instance, some molecules that are candidates for use in a cell wall may produce a wall that is strong, but not easily repairable or amenable to the process of cell division. Further, the availability of materials and their ease of manufacture by the cell are also important considerations in selecting molecular building blocks and in designing the cellular structure. Lastly, all of these conflicting interests must be resolved so that the organism is sufficiently robust and long-lived to compete in its environment.

What we have done in this section is search for common architectural themes in the designs of boats and balloons, and the plans of towns and cities. We find that designs making effective use of available materials often employ specialized structural elements that must act cooperatively in order to function: thin membranes for boundaries, ropes for tensile strength, and walls to balance internal pressure. The choice of construction materials for a given structural element is determined by many factors, such as availability or ease of assembly and repair, with the overall aim of producing a structure with an acceptable lifetime. As we will see in the following sections, evolution has selected many of the same effective design principles as human engineering to produce cells that are adaptable, repairable and functional in a wide range of environments. As we better understand Nature's building code, we will discover subtle features that may have application beyond the cellular world.

1.2 Cell shapes, sizes and structures

Despite their immense variety of shapes and sizes, cells display common architectural themes reflecting the similarity of their basic functions. For instance, all cells have a semi-permeable boundary that selectively segregates the cell's contents from its environment. Frequently, cells adopt similar strategies to cope with mechanical stress, such as the reinforced membrane strategy of boats and balloons discussed in Section 1.1. Further, the chemical similarities among the structural elements of different cells are remarkably strong. In this section, we first review some of the basic mechanical necessities of all cells and then provide a general overview of the construction of several representative cells, namely simple cells such as bacteria, as well as complex plant or animal cells. A longer introduction to cell structure can be found in Appendix A or textbooks such as Alberts *et al.* (2008) or Prescott *et al.* (2004).

As described in Section 1.1, the outer boundaries of boats and balloons are fairly thin compared with the linear size of the vessels themselves. Modern skyscrapers also display this design: the weight of the building is carried by an interior steel skeleton, and the exterior wall is often just glass cladding. A cell follows this strategy as well, by using for its boundary a thin membrane whose tensile strength is less important than its impermeability to water and its capability for self-assembly and repair. By using thin, flexible membranes, the cell can easily adjust its shape as it responds to its changing environment or reproduces through division.

Whether this membrane needs reinforcement depends upon the stresses it must bear. Some proteins embedded in the membrane function as mechanical pumps, allowing the cell to accumulate ions and molecules in its interior. If the ion concentrations differ across its membrane, the cell may operate at an elevated osmotic pressure, which may be an order of magnitude larger than atmospheric pressure in some bacteria (bicycle tires are commonly inflated to about double atmospheric pressure). The cell may accommodate such pressure by reinforcing its membrane with a network of strings and ropes or by building a rigid wall. Even if the membrane bears little tension, networks may be present to help maintain a cell's shape.

What other mechanical attributes does a cell have? Some cells can locomote or actively change shape, permitting them to pursue foes. For example, our bodies have specialized cells that can remove dead cells or force their way through tissues to attack foreign invaders. One way for a cell to change its shape is to possess a network of stiff internal poles that push the cell's surface in the desired manner. Consequently, several types of structural filament, each with differing stiffness, are present in the cell: some filaments are part of reinforcing networks while others are associated with locomotion or internal transportation. Of course, conservation of momentum tells us that there is more to cell motion than pushing on its boundary: to generate relative motion, the cell must adhere to a substratum or otherwise take advantage of the inertial properties of its environment.

The operative length scale for cells is the micron or micrometer (μm), a millionth of a meter. The smallest cells are a third of a micron in diameter, while the largest ones may be more than a hundred microns across. Nerve cells have particularly long sections called axons running up to a meter from end to end, although the diameter of an axon is in the micron range. Structural elements of a cell, such as its filaments and sheets, generally have a transverse dimension within a factor of two of 10^{-2} μm, which is equal to 10 nm (a nanometer is 10^{-9} m); that is, they are very thin in at least one direction. For comparison, a human hair has a diameter of order 10^2 μm.

Let's now examine a few representative cells to see how membranes, networks and filaments appear in their construction. The two principal categories of cells are prokaryotes (without a nucleus) such as mycoplasmas and bacteria, and eukaryotes (with a nucleus) such as plant and animal cells. Having few internal mechanical elements, some of today's prokaryotic cells are structural cousins of the earliest cells, which emerged more than 3.5 billion years ago. Later in the Earth's history, eukaryotes adopted internal membranes to further segregate their contents and provide additional active surface area within the cell. We begin by discussing generic designs of prokaryotic cells.

1.2.1 Mycoplasmas and bacteria

Mycoplasmas are among the smallest known cells and have diameters of perhaps a third of a micron. As displayed in Fig. 1.6, the cell is bounded by a plasma membrane, which is a two-dimensional fluid sheet composed primarily of lipid molecules. Described further in Appendix B, the principal lipids of the membrane have a polar or charged head group, to which are attached two hydrocarbon chains. The head groups are said to be hydrophilic, reflecting their affinity for polar molecules such as water, whereas the non-polar hydrocarbon chains are hydrophobic, and tend to shun contact with water. In an aqueous environment, some types of lipids can self-assemble into a fluid sheet consisting of two layers, referred to as a lipid bilayer, with a combined thickness of 4–5 nm. Like slices of bread in a sandwich, the polar head groups of the lipids form the two surfaces of the bilayer, while the hydrocarbon chains are tucked inside. The bilayer is a two-dimensional fluid and does not have the same elastic properties as a piece of cloth or paper. For instance, if you wrap an apple with a flat sheet of paper, the paper develops folds to accommodate the shape of the apple;

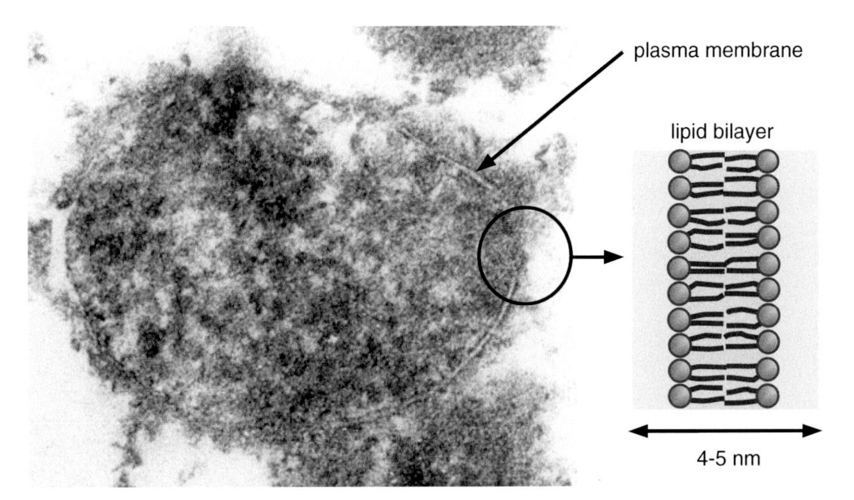

Fig. 1.6 Thin section of *Mycoplasma hominis*, illustrating the plasma membrane isolating the contents of the cell from its surroundings. As shown in the cartoon enlargement, the plasma membrane contains a bilayer of dual-chain lipid molecules, about 4–5 nm thick in its pure form, and somewhat thicker in the presence of other membrane components. Proteins are found in both the plasma membrane and the cell's interior (the fuzzy patches are protein-manufacturing sites called ribosomes). Typical diameter of a mycoplasma is a third of a micron, although the one shown here is 0.6 μm across. Adhering to the external surface of this cell is a precipitate from the animal blood serum in which the cell was grown (courtesy of Dr. Terry Beveridge, University of Guelph).

however, if you dip the apple in thick syrup, the syrup flows and forms a smooth, two-dimensional fluid coating on the apple's surface.

The interior of a mycoplasma contains, among other things, the cell's genetic blueprint, DNA, as well as large numbers of proteins, which also may be embedded in the plasma membrane itself. Although the beautiful world of biochemistry is not the focus of this text, we pause briefly to mention the size and structure of proteins and DNA (more details are provided in Appendix B). Ranging in mass upwards of 10^5 daltons (one dalton, or D, is one-twelfth the mass of a carbon-12 atom), proteins are linear polymers of amino acids containing an amino group ($-NH_3^+$) and an organic acid group ($-COO^-$). Many proteins fold up into compact structures with diameters ranging from a few to tens of nanometers, depending upon the mass of the protein, and some varieties of these globular proteins, in turn, are monomers in still larger structures such as cytoskeletal filaments with diameters ranging up to 25 nm. Like proteins, DNA (deoxyribonucleic acid) and RNA (ribonucleic acid) are polymers with a backbone consisting of a sugar/phosphate repeat unit, to each of which is attached one member of a small set of organic bases, generating the linear pattern of the genetic code. In the form of a double helix, DNA is 2.0 nm in diameter.

Most bacteria both are larger and have a more complex boundary than the simple plasma membrane of the mycoplasma. For instance, the bacterium *Escherichia coli*, an inhabitant of the intestinal tract, has the shape of a cylinder about 1 μm in diameter, and several microns in length, capped by hemispheres at each end. The interior of a bacterium may be under considerable pressure but the presence of a cell wall prevents rupture of its plasma membrane. Except in archaebacteria, the cell wall is composed of layers of a sugar/amino-acid network called peptidoglycan: Gram-positive bacteria have a thick layer of peptidoglycan encapsulating a single plasma membrane (Fig. 1.7(b)), whereas Gram-negative bacteria have just a thin layer of peptidoglycan sandwiched between two lipid bilayers [Fig. 1.7(a)], the inner one of which is the plasma membrane. Parenthetically, this classification reflects the ability of a bacterium to retain Gram's stain. As shown in Fig. 1.7(c), the peptidoglycan network is anisotropic (i.e. does not have the same structure in all directions), with its sugar chains oriented around the girth of the bacterium, the direction which bears the greatest surface stress. The sugar chains are linked transversely with loose strings of amino acids, the latter bearing only half the surface stress that the sugar chains bear. In analogy with a ship's rigging, the sugar chains are like the strong ropes attached to the mast, while the amino-acid strings are weaker cords that link the ropes into a network.

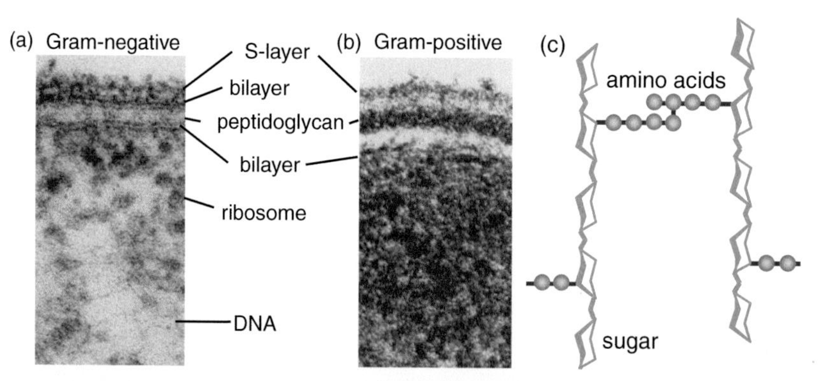

Fig. 1.7 Boundary structure of bacteria. (a) A Gram-negative bacterium, such as *Aermonas salmonicida*, has a very thin layer of peptidoglycan sandwiched between two membranes. Thin strings visible in the cell's cytoplasm are DNA. (b) In Gram-positive bacteria, such as *Bacillus stearothermophilus*, only one bilayer is present, and the peptidoglycan blanket is much thicker. These bacteria have an additional layer of proteinaceous subunits on their surface (S-layer), which is not always present in other bacteria (courtesy of Dr. Terry Beveridge, University of Guelph). (c) The molecular structure of peptidoglycan displays stiff chains of sugar rings, oriented around the girth of the bacterium, cross-linked by floppy strings of amino acids oriented along its axis.

1.2.2 Plant and animal cells

The general layout of a plant cell is displayed in Fig. 1.8(b). Exterior to their plasma membrane, plant cells are bounded by a cell wall, permitting them to withstand higher internal pressures than a wall-less animal cell can support. However, both the thickness and the chemical composition of the wall of a plant cell are different from those of a bacterium. The thickness of the plant cell wall is in the range of 0.1 to 10 μm, which may be larger than the total length of some bacteria, and the plant cell wall is composed of cellulose, rather than the peptidoglycan of bacteria. Also in contrast to bacteria, both plant and animal cells contain many internal membrane-bounded compartments called organelles. Plant cells share most organelles with animal cells, although plant cells alone contain chloroplasts; liquid-filled vacuoles are particularly large in plant cells and may occupy a large fraction of the cell volume.

As illustrated in Fig. 1.8(a), animal cells share a number of common features with plant cells, but their lack of space-filling vacuoles means that animal cells tend to be smaller in linear dimension. Neither do they have a cell wall, relying instead upon the cytoskeleton for much of their mechanical rigidity, although the cytoskeleton does not provide the strength that would allow an animal cell to support a significant internal pressure. Some components of the cytoskeleton meet at the centriole, a cylindrical organelle about 0.4 μm long. The most important organelles of plant and animal cells are the following.

- The nucleus, with a diameter in the range 3 to 10 μm, contains almost all of the cell's DNA and is bounded by a pair of membranes, attached to

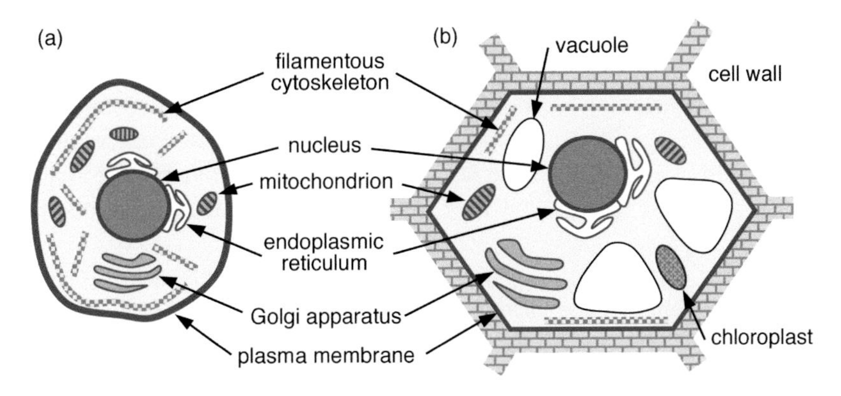

Fig. 1.8 Schematic sections through generic animal (a) and plant (b) cells showing the layout of the organelles and other structural components. Having a more complex internal structure than bacteria, these cells are correspondingly larger, typically 10–100 μm for plants and 10–30 μm for animals.

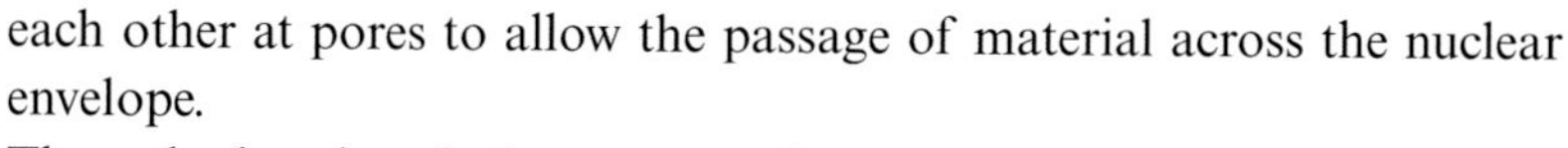

each other at pores to allow the passage of material across the nuclear envelope.

- The endoplasmic reticulum surrounds the nucleus and is continuous with its outer membrane. As a series of folded sheets, the rough endoplasmic reticulum has a small volume compared to its surface, on which proteins are synthesized by strings of ribosomes, like beads on a necklace.
- The membrane-bounded Golgi apparatus is the site of protein sorting, and has the appearance of layers of flattened disks with diameters of a few microns. Small vesicles, with diameters in the range 0.2 to 0.5 μm, pinch off from the Golgi and transport proteins and other material to various regions of the cell.
- Mitochondria and chloroplasts, the latter present only in plant cells, produce the cell's energy currency, ATP (adenosine triphosphate). Shaped roughly like a cylinder with rounded ends (often 0.5 μm in diameter), a mitochondrion is bounded by a double membrane. Also bounded by a double membrane, but containing internal compartments as well, chloroplasts are about 5 μm long and are the site of photosynthesis.

(a)

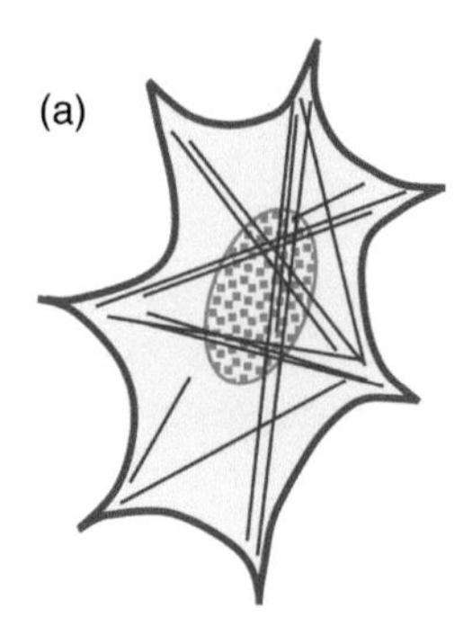

(b)

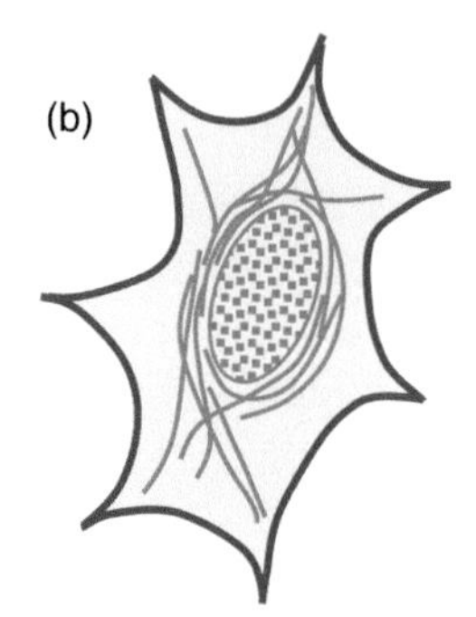

(c)

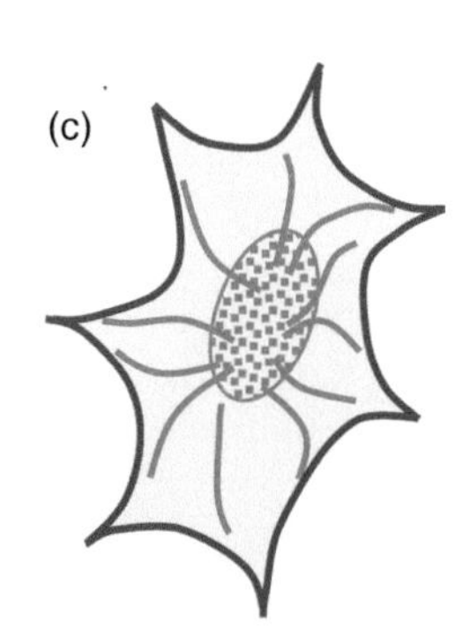

Fig. 1.9

Schematic organization of cytoskeletal filaments in a fibroblast: actin (a), intermediate filaments (b), and microtubules (c).

All of the material within the cell, with the exclusion of its nucleus, is defined as the cytoplasm, which contains organelles as well as the cytosol. The cytoplasm is rich with macromolecules, including DNA and various proteins, both globular and filamentous. For example, the protein content (percent by weight, where 1% = 10 mg/ml) is 20%–32% in bacteria but 35% in red blood cells, neither of which contains organelles. Conversely, the water content of the cytoplasm is just 70%–80% by weight. It's no surprise that the aqueous component of the cytoplasm is well below 90%, for it would be inefficient from the chemical reaction standpoint for reactants to diffuse through a low concentration environment searching out a reaction partner. However, the substantial concentration of proteins increases the viscosity of the cytoplasm and hence reduces the diffusion rates: the effective viscosity of the cytoplasm for diffusion of typical proteins is about three times the viscosity of pure water. Further, the presence of filaments in the cytoplasm may create a meshwork that hinders the motion of large proteins or other objects, further increasing the effective viscosity that they experience (drag forces in viscous environments are treated in Chapter 2).

Permeating the cytosol in some cells, or simply attached to the plasma membrane in others, is the cytoskeleton, a network of filaments of varying size and rigidity. The various filament types of the cytoskeleton may be organized into separate networks with different mechanical properties. In the schematic cell shown in Fig. 1.9, slender filaments of actin are associated with the plasma membrane, while thicker intermediate filaments are connected to cell attachments sites, and stiff microtubules radiate from the microtubule organizing center. In addition to providing the cell with mechanical strength, elements of the cytoskeleton may function as pathways for the transportation of material, much like the roads of a city.

The above survey of cell structure demonstrates several things. First, membranes are ubiquitous components of the cell, providing boundaries for the cell itself and for the cell's organelles and other compartments. Membranes contain both proteins and dual-chain lipids, the latter capable of self-assembly into bilayers under the appropriate conditions, as we will establish in Chapter 5. Being only 4–5 nm thick, bilayers are very flexible and can adapt to the changing shape of the cell as needed. Second, filaments are present in the cell in a variety of forms, sometimes as isolated molecules like DNA or RNA, sometimes as part of a network like the cytoskeleton or cell wall. The filaments range up to 25 nm in diameter; the thickest filaments appear stiff on the length scale of the cell, while the thinnest filaments appear to be highly convoluted. Most individual filaments, networks and membranes of cells are then *soft*, in that they may be easily deformed by forces commonly present in a cell. We now discuss in more detail what is meant by "soft", and describe the thermal fluctuations in the size and shape of filaments and membranes.

1.3 Biomaterials: soft strings and sheets

In a cell, most filaments are not straight like the beams of a skyscraper nor are the membranes flat like the steel sheets in the hull of a boat. For instance, Fig. 1.10(a) is an image of DNA immobilized on a substrate, looking much like long pieces of thread thrown casually onto a table. Similarly, Fig. 1.10(b) demonstrates how membranes, in this case isolated membranes from *Escherichia coli*, can sustain regions of very high curvature. These images hint that the strings and sheets in cells are soft, a hypothesis that is confirmed by observing the ease with which they are deformed by ambient forces within the cell and by those applied externally.

Words like *hard* and *soft* are used in at least two contexts when describing the elastic characteristics of a material. Sometimes the terms are used in a comparative sense to reflect the property of one material relative to another; for example, Moh's scale is a well-known logarithmic measure of hardness (talc has a hardness of 1 on this scale, while diamond has a hardness of 10). In another usage, softness indicates the response of an object to forces routinely present in its environment. In thermal equilibrium, an object can acquire energy from its surroundings, permitting or forcing it to change shape even if energy is needed to do so. The amplitude of these thermal fluctuations in shape depends, of course, on the softness of the material. While rigid objects may not even modestly change shape

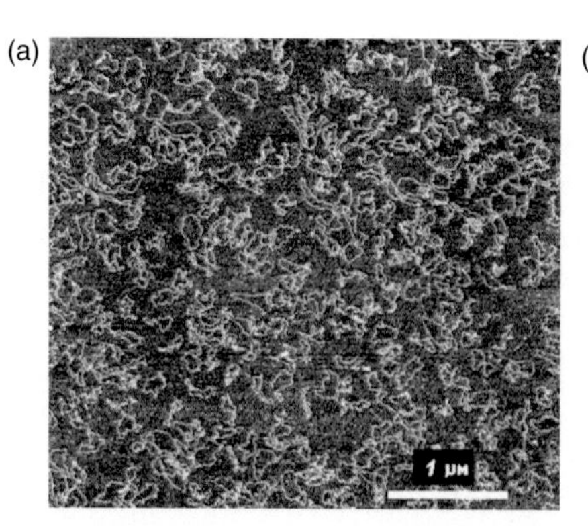

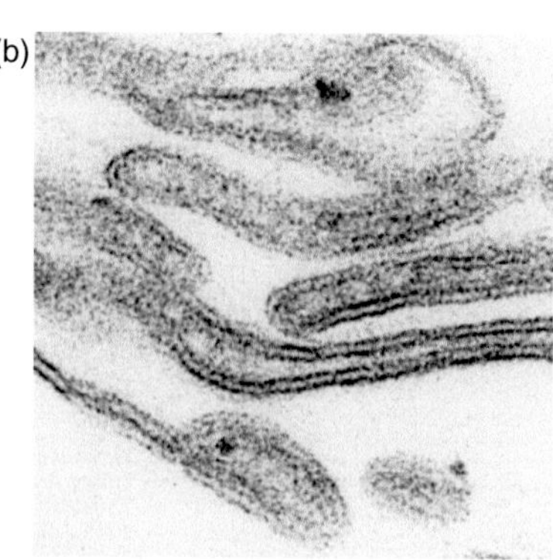

Fig. 1.10 (a) Atomic force microscope image of DNA on a mica substrate (bar is 1 μm; reprinted with permission from Shlyakhtenko *et al.*, 1999; ©1999 by the Biophysical Society). (b) Thin section of outer membranes extracted from *E. coli* after staining with osmium tetroxide and uranyl acetate. The bilayer, clearly visible as parallel black lines, is 7.5 nm thick (courtesy of Dr. Terry Beveridge, University of Guelph).

in response to thermal fluctuations in their energy, flexible filaments may bend from side to side and membranes may undulate at room temperature. We now describe the rigidity and thermally driven shape changes of biological filaments and sheets on cellular length scales; the elastic behavior of macroscopic objects like leaves and feathers can be found in Vogel's very readable book on the subject (Vogel, 1998).

1.3.1 Soft filaments

The resistance that any filament offers to bending depends upon its size and material composition: thick ropes are stiffer than thin strings and steel is more rigid than cooked pasta. To qualitatively understand their bending behavior, consider what happens when a force is applied to the free end of a thin filament, while the other end is fixed, as indicated in Fig. 1.11. The energy required to gently bend a rod into the shape of an arc depends on three factors:

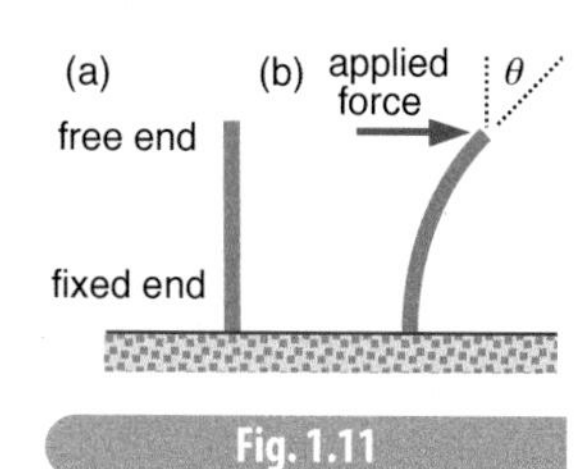

Fig. 1.11 The energy required to gently bend a straight rod (a) is proportional to the square of the bending angle θ, defined in (b).

- the bending angle, θ in Fig. 1.11; the deformation energy increases like the square of the angle, just as the potential energy of a simple spring increases like the square of its displacement from equilibrium.
- the material composition of the rod; for instance, it takes about one hundred times more energy to bend a metallic filament (like copper) than an otherwise identical biofilament, through the same bending angle.
- the diameter (and cross sectional shape) of the filament; for example, the rigidity of a thin, cylindrical rod increases like the fourth power of its diameter.

This last factor means that thick cellular filaments such as microtubules, with diameters of 25 nm, have a bending resistance about two orders of magnitude larger than thin actin filaments, of diameter 8 nm: the fourth power of their diameters has the ratio $(25/8)^4 = 95$. Thus, the cell has at its disposal a selection of filaments spanning a large range of bending rigidity, from floppy threads to stiff molecular ropes, and these filaments bend much more easily than if they were made of metal or similar materials.

What about the thermal motion of a filament as it waves back and forth, exchanging energy with its environment? Very stiff filaments hardly move from their equilibium positions; for them, the value of θ in Fig. 1.11 is usually close to zero. In contrast, highly flexible filaments may sample a bewildering variety of shapes. For instance, at room temperature, an otherwise straight microtubule 10 μm long would be displaced, on average, by about a tenth of a radian (or 6 degrees of arc) because of thermal motion. Were the microtubule made of steel, its mean angular displacement would be less than one degree of arc. In comparison, a 10 μm length of the filamentous protein spectrin, which is much thinner and more flexible than actin, would look like a ball of thread.

(a)

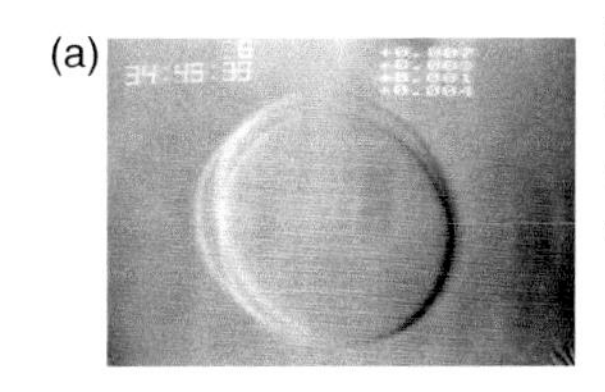

(b)

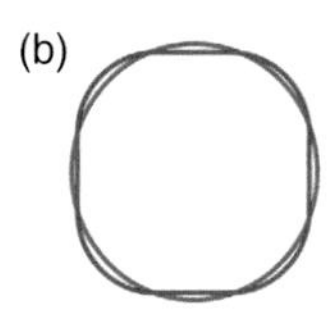

Fig. 1.12

(a) Pure lipid vesicles, whose surface area is slightly more than the minimum needed to contain its aqueous contents, display thermal undulations like those shown in these superimposed video images. (b) Schematic representation of some surface oscillations of a vesicle; a circle representing the mean position of the surface is drawn in grey for comparison (reprinted with permission from Yeung and Evans, 1995; courtesy of Dr. Evan Evans, University of British Columbia, ©1995 by Les Editions de Physique).

1.3.2 Soft sheets

In analogy to flexible filaments, membranes possess a bending resistance that depends on their geometry and material composition. For the membranes of the cell, this resistance covers a much smaller range – perhaps only a factor of ten – than the many orders of magnitude spanned by the bending resistance of cellular filaments. The reason for modest range in bending rigidity is mainly geometrical: reflecting their rather generic lipid composition, viable cell membranes tend to have a thickness of about 4–5 nm, insufficient to take advantage of the power-law dependence of the bending rigidity on membrane thickness. However, these biological membranes do exhibit about two orders of magnitude less resistance to bending than would an otherwise identical membrane made from copper or steel.

The energy required to bend an initially flat bilayer into a closed spherical shape like a cell is neither trivial nor insurmountable, as is demonstrated in Chapter 7. However, gentle undulations of membranes, such as those illustrated in Fig. 1.12(a), can be generated by thermal fluctuations alone. The figure displays images of a pure lipid vesicle, which has a mechanical structure like a water-filled balloon, except that the boundary is a fluid membrane. The images in Fig. 1.12(a) are separated by an elapsed time of 1 s, and have been superimposed to show the amplitude of the undulations. The types of motion executed by the surface are shown schematically in Fig. 1.12(b), where the circle indicates the mean position of the membrane. Clearly, the membrane is sufficiently stiff that the wavelength of the undulations is similar to the size of the vesicle, and not dramatically smaller.

1.3.3 Soft vs. hard: entropy vs. energy

The deformation resistance of a material is quantitatively characterized by its elastic moduli; for instance, a solid has a higher compression modulus than does a gas. Does this mean that the elastic behavior of gases has a different origin or receives different contributions than that of hard solids? The deformation resistance of a solid is primarily energetic: the equilibrium arrangement of the atoms or molecules in a solid at moderate temperatures is determined by energy minimization. Hence, there is an energy cost for displacing these atoms or molecules from their preferred locations when the material is strained.

The situation is different for a gas. J. J. Waterston, a pioneer in the theory of gases, invoked a cloud of flying insects as a metaphor for a gas and we can imagine the effort needed to force a swarm of wasps into a tiny volume (see Section 2–1 of Kauzmann, 1966). During the compression of a dilute gas, there is little change to the interaction energy between its atomic or molecular constituents because they are simply not close to each other, compared to the separation between the constituents of a solid. What changes, then, is not the interaction energy but rather the entropy of the gas. As reviewed in Appendix C, entropy is a measure of the number of configurations that a system can adopt, including spatial configurations as well as molecular shapes. By compressing the volume, the physical space that the system can explore is reduced, and hence its entropy is reduced. According to the second law of thermodynamics, a reduction in entropy cannot occur spontaneously – energy must be added to the system (through work, for example) to decrease its entropy.

The elasticity of soft materials often has both energetic and entropic contributions, depending on its state of strain. Entropic contributions tend to be relatively larger when the system is only modestly deformed, whereas energetic contributions may become important at high deformation. Consider the behavior of a flexible chain in one spatial dimension as a specific example. Suppose the links on the chain are completely flexible so that it takes no energy to introduce a kink by reversing the direction of the chain. The configurations available to such a chain with four links of equal segment length is shown in Fig. 1.13, where the left-hand end of the chain is held fixed and the right-hand end can be pulled by a horizontal force. We assume that the extended end of the chain cannot pass to the left of the fixed end. In Fig. 1.13, there are six configurations with no extension, four with two units of extension, and only one stretched configuration with four units of extension, where "unit" refers to the length of an individual chain element.

In the absence of bending resistance between links on the chain, all configurations in Fig. 1.13 have the same energy. If no force is applied to the right-hand of the chain, it can move freely among the configurations

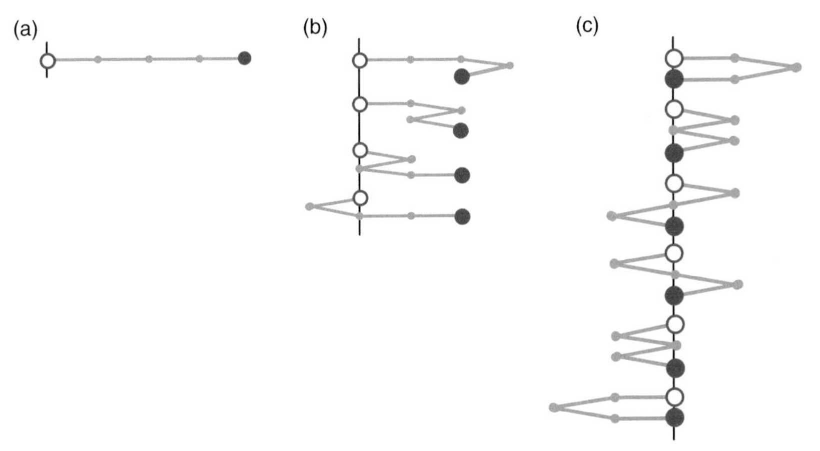

Fig. 1.13 Configurations of a four-segment chain with inequivalent ends in one dimension. The displacement between ends of the chain is 0, 2 and 4 segment lengths in groups (a), (b) and (c), respectively. Switchbacks have been offset for clarity.

without an energy penalty. On the basis of the number of configurations available, the chain is most likely to be found unextended, and least likely to be found fully extended. However, if sufficient force is applied to the right-hand end to pull it by two units of extension to the right, then about half (six out of eleven) of the original configurations of the chain are difficult to access. That is, pulling on the chain reduces its entropy because the system can sample fewer configurations: the chain resists extension for entropic reasons, not energetic ones. Of course, if one attempts to pull the chain beyond four units of extension by stretching an individual link, the resistance will be energetic as well as entropic.

Elasticity arising from the resistance to entropy loss, as we have just described, is well established in polymer physics (Flory, 1953). We will show that entropy also contributes to the elasticity of flexible membranes, not just filaments. From the examples above, we see that the deformation resistance of stiff materials is dominated by energy considerations, while that of soft materials is also influenced by entropy.

1.4 Forces inside and outside the cell

During their lifetime, almost all cells experience forces and stresses that arise as part of the efficient operation of the organism. For instance, the cytoplasm of a bacterium is laden with proteins and ionic compounds that raise its osmotic pressure with respect to the cell's external environment. The high concentration of molecules and ions increases chemical reaction rates in the cytoplasm, but forces the bacterium to surround itself with a

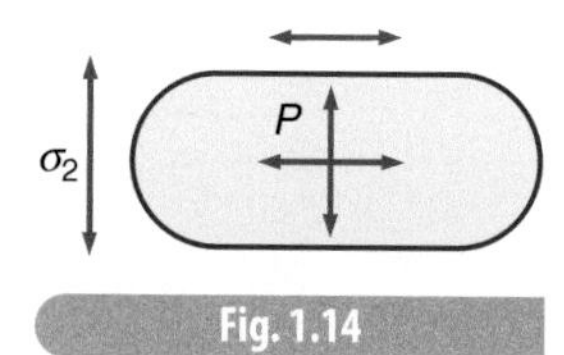

Fig. 1.14

The pressure P within a mechanically simple cell like a bacterium is opposed by the surface stress σ_2 within the cell boundary. As a three-dimensional quantity, P has units of energy per unit volume, whereas the two-dimensional stress has energy per unit area.

cell wall to avoid rupture. Similarly, the cytoplasm is sufficiently viscous that molecular diffusion is slow; consequently, larger cells must find a way of transporting important molecular components through the cytoplasm faster than the time scale of diffusion. Such directed transport of cargo-laden vesicles requires the generation of force to overcome viscous drag on the vesicles. Lastly, during cell division, forces must be exerted in nucleus-bearing cells to segregate sister chromatids.

Consider the bacterium shown in Fig. 1.14. Its cytoplasm is at an elevated pressure P that has units of force per unit area or energy per unit volume; for some bacteria, P can range above 10 atmospheres (10^6 J/m^3). The pressure exerts a force per unit area on the boundary of the cell that is the same at any location, schematically represented by the single-headed arrows in the diagram. The cell wall, which can be viewed as a two-dimensional sheet for our purposes, experiences a surface stress σ_2 having units of energy per unit area, not the energy per unit volume of (three-dimensional) pressure. In Fig. 1.14, surface stress is indicated by the two-headed arrows drawn parallel to the cell boundary. To within some simple numerical factors that will be derived in Chapter 10, σ_2 is proportional to the product RP, where R is the radius of the cell. For a bacterium of radius 1 μm, this shows that the surface stress is in the range of 1 J/m^2. The bacterial cell wall is relatively stiff and can accommodate a surface stress of this magnitude without significant deformation.

Compared to a cell wall, the boundary of the human red blood cell is soft: its plasma membrane easily deforms as the cell passes through narrow capillaries, and its cytoskeleton helps restore the cell to its rest shape once passage is complete. The cytoskeleton of the mature red cell is particularly simple, as it is effectively a two-dimensional mesh attached to the plasma membrane; it does not crisscross through the volume of the cell. The magnitude of the deformation that a red cell can sustain is demonstrated in Fig. 1.15(a), which displays the cell's shape as it is drawn up a micropipette by suction. Part (b) is an image of the density of the membrane-associated cytoskeleton, to which fluorescent molecules have been attached, showing how the cytoskeleton becomes ever more dilute, and hence less visible in the image, as it stretches up the pipette. Part (c) is a computer simulation of a similar deformation, based on a model in which the elasticity of the cytoskeleton arises from the entropy of its thin filaments. The images indicate the magnitude of the deformation to which the materials of a cell can be subjected without failure: the fluid membrane permits the cell to squeeze into a narrow pipette, and the density of the cytoskeleton itself evidently drops by a factor of two along the length of the aspirated segment of the red cell.

As described more fully in Chapter 11, the cell has designed a number of molecular motors for generating force. Consider the origin of Watt's steam engine for a moment: the fundamental motion is the expansion of a

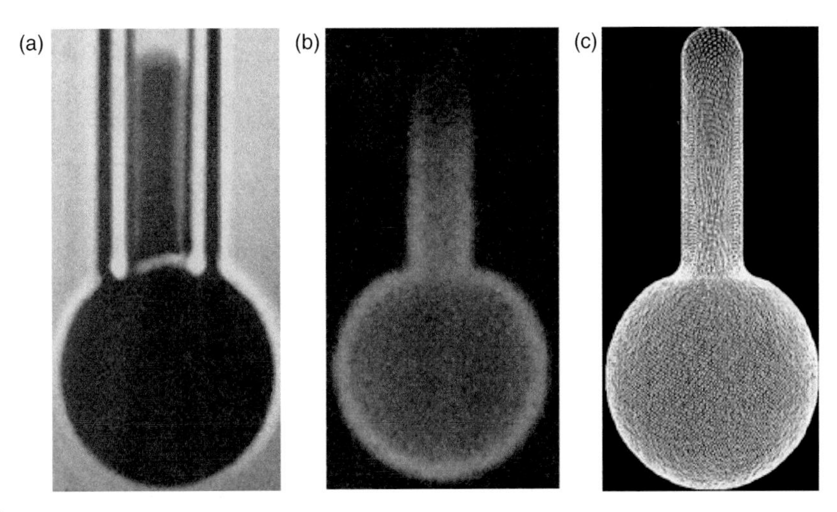

Fig. 1.15 Deformation of a human red blood cell as it is drawn up a pipette approximately 1 μm in diameter. Part (a) is a bright-field microscope image of the cell and the pipette, while (b) is a fluorescence image showing the density of the cytoskeleton. Part (c) is a computer simulation of the experiment, based on the entropic contribution of the cytoskeleton [(a) and (b) reprinted with permission from Discher *et al.* (1994); ©1994 by the American Association for the Advancement of Science; (c) from Discher *et al.* (1998)].

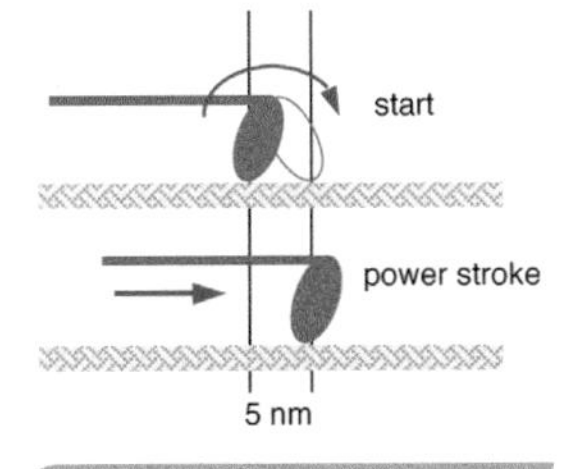

Fig. 1.16 Schematic representation of a myosin motor (magenta) pulling itself along an actin (green) filament. From its initial position, the myosin detaches from the filament, then reattaches at a location about 5 nm away, before the power stoke pulls it. The red arrows indicate the direction of motion. The orientation of the myosin head with respect to its tail is the same at the end of the power stroke as at the initial location.

gas causing the linear motion of a sliding piston within a cylinder. Yet the development of the steam engine to be more than just a poorly controlled means of pumping water from ditches came with the creation of mechanical means of obtaining rotational movement from it (for example, the offset drive of the wheels of a train engine) and the invention of the governor to control the engine's speed. These ideas also apply to a molecular motor in the cell, where the fundamental motion is linear as it *pulls* itself along a filament, a motion that has been adapted to create the push of cell division. An example of the actin/myosin motor is displayed in Fig. 1.16, where the protein myosin undergoes changes in conformation as it pulls itself along an actin filament. Further, the rotor/stator design of modern rotary motors also appears in the multi-component motor that drives the motion of flagella, the whips that propel some types of bacteria.

There are many examples in which the role of molecular motors is to overcome the drag forces arising in the cell's viscous environment, both internally and externally. In fluid mechanics, a quantity called the Reynolds number indicates the relative importance of viscosity in the motion of an object – the Reynolds number is a measure of the inertial force compared to the force from viscous drag. A dimensionless quantity, the Reynolds number depends on the viscosity of the environment and the size of the moving object, among other things. If the Reynolds number of the motion is much less than unity, the motion is dominated by viscous effects. For most situations in the cell, the Reynolds number is remarkably small: one-millionth or less!

Let's estimate the magnitude of drag forces present in the cell. At low speeds, drag is proportional to the product of the speed and width of the moving object, among other things, meaning that drag forces can span a range of values. Taking a representative system that is treated as an example in Chapter 2, the drag force experienced by a spherical bacterium of radius 1 μm moving at a speed of 20 μm/s in water is 0.4 pN, where pN is a piconewton of force, or 10^{-12} N. This is the force that the bacterium's flagella, its propulsion unit, must generate in order to keep the cell moving. Tiny as this force may appear to be, it is substantial on the cellular level and if the flagella were suddenly disabled, the cell would come to a complete stop in a distance much smaller than an atomic diameter. This is the nature of life at low Reynolds number: the lack of inertial effects strongly influences the strategies a cell must employ in order to swim, even in a low viscosity fluid.

Just as with everyday machinery, the motion of molecular motors is repetitive, a series of steps in which chemical energy is converted into mechanical energy. Familiar from introductory physics courses, the work done by the motor in a specific step is equal to the usual product of force and distance for linear motion, or torque and angle for circular motion. Using a step size of 5 nm, and a force of 4 pN as representative values, the corresponding work per step is 2×10^{-20} J. This can be compared to the energy available per hydrolysis of ATP (adenosine triphosphate, the cell's most common energy currency) of 8×10^{-20} J. In other words, the hydrolysis of a single ATP molecule is sufficient to drive a motor through one step.

The remaining concept that we wish to address in this section is the magnitude of the thermal energy scale compared to the typical mechanical energies in the cell. Let's consider the imaginary box of molecules in Fig. 1.17, each molecule of mass m roving at random throughout the box at a density sufficiently low that the molecules form an ideal gas. The molecules are free to collide with each other, exchanging momentum and kinetic energy when they do so, such that they possess a distribution of speeds and kinetic energies. That is, the speed of the particles is not fixed at one particular value, but rather forms a distribution, with few molecules traveling very slowly, few traveling very fast, and many traveling in some intermediate range. Quantities such as the mean speed and kinetic energy of this molecular gas can be calculated *via* statistical mechanics, which shows that the mean kinetic energy of a molecule is $(3/2)k_BT$ in three dimensions, where k_B is Boltzmann's constant (1.38×10^{-23} J). For a gas at room temperature (20 °C), the combination $k_BT = 4 \times 10^{-21}$ J.

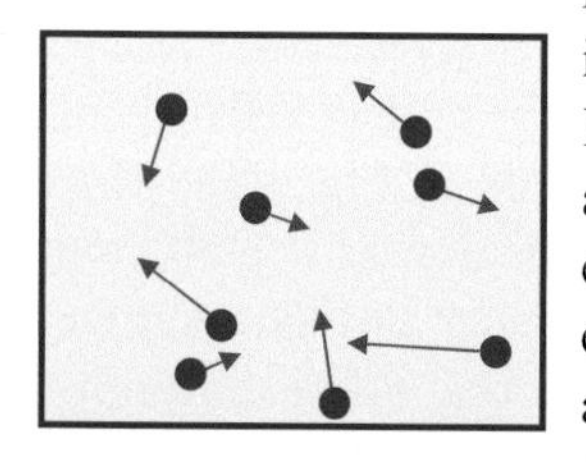

Fig. 1.17

Schematic drawing of an ensemble of particles moving within a confined geometry. The particles exchange energy and momentum by colliding with each other and the walls of the container, which is kept at a fixed temperature T.

How does the thermal energy scale compare to the energy scales in the cell? One benchmark is the work done per step in a typical molecular motor, which we calculated as 2×10^{-20} J, or about five times k_BT. This is not unexpected: if the work done by the motor were less than k_BT, the motor would be ineffective, as random thermal fluctuations would overwhelm its work and the motor would move randomly. In contrast, isolated molecules that are much smaller than proteins are more influenced by thermal effects. For example, we will demonstrate in Chapter 3 that the varying positions of the links of a polymer arising from thermal fluctuations contribute to the elasticity of the polymer. Thus, the thermal energy scale is generally not too far removed from other energy scales in the cell, and effects arising from thermal fluctuations may be important.

Summary

Many of the design principles that have been developed for structures such as buildings and bridges are equally applicable to the architecture of the cell. For example, the efficient usage of available materials may be most readily achieved through composite systems, in which the required functionality is obtained only from the combined properties of the individual structural elements. We see this strategy in the design of the cell, where the plasma membrane provides the necessary barrier for confining the cell's contents and is reinforced as necessary by networks and walls to withstand the stresses of the membrane's environment. We observe that at least one linear dimension (i.e. width or thickness) of most structural elements of the cell is very small, say 4–5 nm for the thickness of a membrane or 8–25 nm for the diameter of many biofilaments, so that most mechanical components of the cell are soft in some respects because of their small dimensions. Further, the resistance of soft biological materials to deformation may be a hundred times less than conventional hard materials such as metals. In thermal equilibrium, soft filaments and sheets oscillate and undulate because the energy required for modest changes in shape is available in the cell. The presence of soft materials is required not only for the everyday tasks of the cell but also for its growth and division, necessitating the development of a much broader building code than what is applicable to human engineering.

In the next several chapters, we investigate the generic characteristics of soft filaments and sheets. The cytoskeleton is one of the primary topics of this text, and Part I is entirely devoted to the behavior of flexible filaments, both in isolation (Chapters 3 and 4), and as they are welded into networks (Chapters 5 and 6). Part II approaches our second principal topic, membranes, in a similar vein: the molecular structure and self-assembly of bilayers are treated in Chapter 7, while membrane undulations and interactions are the subjects of Chapters 8 and 9. Throughout

Parts I and II, many prominent features of soft structures are shown to be rooted in their entropy, and are most easily understood by means of statistical mechanics. An introduction to the statistical concepts needed for the text is given in Chapter 2, while Appendices C and D provide additional background material from statistical mechanics and elasticity that underlie Parts I and II.

We assemble these isolated ropes and sheets into some simple, but complete, cells in Part III. The shapes of biological structures such as vesicles and mammalian red blood cells are interpreted with the aid of mechanical models in Chapter 10. The motion of a cell, including its molecular basis, is the subject of Chapter 11. We then devote the last two chapters of the text to cell growth and division, including mechanisms for the transcription and replication of DNA as well as for the control of the division cycle.

2 Soft materials and fluids

The softness of a material implies that it deforms easily when subjected to a stress. For a cell, an applied stress could arise from the cell's environment, such as the action of a wave on water-borne cells or the pressure from a crowded region in a multicellular organism. The exchange of energy with the cell's environment due to thermal fluctuations can also lead to deformations, although these may be stronger at the molecular level than the mesoscopic length scale of the cell proper. For example, Fig. 1.12 shows the fluctuations in shape of a synthetic vesicle whose membrane is a pure lipid bilayer that has low resistance to out-of-plane undulations because it is so thin. At fixed temperature, flexible systems may sample a variety of shapes, none of which need have the same energy because fixed temperature does not imply fixed energy. In this chapter, the kind of fluctuating ensembles of interest to cell mechanics are introduced in Section 2.1, followed up with a review of viscous fluids and their role in cell dynamics in Section 2.2. Many of the statistical concepts needed for describing fluctuating ensembles are then presented, using as illustrations random walks in Section 2.3 and diffusion in Section 2.4. Lastly, the subject of correlations is presented in Section 2.5, focusing on correlations within the shapes of long, sinuous filaments.

2.1 Fluctuations at the cellular scale

Among the common morphologies found among cyanobacteria (which trace their lineage back billions of years) are filamentous cells, two examples of which are displayed in Fig. 2.1. The images are shown at the same magnification, as indicated by the scale bars. The upper panel is the thin filament *Geitlerinema* PCC 7407, with a diameter of 1.5 ± 0.2 μm, while the lower panel is the much thicker filament *Oscillatoria* PCC 8973, with a diameter of 6.5 ± 0.7 μm (Boal and Ng, 2010). The filaments have been cultured in solution, then mildly stirred before imaging; clearly, the thinner filament has a more sinuous appearance than the thicker filament when seen at the same magnification. This is as expected: the resistance to bending possessed by a uniform solid cylinder grows like the fourth power of its diameter, so the thinner filament should have much less resistance to bending and hence appear more sinuous than the thicker one.

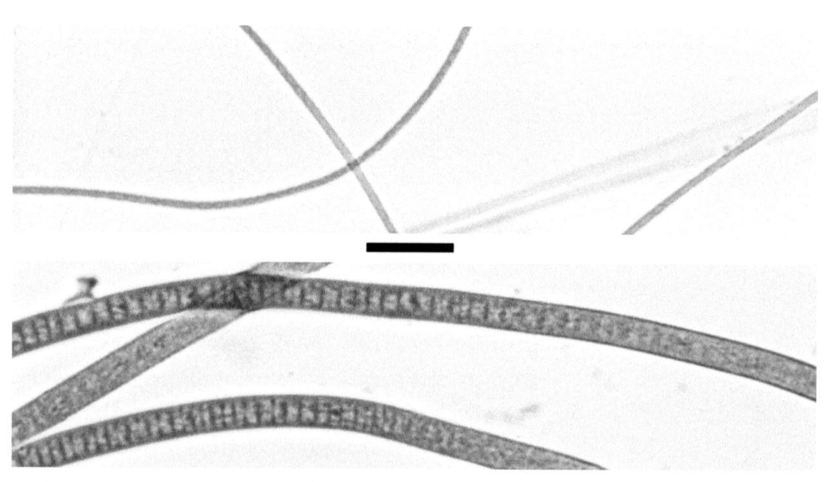

Fig. 2.1 Shape fluctuations exhibited by two species of filamentous cyanobacteria: *Geitlerinema* PCC 7407 (upper, 1.5 ± 0.2 μm) and *Oscillatoria* PCC 8973 (lower, 6.5 ± 0.7 μm), where the mean diameter of the filament is indicated in brackets. Both images are displayed at the same magnification, and the scale bar is 20 μm.

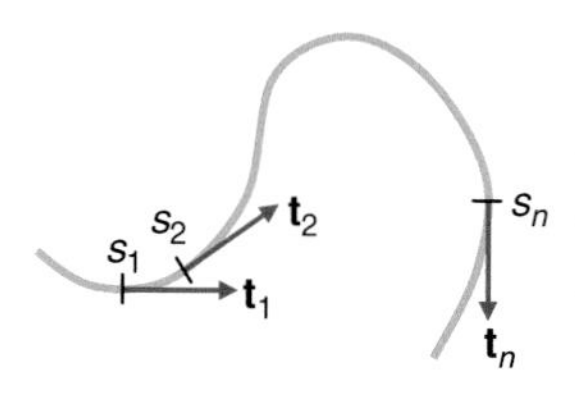

Fig. 2.2

A sinuous curve (green) can be characterized by the variation in the orientation of its tangent vectors (red, $\mathbf{t}_1, \mathbf{t}_2, \ldots \mathbf{t}_n$) at locations along the curve ($s_1, s_2, \ldots s_n$). The separation Δs between locations is equal to the distance along the curve, for example $\Delta s = |s_1 - s_2|$, but is not equal to the displacement between the points.

We can quantify how "sinuous" a curve is by examining the behavior of the tangent vector $\mathbf{t}$ to the curve at different locations along it. It's easiest to work with a unit tangent vector, which means that the length of the vector is unity according to $\mathbf{t} \cdot \mathbf{t} = 1$, where the notation $\mathbf{a} \cdot \mathbf{b}$ represents the "dot" or scalar product of the two vectors $\mathbf{a}$ and $\mathbf{b}$. On Fig. 2.2 is drawn an arbitrarily shaped curve (green), along which several unit tangent vectors have been constructed (red: $\mathbf{t}_1, \mathbf{t}_2, \ldots \mathbf{t}_n$) at three locations along the curve ($s_1, s_2, \ldots s_n$). The separation Δs between locations is equal to the distance (or arc length) along the curve; for example, between locations s_1 and s_2, the separation $\Delta s = |s_1 - s_2|$. Thus, the separation between points s_1 and s_n is much larger than the (vector) displacement between the points: Δs takes into account the path length from s_1 to s_n, whereas the magnitude of the displacement is the distance along a straight line drawn between the two positions.

Let's suppose for a moment that a finite ensemble of vectors $\{\mathbf{t}_i\}$ is selected at N random locations along the curve. The magnitude of the tangent vectors $\mathbf{t}_i$ does not change with location s_i, so that an average that is taken over the ensemble obeys

$$\langle \mathbf{t}_i \cdot \mathbf{t}_i \rangle \equiv (1/N) \Sigma_{i=1}^{N} \mathbf{t}_i \cdot \mathbf{t}_i = (1/N) \Sigma_{i=1}^{N} 1 = (1/N) \cdot N = 1, \tag{2.1}$$

because each contribution $\mathbf{t}_i \cdot \mathbf{t}_i = 1$ in the ensemble average, the latter denoted by $\langle \cdots \rangle$. What happens if we take the scalar product of vectors at different locations? If the curve is a straight line, then $\mathbf{t}_i \cdot \mathbf{t}_j = 1$ even if

$i \neq j$, because all the tangent vectors to a straight line point in the same direction. However, if the tangent vectors point in different directions, then $\mathbf{t}_i \bullet \mathbf{t}_j$ can range between −1 and +1, so we would expect in general

$$-1 \leq \langle \mathbf{t}_i \bullet \mathbf{t}_j \rangle \leq 1, \qquad (i \neq j) \tag{2.2}$$

where the equal signs hold only if all tangent vectors point in the same direction. If the set $\{\mathbf{t}_i\}$ is truly random, then the ensemble average will contain about as many cases where $\mathbf{t}_i \bullet \mathbf{t}_j$ is positive as there are examples where it is negative, in which case

$$\langle \mathbf{t}_i \bullet \mathbf{t}_j \rangle \rightarrow 0 \qquad (i \neq j, \text{ random orientations}) \tag{2.3}$$

in the limit where N is large. Using calculus, it's easy enough to generalize these results to the situation where $\mathbf{t}$ is a continuous function of s, as is done in later chapters; for the time being, all we need is the discrete case. In Section 2.5 of this chapter, we examine how $\mathbf{t}_i \bullet \mathbf{t}_j$ behaves when s_i and s_j are separated by a fixed value, rather than averaged over all values of Δs considered for Eqs. (2.2) and (2.3); we will show that $\langle \mathbf{t}_i \bullet \mathbf{t}_j \rangle_{\Delta s}$ quantitatively characterizes how sinuous the path is.

The shapes of cells in an ensemble provide another example of fluctuations of importance in cell mechanics. Figure 2.3 shows a small collection of the eukaryotic green alga *Stichoccocus S*, a common alga found in freshwater ditches, ponds and similar environments; the green organelle in the cell's interior is a chloroplast. The width of these cells is fairly uniform from one cell to the next, having a mean value of 3.56 ± 0.18 μm, but the length is more variable, with a mean of 6.98 ± 1.29 μm, where the second number of each pair is the standard deviation. The large value of the standard deviation of the cell length relative to its mean value reflects the fact that the length changes by a factor of two during the division cycle.

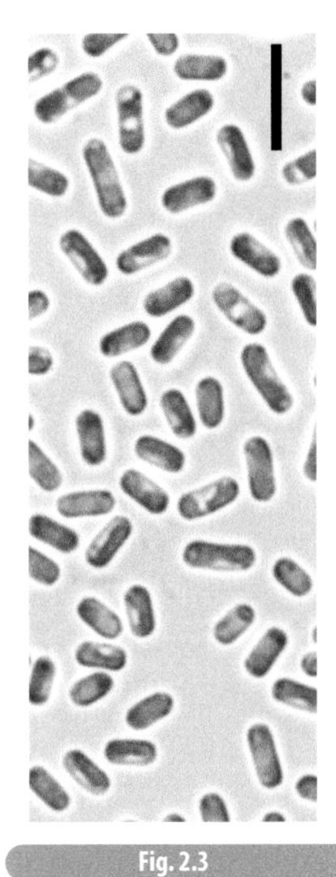

Fig. 2.3

Sample of eukaryotic green alga *Stichococcus S*; this particular strain is from a freshwater ditch in Vancouver, Canada. Scale bar is 10 μm. (Forde and Boal, unpublished).

Although the mean values of cell dimensions are obviously useful for characterizing the species, even more information can be gained from the distribution of cell shapes, as will be established in Chapter 12. For now, we simply wish to describe how to construct and utilize a continuous distribution from an ensemble using data such as the cell length and width. Knowing that *Stichococcus* grows at a fairly constant width, we choose as our geometrical observable the length to width ratio Λ, just to make the observable dimensionless. Each cell has a particular Λ, and from the ensemble one can determine how many cells Δn_Λ there are for a range $\Delta\Lambda$ centered on a given Λ. The total number of cells in the ensemble N is just the sum over all the cells Δn_Λ in each range of Λ.

Now, let's convert *numbers* into *probabilities*; that is, let's determine the probability of finding a cell in a particular range of Λ. This is straightforward: if there are Δn_Λ cells in the range, then the probability of finding a cell in this range out of the total population N is just $\Delta n_\Lambda/N$. Unfortunately, the probability as we have defined it depends on the range $\Delta\Lambda$ that we have used for collecting the individual cell measurements. We can remove this dependence on $\Delta\Lambda$ by constructing a probability density $\mathcal{P}(\Lambda)$ simply by dividing the probability for the range by the magnitude of the range itself, $\Delta\Lambda$. Note that the probability density has units of Λ^{-1} because of the operation of division. Let's put these ideas into equations. Working with the finite range of $\Delta\Lambda$, we said

$$[\textit{probability of finding cell in range } \Delta\Lambda] = \Delta n_\Lambda / N, \tag{2.4}$$

and

$$\begin{aligned}[\textit{probability density around } \Lambda] &= \mathcal{P}(\Lambda) \\ &= [\textit{probability of finding cell in range } \Delta\Lambda] / \Delta\Lambda,\end{aligned} \tag{2.5}$$

so that

$$[\textit{number of cells in range } \Delta\Lambda] = \Delta n_\Lambda = N\,\mathcal{P}(\Lambda)\,\Delta\Lambda. \tag{2.6}$$

If we now make the distribution continuous instead of grouping it into ranges of $\Delta\Lambda$, Eq. (2.6) becomes

$$\mathrm{d}n_\Lambda = N\,\mathcal{P}(\Lambda)\,\mathrm{d}\Lambda. \tag{2.7}$$

By working with probability densities, we have removed the explicit dependence of the distribution on the number of cells in the sample, and at the same time have obtained an easily normalized distribution function:

$$\int \mathcal{P}(\Lambda)\,\mathrm{d}\Lambda = 1, \tag{2.8}$$

from which the mean value of Λ is

$$\langle \Lambda \rangle = \int \Lambda\,\mathcal{P}(\Lambda)\,\mathrm{d}\Lambda. \tag{2.9}$$

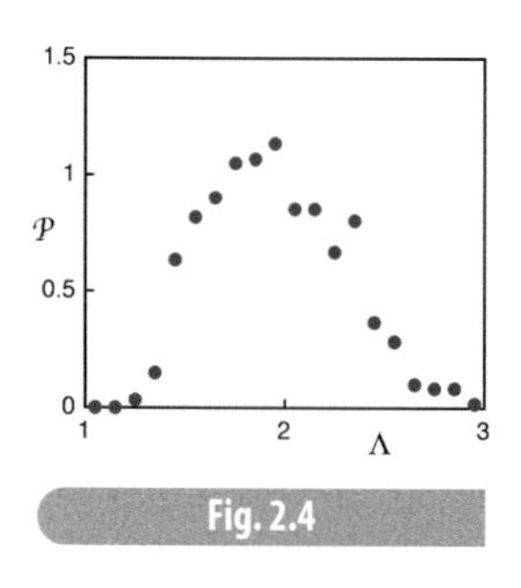

Fig. 2.4

Probability density $\mathcal{P}$ for the length to width ratio Λ of the green alga *Stichococcus S*, a specimen of which is shown in Fig. 2.3 (Forde and Boal, unpublished).

Having done all of this formalism, what do the data themselves look like? Figure 2.4 shows the probability density for the length-to-width ratio of the green alga *Stichococcus S* in Fig. 2.3. The fact that $\mathcal{P}$ is zero at small values of Λ just means that there are no cells in this range: $\Lambda = 1$ corresponds to a spherical cell, and *Stichococcus* is always elongated, like a cylindrical capsule. The peak in $\mathcal{P}$ at the smaller values of Λ observed in the population indicates that the cell grows most slowly during this time. Once the cell begins to grow rapidly, there will be (relatively) fewer examples of it in a steady-state population, and that's what the data indicate is occuring at large values of cell length, near the end of the division cycle.

These first two examples of fluctuations at the cellular level dealt with cell shape, either fluctuations in the local orientation of a single, long biofilament, or the fluctuations in cell length among a population under steady

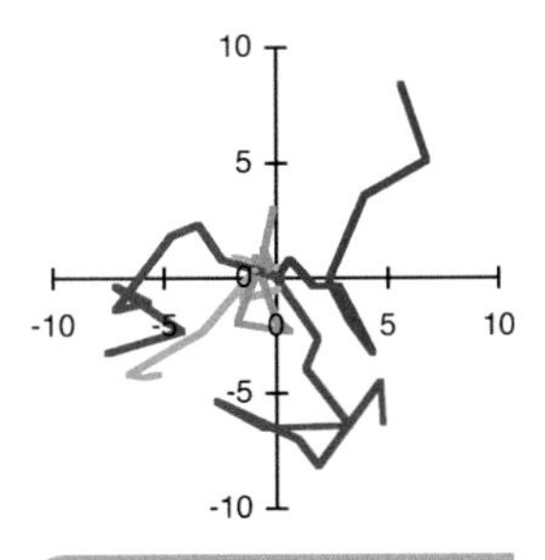

Fig. 2.5

Trajectories of five spherical polystyrene beads in pure water at room temperature, as observed under a microscope. The beads have a diameter of 1 μm, and the tracks were recorded at 5 second intervals.

state growth. As a final illustration, we examine the random motion of objects with sizes in the micron range immersed in a stationary fluid: here, the objects are plastic spheres, and in Chapter 11, they are self-propelled cells. In everyday life, we're familiar with several types of motion within a fluid, even when the fluid as a whole has no overall motion. For instance, convection is often present when there is a density gradient in the fluid, such that less dense regions rise past more dense regions: warm (less dense) air at the surface of the Earth rises through cooler air above it. But even if the fluid has a uniform density and displays no convective motion that we can see with the naked eye, there may be motion at length scales of microns or less.

Figure 2.5 shows the trajectories of small plastic (polystyrene) beads tracked at 5 s intervals as seen under a microscope; the spherical beads have a diameter of 1 μm and they are immersed in pure water in a small chamber on a microscope slide. Several trajectories are displayed, all taken from a region within about a hundred microns. The tracks possess

- no overall drift in a particular direction that might indicate convection or fluid flow,
- no straight line behavior that would indicate motion at a constant velocity.

Rather, the trajectories change speed and direction at random, though on a time scale finer than what appears in the figure, because the time between measurements is a relatively long 5 s. This is an example of Brownian motion, which arises because of the exchange of energy and momentum between the plastic beads and their fluid environment, much like the exchange of energy and momentum among particles in a box described in Section 1.4.

An instantaneous velocity can be assigned to the beads, but it changes constantly in magnitude and direction. A plot of the position of the beads relative to their initial location when their motion began to be recorded, exhibits much scatter from bead to bead, and even the mean value of the (magnitude of the) displacement does not increase linearly with time which would be expected for constant speed. However, the mean value of the squared displacement does rise linearly with time, which is characteristic of random motion as established in Section 2.3. There aren't sufficient trajectories in Fig. 2.5 to obtain an accurate description of the motion, so we must be content with the poorly determined result $\langle r^2 \rangle = (1.1 \pm 0.3\ \mu\text{m}^2/\text{s})\, t$, where r is the magnitude of the displacement from the origin, t is the time, and the ensemble average $\langle \cdots \rangle$ is taken over just five trajectories. Not only can the $\langle r^2 \rangle \propto t$ behavior be explained from random motion, the proportionality constant ($1.1 \pm 0.3\ \mu\text{m}^2/\text{s}$) has its origin in the fluctuations in kinetic energy of particles moving in a viscous fluid.

In Section 1.4, we stated that the mean kinetic energy of a particle as it exchanges energy with its neighbors in an ideal system is equal to 3/2 $k_B T$ in three dimensions, where T is the temperature in Kelvin and k_B is Boltzmann's constant (1.38×10^{-23} J/K). At room temperature, 3/2 $k_B T = 6 \times 10^{-21}$ J. One thing to note is that the mean kinetic energy of the particles in this system is independent of their mass, meaning that lighter particles travel faster than heavier ones. For a hydrogen molecule, with a mass of 3.3×10^{-27} kg, the root mean square (rms) speed is 1930 m/s, found by equating $3k_B T/2 = m\langle v^2 \rangle/2$. The tiny plastic beads of Fig. 2.5 have a much greater mass than a diatomic molecule like H_2, and their mean speed is thus many orders of magnitude smaller. Combining (i) the distribution of speeds in a gas at equilibrium with (ii) the drag force on an object moving in viscous medium, shows that $\langle r^2 \rangle = 6Dt$ in three dimensions, where the diffusion coefficient D is given by $D = k_B T \,/\, 6\pi\eta R$ for spheres of radius R moving in a fluid with viscosity η. We return to this expression, called the Einstein relation, in Section 2.4 and use it to interpret the measurements in Fig. 2.5 in the end-of-chapter problems.

Before undertaking any further analysis of cell motion, we review in Section 2.2 the effects of viscous drag on the movement of objects in a fluid medium. This provides a better preparation and motivation for the discussion of random walks in Section 2.3 and diffusion in Section 2.4. The formalism of correlation functions is presented in Section 2.5, but the material does not involve the properties of fluids so Sections 2.2 and 2.4 need not be read before starting Section 2.5.

2.2 Movement in a viscous fluid

A fluid is a material that can resist compression but cannot resist shear. Passing your hand through air or water demonstrates this, in that the air or water does not restore itself to its initial state once your hand has passed by – rather, there has been mixing and rearranging of the gas or liquid. Yet even if fluids have zero shear resistance, this does not mean that their deformation under shear is instantaneous: there is a characteristic time scale for a fluid to respond to an applied stress. For example, water spreads fairly rapidly when poured into a bowl, whereas salad dressing usually responds more slowly, and sugar-laced molasses slower still. What determines the response time is the strength and nature of the interactions among the fluid's molecular components. For example, the molecules could be long and entangled (as in a polymer) or they could be small, but strongly interacting (as in water or molten glass).

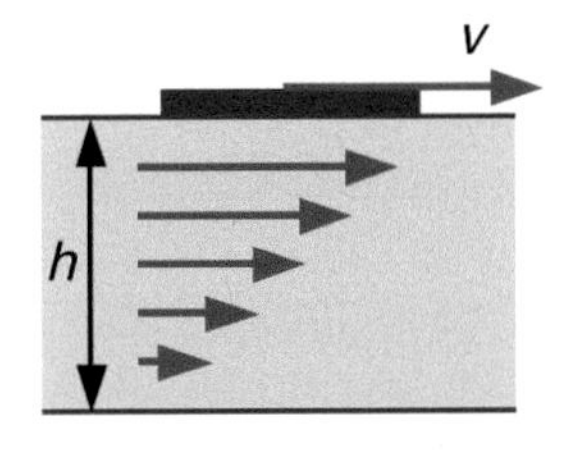

Fig. 2.6

In one measurement of viscosity, a horizontal force F is applied to a flat plate of area A in contact with a liquid, resulting in the plate moving at a velocity v. The height of the liquid in its container is h and the speed of the fluid at the lower boundary is zero.

At low speeds, the response time of a fluid to accommodating an applied stress depends on a physical property called the viscosity, η, among other factors. Unlike an elastic parameter like the compression modulus, which has the same dimensions of stress (force per unit area, or energy per unit volume in three dimensions), the dimensions of η include a reference to time. We illustrate this by considering one means of measuring η, which involves the application of a horizontal force to the surface of an otherwise stationary fluid, as illustrated in Fig. 2.6. In the figure, a flat plate of area A on one side is pulled along the surface of the fluid with a force F, giving a shear stress of F/A. If the material in the figure were a solid, it would resist this stress until it attained a deformed configuration where the applied and reaction forces were in equilibrium. But a fluid doesn't resist shear, and the floating plate continues to move at a speed v as long as the stress is applied. The magnitude of the speed depends inversely on the viscosity: the higher the viscosity the lower the speed that can be achieved with a given stress. The relationship has the form:

$$F/A = \eta\,(v/h), \qquad (2.10)$$

where h is the height of the liquid in its container. Note that the fluid is locally stationary at its boundaries: it is at rest at the bottom of the container and moving with speed v beside the plate.

Elastic quantities such as the bulk modulus or shear modulus appear in Hooke's law expressions of the form [*stress*] = [*elastic modulus*] • [*strain*]. Strain is a dimensionless ratio like the change in volume divided by the undeformed volume, so elastic moduli must have the dimensions of stress. Equation (2.10) is different from this, in that the ratio v/h is not dimensionless but has units of $[time]^{-1}$, so that η has dimensions of [*force/area*] • [*time*], or kg/m • s in the MKSA system. Thus, η provides the time scale for the relaxation, as expected. There are a variety of ways of measuring η; the viscosities of some familiar fluids are given in Table 2.1. Viscosity is often quoted in units of Poise or P, which has the equivalence of kg/m • s ≡ 10 P.

2.2.1 Translational drag

Moving through a viscous fluid, an object experiences a drag force whose magnitude depends on the speed of the object with respect to the fluid. At low speeds where the motion of the object does not induce turbulence in the fluid, the drag force rises linearly with the speed, whereas at high speeds where turbulence is present, the drag force rises like the square of the speed. The detailed relationship between the drag force F_{drag} and the speed v depends on the shape of the object among other things, so for the time being we will simply write the relationship as

Table 2.1 Viscosities of some familiar fluids measured at 20 °C. A commonly quoted unit for viscosity is the Poise; in MKSA system, 1 kg/m • s = 10 P.

Fluid	η (kg/m • s)	η (P)
Air	1.8×10^{-5}	1.8×10^{-4}
Water	1.0×10^{-3}	1.0×10^{-2}
Mercury	1.56×10^{-3}	1.56×10^{-2}
Olive oil	0.084	0.84
Glycerine	1.34	13.4
Glucose	10^{13}	10^{12}
Mixtures: blood	2.7×10^{-3}	2.7×10^{-2}

$$F_{\text{drag}} = c_1 v \quad \text{(low speeds)} \tag{2.11a}$$

$$F_{\text{drag}} = c_2 v^2, \quad \text{(high speeds)} \tag{2.11b}$$

where the constants c_1 and c_2 depend on a variety of terms. This is for linear motion through the fluid, and there are similar relations for rotational motion as will be described later in this section. Note that the power required to overcome the drag force, obtained from [*power*] = Fv, grows at least as fast as v^2 according to Eq. (2.11). Relatively speaking, viscous forces are so important in the cell that we need only be concerned with the low-speed behavior of Eq. (2.11a); the dynamic properties of systems obeying Eq. (2.11b) are treated in the problem set at the end of this chapter.

Let's now solve the motion of an object subject only to linear drag in the horizontal direction – that is, omitting gravity. The object obeys Newton's law $F = ma = m\,(dv/dt)$, so that the drag force from Eq. (2.11a) gives the relation

$$ma = m\,(\mathrm{d}v/\mathrm{d}t) = -c_1 v, \tag{2.12}$$

where the minus sign indicates that the force is in the opposite direction to the velocity. Equation (2.12) can be rearranged to read

$$\mathrm{d}v/\mathrm{d}t = -(c_1/m)\,v, \tag{2.13}$$

which relates a velocity to its rate of change. This equation does not yield a specific number like $v = 5$ m/s; rather, its solution gives the *form* of the function $v(t)$. It's easy to see that the solution is exponential in form, because

$$\mathrm{d}e^x/\mathrm{d}x = e^x. \tag{2.14}$$

That is, the derivative of an exponential is itself an exponential, satisfying Eq. (2.13). One still has to take care of the factor c_1/m in Eq. (2.13), and it's easy to verify by explicit substitution that

Table 2.2 Summary of drag forces for translation and rotation of spheres and ellipsoids at low speeds. For ellipsoids, the drag force, the torque $\mathcal{T}$ and the angular velocity ω are about the major axis; the expressions apply in the limit where the semi-major axis a is much longer than the semi-minor axis b.

	Sphere	Ellipsoid ($a >> b$)
Translational force	$F = 6\pi\eta Rv$	$F = 4\pi\eta av/\{\ln(2a/b) - 1/2\}$
Rotational torque	$\mathcal{T} = 8\pi\eta R^3\omega$	$\mathcal{T} = (16/3)\,\pi\eta ab^2\omega$

$$v(t) = v_o \exp(-c_1 t / m), \tag{2.15}$$

where v_o is the speed of the object at $t = 0$.

The characteristic time scale for the velocity to decay to $1/e$ of its original value is m/c_1. Even though the object is always moving because the velocity goes to zero only in the limit of infinite time, nevertheless, the object reaches a maximum distance mv_o/c_1 from its original location, also at infinite time. The time-dependence of the distance can be found by integrating Eq. (2.15) to yield:

$$\Delta x = (mv_o / c_1) \bullet [1 - \exp(-c_1 t / m)], \tag{2.16}$$

where the limiting value at $t \to \infty$ is obvious.

The strength of the drag force depends not only on the viscosity at low speeds, but also on the cross-sectional shape that is presented to the fluid by the object in its direction of motion. A cigar, for instance, will experience less drag when moving parallel to its long axis than when moving with that axis perpendicular to the direction of motion. Analytical expressions are available for the drag force at low speeds, two examples of which are given in Table 2.2. The most commonly quoted one is Stokes' law for a sphere of radius R:

$$F = 6\pi\eta Rv. \tag{2.17}$$

This expression will be used momentarily in an example. A sphere is a special case of an ellipsoid of revolution where the semi-major axis a and the semi-minor axis b are both equal: $a = b = R$. When $a >> b$, the drag force becomes

$$F = 4\pi\eta av/\{\ln(2a/b) - 1/2\}, \tag{2.18}$$

for motion at low speed parallel to the long axis of the ellipsoid. In this expression, note that if b is fixed, then the drag force increases with the length of the ellipsoid as $a/\ln a$.

At higher speeds when turbulence is present, the drag force for translational motion not only depends on the square of the speed, but it also has a different dependence on the shape of the object:

$$F = (\rho/2)AC_D v^2, \tag{2.19}$$

where ρ is the density of the fluid and A is the cross-sectional area of the object in its direction of motion (πR^2 for a sphere). The dimensionless drag coefficient C_D is often about 0.5 for many shapes of interest, and somewhat less than this for sports cars (0.3). Note that the drag force in Eq. (2.19) depends on the density of the fluid, rather than its viscosity η in Eq. (2.18). Also, note the dependence on the square of the transverse dimension in Eq. (2.19), compared to the linear dependence in Eq. (2.18).

Example 2.1. Consider an idealized bacterium swimming in water, assuming:

- the bacterium is a sphere of radius $R = 1$ μm,
- the fluid medium is water with $\eta = 10^{-3}$ kg / m • s,
- the density of the cell is that of water, $\rho = 1.0 \times 10^3$ kg/m^3,
- the speed of the bacterium is $v = 2 \times 10^{-5}$ m/s.

What is the drag force experienced by the cell? If the cell's propulsion system were turned off, over what distance would it come to a stop (ignoring thermal contributions to the cell's kinetic energy from the its environment)?

First, we calculate the prefactor c_1 in Eq. (2.11a)

$$c_1 = 6\pi\eta R = 6\pi \cdot 10^{-3} \cdot 1 \times 10^{-6} = 1.9 \times 10^{-8} \text{ kg/s},$$

so that the drag force on the cell can then be obtained from Stoke's law:

$$F_{\text{drag}} = c_1 v = 1.9 \times 10^{-8} \cdot 2 \times 10^{-5} = 0.4 \text{ pN} \quad (\text{pN} = 10^{-12} \text{ N}).$$

To determine the maximum distance that the cell can drift without propulsion, we first calculate the mass of the cell m,

$$m = \rho \cdot 4\pi R^3/3 = 10^3 \cdot 4\pi\,(1 \times 10^{-6})^3/3 = 4.2 \times 10^{-15} \text{ kg},$$

from which the stopping distance becomes, using Eq. (2.16)

$$x = mv_o/c_1 = 4.2 \times 10^{-15} \cdot 2 \times 10^{-5} / 1.9 \times 10^{-8}$$
$$= 4.4 \times 10^{-12} \text{ m} = 0.04 \text{ Å}.$$

2.2.2 Rotational drag

The stress experienced by the surface of an object moving through a viscous fluid can retard the rotational motion of the object, as well as its translational motion. The effect of rotational drag is to produce a torque $\mathcal{T}$ that reduces the object's angular speed ω with respect to the fluid. At low angular speed, the torque from drag is linearly proportional to ω, just as the linear relation Eq. (2.11a) governs translational drag:

$$\mathcal{T} = -\chi\omega. \tag{2.20}$$

where the minus sign indicates $\mathcal{T}$ acts to reduce the angular speed. Here, we adopt the usual convention that counter-clockwise rotation corresponds to positive ω. For a sphere of radius R, the drag parameter χ is

$$\chi = 8\pi\eta R^3, \tag{2.21}$$

where η is the viscosity of the medium. The expression for χ for an ellipsoid of revolution is given in Table 2.2. Similar to the expressions for force, the power required to overcome the torque from viscous drag is given by $[power] = \mathcal{T}\omega$, which grows as ω^2 for Eq. (2.20). Confusion can sometimes arise between frequency (revolutions per second) and angular speed (radians per second): ω is equal to 2π times the frequency of rotation. Both quantities have units of $[time]^{-1}$ because radians are dimensionless.

It's straightforward to set up the dynamical equations for rotational motion under drag, and to solve for the functional form $\omega(t)$ of the angular speed and $\theta(t)$ of the angle swept out by the object. For instance, if the rotation is about the longest or shortest symmetry axis of the object, then the torque produces an angular acceleration α that determines $\omega(t)$ via

$$\mathcal{T} = I\alpha = I\,(\mathrm{d}\omega/\mathrm{d}t) = -\chi\omega, \tag{2.22}$$

where I is the moment of inertia about the axis of rotation. For a sphere of radius R, the moment of inertia about all axes through the center of the sphere is $I = mR^2/2$. As in our discussion of translational motion, Eq. (2.22) determines the functional form of $\omega(t)$:

$$\omega(t) = \omega_o \exp(-\chi t \,/\, I), \tag{2.23}$$

where ω_o is the initial value of ω. Equation (2.23) can be integrated to yield the angle traversed during the slowdown, but this is left as an example in the problem set.

Example 2.2. Consider an idealized bacterium swimming in water, assuming:

- the bacterium is a sphere of radius $R = 1$ μm,
- the fluid medium is water with $\eta = 10^{-3}$ kg / m • s,
- the bacterium rotates at a frequency of 10 revolutions per second.

Find the retarding torque from drag experienced by the cell.

First, the frequency of 10 revolutions per second corresponds to an angular frequency of $\omega = 20\pi$ s^{-1}. Next, the prefactor χ in Eq. (2.21) is

$$\chi = 8\pi\eta R^3 = 8\pi \bullet 10^{-3} \bullet (1 \times 10^{-6})^3 = 8\pi \times 10^{-21} \text{ kg-m}^2\text{/s},$$

so that the magnitude of the drag torque on the cell can then be obtained from:

$$\mathcal{T}_{\text{drag}} = \chi\omega = 8\pi \times 10^{-21} \bullet 20\pi = 1.6 \times 10^{-18} \text{ N-m}.$$

As a final caveat, most readers with a physics background are aware that the kinematic quantities ω, α, and $\mathcal{T}$ are vectors and I is a tensor. Thus, the situations we have described are specific to rotations about a particular set of axes through an object. When ω and $\mathcal{T}$ have arbitrary orientations with respect to the symmetry axes, the motion is more complex than what has been described here.

2.2.3 Reynolds number

In Example 2.1 for translational motion, the drag force is so important that it causes a moving cell to stop in less than an atomic diameter once a cell's propulsion unit is turned off. In the problem set, it is shown that rotational motion also ceases abruptly under similar circumstances. (Note that both of these conclusions ignore any contribution to the kinetic energy from thermal fluctuations.) Put another way, the effect of drag easily overwhelms the cell's inertial movement at constant velocity that follows Newton's First Law of mechanics.

In fluid dynamics, a benchmark exists for estimating the importance of the inertial force compared to the drag force. This is the Reynolds number, a dimensionless quantity given by

$$\mathbf{R} = \rho\, v\, \lambda \,/\, \eta, \qquad (2.24)$$

where v and λ are the speed and length of the object, and ρ and η are the density and viscosity of the medium, all respectively. We won't provide a derivation of $\mathbf{R}$ from the ratio of the inertia to drag forces experienced an object (see Nelson, 2003) as $\mathbf{R}$ will not be used elsewhere in this text. The crossover between drag-dominated motion at small $\mathbf{R}$ and inertia-dominated motion at large $\mathbf{R}$ is in the range $\mathbf{R} \sim 10$–100.

Let's collect the terms on the right-hand side of Eq. (2.24) into properties of the fluid (ρ/η) and those of the object ($v\lambda$); for water at room temperature, ρ/η is 10^6 s/m^2. Common objects like fish and boats, with lengths and speeds of meters and meters per second, respectively, have $v\lambda$ in the range of 1–1000 m^2/s. Thus, **R** for everyday objects moving in water is 10^6 or more, and such motion is dominated by inertia, even though viscous effects are present. This conclusion also applies for cars and planes as they travel through air, where ρ/η is 0.5×10^5 s/m^2 under standard conditions. However, for the motion of a cell, the product $v\lambda$ is far smaller: even if $\lambda = 4$ μm and $v = 20$ μm/s, then $v\lambda = 8 \times 10^{-11}$ m^2/s, such that **R** is less than 10^{-4}. Clearly, this value is well below unity so the motion of a typical cell is dominated by viscous drag. In the context of the Reynolds number, the reason for this is the very small size and speed of cells compared to everyday objects.

2.3 Random walks

The motion of microscopic plastic spheres as they interact with their fluid environment was displayed in Fig. 2.5. As shown, the trajectories are just a coarse representation of the motion, in that the positions of the spheres were sampled every five seconds, so that the fine details of the motion were not captured. However, the behavior of each trajectory over long times is correctly represented, permitting the calculation of the ensemble average $\langle r^2 \rangle$ over the suite of positions $\{\mathbf{r}_k\}$ as a function of the elapsed time t, where the index k is a particle label. In discussing Fig. 2.5, it was pointed out that $\langle r^2 \rangle$ does not increase like t^2, as it would for motion at constant velocity; rather, $\langle r^2 \rangle \propto t$, which we now interpret in terms of the behavior of random walks.

Each step of a walk, random or otherwise, can be represented by a vector $\mathbf{b}_i$, where the index i runs over the N steps of the walk. The contour length of the path L is just the scalar sum over the lengths of the individual steps:

$$L = \Sigma_{i=1,N}\, b_i. \tag{2.25}$$

There is no direction dependence to Eq. (2.25) so that no matter how the path twists and turns, the contour length is always the same so long as the average step size is the same. In contrast, the displacement of the path $\mathbf{r}_{ee}$ from one end to the other is a vector sum:

$$\mathbf{r}_{ee} = \Sigma_{i=1,N}\, \mathbf{b}_i. \tag{2.26}$$

The situation is illustrated in Fig. 2.7 for four arbitrary walks of fixed step length, where the magnitude of $\mathbf{r}_{ee}$ for each walk is obviously less than the contour length L.

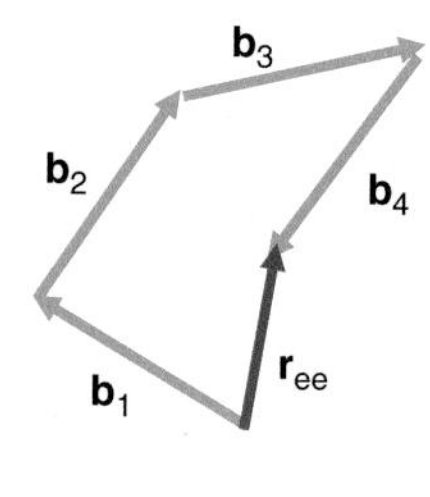

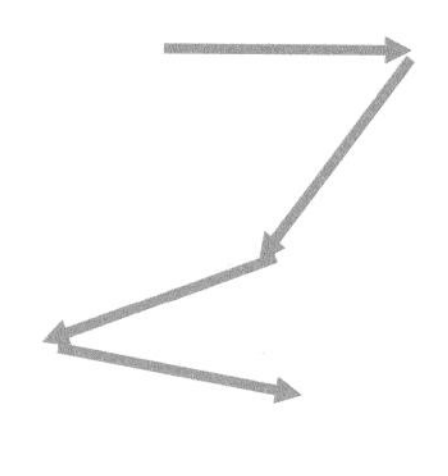

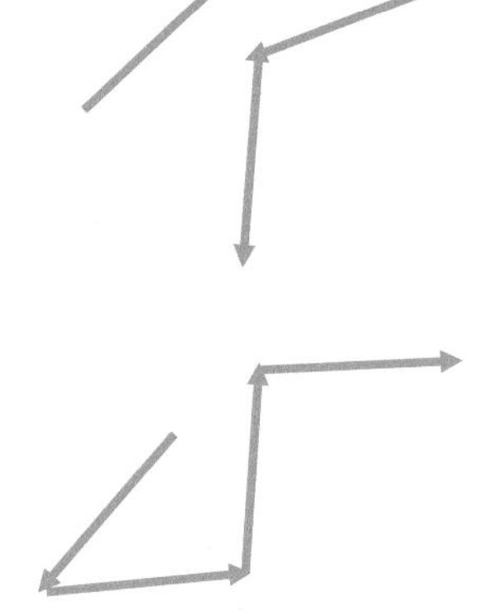

Fig. 2.7

A walk with four steps represented by the vectors $\mathbf{b}_1 \ldots \mathbf{b}_4$ has a contour length $L = b_1 + b_2 + b_3 + b_4$ (scalar sum) and an end-to-end displacement $\mathbf{r}_{ee} = \mathbf{b}_1 + \mathbf{b}_2 + \mathbf{b}_3 + \mathbf{b}_4$ (vector sum). Several walks are shown, each with the same L in this example, but different $\mathbf{r}_{ee}$.

Even though $\mathbf{r}_{ee}{}^2$ may be different for each path, it is easy to calculate the average value of $\mathbf{r}_{ee}{}^2$ when taken over many configurations. For now, each step is assumed to have the same length b, even though the directions are different from step to step. From Eq. (2.26), the general form of the dot product of $\mathbf{r}_{ee}$ with itself for a particular walk is

$$\begin{aligned} \mathbf{r}_{ee} \bullet \mathbf{r}_{ee} &= (\Sigma_{i=1,N}\, \mathbf{b}_i) \bullet (\Sigma_{j=1,N}\, \mathbf{b}_j) \\ &= (\mathbf{b}_1 + \mathbf{b}_2 + \mathbf{b}_3 + \cdots) \bullet (\mathbf{b}_1 + \mathbf{b}_2 + \mathbf{b}_3 + \cdots) \\ &= \mathbf{b}_1{}^2 + \mathbf{b}_2{}^2 + \mathbf{b}_3{}^2 \cdots + 2\mathbf{b}_1 \bullet \mathbf{b}_2 + 2\mathbf{b}_1 \bullet \mathbf{b}_3 + \cdots + 2\mathbf{b}_2 \bullet \mathbf{b}_3 \ldots. \end{aligned} \tag{2.27}$$

In this sum, there are N terms of the form $\mathbf{b}_i{}^2$, each of which is just b^2 if all steps have the same length. Thus, for a given walk

$$\mathbf{r}_{ee}{}^2 = Nb^2 + 2\mathbf{b}_1 \bullet \mathbf{b}_2 + 2\mathbf{b}_1 \bullet \mathbf{b}_3 + 2\mathbf{b}_1 \bullet \mathbf{b}_4 + \cdots + 2\mathbf{b}_2 \bullet \mathbf{b}_3 \ldots. \tag{2.28}$$

But there may be many walks of N steps starting from the same origin, again as illustrated in Fig. 2.7. The average value $\langle \mathbf{r}_{ee}{}^2 \rangle$ is obtained by summing over all these paths

$$\langle \mathbf{r}_{ee}{}^2 \rangle = Nb^2 + 2\langle \mathbf{b}_1 \bullet \mathbf{b}_2 \rangle + 2\langle \mathbf{b}_1 \bullet \mathbf{b}_3 \rangle + 2\langle \mathbf{b}_1 \bullet \mathbf{b}_4 \rangle + \cdots + 2\langle \mathbf{b}_2 \bullet \mathbf{b}_3 \rangle \ldots. \tag{2.29}$$

Each dot product $\mathbf{b}_i \bullet \mathbf{b}_j$ $(i \neq j)$ may have a value between $-b^2$ and $+b^2$. In a large ensemble of random walks, for every configuration with a particular scalar value $\mathbf{b}_i \bullet \mathbf{b}_j = b_{ij}$, there is another configuration with $\mathbf{b}_i \bullet \mathbf{b}_j = -b_{ij}$, so that the average over all available configurations becomes

$$\langle \mathbf{b}_i \bullet \mathbf{b}_j \rangle \to 0. \tag{2.30}$$

Combining Eqs. (2.29) and (2.30) yields the elegant result

$$\langle \mathbf{r}_{ee}{}^2 \rangle = Nb^2. \qquad \text{random walk} \tag{2.31}$$

Example 2.3. Each amino acid in a protein contributes 0.36 nm to its contour length. For example, the protein actin, a major part of our muscles, is 375 amino acids long, giving an overall length of about 135 nm. But the amino acid backbone of a protein does not behave like a stiff rod; rather, it wiggles and sticks to itself at various locations. The random walk gives an approximate value for its size:

$$\langle \mathbf{r}_{ee}{}^2 \rangle = Nb^2 = 375\ (0.36)^2$$

or

$$r_{ee,av} \sim \sqrt{375} \times 0.36 = 7.0 \text{ nm}.$$

In other words, the radius of a random ball of actin (< 10 nm) is *much* less than its length when fully stretched (135 nm).

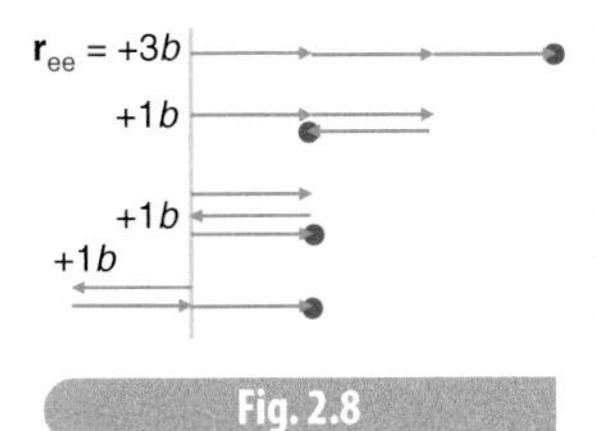

Fig. 2.8

Configurations for a one-dimensional walk with three segments of equal length b; the red dot indicates the end of the path. Only half of the allowed configurations are shown, namely those with displacement $\mathbf{r}_{ee} > 0$.

Random walks were introduced in this section as a description of the thermal motion of microscopic spheres. If each movement of a sphere in a given time step has a fixed length, then Eq. (2.31) establishes that $\langle \mathbf{r}_{ee}^2 \rangle$ should grow linearly with time (i.e. linearly with the number of steps). It will be shown in Chapter 3 that even when the assumption of fixed step size is dropped, $\langle \mathbf{r}_{ee}^2 \rangle$ still rises linearly with time. Yet random walks have greater applicability than just the description of thermal motion. Example 2.3 illustrates the conceptual importance of random walks in understanding the sizes of flexible macromolecules. We now probe the characteristics of random walks more deeply by examining the distribution in $\mathbf{r}_{ee}^2$ within an ensemble of walks, in an effort to understand the distribution of polymer sizes and, in Chapter 3, the importance of entropy in polymer elasticity.

Consider the set of one-dimensional walks with three steps shown in Fig. 2.8: each walk starts off at the origin, and each step can point to the right or the left. Given that each step has 2 possible orientations, there are a total of $2^3 = 8$ possible configurations for the walk as a whole. Using $C(\mathbf{r}_{ee})$ to denote the number of configurations with a particular end-to-end displacement $\mathbf{r}_{ee}$, the eight configurations are distributed according to:

$$C(+3b) = 1 \quad C(+1b) = 3 \quad C(-1b) = 3 \quad C(-3b) = 1. \tag{2.32}$$

The reader will recognize that these values of $C(\mathbf{r}_{ee})$ are equal to the binomial coefficients in the expansion of $(p + q)^3$; i.e. the values are the same as the coefficients $N! \,/\, i! \, j!$ in the expansion

$$(p + q)^N = \Sigma_{i=0,N} \, \{N! \,/\, i! \, j!\} \, p^i q^j, \tag{2.33}$$

where $j = N - i$. Is this fortuitous? Not at all; the different configurations in Fig. 2.8 just reflect the number of ways that the left- and right-pointing vectors can be arranged. So, if there are i vectors pointing left, and j pointing right, such that $N = i + j$, then the total number of ways in which they can be arranged is just the binomial coefficient

$$C(i, j) = N! \,/\, i! \, j!. \tag{2.34}$$

One can think of the configurations in Fig. 2.8 as random walks in which each step (or link) along the walk occurs with probability 1/2. Thus, the probability $P(i, j)$ for there to be a configuration with (i, j) steps to the (left, right) is equal to the product of the total number of configurations ($C(i, j)$ from Eq. (2.34)) with the probability of an individual configuration, which is $(1/2)^i \, (1/2)^j$:

$$P(i, j) = \{N! \,/\, i! \, j!\} \, (1/2)^i \, (1/2)^j. \tag{2.35}$$

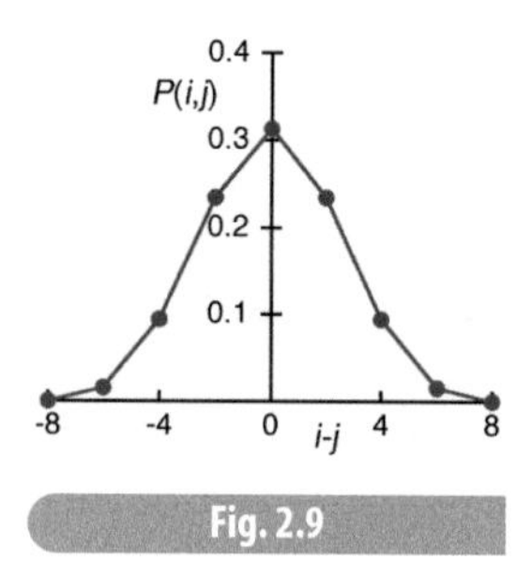

Fig. 2.9

Probability distribution from Eq. (2.35) for a one-dimensional walk with six segments.

Note that the probability in Eq. (2.35) is appropriately normalized to unity, as can be seen by setting $p = q = 1/2$ in Eq. (2.33):

$$\Sigma_{i=0,N}\, P(i,j) = \Sigma_{i=0,N}\, \{N!\,/\,i!\,j!\}\,(1/2)^i\,(1/2)^j = (1/2 + 1/2)^N = 1. \qquad (2.36)$$

What happens to the probability distribution as the number of steps increases and the distribution consequently appears more continuous? The probability distribution for a one-dimensional walk with $N = 6$ is shown in Fig. 2.9, where we note that the end-to-end displacement $\mathbf{r}_{ee} = (j - i) = (2j - N)$ changes by 2 for every unit change in i or j. The distribution is peaked at $\mathbf{r}_{ee} = 0$, as one would expect, and then falls off towards zero at large values of $|\mathbf{r}_{ee}|$ where $i = 0$ or N. As becomes ever more obvious for large N, the shape of the curve in Fig. 2.9 resembles a Gaussian distribution, which has the form

$$\mathcal{P}(x) = (2\pi\sigma^2)^{-1/2} \exp[-(x-\mu)^2\,/\,2\sigma^2]. \qquad (2.37)$$

Normalized to unity, this expression is a probability density (i.e. a probability per unit value of x) such that the probability of finding a state between x and $x + \mathrm{d}x$ is $\mathcal{P}(x)\mathrm{d}x$. The mean value μ of the distribution can be obtained from

$$\mu = \langle x \rangle = \int x\, \mathcal{P}(x)\, \mathrm{d}x, \qquad (2.38)$$

and its variance σ^2 is

$$\sigma^2 = \langle (x - \mu)^2 \rangle = \langle x^2 \rangle - \mu^2, \qquad (2.39)$$

as expected.

Equation (2.37) is the general form of the Gaussian distribution, but the values of μ and σ are specific to the system of interest. As a trivial example, consider a random walk along the x-axis starting from the origin. First, $\mu = 0$ because the vectors $\mathbf{r}_{ee}$ are equally distributed to the left and right about the origin, whence their mean displacement must be zero. Next, $\langle x^2 \rangle = Nb^2$ according to Eq. (2.31), so Eq. (2.39) implies $\sigma^2 = Nb^2$ when $\mu = 0$. Proofs of the equivalence of the Gaussian and binomial distributions at large N can be found in most statistics textbooks. However, the Gaussian distribution provides a surprisingly accurate approximation to the binomial distribution even for modest values of N, as can be seen from Fig. 2.9.

There is more to random or constrained walks and their relation to the properties of polymers than what we have established in this brief introduction. Other topics include the effects of unequal step size or constraints between successive steps, such as the restricted bond angles in a polymeric chain. In addition, the scaling behavior $\langle r_{ee}^{\,2} \rangle \sim N$ may be modified in the presence of attractive interactions between different elements of the walk, useful when the walk is viewed as a polymer chain. These and other properties will be treated in Chapter 3.

2.4 Diffusion

The random walk introduced in Section 2.3 can be applied to a variety of problems and phenomena in physics and biology. In some cases, the trajectory of an object is precisely the linear motion with random forces that we have described in obtaining the generic properties of random walks. The analogous problem of rotational motion with random torques also is a random walk, but in the azimuthal angle of the object about a rotational axis. In other cases, the conformations of a system such as a polymer may be viewed as random walks even though there is no motion of the polymer itself; Example 2.3 illustrates this behavior for the protein actin. In this section of Chapter 2, we apply the properties of random walks to diffusive systems from two different perspectives:

(i) as the translational and rotational motion of a single object in contact with its environment,
(ii) as the collective motion of objects at sufficiently high number density that they can be described by continuous variables such as concentrations.

For the second situation, we will establish how the time evolution of the concentration obeys a relationship like Fick's Law. The inverse problem of the capture of a randomly moving object is treated in Chapter 11.

The trajectory of an individual molecule diffusing through a medium has the form of a random walk, which we characterize by the displacement vector $\mathbf{r}_{ee}$ from the origin of the walk to its end-point. Suppose that the diffusing molecule travels a distance λ before it collides with some other component of the system. Then the random walk tells us that the average end-to-end displacement of the molecule's motion is

$$\langle \mathbf{r}_{ee}^2 \rangle = \lambda^2 N, \tag{2.40}$$

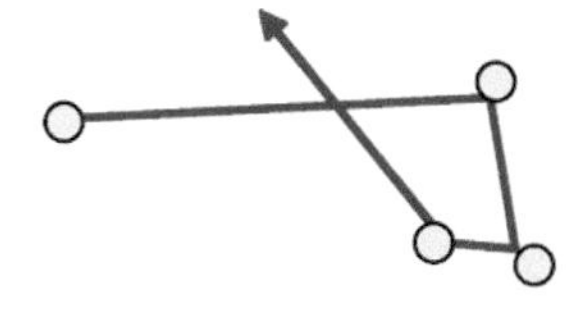

Fig. 2.10

Examples of single-particle diffusion at low (upper panel) and high (lower panel) densities.

where $\langle \cdots \rangle$ indicates an average and where N is the number of steps. How big is λ? As illustrated in Fig. 2.10, λ might be very large for a gas molecule traveling fast in a dilute environment, but λ is rather small for a protein moving in a crowded cell. If there is one step per unit time, then $N = t$ and

$$\langle \mathbf{r}_{ee}^2 \rangle = \lambda^2 t. \tag{2.41}$$

Now, the units of Eq. (2.41) aren't quite correct, in that the left-hand side has units of [*length*^{2}] while the right-hand side has [*length*^{2}]•[*time*]. We accommodate this by writing the displacement as

$$\langle \mathbf{r}_{ee}^2 \rangle \equiv 6Dt, \qquad \text{diffusion in three dimensions} \tag{2.42}$$

where D is defined as the diffusion coefficient. A molecule diffusing in a liquid of like objects has a diffusion coefficient D in the range 10^{-14} to 10^{-10} m²/s, depending on the size of the molecule.

The factor of 6 in Eq. (2.42) is dimension-dependent: for each Cartesian axis, the mean squared displacement is equal to $2Dt$. That is, if an object diffuses in one dimension only (for example, a molecule moves randomly along a track) then

$$\langle \mathbf{r}_{ee}^2 \rangle = 2Dt \qquad \text{diffusion in one dimension} \qquad (2.43)$$

and if it is confined to a plane, such as a protein moving in the lipid bilayer of the cell's plasma membrane, then

$$\begin{aligned} \langle \mathbf{r}_{ee}^2 \rangle &= \langle \mathbf{r}_{ee,x}^2 \rangle + \langle \mathbf{r}_{ee,y}^2 \rangle \qquad \text{diffusion in two dimensions} \\ &= 2Dt + 2Dt = 4Dt. \end{aligned} \qquad (2.44)$$

In all of these cases, D has units of [*length*]² / [*time*].

Example 2.4. How long does it take for a randomly moving protein to travel the distance of a cell diameter, say 10 μm, if its diffusion coefficient is 10^{-12} m²/s?

Inverting Eq. (2.42) yields

$$t = \langle \mathbf{r}_{ee}^2 \rangle / 6D,$$

so that

$$t = (10^{-5})^2 / 6 \bullet 10^{-12} = 16 \text{ s}.$$

Thus, it takes a protein less than a minute to diffuse across a cell at this diffusion coefficient; it would take much longer if the protein were large and $D \sim 10^{-14}$ m²/s.

The diffusion coefficient can be determined analytically for a few specific situations. One case is the random motion of a sphere of radius R subject to Stokes' Law for drag, Eq. (2.17): $F = 6\pi\eta Rv$, where v is the speed of the sphere and η is the viscosity of the fluid. The so-called Einstein relation that governs the diffusion coefficent reads

$$D = k_B T / 6\pi\eta R, \qquad \text{Einstein relation} \qquad (2.45)$$

where k_B is Boltzmann's constant. Now, $k_B T$ is close to the mean kinetic energy of a particle in a thermal environment, so the Einstein equation tells us that:

- the higher the temperature, the greater is an object's kinetic energy and the faster it diffuses,

Table 2.3 Examples of diffusion coefficients, showing the range of values from dilute gases to proteins in water. All measurements are at 25 °C, except xenon gas at 20 °C.

System	D (m^2/s)
Xenon	5760×10^{-9}
Water	2.1×10^{-9}
Sucrose in water	0.52×10^{-9}
Serum albumin in water	0.059×10^{-9}

- the larger an object's size, or the more viscous its environment, the slower it diffuses.

Equation (2.45) permits us to interpret the data presented at the beginning of this chapter for plastic spheres diffusing in water; the calculation is performed in the end-of-chapter problems. Lastly, Table 2.3 provides representative values for the diffusion coefficient for various combinations of solute and solvent. Note that D depends on both of these quantities, as can be seen in Eq. (2.45) where the solute dependence enters through its molecular radius R and the solvent enters through its viscosity η.

Example 2.5. A biological cell contains internal compartments with radii in the range 0.3 to 0.5 μm. Estimate their diffusion coefficient.

Suppose a cellular object like a vesicle has a radius of 0.3 μm and moves in a medium with viscosity $\eta = 2 \times 10^{-3}$ kg / m • s. At room temperature, the Einstein relation predicts

$$D = 4 \times 10^{-21} / (6\pi \bullet 2 \times 10^{-3} \bullet 3 \times 10^{-7}) = 4 \times 10^{-13} \text{ m}^2/\text{s},$$

which has the order of magnitude that we expect.

Although the *translational* motion of an object is the most common example of diffusion, it's not the only one. For example, a molecule like a protein can rotate around its axis at the same time as it travels. Although this rotation could be driven by an external force with a particular angular speed ω, it could also just be random, such that ω changes in both magnitude and direction continuously and randomly. When we talk about a protein docking onto a substrate or receiving site, it may be undergoing rotational diffusion before the optimal orientation is achieved. A random "walk" in angle θ as an object rotates around its axis can be written as

$$\langle \theta^2 \rangle = 2D_r t, \tag{2.46}$$

where D_r is the rotational diffusion coefficient. Once again, the mean change in θ from its original value at $t = 0$ grows like the square root of the elapsed time.

For a sphere rotating in a viscous medium, there is an expression for D_r just like the translational diffusion of Eq. (2.45), namely

$$D_r = k_B T / 8\pi\eta R^3. \qquad \text{rotational diffusion} \qquad (2.47)$$

Note, the units of D_r are [*time* $^{-1}$], whereas D is [*length*^{2}]/[*time*]; hence, there is an extra factor of R^2 in the denominator of Eq. (2.47) compared to Eq. (2.45).

2.4.1 Densities and fluxes

We have approached the phenomenon of diffusion at the microscopic level by considering the trajectories of individual particles, from which ensemble averages can be constructed. This tells us the average behavior of particles moving in a fluid. An alternate approach involves the behavior of macroscopic quantities such as concentrations and fluxes, that themselves represent ensemble averages over the locations of individual particles. Within this description based on local averages, quantities such as the temperature and concentration of the system's components need not be spatially uniform, and their time evolution can be understood using a mathematical formalism that we now develop.

To introduce the concepts behind the mathematics, consider the situation in Fig. 2.11, where a small amount of deep red dye has been placed in a uniform layer at the bottom of a fluid-filled container. The figure shows the appearance of the dye at three different times, starting just after the dye has been introduced (on the left) to after it has diffused through the medium to produce a largely homogenous solution (on the right). We will assume that the concentration of the dye depends only on height (which we will define as the x-direction) and is the same at all locations with the same height at any given time. The concentration $c(x,t)$ then depends on two variables: the location x above the bottom of the container, and the time t from when the dye was introduced. At the microscopic level, we know that dye molecules are moving through the solvent at speeds dictated by the temperature (through $k_B T$), colliding with solvent molecules and slowly moving up through the fluid. At the macroscopic level, we say that the concentration of dye molecules evolves like the schematic representation in Fig. 2.12: at small t, $c(x,t)$ falls rapidly with x while at large times $c(x,t)$ is asymptotically independent of x.

The change in concentration with time is accompanied by a net migration of solute molecules, which is characterized macroscopically by a flux $j(x,t)$. From our microscopic picture, we know that solute molecules are moving in all directions, but that, on average, more of them are moving

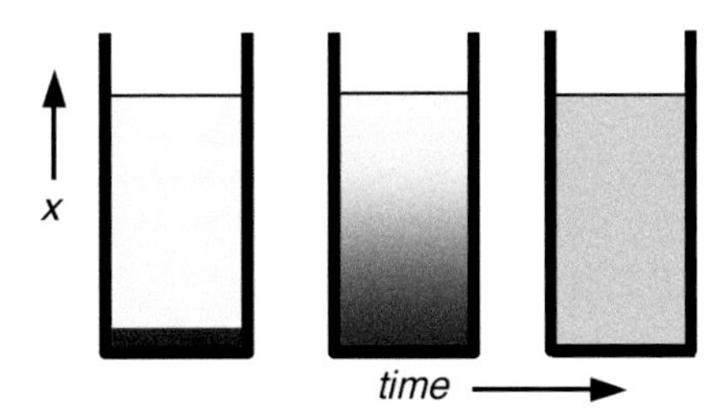

Fig. 2.11 Diffusion of a deep red dye in a light blue solvent. Examples are shown at early, intermediate and late times, from left to right respectively; the x-axis is drawn vertically in this diagram, so the concentration of dye, $c(x,t)$ falls with increasing x. The concentration is uniform in the horizontal plane.

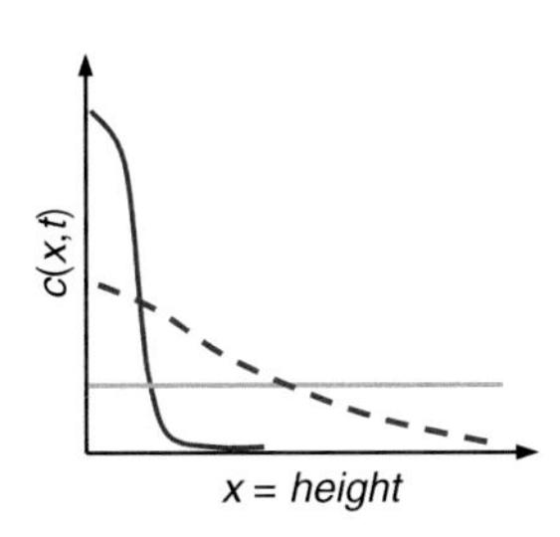

Fig. 2.12 Concentration profiles corresponding to the diffusion of a dye shown in Fig. 2.11.

upward in Fig. 2.11 than are moving downward, giving a net upward drift of solute. The flux is the net number of molecules, per unit area per unit time, crossing an imaginary plane in the yz directions at location x. That is, if the plane has area A and Δn is the net number of molecules crossing it in time Δt, then the flux is $A^{-1}\,\Delta n/\Delta t$, where the infinitesimal limit $A^{-1}\,dn/dt$ is clear. At early times, when the concentration gradient dc/dx is the largest, the flux will also be large. At late times, when the system is almost uniform so that the concentration gradient is very small, the flux is also small. In other words, what drives the flux is the concentration gradient, not the concentration itself: at long times in Fig. 2.11, the concentration may still be large, but the gradient is tiny because the system has become uniform.

Now let's express the previous paragraph in mathematical terms by saying that the flux is proportional to the (negative) gradient of the concentration, as

$$j \propto -dc/dx. \tag{2.48}$$

The minus sign is required in Eq. (2.48) because the flux is positive when the gradient is negative: otherwise, molecules would spontaneously move from regions of low concentration to regions of high concentration, completely counter to our expectations from entropy. The proportionality sign can be removed by introducing the diffusion coefficient D,

$$j = -D\,dc/dx, \qquad \text{Fick's Law} \tag{2.49}$$

an expression known as Fick's First Law of diffusion. At this point, we have not established that D in Eq. (2.49) is the same as the diffusion coefficient appearing in Eqs. (2.42)–(2.44), but we will do so shortly.

Fick's first law is almost the only result for diffusion that appears in the remainder of this book. Nevertheless there are two other important results about diffusion that are easily obtained, and so we present them here. The first is the equation of continuity, which is a conservation law applicable to a variety of situations involving fluids. We consider the diffusion of fluid particles in a cylindrical region of constant cross-sectional

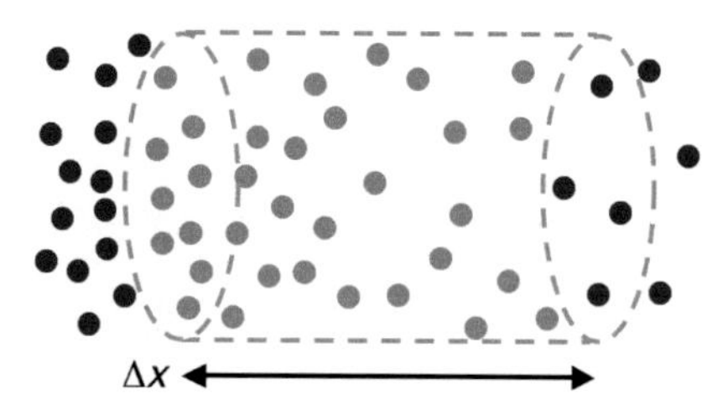

Fig. 2.13 Diffusion of particles along a concentration gradient. The uniform cylinder with cross-sectional area A and length Δx is a strictly mathematical surface: we allow particles to enter and leave this surface only longitudinally through its ends, not laterally through its sides.

area A, as shown in Fig. 2.13. The cylinder is not meant to be a physical object, just a mathematical surface with a defined shape, such that the net motion of particles is along the x-axis. At the left-hand end of the cylinder at $x = 0$, there is a flow of particles into the cylindrical region if $j > 0$, and at the right-hand end at $x = \Delta x$ there is a flow out, again if $j > 0$. If the fluxes at the two ends are not the same, then the number of fluid particles in the cylinder must increase or decrease, where we have imposed the condition that there is no net lateral flow. The total increase in the number of particles ΔN in an elapsed time Δt is then

$$\Delta N = \{[\textit{flux at } x = 0] - [\textit{flux at } x = \Delta x]\} \bullet [\textit{cross-sectional area of cylinder}] \bullet [\textit{elapsed time}].$$

In symbols, this is

$$\Delta N = \{j(0) - j(\Delta x)\}\ A\ \Delta t. \tag{2.50}$$

The change in the concentration Δc of particles in the cylinder arising from diffusion is $\Delta N\ /\ (A\ \Delta x)$, where $A\ \Delta x$ is the volume of the cylinder. Thus, after some rearrangement

$$\Delta c\ /\ \Delta t = \{j(0) - j(\Delta x)\}\ /\ \Delta x. \tag{2.51}$$

By definition, the change in the flux across the cylinder Δj has the opposite sign to the flux difference in Eq. (2.51): $\Delta j = j(\Delta x) - j(0)$. Thus, rewriting Eq. (2.51) in its infinitesimal limit,

$$\mathrm{d}c/\mathrm{d}t = -\mathrm{d}j\ /\ \mathrm{d}x, \qquad \text{continuity equation} \tag{2.52}$$

which is the continuity equation. Note that in Eqs. (2.50)–(2.52), some of the functional dependence of c and j has been suppressed for notational simplicity.

Lastly, we substitute Eq. (2.49) into the right-hand side of Eq. (2.52) to obtain Fick's second law of diffusion (also known as the diffusion equation), namely

$$\partial c/\partial t = D\ \partial^2 c\ /\ \partial x^2, \qquad \text{diffusion equation} \tag{2.53}$$

where we have assumed D does not depend on x. We've had to be slightly less cavalier with our calculus by writing the derivatives as partial derivatives like $\partial/\partial x$, in recognition that c depends on both x and t, so that a derivative with respect to x is taken with t fixed, and vice versa. We can now return to the equivalence of the diffusion coefficient in Eqs. (2.29) and (2.43).

The profiles in Fig. 2.12 schematically represent how the concentration $c(x,t)$ of a layer of dye evolves with time. Equation (2.53) provides us with the means of identifying the appropriate functional form for $c(x,t)$ from an initial configuration. One such solution for a solute initially concentrated at the coordinate origin is

$$c(x,t) = c_o\,(4\pi Dt)^{-1/2}\exp(-x^2/4Dt), \tag{2.54}$$

where c_o is a parameter. The proof is left as an exercise in the end-of-chapter problems. To characterize how the concentration profile changes with time, we evaluate the mean square displacement of the solute particles from the origin using Eq. (2.54). Applying the usual rules for the construction of ensemble averages, we find

$$\begin{aligned}\langle x^2\rangle &= \int x^2 c(x,t)dx \Big/ \int c(x,t)\,dx \\ &= \int_0^\infty x^2\exp(-x^2/4Dt)dx \Big/ \int_0^\infty \exp(-x^2/4Dt)dx.\end{aligned} \tag{2.55}$$

This equation contains definite integrals, whose values are given in the end-of-chapter problems. Substituting,

$$\langle x^2\rangle = (4Dt)\bullet(\sqrt{\pi}/4)/(\sqrt{\pi}/2) = 2Dt. \tag{2.56}$$

This is nothing more than Eq. (2.43), and it shows that we have been using the symbol for the diffusion coefficient correctly.

2.5 Fluctuations and correlations

In the simplest random walk (see Section 2.3), the direction of each step in the walk is completely uncorrelated with its neighbors, and this characteristic gives rise to a particularly simple form for the mean squared end-to-end displacement, $\langle \mathbf{r}_{ee}^2\rangle = Nb^2$, where each walk in the ensemble has N steps of identically the same length b. That neighboring steps are uncorrelated has the mathematical consequence that the mean value of the scalar product of neighboring bond vectors vanishes when the number of walks in the ensemble is large, $\langle \mathbf{b}_i\bullet\mathbf{b}_{i\pm1}\rangle \to 0$, where the average is performed at the same step i on all walks in the ensemble. For the same reason, it is also true that when steps i and j are far from each other on a given walk, $\langle \mathbf{b}_i\bullet\mathbf{b}_j\rangle \to 0$.

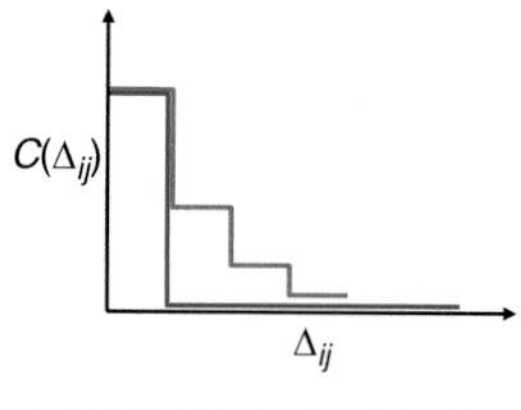

Fig. 2.14

Correlation function $C(\Delta_{ij})$ for the scalar product of bond vectors $\mathbf{b}_i \bullet \mathbf{b}_j$ separated by Δ_{ij} steps on a random walk: the green curve is a schematic representation of a restricted walk while the red curve is for an unrestricted walk.

Suppose now that the directions of neighboring steps are correlated; for instance, suppose that neighboring bond vectors cannot change direction by more than 90°. In terms of scalar products, this implies $0 \leq \mathbf{b}_i \bullet \mathbf{b}_{i\pm1} \leq b^2$, still assuming that all steps have the same length. For bonds pointing in three dimensions, it is straightforward to show that $\langle \mathbf{b}_i \bullet \mathbf{b}_{i\pm1} \rangle = b^2/2$, which is left as an exercise for the interested reader. Going to second-nearest neighbors gives a non-vanishing result for $\langle \mathbf{b}_i \bullet \mathbf{b}_{i\pm2} \rangle$, which is smaller than $b^2/2$, but still larger than zero applicable when the steps are uncorrelated. As the distance along the contour of the walk increases, the mean value $\langle \mathbf{b}_i \bullet \mathbf{b}_j \rangle$ gradually vanishes towards the same limit as walks with uncorrelated steps, as shown schematically in Fig. 2.14. This means that $\langle \mathbf{r}_{ee}{}^2 \rangle$ of the walk with correlated steps still is still proportional to N for long walks, $\langle \mathbf{r}_{ee}{}^2 \rangle \propto N$, but the proportionality constant for correlated walks is no longer b^2; for the particular correlated walk that we have just described, $\langle \mathbf{r}_{ee}{}^2 \rangle > Nb^2$ and the effective "size" of the walk is larger than it is for its uncorrelated cousin.

In describing the behavior of random walks, the quantity $\langle \mathbf{b}_i \bullet \mathbf{b}_j \rangle$ may play the role of a correlation function, a quantity that portrays the magnitude of the correlations in a system as a function of an independent variable, which in the case of random walks or polymers could be the separation $\Delta_{ij} = |i-j|$ along the contour. To illustrate the general behavior of correlation functions, let's start by defining

$$C(\Delta_{ij}) \equiv \langle \mathbf{b}_i \bullet \mathbf{b}_j \rangle_{\Delta(ij)} \,/\, b^2. \tag{2.57}$$

Here, Δ_{ij} may be as small as zero, in which case $C(0) = 1$ because the numerator of the right-hand side of Eq. (2.57) is just the mean squared length of the steps on the walk, b^2 by construction. On the other hand, both types of walks we have discussed above (unrestricted step-to-step bond directions and partly restricted bond directions) display $\langle \mathbf{b}_i \bullet \mathbf{b}_j \rangle_{\Delta(ij)} \to 0$ at large separations. This behavior for $C(\Delta_{ij})$ is plotted schematically in Fig. 2.14: the red curve is the unrestricted walk where the only non-vanishing value of the correlation function is $C(0) = 1$, while the green curve is the restricted walk, where $C(0) = 1$, $C(\Delta_{ij} = 1) = 1/2$, etc. In the language of correlation functions, it is often conventional to define a quantity to be correlated when $C = 1$ and uncorrelated when C vanishes, and we see that this normalization applies to the orientations of the bond directions as expressed in $\langle \mathbf{b}_i \bullet \mathbf{b}_j \rangle$.

The decay of the correlation function provides a quantitative measure of the range of Δ_{ij} where strong correlations are present. In the unrestricted walk, $C(\Delta_{ij})$ falls to zero at $\Delta_{ij} = 1$, whereas in the restricted case, the fall-off is slower because correlations exist to larger values of Δ_{ij} as expected. We will establish in Chapter 3 that $C(x)$ for random walks should decay exponentially with x as

$$C(x) \propto \exp(-x/\xi), \tag{2.58}$$

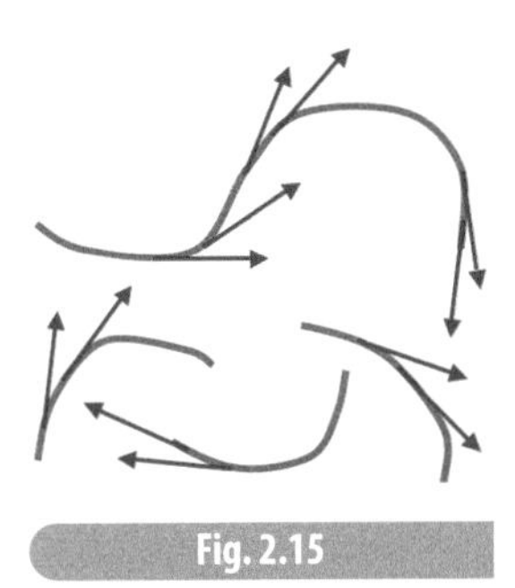

Fig. 2.15

Two approaches of obtaining a tangent-tangent correlation function at modest values of the separation Δs along the curve. In the top panel, a long curve is sampled at many different pairs of locations, each separated by the same Δs. In the bottom panel, a larger number of short paths are sampled to determine the correlation function with the same accuracy, but at only modest values of Δs.

where ξ is a characteristic of the system. For walks and polymers, ξ is the correlation length, while in other situations, ξ could be a correlation time or some other variable. The correlation length is also manifested in the physical "size" of the walk, in that $\langle r_{ee}^2 \rangle$ is linearly proportional to ξ: from Chapter 3, $\langle r_{ee}^2 \rangle = 2\xi L$, where $L = Nb$ is the contour length. For the unrestricted random walk, $\xi = b/2$.

In this text, much of the experimental interest in correlation lengths includes not just random walks and rotations, but also physical systems such as macromolecules and filamentous cells whose size and sinuous shape obey the same mathematical expressions as the random walk. In such cases, it is not always possible to obtain long polymers or filaments that would provide good statistical accuracy for the correlation function at separations large compared to the correlation length. That is, it may not be possible to follow $C(x)$ out to large enough x to see it fall below $1/e$ of its initial value, as in the top panel of Fig. 2.15. When this occurs, an alternate strategy is to measure $C(x)$ accurately at modest values of x by sampling a much larger number of configurations, as in the bottom panel of Fig. 2.15. For this strategy to work, however, the sample must be sufficiently large to yield very accurate values of the correlation function, because the range in x over which $C(x)$ is to be fitted by the exponential function is not large.

Correlation functions describing the orientation of objects, as expressed through tangents to curves or normal vectors to undulating surfaces, have widespread application and appear in many chapters of this text, starting in Chapter 3. However, correlations in time, rather than orientation, are also important, as we illustrate by considering the process of cell division. Consider a species that is capable of forming a robust filament starting from a single cell. After the first doubling time has elapsed, two daughter cells remain attached and form the beginning of a filament. Because the cleavage plane during division is very close to the center of the cell, the daughter cells have almost equal length. These two cells will then grow and divide to create a short filament, with four almost-identical cells. Over an interval of several doubling times, all cells in the filament have very similar lengths, although the average cell length increases continuously throughout the division cycle, as one would expect.

Not only are there are small variations in the length of the daughter cells, there are variations in the growth rates of the cells such that their division times gradually lose synchronization. Thus, as the filament lengthens, there will be regions of 2^4 or 2^5 cells all of which have grown in synchrony, adjoining other regions of where the cell lengths are slightly different. The greatest variation will occur when the cells are near their division point, where a string of very long cells that have grown at the same rate lie beside a string of short cells whose parent cell divided a little earlier than its neighbors. Some sample data are shown in Fig. 2.16 from the rod-shaped alga

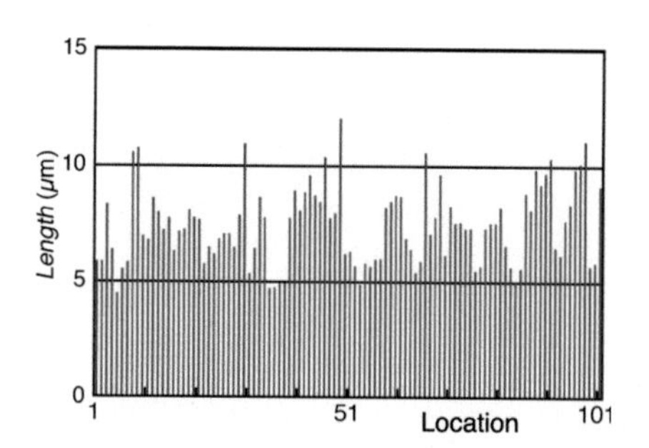

Fig. 2.16 Cell lengths (in μm) for 100 sequential cells as a function of position on a linear filament of the green alga *Stichococcus Y*.

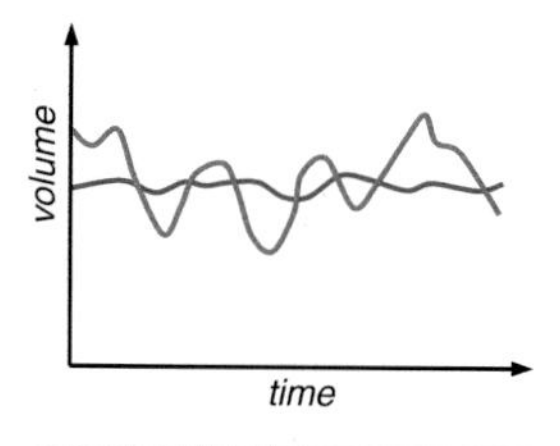

Fig. 2.17 Schematic representation of the fluctuations in volume at fixed temperature for a system with a large (red) and small (green) compression resistance. The fluctuations are recorded as a function of time.

Stichococcus; the plot shows the length of each cell as a function of its position in the linear filament. The cell lengths are locally correlated: domains of high correlation in length between neighboring cells have grown in synchrony from their common parent.

Before closing this chapter, we make a slightly more mathematical pass through a concept introduced in Chapter 1 when discussing the shape fluctuations of soft filaments in thermal equilibrium. Fluctuations arise in the energy of a system owing to the interaction with its environment at fixed temperature. As we described in Chapter 1, the stiffer a filament is, the smaller will be the variations in its shape arising from thermal fluctuations. This is illustrated in Fig. 2.17 for two systems with the same mean volume V_{av} but differing compression resistance. The red curve represents a stiff system with a large compression resistance: as time passes, the volume of the system stays fairly close to its mean value. In contrast, the green curve has low compression resistance, which reflects its ability to explore a larger range of configurations without a large change in its energy.

Put mathematically, the statistical variance of a quantity x about its mean value x_{av}, namely $\langle (x - x_{av})^2 \rangle$, is inversely proportional to a system's resistance to change of that quantity. That is, the variance of the volume of a system is inversely proportional to its compression modulus; similarly, the variance in the energy of a system at fixed temperature is inversely proportional to its heat capacity. Let's now make this exact. The compression modulus of a material is defined as

$$K_V^{-1} = -V^{-1} (\partial V / \partial P)_T, \tag{2.59}$$

where the subscript "T" means that the partial derivative is performed at constant temperature. This definition implies that the larger the volume change (ΔV) in response to a pressure change (ΔP), the *smaller* the compression modulus: a large value for K_V means that the system has a large resistance to change. The minus sign is needed on the right-hand side of the equation because the volume decreases as the pressure increases. Note that K_V has the units of pressure, or energy per unit volume.

Establishing the link between K_V and the variance of V takes a few more concepts from statistical mechanics than we have at our fingertips right now, so the proof is relegated to Appendix D. The relation is

$$K_V^{-1} = \langle (V - V_o)^2 \rangle / (k_B T V_o), \tag{2.60}$$

where V_o is the mean value of the volume at the fixed temperature and pressure of interest. As required, this equation provides K_V with units of energy (through $k_B T$) per unit volume and it demonstrates the inverse relation between K_V and the variance of the volume. Similar expressions can be obtained for other elastic moduli such as the shear modulus and Young's modulus.

Summary

The softness of the cell's mechanical components means that their shapes may fluctuate as the component exchanges energy with its thermal environment. This is particularly true at the molecular level, but may also be seen at larger length scales for biofilaments like actin or membranes like the lipid bilayer. In this chapter, we provide some of the mathematical formalism needed for the description of fluctuating systems, and apply this formalism to linear polymers and diffusion in a viscous medium.

The distribution of values of an observable A is described by the probability density $\mathcal{P}(A)$, which is obtained from the number of times dn_A that the observable is found in the range A to $A + dA$, in a sample of N trials; namely, $\mathcal{P}(A)\, dA = dn_A / N$. Having units of A^{-1}, $\mathcal{P}(A)$ is normalized to unity via $\int \mathcal{P}(A)\, dA = 1$, permitting the calculation of ensemble averages via expressions such as $\langle A^n \rangle \equiv \int A^n\, \mathcal{P}(A)\, dA$. Although the mean value of an observable $\langle A \rangle$ is useful for characterizing the properties of a system, it is often important to understand the behavior of the underlying distribution $\mathcal{P}(A)$ itself. As an example, the probability of a random walk in one dimension, consisting of i steps to the left and j steps to the right, is $P(i, j) = \{N! / i!\, j!\}\, (1/2)^i\, (1/2)^j$, where the probability of stepping in either direction is equal to 1/2 and the total number of steps is N. When N is large, the end-to-end distance of the walk satisfies $\langle r_{ee}^2 \rangle = Nb^2$, where each step in the walk has identically the same length b. The important feature of this result is that the mean value of r_{ee}^2 is proportional to N, rather than N^2; equivalently, $\langle r_{ee}^2 \rangle$ is proportional to the contour length $L = Nb$, rather than the square of the contour length as it would be for a straight line. As N becomes large, the probability distribution $P(i, j)$ looks increasingly like a smooth Gaussian distribution, with a probability density of the form $\mathcal{P}(x) = (2\pi\sigma^2)^{-1/2} \exp[-(x-\mu)^2 / 2\sigma^2]$, where the parameters are given by $\mu = \langle x \rangle$ and $\sigma^2 = \langle x^2 \rangle - \mu^2$.

By means of the fluctuation-dissipation theorem, probability distributions such as $\mathcal{P}(A)$ permit one to determine material characteristics such as elastic moduli or the specific heat. In thermal equilibrium, the energy and other characteristics of a system fluctuate about some mean value, say A_o. The variance of the fluctuations of this characteristic, $\langle (A - A_o)^2 \rangle$, is inversely proportional to the system's resistance to change of that quantity. For example, the compression modulus K_V at constant temperature is related to the change in volume V with pressure via $K_V^{-1} = -V^{-1}(\partial V / \partial P)_T$: the more rapidly the volume changes with pressure, the lower the compression modulus. In terms of fluctuations, the compression modulus is given by $K_V^{-1} = \langle (V - V_o)^2 \rangle \,/\, (k_B T V_o)$, where V_o is the equilibrium value of the volume at fixed T and P.

Probability distributions $\mathcal{P}(A)$ are functions of a single variable, A. Yet within a sequence of measurements, there may be correlations among the values of A as measured at different locations within the system or at the same location but at different times. An example is the local orientation of a unit tangent vector $\mathbf{t}(s)$ measured at location s along a wiggling biofilament. The mean value of $\mathbf{t}$ measured over all locations at the same time, or a specific location as a function of time, vanishes if the filament executes random motion: $\langle \mathbf{t} \bullet \mathbf{t}_o \rangle = 0$, where $\mathbf{t}_o$ is a fixed reference direction, although $\langle t^2 \rangle = 1$ because the vector has unit length. However, the directions of $\mathbf{t}$ at nearby locations s_i and s_j may be correlated such that $\langle \mathbf{t}_i \bullet \mathbf{t}_j \rangle \neq 0$ when averaged over all pairs of locations separated by the same $\Delta s = |s_i - s_j|$. If the local orientations of the filament are uncorrelated at large separations, then $\langle \mathbf{t}_i \bullet \mathbf{t}_j \rangle \rightarrow 0$ as $\Delta s \rightarrow \infty$. This suggests that a correlation length ξ characterizes the decay of the correlations with increasing distance. For many systems, the decay is exponential, and the correlation function $C(\Delta s) \equiv \langle \mathbf{t}_i \bullet \mathbf{t}_j \rangle_{\Delta s}$ can be parametrized as $C(\Delta s) = \exp(-\Delta s / \xi)$. As will be established later, the mean square end-to-end displacement of a random walk can be written as $\langle r_{ee}^{\ 2} \rangle = 2\xi L$, where L is the contour length of the walk as defined above.

The diffusion of the cell's molecular components, and perhaps the random motion of the cell itself in a fluid environment, are formally similar to random walks. In a viscous environment, the motion of an object is affected by the presence of drag forces that depend upon the instantaneous speed of the object: the higher the speed, the larger the force. For slowly moving objects, viscous drag exerts a force $F \propto -v$ for linear motion and a torque of $\mathcal{T} \propto -\omega$ for rotational motion, where ω is the angular speed and the minus sign indicates that drag opposes the motion of the object. Given that force is proportional to the rate of change of velocity according to Newton's Second Law of mechanics, the speed of an object initially moving a low speeds in a viscous environment obeys $\mathrm{d}v/\mathrm{d}t \propto -v$, which means that $v(t)$ must decay exponentially with time (similarly for torque and the rate of change of the angular velocity). The form of the drag force is known analytically for a few simple shapes like a sphere of radius R: $F = -6\pi\eta Rv$ (Stokes' Law) and $\mathcal{T} = -8\pi\eta R^3\omega$, where η is the viscosity of the medium. In fluid mechanics, a useful benchmark for assessing the importance of drag is provided by a dimensionless quantity called the Reynolds number $\mathbf{R} \equiv \rho v\lambda/\eta$, where ρ is the density of the fluid and λ is the length of the object along the direction of motion. When **R** is large, the motion is dominated by inertia, while at small values of **R**, the motion is dominated by drag; the transition between these two domains occurs for **R** around 10–100. A bacterium swimming in water has $\mathbf{R} < 10^{-4}$, meaning that its motion is overwhelmingly dominated by drag.

Strong as the drag forces on a cell may be, this does not mean that cells or their molecular components are motionless unless they have some means of generating movement. The energy provided through the interaction of the cell with its thermal environment causes the cell to move randomly, even if slowly. This is an example of diffusive motion, which is formally equivalent to a random walk in that the end-to-end displacement of the trajectory of a diffusing particle obeys $\langle r_{ee}^{\ 2} \rangle = 2Dt$, $4Dt$ or $6Dt$ in one, two or three dimensions, respectively, where D is called the diffusion coefficient; in these expressions, the elapsed time t takes the place of the number of steps N in a random walk. Einstein analyzed the diffusion of a sphere in a thermal environment, and was able to establish that $D = k_BT / 6\pi\eta R$, where η is the viscosity of the medium, as usual. Objects can also rotate diffusively, with their angular change obeying $\langle \theta^2 \rangle = 2D_rt$, where the rotational diffusion coefficient of a sphere has the form $D_r = k_BT / 8\pi\eta R^3$.

We have introduced diffusion in terms of the motion of a single object or particle, both its translational motion through its environment or its rotational motion about an axis. Another approach to diffusion involves quantities that represent averages over many objects, assumed to have high enough numerical density that the averages apply to a local region of the system. That is, we assume that the concentration of particles, $c(x,t)$ can be determined within a sufficiently small volume that $c(x,t)$ can be assigned a meaningful and accurate value for each location x and time t. For our purposes, we take the concentration to vary only along the x-axis of the system. Changes in the concentration with time usually result from a flux $j(x,t)$, which is the net number of particles per unit area per unit time crossing an imaginary plane in the yz directions at a location x. Fick's first law relates the flux to the gradient of the concentration through $j = -D\, \mathrm{d}c/\mathrm{d}x$. where D is the diffusion coefficient introduced previously. As particles diffuse through the system, the concentration at a specific location changes with time according to the equation of continuity $\mathrm{d}c/\mathrm{d}t = -\mathrm{d}j/\mathrm{d}x$. These two equations for flux can be combined to yield the diffusion equation $\partial c/\partial t = D\, \partial^2 c/\partial x^2$, where partial derivatives are required because $c(x,t)$ depends on both x and t.

Having made a first pass through the mathematical machinery for describing fluctuating systems, the next two chapters deal with the properties of flexible polymers in greater detail. If much of the material in the current chapter is new to the reader, it is probably advisable to solve a selection of the problems in the next section to build a working knowledge of applying the formalism to physical systems.

Problems

Applications

2.1. In a series of experiments, a parameter $\mathcal{A}$ is found to have values between 0 and 1 ($\mathcal{A}$ is measured in fictitious units we will call dils) according to the following distribution.

Range of $\mathcal{A}$ (dils)	Number
0.0 – 0.2	15
0.2 – 0.4	65
0.4 – 0.6	55
0.6 – 0.8	30
0.8 – 1.0	10

Determine the probability density $\mathcal{P}(\mathcal{A})$ for the distribution (including units), and evaluate $\langle \mathcal{A} \rangle$ and $\langle \mathcal{A}^2 \rangle$.

2.2. Calculate the mass of a plastic bead with a diameter of 1 μm and a density of 1.0×10^3 kg/m^3. Find its root mean square speed at room temperature if its mean kinetic energy is equal to $3k_BT/2$.

2.3. The magnitude of the viscous drag force exerted by a stationary fluid on a spherical object of radius R is $F = 6\pi\eta Rv$ at low speeds and $F = (\rho/2)AC_Dv^2$ at high speeds. Apply this to a spherical cell 1 μm in radius, moving in water with $\eta = 10^{-3}$ kg/m • s and $\rho = 10^3$ kg/m^3. Take the cell to have the same density as water, and let its drag coefficient C_D be 0.5.

(a) Plot the two forms of the drag force as a function of cell speed up to 100 μm/s.
(b) Find the speed at which the linear and quadratic drag terms are the same.

2.4. What is the drag force that a molecular motor must overcome to transport a vesicle in a cell? Assume that the vesicle has a radius of

100 nm and travels at 0.5 μm/s. Take the viscosity of the cytoplasm to be one hundred times that of water. What power must the motor generate as it transports the vesicle?

2.5. Some bacteria have the approximate shapes of spherocylinders – a uniform cylinder (length L, radius R) which is capped at each end by hemispheres. Take *E. coli* to have such a shape, with a diameter of 1 μm and an overall length of 4 μm. Find the drag force experienced by a cell of this shape if the drag force is in the turbulent regime where $F_{drag} \sim v^2$. Take the density of the fluid medium to be 10^3 kg/m^3, the drag coefficient to be 0.5 and the cell to be traveling at 20 μm/s.

(a) Evaluate the force for two different orientations of the cell – motion along its symmetry axis and motion transverse to its axis. Quote your answer in pN.
(b) Show that the ratio of the drag forces in these orientations (transverse : longitudinal) is equal to $1 + 12/\pi$.

2.6. Calculate the power required to maintain a spherical bacterium (diameter of 1 μm) rotating at a frequency of 10 Hz when it is immersed in a fluid of viscosity 10^{-3} kg/m•s. If the energy released per ATP hydrolysis is 8×10^{-20} J, how many ATP molecules must be hydrolyzed per second to support this motion?

2.7. The Reynolds number for the motion of cells in water is in the 10^{-5} range. To put this number into everyday context, consider a person swimming in a fluid of viscosity η. Making reasonable assumptions for the length and speed of the swimmer, what value of η corresponds to a Reynolds number of 10^{-5}? Compare your result with the fluids in Table 2.1.

2.8. The sources of household dust include dead skin cells, which we will model as cubes 5 μm to the side having a density of 10^3 kg/m^3.

(a) What is the root mean square speed of this hypothetical dust particle at $T = 20$ °C due to thermal motion alone?
(b) To what height h above the ground could this particle rise at room temperature before its thermal energy is lost to gravitational potential energy? Assume that the gravitational acceleration g is 10 m/s^2.

2.9 (a) A particular insect flies at a constant speed of 1 m/s, and randomly changes direction every 3 s. How long would it take for the end-to-end displacement of its random motion to equal 10 m on average?
(b) Suppose the insect emitted a scent that was detectable by another of its species even at very low concentrations. The molecule of the scent travels at 300 m/s, but changes direction through collisions

every 10^{-11} s. Once the molecule has been released by the insect, how long would it take for the end-to-end displacement of its trajectory to equal 10 m on average?

(c) From your results in parts (a) and (b), what is the better strategy for an insect looking for a mate, which it can detect through an emitted pheromone: (i) be motionless and wait for a scent, or (ii) actively search for a mate by flying?

2.10. An experiment was described in Section 2.1 in which the two-dimensional trajectories of plastic beads in water were found to obey $\langle r_{ee}^2 \rangle = (1.1 \pm 0.3\ \mu m^2/s)t$. What diffusion coefficient D describes these data? Using the Einstein relation, calculate D expected for spheres of diameter 1 μm moving in water, and compare with the measured value.

2.11. A spherical bacterium with a radius of 1 μm moves freely in water at 20 °C.

(a) What is its rotational diffusion coefficient?

(b) What is the root mean square change in angle around a rotational axis over an interval of 1 minute arising from thermal motion?

Be sure to quote your units for parts (a) and (b).

2.12. The plasma membrane plays a pivotal role in maintaining and controlling the cell's contents, as can be seen in the following simplified example. Suppose that the number density of a particular small molecule, which we'll call molly, is 10^{25} m^{-3} higher in the cytoplasm than the medium surrounding the cell. Let molly have a diffusion coefficient in the cytoplasm of 5×10^{-10} m^2/s.

(a) If the drop in concentration of molly from the inside to the outside of the cell occurred over a distance of 5 nm, the thickness of the lipid bilayer, with what flux would molly pass out of the cell?

(b) If the cells in question are spherical with a diameter of 4 μm, how many copies of molly are there per cell?

(c) If molly continued to diffuse out of the cell at the rate found in (a), ignoring the drop in the internal concentration of molly with time, how long would it take for the concentration of molly to be the same on both sides of the membrane?

2.13. Consider three different power-law forms of the drag force with magnitudes:

$$F_{1/2} = a\, v^{1/2} \quad \text{(square root)}$$

$$F_1 = b\, v^1 \quad \text{(linear)}$$

$$F_{3/2} = c\, v^{3/2} \quad \text{(3/2 power).}$$

Traveling horizontally from an initial speed v_o, an object experiencing one of these drag forces would come to rest at

$$x_{max} = (2m/3a)v_o^{3/2} \quad \text{(square root)}$$

$$x_{max} = mv_o / b \quad \text{(linear)}$$

$$x_{max} = 2m\, v_o^{1/2}/c \quad \text{(3/2 power).}$$

(a) Determine the coefficients a, b and c (quote your units) for a cell of mass 1×10^{-14} kg whose drag force is measured to be 5 pN when traveling at 10 μm/s.

(b) Find the maximum displacement that the cell could reach for each force if $v_o = 1$ μm/s.

2.14. In a particularly mountainous, and imaginary, region of the world, the only way to get from A to B is by one of several meandering roads. A group of unsuspecting tourists set out from town A one morning with enough gas in each of their cars to travel 100 km, which is 10 km more than the displacement from A to B. They take different routes, but each car runs out of gas before the reaching the destination, at displacements of 60, 64, 75 and 83 km from town A as the proverbial crow flies. Treating them as an ensemble, what is the persistence length of the roads?

2.15. The persistence length of DNA is measured to be 53 nm.

(a) Over what distance along the contour has its tangent correlation function dropped to 1/10?

(b) Using the result from Chapter 3 that $\langle r_{ee}^2 \rangle = 2\xi_p L$, find the root mean square end-to-end length of a strand of DNA from *E. coli* with a contour length of 1.6 mm. Quote your answer in microns.

Formal development and extensions

Some of the following problems require definite integrals for their solution.

$$\int_0^\infty \exp(-x^2)\,dx = \sqrt{\pi}/2 \qquad \int_0^\infty x^2 \exp(-x^2)\,dx = \sqrt{\pi}/4$$

$$\int_0^\infty x^3 \exp(-x^2)\,dx = 1/2 \qquad \int_0^\infty x^4 \exp(-x^2)\,dx = 3\sqrt{\pi}/8$$

2.16. The probability density for a particular distribution has the form $\mathcal{P}(\Lambda) = A\Lambda^n$, where n is a parameter and Λ has a range $0 \leq \Lambda \leq 1$.

Determine the normalization constant A for this power-law function and calculate $\langle A \rangle$ and $\langle A^2 \rangle$.

2.17. At a temperature T, the distribution of speeds v for an ideal gas of particles has the form $\mathcal{P}(v) = Av^2 \exp(-mv^2/2k_BT)$, where m is the mass of each particle.

(a) Determine A such that $\int_o^\infty \mathcal{P}(v)\, dv = 1$.
(b) Show that the mean kinetic energy is $\langle K \rangle = (3/2)k_BT$.
(c) Show that the most probable kinetic energy of the particles is k_BT.

2.18. (a) In Section 2.2, we stated that the position $x(t)$ of an object of mass m, subject to the drag force $F = c_1 v$, is described by

$$x(t) = (mv_o/c_1) \bullet [1 - \exp(-c_1 t/m)].$$

Differentiate this to establish that the corresponding velocity is

$$v(t) = v_o \exp(-t/t_{visc}),$$

with a characteristic time $t_{visc} = m/c_1$.

(b) The quadratic drag force is parametrized by $F = c_2 v^2$, resulting in

$$x(t) = (v_o/k) \bullet \ln(1 + kt),$$

where

$$k = c_2 v_o/m.$$

Establish that the corresponding velocity is

$$v(t) = v_o / (1 + c_2 v_o t/m).$$

Does x reach a limiting value for quadratic drag?

2.19. Calculate the ratio of the drag force at low speeds for an ellipsoid compared to a sphere; the direction of motion is along the semi-major axis of the ellipsoid. Take the semi-minor axis of the ellipsoid to be R, and plot your results as a function of $\alpha \equiv a/R$ for $2 \le \alpha \le 5$. Comment on how fast the drag force rises with the length of the ellipsoid.

2.20. Consider a power-law form of the drag force with magnitude:

$$F_{1/2} = a\, v^{1/2} \quad \text{(square root)}.$$

Establish that an object traveling horizontally from an initial speed v_o, would come to rest at $x_{max} = (2m/3a)v_o^{3/2}$ for such a force.

2.21. Consider a power-law form of the drag force with magnitude:

$$F_{3/2} = c\, v^{3/2} \quad \text{(3/2 power)}.$$

Establish that an object traveling horizontally from an initial speed v_o, would come to rest at $x_{max} = 2m\, v_o^{1/2}/c$ for this force.

2.22. Consider a rotational drag torque of the form $\mathcal{T} = -\chi\omega$, where $\chi = 8\pi\eta R^3$.

(a) Find $\omega(t)$ and $\theta(t)$ for an object rotating at an initial angular frequency ω_o, experiencing no force other than drag. Take the initial angle to be zero.

(b) Show that the characteristic time scale for the decay of the angular speed is I/χ, where I is the moment of inertia about the axis of rotation. Show that this time scale is $m / (20\pi\eta R)$ for a sphere of mass m and radius R.

(c) Find the value of θ as $t \to \infty$ if the initial frequency is 10 revolutions per second; assume $R = 1$ μm, $\eta = 10^{-3}$ kg/m•s, and the mass of the object is 4×10^{-15} kg.

2.23. It is established in Problem 2.22 that the time scale for the decay of rotational speed of a sphere subject to rotational drag is $m/(20\pi\eta R)$ where m and R are the mass and radius of the sphere.

(a) For a cell with mass 4×10^{-15} kg and radius 1 μm, calculate the time scales for the decay of translational and rotational motion when the cell is immersed in a fluid of viscosity 10^{-3} kg/m•s.

(b) Show analytically that the ratio of the times (rotational: translational) is 3/10.

2.24. Consider a one-dimensional random walk centered on $x = 0$, where the probabilities of stepping to the left or to the right are both equal to 1/2. Find the probability $P(i, j)$ of the discrete walk at $x = 0$ for walks with $N = 4$, 8 and 12 steps. Compare this result with the continuous probability density $\mathcal{P}(x)\Delta x$. [*Hint: you must determine what value of Δx corresponds to the change in end-to-end length of the walk when i and j change by one unit each at fixed N.*] For simplicity, make the step length b equal unity.

2.25. By explicit substitution, show that the following expression for the concentration satisfies Fick's second law of diffusion:

$$c(x,t) = c_o\,(4\pi Dt)^{-1/2} \exp(-x^2/4Dt).$$

2.26. In three dimensions, the concentration of a mobile species spreading from a point at the coordinate origin is given by

$$c(\mathbf{r},t) = c_o\,(4\pi Dt)^{-3/2} \exp(-r^2/4Dt).$$

For this profile, calculate the time dependence of $\langle r^2 \rangle$, just as we calculated $\langle x^2 \rangle$ in Section 2.4. Comment on the factor of 6 in your result. As a three-dimensional average, the volume element in $\langle \cdots \rangle$ in polar coordinates is $d\varphi \sin\theta \, d\theta \, r^2 dr$, where $0 \leq \theta \leq \pi$ and $0 \leq \varphi \leq 2\pi$ as usual.

2.27. Consider a restricted random walk in which neighboring bond vectors, each with length b, have directions that are different by 90° at the most.

(a) Show that $\langle \mathbf{b}_i \bullet \mathbf{b}_{i\pm1} \rangle / b^2 = 1/2$ for walks in three dimensions.
(b) Find $\langle \mathbf{b}_i \bullet \mathbf{b}_{i\pm1} \rangle / b^2$ for walks in two dimensions. Explain in words why $\langle \mathbf{b}_i \bullet \mathbf{b}_{i\pm1} \rangle$ in two dimensions should be larger or smaller than in three dimensions.
(c) What is the minimum value of $\mathbf{b}_i \bullet \mathbf{b}_{i\pm2} / b^2$?

2.28. A variable x fluctuates around its mean value x_{av}. For an ensemble of measurements, show that $\langle \Delta x^2 \rangle = \langle x^2 \rangle - x_{av}^2$, where $\Delta x = x - x_{av}$.

2.29. The exponential is not the only function that decays smoothly to zero with time: there are other algebraic forms that might appear as correlation functions. Suppose that the correlation function $C(x)$ has the form $1/(1 + x/a)$. Equate the derivative of $C(x)$ with that of the exponential $\exp(-x/\xi)$ at small values of x in order to relate the parameter a to the persistence length ξ. With this identification, calculate the ratio of the two functions at $x = \xi$.

2.30. In the NPT ensemble, the temperature T, pressure P and number of particles N are all fixed, but the volume V fluctuates about a mean value determined by the choice of NPT. Consider an ideal gas obeying $PV = Nk_BT$ (where V is really the mean value of V).

(a) Show that the compression modulus K_V is equal to the pressure.
(b) Find how the volume fluctuations $\Delta V^2 / V_{av}^2$ depends on N, where $\Delta V = V - V_{av}$.

PART I

RODS AND ROPES

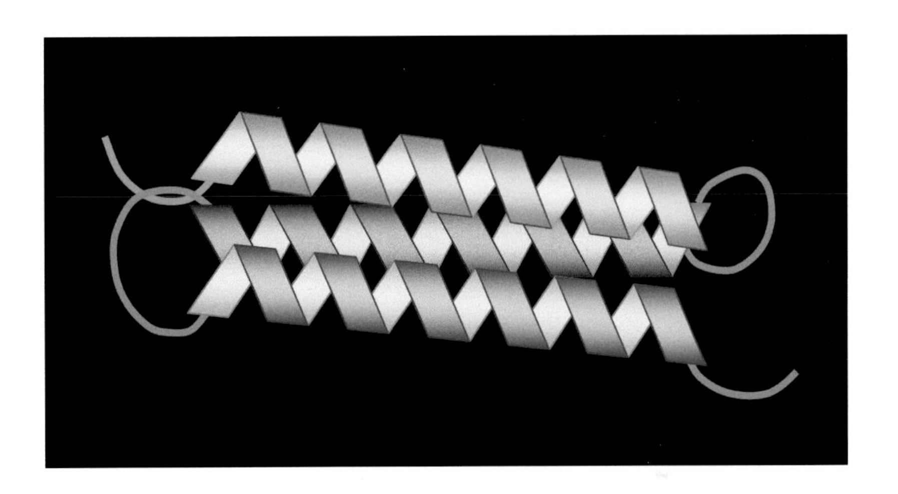

3 Polymers

The structural elements of the cell can be broadly classified as filaments or sheets, where by the term filament, we mean a string-like object whose length is much greater than its width. Some filaments, such as DNA, function as mechanically independent units, but most structural filaments in the cell are linked to form two- or three-dimensional networks. As seen on the cellular length scale of a micron, individual filaments may be relatively straight or highly convoluted, reflecting, in part, their resistance to bending. Part I of this book concentrates on the mechanical properties of biofilaments: Chapter 3 covers the bending and stretching of simple filaments while Chapter 4 explores the structure and torsion resistance of complex filaments. The two chapters making up the remainder of Part I consider how filaments are knitted together to form networks, perhaps closely associated with a membrane as a two-dimensional web (Chapter 5) or perhaps extending though the three-dimensional volume of the cell (Chapter 6).

3.1 Polymers and simple biofilaments

At the molecular level, the cell's ropes and rods are composed of linear polymers, individual monomeric units that are linked together as an unbranched chain. The monomers need not be identical, and may themselves be constructed of more elementary chemical units. For example, the monomeric unit of DNA and RNA is a troika of phosphate, sugar and organic base, with the phosphate and sugar units alternating along the backbone of the polymer (see Chapter 4 and Appendix B). However, the monomers are not completely identical because the base may vary from one monomer to the next. The double helix of DNA contains two such sugar–base–phosphate strands, with a length along the helix of 0.34 nm per pair of organic bases, and a corresponding molecular mass per unit length of about 1.9 kDa/nm (Saenger, 1984).

The principal components of the cytoskeleton – actin, intermediate filaments and microtubules – are themselves composite structures made from protein subunits, each of which is a linear chain hundreds of amino acids long. In addition, some cells contain fine strings of the protein spectrin,

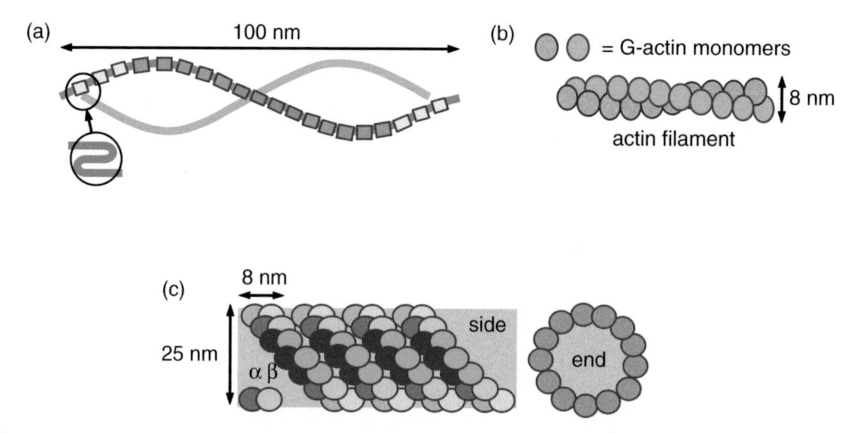

Fig. 3.1 (a) Two spectrin chains intertwine in a filament, where the boxes represent regions in which the protein chain has folded back on itself (as in the inset). The two strings are stretched and separated for clarity. (b) Monomers of G-actin associate to form a filament of F-actin, which superficially appears like two intertwined strands. (c) Microtubules usually contain 13 protofilaments, whose elementary unit is an 8 nm long dimer of the proteins α and β tubulin. Both side and end views of the cylinder are shown.

whose structure we discuss first before considering the thicker filaments of the cytoskeleton. As organized in the human erythrocyte, two pairs of chains, each pair containing two intertwined and inequivalent strings of spectrin (called α and β), are joined end-to-end to form a filament about 200 nm in contour length. The α and β chains have molecular masses of 230 and 220 kDa, respectively, giving a mass per unit length along the tetramer of 4.5 kDa/nm. An individual chain folds back on itself repeatedly like a Z, so that each monomer is a series of 19 or 20 relatively rigid barrels 106 amino acid residues long, as illustrated in Fig. 3.1(a).

Forming somewhat thicker filaments than spectrin, the protein actin is present in many different cell types and plays a variety of roles in the cytoskeleton. The elementary actin building block is the protein G-actin ("G" for globular), a single chain of approximately 375 amino acids having a molecular mass of 42 kDa. G-actin units can assemble into a long string called F-actin ("F" for filamentous), which, as illustrated in Fig. 3.1(b), has the superficial appearance of two strands forming a coil, although the strands are not, in fact, independently stable. The filament has a width of about 8 nm and a mass per unit length of 16 kDa/nm. Typical actin monomer concentrations in the cell are 1–5 mg/ml; as a benchmark, a concentration of 1 mg/ml is 24 μM for a molecular mass of 42 kDa.

The thickest individual filaments are composed of the protein tubulin, present as a heterodimer of α-tubulin and β-tubulin, each with a molecular mass of about 50 kDa. Pairs of α- and β-tubulin form a unit 8 nm in length, and these units can assemble α to β successively into a hollow microtubule consisting of 13 linear protofilaments (in almost all cells), as shown in Fig. 3.1(c). The overall molecular mass per unit length of a microtubule is about 160 kDa/nm, ten times that of actin. Tubulin is present at concentrations of a few milligrams per milliliter in a common cell; given a molecular mass of 100 kDa for a tubulin dimer, a concentration of 1 mg/ml corresponds to 10 μM.

Intermediate filaments lie in diameter between microtubules and F-actin. As will be described further in Chapter 4, intermediate filaments are composed of individual strands with helical shapes that are bundled together to form a composite structure with 32 strands. Depending on the type, an intermediate filament has a roughly cylindrical shape about 10 nm in diameter and a mass per unit length of about 35 kDa/nm, with some variation. Desmin and vimentin are somewhat higher at 40–60 kDa/nm (Herrmann *et al.*, 1999); neurofilaments are also observed to lie in the 50 kDa/nm range (Heins *et al.*, 1993). Further examples of composite filaments are collagen and cellulose, both of which form strong tension-bearing fibers with much larger diameters than microtubules. In the case of collagen, the primary structural element is tropocollagen, a triple helix (of linear polymers) which is about 300 nm long, 1.5 nm in diameter with a mass per unit length of about 1000 Da/nm. In turn, threads of tropocollagen form collagen fibrils, and these fibrils assemble in parallel to form collagen fibers.

The design of cellular filaments has been presented in some detail in order to illustrate both their similarities and differences. Most of the filaments possess a hierarchical organization of threads wound into strings, which then may be wound into ropes. The filaments within a cell are, to an order of magnitude, about 10 nm across, which is less than 1% of the diameters of the cells themselves. As one might expect, the visual appearance of the cytoskeletal filaments on cellular length scales varies with their thickness. The thickest filaments, microtubules, are stiff on the length scale of a micron, such that isolated filaments are only gently curved. In contrast, intertwined strings of spectrin are relatively flexible: at ambient temperatures, a 200 nm filament of spectrin adopts such convoluted shapes that the distance between its end-points is only 75 nm on average (for spectrin filaments that are part of a network).

The biological rods and ropes of a cell may undergo a variety of deformations, depending upon the nature of the applied forces and the mechanical properties of the filament. Analogous to the tension and compression experienced by the rigging and masts of a sailing ship, some forces lie along the length of the filament, causing it to stretch, shorten or perhaps buckle. In other cases, the forces are transverse to the filament, causing it to bend or twist. Whatever the deformation mode, energy may be required to distort the filament from its "natural" shape, by which we mean its shape at zero temperature and zero stress. Consider, for example, a uniform straight rod of length L bent into an arc of a circle of

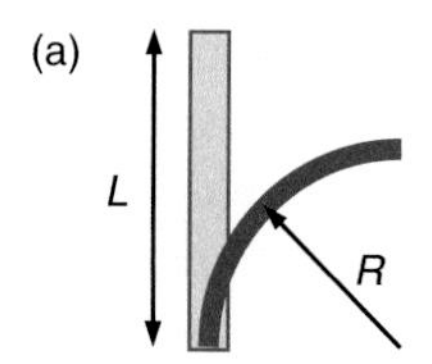

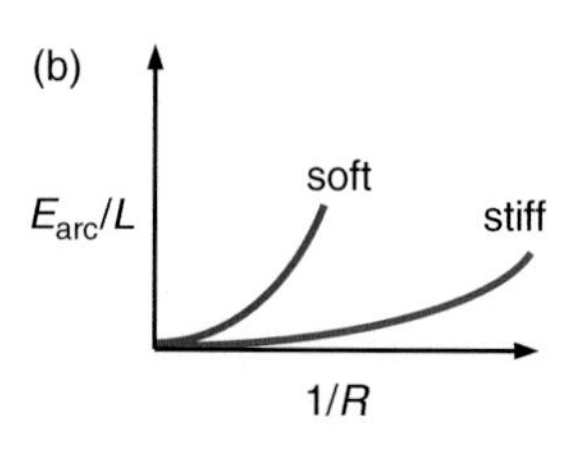

Fig. 3.2

Bending a rod of length L into the shape of an arc of radius R (in part (a)) requires an input of energy E_{arc} whose magnitude depends upon the severity of the deformation and the stiffness of the rod (b).

radius R, as illustrated in Fig. 3.2(a). Within a simple model introduced in Section 3.2 for the bending of rods, the energy E_{arc} required to perform this deformation is given by

$$E_{arc} = \kappa_f L / 2R^2, \tag{3.1}$$

where κ_f is called the flexural rigidity of the rod: large κ_f corresponds to stiff rods. Figure 3.2(b) displays how the bending energy behaves according to Eq. (3.1): a straight rod has $R = \infty$ (an infinite radius of curvature means that the rod is straight) and hence $E_{arc} = 0$, while a strongly curved rod might have L/R near unity, and hence $E_{arc} >> 0$, depending on the magnitude of κ_f.

The flexural rigidity of a uniform rod can be written as the product of its Young's modulus Y and the moment of inertia of its cross section $\mathcal{I}$,

$$\kappa_f = \mathcal{I}Y, \tag{3.2}$$

where Y and $\mathcal{I}$ reflect the composition and geometry of the rod, respectively. Stiff materials, such as steel, have $Y \sim 2 \times 10^{11}$ J/m^3, while softer materials, such as plastics, have $Y \sim 10^9$ J/m^3. The moment of inertia of the cross section, $\mathcal{I}$ (not to be confused with the moment of inertia of the mass, familiar from rotational motion), depends upon the shape of the rod; for instance, a cylindrical rod of constant density has $\mathcal{I} = \pi R^4/4$. Owing to its power-law dependence on filament radius, the flexural rigidity of filaments in the cell spans nearly five orders of magnitude.

We know from Chapters 1 and 2 that the energy of an object in thermal equilibrium fluctuates with time with an energy scale set by $k_B T$, such that an otherwise straight rod bends as it exchanges energy with its environment (see Fig. 3.3). The fluctuations in the local orientation of a sinuous filament can be characterized by the persistence length ξ_p that appears in the tangent correlation function introduced in Section 2.5: the larger the persistence length, the straighter a section of a rod will appear at a fixed

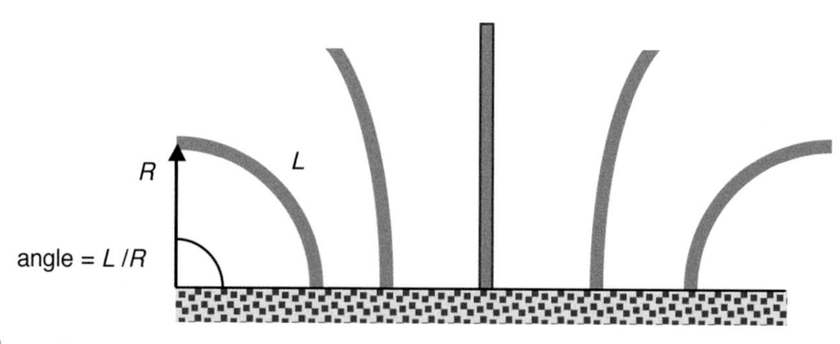

Fig. 3.3

Sample configurations of a very flexible rod at non-zero temperature as it exchanges energy with its surroundings. The base of the filament in the diagram is fixed. The configuration at the far left is an arc of a circle subtending an angle of L/R radians.

viewing distance. Intuitively, we expect that ξ_p should be directly proportional to the flexural rigidity κ_f (stiffer filaments are straighter) and inversely proportional to temperature $k_B T$ (colder filaments are straighter). A mechanical analysis shows that the combination $\kappa_f / k_B T$ is in fact the persistence length ξ_p of the filament

$$\xi_p = \kappa_f / k_B T = Y\mathcal{I} / k_B T, \tag{3.3}$$

as established in the treatment of shape fluctuations in Section 3.3.

If its persistence length is large compared to its contour length, i.e. $\xi_p >> L$, a filament appears relatively straight and recognizable as a rod. However, if $\xi_p << L$, the filament adopts more convoluted shapes, such as that in Fig. 3.4(a). What is the likelihood that a particular filament will be observed in one of its contorted shapes, as opposed to a rod-like one? Using the end-to-end displacement r_{ee} as a measure, there are many contorted shapes with r_{ee} close to zero, but very few extended ones with r_{ee} ~ L, as illustrated in Fig. 1.13 and Fig. 2.8. If there is little or no energy difference between these shapes compared to $k_B T$, then any specific configuration is as likely as any other and the filament will adopt a convoluted shape more frequently than a straight one. We can also view this conclusion in terms of entropy, which is proportional to the logarithm of the number of configurations available to a system (see Appendix C). The large family of shapes with $r_{ee}/L \approx 0$ contributes significantly to the system's entropy, while $r_{ee}/L \approx 1$ contributes much less.

Now consider what happens as we stretch a flexible filament by pulling on its ends, as in Fig. 3.4(b). Stretching the filament reduces the number of configurations available to it, thus lowering its entropy; thermodynamics tells us that this is not a desirable situation – systems do not spontaneously lower their entropy, all other things being equal. Because of this, a force must be applied to the ends of the filament to pull it straight and the filament is elastic by virtue of its entropy, as explained in Section 1.3.

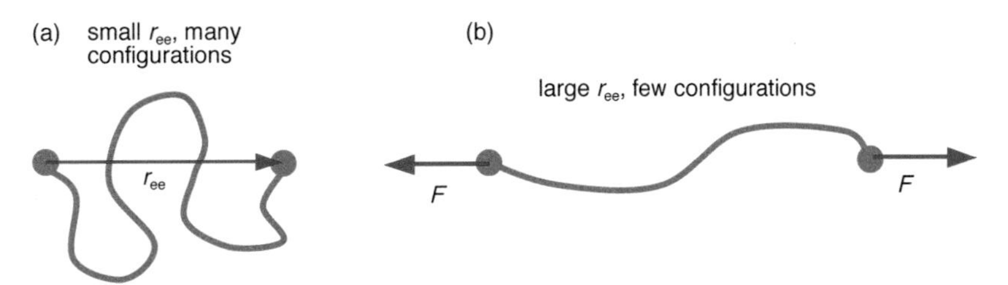

Fig. 3.4 Two samples from the set of configurations available to a highly flexible filament. The end-to-end displacement vector r_{ee} is indicated by the arrow in part (a). The number of configurations available at a given end-to-end distance is reduced as a force F is applied to the ends of the filament in (a) to stretch it out like that in (b).

For small extensions, this force is proportional to the change in r_{ee} from its equilibrium value, just like Hooke's Law for springs. In fact, the elastic behavior of convoluted filaments can be represented by an effective spring constant k_{sp} given by

$$k_{sp} = 3k_B T / 2L\xi_p, \tag{3.4}$$

which is valid for a filament in three dimensions near equilibrium (see Section 3.4).

We have emphasized the role of entropy in the structure and elastic properties of the cell's mechanical components simply because soft materials are so common. However, deformations of the relatively stiff components of the cell are dominated by energetic considerations familiar from continuum mechanics. Under a tensile force F, rods of length L, cross-sectional area A and uniform composition, stretch according to Hooke's Law $F = k_{sp}\,\Delta L$, where the spring constant k_{sp} is given by $k_{sp} = YA / L$. Under a compressive force, a rod first compresses according to Hooke's Law, but then buckles once a particular threshold F_{buckle} has been reached: $F_{buckle} = \pi^2 \kappa_f / L^2$. Microtubules may exhibit buckling during the cell division process of eukaryotic cells.

Thus, we see that filaments exhibit elastic behavior with differing microscopic origins. At low temperatures, a filament may resist stretching and bending for purely energetic reasons associated with displacing atoms from their most energetically favored positions. On the other hand, at high temperatures, the shape of a very flexible filament may fluctuate strongly, and entropy discourages such filaments from straightening out. In Sections 3.3 and 3.4, we investigate these situations using the formalism of statistical mechanics, but not until the bending of rods is expressed mathematically in Section 3.2. The buckling of rods under a compressive load is studied in Section 3.5, following which our formal results are applied to the analysis of biological polymers in Section 3.6. More details on the structure of filaments in the cytoskeleton can be found in Chapters 7–9 of Howard (2001).

3.2 Mathematical description of flexible rods

The various polymers and filaments in the cell display bending resistances whose numerical values span six orders of magnitude, from highly flexible alkanes through somewhat stiffer protein polymers such as F-actin, to moderately rigid microtubules. Viewed on micron length scales, these filaments may appear to be erratic, rambunctious chains or gently curved rods, and their elastic properties may be dominated by entropic or energetic effects. In selecting a framework for interpreting the characteristics of cellular filaments, one can choose among several simple pictures of linear polymers, each picture emphasizing different aspects of the polymer. In this section, we view the filament as a smoothly curving rod in contrast to the wiggly segmented chain represented by a random walk (introduced in Section 2.3). These two pictures of linear polymers overlap, of course, and there are links between their parametrizations.

3.2.1 Arc length and curvature

Our primary interest at cellular length scales are filaments whose local orientation changes smoothly along their length. For the moment, the cross-sectional shape and material composition of the filament will be ignored so that it can be described as a continuous curve with no kinks or discontinuities. As displayed in Fig. 3.5(a), each point on the curve corresponds to a position vector **r**, represented by the familiar Cartesian triplet (x, y, z). In Newtonian mechanics, we're already familiar with the idea of describing the trajectory of a projectile by writing its coordinates as a function of time, $\mathbf{r}(t)$, where t appears as a parameter. For the curve that represents a filament, we do something similar except **r** is written as a function of the arc length s [$\mathbf{r}(s)$ or the triplet $x(s)$, $y(s)$, $z(s)$], where s follows along the contour of the curve, running from 0 at one end to the full contour length L_c at the other. To illustrate how this works, consider a circle of radius R lying in the xy plane; the x and y coordinates of the circle are related to each other through the familiar equation $x^2 + y^2 = R^2$. In a parametric approach where the arc length s is used as a parameter, the coordinates are written as $x(s) = R\cos(s/R)$ and $y(s) = R\sin(s/R)$, where s is zero at $(x,y) = (R,0)$ and increases in a counter-clockwise fashion along the perimeter of the circle.

The function $\mathbf{r}(s)$ contains all the information needed to describe a sinuous curve, so that $\mathbf{r}(s)$ can be used to generate other characteristics of the

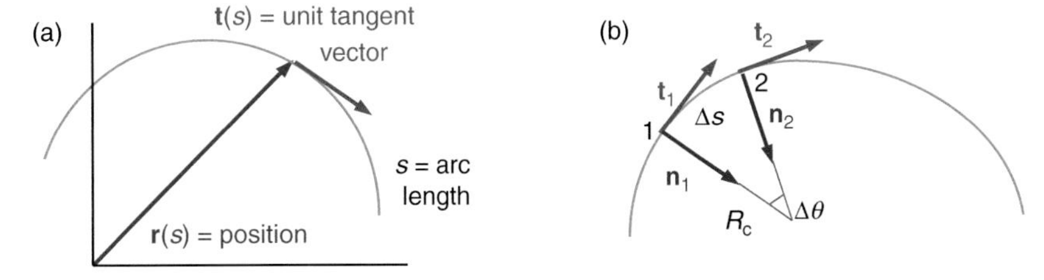

Fig. 3.5 (a) A point on the curve at arc length s is described by a position vector $\mathbf{r}(s)$ and a unit tangent vector $\mathbf{t}(s) = \partial\mathbf{r}/\partial s$. (b) Two locations are separated by an arc length Δs subtending an angle $\Delta\theta$ at a vertex formed by extensions of the unit normals $\mathbf{n}_1$ and $\mathbf{n}_2$. Extensions of $\mathbf{n}_1$ and $\mathbf{n}_2$ intersect at a distance R_c from the curve.

curve such as its local orientation. One of these is the unit tangent vector **t** that follows the direction of the curve as it winds its way through space, as shown in Fig. 3.5. In two dimensions, the (x,y) components of **t** are $(\cos\theta, \sin\theta)$, where θ is the angle between **t** and the x-axis. For a short section of arc Δs, over which the curve appears straight, the pair $(\cos\theta, \sin\theta)$ can be replaced by $(\Delta r_x/\Delta s, \Delta r_y/\Delta s)$, which becomes $(\partial r_x/\partial s, \partial r_y/\partial s)$ in the infinitesimal limit, or

$$\mathbf{t}(s) = \partial\mathbf{r}/\partial s. \tag{3.5}$$

Although Eq. (3.5) was derived as a two-dimensional example, it is valid in three dimensions as well.

How sinuous a curve appears depends on how rapidly **t** changes with s. Consider two nearby positions on a curve, which are labeled 1 and 2 on the curve illustrated in Fig. 3.5(b). If the curve were a straight line, the unit tangent vectors $\mathbf{t}_1$ and $\mathbf{t}_2$ at points 1 and 2 would be parallel; in other words, the orientations of the unit tangent vectors to a straight line are independent of position. However, such is not the case with curved lines. As we recall from introductory mechanics, the vector $\Delta\mathbf{t} = \mathbf{t}_2 - \mathbf{t}_1$ is perpendicular to the curve in the limit where positions 1 and 2 are infinitesimally close. Thus, the rate of change of **t** with s is proportional to the unit normal vector to the curve **n**, and we define the proportionality constant to be the curvature C

$$\partial\mathbf{t}/\partial s = C\mathbf{n}, \tag{3.6}$$

where C has units of inverse length. We can substitute Eq. (3.5) into (3.6) to obtain

$$C\mathbf{n} = \partial^2\mathbf{r}/\partial s^2. \tag{3.7}$$

Some care must be taken about the direction of **n**. For example, consider the arc drawn in Fig. 3.5(b). Proceeding along the arc, one can see that $\Delta\mathbf{t}$ from location to location points to the "inside" or concave side of the arc, not the convex side.

The reciprocal of C is the local radius of curvature of the arc, as can be proven by extrapolating nearby unit normal vectors $\mathbf{n}_1$ and $\mathbf{n}_2$ to their point of intersection. In Fig. 3.5(b), positions 1 and 2 are close by on the contour and they define a segment that is approximately an arc of a circle with radius R_c. The segment has length Δs along the contour and subtends an angle $\Delta\theta = \Delta s\, / R_c$ with respect to the location where $\mathbf{n}_1$ and $\mathbf{n}_2$ intersect. However, $\Delta\theta$ is also the angle between $\mathbf{t}_1$ and $\mathbf{t}_2$; that is, $\Delta\theta = |\Delta\mathbf{t}|\, / t = |\Delta\mathbf{t}|$, the second equality following from $|\mathbf{t}| = t = 1$. Equating these two expressions for $\Delta\theta$ yields $|\Delta\mathbf{t}|\, / \Delta s = 1/R_c$, which can be compared with Eq. (3.6) to give

$$C = 1/R_c. \tag{3.8}$$

Lastly, the unit normal vector **n**, which is $\Delta\mathbf{t} / |\Delta\mathbf{t}|$, can itself be rewritten using $|\Delta\mathbf{t}| = \Delta\theta$

$$\mathbf{n} = \partial\mathbf{t}/\partial\theta, \tag{3.9}$$

in the limit where $\Delta\theta \to 0$.

3.2.2 Bending energy of a thin rod

The quantities $\mathbf{t}(s)$, $\mathbf{n}(s)$ and the local curvature C describe the shape of a flexible rod or rope, but they do not tell us the filament's dynamics. To understand the latter, we must find the forces or energies involved in deforming a filament by bending or twisting it. Suppose that we take a straight rod of length L_c with uniform density and cross section, and bend it into an arc with radius of curvature R_c, as in Fig. 3.3. The energy E_{arc} associated with this deformation is determined in many texts on continuum mechanics, and has the form (Landau and Lifshitz, 1986)

$$E_{arc} / L_c = \kappa_f / 2R_c^2 = Y\mathcal{I} / 2R_c^2. \tag{3.10}$$

Recalling Eq. (3.2), the flexural rigidity κ_f is equal to the product $Y\mathcal{I}$, where Y is Young's modulus of the material, and $\mathcal{I}$ is the moment of inertia of the cross section (see Fig. 3.6). Young's modulus appears in expressions of the form [*stress*] = Y [*strain*], and has the same units as stress, since strain is dimensionless (see Appendix D for a review of elasticity theory). For three-dimensional materials, Y has units of energy density, and typically ranges from 10^9 J/m^3 for plastics to 10^{11} J/m^3 for metals.

The moment of inertia of the cross section $\mathcal{I}$ is defined somewhat similarly to the moment of inertia of the mass: it is an area-weighted integral of the squared distance from an axis

$$\mathcal{I}_y = \int x^2 \, dA, \tag{3.11}$$

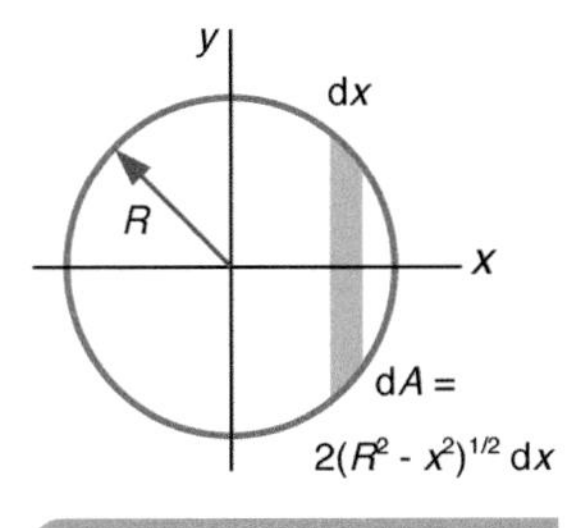

Fig. 3.6

Section through a cylindrical rod showing the *xy* axes used to evaluate the moment of inertia of the cross section $\mathcal{I}_y$ in Eq. (3.11).

where the xy plane defined by the integration axes is a cross section perpendicular to the length of the rod, and dA is an element of surface area in that plane. The subscript y on the moment $\mathcal{I}$ indicates that the bending deformation occurs around the y-axis. It is generally advantageous to perform the integration in strips parallel to the y-axis so that the strips have constant values of x. For example, if the rod is a cylinder of radius R, the cross section has the shape of a solid disk with an area element dA at position x given by $dA = 2(R^2 - x^2)^{1/2}dx$, as shown in Fig. 3.6. Hence, for a solid cylinder

$$\mathcal{I}_y = 4 \int_0^R x^2 \left(R^2 - x^2\right)^{1/2} dx = \pi R^4 / 4. \qquad \text{solid cylinder} \tag{3.12}$$

Should the rod have a hollow core of radius R_i, like a microtubule, then the moment of inertia $\pi R^4/4$ of a solid cylinder would be reduced by the moment of inertia $\pi R_i^4/4$ of the core:

$$\mathcal{I}_y = \pi(R^4 - R_i^4)/4. \quad \text{hollow cylinder} \tag{3.13}$$

Other rods of varying cross-sectional shape are treated in the end-of-chapter problems.

The deformation energy per unit length of the arc in Eq. (3.10) is inversely proportional to the square of the radius of curvature, or, equivalently, is proportional to the square of the curvature C from Eq. (3.8). In fact, one would expect on general grounds that the leading order contribution to the energy per unit length must be C^2, just as the potential energy of an ideal spring is proportional to the square of the displacement from equilibrium. Alternatively, then, the energy per unit length could be written as $E_{arc}/L = \kappa_f (\partial\mathbf{t}/\partial s)^2/2$ by using Eq. (3.6), an expression that is slightly closer mathematically to the functions representing the shape of the curve. For example, a straight line obeys $\partial\mathbf{t}/\partial s = 0$, for which the bending energy obviously vanishes. Further, there is no need for the curvature to be constant along the length of the filament, and the general expression for the total energy of deformation E_{bend} is, to lowest order,

$$E_{bend} = (\kappa_f / 2) \int_0^{L_c} (\partial\mathbf{t} / \partial s)^2 \, ds, \tag{3.14}$$

where the integral runs along the length of the filament. This form for E_{bend} is called the Kratky–Porod model; it can be trivially modified to represent a rod that is intrinsically curved even when it is not under stress.

3.2.3 Directional fluctuations and persistence length

At zero temperature, a filament adopts a shape that minimizes its energy, which corresponds to a straight rod if the energy is governed by Eq. (3.14). At non-zero temperature, the filament exchanges energy with its environment, permitting the shape to fluctuate, as illustrated in Fig. 3.7(a). According to Eq. (3.14), the bending energy of a filament rises as its shape becomes more contorted and the local curvature along the filament grows; hence, the bending energy of the configurations increases from left to right in Fig. 3.7(a). Now, the probability $\mathcal{P}(E)$ of the filament being found in a specific configuration with energy E is proportional to the Boltzmann factor $\exp(-\beta E)$, where β is the inverse temperature $\beta = 1/k_BT$ (see Appendix C for a review). The Boltzmann factor tells us that the larger the energy required to deform the filament into a specific shape, the lower the probability that the filament will have that shape, all other things being equal. Thus, a filament will adopt configurations with small average curvature if its flexural rigidity is high or the temperature is low; the shapes will resemble sections of circles, becoming contorted only at high temperatures.

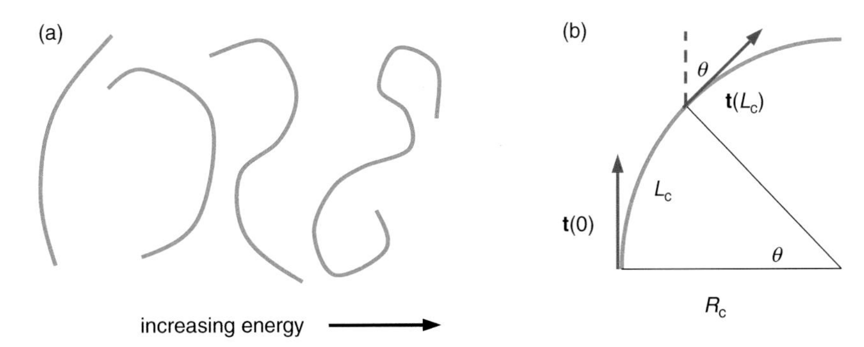

Fig. 3.7 (a) Sample of configurations available to a filament; for a given κ_f the bending energy of the filament rises as its shape becomes more contorted. (b) If the filament is a section of a circle, the angle subtended by the contour length L_c is the same as the change in the direction of the unit tangent vector **t** along the arc.

Let's now examine the bending energy of a specific filament that can sustain only gentle curves; its contour length is sufficiently short that the curvature of the bend is constant. The shape can then be uniquely parametrized by the angle θ between the unit tangent vectors $\mathbf{t}(0)$ and $\mathbf{t}(L_c)$ at the two ends of the filament as in Fig. 3.7(b). This angle is the same as that subtended by the contour length (i.e. $\theta = L_c/R_c$) because we have taken the shape to be an arc of a circle with a radius R_c. Thus, each configuration has a bending energy determined by the value of θ

$$E_{\text{arc}} = \kappa_f \theta^2 / 2L_c, \tag{3.15}$$

where we have removed R_c from Eq. (3.10) in favor of θ by using $R_c = L_c/\theta$.

At non-zero temperature, the angle θ changes as the filament waves back and forth exchanging energy with its thermal environment: at higher temperatures, the oscillations have a larger amplitude and the filament samples larger values of θ than at lower temperatures. To characterize the magnitude of the oscillations, we evaluate the mean value of θ^2, denoted by the conventional $\langle \theta^2 \rangle$. If the filament has a fixed length, $\langle \theta^2 \rangle$ involves a weighted average of the three-dimensional position sampled by the end of the filament. That is, with one end of the filament defining the direction of a coordinate axis (say the z-axis), the other end is described by the polar angle θ and the azimuthal angle ϕ, such that the ensemble average is

$$\langle \theta^2 \rangle = \int \theta^2 \, \mathcal{P}(E_{\text{arc}}) \mathrm{d}\Omega \, / \int \mathcal{P}(E_{\text{arc}}) \mathrm{d}\Omega, \tag{3.16}$$

where E_{arc} is given by Eq. (3.15) and where the integral must be performed over the solid angle $\mathrm{d}\Omega = \sin\theta \, d\theta \, \mathrm{d}\phi$. The bending energy E_{arc} is independent of ϕ, allowing the azimuthal angle to be integrated out, leaving

$$\langle\theta^2\rangle = \int\theta^2\, \mathcal{P}(E_{arc}) \sin\theta \, d\theta \,/ \int \mathcal{P}(E_{arc}) \sin\theta \, d\theta, \tag{3.17}$$

where the range of θ in the integrals is 0 to 2π.

In thermal equilibrium, the probability $\mathcal{P}(E_{arc})$ of finding a configuration with a given bending energy E_{arc} is given by the Boltzmann factor $\exp(-\beta E_{arc})$, so that Eq. (3.17) becomes

$$\langle\theta^2\rangle = \int \theta^2 \exp(-\beta E_{arc}) \sin\theta \, d\theta \,/ \int \exp(-\beta E_{arc}) \sin\theta \, d\theta. \tag{3.18}$$

According to Eq. (3.15), the bending energy E_{arc} rises quadratically with θ, with the result that the Boltzmann factor decays rapidly with θ. Conseqently, the $\sin\theta$ factors in Eq. (3.18) are sampled only at small θ, and can be replaced by the small angle approximation $\sin\theta \sim \theta$. Hence, Eq. (3.18) becomes

$$\langle\theta^2\rangle = (2L_c \,/\, \beta\kappa_f) \int x^3 \exp(-x^2) \, dx \,/ \int x \exp(-x^2) \, dx, \tag{3.19}$$

after substituting Eq. (3.15) for E_{arc} and changing variables to $x = (\beta\kappa_f \,/\, 2L_c)^{1/2}\theta$. In the small oscillation approximation, the upper limits of the integrals in Eq. (3.19) can be extended to infinity with little error, whence both integrals are equal to 1/2 and cancel out. Thus, the expression for the mean square value of θ is

$$\langle\theta^2\rangle \cong 2s \,/\, \beta\kappa_f, \qquad \text{small oscillations} \tag{3.20}$$

where we have replaced L_c by the arc length s in anticipation of making the contour length a variable. The combination $\beta\kappa_f$ has the units of length, and is defined as the persistence length ξ_p of the filament:

$$\xi_p \equiv \beta\kappa_f. \tag{3.21}$$

Note that, for thermal systems, ξ_p decreases with increasing temperature.

A directional persistence length ξ was introduced in Section 2.5 by means of the tangent correlation function $\langle \mathbf{t}(0) \cdot \mathbf{t}(s) \rangle$, and we will now show that ξ and ξ_p are one and the same. Still assuming that the shapes of the filament are arcs of circles, the ensemble average $\langle \mathbf{t}(0) \cdot \mathbf{t}(s) \rangle = \langle \cos\theta \rangle$, which has a maximum absolute value of unity because $|\mathbf{t}| = 1$. At low temperatures where θ is usually small, we again invoke the small approximation that leads from Eq. (3.18) to Eq. (3.19): $\cos\theta \sim 1 - \theta^2/2$, permitting the correlation function to be written as

$$\langle \mathbf{t}(0) \cdot \mathbf{t}(s) \rangle \sim 1 - \langle\theta^2\rangle/2. \tag{3.22}$$

The variance in θ in this small oscillation limit is given by Eq. (3.20), so that

$$\langle \mathbf{t}(0) \cdot \mathbf{t}(s) \rangle \sim 1 - s/\xi_p \qquad (s/\xi_p << 1), \tag{3.23}$$

where the arc length s now appears as a parameter: the equation is valid for filaments of varying length. Equation (3.23) can be used to obtain the mean squared difference in the tangent vectors

$$\langle [\mathbf{t}(s) - \mathbf{t}(0)]^2 \rangle = 2 - 2\langle \mathbf{t}(0) \cdot \mathbf{t}(s) \rangle \sim 2s/\xi_p \qquad (s/\xi_p << 1). \tag{3.24}$$

When a filament's contour length is short compared with ξ_p, Eq. (3.23) correctly predicts that $\langle \mathbf{t}(0) \cdot \mathbf{t}(L_c) \rangle$ initially dies off linearly as L_c grows. However, if $L_c >> \xi_p$, the filament appears floppy and $\langle \mathbf{t}(0) \cdot \mathbf{t}(L_c) \rangle$ should vanish as the tangent vectors at the ends of the filament become uncorrelated, a regime not included in Eq. (3.23) because it was derived in the limit of small oscillations. Rather, the correct expression for the tangent correlation function applicable at short and long distances is

$$\langle \mathbf{t}(0) \cdot \mathbf{t}(s) \rangle = \exp(-s/\xi_p), \tag{3.25}$$

from which we see that Eq. (3.23) is the leading order approximation via $\exp(-x) \sim 1 - x$ at small x. Intuitively, one would expect to obtain an expression like Eq. (3.25) by applying Eq. (3.23) repeatedly to successive sections of the filament; a more detailed derivation can be found in Doi and Edwards (1986).

3.3 Sizes of polymer chains

A function of both temperature and bending resistance, the persistence length of a filament sets the length scale of its thermal undulations. If the contour length of the filament is much smaller than its persistence length, the filament can be viewed as a relatively stiff rod undergoing only limited excursions from its equilibrium shape. In contrast, a very flexible polymer samples an extensive collection of contorted shapes with erratically changing directions. Do the configurations in this collection have any large scale characteristics, or are they just an unruly mob of rapidly changing tangents and curvatures? If the ensemble of configurations do have common or universal features, upon what properties of the filament do they depend? Here, we study several polymer families, characterized by their connectivity and interactions, to answer these questions.

3.3.1 Ideal chains and filaments

In Section 2.3, we derive several properties of ideal random walks, and establish that the mean square value of the end-to-end displacement vector $\mathbf{r}_{ee}$ obeys $\langle r_{ee}^2 \rangle = Nb^2$, where N is the number of steps in the walk and b is the length of each step (assumed identical for all steps). We argue that flexible polymers might be described by such walks, and apply the expression for $\langle r_{ee}^2 \rangle$ to floppy proteins to demonstrate how the radius of its folded state should be much less than its contour length. We now perform the same kind of analysis to the continuous representation of flexible filaments introduced in Section 3.2, rather than the segmented configurations of Section 2.3.

We start with the conventional end-to-end displacement vector $\mathbf{r}_{ee} \equiv \mathbf{r}(L_c) - \mathbf{r}(0)$, where $\mathbf{r}(s)$ is the continuous function that denotes the position of the filament at arc length s. The mean square value of $\mathbf{r}_{ee}$ is then

$$\langle \mathbf{r}_{ee}^2 \rangle = \langle [\mathbf{r}(L_c) - \mathbf{r}(0)]^2 \rangle. \tag{3.26}$$

The unit tangent vector $\mathbf{t}(s)$ was introduced in Eq. (3.5) as a derivative of the position $\mathbf{r}(s)$, which means that $\mathbf{r}(s)$ at any location can be found by integrating $\mathbf{t}(s)$, as in

$$\mathbf{r}(s) = \mathbf{r}(0) + \int_0^s \mathrm{d}u \ \mathbf{t}(u). \tag{3.27}$$

The representation of $\mathbf{r}(s)$ in Eq. (3.27) can be substituted into Eq. (3.26) to yield

$$\langle \mathbf{r}_{ee}^2 \rangle = \int_0^{L_c} \mathrm{d}u \int_0^{L_c} \mathrm{d}v \langle \mathbf{t}(u) \bullet \mathbf{t}(v) \rangle, \tag{3.28}$$

after moving the ensemble average inside the integral. According to Eq. (3.25), the correlation function $\langle \mathbf{t}(s) \bullet \mathbf{t}(0) \rangle$ decays exponentially as $\exp(-s/\xi_p)$, which means that Eq. (3.28) can be rewritten as

$$\langle \mathbf{r}_{ee}^2 \rangle = \int_0^{L_c} \mathrm{d}u \int_0^{L_c} \mathrm{d}v \ \exp(-|u-v| / \xi_p). \tag{3.29}$$

The absolute value operation in the exponential looks slightly awkward, but it can be handled by breaking the integral into two identical pieces where one integration variable is forced to have a value less than the other during integration:

$$\langle \mathbf{r}_{ee}^2 \rangle = 2 \int_0^{L_c} \mathrm{d}u \int_0^{u} \mathrm{d}v \ \exp(-[u-v] / \xi_p). \tag{3.30}$$

It is straightforward to solve this integral using a few changes of variables

$$\begin{aligned} 2 \int_0^{L_c} \exp(-u/\xi_p) \ \mathrm{d}u \int_0^{u} \mathrm{d}v \ \exp(v/\xi_p) \\ &= 2 \int_0^{L_c} \mathrm{d}u \ \exp(-u/\xi_p) \bullet \xi_p \bullet \left[\exp(u/\xi_p) - 1\right] \\ &= 2\xi_p^2 \int_0^{L_c/\xi_p} \mathrm{d}w \ [1 - \exp(-w)]. \end{aligned} \tag{3.31}$$

Evaluating the last integral gives

$$\langle \mathbf{r}_{ee}{}^2 \rangle = 2\xi_p L_c - 2\xi_p{}^2 [1 - \exp(-L_c / \xi_p)]. \quad \text{continuous curve} \quad (3.32)$$

This result simplifies in two limits. If $\xi_p >> L_c$, Eq. (3.32) reduces to $\langle \mathbf{r}_{ee}{}^2 \rangle^{1/2} = L_c$ using the approximation $\exp(-x) \sim 1 - x + x^2/2\ldots$ valid at small x; in this limit, the filament appears rather rod-like with an end-to-end displacement close to its contour length. At the other extreme where $\xi_p << L_c$, Eq. (3.32) is approximately

$$\langle \mathbf{r}_{ee}{}^2 \rangle \cong 2\xi_p L_c \quad (\text{if } L_c >> \xi_p), \quad (3.33)$$

implying that, over long distances compared to the persistence length, $\langle \mathbf{r}_{ee}{}^2 \rangle^{1/2}$ grows like the square root of the contour length, not as the contour length itself. In other words, long polymers appear convoluted, and their average linear dimension increases much more slowly than their contour length.

The behavior of $\langle \mathbf{r}_{ee}{}^2 \rangle$ for continuous filaments with $\xi_p << L_c$ is the same as that of ideal segmented chains $\langle r_{ee}{}^2 \rangle = Nb^2$ once Nb is replaced by the contour length L_c and the step size is identified with $2\xi_p$ in Eq. (3.33) such that

$$\xi_p = b/2. \quad \text{ideal chains} \quad (3.34)$$

In other words, both descriptions show that the linear dimension of very flexible filaments increases as the square root of the contour length. The scaling behavior $\langle \mathbf{r}_{ee}{}^2 \rangle^{1/2} \sim N^{1/2}$ or $L_c{}^{1/2}$ in Eqs. (2.31) and (3.33) is referred to as *ideal* scaling. Note that our determination of the ideal scaling exponent does not depend on the dimension of the space in which the chain resides: random chains in two dimensions (i.e. confined to a plane) or three dimensions both exhibit the same scaling behavior.

Ideal scaling of polymer chains can appear even if the orientations of neighboring segments are not completely random, although the persistence length of the polymer will not be $b/2$. As an example, consider the usual set of bond vectors $\{\mathbf{b}_i\}$ from which the end-to-end displacement vector is constructed via $\mathbf{r}_{ee} = \Sigma_{i=1,N}\, \mathbf{b}_i$. In the freely rotating chain model, successive chain elements $\mathbf{b}_i$ and $\mathbf{b}_{i+1}$ are forced to have the same polar angle α, although the bonds may swivel around each other and each bond has the same length b. As usual, the ensemble average $\langle \mathbf{r}_{ee}{}^2 \rangle$ has the formal representation

$$\langle \mathbf{r}_{ee}{}^2 \rangle = \Sigma_i \, \Sigma_j \, \langle \mathbf{b}_i \bullet \mathbf{b}_j \rangle,$$

but there are now restrictions present within $\langle \cdots \rangle$. This model is solved in the end-of-chapter problems (see also Flory (1953), p. 414), and yields

$$\langle \mathbf{r}_{ee}{}^2 \rangle = Nb^2 (1 - \cos\alpha) \,/\, (1 + \cos\alpha), \quad (3.35)$$

in the large N limit.

Now, $\langle \mathbf{r}_{ee}{}^2 \rangle^{1/2}$ in Eq. (3.35) obeys the scaling exponent $N^{1/2}$, demonstrating that self-intersecting freely rotating chains are ideal. Further, Eq. (3.35)

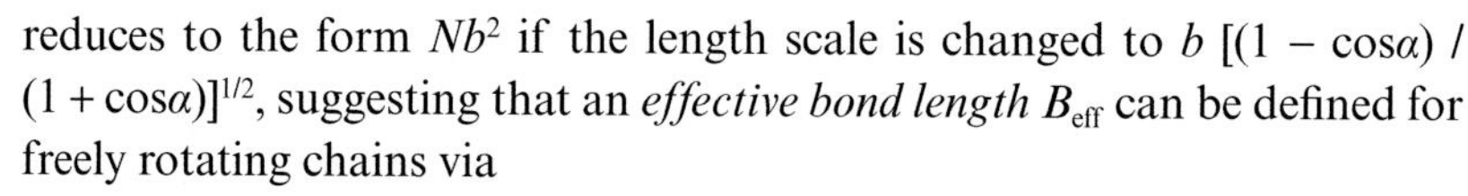

reduces to the form Nb^2 if the length scale is changed to $b\,[(1 - \cos\alpha) / (1 + \cos\alpha)]^{1/2}$, suggesting that an *effective bond length* B_{eff} can be defined for freely rotating chains via

$$B_{\mathrm{eff}} = b\,[(1 - \cos\alpha) / (1 + \cos\alpha)]^{1/2}, \tag{3.36}$$

and $\langle \mathbf{r}_{ee}^{\,2} \rangle$ is expressed as $NB_{\mathrm{eff}}^{\,2}$. The effective bond length is only one of the parametrizations commonly employed for ideal chains with N segments:

$$\langle \mathbf{r}_{ee}^{\,2} \rangle = \begin{cases} NB_{\mathrm{eff}}^{\,2} \\ L_c \mathcal{L}_K \\ 2L_c \xi_p. \end{cases} \tag{3.37}$$

Another parametrization is the Kuhn length, $\mathcal{L}_K$, defined in analogy with the monomer length: $\langle \mathbf{r}_{ee}^{\,2} \rangle = N_K \mathcal{L}_K^{\,2}$ and $L_c = N_K \mathcal{L}_K$, with N_K the number of Kuhn lengths in the contour length.

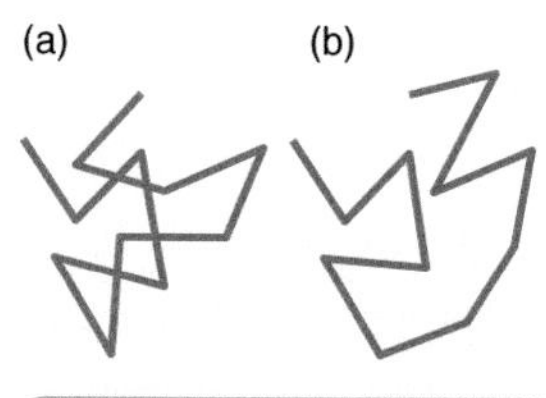

Fig. 3.8

Self-avoidance changes the scaling properties of chains in one-, two- and three-dimensional systems. In the two-dimensional configurations displayed here, (a) is a random chain and (b) is a self-avoiding chain.

3.3.2 Self-avoiding linear chains

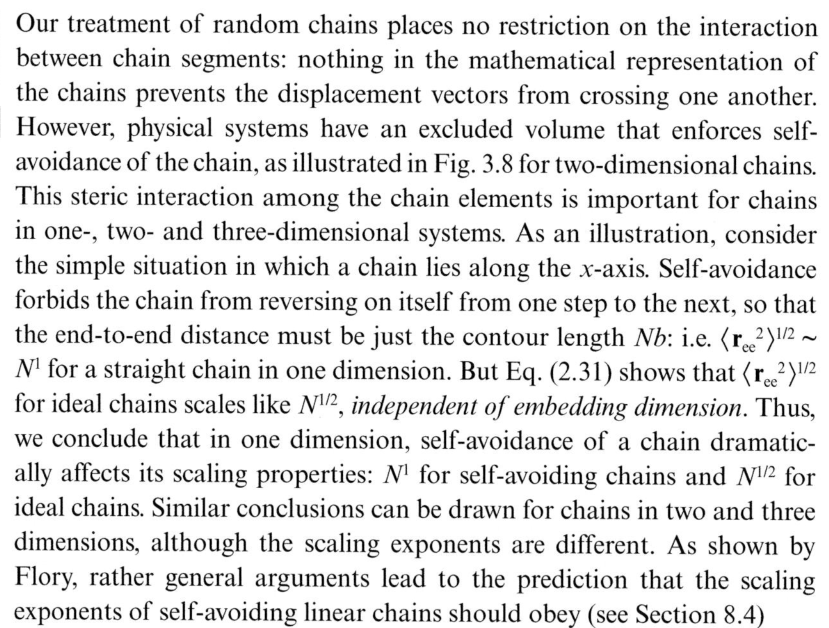

Our treatment of random chains places no restriction on the interaction between chain segments: nothing in the mathematical representation of the chains prevents the displacement vectors from crossing one another. However, physical systems have an excluded volume that enforces self-avoidance of the chain, as illustrated in Fig. 3.8 for two-dimensional chains. This steric interaction among the chain elements is important for chains in one-, two- and three-dimensional systems. As an illustration, consider the simple situation in which a chain lies along the x-axis. Self-avoidance forbids the chain from reversing on itself from one step to the next, so that the end-to-end distance must be just the contour length Nb: i.e. $\langle \mathbf{r}_{ee}^{\,2} \rangle^{1/2} \sim N^1$ for a straight chain in one dimension. But Eq. (2.31) shows that $\langle \mathbf{r}_{ee}^{\,2} \rangle^{1/2}$ for ideal chains scales like $N^{1/2}$, *independent of embedding dimension*. Thus, we conclude that in one dimension, self-avoidance of a chain dramatically affects its scaling properties: N^1 for self-avoiding chains and $N^{1/2}$ for ideal chains. Similar conclusions can be drawn for chains in two and three dimensions, although the scaling exponents are different. As shown by Flory, rather general arguments lead to the prediction that the scaling exponents of self-avoiding linear chains should obey (see Section 8.4)

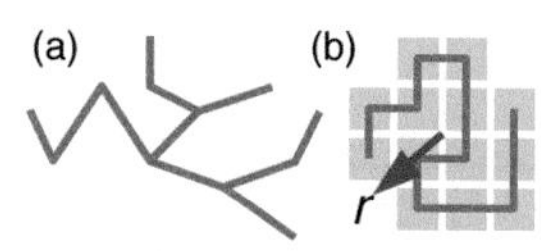

Fig. 3.9

Sample configurations of a branched polymer (a) and a dense chain (b) in two dimensions. To aid the argument in the text, the chain in (b) consists of linked squares, which, when packed tightly together, cover an area $\sim r^2$ in two dimensions.

$$\nu_{\mathrm{FL}} = 3 / (2+d), \tag{3.38}$$

where d is the embedding dimension. Equation (3.38) gives $\nu_{\mathrm{FL}} = 1$, 3/4, 3/5 and 1/2, in one to four dimensions, respectively, predictions which have been shown to be exact or nearly so. As the ideal scaling exponent cannot be less than 1/2, Eq. (3.38) is not valid in more than four dimensions; hence, the effects of self-avoidance are irrelevant in four or more dimensions and the scaling is always ideal.

3.3.3 Branched polymers

The polymers discussed in most of this text are linear chains; however, there are many examples of polymers with extensive side branches. The scaling behavior of such *branched polymers* should not be the same as single chains, since branching adds more monomers along the chain length as illustrated in Fig. 3.9(a). Because a branched polymer has more than two ends, the end-to-end displacement has to be replaced by a different measure of the polymer size, such as the radius of gyration, R_g (see end-of-chapter problems). The radius of gyration for branched polymers is found to have a scaling form

$$\langle R_g^{\,2} \rangle^{1/2} \sim N^{\nu}, \tag{3.39}$$

where N is the number of polymer segments and $\nu = 0.64$ and 0.5 in two and three dimensions, respectively (see Section 8.4). In comparison, self-avoiding linear chains have scaling exponents of 3/4 and 0.59, respectively (see Eq. (3.38)), meaning that the spatial region occupied by branched polymers grows more slowly with N than does that of linear chains; i.e. linear chains are less dense than branched polymers. Fluid membranes also behave like branched polymers at large length scales (see Section 8.4).

3.3.4 Collapsed chains

None of the chain configurations described so far in this section is as compact as it could be. Consider a system of identical objects, say squares in two dimensions or cubes in three dimensions, having a length b to the side such that each object has a "volume" of b^d in d dimensions, and N of these objects have a volume Nb^d. The configuration of the N objects with the smallest surface area is the most compact or the most *dense* configuration, as illustrated in Fig. 3.9(b), and we denote by r the linear dimension of this configuration. Ignoring factors of π and the like, the total volume Nb^d of the most compact configuration is proportional to r^d, so that r itself scales like

$$r \sim N^{1/d} \text{ (dense)}. \tag{3.40}$$

Polymers can be made to collapse into their most dense configurations by a variety of experimental means, including changes in the solvent, and it is observed that the collapse of the chains occurs at a well-defined phase transition.

The scaling exponents of all the systems that we have considered in this section are summarized in Table 3.1. If the chains are self-avoiding, $1/d$ represents the lower bound on the possible values of the scaling exponents for the "size" of the configurations, and the straight rod scaling of $\langle R_g^{\,2} \rangle^{1/2} \sim N^1$ represents the upper bound. One can see from the table that random or

Table 3.1 Exponents for the scaling law $\langle R_g^2 \rangle^{1/2} \sim N^\nu$ for ideal (or random) chains, self-avoiding chains and branched polymers, as a function of embedding dimension d. Collapsed chains have the highest density and obey $\langle R_g^2 \rangle^{1/2} \sim N^{1/d}$.

Configuration	$d = 2$	$d = 3$	$d = 4$
Ideal chains	1/2	1/2	1/2
Self-avoiding chains	3/4	0.59	1/2
Branched polymers	0.64	1/2	
Collapsed chains	1/2	1/3	1/4

self-avoiding chains, as well as branched polymers, exhibit scaling behavior that lies between these extremes.

3.4 Entropic elasticity

The distribution of end-to-end displacements $\mathbf{r}_{ee}$ for random walks in one dimension was derived in Section 2.3; viewing the walks as one-dimensional linear polymers, it was argued that entropy favored polymer configurations that were convoluted rather than straight. In this section, the analysis is extended to walks or polymer chains in three dimensions, which allows for a larger variety of configurations. The three-dimensional distributions confirm that it is highly unlikely for a random chain to be found in a fully stretched configuration: the most likely value of r_{ee}^2 for a freely jointed chain is not far from its mean value of Nb^2, for chains with N links of length b. Consequently, as a polymer chain is made to straighten out by an external force, its entropy decreases and work must be done on the chain to stretch it: in other words, the polymer behaves elastically because of its entropy. Here, we will establish that the Hooke's law spring constant associated with a polymer's entropic resistance to stretching increases with temperature as $3k_BT / Nb^2$ for three-dimensional chains.

3.4.1 Random chain in three dimensions

Let us briefly revisit the results from Section 2.3 for ideal random walks: the mean squared end-to-end displacement obeys $\langle r_{ee}^2 \rangle = Nb^2$ in any dimension, while the probability distribution for $r_{ee,x}$ in one dimension obeys $\mathcal{P}(x) \propto \exp(-r_{ee,x}^2 / 2\sigma^2)$, where $\sigma^2 \equiv Nb^2$. How does the probability change in three dimensions? By projecting their configurations onto a set of Cartesian axes, as illustrated in Fig. 3.10, three-dimensional random

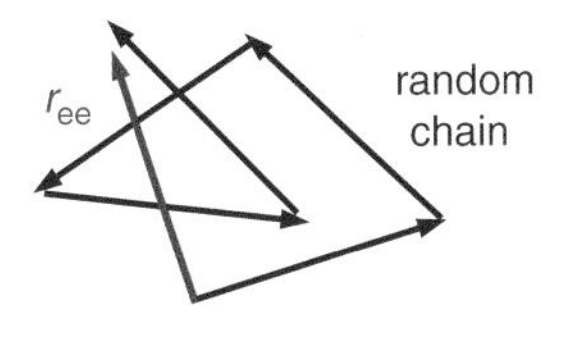

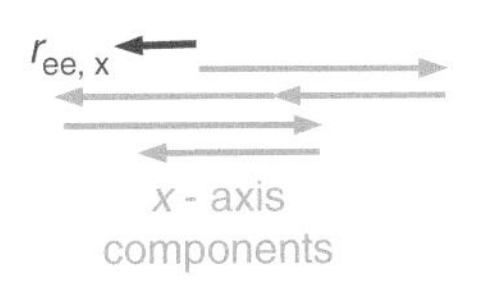

Fig. 3.10

Projection of the segments of a two-dimensional chain onto the x-axis.

chains can be treated as three separate one-dimensional systems. For example, the x-component of the end-to-end displacement vector, $r_{ee,x}$, is just the sum of the individual monomer vectors as projected onto the x-axis:

$$r_{ee,x} = \Sigma_i \, b_{i,x}, \tag{3.41}$$

where $b_{i,x}$ is the x-projection of the monomer vector $\mathbf{b}_i$. For freely jointed chains, the component $b_{i,x}$ is independent of the component $b_{i+1,x}$, so the projections form a random walk in one dimension, although the x-axis projections are of variable length even if all monomers have the same b in three dimensions. If the number of segments is large, the probability distribution with variable segment length has the same form as the distribution with uniform segment length (Chapter 1 of Reif, 1965),

$$\mathcal{P}(x) = (2\pi\sigma_x^{\,2})^{-1/2} \exp(-r_{ee,x}^{\,2} \,/\, 2\sigma_x^{\,2}), \tag{3.42}$$

except that the variance is given by

$$\sigma_x^{\,2} = N\langle b_x^{\,2} \rangle. \tag{3.43}$$

In this variance, b^2 of the strictly one-dimensional walk with fixed step size has been replaced by $\langle b_x^{\,2} \rangle \leq b^2$ for variable step size. Of course, one could still say that Eq. (3.43) incorporates the strictly one-dimensional case, in that $\langle b_x^{\,2} \rangle = b^2$ if the step size is constant.

It is straightforward to determine $\langle b_x^{\,2} \rangle$ even when the projected steps are of unequal length. The expectation of the step length in three dimensions must have the form

$$\langle b^2 \rangle = \langle b_x^{\,2} + b_y^{\,2} + b_z^{\,2} \rangle = \langle b_x^{\,2} \rangle + \langle b_y^{\,2} \rangle + \langle b_z^{\,2} \rangle. \tag{3.44}$$

Because of symmetry, we anticipate that the mean projections must be the same along each of the Cartesian axes, so

$$\langle b_x^{\,2} \rangle = \langle b_y^{\,2} \rangle = \langle b_z^{\,2} \rangle = b^2/3. \tag{3.45}$$

Hence, the variance in Eq. (3.43) is

$$\sigma_3^{\,2} \equiv \sigma_x^{\,2} = Nb^2/3, \qquad \sigma_3^{\,2} \text{ in three dimensions} \tag{3.46}$$

where we have introduced the new symbol σ_3 just to avoid notational confusion between one- and three-dimensional walks.

Equation (3.42) is the probability density for $\mathbf{r}_{ee}$ as projected onto the x-axis. By symmetry, similar expressions exist for the y- and z-axis projections. These three distributions can be combined to give the probability density for finding $\mathbf{r}_{ee}$ in a volume $dx\,dy\,dz$ centered on the specific position (x,y,z), namely $\mathcal{P}(x,y,z)\,dx\,dy\,dz$. Thus, $\mathcal{P}(x,y,z)$ is the product of the probability distributions in each of the Cartesian directions

$$\mathcal{P}(x,y,z) = \mathcal{P}(x)\,\mathcal{P}(y)\,\mathcal{P}(z) = (2\pi\sigma_3^{\,2})^{-3/2} \exp[-(x^2+y^2+z^2) \,/\, 2\sigma_3^{\,2}], \tag{3.47}$$

where σ_3^2 is still given by Eq. (3.46), and where $x \equiv r_{ee,x}$, etc. Equation (3.47) says that, of all possible chain configurations, the most likely set of coordinates for the tip of the chain is (0,0,0), which is the coordinate origin of the chain or random walk; it does *not* say that the most likely value of r_{ee} is zero. Indeed, the distribution of the magnitude of $\mathbf{r}_{ee}$ must reflect the fact that many different coordinate positions (x,y,z) have the same r, although $\mathbf{r}_{ee}$ may point in various directions at that value of r. The probability for the chain to have a radial end-to-end distance between r and $r + \mathrm{d}r$ is $\mathcal{P}_{rad}(r)\,\mathrm{d}r$, where $\mathcal{P}_{rad}(r)$ is the probability per unit length obtained from

$$\int_{angle} \mathcal{P}(x,y,z)\,\mathrm{d}x\,\mathrm{d}y\,\mathrm{d}z = \mathcal{P}_{rad}(r)\,\mathrm{d}r. \tag{3.48}$$

Replacing $\mathrm{d}x\,\mathrm{d}y\,\mathrm{d}z$ by the angular expression $r^2\,\mathrm{d}r\,\sin\theta\,\mathrm{d}\theta\,\mathrm{d}\phi$, the θ and ϕ integrals in Eq. (3.48) can be done immediately, as $x^2+y^2+z^2 = r^2$ so that there is no angular dependence on the right-hand side of Eq. (3.47). Thus,

$$\mathcal{P}_{rad}(r) = 4\pi r^2\,(2\pi\sigma_3^2)^{-3/2}\exp(-r^2 / 2\sigma_3^2). \tag{3.49}$$

It's the extra factor of r^2 outside of the exponential that shifts the most likely value of r_{ee} away from zero.

Figure 3.11 shows the behavior of Eq. (3.49), as well as the projection of the chain on the x-axis. We can equate to zero the derivative of $\mathcal{P}_{rad}(r)$ with respect to r to find the most likely value of r_{ee}. A summary of the results for ideal chains in three dimensions is:

$$r_{ee,\,most\ likely} = (2/3)^{1/2}\,N^{1/2}b, \tag{3.50}$$

$$\langle r_{ee}\rangle = (8/3\pi)^{1/2}\,N^{1/2}b, \tag{3.51}$$

and, of course,

$$\langle \mathbf{r}_{ee}^2\rangle = Nb^2. \tag{3.52}$$

Note that r_{ee} in Eqs. (3.50) and (3.51) is the scalar radius $r_{ee} = (\mathbf{r}_{ee}^2)^{1/2}$.

3.4.2 Entropic elasticity

The probability distribution functions, as illustrated in Fig. 3.11, confirm our intuition that far more chain configurations have end-to-end displacements close to the mean value of r_{ee} than to the chain contour length L_c. Being proportional to the logarithm of the number of configurations, the entropy S of a polymer chain must decrease as the chain is stretched from its equilibrium length. Now the free energy of an ensemble of chains at a temperature T is $F = E - TS$, which is simply $F = -TS$ for freely jointed

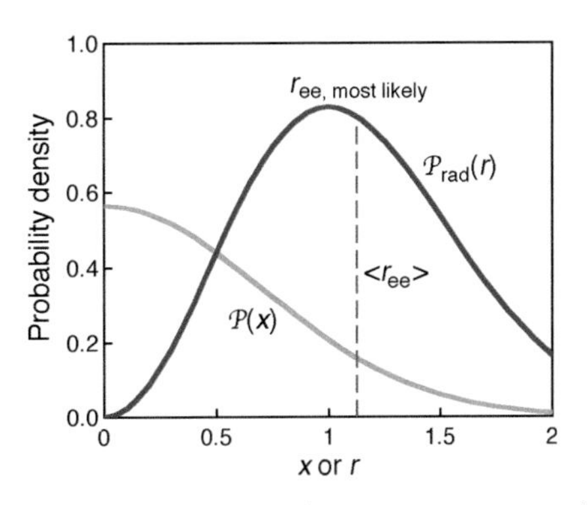

Fig. 3.11 Probability distributions for random chains in three dimensions. Two cases are shown: the three-dimensional distribution (red curve) as a function of $r = r_{ee}$, and the x-axis projection (blue curve) as a function of $x = r_{ee,x}$ ($\sigma^2 = 1/2$ in both distributions). The dashed vertical line is $\langle r_{ee} \rangle$ in three dimensions.

chains, since their configurations all have vanishing energy E. Thus, S decreases and F increases as the chain is stretched at non-zero temperature; in other words, work must be done to stretch the chain, and the chain is elastic by virtue of its entropy.

Viewed as a spring obeying Hooke's Law, the effective force constant of a polymer chain can be extracted by comparing the distributions for the end-to-end displacement of the chain with that of an ideal spring, whose fluctuations can be calculated using statistical mechanics. Now, a Hookean spring has a potential energy $V(x)$ equal to $k_{sp} \cdot x^2/2$, where x is the displacement from equilibrium and k_{sp} is the force constant of the spring. Aside from an overall normalization factor, the probability distribution $\mathcal{P}(x)$ for the spring displacement x is proportional to the usual Boltzmann factor $\exp(-E/k_B T)$, which becomes, for the Hooke's Law potential

$$\mathcal{P}(x) \sim \exp(-k_{sp} \cdot x^2 / 2k_B T). \tag{3.53}$$

The probability distribution for the displacement of an ideal chain according to Eq. (3.42) is $\mathcal{P}(x) \sim \exp(-x^2/2\sigma_d^2)$, again aside from an overall normalization factor. Comparing the functional form of the two distributions at large x gives $k_{sp} = k_B T / \sigma_d^2$, where $\sigma_d^2 = Nb^2/d$ for ideal chains embedded in d dimensions [the dimensionality can be seen from Eq. (3.46)]. Hence, in three dimensions, we expect

$$k_{sp} = 3k_B T / Nb^2 = 3k_B T / 2\xi_p L_c, \qquad k_{sp} \text{ in three dimensions} \tag{3.54}$$

using $L_c = Nb$ and $\xi_p = b/2$ for an ideal chain. Observe that k_{sp} increases with temperature, which is readily demonstrated experimentally by hanging a weight from an elastic band, and then using a device (like a hair dryer) to heat the elastic. The weight will be seen to rise as the elastic heats up, since k_{sp} increases simultaneously and provides greater resistance to the stretching of the elastic band by the weight.

3.4.3 Highly stretched chains

The Gaussian probability distribution, Eq. (3.42), gives a good description of chain behavior at small displacements from equilibrium. It predicts, from Eq. (3.54), that the force f required to produce an extension x in the end-to-end displacement is $f = (3k_BT / 2\xi_pL_c)x$, which can be rewritten as

$$x/L_c = (2\xi_p / 3k_BT)\,f. \tag{3.55}$$

If the chain segments are individually inextensible, the force required to extend the chain should diverge as the chain approaches its maximal extension, $x/L_c \rightarrow 1$. Such a divergence is not present in Eq. (3.55), indicating that the Gaussian distribution must be increasingly inaccurate and ultimately invalid as an inextensible chain is stretched towards its contour length.

Of course, the Gaussian distribution is only an approximate representation of freely jointed chains; fortunately, the force–extension relation of rigid, freely jointed rods can be obtained analytically. For those familiar with the example, the problem is analogous to the alignment of spin vectors in an external field, where the spin vectors represent the projection of the polymer segments along the direction of the applied field. It is straightforward to show (Kuhn and Grün, 1942; James and Guth, 1943; see also Flory, 1953, p. 427) that the solution has the form

$$x/L_c = \mathcal{L}(2\xi_p f / k_BT), \tag{3.56}$$

where $\mathcal{L}(y)$ is the Langevin function

$$\mathcal{L}(y) = \coth(y) - 1/y. \tag{3.57}$$

Note that x in Eq. (3.56) is the projection of the end-to-end displacement along the direction of the applied force. For small values of f, Eq. (3.56) reduces to the Gaussian expression Eq. (3.55); for very large values of f, the Langevin function tends to 1 so that x asymptotically approaches L_c in Eq. (3.56), as desired.

The force–extension relation of freely jointed rods provides a reasonably accurate description of biopolymers. Its weakness lies in viewing the polymer as a chain of rigid segments: thick filaments such as microtubules and DNA surely look more like continuously flexible ropes than chains of rigid rods. A more appropriate representation of flexible filaments can be derived from the Kratky–Porod energy expression, Eq. (3.14), and is referred to as the worm-like chain (WLC). Although the general form of the WLC force–extension relationship is numerical, an accurate interpolation formula has been obtained by Marko and Siggia (1995):

$$\xi_p f / k_B T = (1/4)(1 - x/L_c)^{-2} - 1/4 + x/L_c. \tag{3.58}$$

Again, the force diverges in this expression as $x/L_c \rightarrow 1$, as desired. Equation (3.58) and the freely jointed chain display the same behavior at both large and small forces, although their force–extension curves may disagree by as much as 15% for intermediate forces.

3.5 Buckling

The filaments and sheets of a cell are subject to stress from a variety of sources. For example:

- the membrane and its associated networks may be under tension if the cell has an elevated osmotic pressure,
- components within the cell such as vesicles and filaments experience a variety of forces as they are dragged by molecular motors,
- inequivalent elements of the cytoskeleton may bear differentially the compressive and tensile stresses of a deformation.

As described earlier, such forces in the cell generally have magnitudes in the piconewton range. For a comparison at a macroscopic scale, we calculate the force required to bend a strand of hair. The flexural rigidity κ_f of a solid cylindrical filament of radius R is equal to $\pi Y R^4/4$, where Y is the Young's modulus of the material; with $R = 0.05$ mm and $Y = 10^9$ J/m^3 (typical of biomaterials), we expect $\kappa_f = 5 \times 10^{-9}$ J•m for a strand of hair. With one end of a filament of length L held fixed, the free end moves a distance $z = FL^3/(3\kappa_f)$ when subjected to a transverse force F (see end-of-chapter problems). Thus, a force of 1.5×10^5 pN is required to move the free end of a 10 cm strand through a distance of 1 cm. In other words, even this imperceptibly small force on our finger tip is five orders of magnitude larger than the typical force on a filament in the cell.

Newton's Third Law of mechanics tells us that a tensile stress on one component of a cell in equilibrium must be balanced by a compressive stress on another. In the design of bridges and houses, one often sees rigid beams and bars carrying either tension or compression. The simple truss in Fig. 3.12(a) demonstrates how a vertical load is distributed across three beams in a triangle: the two thick elements are under compression while the thin element at the base is subject only to tension. As a design, these

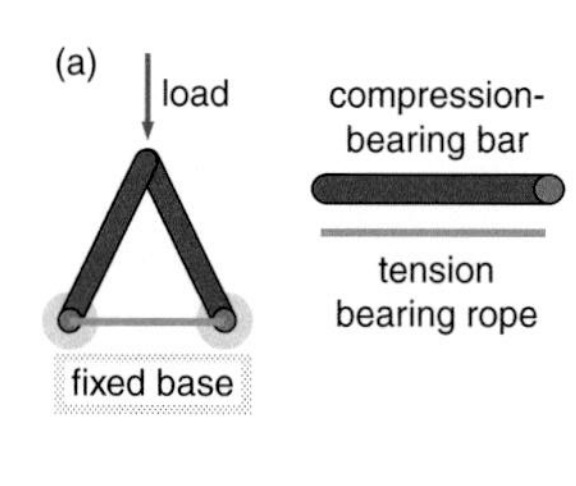

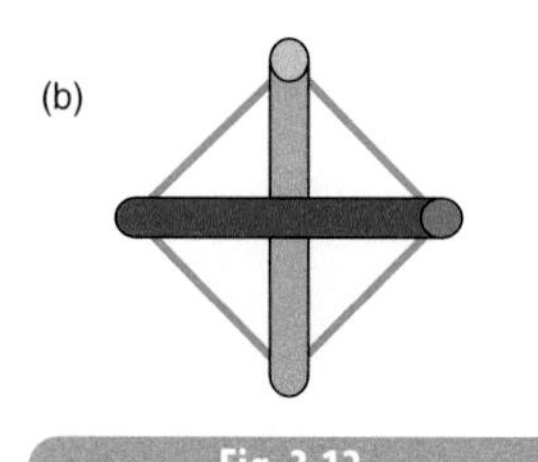

Fig. 3.12

(a) Three elements linked in a triangle bearing a vertical load; two bars are under compression while the rope is under tension. (b) A two-dimensional tensegrity structure of ropes and bars: no two compression-bearing bars are attached.

couplets may make efficient use of materials because a tensile element, in general, needs only a fraction of the cross-sectional area of a compressive element to do its job properly. Space-filling structures built from components that individually bear only tension or only compression include the so-called tensegrities, a two-dimensional example of which is drawn in Fig. 3.12(b). Coined by R. Buckminster Fuller as *tensile-integrity structures* in 1962, tensegrities are intriguing in that rigid compressive elements are often linked only by tension-bearing flexible ropes. The possibility that tensegrities can provide cells with rigidity at an economical cost of materials has been raised by Ingber (see Ingber, 1997, and references therein; Maniotis *et al.*, 1997). Certainly, the filaments of the cell do span a remarkable range of bending stiffness – a microtubule is a million times stiffer than a spectrin tetramer – and these filaments may be capable of forming a delicately balanced network if the cell could direct its assembly.

The importance of compression-bearing rods in the cell's architecture depends upon their buckling resistance; in Fig. 3.12(b), the two compressive elements will buckle if the tension sustained by the ropes is too great. Buckling occurs when a force applied longitudinally to a bar exceeds a specific threshold value, which depends upon the length of the bar and its rigidity. We calculate this buckling threshold in two steps. First, we describe the bending of a beam or rod in response to an applied torque (leading to Eq. (3.63)), then we apply this equation to the specific problem of buckling. The calculation follows that of Chapter 38 of Feynmann *et al.* (1964); a more general treatment can be found in Section 21 of Landau and Lifshitz (1986). Readers not interested in the derivation may skip to Eq. (3.69) to see the application to microtubules.

Suppose that we gently bend an otherwise straight bar by applying a torque about its ends. A small segment of the now-curved bar would look something like Fig. 3.13(a), where the top surface of the bar is stretched and the bottom surface is compressed. Near the middle of the bar (depending in part on its cross-sectional shape) lies what is called the neutral surface, within which there is no lateral strain with respect to the original shape. Let's assume that the bend is very gentle and that the neutral surface runs through the midplane of the bar. Measured from the neutral surface, the radius of curvature R is taken to be constant on the small segment in the figure.

The segment has an arc length s along the neutral surface and a length $s + \Delta s$ at a vertical displacement y, where $\Delta s > 0$ when $y > 0$. Because the arcs in Fig. 3.13 have a common center of curvature, then by simple geometry $(s + \Delta s)/s = (R + y)/R$, or

$$\Delta s\,/s = y/R. \tag{3.59}$$

However, $\Delta s\,/s$ is the strain in the longitudinal direction (the strain is the relative change in the length; see Appendix D), telling us that the

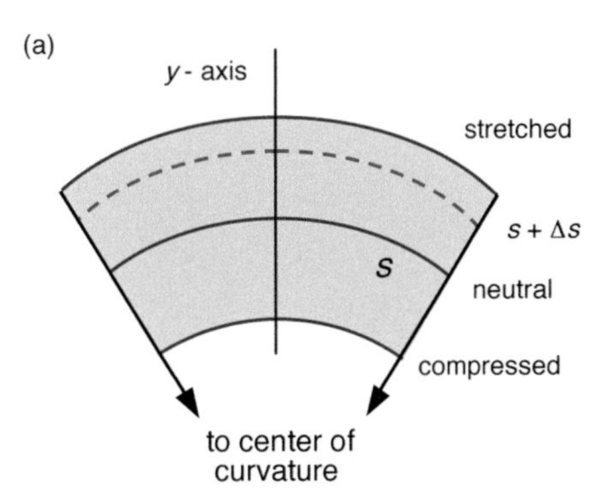

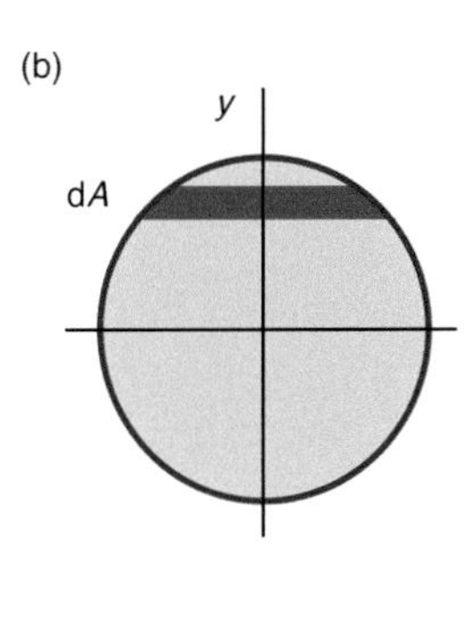

Fig. 3.13 (a) An exaggerated view of a curved rod lying in the plane of the drawing; (b) the (solid) cylindrical rod in cross section. The arc length along the neutral surface is s, which changes to $s + \Delta s$ at a vertical displacement y.

longitudinal strain at y is equal to y/R. The stress that produces this strain is the force per unit area at y, which we write as $\mathrm{d}F/\mathrm{d}A$, where $\mathrm{d}A$ is the unshaded region at coordinate y in the cross section displayed in Fig. 3.13(b). Stress and strain are related through Young's modulus Y by

$$[stress] = Y\,[strain] \tag{3.60}$$

which becomes $\mathrm{d}F/\mathrm{d}A = Yy/R$, or

$$\mathrm{d}F = (yY/R)\,\mathrm{d}A. \tag{3.61}$$

This element of force results in a torque around the mid-plane. Recalling from introductory mechanics that torque is the cross product of force and displacement, the torque must be equal to $y\,\mathrm{d}F$, which can be integrated to give the bending moment $\mathcal{M}$:

$$\mathcal{M} = \int y\,\mathrm{d}F = (Y/R)\int y^2\,\mathrm{d}A, \tag{3.62}$$

after substituting Eq. (3.61) for the force; equivalently

$$\mathcal{M} = Y\mathcal{I}/R. \tag{3.63}$$

The quantity $\mathcal{I}$ made its debut in Section 3.2 as the moment of inertia of the cross section, and has the form

$$\mathcal{I} = \int_{\text{cross section}} y^2\,\mathrm{d}A,$$

where the integration is performed only over the cross section of the bar.

We now apply Eq. (3.63) to the buckling problem, specifically the forces applied to the bar in Fig. 3.14. The coordinate system is defined with $x = 0$ at one end of the bar, whose contour length is L_c. Then, at any given height $h(x)$, the bending moment $\mathcal{M}$ arising from the force F applied to the ends of the bar is equal to

$$\mathcal{M}(x) = Fh(x), \tag{3.64}$$

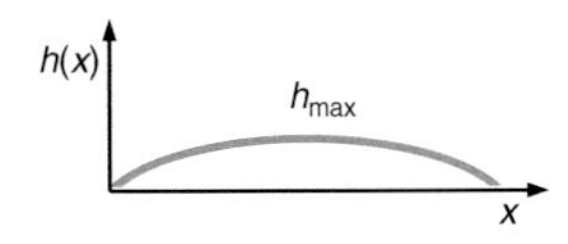

Fig. 3.14 A rod subject to a sufficiently large compressive force F in its longitudinal direction will buckle. The deformed shape can be characterized by a function $h(x)$.

as expected from the definition of torque ($\mathbf{r} \times \mathbf{F}$). We replace the moment using Eq. (3.63) to obtain

$$Y\mathcal{I} / R(x) = F\,h(x), \tag{3.65}$$

where we emphasize that the radius of curvature R is a function of position by writing it as $R(x)$. From Section 3.2, the radius of curvature at a position $\mathbf{r}$ is defined by $\mathrm{d}^2\mathbf{r}/\mathrm{d}s^2 = \mathbf{n}/R$, where s is the arc length along the curve and $\mathbf{n}$ is a unit normal at $\mathbf{r}$. For gently curved surfaces, $\mathrm{d}^2\mathbf{r}/\mathrm{d}s^2$ can be replaced by $\mathrm{d}^2h/\mathrm{d}x^2$, so that $1/R = -\mathrm{d}^2h/\mathrm{d}x^2$ (the minus sign is needed because $\mathrm{d}^2h/\mathrm{d}x^2$ is negative for our bent rod as drawn). Thus, Eq. (3.65) becomes

$$\mathrm{d}^2h/\mathrm{d}x^2 = -(F / Y\mathcal{I})h(x), \tag{3.66}$$

which is a differential equation for $h(x)$, showing that the second derivative of the height is proportional to the height itself.

From first-year mechanics courses, we recognize this equation as having the same functional form as simple harmonic motion of a spring ($\mathrm{d}^2x/\mathrm{d}t^2 \propto -x(t)$), which we know has a sine or cosine function as its solution. For the specific situation in Fig. 3.14, the solution must be

$$h(x) = h_{\mathrm{max}} \sin(\pi x / L_{\mathrm{c}}), \tag{3.67}$$

where h_{max} is the maximum displacement of the bend, occurring at $x = L_{\mathrm{c}}/2$ in this approximation. As required, Eq. (3.67) has the property that $h(0) = h(L_{\mathrm{c}}) = 0$. What we are interested in is the allowed range of forces under which buckling will occur, and this can be found by manipulating the solution given by Eq. (3.67). Taking the second derivative of this solution

$$\mathrm{d}^2h/\mathrm{d}x^2 = -(\pi/L_{\mathrm{c}})^2\, h_{\mathrm{max}} \sin(\pi x / L_{\mathrm{c}}) = -(\pi/L_{\mathrm{c}})^2 h(x). \tag{3.68}$$

The proportionality constant $(\pi/L_{\mathrm{c}})^2$ in Eq. (3.68) must be equal to the proportionality constant $F / Y\mathcal{I}$ in Eq. (3.66). This yields

$$F_{\mathrm{buckle}} = \pi^2 Y\mathcal{I} / L_{\mathrm{c}}^2 = \pi^2\kappa_{\mathrm{f}} / L_{\mathrm{c}}^2. \tag{3.69}$$

Now, this expression for the force is independent of h_{max}. What does this mean physically? If the applied force is less than F_{buckle}, the beam will not bend at all, simply compress. However, if $F > F_{\mathrm{buckle}}$, the rod buckles as its ends are driven towards each other. To find out what happens at larger displacements, greater care must be taken with the expression for the curvature. This type of analysis can be applied to the buckling of membranes as well, and has been used to determine the bending rigidity of bilayers (Evans, 1983).

The simple fact that their persistence lengths are comparable to, or less than, cellular dimensions tells us that single actin and spectrin filaments do not behave like rigid rods in the cell. On the other hand, a microtubule appears to be gently curved, because its persistence length is ten to several hundred times the width of a typical cell (see Table 3.2). Can a microtubule

Table 3.2 Linear density λ_p (mass per unit length) and persistence length ξ_p of some biologically important polymers. For the proteins ubiquitin, tenascin and titin, λ_p refers to the unraveled polypeptide.

Polymer	Configuration	λ_p (Da/nm)	ξ_p (nm)
Long alkanes	linear polymer	~110	~0.5
Ubiquitin	linear filament	~300	0.4
Tenascin	linear filament	~300	0.42 ± 0.22
Titin	linear filament	~300	0.4
Procollagen	triple helix	~380	15
Spectrin	two-strand filament	4500	10–20
DNA	double helix	1900	53 ± 2
F-actin	filament	16 000	$(10–20) \times 10^3$
Intermediate filaments	32 strand filament	~50 000	$(0.1–1) \times 10^3$
Tobacco mosaic virus		~140 000	$\sim 1 \times 10^6$
Microtubules	13 protofilaments	160 000	$4–6 \times 10^6$

withstand the typical forces in a cell without buckling? Taking its persistence length ξ_p to be 3 mm, in the mid-range of experimental observation, the flexural rigidity of a microtubule is $\kappa_f = k_B T \xi_p = 1.2 \times 10^{-23}$ J•m. Assuming 5 pN to be a commonly available force in the cell, Eq. (3.69) tells us that a microtubule will buckle if its length exceeds about 5 μm, not a very long filament compared to the width of some cells. If 10–20 μm long microtubules were required to withstand compressive forces in excess of 5 pN, they would have to be bundled to provide extra rigidity, as they are in flagella.

Both of these expectations for the bucking of microtubules have been observed experimentally (Elbaum *et al.*, 1996). When a single microtubule of sufficient length resides within a floppy phospholipid vesicle, the vesicle has the appearance of an American football, whose pointed ends demarcate the ends of the filament. As tension is applied to the membrane by means of aspirating the vesicle, the microtubule ultimately buckles and the vesicle appears spherical. In a specific experiment, a microtubule of length 9.2 μm buckled at a force of 10 pN, consistent with our estimates above. When a long bundle of microtubules was present in the vesicle, the external appearance of the vesicle resembled the Greek letter ϕ, with the diagonal stroke representing the rigid bundle and the circle representing the bilayer of the vesicle. In other words, although they are not far from their buckling point, microtubules are capable of forming tension–compression couplets with membranes or other filaments. However, more flexible filaments such as actin and spectrin are most likely restricted to be tension-bearing elements.

3.6 Measurements of bending resistance

The bending deformation energy of a filament can be characterized by its flexural rigidity κ_f. Having units of [*energy* • *length*], the flexural rigidity of uniform rods can be written as a product of the Young's modulus Y (units of [*energy* • *length*^{3}]) and the moment of inertia of the cross section $\mathcal{I}$ (units of [*length*^{4}]): $\kappa_f = Y\mathcal{I}$. At finite temperature T, the rod's shape fluctuates, with the local orientation of the rod changing strongly over length scales characterized by the persistence length $\xi_p = \kappa_f / k_B T$, where k_B is Boltzmann's constant. We now review the experimental measurements of κ_f or ξ_p for a number of biological filaments, and then interpret them using results from Sections 3.2–3.5.

3.6.1 Measurements of persistence length

Mechanical properties of the principal structural filaments of the cytoskeleton – spectrin, actin, intermediate filaments and microtubules – have been obtained through a variety of methods. In first determining the persistence length of spectrin, Stokke *et al.* (1985a) related the intrinsic viscosity of a spectrin dimer to its root-mean-square radius, from which the persistence length could be extracted via a relationship like Eq. (3.33). The resulting values of ξ_p covered a range of 15–25 nm, depending upon temperature. Another approach (Svoboda *et al.*, 1992) employed optical tweezers to hold a complete erythrocyte cytoskeleton in a flow chamber while the appearance of the cytoskeleton was observed as a function of the salt concentration of the medium. It was found that a persistence length of 10 nm is consistent with the measured mean squared end-to-end displacement $\langle r_{ee}^2 \rangle$ of the spectrin tetramer and with the dependence of the skeleton's diameter on salt concentration. Both measurements comfortably exceed the lower bound of 2.5 nm placed on the persistence length of a spectrin *monomer* (as opposed to the intertwined helix in the cytoskeleton) by viewing it as a freely jointed chain of segment length b = 5 nm and invoking $\xi_p = b/2$ from Eq. (3.34) (5 nm is the approximate length of each of approximately 20 barrel-like subunits in a spectrin monomer of contour length 100 nm; see Fig. 3.1(a)).

The persistence length of F-actin has been extracted from the analysis of more than a dozen experiments, although we cite here only a few works as an introduction to the literature. The measurements involve both native and

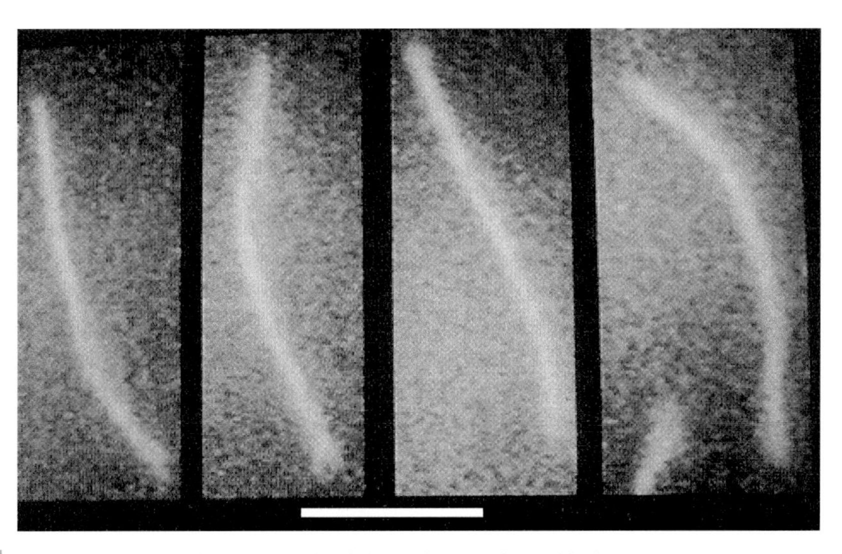

Fig. 3.15 Thermal fluctuations of a rhodamine-labeled actin filament observed by fluorescence microscopy at intervals of 6 s (bar is 5 μm in length; reprinted with permission from Isambert *et al.*, 1995; ©1995 by the American Society for Biochemistry and Molecular Biology).

fluorescently labeled actin filaments, which may account for some of the variation in the reported values of ξ_p. The principal techniques include:

(i) dynamic light scattering, which has given a rather broad range of results, converging on $\xi_p \sim 16$ μm (Janmey *et al.*, 1994);
(ii) direct microscopic observation of the thermal fluctuations of fluorescently labeled actin filaments, as illustrated in Fig. 3.15. Actin filaments stabilized by phalloidin are observed to have $\xi_p = 17$–19 μm (Gittes *et al.*, 1993; Isambert *et al.*, 1995; Brangwynne *et al.*, 2007), while unstabilized actin filaments are more flexible, at $\xi_p = 9 \pm 0.5$ μm (Isambert *et al.*, 1995);
(iii) direct microscopic observation of the driven oscillation of labeled actin filaments give $\xi_p = 7.4 \pm 0.2$ μm (Riveline *et al.*, 1997).

Taken together, these experiments and others indicate that the persistence length of F-actin lies in the 10–20 μm range, about a thousand times larger than spectrin dimers.

Microtubules have been measured with several of the same techniques as employed for extracting the persistence length of actin filaments. Again, both pure and treated (in this case, taxol-stabilized) microtubules have been examined by means of:

(i) direct microscopic observation of the bending of microtubules as they move within a fluid medium, yielding ξ_p in the range of 1–8 mm (Venier *et al.*, 1994; Kurz and Williams, 1995; Felgner *et al.*, 1996);

(ii) direct microscopic observation of the thermal fluctuations of microtubules. Most measurements (Gittes *et al.*, 1993; Venier *et al.*, 1994; Kurz and Williams, 1995; Brangwynne *et al.*, 2007) give a range of 1–6 mm, and up to 15 mm in the presence of stabilizing agents (Mickey and Howard, 1995). More recent work which examines the dependence of ξ_p on the microtubule growth rate confirms the 4–6 mm range (Janson and Dogterom, 2004);

(iii) direct microscopic observation of the buckling of a single, long microtubule confined within a vesicle under controlled conditions, leading to $\xi_p = 6.3$ mm (Elbaum *et al.*, 1996), although one experiment gives notably lower values (Kikumoto *et al.*, 2006).

Thus, the persistence length of microtubules is more than an order of magnitude larger than a typical cell diameter, with many measurements concentrated in the 4–6 mm range. Some experiments have reported that the flexural rigidities of microtubules appears to depend on the length of the filament (Kis *et al.*, 2002; Pampaloni, 2006), a situation that can arise when the shear modulus of a filament is much lower than its longitudinal Young's modulus (Li *et al.*, 2006a).

The filaments of the cytoskeleton are not the only polymers whose mechanical properties are important to the operation of the cell. For example, the packing of DNA into the restricted volume of the cell is a significant challenge, given both the contour length and persistence length of a DNA molecule. Measurements of the DNA persistence length date back at least two decades to the work of Taylor and Hagerman (1990), who found $\xi_p = 45 \pm 1.5$ nm by observing the rate at which a linear strand of DNA closes into a circle. Other experiments directly manipulate a single DNA molecule by attaching a magnetic bead to one end of the filament (while the other is held fixed) and applying a force by means of an external magnetic field. The resulting force–extension relation for DNA from bacteriophage lambda (a virus that attacks bacteria such as *E. coli*) is found to be well-described by the worm-like chain model, Eq. (3.58), which involves only two parameters – the contour length and the persistence length (Bustamante *et al.*, 1994). As well as yielding a fitted contour length in agreement with the crystallographic value, the procedure gives a fitted persistence length of 53 ± 2 nm, in the same range as found earlier by Taylor and Hagerman. Yet another approach records the motion of a fluorescently labeled DNA molecule in a fluid (Perkins *et al.*, 1995), the analysis of which gives $\xi_p \sim 68$ nm (Stigter and Bustamante, 1998), a slightly higher value than that of unlabelled DNA. The torsion resistance of DNA is described in Section 4.3; note that ξ_p may depend strongly on experimental conditions (see, for example, Amit *et al.*, 2003).

The above measurements are summarized in Table 3.2, which also displays the mass per unit length of the filament (from Section 3.1). For comparison, the table includes very flexible alkanes (from Flory, 1969) as well as the proteins procollagen (Sun *et al.*, 2002), tenascin (Oberhauser *et al.*, 1998), titin (Rief *et al.*, 1997) and ubiquitin (Chyan *et al.*, 2004). Single chains of the polysaccharide cellulose exhibit a range of persistence lengths of 5–10 nm, depending on conditions (Muroga *et al.*, 1987). The tobacco mosaic virus is a hollow rod-like structure with a linear density similar to that of a microtubule and a persistence length to match. Experimentally, the flexural rigidity of the virus on a substrate is obtained by observing the response of the virus when probed by the tip of an atomic force microscope (Falvo *et al.*, 1997). The persistence lengths of intermediate filaments are less well understood, with ξ_p of desmin lying in the range 0.1 to 1 μm as measured by dynamic light scattering (Hohenadl *et al.*, 1999). In this case and several others, the persistence length quoted in Table 3.2 is found from the flexural rigidity via Eq. (3.21). The persistence length of fibrin protofilaments with a radius of 10 nm has been measured to be 0.5 μm (Storm *et al.*, 2005).

3.6.2 ξ_p and Young's modulus

The measured persistence lengths in Table 3.2 span more than six orders of magnitude, a much larger range than the linear density, which covers about three orders of magnitude. We can understand this behavior by viewing the polymers as flexible rods, whose flexural rigidity from Eq. (3.2) is

$$\kappa_f = Y\mathcal{I}, \tag{3.70}$$

and whose corresponding persistence length, according to Eq. (3.21), is

$$\xi_p = Y\mathcal{I} / k_B T, \tag{3.71}$$

where the moment of inertia of the cross section for hollow rods of inner radius R_i and outer radius R is (from Eq. (3.13))

$$\mathcal{I} = \pi(R^4 - R_i^4)/4.$$

For some hollow biofilaments like the tobacco mosaic virus, for which $R/R_i \sim 4.5$, only a small error is introduced by neglecting R_i^4 in the expression for $\mathcal{I}$, so that

$$\xi_p \cong \pi Y R^4 / 4k_B T, \tag{3.72}$$

although we note that this expression is in error by a factor of two for microtubules ($R \sim 14$ nm and $R_i \sim 11.5$ nm; see Amos and Amos, 1991). Being raised to the fourth power, R must be known relatively well to make an accurate prediction with Eq. (3.72). Such is not always the case, and a somewhat gentler approach which, in some sense averages over the bumpy atomic boundary of a molecule, replaces R^2 by the mass per unit length λ_p

using the relationship $\lambda_p = \rho_m \pi R^2$ for a cylinder, where ρ_m is the mass per unit volume. Thus, we obtain

$$\xi_p \cong (Y / 4\pi k_B T \rho_m^2) \lambda_p^2, \tag{3.73}$$

which implies that the persistence length should be proportional to the square of the mass per unit length, if Y and ρ_m are relatively constant from one filament to the next.

Data from Table 3.2 are plotted logarithmically in Fig. 3.16 as a test of the quadratic dependence of ξ_p on λ_p suggested by Eq. (3.73). With the exception of spectrin, which is a loosely intertwined pair of filaments, the data are consistent with the fitted functional form $\xi_p = 2.5 \times 10^{-5} \lambda_p^2$, where ξ_p is in nm and λ_p is in Da/nm. Because the data span so many orders of magnitude, the approximations (such as constant Y and ρ_m) behind the scaling law are supportable, and the graph can be used to find the Young's modulus of a generic biofilament. Equating the fitted numerical factor 2.5×10^{-5} nm^3/Da^2 with $Y / 4\pi k_B T \rho_m^2$ gives $Y = 0.5 \times 10^9$ J/m^3 for $k_B T = 4 \times 10^{-21}$ J and $\rho_m = 10^3$ kg/m^3 (which is the density of water, or roughly the density of many hydrocarbons). Although no more accurate than a factor of two, this value of Y is in the same range as $Y = 1$–2×10^9 J/m^3 found for collagen (linear density of 1000 Da/nm; for collagen fibrils, see Shen *et al.*, 2008), but much smaller than that of dry cellulose (8×10^{10} J/m^3) or steel (2×10^{11} J/m^3). Of course, Eq. (3.73) can be applied to individual filaments if their radii and persistence lengths are sufficiently well known, yielding $Y \sim (0.5$–$1.5) \times 10^9$ J/m^3 for most filaments. An analysis of the indentation of microtubules using an AFM tip yields 0.6×10^9 J/m^3 (Schaap *et al.*, 2006). Note that the Young's modulus of individual

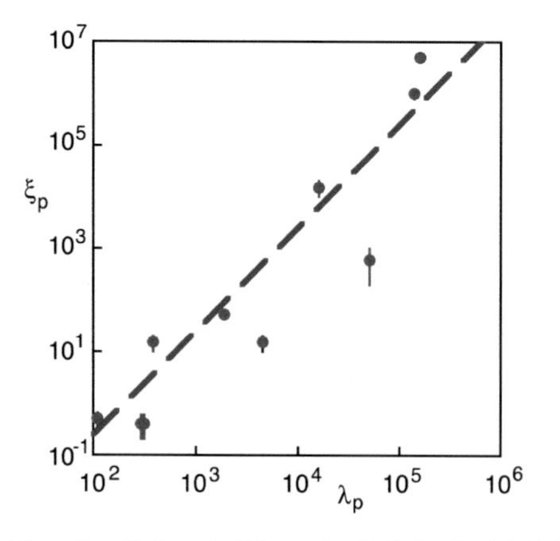

Fig. 3.16 Logarithmic plot of persistence length ξ_p against linear density λ_p for the data in Table 3.2. The straight line through the data is the function $\xi_p = 2.5 \times 10^{-5} \lambda_p^2$, where ξ_p is in nm and λ_p is in Da/nm.

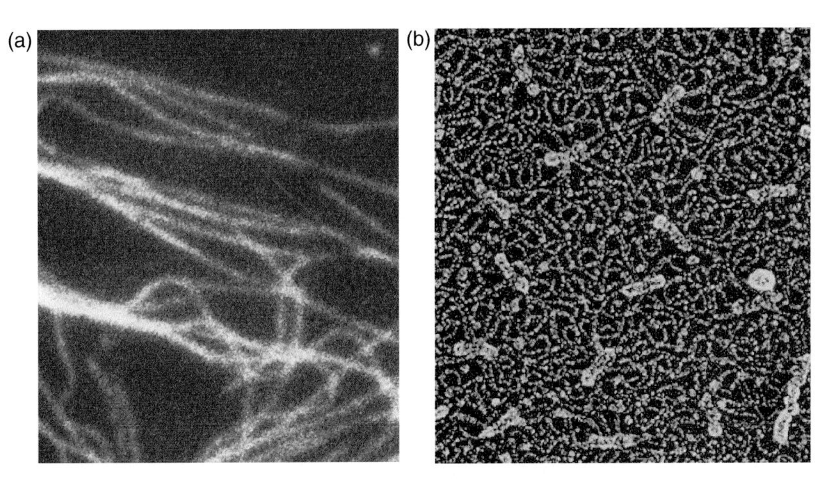

Fig. 3.17 (a) Microtubules, with a persistence length in the millimeter range, are relatively stiff on cellular length scales (image is about 10 μm across, reprinted with permission from Osborn *et al.*, 1978; ©1978 by the Rockefeller University Press), in contrast to spectrin (b), which has a persistence length five orders of magnitude smaller as seen in the human erythrocyte (distance between actin nuggets is about 100 nm; reprinted with permission from Coleman *et al.* 1989; ©1989 by Wiley-Liss; image prepared by John Heuser, Washington University; see also Heuser, 1983).

filaments may be strongly hydration dependent; for example, intermediate filaments from hagfish slime possess a Young's modulus that drops from 3.6×10^9 J/m^3 when dry to just 0.006×10^9 J/m^3 when hydrated (Fudge and Gosline, 2004). The decrease of Y upon hydration of a filament is observed for collagen fibrils as well (van der Rijt *et al.*, 2006).

3.6.3 Filament configurations in the cell

Knowing their flexural rigidities, how do we expect cytoskeletal filaments to behave in the cell? As one representative situation, we examine microtubules, which can easily be as long as a typical cell is wide (say 10 μm) and have a persistence length one hundred times the cell diameter. Microtubules should not display very strong thermal oscillations, and indeed Eq. (3.20) demonstrates that the root mean square angle of oscillation $\langle \theta^2 \rangle^{1/2}$ is about a tenth of a radian (or about 6°) for a sample microtubule with $L_c = 10$ μm if the persistence length is in the mid-range of the experimental values, say $\xi_p = 2 \times 10^3$ μm. This doesn't mean that microtubules in the cell behave quite like steel rods in a plastic bag, as can be seen by the image in Fig. 3.17(a), but a generous amount of energy is required to give our sample microtubule a substantial curvature: for example, Eq. (3.10) for E_{arc} becomes $E_{arc} = (\xi_p / 2L_c)\, k_B T$ when $R_c = L_c$, and this yields $E_{arc} = 100\, k_B T$ for $R_c = L_c =$

10 μm. Indeed, Elbaum *et al.* (1996) observe that a single microtubule with a length longer than the mean diameter of an artificial vesicle can cause the vesicle to deform into an ovoid.

Having a persistence length about one-tenth of its contour length, a spectrin tetramer should appear contorted on cellular length scales. In fact, its mean end-to-end displacement $\langle r_{ee}^2 \rangle^{1/2}$ is just 75 nm, which is only one-third of its contour length (200 nm), so the tetramers form a sinuous web when joined to form a network, as shown in Fig. 3.17(b). Thus, the network of spectrin tetramers in the human erythrocyte cytoskeleton can be stretched considerably to achieve a maximum area that is $(200/75)^2 \sim 7$ times its equilibrium area. As discussed in Section 3.4, highly convoluted chains such as spectrin resist extension, behaving like entropic springs with a spring constant $k_{sp} = 3k_BT / 2\xi_pL_c$ in three dimensions, from Eq. (3.54). Our spectrin tetramer, then, has a spring constant of about 2×10^{-6} J/m^2, which, although not a huge number, helps provide the cytoskeleton with enough shear resistance to restore a red cell to its equilibrium shape after passage through a narrow capillary. Lastly, we recall from the definition $\xi_p = \kappa_f / k_BT$ that the persistence length should decrease with temperature, if the flexural rigidity is temperature-independent. The temperature-dependence of the elasticity of biofilaments has not been as extensively studied as that of conventional polymers, but Stokke *et al.* (1985a) do find that the persistence length decreases with temperature roughly as expected for an entropic spring.

3.6.4 Filamentous cells

Many genera of cells form long filaments whose appearance ranges from beads on a string to relatively rigid rods. Two examples from the world of cyanobacteria are displayed in Fig. 2.1. The mechanical deformation of these filaments can be analyzed within the same formalism introduced earlier in this chapter: each filament possesses an intrinsic resistance to bending, twisting, etc. Yet there are few direct measurements of filament elasticity based on conventional stress–strain curves, in which the deformation of a filament is observed in response to a known applied stress. Further, the thermal fluctuations in overall filament shape are relatively tiny compared to what they are on a molecular level (see end-of-chapter problems), eliminating this approach as an appropriate technique for extracting elastic parameters. However, the scaling behavior of the flexural rigidity can be probed by constructing a tangent correlation function for randomly stirred filaments in a fluid, and one finds that κ_f increases like R^4 or perhaps R^3 with increasing filament radius R for filamentous cyanobacteria within a given genus.

Summary

Sometimes criss-crossing the interior of a cell, sometimes forming a mat or wall around it, the biological chains and filaments of the cell range in diameter up to 25 nm and have a mass per unit length covering more than three orders of magnitude, from ~100 Da/nm for alkanes to 160 kDa/nm for microtubules. A simple mathematical representation of these filaments views them as structureless lines characterized by a position $\mathbf{r}(s)$ and tangent vector $\mathbf{t}(s) = \partial\mathbf{r}(s)/\partial s$, where s is the arc length along the line. The lowest order expression for the energy per unit length of deforming a filament from its straight-line configuration is $(\kappa_f/2)(\partial\mathbf{t}(s)/\partial s)^2$, where κ_f is the flexural rigidity of the filament.

At non-zero temperature, filaments can exchange energy with their surroundings, permitting their shapes to fluctuate as they bend and twist. Their orientation changes direction with both position and time, such that the direction of the tangent vectors to the filament decays as $\langle \mathbf{t}(0) \cdot \mathbf{t}(s) \rangle = \exp(-s/\xi_p)$ due to thermal motion, where the persistence length $\xi_p = \kappa_f / k_B T$ depends upon the temperature T (k_B is Boltzmann's constant). Also, because of fluctuations, the squared end-to-end displacement $\mathbf{r}_{ee}$ of a filament is less than its contour length L_c, having the form $\langle r_{ee}^2 \rangle = 2\xi_p L_c - 2\xi_p^2 [1 - \exp(-L_c/\xi_p)]$, which approaches the rigid rod limit $\langle r_{ee}^2 \rangle^{1/2} \sim L_c$ only when $\xi_p >> L_c$. In comparison, long filaments with relatively short persistence lengths obey the form $\langle r_{ee}^2 \rangle = 2\xi_p L_c$, showing that the linear size of sinuous filaments grows only like the square root of the contour length, a property of all linear chains in which self-avoidance is neglected. When self-avoidance is enforced, the exponent n in the scaling relation $\langle r_{ee}^2 \rangle^{1/2} \propto L_c^n$ is dimension-dependent, achieving ideal behavior $n = 1/2$ only in four dimensions. Branched polymers or linear chains with attractive interactions display still different scaling behavior.

At finite temperature, $\mathbf{r}_{ee}$ does not have a unique value but rather is distributed according to a probability per unit length of the form $\mathcal{P}(x) = (2\pi\sigma^2)^{-1/2} \exp(-x^2 / 2\sigma^2)$ for filaments in one dimension, where x is the displacement from one end of the polymer chain to the other. For freely jointed chains with N identical segments of length b, the variance is given by $\sigma^2 = Nb^2/d$, where d is the spatial dimension of the chain. This distribution demonstrates that a chain is not likely to be found in its fully stretched configuration $r_{ee} = L_c$ because such a configuration is strongly disfavored by entropy. Thus, the free energy of a flexible chain rises as the chain is stretched from its equilibrium value of r_{ee}, and the chain behaves like an entropic spring with a force constant $k_{sp} = 3k_B T / 2\xi_p L_c = 3k_B T / \langle r_{ee}^2 \rangle$ in three dimensions. The ideal spring behavior $f = k_{sp} x$ is valid only at small extensions and the probability distribution function $\mathcal{P}(x)$ does not take into account the fact that $\mathcal{P}(x>L_c) = 0$: the end-to-end displacement of the chain cannot exceed its contour length. As the chain becomes increasingly stretched, the relation between applied force and extension is much better described by the worm-like chain model, where $\xi_p f / k_B T = (2(1 - x/L_c))^{-2} - 1/4 + x/L_c$.

The persistence lengths of a variety of the cell's polymers and filaments have been measured, and they span the enormous range of 0.5 nm for alkanes up to a few millimeters for microtubules. This behavior can be understood by viewing the filament as a flexible rod of uniform density and cross section, permitting the flexural rigidity to be written as $\kappa_f = Y\mathcal{I}$, where Y is the material's Young's modulus. The moment of inertia of the cross section $\mathcal{I}$ has the form $\mathcal{I} = \pi(R^4 - R_i^4)/4$ for a hollow tube of inner and outer radii R_i and R, respectively, predicting that the persistence length has the form $\xi_p = \pi Y(R^4 - R_i^4) / 4k_B T$. Treated as uniform rods, the filaments of the cell have Young's moduli in the range $(0.5 - 1.5) \times 10^9$ J/m^3, which is about two orders of magnitude lower than the moduli of conventionally "hard" materials such as wood or steel, but comparable to plastics.

Problems

Applications

3.1. The interiors of some cylindrically shaped bacteria are known to contain filaments that wind around the inside of the membrane, adopting the shape of a helix as they travel along the cylindrical part of the bacterium. Suppose the filament advances 3 μm along the cylinder while it executes one complete turn, for a cylinder of diameter 1 μm.

(a) What is the length of this section of the filament?

(b) What angle α does it make with respect to a plane perpendicular to the cylindrical axis?

3.2. Consider a large motor neuron running from the brain to the arm containing a core bundle of microtubules. Taking the persistence length of a microtubule to be 2 mm, what energy is required (in $k_B T$ at 300 K) to bend a microtubule of length 20 cm into an arc of radius 10 cm?

3.3. Let θ be the angle characterizing the change in direction of a filament along its length. For a tobacco mosaic virus of contour length 250 nm, determine the value of $\langle \theta^2 \rangle^{1/2}$ arising from thermal fluctuations. Quote your answer in degrees.

3.4. Consider a piece of spaghetti 2 mm in diameter. If the Young's modulus Y of this material is 1×10^8 J/m^3, what is the persistence length of the spaghetti at $T = 300$ K? Is the result consistent with your everyday observations?

3.5. Flagella are whip-like structures typically about 10 μm long whose bending resistance arises from a microtubule core. Treating the flagellum as a hollow rod of inner radius 0.07 μm and outer radius 0.1 μm, find its persistence length at $T = 300$ K if its Young's modulus is 1×10^8 J/m^3. Compare your result with the persistence length of a single microtubule. (*Note: this approximation is not especially trustworthy.*)

3.6. What are the structural advantages for a microtubule to be hollow? Calculate the mass ratio and the flexural rigidity ratio for a hollow microtubule with inner and outer radii 11.5 nm and 14 nm, respectively, compared to a solid microtubule with the same outer radius. What is the most efficient use of construction materials such as proteins to gain rigidity: one solid microtubule or several hollow ones?

3.7. The virus bacteriophage-λ contains a string of 97 000 base pairs in its DNA. (a) Find the contour length of this DNA strand at 0.34 nm/base-pair and compare it with the DNA persistence length. (b) If the DNA is 2 nm in diameter, what is the radius of the smallest spherical volume into which it can be packed? (c) If this DNA becomes a random chain once released into a host cell, what is its root mean square end-to-end displacement? Compare your answers to parts (b) and (c) with the size of a typical bacterium.

3.8. (a) Compare the root mean square end-to-end distance $\langle r_{ee}^2 \rangle^{1/2}$ of strands of spectrin, actin and microtubules 200 nm in contour length, using both the exact and approximate expressions from Section 3.3. Comment on the difference between the results. (b) What is the effective spring constant (in N/m) in three dimensions for each type of protein, for filaments with a contour length of 1 cm at 300 K? Use $\xi_p = 15$, 15×10^3, and 2×10^6 nm for spectrin, actin and microtubules, respectively.

3.9. Consider a 30 μm length of DNA, such as might be found in a virus. What force is required to stretch the DNA to an end-to-end displacement $x = 10$, 20 and 25 μm, according to (a) the Gaussian approximation and (b) the worm-like chain model of Section 3.4? Quote your answer in N, and assume the temperature is 300 K. Comment on the accuracy of the Gaussian approximation.

3.10. Eukaryotic cells package their DNA by wrapping short stretches of it around the rims of disk-shaped proteins called histones. In a typical situation, a 150 base-pair segment of DNA wraps itself 1.7 times around the histone core. (a) What is the curvature of the DNA? (b) What is the bending energy (in units of k_BT) associated with this deformation if the DNA persistence length is 53 nm?

3.11. Compare the two force–extension relations in Eqs. (3.55) and (3.58) by plotting $\xi_p f / k_B T$ against x / L_c. At what value of x / L_c is the difference between these curves the largest?

3.12. *Oscillatoria* is a genus of filamentous cyanobacteria that execute slow side-to-side movement with a typical angular range of $\langle \theta^2 \rangle^{1/2} \sim 0.1$. For the sake of illustration, take the length and diameter of an *Oscillatoria* filament to be 100 μm and 5 μm respectively, and its Young's modulus to be 10^9 J/m^3. (a) Determine $\langle \theta^2 \rangle^{1/2}$ arising from thermal oscillations and compare your result to the observed variation. (b) Find the buckling force applicable to this specimen.

3.13. One end of a microtubule of length $L = 5$ μm is subject to a lateral force of 0.5 pN while the other end is held fixed. Assuming this filament to have a persistence length ξ_p of 2 mm at room temperature, do the following.

(a) Find the lateral displacement z of the free end of the microtubule; use the result from Problem 3.31 without proof and quote your answer in μm.
(b) Estimate the angular displacement θ of the free end by using z from part (a).
(c) Compare this result with $\langle \theta^2 \rangle^{1/2}$ from thermal fluctuations.

3.14. In a hypothetical system, two parallel plates are linked by a large number of identical polymer chains in parallel with each other. The polymers have a length of 50 nm and a persistence length of 0.5 nm.

(a) What is the entropic spring constant for each polymer at a temperature of $T = 300$ K. Quote your answer in N/m.
(b) If the polymers uniformly cover an area of 1 cm^2 on each plate, how many of them are required to generate a macroscopic spring constant of 10^7 N/m between the plates?
(c) What is the rough spacing between polymers in part (b)? For simplicity, assume that they are arranged in a square pattern on each plate. Quote your answer in nm and compare it with the contour length of the polymer.

3.15. Find the ratio of the flexural rigidities of the mast and rigging of an old sailboat. Use the following values for Young's modulus Y and diameter of the components:

- mast: $Y = 10^{10}$ J/m^3, diameter = 30 cm,
- ropes: $Y = 2 \times 10^9$ J/m^3, diameter = 5 cm.

Compare this ratio to that for microtubules and actin, using data from Table 3.2.

Formal development and extensions

Some of the following problems require definite integrals for their solution.

$$\int_0^\infty \exp(-x^2)\,dx = \sqrt{\pi}/2 \qquad \int_0^\infty x^2 \exp(-x^2)\,dx = \sqrt{\pi}/4$$

$$\int_0^\infty x^3 \exp(-x^2)\,dx = 1/2 \qquad \int_0^\infty x^4 \exp(-x^2)\,dx = 3\sqrt{\pi}/8$$

3.16. The curvature $C(s)$ of a particular deformed filament of contour length L is described by the function $C(s) = C_o s/L$, where s is the arc length along the filament.

(a) Draw the shape of the filament over the range $0 \leq s \leq L$ when $C_o = 1/L$.
(b) For a flexural rigidity κ_f, what is the bending energy of the deformation?

3.17. The path of a uniform helix is described by the Cartesian coordinates:

$$x(s) = a\cos(2\pi s/s_o)$$
$$y(s) = a\sin(2\pi s/s_o)$$
$$z(s) = ps/s_o,$$

where s is the arc length, $s_o = (p^2 + (2\pi a)^2)^{1/2}$, and the parameters a and p have the dimensions of length (the radius and pitch of the helix, respectively).

(a) Find the unit tangent vector and unit normal vector to the curve at arbitrary s.
(b) Obtain an expression for the local curvature C.
(c) Find the behavior of C in the two limits $a >> p$ and $a << p$; interpret your results.

3.18. Show that the curvature C of the trajectory of a particle moving with velocity $\mathbf{v}$ and acceleration $\mathbf{a}$ can be found from the cross product $|\mathbf{v} \times \mathbf{a}| = Cv^3$.

3.19. Consider a polymer such as a linear alkane, where the bond angle between successive carbon atoms is a fixed value α, although the bonds are free to rotate around one another.

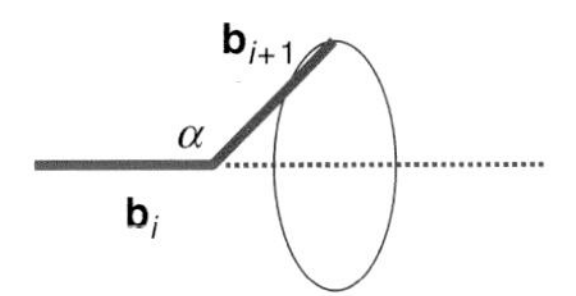

The length and orientation of the bond between atom i and atom $i+1$ defines a bond vector $\mathbf{b}_i$. Assume all bond lengths are the same, and that remote bonds can intersect.

(a) Show that the average projection of $\mathbf{b}_{i+k}$ on $\mathbf{b}_i$ is

$$\langle \mathbf{b}_i \bullet \mathbf{b}_{i+k} \rangle = b^2(-\cos\alpha)^k. \qquad (k \geq 0)$$

[*Hint: start with* $\langle \mathbf{b}_i \bullet \mathbf{b}_{i+1} \rangle$ *and iterate.*]

(b) Write $\langle \mathbf{r}_{ee}{}^2 \rangle$ in terms of $\langle \mathbf{b}_i \bullet \mathbf{b}_j \rangle$ to obtain

$$\langle \mathbf{r}_{ee}{}^2 \rangle / b^2 = N\,[1 + (2\text{–}2/N)(-\cos\alpha) + (2\text{–}4/N)(-\cos\alpha)^2 + \ldots].$$

(c) Use your result from (b) to establish that, in the large N limit

$$\langle \mathbf{r}_{ee}{}^2 \rangle = Nb^2\,(1 - \cos\alpha) \,/\, (1 + \cos\alpha).$$

(d) What is the effective bond length (in units of b) in this model at the tetrahedral value of 109.5°?

3.20. The backbones of polymers such as the polysiloxanes have alternating unequal bond angles, even though the bond lengths are all equal.

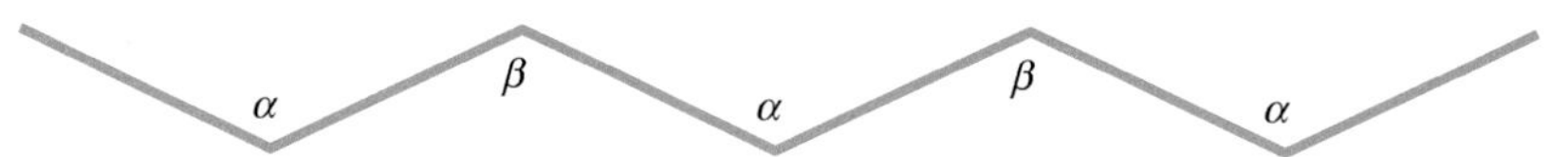

(a) Show that, if self-intersections of this type of chain are permitted, its effective bond length is $B_{\text{eff}}{}^2 = b^2\,(1 - \cos\alpha)\bullet(1 - \cos\beta) \,/\, (1 - \cos\alpha\,\cos\beta)$. [*Hint: follow the same steps as in Problem 3.19.*]

(b) Confirm that this expression reduces to the fixed-angle rotating chain expression in Problem 3.19(c) when $\alpha = \beta$.

(c) Evaluate B_{eff} when $\alpha = 109.5°$, $\beta = 130°$ and $b = 0.17$ nm.

3.21. Show that $\langle |\mathbf{r}_{ee}| \rangle = (8/3\pi)^{1/2} N^{1/2} b$ for ideal chains in three dimensions.

3.22. The radius of gyration, or root mean square radius, R_g of the N+1 vertices in a linear chain, is defined by

$$R_g{}^2 = [\Sigma_{i=1,N+1}\,(\mathbf{r}_i - \mathbf{r}_{cm})^2] \,/\, (N+1),$$

where $\mathbf{r}_i$ is the position vector of each of the N+1 vertices and $\mathbf{r}_{cm}$ is the center-of-mass position $\mathbf{r}_{cm} = \Sigma_i \mathbf{r}_i \,/\, (N+1)$. Show that $\langle R_g{}^2 \rangle = \langle r_{ee}{}^2 \rangle / 6$ for ideal chains. [*Hint: recast the problem to read* $R_g{}^2 \propto \Sigma\Sigma\, \mathbf{r}_{ij}{}^2$ *and then use* $\langle r_{ij}{}^2 \rangle = |j - i| b^2$ *where* $\mathbf{r}_{ij}$ *is the displacement between vertices i and j. Justify!*]

3.23. Find $r_{\text{ee, most likely}}$ and $\langle |\mathbf{r}_{ee}| \rangle$ for ideal chains in two dimensions.

3.24. Suppose that a particle moves only in one direction at a constant speed v but changes direction randomly at the end of every time interval Δt.

(a) Find the diffusion coefficient D of the motion as a function of v and Δt, given the diffusion equation $\langle x^2 \rangle = 2Dt$.
(b) Find the temperature dependence of D if the kinetic energy of the particle is $k_B T/2$.

3.25. Consider a three-dimensional ideal chain of 50 segments, each with length 10 nm.

(a) What are $\langle |\mathbf{r}_{ee}| \rangle$ and $\langle \mathbf{r}_{ee}^2 \rangle^{1/2}$?
(b) What is the effective spring constant (in N/m) at 300 K?
(c) If the chain has charges +/–e on each end and is placed in a field of 10^6 V/m, what is the change in the end-to-end distance?

3.26. The results in the text for the distribution of $\mathbf{r}_{ee}$ for random chains in three dimensions can be generalized easily to random chains in d dimensions. For chains whose N segments have a uniform length b, show that:

(a) the distribution of end-to-end distances has the conventional Gaussian form, but with $\sigma_d^2 = Nb^2/d$,
(b) the effective spring constant is $k_{sp} = dk_B T / Nb^2$.

3.27. Show that the flexural rigidity κ_f of a solid beam having a rectangular cross section of width w and thickness t is given by $\kappa_f = Ywt^3/12$, where the axis of the bending deformation is through the center of the rectangle as shown.

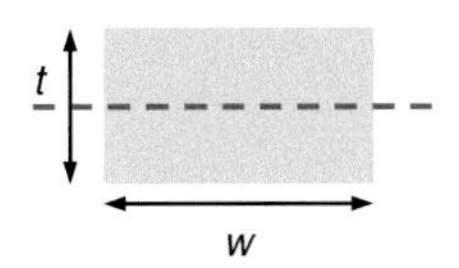

3.28. Consider a hollow rod with the cross section of a square as shown, where the length of a side is a and the thickness is t. Determine the moment of inertia of the cross section of this shape in the limit where $t << a$. Place the axis of the bending motion through the center of the square.

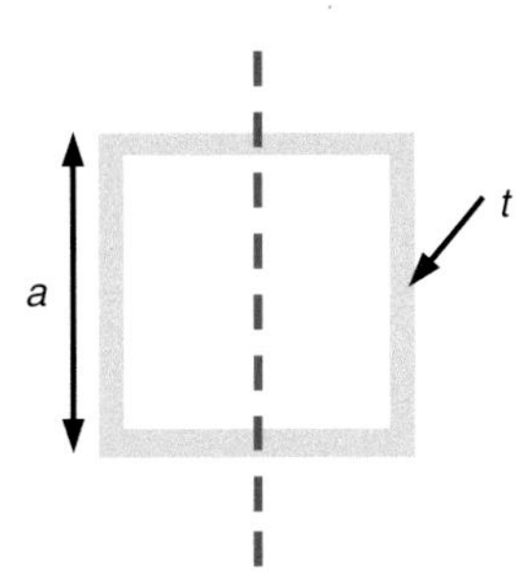

3.29. Determine the moment of inertia of the cross section $\mathcal{I}$ for the three regular-shaped rods with cross sections (S_1, S_2, H) as shown (the dashed line indicates the axis around which the rod is to bend).

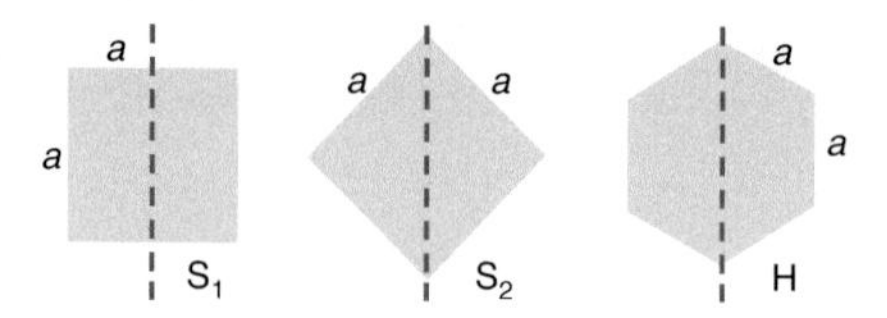

Take the length of all sides to be a. Find the ratio of these moments to that of a cylinder of radius R ($\mathcal{I} = \pi R^4/4$), imposing the condition that all shapes have the same cross-sectional area πR^2 to express a in terms of R.

3.30. Two uniform cylindrical rods, each of radius R, are joined side by side as shown. What is the moment of inertia of the cross section $\mathcal{I}$ of the combined pair? Compared to the flexural rigidity of a single rod (with a bending axis through its center), what is the rigidity of the pair? You may NOT use the parallel axis theorem.

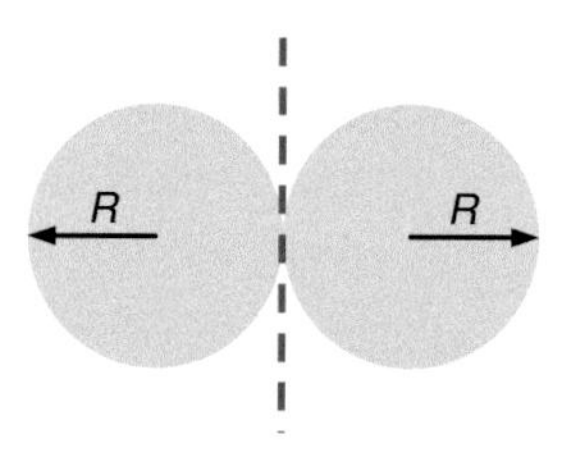

3.31. A massless rod of length L lies horizontally with one end free and one end held so that it can neither translate nor rotate. A force F is applied in the upward direction to the free end.

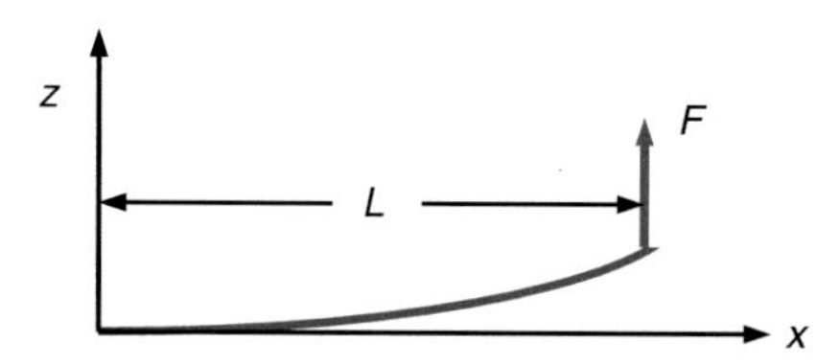

For gentle bends, show that the displacement $z(x)$ of the rod at any horizontal position x is given by

$$z(x) = (F / Y\mathcal{I}) \bullet (Lx^2/2 - x^3/6),$$

where Y and $\mathcal{I}$ are the beam's Young's modulus and moment of inertia of the cross section. What is the displacement of the rod at $x = L$? [*Hint: find the analog of Eq. (3.65) for a force perpendicular to the filament; after Chapter 38 of Feynmann et al.*, 1964].

4 Complex filaments

The polymers and biofilaments described in Chapter 3 are structurally simple, such that the deformation energy of their conformations depends mainly on their resistance to bending. Consequently, the suite of shapes adopted by such filaments represents a balance between bending energy and entropy, and unless the filaments possess permanent kinks, then they are straight at low temperature. In this chapter, we consider filaments with more complex structures that arise because of their torsion resistance or because of the attraction between non-adjacent locations of the filaments. Torsion resistance can occur even in polymers because of constraints on the rotational freedom of the polymer's backbone arising from torsion-resistant bonds or from steric constraints among side-groups to the backbone. However, in this chapter we are mainly concerned with the classic torsion resistance of thick filaments associated with shear resistance of their construction material.

This chapter begins with a descriptive overview of structures such as single and multiple helices that are observed in macromolecules and cellular filaments on length scales of one to ten nanometers. The mathematical description of helical shapes begins in Section 4.2, including the topological characteristics of twist, writhe and linking number. Then in Section 4.3, the deformation energy of a system subject to torsion is formally presented, permitting the torsional rigidity to be expressed in terms of elastic moduli such as the shear modulus and Young's modulus. Experimental measurements of the torsional rigidity are also summarized in this section. The complex architecture of globular proteins appears in both their secondary structure (including helical segments) as well their tertiary structure (overall shape). Experimental measurements of the unfolding of protein segments with secondary structure is studied in Section 4.4, while the qualitative features of RNA folding and protein folding into tertiary structure are laid out in Section 4.5. The origin of these folded structures is the effective attraction of sections of the molecule that are not immediately adjacent to each other. To quantitatively understand what type of structures arise from such interactions, a version of the HP model (for Hydrophobic/Polar model) is studied in Section 4.6. The chapter is summarized before the end-of-chapter problems.

4.1 The structure of complex filaments

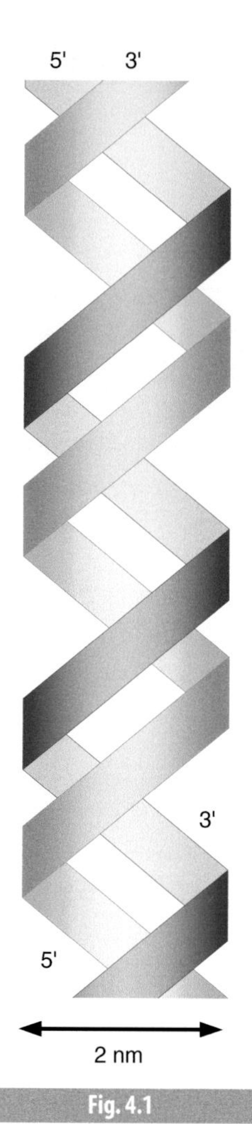

Fig. 4.1

Schematic representation of the B form of DNA. Each DNA strand has a backbone of alternating sugar–phosphate pairs, with an organic base attached to the sugar. Hydrogen bonds between the bases on opposite strands are the basis for creating the double helix. The 3′ and 5′ labels are explained in Appendix B.

The rods and ropes of importance in the cell range from macromolecules with widths of a nanometer or less, to composite filaments with widths greater than ten nanometers. We begin by reviewing the helical structure of the cell's genetic blueprint, DNA. The chemical composition of DNA is described in Appendix B: each linear DNA string is composed of nucleotide monomers (sugar–base–phosphate) linked together by a covalent bond between a sugar on one nucleotide and a phosphate on its neighbor. The organic bases lie to the side of the backbone, but enable two strings to hydrogen-bond to each other to form a double helix, as displayed in Fig. 4.1. Note the asymmetric displacement of the two helices in the figure. The B and A forms of DNA are both right-handed helices with 10 phosphate groups per turn of the helix for B-DNA and 11 phosphates for A-DNA. In contrast, the Z form of DNA is a left-handed double helix. The diameters of the A and B forms are about 2.0 nm.

Individual proteins often have substructures based on repeated sequences of residues. In some cases, the pathway through space followed by the residues in a substructure is relatively random, although the structure itself is not loose or floppy. In other cases, regular patterns such as α-helices or β-sheets are formed, sometimes as isolated units or sometimes in association with other patterns of the same kind. The Z-barrels of spectrin that were described in Chapter 3 are examples of a linked series of α-helices shown schematically in Fig. 4.2. The barrel is a folded domain within what is otherwise a filament, and the barrels will unfold when the chain is subject to a longitudinal tension. There are examples of other repeated patterns that unfold under tension, as will be described in detail in Section 4.4.

The most common structural filaments within cells are actin, microtubules and intermediate filaments; the first two of these are described in Chapter 3. Intermediate filaments have a complex hierarchical structure, as illustrated by the model in Fig. 4.3. The basic building blocks of the filament are two protein chains intertwined as a helix. Pairs of helices lie side-by-side to form a linear protofilament some 2–3 nm in width. The intermediate filament itself is a bundle of eight protofilaments in a roughly cylindrical shape about 10 nm in diameter. Many intermediate-filament monomers have masses in the 40–70 kDa range and lengths of the order 50 nm, such that the mass per unit length of a four-strand protofilament is about 4.5 kDa/nm, and that of a 32-strand filament is about 35 kDa/nm, with some variation.

Biological filaments can also be found outside of the cell. Frequently present in connective tissue such as tendon, the many types of collagen are fibrous proteins that are organized into hierarchical structures such as

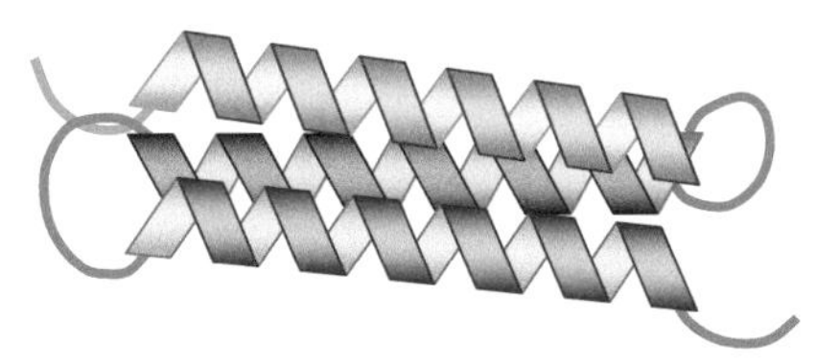

Fig. 4.2 Schematic representation of three α-helices folded to form a barrel as part of a polypeptide chain.

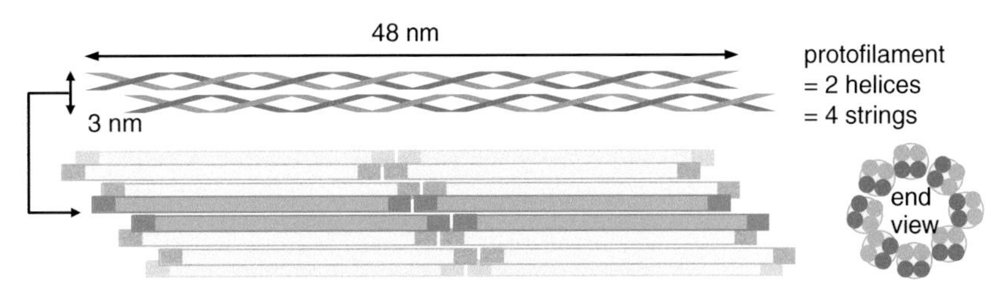

Fig. 4.3 Model of an intermediate filament, consisting of eight protofilaments, each containing four protein strings, intertwined as two helical doublets.

ropes and mats. The string of amino acids in type I collagen, for example, is arranged in a left-handed helix with 3.3 amino-acid residues per turn, and three such strands associate to form a larger molecule called tropocollagen. Displayed in Fig. 4.4, tropocollagen is about 300 nm long and 1.5 nm in diameter, with a mass per unit length of about 1000 Da/nm. Many threads of tropocollagen are organized into a collagen fibril, which may have a diameter of 10–300 nm. The collagen fibrils themselves can assemble in parallel formation into a collagen fiber.

As a last example of cellular filaments, we mention cellulose, which, as the principal tension-bearing component of the plant cell, is a linear polysaccharide made from the sugar glucose (see Appendix B for more details). A bundle of 60–70 molecular strands of cellulose lie parallel to each other to form a linear microfibril that typically has a diameter of about 10 μm. In the plant cell wall, microfibrils are themselves linked by hemicellulose molecules to form a sheet-like structure in which the fibrils are approximately aligned with each other. The cell wall then consists of multiple sheets of microfibrils, with successive sheets having different orientations of the fibrils, as in Fig. 4.5.

As a simplified model for describing the torsion response of complex filaments in the cell, let's consider a uniform cylinder that is perfectly straight in its unstressed state. When a torque $\mathcal{T}$ around the symmetry axis of the cylinder is applied to its ends, the cylinder remains straight but the ends rotate with respect to each other through an angle per unit length α, which is the total angle of rotation of the ends divided by the length of the cylinder. The relation between α and $\mathcal{T}$ has the same form as Hooke's

(a)

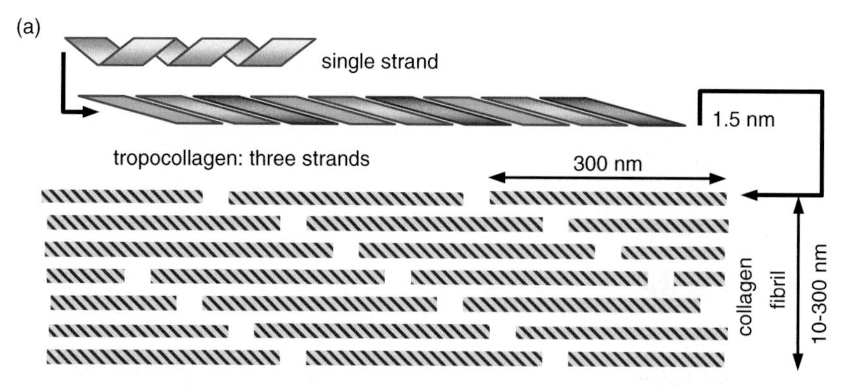

(b)

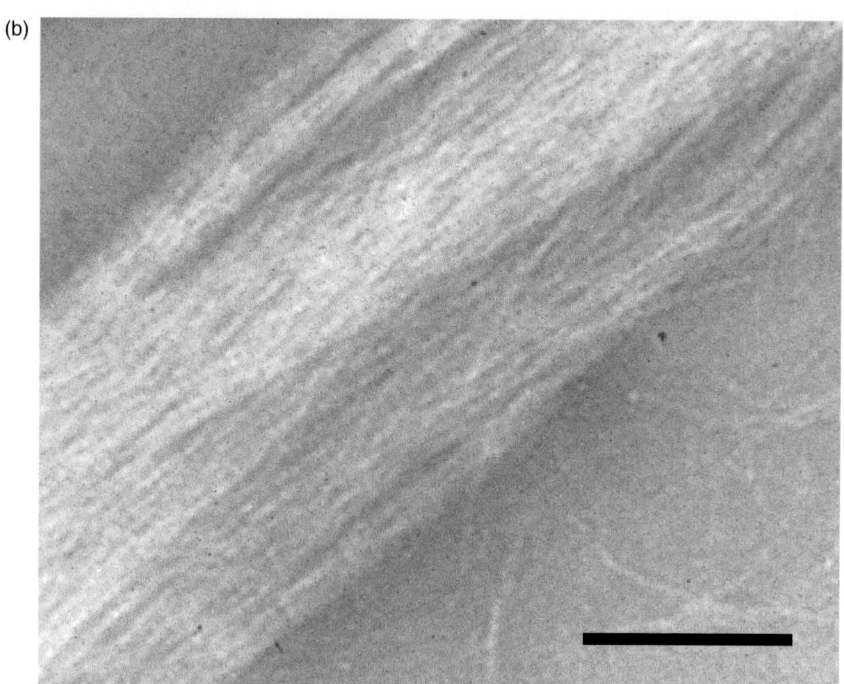

Fig. 4.4 (a) Schematic representation of the hierarchical structure of type I collagen. A single collagen molecule is a left-handed helix; three such strands form a right-handed helix of tropocollagen, many threads of which form a collagen fibril. (b) TEM image of a section of a fibril formed *in vitro* from rat type I collagen. Scale bar = 200 nm. (Courtesy of Dr. Clara Chan, Dr. Andrew Wieczorek and Dr. Nancy Forde, Simon Fraser University.)

Law for the extension of a spring, where the extension is proportional to the applied force; under torsion, the deformation angle per unit length is proportional to the applied torque for small deformations. In symbols, $\mathcal{T} = \kappa_{tor}\alpha$, where the torsional rigidity κ_{tor} is the angular analog of the spring constant in Hooke's Law.

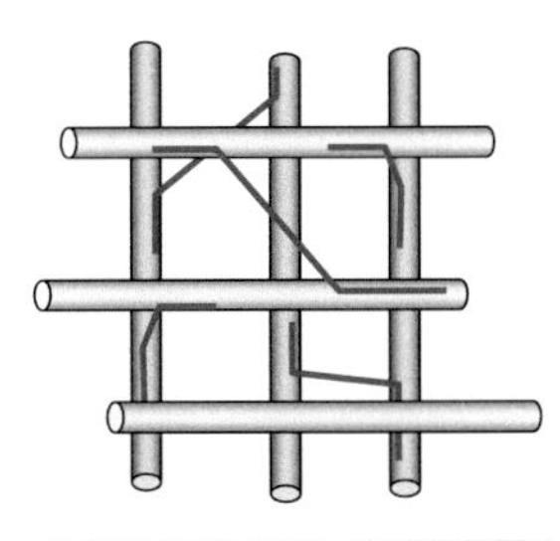

Fig. 4.5

In the plant cell wall, 60–70 individual cellulose polymers are bundled together to form a microfibril. Hemicellulose (red) links the microfibrils at 20–40 nm intervals to form oriented sheets (two orthogonal sheets are shown). The pectin network of the plant cell wall is not displayed.

For larger values of the torque, the cylinder no longer remains straight and ultimately forms a supercoil. This behavior can be seen in everyday life if one twists an electrical cord or a telephone cord: for small torsion the cord simply twists without forming a helix while at large torsion, a coil forms perpendicular to the original symmetry axis of the cord. This is often an annoying feature of telephone cords. The topology of the coiled conformations is described by the use of two characteristics, twist (Tw) and writhe (Wr); when the cord spontaneously forms a supercoil at fixed torque, the sum of the twist and writhe does not change. The conserved quantity is called the linking number Lk, where $Lk = Tw + Wr$. Twist is the number of rotations of the cord around its axis, while writhe is related to the number of times the plane of a coiled state rotates around the axis of the coil. There are other subtleties to these quantities, particularly to writhe, and these are explained in more detail in Section 4.2.

The torsional rigidity has been measured for two cellular filaments in particular – DNA and F-actin, although it may be possible to extend the current experimental techniques to other biofilaments of importance in the cell. The measurements for κ_{tor} yield (2–4) $\times 10^{-28}$ J•m for DNA and (3–8) $\times 10^{-26}$ J•m for F-actin, depending on the experimental conditions. Expressed as a length scale, the combination $\kappa_{tor}/k_B T$ is 50–100 nm for DNA and 7.5–20 μm for F-actin. These numbers are in the same range as the persistence lengths for DNA and F-actin, which are equal to $\kappa_f/k_B T$, where κ_f is the flexural rigidity. The reason why κ_f and κ_{tor} are so similar can be seen through a continuum mechanics analysis of torsional deformations, which shows that for a uniform cylinder of radius R, the torsional rigidity is given by $\mu \pi R^4/2$, where μ is the shear modulus. This expression is exactly the same as that for the flexural rigidity if $\mu = Y/2$, a relation that is approximately obeyed for many materials. The continuum mechanics prediction that κ_{tor} should scale like the fourth power of the radius explains why κ_{tor} for F-actin is three orders of magnitude larger than that of DNA.

Flexible filaments with very simple architecture, such as linear alkanes, possess elasticity by virtue of their entropy, as described in Chapter 3. At low extension, their force–extension relation is linear as in Hooke's Law, becoming smoothly non-linear as the filament becomes increasingly stretched. Complex filaments such as spectrin, with its series of folded regions in the shape of the letter Z, unfold sequentially under increasing tension. As a result, their force–extension curves exhibit a sawtooth pattern in which the filament stretches like a simple polymer until a threshold force is reached where one of the folded sections open up and the force drops substantially; this pattern is repeated for each folded section in turn. The peak force at which the filament unfolds depends on the rate at which the filament is stretched – the more gently a filament is pulled, the lower the threshold for unfolding. An analysis of the sawtooth pattern permits the bending rigidity of the filament to be determined, as described in Section 4.4.

The rigid segments that make up the folded regions of proteins such as spectrin are often helical in construction, which is one of two common patterns found in proteins. Another pattern is a two-dimensional sheet consisting of straight sections of the polypeptide chain lying in parallel; individual strands of the sheet are not by themselves rigid, they only form a regular structure by hydrogen bonding with a neighboring strand. These two patterns, α-helices and β-sheets, are the most common ones in proteins when folded into their native conformation: the helices and sheets are linked together by other segments of the same linear polypeptide that makes up the entire protein. Although the linking regions have no regular shape, the overall structure of a globular protein is fairly robust, which in most cases is a necessity for the functionality of the protein.

General features of protein folding can be studied within mathematically simple models in which the linear polypeptide is placed on a lattice and the interactions among the twenty different amino acids is greatly simplified. One of the simplest models is the HP model (for Hydrophobic/Polar model), that can have as little as just one parameter in its simplest form. Although the folded structures may just be creatures of the model, nevertheless one can see what factors are important in making a particular conformation *encodable*, meaning that there is a sequence that has that conformation as its native state. By numerical evaluation of all combinations of sequences and conformations for a model polypeptide of a given length, one can also assess which structures are the most *designable*, meaning that there are many sequences that encode for it. Computational models are also useful for studying the kinetics of folding, and the thermodynamics of the denaturation of folded states.

4.2 Torsion, twist and writhe

One topic often discussed in introductory physics courses is the torsion pendulum: a torque is applied to an object suspended on a thin wire causing the object to rotate around the wire through an initial angle ϕ. The object is then released, and it subsequently oscillates about its equilibrium position. At small displacements, ϕ is linearly proportional to the magnitude of the applied torque $\mathcal{T}$,

$$\mathcal{T} \propto \phi. \tag{4.1}$$

As we will now show, the proportionality constant in Eq. (4.1) can be written in terms of the elasticity and shape of the wire. The resulting expressions will be used in subsequent sections of this chapter to analyze the torsion resistance of biofilaments.

4.2.1 Twist without curvature

We begin by considering a rod in the shape of a uniform cylinder with length L, radius R and shear modulus μ, as in Fig. 4.6. For this subsection, we study only rods with no curvature in either their equilibrium or deformed state. Fixing the bottom end of the cylinder, a torque is applied around its top end, causing the top to rotate through an angle ϕ. Applying this same torque to a cylinder of length $2L$ would cause the end to rotate through an angle 2ϕ. This is the same effect as seen in the stretching of linear springs: when a mass m is hung from a spring of length $2L$ the end of the spring moves through twice the distance as when the mass is hung from a spring (with the same construction) of length L. Consequently, rather than use ϕ as the measure of angular deformation, it is better to use the *twist* of the cylinder α, which is the rate of change of the rotational angle ϕ as a function of the length of the cylinder, or

$$\alpha \equiv \phi / L. \qquad \text{twist, no curvature} \qquad (4.2)$$

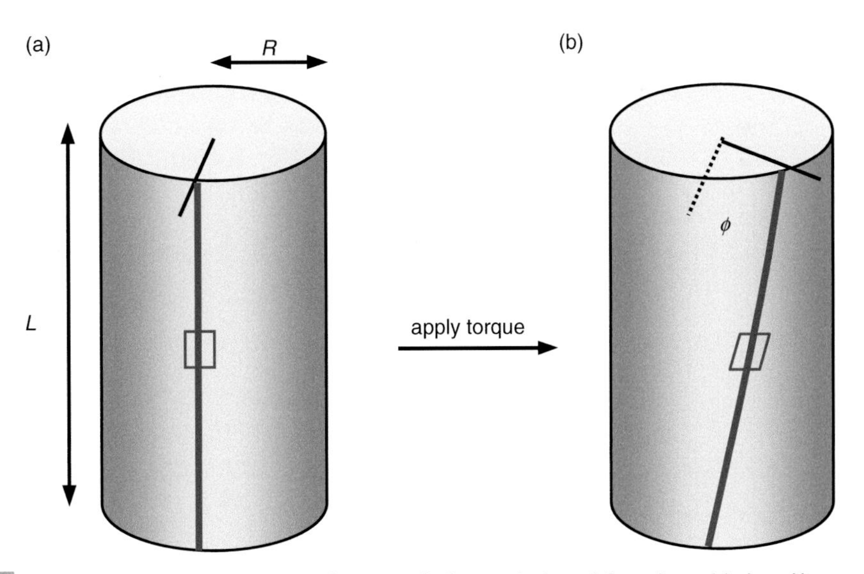

Fig. 4.6 Twisting of a uniform cylinder arising from an applied torque. In the undeformed state (a), the red line runs parallel to the axis of the cylinder, while in the deformed state (b) the top of the line is rotated by an angle $\phi = \alpha L$ with respect to the bottom, where α is the angular change per unit length along the cylinder. In this diagram, the deformation is right-handed ($\alpha > 0$); a left-handed deformation would have the opposite rotational sense.

Note that twist is not an angle, it has units of inverse length and is constant along the length of the cylinder. Twist can be either positive or negative: $\alpha > 0$ implies that $\phi > 0$ and the deformation is counter-clockwise or right-handed, while $\alpha < 0$ implies that $\phi < 0$ and the deformation is clockwise or left-handed. Equation (4.1) can be written in terms of twist as

$$\mathcal{T} = \kappa_{tor}\alpha, \qquad (4.3)$$

where κ_{tor} is the torsional rigidity, the analog of the flexural rigidity κ_f introduced in previous chapters.

Imagine that a small rectangle is drawn on the side of the undeformed cylinder in Fig. 4.6(a). When the cylinder is deformed as in Fig. 4.6(b), the top and bottom edges of this rectangle remain horizontal, whereas its sides are slanted at the same angle per unit length α as the cylinder as a whole. Thus, the initially rectangular shape becomes a parallelogram under deformation; that is, the rectangle undergoes shear. By implication, the torsional rigidity κ_{tor} should be proportional to the shear modulus μ of the cylinder just as the flexural rigidity κ_f of a filament is proportional to its Young's modulus Y. In fact, it can be shown that the torsional rigidity of a uniform cylinder is

$$\kappa_{tor} = \mu\,\pi R^4/2 = [Y/(1+\sigma_p)] \bullet (\pi R^4/4), \qquad \text{solid cylinder} \qquad (4.4)$$

which has a very similar form as the flexural rigidity $\kappa_f = Y\pi R^4/4$; both κ_{tor} and κ_f have units of [*energy*]•[*length*]. In Eq. (4.4), σ_p is the Poisson ratio.

The proof of Eq. (4.4) involves an analysis of stresses and strains that we are not quite ready to perform in this chapter; details can be found in such standard sources as Section 12.5 of Fung (1994) and Section 17 of Landau and Lifshitz (1986). From Eq. (4.4), it's easy to prove that the torsional rigidity of a hollow tube must have the form

$$\kappa_{tor} = \mu\pi(R_{outer}{}^4 - R_{inner}{}^4)/2. \qquad \text{hollow cylinder} \qquad (4.5)$$

Further, a solid rod with the cross-sectional shape of an ellipse obeys

$$\kappa_{tor} = \mu\pi a^3 b^3 \,/\, (a^2 + b^2), \qquad \text{elliptical rod} \qquad (4.6)$$

where a and b are the semi-major and semi-minor axes of the ellipse. Clearly Eq. (4.6) reduces to Eq. (4.4) when the ellipse becomes a circle ($a = b = R$).

The energy density per unit length $\mathcal{E}$ of the twist deformation can be obtained by integrating Eq. (4.3) over the twist angle per unit length to obtain the expected quadratic expression

$$\mathcal{E} = \kappa_{tor}\alpha^2/2. \qquad (4.7)$$

As explained further in Chapter 6, the Young's modulus of many solids is about two or three times the shear modulus, so the torsional rigidity ($\mu\pi R^4/2$) is similar in magnitude to the flexural rigidity ($Y\pi R^4/4$). Indeed

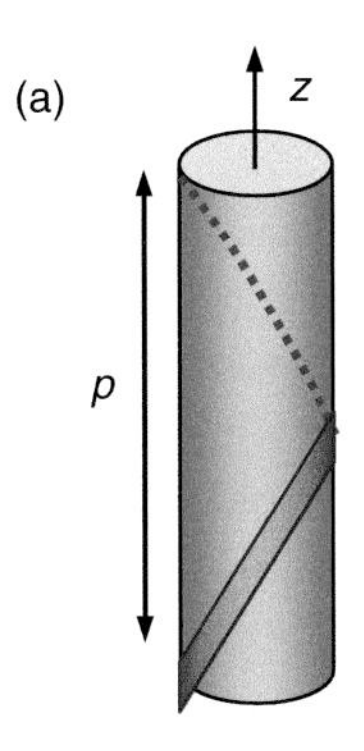

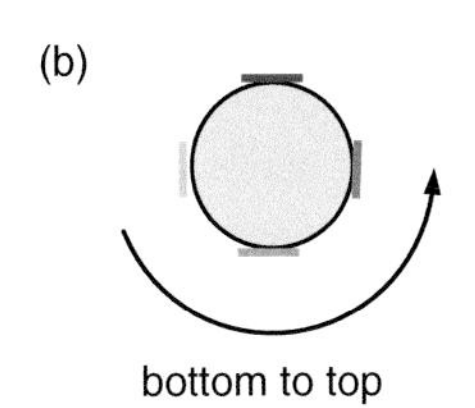

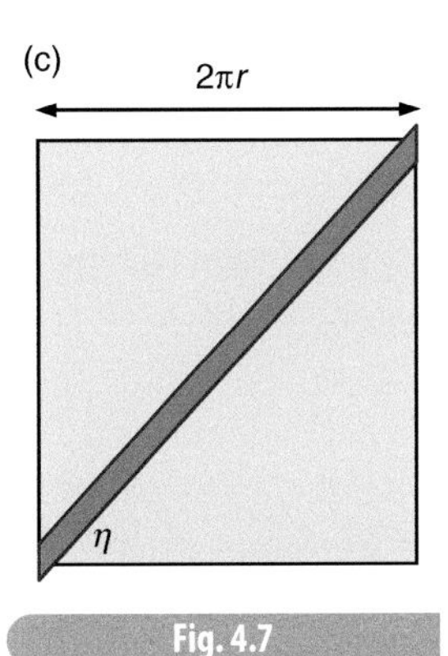

Fig. 4.7

(a) Side view of a beam with rectangular cross section as it winds around an imaginary cylinder of radius r, completing one turn in a distance p along the axis of the cylinder. (b) Top view of the beam, showing the rotation of its cross-sectional shape at intervals; the beam rotates counter-clockwise around the cylinder. (c) Location of the beam on the surface if the latter were "unrolled" and made flat.

if $Y = 2\mu$, then $\kappa_{tor} = \kappa_f$. Consequently, the energy per unit length required to twist a cylindrical rod (while keeping it straight) is not dissimilar to the energy required to bend the rod into a helix as it twists. Which configuration is favored energetically depends on the applied torque and the composition of the rod, but we now understand why telephone cords assume the complex shapes that they do as the telephone handset is rotated (with the proliferation of cell phones, this analogy will be familiar to fewer and fewer readers).

Example 4.1. A hypothetical cellular filament of diameter 10 nm and length 5 μm is held with one end fixed while the other experiences a force couplet of 1 pN that generates a torque of 10^{-20} N-m around its symmetry axis. Assuming the filament has a shear modulus of 10^8 J/m^3, find the twist angle per unit length α and the total energy of the deformation.

The first step is the numerical evaluation of the torsional rigidity from Eq. (4.4):

$$\kappa_{tor} = \mu\pi R^4/2 = 10^8 \bullet \pi \bullet (5 \times 10^{-9})^4 / 2 \cong 10^{-25} \text{ J-m}.$$

Then from Eq. (4.3) the twist α is

$$\alpha = \mathcal{T}/\kappa_{tor} = 10^{-20} / 10^{-25} = 10^5 \text{ m}^{-1}.$$

Lastly, from Eq. (4.7) the deformation energy per unit length is

$$\mathcal{E} = \kappa_{tor}\alpha^2/2 = 10^{-25} \bullet (10^5)^2 / 2 = 5 \times 10^{-16} \text{ J/m},$$

so the total energy is

$$E = \mathcal{E}L = 5 \times 10^{-16} \bullet 5 \times 10^{-6} = 2.5 \times 10^{-21} \text{ J}.$$

Thus, the deformation energy is close to $k_BT = 4 \times 10^{-21}$ J at room temperature.

4.2.2 Twist with curvature

From the similar values of the torsional and flexural rigidities of solid cylinders, we expect that the general deformation of a rod or beam under torsion may involve both twisting of the rod along its symmetry axis as well as twisting of the axis itself as it bends into a helical shape. To see how the twisting and bending modes may be coupled, we take a beam of rectangular cross section and wind it around an imaginary cylinder of radius r to form a helix as in Fig. 4.7. Panel (a) shows a side view of the beam while the panel (b) is a view down the symmetry axis of the helix. To make the twist of the beam a little clearer, the center panel only show sections through the beam at intervals along its length. Clearly, the long axis of the beam executes a spiral at the same time as the plane of the beam rotates around the imaginary cylinder; the interested reader can see the same effect by winding a ribbon around a cardboard tube.

The *pitch* (p) of the beam is defined as the length along the helical axis (not the beam axis) during which the beam completes one rotation as a helix; in Fig. 4.7, the plane of the rectangular beam also completes one rotation in this interval in this situation. If the beam were straight, then the pitch would be given by $p = 2\pi/\alpha$ from Eq. (4.2) when ϕ is set equal to 2π and L is set equal to p. However, the general relation between p and α is made complicated in the presence of curvature. First, let's understand the relation between the arc length of the beam and its pitch by "unrolling" the surface of the imaginary cylinder without changing the location of the beam on it. In the helical state, the "height" of the beam's path (its projection on the z-axis) increases linearly with the rotational angle around the axis because α is constant; consequently the beam executes a diagonal on the unrolled surface in Fig. 4.7(c). Now the base of this rectangular surface is $2\pi r$, and the overall height is just the pitch p, so the length s_{helix} of the beam in one circuit around the helix must be

$$s_{helix} = [(2\pi r)^2 + p^2]^{1/2}. \tag{4.8}$$

From Fig. 4.7(c), it can also be shown that the angle η between the beam and the plane perpendicular to the helical axis is given by

$$\tan\eta = p/2\pi r. \tag{4.9}$$

Compared to the simple twisting of a beam about its own (straight) axis in Fig. 4.6, what makes the geometry of Fig. 4.7 interesting is that the beam twists around its axis at the same time as that axis follows a helical path through space. Although the mathematical surface that the beam hugs in Fig. 4.7 is referred to as imaginary, a physical realization of it would be the winding of DNA (itself a helix) around a histone barrel to form another helix of unequal radius. Consequently, there are two contributions to the orientation of the beam from one location to the next, as illustrated in Fig. 4.8. Panel (a) shows a series of rectangles as their orientation twists around a straight-line path. The unit tangent vector **t** to the path remains fixed along the straight line, so the change Δ**t** from location to location is zero. However, a vector **n** normal to the path and attached to the rectangle rotates around the axis of the path, so that Δ**n** is not zero. Note that both **n** and Δ**n** are perpendicular to **t** here. Panel (b) shows a different situation in which the normal vectors to the path do not change orientation, so now $\Delta\mathbf{n} = 0$, while the tangent vectors to the path rotate along the path with $\Delta\mathbf{t} \neq 0$. In panel (b), both **t** and Δ**t** are perpendicular to **n**.

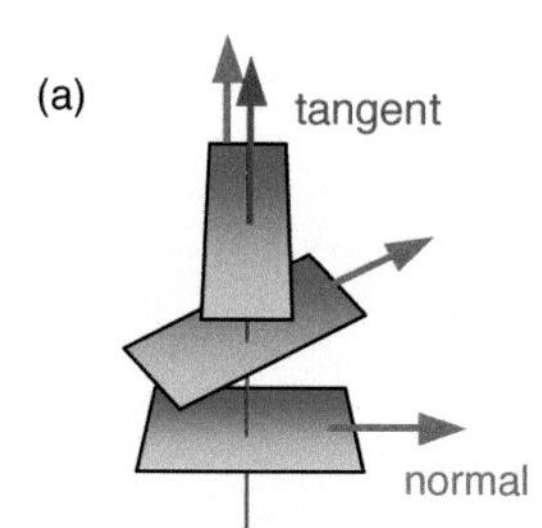

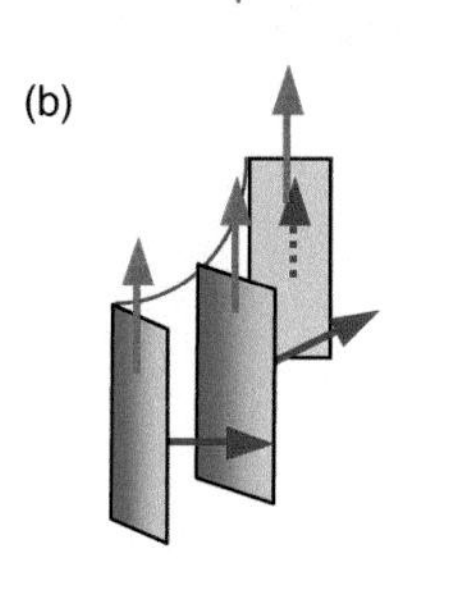

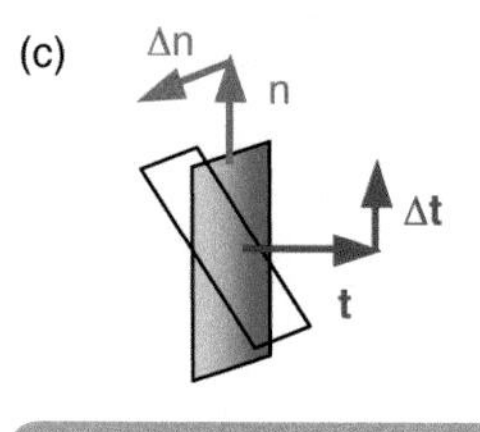

Fig. 4.8

(a) Positions of a series of rectangles following a straight axis: the normal vectors **n** rotate but the tangents **t** do not. (b) Positions of a series of rectangles following a circular path: the tangent vectors **t** rotate but the normals **n** do not. (c) Two orientations of a rectangle where the difference in the tangent vector Δ**t** and the normal vector Δ**n** are orthogonal.

As shown in Fig. 4.8(c), a change in the orientation of the cross section through a beam involves both a change $\Delta\mathbf{t}$ in the tangent vector as well as a rotation $\Delta\mathbf{n}$ of the normal in the plane of the cross section. As drawn, these changes are themselves orthogonal, so the magnitude of the total change is found from the vector sum to be $(\Delta\mathbf{t}^2 + \Delta\mathbf{n}^2)^{1/2}$. Let's find $\Delta\mathbf{t}$ and $\Delta\mathbf{n}$ over a small distance Δs along the path of the helix. The change in the direction of the tangent is the product of the curvature C and Δs, or $|\Delta\mathbf{t}| = C \bullet \Delta s$. Similarly, the change in the orientation of the beam's rectangular shape with respect to the beam's axis is the product of the twist α and Δs, or $|\Delta\mathbf{n}| = \alpha \bullet \Delta s$. Thus, $|\Delta\mathbf{n}| \,/\, |\Delta\mathbf{t}| = \alpha / C$. Given that the total change in orientation is just 2π over the arc length s_{helix}, then the orientational change that occurs over the small element of arc Δs must be $2\pi\ (\Delta s \,/\, s_{\text{helix}})$. Hence,

$$[2\pi\ (\Delta s \,/\, s_{\text{helix}})]^2 = \Delta\mathbf{t}^2 + \Delta\mathbf{n}^2 = (C\ \Delta s)^2 + (\alpha\ \Delta s)^2, \tag{4.10}$$

or

$$(2\pi \,/\, s_{\text{helix}})^2 = C^{\,2} + \alpha^2. \tag{4.11}$$

Although the result will look a little messy, let's use Eq. (4.8) to eliminate s_{helix} from Eq. (4.11), leaving

$$(2\pi)^2 \,/\, (C^{\,2} + \alpha^2) = (2\pi r)^2 + p^2. \tag{4.12}$$

This is a first step in expressing C and α in terms of r and p.

At the angle η in Fig. 4.7, the tangent vector $\mathbf{t}$ rotates around the helical axis, covering a distance of $2\pi\ \cos\eta$ in one complete revolution. Thus, for a given Δs, $|\Delta\mathbf{t}| = 2\pi\ \cos\eta\ (\Delta s \,/\, s_{\text{helix}})$. Similarly, $\mathbf{n}$ rotates around the helical axis, but at a different radius, covering a distance of $2\pi\ \sin\eta$ in one complete revolution. Thus, $|\Delta\mathbf{n}| = 2\pi\ \sin\eta\ (\Delta s \,/\, s_{\text{helix}})$, which can be combined with the expression for $|\Delta\mathbf{t}|$ to yield $|\Delta\mathbf{n}| \,/\, |\Delta\mathbf{t}| = \tan\eta = p \,/\, 2\pi r$, where the second equality follows from Eq. (4.9). Now we showed in the previous paragraph the additional relation $|\Delta\mathbf{n}| \,/\, |\Delta\mathbf{t}| = \alpha \,/\, C$. Combining both expressions for $|\Delta\mathbf{n}| \,/\, |\Delta\mathbf{t}|$ we obtain

$$p \,/\, 2\pi r = \alpha \,/\, C. \tag{4.13}$$

Equations (4.12) and (4.13) can be solved to yield

$$r = C \,/\, (C^{\,2} + \alpha^2), \tag{4.14a}$$

$$p = 2\pi\alpha \,/\, (C^{\,2} + \alpha^2) \tag{4.14b}$$

and

$$\alpha = 2\pi p \,/\, (4\pi^2 r^2 + p^2) \tag{4.15a}$$

$$C = 4\pi^2 r \,/\, (4\pi^2 r^2 + p^2). \tag{4.15b}$$

Before leaving these two expressions, we check their limiting values. If the beam just forms a circle (in a plane) and does not rise to form a helix, then the twist angle per unit length α vanishes and Eq. (4.14a) gives $r = 1/C$ as it should: the curvature is just the reciprocal of the radius of the circle. Similarly, if the beam twists around a straight line with no curvature ($C=0$), then Eq. (4.14b) gives $p = 2\pi/\alpha$ as it should: the pitch of the twisting beam is equal to 2π divided by the change in angle per unit length.

4.2.3 Twist and writhe

The shape that a rod assumes when it is subject to torsion is the result of a competition between bending and torsion resistance, which in turn depends on the cross sectional shape of the rod. Let's consider a rod with a rectangular cross section (thinner than it is wide), choosing a belt as an example. A belt can be twisted fairly easily into the form displayed in Fig. 4.9(a): one end of the belt in the diagram has been twisted through two complete rotations about its long axis as shown by the alternating colors, one for each side of the belt. The semi-circular arrow in the diagram indicates the direction of the twist to form a right-handed spiral; large enough longitudinal forces are applied to the opposite ends of the belt to keep its axis straight. If the forces are reduced, we know from experience that the belt may untwist slowly as at first one, and then two, loops appear; this topology is shown in Fig. 4.9(b). The thinner the belt, the more likely that loops will appear, an effect all too familiar with the very thin paper decorations used at parties.

The reason for the change in the appearance of the belt lies in the dependence of the torsional and bending rigidities on cross-sectional shape. For a solid cylinder, we know that ratio of the flexural to torsional rigidities is given by $\kappa_f / \kappa_{tor} = (Y/\mu)/2$, after canceling a common factor of πR^4 in

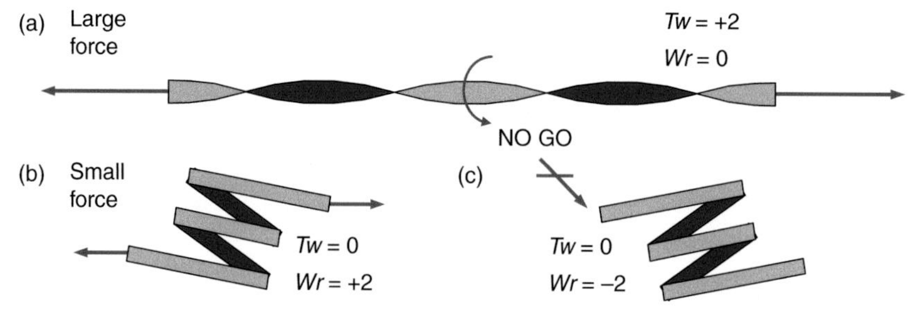

Fig. 4.9 Sample configurations available to a belt subject to torsion. The configurations (a) and (b) involve right-handed spirals and can be smoothly deformed into one another. In contrast, configuration (c) is left-handed, and cannot be obtained from (a) or (b) without cutting the belt and rejoining it, assuming that the ends of the belt are fixed.

the expressions for κ_f and κ_{tor}. However, a different ratio is obtained for rectangular beams having width w and thickness t. The flexural rigidity of a beam with this cross section is exactly $\kappa_f = Ywt^3/12$, while the torsional rigidity is $\kappa_f = \mu wt^3/3$ when $t << w$. Thus, for thin beams, $\kappa_f / \kappa_{tor} = (Y/\mu)/4$, which is half the value of this ratio for cylinders. In other words, it becomes relatively easier to bend a rod than twist it (around a straight axis) as one moves from a symmetric shape like a cylinder to a highly asymmetric shape like a thin beam or ribbon.

The two shapes in Fig. 4.9 parts (a) and (b) are a warning that a given configuration may be deformable into many other shapes without having to cut the belt and glue it back together. An example of a shape that is not available to a continuous deformation of Fig. 4.9(a) is displayed in Fig. 4.9(c): if the ends of the belt are fixed, then the right-handed spiral of part (a) cannot be deformed into the left-handed loops of part (c) without cutting the belt. Mathematicians use two terms, twist and writhe, to describe the overall topology of rods such as those in Fig. 4.9. Twist, given the symbol Tw, is the number of complete turns made by a vector normal to the axis of the rod (and in its plane) as it is propagated from one end to the other. In Fig. 4.9(a), the plane of the rod rotates about the axis twice (not four times) from the bottom to the top, in a right-handed spiral. Thus, Fig. 4.9(a) has $Tw = +2$, where the plus sign indicates right-handedness. In panels (b) and (c) of Fig. 4.9, a normal to rod's axis (and lying in the plane of the rod) does not change direction around the loop, and so both of these configurations have $Tw = 0$.

The difference between the configurations in panels (b) and (c) lies in their writhe, given the symbol Wr. In the context of a beam or ribbon with free ends (i.e. ends that are not attached to each other), the writhe is the number of loops, taking into account the handedness of the spiral. In panel (b), the writhe is +2 because there are two complete loops and the spiral is right-handed, while the mirror image of this configuration is left-handed with $Wr = -2$, as shown in panel (c). If the ends of the beam or ribbon are linked to form a circle, the writhe may be found in a similar way. For illustration, several closed-loop configurations with their corresponding writhe are displayed in Fig. 4.10; all of the configurations are right-handed, but their left-handed counterparts can be found by taking their mirror images. Looking down the long axis of configuration (b) in the plane of the diagram, the plane of the loop does half a turn (counter-clockwise) as the path of the rod is followed from top to bottom, then another half as the path is followed back to the top, for a total writhe of +1. Similarly, in configuration (c), the plane executes one complete right-handed (counter-clockwise) turn on the way down, and a second one as the path returns to the top. This figure does not represent the most general situation in that no consideration is given to the twist of the rod; closed loops may have non-zero values for both Tw and Wr.

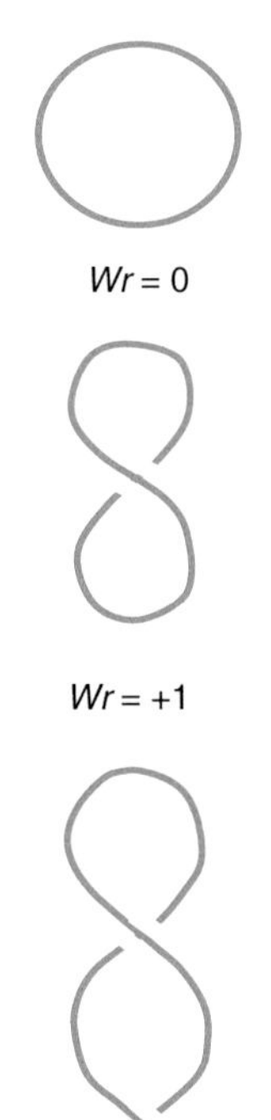

Fig. 4.10

Sample configurations available to a filament joined end-to-end as a closed loop, with their corresponding writhe as indicated. No reference is made to the twist of the filament, which may also be non-zero. None of these configurations is left-handed.

Configurations (a) and (b) in Fig. 4.9 can be continuously deformed into one another without cutting the belt and reattaching it after increasing or decreasing its twist. Yet, both the twist and writhe of configurations (a) and (b) are different, which tells us that Tw and Wr each can change as the path of the belt changes. This does not mean that Tw and Wr can change continuously to any arbitrary value, however, as configuration (c) with $Tw = 0$ and $Wr = -2$ cannot be reached without cutting the belt, twisting it and then rejoining it. The rule governing what values of Tw and Wr are permissible in a continuous deformation is that their algebraic sum must be constant. The sum is called the linking number, Lk:

$$Lk = Tw + Wr, \tag{4.16}$$

where the signs of Tw and Wr must be taken into account when performing the sum. For example, configurations (a) and (b) in Fig. 4.9 both have $Lk = +2$, and are continuously deformable into each other, while configuration (c) has $Lk = -2$ and consequently is not accessible from the other configurations. The fact that Lk does not change under a continuous deformation ($\Delta Lk = 0$) means that ΔTw and ΔWr are correlated: $\Delta Tw = -\Delta Wr$ according to Eq. (4.16). There are other mathematical properties of twist and writhe which are important to some of their applications to DNA, and the interested reader is referred to standard texts for further reading.

4.3 Measurements of torsional rigidity

There are rather fewer measurements of the torsional rigidity κ_{tor} of filaments in the cell than there are of the flexural rigidity κ_f as compiled in Section 3.5. This should not surprise us: the bending resistance can be extracted by means of several experimental approaches (stress–strain relationship, thermal fluctuations of shape, filament buckling), and some of these are challenging to implement in the context of torsion. Only the response of DNA and actin to torsion has been well studied so far, although efforts are underway to measure other important biofilaments.

The magnitude of the side-to-side motion of a filament arising from thermal fluctuations in its energy, as introduced in Chapter 1, is inversely proportional to its bending resistance. The torsional analog to this is the random rotation of a filament about its axis of symmetry; again, the larger the fluctuations in the rotational angle, the smaller the resistance to torsion. Several experiments have been performed using this approach, such as that developed by Tsuda *et al.* (1996) to obtain the torsional rigidity of F-actin. The idea is to fix one end of a filament while a micron-scale sphere attached to the other end is held in an optical trap, free to rotate about

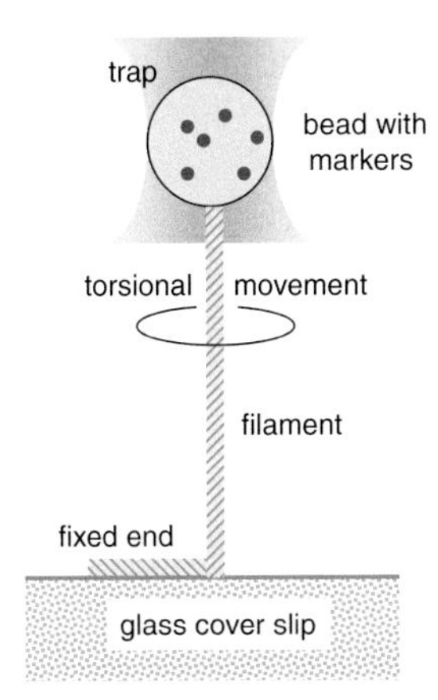

Fig. 4.11

Measurement of the rotation of a filament associated with thermal fluctuations in its torsional energy. The bottom end of the filament is fixed to a substrate while the top is attached to a bead held in an optical trap, which is adjusted to pull the biofilament taut. Seen through a microscope with a viewing axis parallel to the filament, tiny fluorescent markers on the bead indicate the rotational angle of the bead around the filament axis. The diagram is not to scale: a typical bead diameter is 2 μm, two orders of magnitude larger than the diameters of cellular filaments like F-actin or microtubules.

the mean location of the symmetry axis of the filament, as in Fig. 4.11. Placing fluorescent markers on the surface of the sphere permits the measurement of its rotation in the plane perpendicular to the axis of the filament. Defining ϕ as the rotational angle with respect to the mean azimuthal orientation of the sphere, then the distribution in measured values $N(\phi)$ of ϕ should be governed by the Boltzmann factor:

$$N(\phi) \propto \exp(-\kappa_{\text{tor}}\phi^2 / 2k_BTL), \tag{4.17}$$

where L is the length of the filament being studied. The variance $\langle \phi^2 \rangle$ of the rotational angle can be shown to obey

$$\langle \phi^2 \rangle = k_BTL / \kappa_{\text{tor}}. \tag{4.18}$$

The torsional rigidity can be obtained from the observed distribution $N(\phi)$ as well as the measured values of $\langle \phi^2 \rangle$ as a function of L.

Another possibility for extracting κ_{tor} is the buckling of a filament subject to a torque about its symmetry axis, which is the angular analog of the buckling of a filament under longitudinal compression. As described in Chapter 3, compressive buckling was used successfully to obtain κ_f of microtubules. In the torsional case, the deformation of a uniform straight cylinder under a small torsional stress is a uniform twist about the symmetry axis, as in Fig. 4.6. But as the stress increases, the cylinder will buckle and become a helix or some other supercoil configuration, depending on the magnitude of the stress and boundary conditions on the filament (for buckling of cylindrical tubes, see Flügge 1973, p. 463; Yi *et al.*, 2008).

The properties of supercoils in DNA have been well studied from both theoretical and experimental perspectives because of their importance in DNA replication and packaging. In its relaxed state, a segment of a DNA double helix is characterized by a linking number Lk_o equal to the number of helical turns Tw_o (the length of the segment divided by the helical repeat distance of 3.4 nm for B-form DNA). Both in the cell and in the lab, the linking number of a DNA segment need not be Lk_o, and it is conventional to characterize the segment by the relative change in linking number σ_{Lk}, defined by

$$\sigma_{\text{Lk}} \equiv (Lk / Lk_o) - 1. \tag{4.19}$$

For most cells, σ_{Lk} is negative, with a range of about −0.05 to −0.07; however, hyperthermophilic archaea may display positive σ_{Lk} as high as +0.03 (López-García, 1998).

The conformations adopted by a filament such as DNA depend on both σ_{Lk} and the tensile force applied along the filament. At high tensile force, the filament will be extended along its length, adopting a helical shape with $Tw \neq 0$ for $\sigma_{\text{Lk}} \neq 0$. As the stretching force is reduced, twist will be converted into writhe if the linking number is fixed, and shapes called

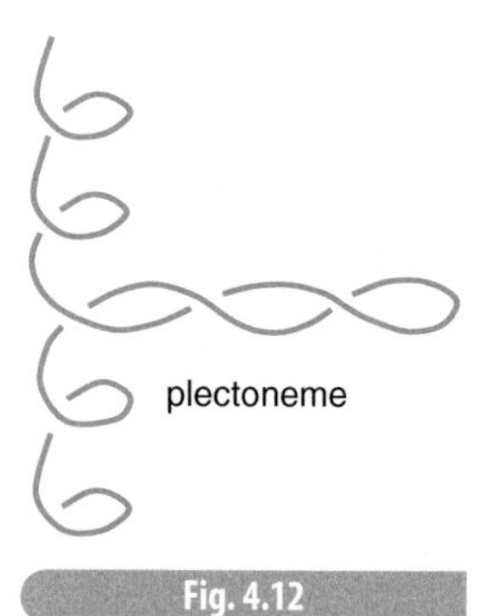

Fig. 4.12

Formation of a plectoneme from an overwound coil.

plectonemes will emerge, as in Fig. 4.12. Although the calculation of the energy-minimizing shape of the filament generally must be done numerically (for example Yang *et al.*, 1993), the problem has been solved by approximation by Marko and Siggia (1994) to reveal the relative importance of various contributions to the filament energy and entropy. The appearance of plectomenes in DNA has been investigated experimentally (Strick *et al.*, 1996; Brutzer *et al.*, 2010).

The mathematical theory of buckling under torsion and the formation of supercoils is more complex than buckling under compression as presented in Section 3.5. To begin with, the formal description of a coiled filament involves more than just following the tangent vector $\mathbf{t}$ to the filament as a function of arc length. For torsion, one also follows the orientation of two vectors normal to $\mathbf{t}$ that rotate as the filament twists (for a more detailed picture, see Yang *et al.*, 1993). Further, the cross-sectional shape of a filament like DNA is not circular, which may be important in some situations (see, for example, Kessler and Rabin, 2003, and references therein). The presence of torsion resistance will also affect some results obtained for polymers in Chapter 3; for example, the force–extension relation of a linear polymer under tension will be modified when $\kappa_{tor} \neq 0$ (Moroz and Nelson, 1997b; Bouchiat and Mézard, 1998). While a limited number of classic deformation problems have been solved analytically with $\kappa_{tor} \neq 0$ (see Love, 1944), others have only yielded to numerical methods. The properties of filaments with intrinsic curvature and torsion are described in Zhou *et al.*, (2005); the spontaneous coiling of membrane tubes has been observed by Tsafrir *et al.* (2001).

We now review some of the measurements of κ_{tor} for filaments of importance in the cell, organizing the presentation of the results according to the composition of the filament.

4.3.1 DNA

The earliest determinations of κ_{tor} for DNA were based on spectroscopic techniques such as fluorescence polarization anisotropy (for an overview, see Fujimoto *et al.*, 2006). The development of single molecule manipulation through optical or magnetic traps permitted the direct observation of the response of a DNA helix to an applied torque. For example, the approach of Bryant *et al.* (2003) uses a variant of the experimental configuration in Fig. 4.11: one end of a filament under tension is held fixed while the other is rotated by means of an attached micropipette. The measurement consists of tracking the rotation of a marker bead placed partway along the DNA segment as the wound DNA is allowed to relax toward its stress-free configuration. The measured velocity of the marker (around the axis of the DNA) permits the calculation of the drag force it experiences, which in turn generates a torque vs. displacement graph and κ_{tor}.

The torsional rigidity may depend on conditions at the molecular level, such as the length of the DNA segment (Horowitz and Wang, 1984) or the presence of bound molecules (Wu *et al.*, 1988) or some aspects of filament curvature (Kahn *et al.*, 1994). However, the variation of κ_{tor} with the local base-pair sequence (Fujimoto and Schurr, 1990) is not observed to be strong. Quoted values for κ_{tor} lie mostly within $(2-4) \times 10^{-27}$ J•m, although this variation is outside the statistical accuracy of each individual experiment. Fujimoto *et al.* (2006) argue that the discrepancy among the values of κ_{tor} found by different experiments may reflect the stress state of the DNA being examined: states of higher stress (bending or stretching) tend to display higher κ_{tor}.

4.3.2 F-actin

Some of the pioneering measurements of κ_{tor} for filamentous actin were performed using spectroscopic techniques such as transient absorption anisotropy or related methods. In these experiments, the physical rotation of the filament is not observed directly, but the torsion resistance is obtained from an analysis of the decay of a spectroscopic signal. Two such experiments yield values for κ_{tor} in the range of $(1-2) \times 10^{-27}$ J•m (Yoshimura *et al.*, 1984; Prochniewicz *et al.*, 1996). As will be described below, these values are more than an order of magnitude lower than what one would expect for the measured flexural rigidity κ_f if actin behaves like a uniform cylinder.

The advent of optical traps led to the development of experimental configurations in which the rotation of filamentous actin could be imaged directly by microscope (for a review of the application of optical traps to single molecule measurements, see Neuman *et al.*, 2007). Two such experiments for measuring thermal fluctuations in rotational angle involve:

- attaching a pair of beads part-way along a taut filament and determining the rotational angle of the filament from the relative location of the beads with respect to each other (Yasuda *et al.* 1996),
- attaching to the free end of a taut filament a single bead coated with fluorescent markers to display the rotation (Tsuda *et al.*, 1996).

The "two-bead" experiment yields values for κ_{tor} that depend on the choice of ion bound to the actin monomers: $\kappa_{tor} = (2.8 \pm 0.3) \times 10^{-26}$ J•m for bound Mg^{2+} ions and $\kappa_{tor} = (8.5 \pm 1.3) \times 10^{-26}$ J•m for bound Ca^{2+} ions (Yasuda *et al.*, 1996). Within the same experiment, the flexural rigidity is found to be $\kappa_f = (6.0 \pm 0.2) \times 10^{-26}$ J•m, in the middle of the range of κ_{tor} values and largely independent of the nature of the bound cation.

The "single-bead" experiment yields a value at the high end of this range, namely $(8.0 \pm 1.2) \times 10^{-26}$ J • m (Tsuda *et al.*, 1996).

4.3.3 Microtubules

Microtubules are moderately stiff by cellular standards, and this may limit the accuracy of measurements that rely on thermal fluctuations, as we now show. The flexural rigidity and torsional rigidity of uniform cylinders both scale with the radius of the cylinder in the same way, namely R^4. To a first approximation, then, one would expect that the ratio of κ_{tor} for microtubules compared to actin should be the same as the ratio of κ_f for microtubules compared to actin, or about three orders of magnitude according to Table 3.2. That is, continuum mechanics predicts $\kappa_{tor}(\text{MT}) \sim 10^3\ \kappa_{tor}(\text{actin})$. For thermal fluctuations, the mean square angular displacement $\langle \phi^2 \rangle$ of a filament from its equilibrium configuration is inversely proportional to κ_{tor}, so $\langle \phi^2 \rangle$ should be three orders of magnitude smaller for a microtubule compared to an actin filament of the same length. Experimentally, $\langle \phi^2 \rangle$ is in the range 10^3 square degrees for actin, so one would expect $\langle \phi^2 \rangle$ to be just one square degree for microtubules. Such a small value will be comparable to the experimental uncertainty under many conditions, so that the thermal approach is much less promising for microtubules, and it may be more useful to explore conventional stress–strain curves, or filament buckling under torsion.

4.3.4 Other filaments

We close this discussion by mentioning two other proteins of biological importance – kinesin and collagen.

- *Kinesin* is a motor protein that transports cargo along microtubules within the cell. As described in more detail in Chapter 11, the main body of kinesin is a helical segment more than 40 nm long, at one end of which are two short "legs" that move along a microtubule (the other end is attached to its cargo). The torsion resistance of the region near the microtubule has been measured, yielding $\kappa_{tor}/L = (117 \pm 19) \times 10^{-24}$ J, where L is the effective length of the filament involved in the rotational motion (Hunt and Howard, 1993). Based on Eq. (4.18), the magnitude of the rotational motion of this section of kinesin due to thermal fluctuations at room temperature is $\langle \phi^2 \rangle^{1/2} \sim 2\pi$.
- *Collagen* fibers have a hierarchical structure based on tropocollagen, a triple helix of three single strands of the elementary protein. With a diameter of 1.5 nm, tropocollagen should exhibit thermal fluctuations in rotation about its axis that are comparable to DNA and much larger than actin, which has a diameter of approximately 8 nm. Thus, the

Table 4.1 Measured values of the torsional rigidity κ_{tor} for two filaments of importance in the cell. The values for the flexural rigidity κ_f are taken from the persistence lengths ξ_p in Table 3.2 using the conversion $\xi_p = \kappa_f / k_B T$.

Filament	κ_{tor} (J • m)	$\kappa_{tor} / k_B T$ (m)	$\kappa_f / k_B T$ (m)
DNA	$(2–4) \times 10^{-28}$	$(50–100) \times 10^{-9}$	53×10^{-9}
F-actin	$(3–8) \times 10^{-26}$	$(7.5–20) \times 10^{-6}$	$(10–20) \times 10^{-6}$

fluctuation approach used for extracting κ_{tor} for DNA and actin, should also work for collagen (N. Forde, private communication).

The measurements of κ_{tor} summarized in Table 4.1 can be interpreted using standard results from continuum mechanics. Comparing Eq. (4.4) for torsional rigidity with Eqs. (3.2) and (3.12) for flexural rigidity shows that

$$\kappa_{tor} = \kappa_f \, / \, (1 + \sigma_p) \tag{4.20}$$

for a solid cylinder, where σ_p is the Poisson ratio. Given that the Poisson ratio has a value of 0.5 or less for many materials, we would expect κ_{tor} to be more than 2/3 of κ_f if the filaments were uniform cylinders. Of course, some cellular filaments have elliptical cross sections and some are hollow, so Eq. (4.20) is only a benchmark. Further, the measured values of κ_{tor} and κ_f aren't known well enough to accurately test this expectation. However, a comparison of the data in columns 3 and 4 of Table 4.1 confirms that κ_{tor} and κ_f most likely obey Eq. (4.20) to within at least a factor of two.

4.4 Stretching of folded polymers

Many filaments of importance in the cell have relatively uniform structures, permitting us to represent them either as freely jointed chains when their contour length L_c is much longer than their persistence length ξ_p, or as gently bending or twisting rods when $L_c << \xi_p$. In either of these regimes, the elastic response is linear in the sense that it is described by Hooke's Law: stress is linearly proportional to strain. However, from the mere fact that the filaments have a finite length, we know that linear response must break down for long chains ($L_c >> \xi_p$) when their extended length x approaches their contour length. The Langevin description of Eq. (3.56) and the worm-like chain model of Eq. (3.58) were developed for application across the full range of x.

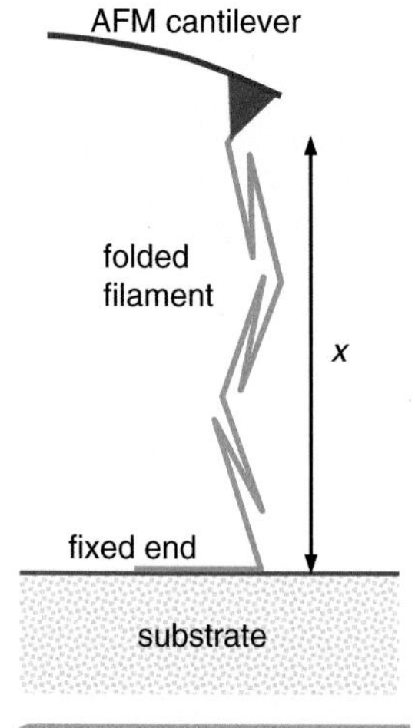

Fig. 4.13

Measurement of force–extension relationship for a folded polymer using an atomic force microscope (AFM). The "sensor" end of the polymer (top) is attached to the tip of an AFM that can be moved to a location x; the force experienced by the polymer is determined from the deflection of the AFM cantilever. The experiments can also be done by replacing the AFM with an optical trap. The diagram is not to scale.

The force–extension behavior of a filament may display non-linear characteristics well before the large extension limit is reached. For example, the cytoskeletal filament spectrin introduced in Chapter 3 can be described as a series of linked barrels, within each of which the linear protein molecule is folded back upon itself in the shape of the letter Z. As the filament is increasingly extended by a tensile force, each barrel may unfold in turn, yielding an overall contour length much larger than the 100 nm length of a single molecule at zero stress. This type of unfolding behavior has been observed and quantified for a number of proteins, and includes not just the back-folded Z-like structures of spectrin.

Many of the single-molecule experiments that will be summarized here are performed at fixed strain: one end of the molecule is attached to a rigid support while the other end is attached to force sensor such as the tip of an atomic force microscope (AFM) or a bead held in an optical trap, as in Fig. 4.13 (most measurements reported so far are based on the AFM approach). The sensor end can be moved through a series of positions to change the extension x of the molecule. For each value of x, the mean reaction force $\langle F \rangle$ exerted by the stretched molecule can be obtained from the deflection of the AFM cantilever, for example. An alternate approach to the fixed strain ensemble is the fixed stress ensemble in which F is fixed and x fluctuates by an amount Δx around the mean value $\langle x \rangle$, where the fluctuations are related to F by the equipartition theorem $F/x = k_B T / \langle \Delta x^2 \rangle$ (Strick *et al.*, 1996). The fixed stress and fixed strain ensembles are inequivalent, just as are the NVT and NPT ensembles for gases described in Appendices C and D. As applied to single-molecule experiments, the differences between the two ensembles are discussed by Keller *et al.* (2003) and Neumann (2003).

At the equilibrium location $x = 0$, F vanishes; but as the sensor end of the protein is moved so as to extend the molecule, F rises with x. At a particular value of x, one of the folded regions of the molecule opens up, increasing its overall contour length. If the added length is large, then the reaction force from the protein may drop precipitously. For most of the proteins discussed in this section, a Z-shaped folded region adds more than 20 nm to the protein contour length when it unfolds. As x increases further by movement of the sensor end of the protein, F once again rises until the next unfolding event occurs. The resulting force vs. extension graph has a saw-tooth form as illustrated schematically in Fig. 4.14. As a technical note, when the applied force is delivered through an AFM tip, the reduced force on the tip from unfolding permits the tip to relax by a few nanometers, but this change is still much less than the simultaneous change in the contour length (Rief *et al.*, 1997).

Each domain of the sawtooth pattern shows generally similar behavior: when the applied force is small, F rises linearly with x like an ideal spring. But as the molecule becomes increasingly stretched, the rise in F is

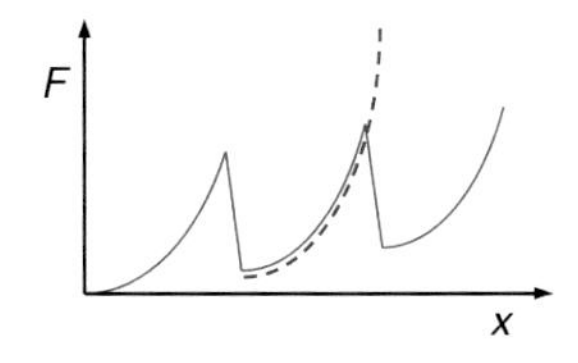

Fig. 4.14

Schematic force–extension curves (F as a function of x) for a folded polymer under tension. Each sawtooth domain can be approximately described by the worm-like chain model (dashed line); however, the contour length of the polymer for a given number of folds is longer than the extension at the peak force, as shown.

faster than linear, as one would expect from the Langevin description or the worm-like chain model. If the protein behaved like an ideal entropic spring between each unfolding event, then the initial spring constant k_{sp} in the linear regime could be different from one domain to the next: according to Eq. (3.46), the entropic spring constant of a freely jointed chain with N segments of equal length b is given by

$$k_{sp} = dk_B T / Nb^2, \tag{4.21}$$

where d is the spatial dimension in which the chain fluctuates. If an unfolding event adds length to the protein by increasing the number of links N, then k_{sp} should decrease, although the size of the decrease may be small if N is large. However, the increased configuration space available to the protein caused by an unfolding event is somewhat offset as the ends of the molecule move further apart, which acts to decrease the available configuration space. Experimentally, k_{sp} in the linear regime is observed to decrease from domain to domain as the unfolding process proceeds (Rief *et al.*, 1997).

Experimentally, the force–extension curves show a strong upward trend in the force as the molecule becomes increasingly stretched. However, the peak force does not correspond to a fully extended chain. In most situations, a domain in the chain unfolds before the completely extended state is reached, and the force quickly returns to a small value as the molecule is stretched further after a domain unfolds. Although it is not directly measured in the single-molecule stretching experiments described here, the contour length of the molecule can be obtained as a parameter in a model for the extension dynamics.

We now turn to measurements of folded proteins (for a review, see Deniz *et al.*, 2007). In all of these studies, the analysis is based on fitting the force–extension data with the worm-like chain model, which yields the filament's contour length and persistence length.

(1) Spectrin

The cytoskeletal network of the red blood cell is composed of tetramers of the filamentous protein spectrin: a pair of dimers each of which has one α-spectrin and one β-spectrin subunit joined end-to-end, as described in Chapter 3. Studies of the spectrin repeat units in a single filament have been done by Rief *et al.* (1999). An analysis of the sawtooth shapes of the force–extension curves shows that unfolding the Z-shaped units adds 31.3 ± 0.3 nm to the filament's length and that the persistence length of the single filament lies at the high end of the 0.4–0.8 nm range. Further work has explored the unfolding steps in more detail (Law *et al.*, 2003a), as well as the temperature dependence of the force required for unfolding

(Law *et al.*, 2003b). Batey *et al.* (2006) place the change in free energy of a spectrin unfolding event at 6–10 kcal/mol or 10–15 $k_B T$.

(2) Ubiquitin

Each monomer of this small protein has a molecular weight of 8433 Da and a contour length of 28 nm, giving a mass per unit length of 300 Da/nm. Using eight globular monomers linked as a linear chain, Chyan *et al.* (2004) measure force–extension curves of the globular form of ubiquitin as it unfolds, obtaining a persistence length of 0.4 nm for the protein and find that each unfolding event adds 25 nm to the contour length.

(3) Tenascin

A protein found in the extracellular matrix, tenascin has folded domains whose force–extension behavior has been studied using the worm-like chain model. Oberhauser *et al.* (1998) assign a persistence length of 0.42 ± 0.22 nm to tenascin, while Rief *et al.* (1998) find a somewhat broader range around 0.4 nm. An unfolding event adds 31 nm to the length of the protein; unfolded, the mass per unit length of tenascin is approximately 300 Da/nm.

(4) Titin

Fragments of titin were among the earlier proteins whose mechanical unfolding was studied with atomic force microscopy (Rief *et al.*, 1997, 1998). As with several proteins described in this section, the persistence length of the unfolding segment was found to be about 0.4 nm, with an unfolding event adding 28–29 nm to the filament contour length. Since the original measurements, the unfolding process and its dependence on local composition has been characterized in much more detail (Leake *et al.*, 2004; Di Cola *et al.*, 2005; Nagy *et al.*, 2005; Higgins *et al.*, 2006).

Other molecules or units that are known to exhibit unfolding under tension include ankyrin (Li *et al.*, 2006b; Serquera *et al.*, 2010), dystrophin (Bhasin *et al.*, 2005), heparin (Marszalek *et al.*, 2003), myosin (Root *et al.*, 2006) and Von Willebrand factor (Ying *et al.*, 2010); the unbinding of DNA in chromatin has also been measured with the force–extension approach (Pope *et al.*, 2005). More structurally complex biofilaments whose mechanical strength has been investigated by similar techniques include collagen (Gutsmann *et al.*, 2004; Shen *et al.*, 2008; Yang *et al.*, 2008; Gevorkian *et al.*, 2009).

In most of the experiments summarized above, the magnitudes of the forces involved at the unfolding point are generally in the domain of tens of piconewtons or more. However, the force at a rupture or unfolding event

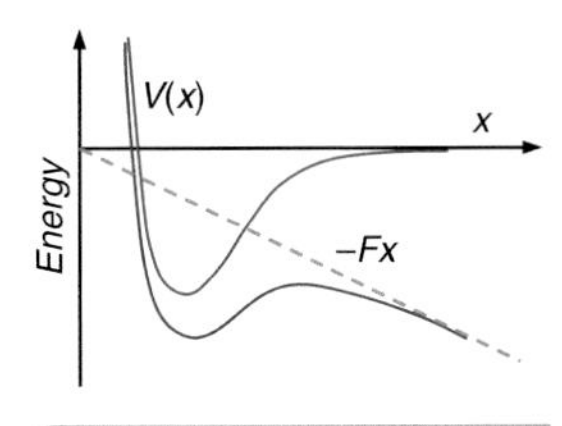

Fig. 4.15

Effect of a tensile force of constant magnitude F on the total energy of a bound system. The red curve is the potential energy $V(x)$ of the system when F vanishes, while the blue curve shows the sum of $V(x)$ and the work done by the force on the system, $-Fx$. The work is the green dashed line.

rises logarithmically with the rate at which the tensile force is increased – the faster the force is applied, the higher the apparent rupture force. This is a thermal effect described more fully in a moment. Not all experiments examine the rupture force as a function of loading rate, so we have chosen not to report the unfolding force rather than present data that may not be comparable from one experiment to the next.

The reason that the force for bond-breaking depends on the rate at which the force is applied can be seen by considering the generic binding potential $V(x)$ of Fig. 4.15. In the absence of an external tensile force, the potential energy has a minimum value at a physical location x_{min} where the derivative of the potential vanishes: $dV/dx = 0$ at x_{min}. However, when the system experiences an external tensile force, the combined potential energy plus the work done by the force ($\int F \cdot dx$) decreases without limit as $x \rightarrow \infty$. This behavior is illustrated in Fig. 4.15 for the case where F is constant. Whereas the energy minimum is at x_{min} when $F = 0$, the minimum is at $x \rightarrow \infty$ when $F > 0$. Under tension at zero temperature, a system can be trapped near x_{min}, but at finite temperature, thermal fluctuations will always permit the system to hop out of the metastable location at escape to $x \rightarrow \infty$.

Mathematical formalisms for interpreting the apparent force for rupture or unbinding are developed in Bell (1978) and Evans and Ritchie (1997, 1999); see also Shillcock and Seifert (1998b). The properties of the energy landscape in the presence of a tensile force are obtained for a power-law potential energy in the end-of-chapter problems. The logarithmic dependence of the nominal unbinding force on the force loading rate has been observed for a number of proteins, including titin (Evans and Ritchie, 1999) and ubiquitin (Chyan *et al.*, 2004). Lastly, there exist methods for obtaining free energy differences in nonequilibrium systems (Jarzynski, 1997).

4.5 Protein and RNA folding – concepts

Proteins, RNA and DNA are all linear polymers: depending on their form, RNA and proteins may be built from hundreds of monomers, while DNA may have many orders of magnitude more than that. Yet even when polymerized into a chain, the monomers of these macromolecules still have modest binding affinities that enable them to bond to other molecules or to other segments of their own polymer. With DNA, the double-helical structure prevents monomers on the same chain from binding with each other, but this is not true for proteins or RNA. Proteins form globular structures within which are local substructures like α-helices and β-sheets. A single strand of RNA possesses local regions that stick to each other like the parallel sides of the letter U. In this section, we describe the energetics and structures of protein and RNA binding; in Section 4.6, we analyze protein conformations within the context of a simple model for interpreting their most important features.

Proteins commonly have about 300–400 amino-acid residues linked together to form an unbranched polymer, although there are many examples of proteins whose composition lies outside this range. Each residue contributes 0.38 nm to the contour length of the polymer, a number that drops to 0.32–0.34 nm for residues that are part of a β-sheet. For example, the protein actin, which has 374 residues in rabbit skeletal muscle, has a contour length of 142 nm, which is enormous compared to its width as a globular protein. The peptide bond between adjacent residues is too flexible to keep the polymer straight, so that a protein may have a width that resembles that of a self-avoiding random walk. Starting with an ideal random walk as a benchmark, the root mean square end-to-end displacement $\langle r_{ee}^2 \rangle^{1/2}$ of actin should be in the range 7 nm from $\langle r_{ee}^2 \rangle = Nb^2$ if $N =$ 374 and $b = 0.38$ nm (where "ideal" walks can self-intersect). This value is much closer to the observed dimensions of globular actin, which are about 5–7 nm.

Let's now examine the width of globular proteins with a finer eye. First, we extract the root mean square radius of gyration according to the expression $\langle R_g^2 \rangle = \langle r_{ee}^2 \rangle / 6$ for ideal random walks, finding $\langle R_g^2 \rangle^{1/2} = 3$ nm treating actin as a random walk. Knowing that a sphere of radius R has $\langle R_g^2 \rangle =$ $(3/5)R^2$, we would expect that the diameter of an actin polymer should be close to 8 nm if it had a spherical shape. Again, this is comparable to the experimental width, but it ignores two effects that act in opposite ways:

- random chains are very floppy, unlike a functioning protein with its condensed local structure, including α-helices and β-sheets;
- self-avoidance of a polymer will increase its size above the ideal chain value.

Consequently the true description of a condensed protein involves more than a random walk. An indication of the appropriate physics can be found by looking at the scaling properties of the measured radii as a function of the contour length of the polymer. As shown in Chapter 3, the width (or linear dimension) of a random walk scales with the number of segments N as $N^{1/2}$ for ideal walks and $N^{3/5}$ for self-avoiding walks in three dimensions. In contrast, the width of a dense polymer or space-filling walk scales with the number of segments like $N^{1/3}$ in three dimensions. The observed increase in the transverse dimension of proteins with the number of residues is consistent with $N^{1/3}$ scaling, although there is scatter in the data (Dewey, 1993).

The fact that protein size roughly obeys $N^{1/3}$ scaling should not be interpreted to mean that the polypeptide itself is densely packed. For example, if each residue had a certain average effective volume surrounding it,

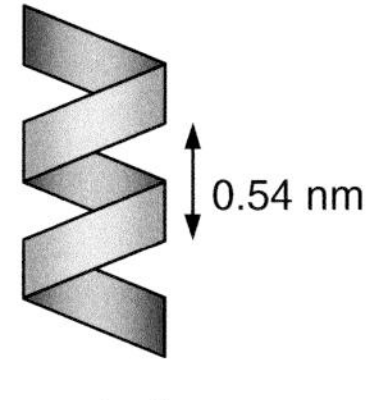

α-helix
right-handed

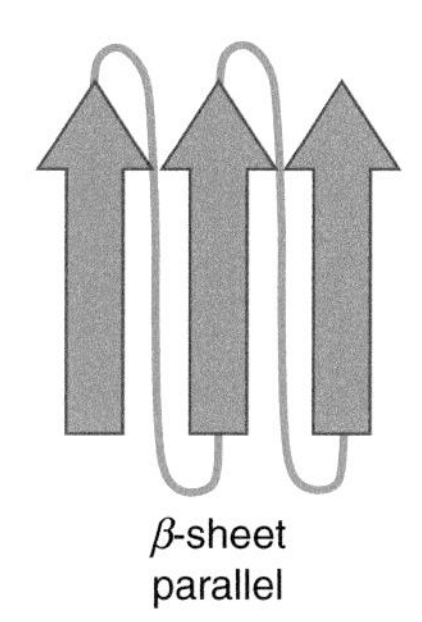

β-sheet
parallel

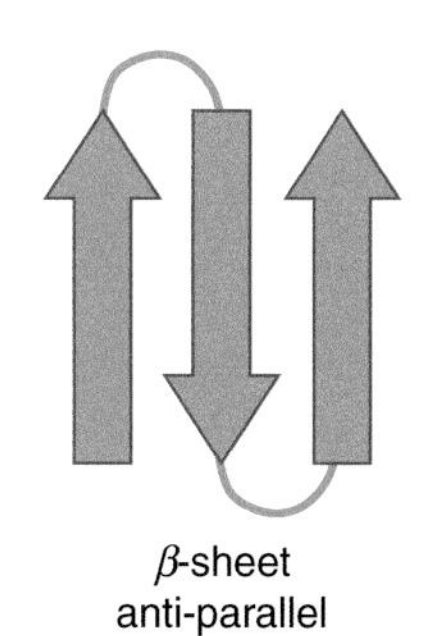

β-sheet
anti-parallel

Fig. 4.16

Schematic representations of α-helices and β-sheets observed in proteins. In a β-sheet, the individual β-strands can adopt either parallel or antiparallel orientations.

including solvent, then $N^{1/3}$ scaling would still apply so long as the effective volumes of the chain were space-filling. Of course, the radius of gyration of the protein would necessarily be larger, reflecting the presence of the solvent around the polypeptide backbone. The presence of solvent in a globular protein is probed further in the end-of-chapter problems.

Scaling analysis describes gross or average features of a protein as a whole, but is silent about condensed substructures that might be present. Several regular patterns appear in proteins (see Creighton, 1993), the two most common being α-helices and β-sheets.

(1) α-helices

Of the different helical forms are known to be present in proteins, the most common one is the α-helix. This is a right-handed structure with 3.6 residues per turn of the helix and a translation of 0.541 nm per turn, corresponding to a translation of 0.15 nm per residue as in Fig. 4.16. The formation of the helix is driven by hydrogen bonding within the polypeptide between the hydrogen of an NH group and the oxygen of the carbonyl group on the fourth residue away from the NH group along the backbone of the chain. The side chains of the residues point outward from the helix.

(2) β-sheets

This regular secondary structure consists of straight strands of the polypeptide (β-strands) lying side by side. From one strand to the next, the arrangement can be either parallel (all strands having the same orientation with respect to each other) or antiparallel (the orientation reverses from strand to strand), as in Fig. 4.16. The conformation of a free β-strand is unsupportable, in the sense that the peptide bonds are too flexible to keep the strand in a regular zig-zag pattern against the effects of entropy. However, when associated in a sheet, the attractive interactions between strands stabilize the individual conformation. The binding arises from hydrogen bonds between peptides on adjacent β-strands: the hydrogen on an NH element of one strand is attracted to the oxygen of a CO group on another. The side chains of the residues lie out of the plane of the sheet. Each amino-acid residue along the polypeptide of a β-strand contributes 0.32 nm to the projected length along the strand in the parallel arrangement, and 0.34 nm in the antiparallel arrangement.

Protein structure and substructure is generally organized into four hierarchical categories. The *primary* structure is just the amino-acid sequence itself, without reference to the shape adopted by the macromolecule. *Secondary* structures are regular conformations like α-helices and β-sheets, as well as the less common 3_{10}-helix and π-helix. Its *tertiary* structure is the organization of the protein as a whole, for example into a condensed form. Lastly, some proteins are members of multi-protein complexes, and their arrangement within such complexes is referred to as their *quaternary* structure.

Before examining the general features of how macromolecules in the cell fold under various conditions, we briefly describe the structure of small RNA molecules. As described in Appendix B, RNA is composed of similar nucleotides as DNA is, except that the sugar is slightly different and the four-letter genetic code is ACGU instead of DNA's ACGT, where uracil of RNA replaces thymine of DNA. In the steps leading up to the manufacture of proteins, the sequence of nucleotides that forms a gene on a single strand of DNA is *transcribed* into messenger RNA or mRNA. Because of base-pair complementarity, the sequence of bases on mRNA is the inverse of that on DNA; that is, the specific sequence CGAT on DNA becomes GCUA on RNA. With three bases on DNA coding for a single amino-acid residue, a typical protein of 300–400 residues requires a functional mRNA molecule of a thousand bases. In eukaryotic cells, the gene containing these 1000 bases of genetic information may be greatly lengthened by the presence of large regions that are non-coding: the coding region of the average human gene is around 1300 bases, yet the average length of a human gene is 20 times that. After a process that removes non-coding regions, etc., the finished mRNA molecule is transported to a ribosome, where it becomes the genetic template for protein production.

The assembly of amino-acid monomers into a protein, or *translation*, involves many other macromolecules and types of RNA, among which is transfer RNA or tRNA. This is a much smaller molecule than mRNA on average, and its job is to carry an amino acid to an mRNA that is undergoing translation, where it is captured by the protein construction machinery when the appropriate codon appears. There must be many forms of tRNA molecule, at least one form for each amino acid, so that a particular codon is identified with a particular residue; more than one codon may correspond to a given residue. Structurally, the residue is carried at one end of the tRNA while the codon-detection apparatus is at the other end, crudely speaking. So far, close to a dozen types of RNA are known to play a role in the cell.

The monomeric units of RNA have more specific binding than the hydrophobic forces that protein residues possess. In RNA, the CG, AU pairs can be hydrogen bonded, permitting different segments of a single strand of RNA to bind to each other as in Fig. 4.17. However, only specific segments will bind because of the specific CG–AU binding affinities: the codes on the two sections of the strand must be complementary in order for the two segments to bind. This feature is evident in Fig. 4.17: the spheres represent nucleotides, where red spheres bind to each other by hydrogen bonds (red) while magenta spheres do not. The blue line indicates the backbone of this 76-nucleotide example of transfer RNA. The binding regions are relatively

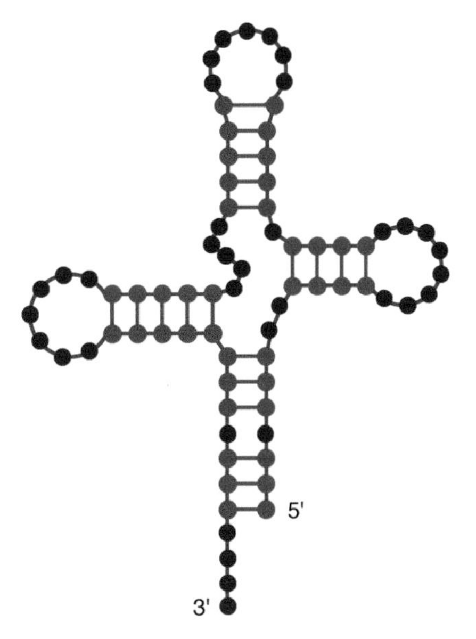

Fig. 4.17 Schematic example of a transfer RNA molecule. The nucleotides are indicated by red and magenta spheres: red spheres are linked to each other by hydrogen bonds between bases while magenta spheres are not. The backbone of the polymer is indicated by the blue line; the 3′ and 5′ labels are explained in Appendix B.

nearby on the linear strand, separated from each other in the example by a short portion of less than ten nucleotides that form a small loop.

The bound or folded structures that can be adopted by a short strand of tRNA are fairly limited. In Fig. 4.17, there are four locally bound regions defined by four to six paired nucleotides, and the sequences can be set so that only these regions are well matched for binding. In contrast, there is a hierarchy of binding interactions in proteins, of which the hydrophobic effect is both an important and a relatively non-specific contributor. Even for a short protein segment, large numbers of partially folded states lie in the same energy range, such that the ground state or *native* conformation may not be readily apparent. As far as energetics is concerned, the hydrogen bonding that appears in RNA folding, and in the α-helix and β-sheets of proteins, generally lies in the range of 3 to 10 k_BT, well below the energies of covalent bonds. For example, the unzipping of 22 H-bonded base pairs in a particular section of RNA corresponded to a free energy change of ~150 kJ/mol, or 62 k_BT for the segment (Liphardt *et al.*, 2001; for DNA, see Bockelmann *et al.*, 2002).

For a sequence of residues to lead to a functional protein, the conformational route to the folded state should not be too tortuous. The concepts at play are illustrated in Fig. 4.18, which shows the energy as a function of a spatial coordinate for a series of hypothetical situations. The spatial

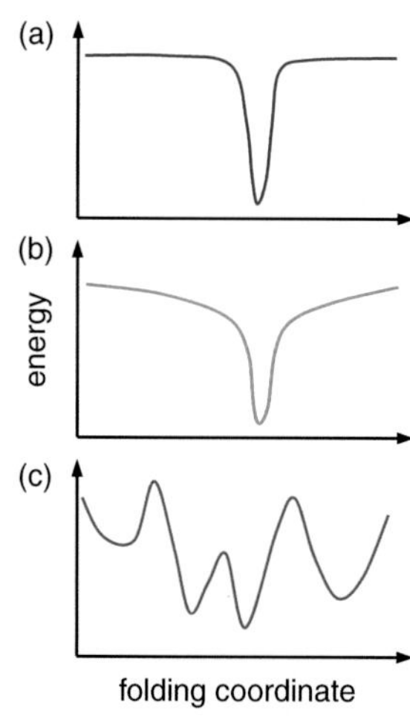

Fig. 4.18

Schematic representations of energy landscapes in protein folding, as a function of the folding coordinate. The folding region in panel (a) is highly localized, while in panel (b) it resides at the center of a shallow funnel. The landscape in panel (c) is rough, with several metastable conformations.

coordinate represents a continuous change in the conformation as it folds; in a protein with several hundred residues, there are many degrees of conformational freedom such that the one-dimensional situation plotted here is obviously a gross oversimplification. Nevertheless, we imagine for the sake of argument that there is one coordinate that captures the most important aspect of the folded state; for example, if the folded state were a simple V-shape, then the coordinate might be the angle at the vertex of the V.

Panel (a) of Fig. 4.18 shows a situation where there is a deeply bound native state to be found in a very confined region of the folding coordinate. Although this might seem to be the ideal situation for a folded protein, an unfolded state at some arbitrary location in the coordinate could wander randomly through space for a long time before stumbling upon the deep well representing the folded state. Borrowing a sports analogy from golf, it's like trying to find the cup in a flat green by putting in random directions. Panel (b) shows a situation that is more time-efficient for finding the folded state: the energy landscape has a gradual slope toward the folded state, which itself is still very well localized. Thus, an unfolded conformation is guided toward the folded state by the decrease in energy as the coordinate changes. Lastly, panel (c) represents a challenge to finding the folded state at all: the coordinate passes through many local energy minima where the protein can become trapped in a metastable state. If these metastable states are long-lived and functionally useless compared to the native state, then the cell may as well not have bothered making the protein in the first place.

The inverse process of folding is the denaturing of a protein by thermal or chemical means (for an overview, see Section 7.4 of Creighton, 1993). In this situation, a macromolecule escapes from its native folded state and explores a variety of partly or completely unfolded conformations. This situation is familiar to us in the kitchen through the denaturing of the protein albumin when an egg is boiled. In the case of thermal denaturation, the transition region over which the unfolding occurs (where likelihood that a protein is in its native state drops from about 95% to 5%) spans a temperature range of about 20 celsius degrees, depending on the protein.

By considering the energy landscape of the folding process, we see that it is important for a protein to fold readily into a unique and stable ground state: if several conformations exist with energies very close to the ground state energy, a protein may not fold to the correct conformation when needed. Whether a polypeptide can fold into an appropriate native state depends of course on the sequence of residues it possesses. A particular native state conformation is said to be *encodable* if there exists at least one sequence for which the conformation is the ground state. Now it may be that more than one sequence folds to the same ground state conformation, and the term *designable* indicates qualitatively how many sequences lead to the same native state: the more sequences, the more designable the native state. Many features of protein folding such as designability, the energy landscape, and folding rates, have been studied within a variety of computational models. One such model that is particularly tractable for small proteins is the HP model (for Hydrophobic/Polar), and we explore its properties in Section 4.6.

4.6 Models for polymer folding

Some conceptual characteristics of proteins introduced in Section 4.5, such as the existence of designable structures, can be viewed quantitatively through the use of one or two very simple computational models. The sequence and structure space of even the barest of models is immense, so the hypothetical interactions between amino acids within such models are forced to be unrealistically simple to keep the models numerically tractable. Nevertheless, the models possess many of the essential features of protein design and permit us to probe the factors that influence the suitability of different designs.

The HP model (Hydrophobic/Polar) is a simple one-parameter model that provides a way of understanding the ground states of folded polymers and how they unfold with increasing temperature (Lau and Dill, 1989; Chan and Dill, 1989; for a thorough review, see Dill *et al.*, 1995). In this model, every amino acid is classified as either hydrophobic (H) or polar (P): the H group has nine amino acids (alanine, cysteine, isoleucine, leucine, methionine, phenylalanine, tryptophan, tyrosine, valine) while the P group has eleven (arginine, asparagine, aspartic acid, glutamic acid, glutamine, glycine, histidine, lysine, proline, serine, threonine). Structures of these acids are given in Appendix B; see Thomas and Dill (1996) for a hierarchy of amino-acid interactions. Within the generic HP model, the interaction energies between the simplified elements of the model protein can be assigned in a variety of ways. For example, one could imagine a hierarchy of energy scales, a different scale for each of the H–H, H–P, P–P contacts, as well as separate scales for the interaction of an H or P site with its aqueous environment (Dill *et al.*, 1995; as an example, see Li *et al.*, 1996).

For the version studied in this chapter, all interactions between one amino acid and another or its (polar) aqueous environment are zero with the exception that hydrophobic species have a repulsive energy $+\varepsilon_H$ with any polar species or the environment; nearest neighbor covalent bonds are not part of this accounting. To distinguish this approach from other forms, we refer to it as "our" HP model, although it has been used elsewhere

(Phillips *et al.*, 2009). This is a slightly different model from the original form in which all interactions are zero except for an attractive term $-\varepsilon$ for H–H contacts, although the two parametrizations yield very similar results (Lau and Dill, 1989); the original model is computationally very efficient and its properties have been heavily studied. In this section, we make the further assumption that the configurations of the polypeptide live on a square lattice in two dimensions or a cubic lattice in three dimensions; lattice sites not occupied by the polymer are occupied by molecules of the aqueous (polar) environment. The HP model is neither the only nor the first lattice model for proteins; an earlier model that has also been extensively studied is the so-called Gō model (Taketomi *et al.*, 1975; Gō and Taketomi, 1978).

By classifying the amino acids into two groups for the purposes of evaluating their interaction energies, the number of protein sequences is greatly reduced. Within the HP model, the number of sequences available to N monomers is 2^N, which is certainly a large number but still very much smaller than 20^N for 20 different amino acids! Further, by placing the monomers onto a lattice, the physical conformation space of the polymer is greatly reduced as well and the lattice symmetry relates groups of these conformations to each other by symmetry operations. Consider, for example, a lattice polymer with just three sites. The allowed conformations are either a straight line or an L-shaped conformation with each arm having the same length. Among the lattice shapes, there are two straight lines (vertical or horizontal) and four orientations of the L-shape (where the L is flipped horizontally, flipped vertically or flipped horizontally and vertically). Yet these six conformations are not independent and can be obtained by rotations of just two basic shapes, a straight line and an L-shape. In other words, the combined sequence/structure space of the HP model is tremendously reduced from the space available to 20 different types of monomers linked to form random chains in physical space.

To see how the model works, consider a hypothetical polypeptide with just four amino acids. The lattice conformations available to this model polymer are:

- a straight line with four consecutive lattice points,
- an L-shape with three lattice points on a line and one point off the axis,
- a zig-zag shape with the two end-points of the polymer lying on opposite sides of the central bond,
- a U-shape (with sharp corners).

We will return to the frequency of appearance of these shapes in an equilibrium ensemble later in this subsection. Onto these structures are placed $2^4 = 16$ sequences of various combinations of H and P amino acids, for example HHPP, HHHP, HPPH. Within our version of the HP model, the energy of each structure/sequence combination is easily calculated. For example, if the structure is a straight line, then the HHHH sequence has an energy of $(3+2+2+3)\varepsilon_H = 10\varepsilon_H$: the H sites at the ends have energy $+3\varepsilon_H$ each from their exposure to the environment while the two H sites in the middle have energy $+2\varepsilon_H$ each. In contrast, the PPPP sequence has zero energy. Not all sequences have the same energy when placed on a given structure even if the number of H and P sites is numerically the same: for the straight line, HHHP has $7\varepsilon_H$ and HHPH has $8\varepsilon_H$ (both sequences have three H and one P sites).

One can see that as the polymer chain folds from a straight line into a U, the H sites have less exposure to the aqueous environment so that the energy of the chain drops as a result. For example, in the HHHH sequence, the L-shape and zig-zag have the same energy as the straight line: $10\varepsilon_H$ as we just calculated. However, the HHHH sequence on the U-shape has a lower energy of $+8\varepsilon_H$ because the two H sites at the ends of the chain have each other as adjacent sites, reducing the interaction with the polar environment. Thus, the hydrophobic effect favors folded states, so we will present the energies of only the U-shaped structure in detail. Figure 4.19 displays the energy of the folded state for all applicable HP sequences; other configurations that can be obtained from these states by rotations or reflections are not independent states and, thus, are not shown. In the figure, each of the amino acids occupies a site on a square lattice and the grey lines represent the covalent ("permanent") polypeptide bonds. Polar amino acids are colored blue, while hydrophobic ones are red. The short red lines emanating from the hydrophobic sites represent repulsive interactions, each with an energy cost ε_H; the total number of these lines is the repulsive energy of the configuration in units of ε_H.

It's tempting to scan Fig. 4.19 for the state with the lowest energy – a task that is done all the time in mechanics. This would imply, however, that we have the freedom to vary the sequence in a protein while its spatial structure remains fixed, whereas the opposite is true for a protein in solution: the sequence is fixed and the protein wiggles about, changing from one structure to another, unless it falls into a deeply bound state. Thus, our task is to search for the spatial structure that minimizes the energy for a given HP sequence. Let's consider two specific sequences PHHP and HPPH that have the same overall composition: two H sites and two P sites each. Shown in Fig. 4.20 are the lattice structures available to these two sequences. We know already from Fig. 4.19 that both sequences have the same energy in the U-shape structure, and Fig. 4.20 yields the further information that the U-shape is an energy minimum for HPPH, but not for PHHP. Thus, a model polypeptide with the PHHP sequence does not have a unique ground state whereas the HPPH sequence does, and it is the latter sequence that is "protein-like" in that it has a unique and folded ground state structure.

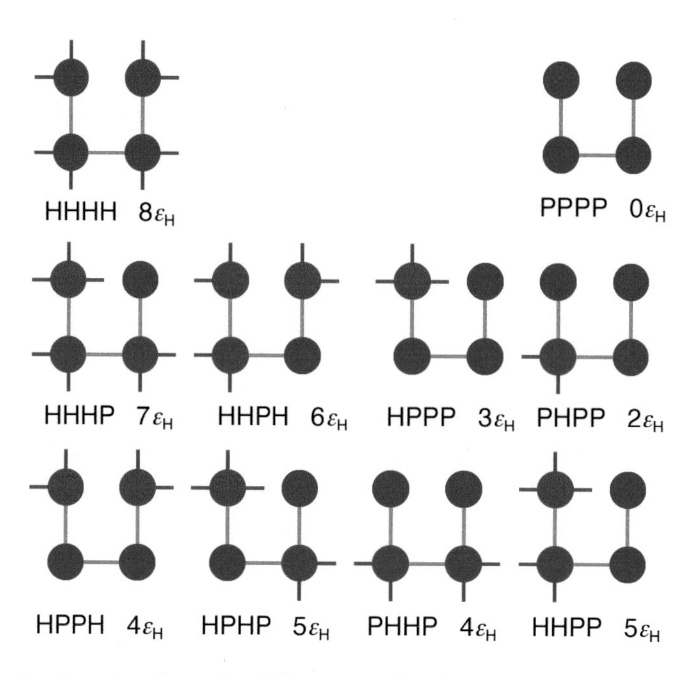

Fig. 4.19 U-shaped configurations of a protein with four amino acids in the HP model. Linked by covalent bonds (grey) occupied sites on the square lattice may be polar (blue) or hydrophobic (red). The repulsive interactions experienced by H sites are indicated by the short red lines; the repulsive energy of the configuration is equal to the number of these lines times the energy parameter ε_H in the version of the HP model considered here. The sequence on each configuration is indicated, starting from the upper left site on the U-shape conformation.

Of all the amino-acid sequences permitted in the HP model for this polypeptide, only HHHH, HPPH and HHPH have unique ground states. All of the other sequences are degenerate, in the sense that all structures of the sequence have the same energy (see end-of-chapter problems). By inspection, the characteristic of the sequence that is associated with its having a unique folded ground state is that the two ends of the polymer are both H sites: that is, when the polymer folds up, the two end-sites that are completely exposed to the aqueous environment become nearest neighbors and the repulsive energy is reduced if both sites are occupied by H species.

In the four-site system, the U-shaped folded structure is the only one that possesses interactions between different sites on the polymer: sites on the straight line, L-shape and zig-zag interact only with their environment. Thus, if there is a ground state associated with polymer folding, it must necessarily be the U-shape, the only shape that is fully folded. However, for longer polymers, this is not the case and there are many

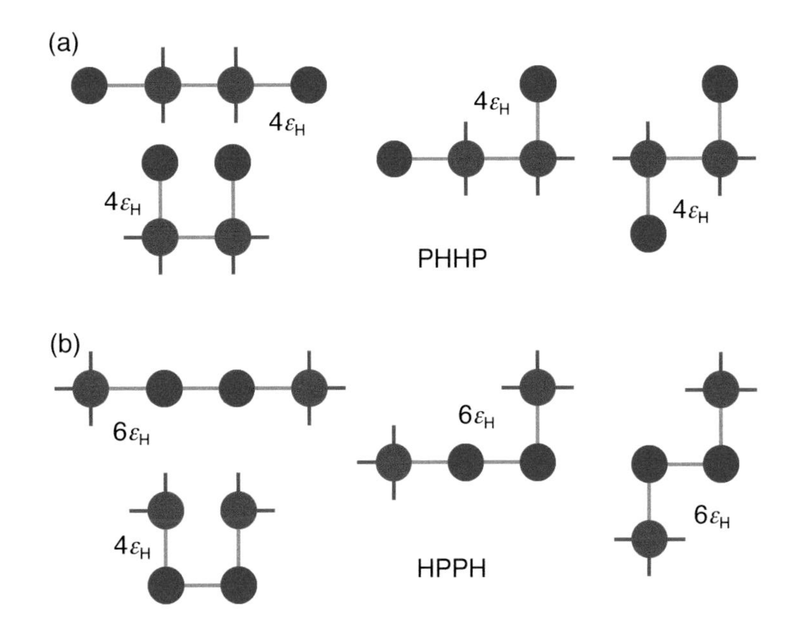

Fig. 4.20 Four structures of the sequences PHHP [panel (a)] and HPPH [panel (b)]; the energy of each configuration within our version of the HP model is given in terms of the energy parameter ε. Other aspects of the figure are the same as in Fig. 4.19.

different folded states, only a subset of which may contain ground states for various sequences. This property can be illustrated through the use of another very simple system: six linked sites on a square lattice. As shown in Fig. 4.21, this system has three different fully folded states: a U-shape, an S-shape and a G-shape, where the six-site U-shape has longer arms than the four-site U. Note that one of these structures, the G-shape, has no internal symmetries, with the result that the order of the sequence must be specified with respect to a particular end of the chain. This is also true of the L-shape of the four-site system; indeed, for longer chains there are ever more structures with no internal symmetries. In Fig. 4.21, the starting point of the sequence is indicated by an asterisk.

For the six-site system, there are $2^6 = 64$ distinct sequences in the HP model. Each of the sequences in Fig. 4.21 has a unique ground state structure as shown: HHPPHH has a U-shape, PHHHPH has an S-shape and HHPHPH has a G-shape, as confirmed in the end-of-chapter problems. Based on only the three folded configurations in the figure, it is not difficult to establish that there are nine sequences having a U-shaped ground state structure, six with an S-shape and three with a G-shape. Thus, 18 out of the 64 sequences possess unique folded ground states,

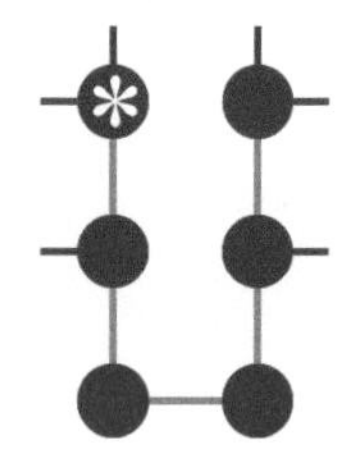
HHPPHH g.s. $6\varepsilon_H$

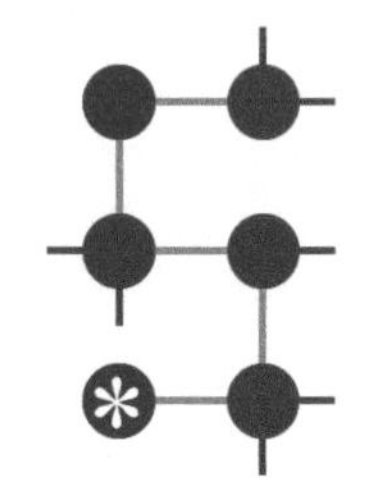
PHHHPH g.s. $7\varepsilon_H$

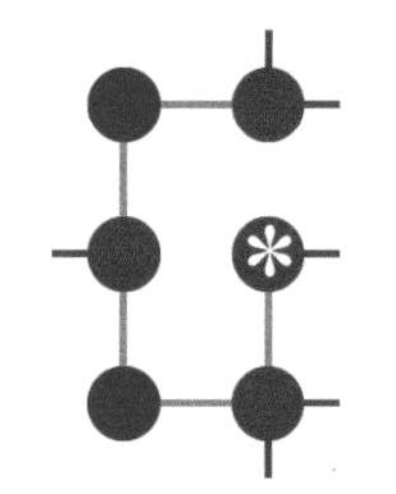
HHPHPH g.s. $6\varepsilon_H$

Fig. 4.21

The three fully folded structures of a six-site HP model polypeptide on a square lattice in two dimensions. Each of the structures emerges as a unique ground state for more than one sequence: each diagram shows a sequence for which the structure is a unique ground state. The starting site of the sequence is indicated by an asterisk; other aspects of the figure are the same as Fig. 4.19.

and all three of the compact structures can be a ground state for at least one sequence. The larger the number of sequences having a particular structure for a ground state, the more *designable* that structure is. In the six-site system, the U-shaped structure is the most designable. Another attribute assigned to a structure is whether it is *encodable*, which means that there is at least one sequence that has the structure as its ground state configuration. The U-, S- and G-shaped structures in Fig. 4.21 are all encodable: nine sequences encode for the U-shaped structure, six encode for the S-shape and three encode for the G-shape. In the four-site system on a lattice, only the U-shape is encodable and any sequence with an H at each end encodes for it.

The number of structures and sequences rises very rapidly with the number of sites in the model polypeptide, even with all the simplifications of the HP model. For example, the number of sequences increases from 64 for six occupied sites, to $2^{10} = 1024$ for ten, and $2^{18} = 262\,144$ for eighteen. The number of structures is truly impressive: even in two dimensions, there are 5.8 million conformations for polymers with eighteen sites on a square lattice although the number of compact conformations is considerably lower. To within an order of magnitude, the latter rises approximately exponentially as $\exp(0.39^*N)$ where N is the number of links on the polymer chain (Chan and Dill, 1989). With existing computing resources, it is difficult to scan all combinations of sequences and structures in a search for encodable structures when the number of monomers is much beyond 50. In contrast, real proteins typically have 300 residues, not 50, and an alphabet of 20 amino acids, not 2.

Challenging as the computational issues might be, there are more than a few general results that have emerged from the study of small chains. For example, about 2.1%–2.4% of sequences on a two-dimensional square lattice have unique ground states, an observation based on chains with up to 18 segments (Chan and Dill, 1991); this number may be higher in three dimensions: 4.75% of sequences on a $3 \times 3 \times 3$ cubic lattice have unique ground states (Li *et al.*, 1996). Also, the model proteins have emphasized the problem of negative design: it's one thing for a sequence to encode for a particular shape (positive design) but quite another to prevent this sequence from adopting one of the huge number of other conformations available to it (negative design). This problem has been studied computationally within the HP model by taking an HP sequence which is known to encode for a particular structure and then searching for other structures for which it encodes and which have the same or lower energy (Yue *et al.*, 1995).

The lowest energy configuration of a system such as our model polymers is the one it adopts at zero temperature. If this configuration is not unique, then members of an ensemble of identically prepared polymers will be distributed among these energy-equivalent configurations even at vanishing

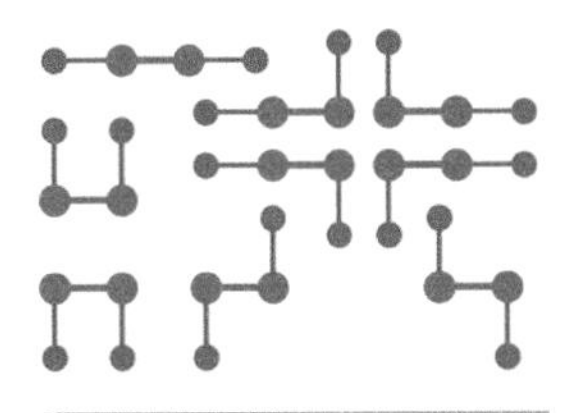

Fig. 4.22

Conformations available to a 4-site polymer on a square lattice in two dimensions. A second set of conformations with the same relative populations is obtained from these nine by rotating the central bond through 90 degrees.

temperature. However, at finite temperature any configuration with finite energy relative to the ground state can be present in the ensemble, thus reducing the population of polymers in the ground state. As usual, if a system can fluctuate among more than one conformation, the probability of finding it in a given conformation involves the Boltzmann weight $\exp(-\Delta E/k_B T)$ and the degeneracy of the state, where ΔE is the energy of the state with respect to the ground state and T is the temperature. The degeneracy g is the number of states with the same energy.

As an example, consider a 4-site HP system with the sequence HPPH, as displayed in panel (b) of Fig. 4.20. The figure does not represent correctly the degeneracy of the conformations in an ensemble, but it's easy to calculate the ensemble of conformations on a lattice. Suppose the two middle sites have a fixed location on a horizontal line, then the two end sites each can independently assume three orientations for a total of $3 \times 3 = 9$ conformations as shown in Fig. 4.22: 2 U-shapes (up and down), 4 L-shapes, 2 zig-zags and 1 straight line. Put another way, the straight line is unique, the U-shape and zig-zag are both two-fold degenerate while the L-shape is four-fold degenerate. Making the bond between the two middle sites vertical instead of horizontal gives the same population of conformations. All the shapes have energy $\Delta E = +2\varepsilon_H$ with respect to the U-shape, so the likelihood P_{gs} of finding the system in the ground state is

$$P_{gs} = 2 / [2 + 7 \cdot \exp(-2\varepsilon_H/k_B T)]. \tag{4.22}$$

The denominator in this expression is the sum over the combined Boltzmann factors and degeneracies of all allowed configurations, also known as the partition function. At zero temperature, $P_{gs} = 1$ as expected, while at high temperature, P_{gs} approaches 2/9. The temperature at which there is a 50% probability of finding the model protein in the ground state is $k_B T = 1.6\varepsilon_H$, which is less than the energy difference between the ground state and all the other configurations ($2\varepsilon_H$). Other approaches to the counting of accessible states are described in Chan and Dill (1989), particularly footnote 23.

By counting the accessible conformations, the thermal properties of the model proteins can be calculated analytically for the shortest systems. Continuing with the 4-site HPPH system, the mean interaction energy $\langle U \rangle$ of an ensemble of polymers is

$$\langle U \rangle = 14\varepsilon_H \exp(-2\beta\varepsilon_H) / [2 + 7 \cdot \exp(-2\beta\varepsilon_H)]. \tag{4.23}$$

if the interaction energy uses the ground state energy as a reference point. This follows because there are two states with energy zero, and seven states with energy $2\varepsilon_H$; the rest of the equation are just the Boltzmann factors and degeneracies of the states. In the zero temperature limit, $\beta \to \infty$ and $\langle U \rangle = 0$ according to Eq. (4.23), as it should be. In the high temperature limit, $\langle U \rangle = (14/9)\varepsilon_H$, which is less than the energy difference of $2\varepsilon_H$ between the folded and unfolded conformations of HPPH. As a function of temperature, $\langle U \rangle$ initially rises slowly from $T = 0$, then more rapidly as the population shifts to unfolded conformations, before slowly approaching its asymptotic value of (14/9) ε_H.

The increase of $\langle U \rangle$ with temperature – slower, faster, slower – suggests that the model protein's specific heat will have a peak value in the temperature region where unfolding occurs. From Eq. (4.23), it is straightforward to obtain the specific heat C_V analytically from the derivative of $\langle U \rangle$ with respect to temperature,

$$C_V = (\partial U / \partial T)_V. \tag{4.24}$$

As we have implemented it, the lattice has a fixed spacing so the heat capacity is determined at constant volume, C_V, rather than constant pressure. As derived in the end-of-chapter problems, the specific heat corresponding to Eq. (4.24) has the form

$$C_V = 56\varepsilon_H^2 k_B \beta^2 \exp(-2\beta\varepsilon_H) / [2 + 7 \bullet \exp(-2\beta\varepsilon_H)]^2. \tag{4.25}$$

The peak in C_V occurs at a temperature $k_B T$ of $0.685\varepsilon_H$, which is about 1/3 of the energy difference $2\varepsilon_H$ between the ground state and all other states.

The effect of making the proteins longer is to increase the importance of the unfolded configurations compared to the ground state. This means that the ground state population will be lower compared to the unfolded states, and the peak in C_V will be shifted to lower temperatures. An easy system that illustrates the magnitude of the effects is the 5-site HPPHP sequence, whose properties can be easily obtained from the 4-site HPPH that we have already examined. By adding the extra P site, the total number of conformations almost triples, and the ground state configuration becomes degenerate as there are two distinct ways of attaching the extra P site to the end of the U-shape conformation that is the ground state of the HPPH sequence. As obtained in several end-of-chapter problems, an immediate consequence of the additional unfolded states is that the temperature where the ground state is only half occupied drops by about 25% compared to HPPH. Correspondingly, the temperature of the peak in C_V associated with unfolding drops by 8%. These two changes are not entirely small, yet they arise from the addition of just one more residue to the model protein.

The kinetics of the folding process can be investigated at least qualitatively within lattice models through the use of computer simulations. For example, the evolution of an unfolded conformation can be followed as it explores partially folded states, reducing its energy in the process. Within a simulation, the polymer evolves from one state to another according to a set of "allowed" moves: a given link on the polymer chain may rotate to a new location if it is at the end of the chain, or two links may change simultaneously so that a vertex can "flip" across the chain itself (like inverting the letter V, where the arms of the V represent links on the chain and the tip represents an occupied site). There are challenges in relating the physically elapsed time for a process with a simulation quantity such as the number of vertex moves needed to go from one specific conformation to another, but these two quantities are nevertheless correlated.

On its way through a suite of conformations, the model protein may become "trapped" in a metastable configuration: a conformation that is not the true ground state of the sequence but has a lower energy than any conformation which can be reached from it by just a few moves of a bond or vertex. An example of this is shown in Fig. 4.23 (from Chan and Dill, 1994): the straight line is a completely unfolded configuration while the two other states are compact. One of these compact states is metastable, with 5 HH contacts (repulsive energy = $+11\varepsilon_H$) while the other is the true ground state with 6 HH contacts (energy = $+9\varepsilon_H$). The metastable configuration can only reach the ground state through a series of moves that take it through states with significantly higher energy. At zero temperature, the protein cannot escape from such a metastable state, but at finite temperature, thermal fluctuations will allow it to wiggle away, even if very slowly.

Simulation studies describe the appearance of the energy landscape for protein folding, answering questions such as "how common are metastable states and what is their energy?". They also probe the large-scale nature of the folding process itself by distinguishing among competing views of the time sequence for folding. For example, a commonly held view is that local structures such as helices and sheets form first, followed by condensation of these components into the globular structure itself. However, some

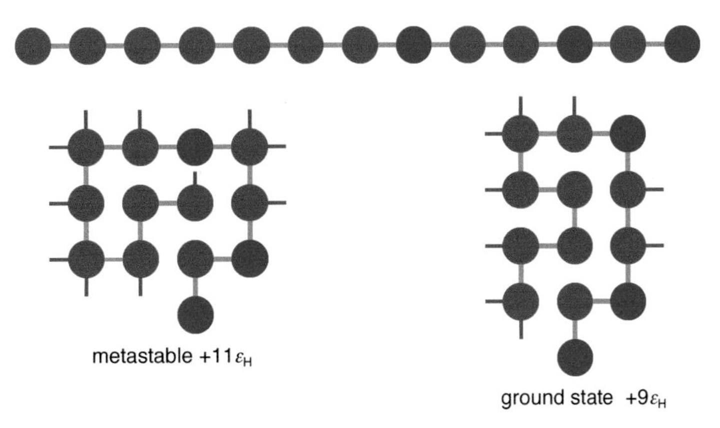

Fig. 4.23 Conformation and energy of the ground state and a metastable state for the sequence HHHHHHHPHHPHP of a 13-site polymer in the HP model (after Chan and Dill, 1994). The linear conformation is displayed at the top of the figure to make the sequence more apparent (blue sites are polar and red sites are hydrophobic).

simulation studies paint a different picture in which there is a fast collapse of the protein on a large scale, followed by slow rearrangement at the local level to form other structures (Chan and Dill, 1994). Simulations have also revealed that certain local arrangements are preferred in both two and three dimensions (Li *et al.*, 1996; other generic results are given in Dill *et al.*, 1995). Simulations of RNA folding have related the stability of folded structures to the size of the genetic alphabet, a significant question in evolution (Mukhopadhyay *et al.*, 2003).

Summary

Many of the important filaments in the cell have properties beyond those of the ideal chains described in Chapter 3. The cell's macromolecules often are helical or are folded into condensed systems with both irregular and patterned substructures within them. Composite filaments assembled from proteins possess torsion resistance that rivals their bending resistance and leads to the formation of supercoiled conformations when subject to sufficiently large torques. The mathematical description of helical structures is slightly more involved than that for simple bending. For example, when subject to a relative torque $\mathcal{T}'$, the ends of a uniform beam rotate with respect to each other, generating a rotational angle α per unit length along the beam if the axis of the beam remains straight. However, for larger $\mathcal{T}'$, the axis of the beam may deform into the shape of a helix, meaning that the beam is simultaneously twisted and curved. If p is the pitch of the helix (the length along the symmetry axis over which the helix completes one turn), the curvature C of the deformed beam is governed by $p = 2\pi\alpha / (C^2 + \alpha^2)$ and the radius r of the helix is $r = C/(C^2 + \alpha^2)$.

More complex structures than helices are possible for filaments subject to torsional stress. A familiar example is the behavior of an electrical cord or telephone handset cord that is repeatedly wound in one direction: at some point, the cord dramatically changes appearance through the creation of a plectoneme, a coil that emerges at right angles to the original symmetry axis of the cord. The formation of conformations like plectonemes is governed by a quantity called the linking number Lk, the role of which can be illustrated through the deformations of a ribbon. If the axis of the ribbon is held straight by applying sufficient tension along its length, then its twist Tw is the number of turns it executes as one end is wound with respect to the other. If the tension is released but the ends are held fixed, the ribbon may reduce its twist through the formation of loops, which have the appearance of a thin ribbon wound around a finger. The number of loops is defined as the withe Wr of the ribbon. The linking number is the sum of the twist and the writhe: $Lk = Tw + Wr$, and it is conserved in the formation of supercoils under continuous deformation.

For modest deformations, the relation between torque and the rotational angle per unit length is governed by the standard Hooke's Law form $\mathcal{T}' = \kappa_{tor}\alpha$, where κ_{tor} is the torsional rigidity. This expression can be integrated over α to obtain the energy density per unit length of the deformation, $\mathcal{E} = \kappa_{tor}\alpha^2/2$, that can itself be integrated along the length of the beam to yield the total energy of the deformed conformation. Torsional deformations can be analyzed within the formalism of continuum mechanics to relate κ_{tor} to elastic moduli such as the shear modulus μ and the Young's modulus Y. For example, for a uniform solid cylinder of radius R, one finds $\kappa_{tor} = \mu\pi R^4/2 = [Y/(1 + \sigma_p)] \cdot \pi R^4/4$, where σ_p is the Poisson ratio. Except for the factor of $(1 + \sigma_p)$, this expression is identical to that for the flexural rigidity κ_f for a solid cylinder, namely $\kappa_f = Y\pi R^4/4$, although beams with other cross sections may exhibit different relationships. Given the generally small values of σ_p, the continuum mechanics analysis predicts that the torsional rigidity and flexural rigidity should be within a factor of two of each other, and this is found to be case experimentally. For example, DNA is found to have $\kappa_f / k_BT \cong 53$ nm and $\kappa_{tor} / k_BT = 50$–100 nm, depending on conditions.

Hydrogen bonding between non-adjacent locations on a polymer may result in local substructures that add complexity to the polymer's mechanical properties. For example, the α-helices and β-sheets within proteins are a result of hydrogen bonding, as is the formation of the double helix of DNA. The unfolding and refolding of these structures can be probed through the measurement of force vs. extension curves at the single-molecule level. For most proteins with folded substructures, the force rises with extension in a manner quantitatively similar to the worm-like chain model for polymers, until a threshold force is reached where the substructure unfolds and the force drops dramatically. Under most experimental conditions, the threshold force is tens of piconewtons, depending on the rate at which the force is applied. Fortunately, there are techniques for taking into account the loading-rate dependence to the force.

The formation of patterns such as helices and sheets in a protein are part of the more general problem of the condensation of a polypeptide into its tertiary structure. Not all of the available sequences of amino-acid residues lead to an appropriate native state. In some cases, the sequence does not produce a condensed state at all, while in others, the ground state may be difficult to achieve because of the presence of metastable states along the folding pathway. These issues can be studied systematically through the use of models that greatly simplify both the interaction between residues and the spatial configurations that they can adopt. Within the context of this parameter and conformation space, the models can probe the most important contributions to whether a conformation is *encodable* (i.e. it is the native state of at least one sequence), or to which conformations are the most designable (i.e. it is encodable for the largest number of sequences). The models also examine the kinetics of folding and how a protein denatures with increasing temperature.

Problems

Applications

4.1. A ribbon is wrapped into the shape of a helix with pitch 5 μm and radius 0.5 μm.

(a) Find the curvature and twist angle per unit length of the helix.
(b) Is your value for α different than $2\pi/5$? If so, explain why.

4.2. Actin filaments can be modeled as a solid rod having an elliptical cross section with major and minor axes of 9 nm and 5 nm, respectively. The axes of the ellipse complete one full rotation around the symmetry axis of the filament every 74 nm.

(a) Assigning a generic value of 10^8 J/m^3 to the shear modulus of the filament, what is its torsional rigidity? Quote your answer in J-m.
(b) What distance does a point at the end of the semi-major axis (the "ridge" of the filment) sweep out in one revolution of the ellipse? (Diagram (b) here shows the location of a point on the ridge.)
(c) What distance does a point at the end of the semi-minor axis (the "trough" of the filament) sweep out in one revolution of the ellipse? (Diagram (c) shows the location of a point on the trough.)

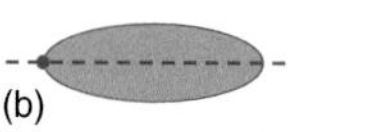
(b)

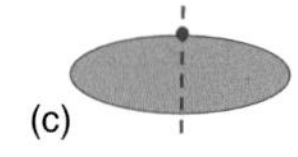
(c)

4.3. To illustrate the magnitude of the torsional deformations expected for cells, supposed that a torque of 10^{-20} N-m is applied to the free end of a 10 μm long filament while the other end is held fixed.

(a) If the filament is cylindrical with a diameter of 0.01 μm and a shear modulus of 10^8 J/m^3, what is its torsional rigidity?
(b) Through what angle will the free end of the filament rotate, assuming that the filament remains straight?

4.4. Suppose that an attempt is made to measure the torsional rigidity of a microtubule by observing thermal fluctuations in the filament's rotation about its axis. If the experiment has an uncertainty of $\pm 5^\circ$, how long must the microtubule be so that the measured $\langle \phi^2 \rangle^{1/2}$ be three times the uncertainty? For κ_{tor} of the microtubule, use a value one thousand times larger than κ_{tor} of F-actin.

4.5. Here, we examine the measurement of the torsional rigidity of DNA, as described in Section 4.3. In the experiment, the DNA is held taut, with one end fixed and the other free to rotate; the length of this segment is 14 800 base pairs. A sphere of diameter 1.0 μm is attached on the side of the DNA strand at its free end; the DNA is overwound by 75 turns and, upon release, the sphere follows the unwinding of the DNA by rotating about an axis through its edge. The drag torque experienced by the sphere is equal to $14\pi R^3 \eta\omega$, where η is the viscosity (10^{-3} kg/m • s) and ω is the angular velocity. If ω is observed to be 7 radians per second, what is κ_{tor}?

4.6. A short length of a particular protein has 24 rigid segments of length 15 nm each. In its native state, the segments fold back on one another to form a series of Z-like barrels as in Fig. 4.2; assume each barrel can pivot independently of its neighbor.

(a) What is the effective contour length of the protein as each of the barrels unfolds?
(b) Plot the three-dimensional entropic spring constant as a function of the effective contour length.
(c) Using Hooke's Law with the spring constants from part (b), calculate the values of the polymer extension x when the force reaches 50 pN for the first three unfolding events; quote your answer in microns. Compare the calculated extensions with the contour length of the completely unfolded protein – how do you interpret your results?

4.7. Suppose that two polymers composed of rigid links (each with length b) lie side by side, and there is a contact energy ε between each pair of vertices in contact with each other on neighboring chains (i.e. ε does not apply to vertices on the same polymer). The chains are unzipped by applying a couplet of forces to one end of each chain as in the diagram.

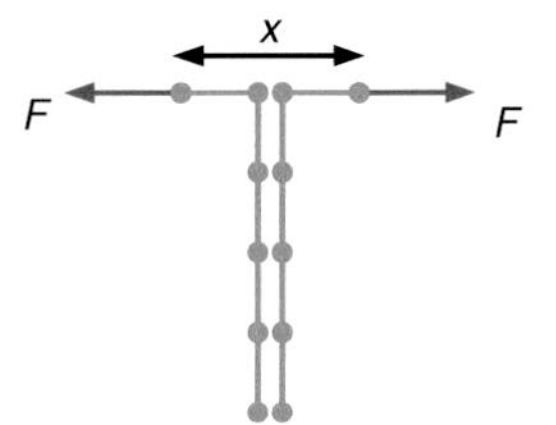

(a) Neglecting thermal fluctuations, find the force needed to break a hydrogen bond between vertices. If $b = 0.34$ nm and $\varepsilon = 4$ kcal/mol, what is this force in pN?

(b) Again, neglecting thermal fluctuations, find the force needed to break a covalent bond between vertices. If $b = 0.15$ nm and $\varepsilon = 90$ kcal/mol, what is this force in pN?

4.8. Calculate the energies of the sequences HHHH, HHHP, HHPH, HPPP, PHPP, HPHP and HHPP of a 4-site polypeptide within the HP model, as applied to the four independent conformations available on a square lattice (i.e. straight, L, zig-zag and U). Which of the sequences have unique folded ground states?

4.9. Calculate the energies of the sequences HHPPHH, PHHHPH and HHPHPH of a 6-site polypeptide within the HP model, as applied to the three folded structures available on a square lattice (i.e. U-, S- and G-shapes). Find the ground-state structure for each sequence, if it exists.

4.10. Calculate the energies of the sequences PHPHHH, HPPHPH, PHPHHP, PHPPHH, PHPPHP, PPHHHH, PPHHPH, HPPHHH, PPHPHH, PHHHHP, PPHPPH, PHHPPH and PHHPHP of a 6-site polypeptide within the HP model, as applied to the three folded structures available on a square lattice (i.e. U-, S- and G-shapes). Find the ground state structure for each sequence, if it exists.

4.11. Draw the conformations for a five-site model protein on a two-dimensional square lattice, with the site labels starting from the left as in the diagram.

Start with the nine conformations of the four-site protein, then attach the fifth site to the fourth at all allowed locations. How many conformations are there in total? For the sequence HPPHP, what are the energies of the conformations in the HP model [*you can use results from the four-site problem without proof to greatly simplify your solution*]?

4.12. For the conformations described in Problem 4.11, show that the probability of finding a polymer in the ground state configuration is $P_{gs} = 4 / [4 + 21 \cdot \exp(-2\varepsilon_H/k_B T)]$. In terms of the energy parameter ε_H, at what temperature is the ground state only 50% occupied? Compare your calculated temperature with that of the four-site model given in the text, and comment on the origin of the difference.

4.13. The protein actin has a molecular mass of 42 kDa and assumes a globular form that is roughly the shape of a spherocylinder 5 nm across and 7 nm long. (a) Based on the actin monomer alone, what is the density of actin in this form? (b) Bulk proteins, hydrocarbons and water should have similar densities, given their atomic compositions. Compare your answer from part (a) with that of water. What do you conclude about the packing of actin in its globular structure?

4.14. The typical volume occupied by an amino-acid residue in the interior of a folded protein is 0.15 nm^3. (a) Use this value to calculate the volume of globular actin with 374 residues. (b) Compare your result with the approximate expression applicable to proteins [*volume*] = $1.27 \times 10^{-3} \bullet$ MW (nm^3/Da), where MW is the molecular mass in Da (p. 229 of Creighton, 1993). (c) Use your result from part (a) to calculate the radius of actin if it were a sphere, and compare it to the dimensions quoted in Problem 4.13. Qualitatively, how much of the interior of this globular form is occupied by the polypeptide?

4.15. Calculate the temperature range over which the ground state occupancy of a polymer drops from 90% to 50% for the 5-site HP model in Problem 4.12. As a specific example, assign the hydrophobic energy parameter ε_H = 1.5 kcal/mol. What is the lowest value of P_{gs} in this system? You may use the probability function in Problem 4.12 without proof.

Formal development and extensions

4.16. The mass per unit length of an initially uniform solid cylinder of radius R and density ρ is reduced by half by hollowing out its core to a radius R_i. What is the resulting decrease in the torsional resistance of the cylinder (quote your answer in percent)?

4.17. Consider two configurations of a linear ribbon of length L (i.e. the ends are not attached to each other): one configuration is pure twist ($Wr = 0$) and other is pure writhe ($Tw = 0$) and both have linking number $Lk = 2$. The deformation of the ribbon is smooth in the sense that both the curvature and the twist angle per unit length are constant along the ribbon. The pure twist state has exactly two turns and the pure writhe state has exactly two loops. Assuming that the flexural and torsional rigidities of this ribbon are identical, what is the ratio of the deformation energies of the two configurations?

4.18. A uniform filament of length L is wrapped into a simple circular loop, with no twist along its path.

(a) What is the bending energy of this configuration?
(b) Holding one side of the loop fixed, the opposite side is twisted to form a figure-8 with constant curvature along its length, as in the diagram.

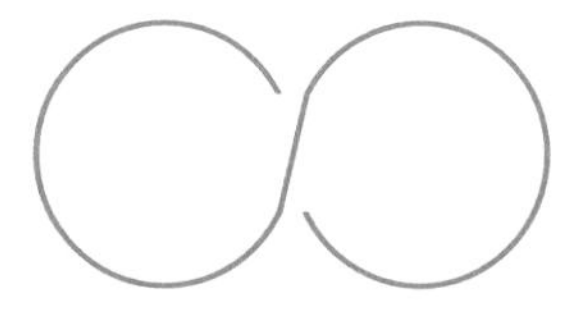

(b) What are Tw and Wr of this configuration?
(c) Find the bending energy, torsional energy and total energy of the new configuration.

4.19. (a) Confirm that if the thermal fluctuations in the rotational angle ϕ are governed by the expression $N(\phi) \propto \exp(-\kappa_{tor}\phi^2 / 2k_BTL)$, the variance of ϕ is given by $\langle \phi^2 \rangle = k_BTL / \kappa_{tor}$.
(b) How long must an actin filament be such that $\langle \phi^2 \rangle^{1/2} = \pi$? Choose its value of κ_{tor} to lie in the mid-range of Table 4.1.

4.20. Consider a piece of ribbon (or belt) lying on a table, making n complete loops as shown. In this configuration, the ribbon bends along its length, but it does not twist.

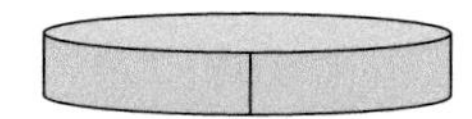

(a) What are the values of Tw and Wr of this configuration?
(b) Holding the two ends of the ribbon such that their planes remain parallel, lift one end until the axis of the ribbon is straight, as in Fig. 4.9. What are Tw and Wr now?
(c) At any intermediate point as you lifted one end, show that Tw and Wr need not be integers, and are given by

$$Tw = n \sin\eta$$

$$Wr = n\,(1 - \sin\eta),$$

where η is the opening angle of the supercoil. [*This problem is not mathematically difficult; show lots of diagrams to demonstrate that you understand the spatial configurations. See Marko and Siggia* (1994) *for plectonemes.*]

4.21. A filament subjected to a stretching force f will undergo torsional buckling only if the torque applied to the filament exceeds a critical value $\tau_{crit} = 2(\kappa_f f)^{1/2}$, where κ_f is the flexural rigidity [*this apples only at zero temperature; see Moroz and Nelson*, 1997b]. Let's apply this idea to a 2000 base-pair segment of DNA. Find the tensile force needed to suppress buckling as a function of the number of extra turns n applied to the DNA. Plot f as a function of n up to 200 turns. How many extra turns can a force of 1 pN sustain? Use $\kappa_f = 2 \times 10^{-28}$ J•m and $\kappa_{tor} = 3 \times 10^{-28}$ J•m.

4.22. Find the work needed to extend a polymer from $x = 0$ to $x = L/2$ in the worm-like chain model, where L is the contour length.

4.23. To illustrate the effects of tensile force on the effective potential energy $V(x)$ of a polymer, consider the hypothetical form $V(x) = \sigma_2/x^2 - \sigma_1/x$, where σ_1 and σ_2 are constants with units of energy if we treat x as dimensionless.

(a) Sketch the potential energy and find the location x_o where $V(x)$ is minimized.

(b) A tensile force is applied to the system, contributing the work $-fx$ to the total energy, where f has the units of energy when x is dimensionless. By what amount $\delta \equiv x - x_o$ is the location of the total energy minimum shifted when a small force is applied (i.e. solve the problem in the limit of small δ)? Based on the discussion of this problem in the text, what sign should δ have?

4.24. Calculate the force–extension relationship for a freely jointed chain in one dimension, subject to a tensile force f applied to its ends. Let the chain have N links, which can be viewed as vectors with length b having n_L links pointing left and n_R links pointing right. The number of configurations for this chain is given by $C(n_L,n_R) = N! \,/\, n_L!n_R!$ as derived in Chapter 3.

(a) Use Stirling's approximation $\ln N! = N \ln N - N$ valid at large N to establish that the entropic contribution $k_BT \ln C(n_L,n_R)$ to the free energy G_T is given by:

$$G_T = k_BT\,[n_R \ln n_R + (N - n_R)\ln(N - n_R)].$$

(b) To your result in part (a) add the work done by the tensile force to obtain the total free energy

$$G = -2f\,bn_R + k_BT\,[n_R \ln n_R + (N - n_R)\ln(N - n_R)].$$

(c) Minimize the free energy with respect to n_R to show $n_R/n_L = \exp(2fb \,/\, k_BT)$.

(d) Expressing the end-to-end length of the chain x and the contour length L_c in terms of n_R and n_L, show that $x/L_c = \tanh(fb \,/\, k_BT)$.

(e) Find the limiting value of x as $f \to \infty$. [*Note that constant terms have been discarded from parts* (*a*) and (*b*).]

4.25. The mean interaction energy for the HPPH sequence of the four-site protein is given by

$$\langle U \rangle = 14\varepsilon_H \exp(-2\beta\varepsilon_H) \,/\, [2 + 7 \bullet \exp(-2\beta\varepsilon_H)],$$

where $\beta = 1/\,k_BT$. Show that the specific heat at constant volume C_V for this system is

$$C_V = 56\varepsilon_H^2 k_B \beta^2 \chi \,/\, (2 + 7\chi)^2,$$

where the function χ is defined by $\chi = \exp(-2\beta\varepsilon_H)$.

4.26. Find the maximum value of the specific heat for the HPPH sequence in the HP model, starting from $C_V = 56\varepsilon_H^2 k_B \beta^2 \chi / (2 + 7\chi)^2$, where the function χ is defined by $\chi = \exp(-2\beta\varepsilon_H)$.

4.27. Problems 4.11 and 4.12 probe the HPPHP sequence on a square lattice in two dimensions. Establish that the mean energy $\langle U \rangle$ (with respect to the ground state) of this model protein is

$$\langle U \rangle = 42\varepsilon_H \exp(-2\beta\varepsilon_H) \,/\, [4 + 21 \bullet \exp(-2\beta\varepsilon_H)].$$

Show that its specific heat at constant volume is given by $C_V = 336\varepsilon_H^2 k_B \beta^2 \chi / (4 + 21\chi)^2$, where $\chi = \exp(-2\beta\varepsilon_H)$.

4.28. For the HPPHP sequence in our HP model, find the temperature at which the specific heat has its maximum value. You may assume without proof, the result from the previous problem that $C_V = 336\varepsilon_H^2 k_B \beta^2 \chi / (4 + 21\chi)^2$, where $(\beta) = \exp(-2\beta\varepsilon_H)$. Quote your temperature $k_B T$ in terms of ε_H.

4.29. Suppose that the potential energy of a system is described by the function $V(x) = -C\cos(\alpha x)$ where α, $C > 0$.

(a) At what location is $V(x)$ a minimum? Is there more than one minimum?

(b) Add a quadratic piece bx^2 to yield $V(x) = -C\cos(\alpha x) + bx^2$ (α, b, $C > 0$). Where is the minimum in $V(x)$ now? What equation determines the location of the lowest-energy metastable states? Does this equation always have solutions for arbitrary values of α, b, C?

4.30. Consider the one-dimensional potential energy function $V(x) = cx^4$, where the coordinate x runs from $-\infty$ to $+\infty$ and where c is a positive constant.

(a) At what location is $V(x)$ a minimum? Is there more than one minimum?

(b) Now add a term ax, to yield $V(x) = ax + cx^4$ (a, $c > 0$). Where is the minimum in $V(x)$? Is there more than one local minimum?

(c) Lastly, add a quadratic term bx^2 so that $V(x) = ax - bx^2 + cx^4$ (a, $c > 0$; $b < 0$). Sketch the resulting function for small values of a. Is there more than one local minimum? If so, which state is stable and which is metastable for small values of a?

5 Two-dimensional networks

With some notable exceptions, such as DNA, most filaments in the cell are linked together as part of a network, which may extend throughout the interior of the cell, or be associated with a membrane as an effectively two-dimensional structure such as shown in Fig. 5.1. This chapter concentrates on the properties of planar networks, particularly the temperature- and stress-dependence of their geometry and elasticity. To allow sufficient time to develop the theoretical framework of elasticity, we focus in this chapter only on networks having uniform connectivity, namely those with the four-fold and six-fold connectivity found in a number of biologically important cells.

5.1 Soft networks in the cell

Two-dimensional networks arise in a variety of situations in the cell; they may be attached to its plasma or nuclear membrane, or be wrapped around a cell as its wall. Containing neither a nucleus nor other cytoskeletal components such as microtubules, for example, the human red blood cell possesses only a membrane-associated cytoskeleton. Composed of tetramers of the protein spectrin, the erythrocyte cytoskeleton is highly convoluted *in vivo* (see Fig. 3.17(b)), but can be stretched by about a factor of seven in area to reveal its relatively uniform four- to six-fold connectivity, as shown in Fig. 5.1(a) (Byers and Branton, 1985; Liu *et al.*, 1987; Takeuchi *et al.*, 1998). Roughly midway along their 200 nm contour length, the spectrin tetramers are attached to the plasma membrane by the protein ankyrin (using another protein called band 3 as an intermediary). About 120 000 tetramers cover the 140 μm^2 membrane area of a typical erythrocyte, corresponding to a tetramer density of about 800 μm^{-2}. The tetramers are attached to one another at junction complexes containing actin segments 33–37 nm long; perhaps 35 000 junction complexes, with an average separation of about 75 nm, cover the erythrocyte.

An example of a two-dimensional biological network with lower connectivity than the red cell is the lateral cortex of the auditory outer hair cell, where the network junctions are frequently connected to just four of their neighbors. In guinea pigs, outer hair cells are roughly cylindrical in shape

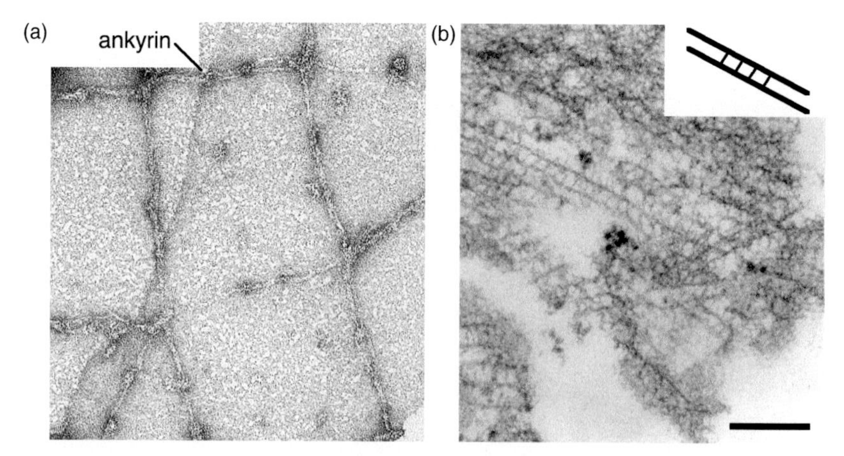

Fig. 5.1 (a) Membrane-associated cytoskeleton of the human erythrocyte. To show its connectivity, the cytoskeleton has been stretched, with a separation between junction complexes of about 200 nm (from Byers and Branton, 1985; courtesy of Dr. Daniel Branton, Harvard University). (b) Section from the cortical lattice of an auditory outer hair cell; inset illustrates how the long, regularly spaced circumferential filaments are cross-linked (bar = 200 nm; reprinted with permission from Holley and Ashmore, 1990; ©1990 by the Company of Biologists; courtesy of Dr. Matthew Holley, University of Bristol).

and about 10 μm in diameter, with the lateral cortex lying inside of, and attached to, the plasma membrane. As displayed in Fig. 5.1(b), the cortex consists of parallel filaments ~5–7 nm thick, spaced 50–80 nm apart and wound around the axis of the cylinder. These filaments are cross-linked at intervals of ~30 nm (with a range of 10–50 nm) by thinner filaments just 2–3 nm thick (from Tolomeo *et al.*, 1996). Morphological and immunological studies suggest that the circumferential filaments are actin, and the crosslinks are spectrin, implying that the network is elastically anisotropic, being stiffer around the cylinder and more flexible along its axis.

The nuclear lamina is a further example of a network with extensive regions of four-fold connectivity (Aebi *et al.*, 1986; McKeon *et al.*, 1986). The nucleus is bounded by two membranes, and the lamina lies in the interior of the nucleus, adjacent to the inner nuclear membrane as shown in Fig. 5.2. Assembled into a 10–20 nm thick meshwork, the intermediate filaments of the lamina are 10.5 ± 1.5 nm in diameter and are typically separated by about 50 nm in the network. Composed of varieties of the protein lamin, the intermediate filaments form a dynamic network that can disassemble during cell division.

On a much shorter length scale than the inter-vertex separation of the networks described above, the few layers of peptidoglycan in the cell wall of Gram-negative bacteria form a tightly knit structure with three-fold coordinated junctions, having the approximate appearance

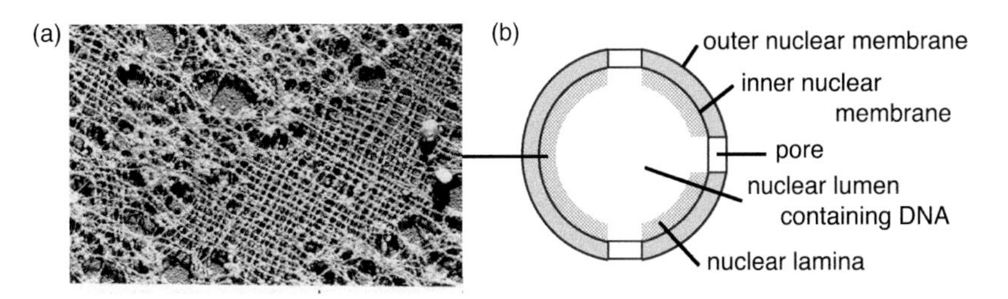

Fig. 5.2

(a) Electron micrograph of a region about 2.5 μm in length from the two-dimensional nuclear lamina in a *Xenopus* ooctyte, showing its square lattice of intermediate filaments (reprinted with permission from Aebi *et al.*, 1986; ©1986 by Macmillan Magazines Limited). Part (b) illustrates the approximate placement of the lamina adjacent to the inner nuclear membrane. For clarity, the endoplasmic reticulum has been omitted from the drawing. Although the pore is shown as a simple hole, in fact it is a 100 nm wide complex of proteins.

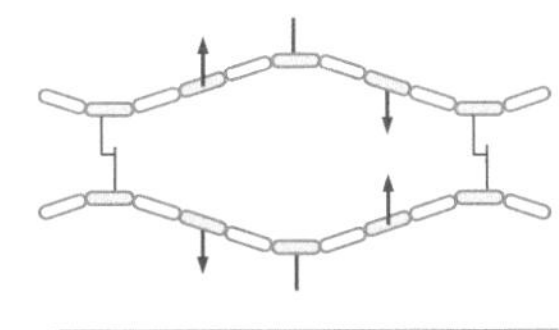

Fig. 5.3

Model for the connectivity of the peptidoglycan network of bacterial cell walls. The stiff chains of sugar molecules, represented by cylinders, are cross-linked both in- and out-of-plane by flexible chains of amino acids. Out-of-plane links are indicated by arrows (reprinted with permission from Koch and Woeste, 1992; ©1992 by the American Society for Microbiology).

of a planar honeycomb. As illustrated in Fig. 5.3, peptidoglycan consists of stiff chains of sugar molecules running in one direction, linked transversely by flexible chains of amino acids. However, the network connectivity may not be entirely two-dimensional: there are thought to be covalent links out of the plane formed by adjacent sugar chains, as indicated by the arrows in the figure. As with the erythrocyte cytoskeleton, the flexible amino acid chains allow a peptidoglycan network to be stretched under stress, and the network area may increase by up to a factor of three compared to the *in vivo* area. The orientation of this anisotropic network is similar to that of the cortex of the auditory outer hair cell: the stiffer elements (sugars) run around the girth of the bacterium's cylindrical shape, while the softer elements lie parallel to its axis of symmetry.

The two-dimensional cellular structures described above have been presented as networks, rather than solid sheets, although these two morphologies display the same mechanical properties when measured on the appropriate length scale, as will be described in Chapter 8. Appropriately, the diameter of the filaments in our cellular networks is much less than the junction spacing, which ranges from a few nanometers in the bacterial cell wall to more than 50 nm in the erythrocyte cytoskeleton. The elastic properties of network filaments, both their resistance to bending and, for sinuous filaments, their resistance to stretching, are described in Chapter 3, where we emphasize the entropic contribution to their elasticity. What happens when we link filaments, both soft and stiff, into a network? How much of the network's behavior faithfully mimics that of its individual components, and how much represents collective effects arising from interactions among filaments?

To address these questions, we need to determine how a network changes in response to stress, in the same way as we characterized, in Chapters 3

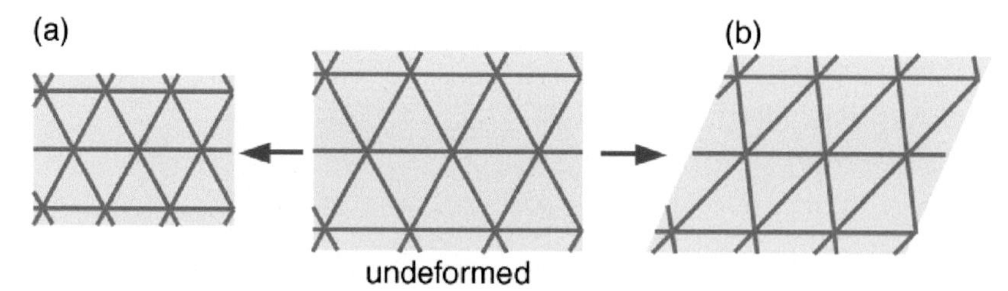

Fig. 5.4 Fundamental deformation modes of a triangular network in two dimensions: (a) a compression mode which changes the network area without affecting its internal angles; (b) a shear mode which changes the angles, but leaves the area untouched.

and 4, the bending and twisting deformations of flexible filaments. In principle, two-dimensional networks, like their three-dimensional cousins, have many deformation modes, although not all of the modes are independent. In two dimensions, isotropic materials (i.e. having an internal structure which is independent of direction), or networks with uniform six-fold connectivity, possess in-plane deformations involving only two fundamental modes – compression and shear – as illustrated in Fig. 5.4. The compression mode displayed in Fig. 5.4(a) preserves the internal angles of the network by uniformly scaling its linear dimensions, in contrast to the shear mode of Fig. 5.4(b), which preserves the area of the network, but not its internal angles. The resistance of two-dimensional, isotropic networks against these deformation modes is parametrized by just two elastic moduli, the area compression modulus K_A and the shear modulus μ, although we emphasize again that networks without full rotational symmetry possess a larger number of independent deformation modes, and hence require a larger number of moduli.

Reflecting the large-scale behavior of a system, elastic moduli can be related to the properties of its individual elements; for example, a uniform network of springs with force constant k_{sp} obeys

$$K_A = \sqrt{3}\, k_{sp} / 2 \quad \mu = \sqrt{3}\, k_{sp} / 4, \quad \text{(zero temperature and stress)} \tag{5.1}$$

if the springs are connected at six-fold coordinated junctions as illustrated in Fig. 5.4. Note that, in two dimensions, the moduli have units of energy per unit area, the same as a spring constant. Similar relations to Eq. (5.1) can be obtained for the elastic moduli of networks with other symmetries, such as four-fold connectivity. Now, Eq. (5.1) applies only for small deformations of an unstressed network at zero temperature; as the temperature increases, or the network is subject to an external (two-dimensional) stress, the network may stray from its uniform appearance, as in Fig. 5.5, such that the elastic moduli change, and Eq. (5.1) becomes increasingly inaccurate. Approximate expressions for the temperature- and stress-dependence of the elastic moduli are available for several network structures of biological

(a)

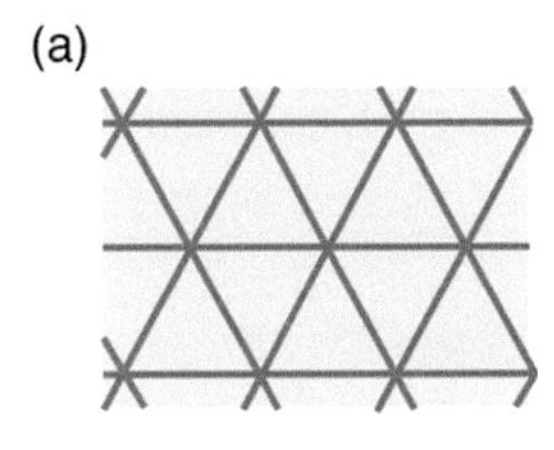

(b)

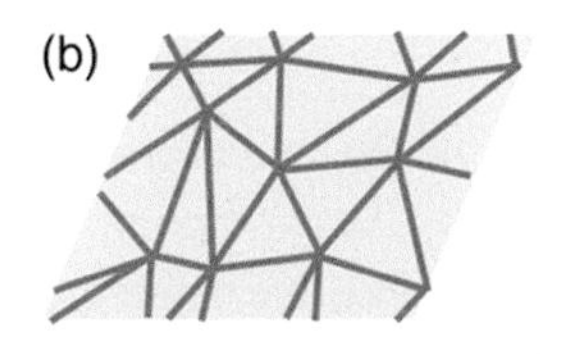

Fig. 5.5

Thermal fluctuations change a zero-temperature network, such as in part (a), into a more erratic configuration like (b), which may also be under a two-dimensional stress applied externally.

importance, and these can be used for interpreting the observed deformation of cellular networks.

We saw in Section 3.4 that a sinuous polymer can be approximately represented by a spring, so long as the polymer is not displaced too far from its equilibrium position. However, at large extensions, the simple spring description of a polymer fails, because a Hookean spring can be extended indefinitely, whereas real protein filaments cannot. Similar difficulties beset pure spring representations of networks: at zero temperature, for example, spring networks with uniform six-fold connectivity are predicted to expand without limit when subjected to a tensile stress exceeding $\surd 3\, k_{sp}$, or collapse under moderate compression. Thus, in these large stress regimes, a successful description of a biological network requires a polymer representation that incorporates steric repulsion among the chain elements, and recognizes the maximal extension of the filament.

The mean shape adopted by a network under stress depends upon the elastic moduli as one might expect: a network with a large compression modulus undergoes only small deformations for a particular stress compared to the deformations experienced by a system with a small K_A. Equally interesting is the directional response of a network to a tensile stress applied along one direction only (referred to as a uniaxial stress). As illustrated in Fig. 5.6, when most materials are stretched in one direction, they shrink in the transverse direction. These changes in shape are characterized by the Poisson ratio, which is the fractional decrease in length along the transverse direction divided by the fractional increase in length along the longitudinal direction. Generally, the Poisson ratio is observed to be +1/3, meaning that the fractional shrinkage in the transverse direction is about 30% of the fractional extension along the direction of the stress. However, networks under certain ranges of stress, or two-dimensional networks fluctuating in three dimensions, may undergo little shrinkage in the transverse direction, or even expand transversely, when stretched longitudinally.

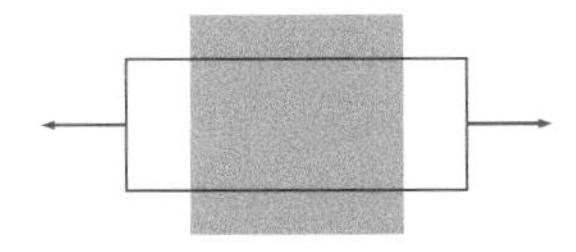

Fig. 5.6

Behavior of conventional materials when subject to a uniaxial stress: the object, whose original position is indicated by the shaded region, stretches in the direction of the stress, but contracts in the transverse direction.

In the following three sections of this chapter, we review and apply the theory of elasticity to networks in two dimensions. Section 5.2 begins with an introduction to elastic moduli in two dimensions as applied to two networks in particular, namely those with uniform six-fold coordination (triangular networks) and four-fold coordination (square networks). The elastic moduli are obtained for spring networks both as an approximation useful for modest deformations, and as an illustration of the relationship between the macroscopic properties of a system and their microscopic components; networks with uniform six-fold coordination are treated in Section 5.3, while those with four-fold coordination are the subject of Section 5.4. Lastly, the formal results of Sections 5.2 to 5.4 are applied to a number of cellular networks in Section 5.5. Students unfamiliar with the tensor description of elastic deformations might consider reading part of Appendix D before tackling the proofs in the following three sections.

5.2 Elasticity in two dimensions

The mechanical behavior of a membrane or a two-dimensional network depends upon whether it can undulate or is forced to lie flat in a plane. In this chapter, we consider only networks that are globally flat, and delay to Chapter 8 the more general question of out-of-plane motion. We begin our treatment of networks by introducing the strain tensor, which can be used for describing the deformation of two- or three-dimensional objects, similar to the way that the displacement from equilibrium, Δx, describes the deformation of a spring in one dimension. Unlike a simple spring where the applied force and the deformation are parallel, the general situation for materials is more complex: a force applied in one direction can result in a deformation in a different direction, as seen when one squeezes Jello, for example. The resistance of a material against deformation is reflected in its elastic moduli, which are the analogs of the spring constant in Hooke's law: the larger the modulus, the smaller the deformation for a given force. For any material, there may be many different elastic moduli linking the magnitude and direction of the applied force with those of the resulting deformation, although in practice the number of independent moduli is strongly reduced because of symmetry considerations; note that the units of the moduli are energy density and therefore depend on the dimensionality of the system. Lastly, we provide a link between the macroscopic elastic properties of a network and its microscopic structure by determining the elastic moduli of a six-fold network of springs at zero temperature.

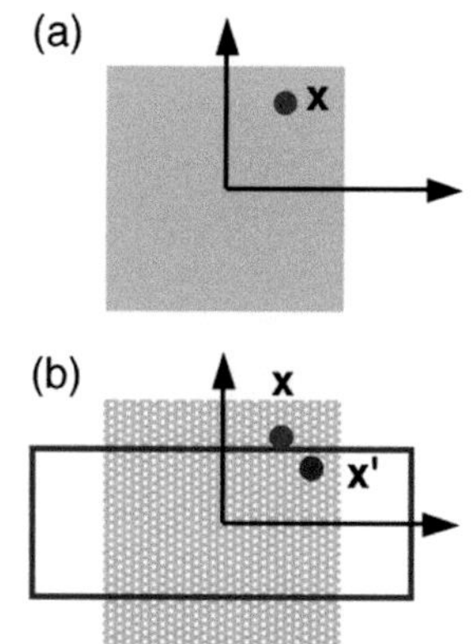

Fig. 5.7

Deformation of a square. The object before (a) and after (b) deformation is shown as the wire outline; for reference, the shaded region of part (b) also indicates the shape and labels of the undeformed object. The position of a specific element of the object shifts from $\mathbf{x}$ to $\mathbf{x}'$ under strain.

5.2.1 Deformations and the strain tensor

How do we describe a deformation mathematically? Consider what happens to a square when it is deformed into a rectangle, as represented by the shaded region and wire outline in Fig. 5.7, respectively. Under the deformation, a mark on the square moves from its original position $\mathbf{x}$ to a new position $\mathbf{x}'$, according to a coordinate system that does *not* follow the object during deformation. The difference in position vectors $\mathbf{x}$ and $\mathbf{x}'$ is illustrated by Fig. 5.7(b), resulting in a displacement vector $\mathbf{u}$ defined by

$$\mathbf{u} \equiv \mathbf{x}' - \mathbf{x}. \tag{5.2}$$

In general, $\mathbf{u}$ is not uniform over the surface of the object because a constant displacement $\mathbf{u}$ of all positions $\mathbf{x}$ is simply a translation of the object.

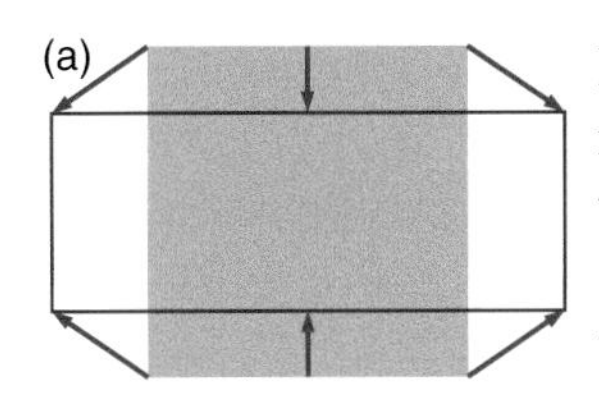

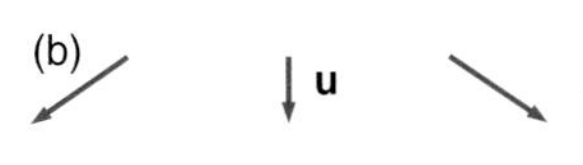

Fig. 5.8

(a) Movement of positions on a square (shaded region) as it changes to a rectangle (wire outline). (b) The deformation is described by the displacement vector **u**, which varies locally on the object.

Rather, the direction and magnitude of **u** varies locally with **x**, as shown in Fig. 5.8, and this variation exposes how the elements of the object move with respect to each other during the deformation.

Referring to Fig. 5.9(a), consider two locations $\mathbf{x}^a$ and $\mathbf{x}^b$ that are initially separated by a vector d**x**,

$$\mathrm{d}\mathbf{x} = \mathbf{x}^b - \mathbf{x}^a. \tag{5.3}$$

Denoting by $\mathbf{u}^a$ and $\mathbf{u}^b$ the change in the positions of the marks at $\mathbf{x}^a$ and $\mathbf{x}^b$ arising from the deformation, the separation $\mathrm{d}\mathbf{x}' = \mathbf{x}'^b - \mathbf{x}'^a$ is given by

$$\mathrm{d}\mathbf{x}' = \mathrm{d}\mathbf{x} + \mathbf{u}^b - \mathbf{u}^a. \tag{5.4}$$

When points a and b are close to each other, Eq. (5.4) can be rewritten as

$$\mathrm{d}x_i' = \mathrm{d}x_i + \Sigma_j (\partial u_i/\partial x_j)\mathrm{d}x_j, \tag{5.5}$$

where the components of each vector are indicated by the subscripts i and j. The distance between nearby points can be evaluated using Eq. (5.5). Before the deformation, the squared distance $\mathrm{d}\ell^2$ between neighboring points is just

$$\mathrm{d}\ell^2 = \Sigma_i\, \mathrm{d}x_i^2, \tag{5.6}$$

which changes to

$$\begin{aligned} \mathrm{d}\ell'^2 &= \Sigma_i\, \mathrm{d}x_i'^2 = \Sigma_i\, [\mathrm{d}x_i + \Sigma_j (\partial u_i / \partial x_j)\mathrm{d}x_j]^2 \\ &= \mathrm{d}\ell^2 + 2\, \Sigma_{i,j} (\partial u_i / \partial x_j)\, \mathrm{d}x_i\, \mathrm{d}x_j \\ &\quad + \Sigma_{i,j,k} (\partial u_i / \partial x_j)(\partial u_i / \partial x_k)\, \mathrm{d}x_j\, \mathrm{d}x_k, \end{aligned} \tag{5.7}$$

after the deformation. Rearranging the summation indices, Eq. (5.7) can be written as

$$\mathrm{d}\ell'^2 = \mathrm{d}\ell^2 + 2\, \Sigma_{i,j}\, u_{ij}\, \mathrm{d}x_i\, \mathrm{d}x_j, \tag{5.8}$$

where

$$u_{ij} \equiv 1/2\, [\partial u_i / \partial x_j + \partial u_j / \partial x_i + \Sigma_k (\partial u_k / \partial x_i)(\partial u_k / \partial x_j)]. \tag{5.9}$$

The quantity u_{ij} is called the *strain tensor*, which is clearly symmetric in indices i and j. There are four components of u_{ij} in two dimensions and nine components in three dimensions. Note that u_{ij} is unitless and is symmetric in indices i and j.

For small deformations, the last term in Eq. (5.9) may be neglected, yielding

$$u_{ij} \cong 1/2\, (\partial u_i / \partial x_j + \partial u_j / \partial x_i). \tag{5.10}$$

If the object is simply translated by an applied force, then **u** is everywhere the same and all derivatives of the displacement vector vanish; hence, $u_{ij} = 0$. In other words, if the force is constant across a body, it accelerates but

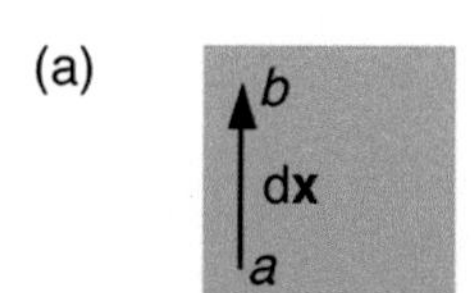

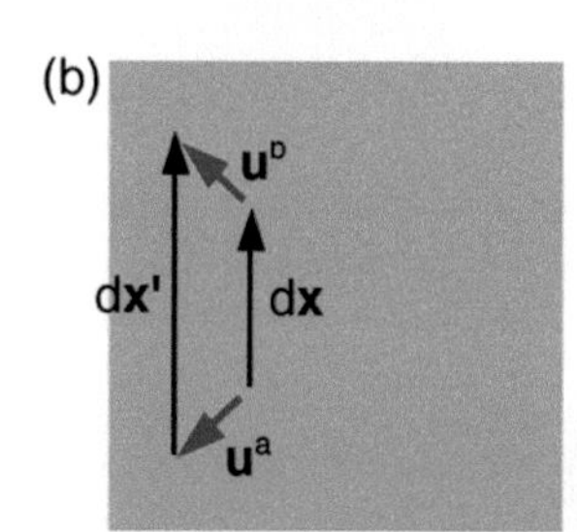

Fig. 5.9

Change in vector **dx** to **dx′** during a shape-preserving deformation, where **dx** is the displacement between locations *a* and *b*. Parts (a) and (b) show the object before and after the deformation, respectively. The initial displacement vector **dx** is superimposed on part (b) to show the changes more clearly.

does not deform. As established in the problem set, one important feature of the strain tensor is that the sum of its diagonal elements (the *trace* of the tensor, or tr$u$) is equal to the fractional change in area (two dimensions) or volume (three dimensions):

$$(dA' - dA)/dA = \mathrm{tr}u. \qquad \text{(two dimensions)} \qquad (5.11)$$

Further properties of the strain tensor can be found in Appendix D; alternate definitions of the strain tensor are described in Fung (1994).

Example 5.1. Suppose that the deformation is a modest version of Fig. 5.8, in which all positions are uniformly scaled, with their x-coordinates increasing from x to $1.1x$, and their y-coordinates decreasing from y to $0.9y$. Find the strain tensor and the fractional change in area of the deformation in the small deformation limit.

The elements of the strain tensor are found from the fractional changes in position:

$$u_{xx} = (1.1-1.0)x / x = 0.1$$

$$u_{yy} = (0.9-1.0)y / y = -0.1.$$

Since the magnitude of the scaling in a given Cartesian direction is independent of the transverse direction in this example, then

$$u_{xy} = u_{yx} = 0.$$

From these elements, the trace of the strain tensor is

$$\mathrm{tr}u = u_{xx} + u_{yy} = 0.1 - 0.1 = 0$$

and there is no change in area

$$(dA' - dA)/dA = \mathrm{tr}u = 0.$$

Note the important result here that u_{ij} is the same everywhere on the object in Fig. 5.8, even though **u** varies locally, so that u_{ij} describes the global behavior of this particular deformation. Note also that these results are not exact, but apply in the small deformation limit.

5.2.2 Forces and the stress tensor

An object deforms in response to external forces, which may have arbitrary orientations with respect to the object. For instance, in Fig. 5.10, a cube is subject to pairs of forces in the x direction: in part (a), the forces are applied perpendicular to the boundary of the object facing the x-axis,

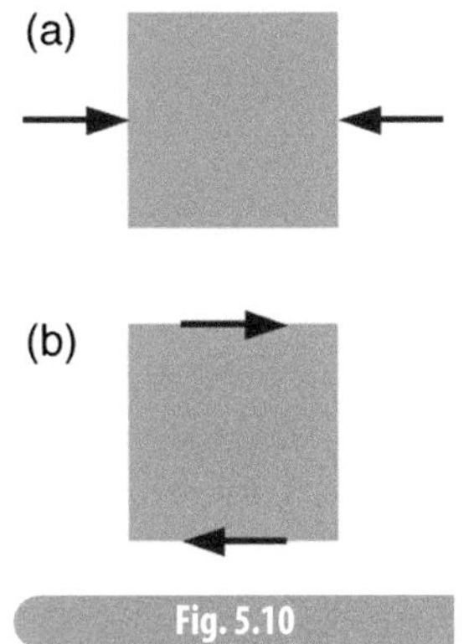

Fig. 5.10

Pairs of forces applied to a cube, seen in cross section. In (a), the forces are applied to the surfaces facing the x-axis, while in (b), the forces are applied parallel to surfaces facing the y-axis.

resulting in a compression, while in part (b), the forces are applied parallel to the boundaries facing the y-axis, resulting in a shear. Thus, to determine how a force deforms an object, one must take into account its orientation with respect to the boundary of the object. Further, if the force is spread over a large surface it will not deform the object as much as if it is applied to a small surface, where, in this chapter, we use the term *surface* to represent a perimeter in two dimensions. For the deformation of extended objects the analog of the force is the stress tensor σ_{ij}, which has units of energy density.

The stress tensor can be found in terms of forces and lengths, as will be done in Chapter 6. However, the physical units of the stress tensor depend on the dimensionality of the system, so we choose not to work with σ_{ij} in this chapter in order to avoid notational confusion. Rather, we define the stress tensor in terms of the change in the free energy density $\mathcal{F}$:

$$\sigma_{ij} = \partial\mathcal{F} / \partial u_{ij}. \tag{5.12}$$

This equation, which is valid in both two and three dimensions, is analogous to the relation $F = -\mathrm{d}V / \mathrm{d}x$ between the force F and the potential energy V in one dimension. Just as with the strain tensor, the stress tensor is symmetric under exchange of indices i and j

$$\sigma_{ij} = \sigma_{ji}. \tag{5.13}$$

The stress tensor will be treated in greater detail in Chapter 6.

5.2.3 Elastic moduli

Just as the potential energy of a Hooke's law spring is quadratic in the square of the displacement, the change in the free energy density $\Delta\mathcal{F}$ of a continuous object under deformation is quadratic in the strain tensor u_{ij}:

$$\Delta\mathcal{F} = 1/2\ \Sigma_{i,j,k,l}\ C_{ijkl}\ u_{ij}\ u_{kl}. \tag{5.14}$$

The constants C_{ijkl} in Eq. (5.14) are called the elastic moduli and they look intimidating – there are four indices $ijkl$ each of which can have values of x or y in two dimensions (or linear combinations of x and y). However, symmetry principles come to the rescue in reducing the number of truly independent moduli. For example, because u_{ij} is symmetric under the exchange of i and j, then so must the moduli: $C_{ijkl} = C_{jikl}$ and also $C_{ijkl} = C_{ijlk}$ for all values of x and y. Further, the combination $u_{ij}u_{kl}$ in Eq. (5.14) is symmetric under the exchange of the ij pair of indices with the kl pair. These symmetry considerations alone reduce the number of independent moduli from $2^4 = 16$ in two dimensions down to just 6. Additional symmetry conditions imposed by the material itself decreases the number of independent moduli even further (see Landau and Lifshitz, 1986). Let's now examine two different systems to see the implications.

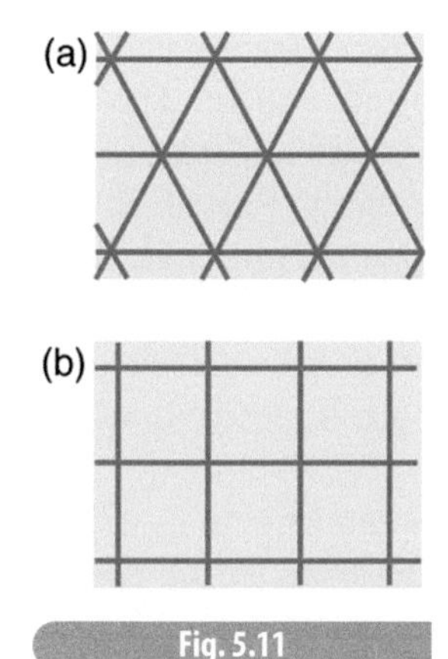

Fig. 5.11

Two-dimensional networks with six-fold (a) and four-fold symmetry (b).

5.2.4 Six-fold networks in 2D

The network displayed in Fig. 5.11(a) is an example of a system with six-fold symmetry: each vertex in the network is attached to six others, and the links between the vertices have equal lengths. Thus, the network can be rotated in steps of $2\pi/6$ radians and its appearance is unchanged locally, ignoring the overall shape of the object. This rotational symmetry can be imposed on the elastic moduli to generate more relationships among its components, as is shown in detail in Appendix D. As a result, there are only two independent moduli, not $2^4 = 16$ that one might naïvely expect. The two moduli can be chosen to be the area compression modulus K_A and the shear modulus μ, both in two dimensions, and these two moduli are expressed in terms of C_{ijkl} in Appendix D. In terms of K_A and μ, Eq. (5.14) becomes

$$\Delta\mathcal{F} = (K_A/2)(u_{xx}+u_{yy})^2 + \mu\ \{(u_{xx}-u_{yy})^2/2 + 2u_{xy}^{\ 2}\}. \quad \text{(six-fold symmetry)} \tag{5.15}$$

It turns out that isotropic materials also obey this expression; that is, in two dimensions, isotropic materials and those with six-fold symmetry are both described by just two elastic moduli at small deformations. To confirm that K_A and μ play their appropriate role in Eq. (5.15), we work through the following example.

Example 5.2. Find the expression for the free energy of deformation $\Delta\mathcal{F}$ in Eq. (5.15) under the following deformations:

(a) pure compression with $u_{xx} = u_{yy} \neq 0$ and $u_{xy} = 0$.
(b) simple shear with $u_{xx} = u_{yy} = 0$ and $u_{xy} \neq 0$.

(a) The elements of the strain tensor were determined for a pure compression in Example 5.1 and are seen to correspond to $u_{xx} = u_{yy} \neq 0$ and $u_{xy} = 0$. Substituting these into Eq. (5.15) yields

$$\Delta\mathcal{F} = (K_A/2)(u_{xx} + u_{yy})^2 + \mu\ \{(u_{xx} - u_{xx})^2/2 + 2\bullet 0\} = (K_A/2)(u_{xx} + u_{yy})^2.$$

But $u_{xx} + u_{yy}$ is just the trace of the strain tensor $\mathrm{tr}u$, and we know from Eq. (5.11) that $\mathrm{tr}u$ is the fractional change in the area: $\mathrm{tr}u = \Delta A\ /\ A$. Thus,

$$\Delta\mathcal{F} = (K_A/2)(\Delta A\ /\ A)^2.$$

Clearly, K_A plays the role of the compression modulus in this equation.

(b) In this deformation mode, the extensions along the x- and y-axes don't change ($u_{xx} = u_{yy} = 0$), and the network deforms like the one in Fig. 5.4(b), which is a simple shear. Then, Eq. (5.15) becomes

$$\Delta\mathcal{F} = (K_A/2)\bullet 0 + \mu\ \{0 + 2u_{xy}^{\ 2}\} = 2\mu u_{xy}^{\ 2}.$$

Here, the free energy depends only on μ, which must be the shear modulus.

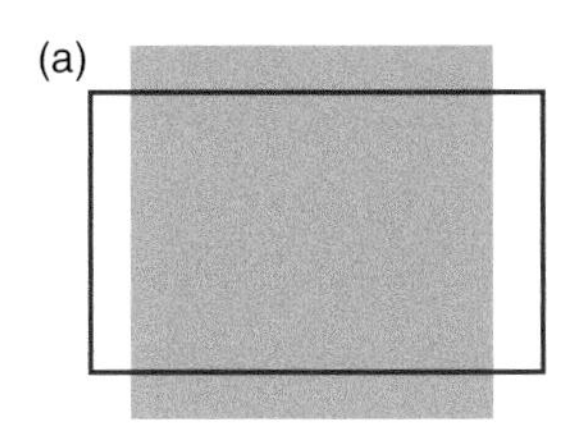

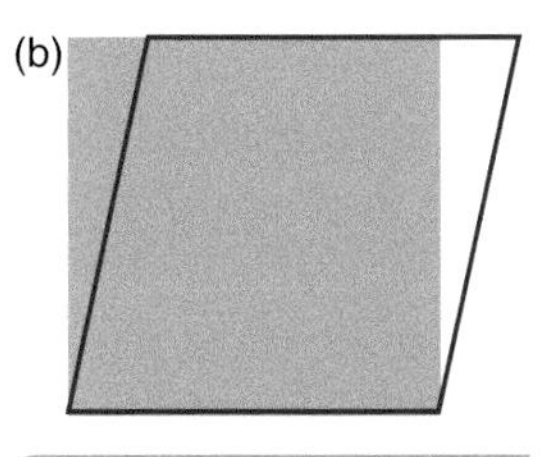

Fig. 5.12

Two potentially inequivalent shear modes in four-fold systems: pure shear (a) and simple shear (b). The shaded region shows the original configuration, while the wire outline is the deformed object.

The Poisson ratio σ_p is a measure of how much a material *contracts* laterally when it is stretched longitudinally. If the stretching force is applied along the x-axis,

$$\sigma_p = -u_{yy} / u_{xx}. \tag{5.16}$$

For objects in two dimensions, σ_p is

$$\sigma_p = (K_A - \mu)/(K_A + \mu). \qquad \text{(two dimensions)} \tag{5.17}$$

The proof is given in the problem set.

5.2.5 Four-fold networks in 2D

The system represented in Fig. 5.11(b) is an example of a network with four-fold symmetry: each of the vertices is connected to four nearest neighbours. The symmetries of this network are different than those of the six-fold network, and as a result, the free energy density has a different form. The steps in determining the relationships among the moduli are described in more detail in Appendix D, and lead to the expression for the free energy density $\Delta\mathcal{F}$ of

$$\Delta\mathcal{F} = (K_A/2)(u_{xx} + u_{yy})^2 + (\mu_p/2)(u_{xx} - u_{yy})^2 + 2\mu_s u_{xy}^2. \qquad \text{(four-fold symmetry)} \tag{5.18}$$

There are two shear moduli present in Eq. (5.18), μ_p and μ_s, reflecting the resistance against the two shear modes (pure shear and simple shear, respectively) shown in Fig. 5.12. One can see the nature of the modes by examining Eq. (5.18).

(i) The pure shear mode in Fig. 5.12(a) involves compression in one direction and expansion in the orthogonal direction. If the area is preserved in this mode, then the trace of the strain tensor must vanish and consequently $u_{xx} = -u_{yy}$. The only contribution to $\Delta\mathcal{F}$ then comes from the μ_p term.
(ii) The simple shear mode in Fig. 5.12(b) involves a change in angle between the horizontal and vertical elements, but not a change in the horizontal or vertical lengths if the area is preserved. Thus, only $u_{xy} \neq 0$, and the single contribution to $\Delta\mathcal{F}$ arises from the μ_s term.

Both of these modes can be related to each other through a coordinate transformation (see Fung, 1994, p. 132). Viewing the network bonds as springs, for example, one can see that that pure and simple shear modes have different energies, and hence there is no reason why μ_p and μ_s should be the same.

5.2.6 Networks of springs

Our next step in exploring elasticity is to relate the elastic moduli, which describe a system at large length scales, to the properties of the individual

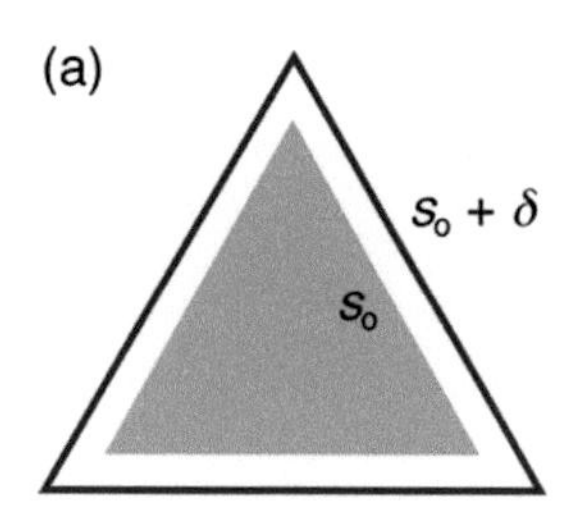

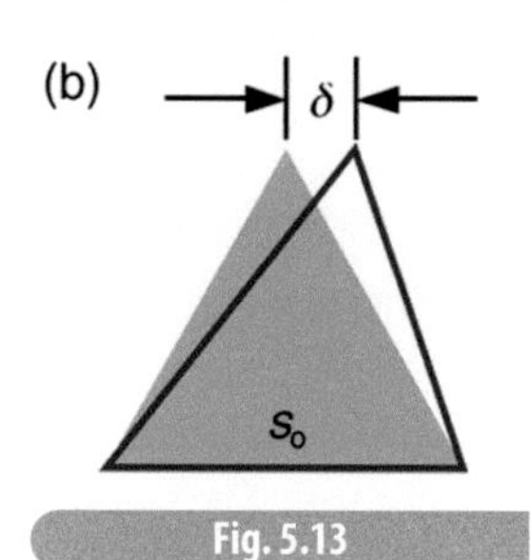

Fig. 5.13

Infinitesimal deformations (exaggerated for effect) used for calculating the compression modulus (a) and shear modulus (b) of a triangular network at zero temperature. In (a), all springs are stretched in length from s_o to $s_o + \delta$, while in (b), the top vertex is moved a distance δ to the right.

components of the system. To show how the calculation is done, we examine a specific situation, namely a six-fold network of springs. Our choice isn't arbitrary, as the results from spring networks provide a useful benchmark for analyzing two-dimensional networks in the cell. Each spring has an unstretched length of s_o, and a potential energy V_{sp} of

$$V_{sp} = k_{sp}(s - s_o)^2 / 2, \tag{5.19}$$

where k_{sp} is the force constant. The relationship between microscopic and macroscopic descriptions is found by comparing two expressions for the change in free energy density under an infinitesimal deformation – one in terms of elastic constants (K_A and μ), and the other in terms of spring characteristics (k_{sp} and s_o).

We consider first the pure compression mode of Fig. 5.13(a). If each spring is stretched a small amount $\delta \equiv s - s_o$ away from its equilibrium length s_o, the change in potential energy ΔU per vertex is, from the energy of a single spring in Eq. (5.19),

$$\Delta U = 3\,\Delta V_{sp} = 3k_{sp}\delta^2/2, \tag{5.20}$$

where the factor of three arises because there are three times as many springs as there are vertices. Dividing Eq. (5.20) by the network area per vertex of $A_v = \surd 3\, s_o^2/2$, for triangular networks, shows that the change in the free energy density $\Delta\mathcal{F}$ for small deformations is

$$\Delta\mathcal{F} = \Delta U / A_v = \surd 3\, k_{sp}(\delta/s_o)^2. \tag{5.21}$$

Next, we write the strain tensor in terms of s_o and δ. Because the deformations are uniform in the x and y directions, it follows that $u_{xx} = u_{yy} = \delta/s_o$. Further, the off-diagonal elements vanish, $u_{xy} = 0$, because the displacement in the y-direction is independent of the position of the triangle in the x-direction. Substituting these elements of the strain tensor into Eq. (5.18) gives

$$\Delta\mathcal{F} = 2K_A(\delta/s_o)^2, \tag{5.22}$$

for pure compression. Comparing Eqs. (5.21) and (5.22) yields

$$K_A = \surd 3\, k_{sp}/2. \qquad \text{(six-fold network)} \tag{5.23}$$

An analogous method reveals an expression for the shear modulus, based upon the simple shear mode of Fig. 5.13(b). The steps are calculated as an end-of-chapter problem, and yield

$$\mu = \surd 3\, k_{sp}/4. \qquad \text{(six-fold network)} \tag{5.24}$$

From Eqs. (5.23) and (5.24) we see that the ratio of K_A to μ is 2 for six-fold spring networks at zero temperature. Note that Eqs. (5.23) and (5.24) do not depend on s_o, and that thermal fluctuations in the spring length have been ignored.

5.3 Isotropic networks

Not unexpectedly, the engineering specifications for a biological network vary from cell to cell. For example, a red blood cell needs a flexible network that permits it easily to deform in a capillary while a bacterium needs a strong envelope to resist its internal pressure. Thus, a cell may experience a variety of stresses during its everyday operation and adopts a network design to respond to these stresses in an appropriate way. How do the elastic properties of these networks depend upon quantities such as stress or temperature? Considering the stress-dependence for a moment, we know that the compression modulus of an ideal gas, for instance, is equal to its pressure (see end-of-chapter problems), indicating that a gas becomes more resistant to compression as its pressure rises. Do similar relationships exist for networks?

In this section, we determine the stress- and temperature-dependence of the elasticity and shape of a two-dimensional spring network with six-fold connectivity, similar to the cytoskeleton of the human erythrocyte. The properties of these networks are known from both analytical and computational treatments (Boal *et al.*, 1993; Discher *et al.*, 1997; Lammert and Discher, 1998; Wintz *et al.*, 1997). Four-fold networks are examined separately in Section 5.4, permitting us to observe the dependence of elastic properties on network connectivity. Although our primary interest is the behavior of networks under tension, there are situations in which a two-dimensional network can be compressed without buckling, encouraging us to consider both compression and tension.

(a)

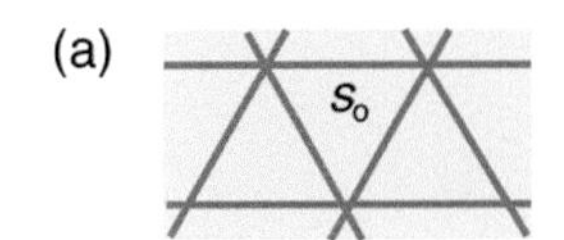

(b)

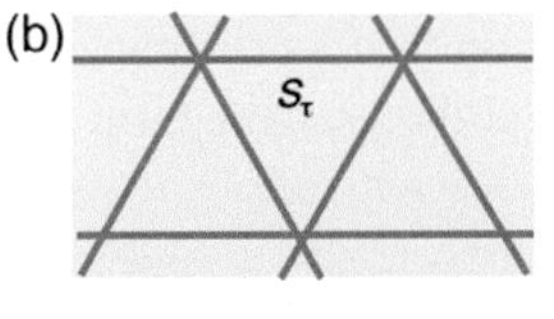

Fig. 5.14

The elementary plaquette in a planar network with six-fold connectivity has the shape of an equilateral triangle if all elements have the same length s. When the network is placed under a two-dimensional tension, s changes from its unstressed value s_o in (a) to a new value $s\tau$ in (b).

5.3.1 Six-fold spring networks under stress

A network with six-fold connectivity, as illustrated in Fig. 5.14, can be viewed as a plane of triangular plaquettes, each plaquette bounded by three mechanical elements attached to neighboring vertices. In a network of springs at zero temperature and stress, all elements have the same length s_o and the plaquettes have the shape of equilateral triangles. The spring length changes from s_o to s_τ when the network is placed under a two-dimensional isotropic tension τ, but the plaquettes remain equilateral triangles, albeit with different area. For such a network, s_τ can be calculated by minimizing the enthalpy H

$$H = E - \tau A, \tag{5.25}$$

where E is the energy of the springs and A is the area of the network. Here we adopt the sign convention that $\tau > 0$ corresponds to tension and $\tau < 0$ to compression. The quantities we calculate are normalized per network vertex, where each triangular plaquette corresponds to 1.5 springs and 0.5 vertices. As there are three springs for each vertex in the network, the energy per vertex for springs of equal length s is $(3/2)k_{sp}(s - s_o)^2$ and the area per vertex A_v is

$$A_v = \sqrt{3}\, s^2/2. \tag{5.26}$$

Hence, the enthalpy per vertex H_v is

$$H_v = (3/2)k_{sp}(s - s_o)^2 - \sqrt{3}\, \tau s^2/2, \tag{5.27}$$

for a six-fold spring network with equilateral plaquettes. The spring length s_τ that minimizes H_v for a particular stress can be found by equating the derivative $\partial H_v / \partial s$ to zero, yielding

$$s_\tau = s_o / (1 - \tau / [\sqrt{3}\, k_{sp}]), \qquad \text{(six-fold symmetry)} \tag{5.28}$$

such that the minimum value of the enthalpy per vertex is

$$H_{v,min} = -(\sqrt{3}/2)\tau s_o^2 / (1 - \tau / [\sqrt{3}\, k_{sp}]). \qquad \text{(equilateral plaquettes)} \tag{5.29}$$

Equation (5.28) predicts that the network expands under tension ($\tau > 0$) and shrinks under compression ($\tau < 0$).

Fig. 5.15 displays the area per vertex as obtained from s_τ in Eq. (5.28), which assumes all plaquettes are equilateral triangles. First, observe that the area increases without bound as the tension approaches a critical value τ_{exp},

$$\tau_{exp} = \sqrt{3}\, k_{sp}, \qquad \text{(six-fold symmetry)} \tag{5.30}$$

as can be obtained from Eq. (5.28) by inspection. The network area blows up at large tension because the tension contribution to H_v dominates at $\tau >> 0$. Although both the energy of the springs and the pressure term τA_v scale like s^2 at large extensions, they enter the enthalpy with opposite signs

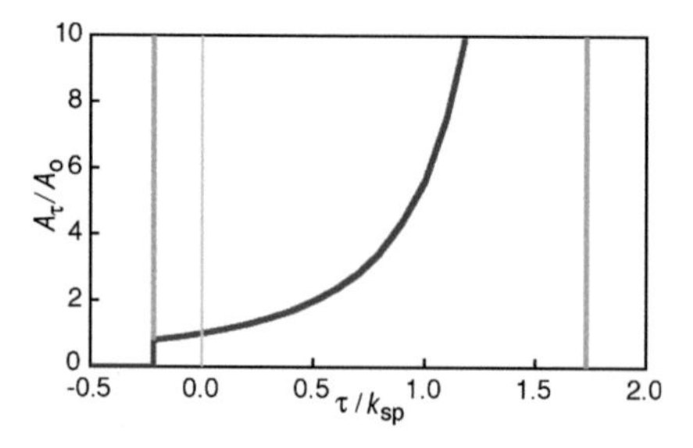

Fig. 5.15 Area per vertex expected for a spring network at zero temperature, shown as function of the reduced tension τ/k_{sp}, and normalized to $A_o = \sqrt{3}\, s_o^2/2$. Under a tensile stress, the area expands without bound if $\tau < \sqrt{3}\, k_{sp}$; under compression, the network collapses if $\tau < \tau_{coll} = -(\sqrt{3}/8)\, k_{sp}$ (after Discher *et al.*, 1997).

if $\tau > 0$. Hence, if τ is sufficiently large, the tension term must dominate over the spring energy, and the network expands without limit.

A second feature to note in Fig. 5.15 is the collapse of the network under compression, a behavior *not* present in networks of equilateral plaquettes. We have assumed, in deriving Eq. (5.29), that the enthalpy is minimized when its derivative with respect to s vanishes. However, Eq. (5.29) may not be the global minimum of H if shapes other than equilateral plaquettes are considered. Indeed, equilateral triangles enclose the largest area for a given perimeter, and therefore may not be the optimal plaquette shape under compression, where the tension term τA drives the system towards small areas. Clearly, the smallest τA contribution at $\tau < 0$ is given by plaquettes with zero area, of which those shaped like isosceles triangles have the lowest spring energy, namely $(k_{sp}/2) \bullet [(2s_{Iso} - s_o)^2 + 2(s_{Iso} - s_o)^2]$, where a zero-area triangle has two short sides of length s_{Iso} and a long side of length $2s_{Iso}$. The minimum value of this energy as a function of s_{Iso} is $k_{sp}s_o^2/6$ and occurs for $s_{Iso} = 2s_o/3$. Thus, the enthalpy per vertex of the equilateral network rises with pressure ($\tau < 0$) according to Eq. (5.29) until it exceeds the enthalpy of a network of zero-area isosceles plaquettes ($k_{sp}s_o^2/6$) at a collapse tension τ_{coll} given by

$$\tau_{coll} = -(\sqrt{3}/8)\, k_{sp}, \qquad \text{(six-fold symmetry)} \qquad (5.31)$$

or, equivalently, at a spring length $s_\tau/s_o = 8/9$. That is, the network collapses when subject to a two-dimensional pressure exceeding $|\tau_{coll}|$.

The elastic moduli of equilateral spring networks under stress can be related to the component characteristics k_{sp} and s_o using the same technique as was employed in Section 5.2 for unstressed networks. For a known deformation, two expressions are obtained for the change in the free energy density: one expression is in terms of the spring variables (k_{sp} and s_o) while the other is in terms of the elastic moduli and the strain tensor. As shown in Fig. 5.16, the deformation modes of choice are the same as those used in Section 5.2 except that, here, the small perturbations are performed on a state that is already under tension with sides of length s_τ.

Because the extraction of the elastic moduli from the deformation modes in Fig. 5.16 follows Section 5.2 very closely, the calculations are not repeated here. Rather, Table 5.1 summarizes the strain tensor and the changes in free energy density expected for the two deformation modes. Note that the deformations in Fig. 5.16 are with respect to the stressed state with spring length s_τ; i.e. the strain tensor measures a perturbation about a reference state which is itself deformed with respect to the unstressed state with spring length s_o. The elastic moduli can be obtained by equating columns 2 and 4 in Table 5.1, leading to the zero-temperature relations

$$K_A = (\sqrt{3}\, k_{sp}/2) \bullet (1 - \tau / [\sqrt{3}\, k_{sp}]) \qquad \text{(six-fold symmetry, } T = 0) \qquad (5.32a)$$

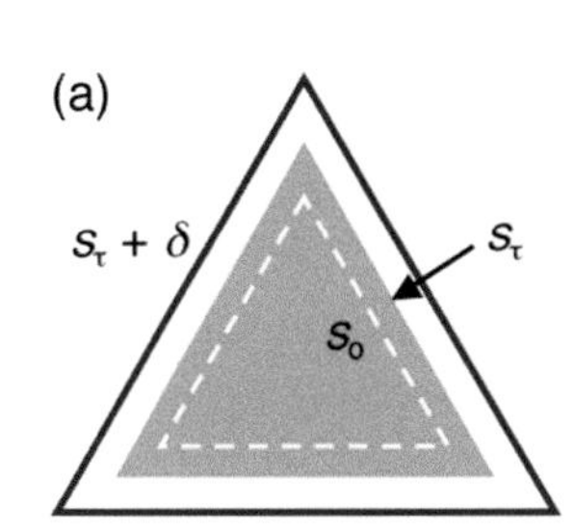

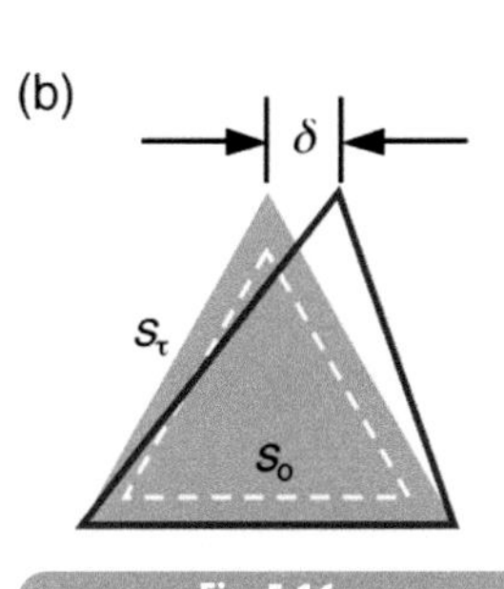

Fig. 5.16

An unstressed equilateral triangle has sides of length s_o, indicated by the dashed line. An *equilateral* triangle under tension, with sides of length s_τ, is shown as the shaded region; this triangle is given an infinitesimal deformation as shown in outline. Mode (a) measures the area compression modulus, while mode (b) measures the shear modulus.

Table 5.1 Change in free energy density $\Delta\mathcal{F}$ for deformation modes (a) and (b) in Fig. 5.16 of a triangulated spring network under stress. Column 2 (microscopic) refers to the enthalpy change expressed in spring variables, while column 4 (continuum) is the free energy change using Eq. (5.16) and the strain tensor in column 3.

Mode	$\Delta\mathcal{F}$ (*microscopic*)	Strain	$\Delta\mathcal{F}$ (*continuum*)
(a)	$\sqrt{3}\, k_{sp}\, (1 - \tau / [\sqrt{3}\, k_{sp}]) \bullet (\delta / s_\tau)^2$	$u_{xx} = u_{yy} = \delta / s_\tau$ $u_{xy} = 0$	$2K_A(\delta / s_\tau)^2$
(b)	$(k_{sp} / 2\sqrt{3}) \bullet (1 + \sqrt{3}\, \tau / k_{sp}) \bullet (\delta / s_\tau)^2$	$u_{xx} = u_{yy} = 0$ $u_{xy} = (\delta / s_\tau) / \sqrt{3}$	$(2/3)\mu(\delta / s_\tau)^2$

$$\mu = (\sqrt{3}\, k_{sp}/4) \bullet (1 + \sqrt{3}\, \tau / k_{sp}). \tag{5.32b}$$

These expressions reduce to Eqs. (5.23) and (5.24) in the zero-stress limit. We remark that a more direct way of obtaining Eq. (5.32a) is to simply differentiate the area per vertex $\sqrt{3}\, s_\tau^2/2$ with respect to τ, and substitute the result into the definition of the areal compression modulus. Equations (5.32) show that the moduli do depend on the tension τ, becoming more resistant to shear as the network is stretched. The fact that K_A vanishes at $\tau = \sqrt{3}\, k_{sp}$ has already been anticipated, since the area per vertex grows without bound at this tension. Note that Eqs. (5.32) are not appropriate for collapsed networks.

The stress-dependence of the moduli has interesting implications for the Poisson ratio, which is a measure of how a material contracts in the transverse direction when it is stretched longitudinally. Specifically, the Poisson ratio σ_p is defined in Eq. (5.16) as the ratio of the strains $\sigma_p = -u_{yy} / u_{xx}$, where the stress is applied along the x-axis (not isotropically, as in Table 5.1). The negative sign is chosen so that conventional materials have a positive Poisson ratio: they shrink transversely when stretched longitudinally. In the end-of-chapter problems, we show how to relate the Poisson ratio to the elastic moduli, yielding $\sigma_p = (K_A - \mu) / (K_A + \mu)$, which is Eq. (5.17). Even though K_A and μ are both required to be positive (otherwise the material will spontaneously deform by compression or shear), there is no requirement that $K_A > \mu$ to ensure $\sigma_p > 0$. From Eqs. (5.23) and (5.24), a triangular network at zero temperature and stress has $K_A/\mu = 2$ and consequently $\sigma_p = 1/3$. However, the stress-dependence of the elastic moduli in Eq. (5.32) shows that the compression modulus decreases with tension, while the shear modulus increases with tension, leading to

$$\sigma_p = [1 - 5\tau / (\sqrt{3}\, k_{sp})] / [3 + \tau / (\sqrt{3}\, k_{sp})]. \qquad \text{(six-fold symmetry)} \tag{5.33}$$

This expression for σ_p becomes negative for a six-fold spring network under tension for $\tau / k_{sp} > \surd 3 /5$, indicating that such a network should expand laterally when stretched longitudinally in this range of stress.

Just how well do our zero-temperature calculations reproduce the stress-dependence of a spring network at non-zero temperature? From the existence of the collapse transition alone, we know to be wary of severe approximations: the collapse transition only appears once we drop the assumption that all springs have the same length. The calculations presented here should be accurate for triangulated networks at low temperature and/or high tension, where the springs are expected to be of uniform length even if it is different from s_o. A Monte Carlo simulation has shown Eq. (5.28) to be accurate at the 90% level or better over the temperature range $k_B T = k_{sp} s_o^2/8$ to $k_{sp} s_o^2$ (Boal *et al.*, 1993).

5.3.2 Six-fold spring networks at non-zero temperature

Our calculation of the stress-dependence of spring networks assumed that all springs have a common length, an assumption valid at low temperatures or high tensions relative to the scales set by $k_{sp} s_o^2$ and k_{sp}, respectively. However, as the temperature rises, the springs increasingly fluctuate in length. Consider an isolated spring, for which the probability of the displacement from equilibrium, $x \equiv s - s_o$, is proportional to the Boltzmann factor $\exp(-\Delta U / k_B T) = \exp(-k_{sp} x^2 / 2 k_B T)$. Properly normalized, the probability $\mathcal{P}(x)\mathrm{d}x$ that x lies between x and $x + \mathrm{d}x$ is given by

$$\mathcal{P}(x)\mathrm{d}x = \mathrm{d}x \bullet \exp(-\beta k_{sp} x^2/2) \,/ \int_{-\infty}^{\infty} \exp(-\beta k_{sp} x^2/2)\, \mathrm{d}x, \qquad (5.34)$$

where β is the inverse temperature $(k_B T)^{-1}$. Derived as an example in Appendix C, the mean square value of x obtained from this probability distribution is

$$\langle x^2 \rangle = \int_{-\infty}^{\infty} x^2\, \mathcal{P}(x)\, \mathrm{d}x = k_B T / k_{sp}, \qquad (5.35)$$

Fig. 5.17

Snapshot of a triangular network of springs as seen in a computer simulation. The network is held at zero tension and the relatively modest temperature of $k_B T = k_{sp} s_o^2/4$. The system is subject to periodic boundary conditions in the shape of a rectangle, whose location is indicated by the black background (after Boal *et al.*, 1993).

confirming that the fluctuation in the spring length increases with temperature. Thus, we expect spring networks to lose their regular appearance as the temperature grows, as can be seen in the simulated configuration displayed in Fig. 5.17.

It is not immediately clear from Fig. 5.17 how the network area should change with temperature, as there is considerable variation in the spring length and plaquette area. In fact, the thermal behavior of a network depends on its connectivity, as will be shown in Section 5.4 by comparing networks with four- and six-fold connectivity. As determined by computer simulation, the

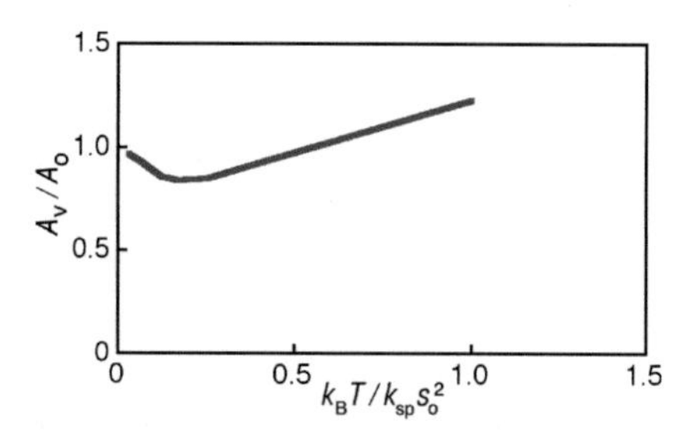

Fig. 5.18 Area per vertex $\langle A_v \rangle$ for a six-fold network of springs, shown as a function of the reduced temperature $k_BT/k_{sp}s_o^2$. The area is normalized to the area per vertex at zero stress and temperature, namely $A_o = \sqrt{3}\, s_o^2/2$, and is obtained by simulation.

temperature-dependence of the area per vertex for the six-fold coordinated network is displayed in Fig. 5.18. Although one might expect the network to expand with temperature, as most conventional materials do, the area per vertex in the six-fold network surprisingly declines with temperature for $k_BT/k_{sp}s_o^2$ between 0 and 0.2. That is, the network has a negative coefficient of thermal expansion at low temperatures (Lammert and Discher, 1998). At higher temperatures, the area increases linearly with temperature with a slope close to 1/2. In contrast to the behavior of the area per vertex, the elastic moduli vary only weakly with temperature: both K_A and μ are within 10%–20% of their zero-temperature values ($\sqrt{3}\, k_{sp}/2$ and $\sqrt{3}\, k_{sp}/4$, respectively) up to a reduced temperature of $k_BT/k_{sp}s_o^2 = 0.3$. We return to the temperature-dependence of two-dimensional networks in Section 5.4.

5.4 Networks with low coordination number

Included among the simple two-dimensional networks of biological significance are those with uniform four-fold connectivity: for example, the nuclear lamina and the cortex of the auditory outer hair cell. In this section, we examine four-fold networks of identical springs, although we recognize that this system may represent only four-fold cellular networks of uniform composition. The free energy density $\Delta\mathcal{F}$ associated with the deformation of four-fold materials was shown in Section 5.2 to have a slightly more complicated form than that of six-fold or isotropic materials, namely, from Eq. (5.18)

$$\Delta\mathcal{F} = (K_A/2)(u_{xx} + u_{yy})^2 + (\mu_p/2)(u_{xx} - u_{yy})^2 + 2\mu_s u_{xy}^2, \qquad \text{(four-fold symmetry)}$$

an expression containing two shear moduli, μ_p and μ_s. If there is no restoring force between adjacent network elements, i.e. if the energy does not depend even implicitly upon the angle between neighboring bonds, the ground state is not unique and is said to be degenerate. For example, in the

Fig. 5.19

With no bending resistance, each of these spring configurations has the same energy, demonstrating that there is no energy penalty against simple shear deformations, to lowest order in the strain tensor.

absence of angle-dependent forces, all of the plaquettes in Fig. 5.19 have the same energy, permitting a network to move among these configurations with no energy penalty. As a consequence, the simple shear modulus of this system vanishes at zero temperature and stress, a behavior not seen in six-fold networks.

Rather than immediately address the implications of ground state degeneracy, we first describe an approximation for the network that should be valid at simultaneously large tension and low temperature, namely that all plaquettes have the shape of squares. This is like the equilateral plaquette approximation for the six-fold network in Section 5.2, which describes the low temperature behavior of that network very well. For square plaquettes, all springs in the network have a length s_o when unstressed, and s_τ when subject to a two-dimensional tension τ. As there are two springs and one vertex per plaquette, the enthalpy per vertex in this approximation is $H_v = k_{sp}(s - s_o)^2 - \tau s^2$. The spring length s_τ that minimizes H_v for square plaquettes is found to be

$$s_\tau = s_o / (1 - \tau / k_{sp}). \qquad \text{(square plaquettes)} \qquad (5.36)$$

From Eq. (5.36), we see that the spring length and, consequently, the area per vertex of square plaquettes both grow without bound at a tension given by

$$\tau_{exp} = k_{sp}. \qquad \text{(square plaquettes)} \qquad (5.37)$$

This behavior occurs for the same reason that it does for six-fold spring networks: the energy $[k_{sp}(s - s_o)^2]$ and tension (τs^2) terms are both proportional to s^2 at large extension, and the tension dominates if τ/k_{sp} is sufficiently large. Under any positive pressure (negative tension) at zero temperature, four-fold networks collapse because all plaquettes have the same energy, including the plaquette with zero area:

$$\tau_{coll} < 0. \qquad \text{(four-fold symmetry)} \qquad (5.38)$$

The elastic moduli of the square plaquette network can be found by performing small deformations on a plaquette of length s_τ to the side, as illustrated by the three deformation modes in Fig. 5.20. The calculation of the moduli follows the same procedure as that used for six-fold networks in Section 5.3, namely two expressions are generated for the free energy change of each deformation mode – one involving the elastic moduli and one involving spring variables – and the moduli are determined by equating the expressions for the free energy change. The method follows Section 5.3 so closely that we will not repeat it here; rather, we quote the results

$$K_A = (k_{sp} - \tau)/2$$
$$\mu_p = (k_{sp} + \tau)/2 \qquad \text{(square plaquettes)} \qquad (5.39)$$
$$\mu_s = \tau,$$

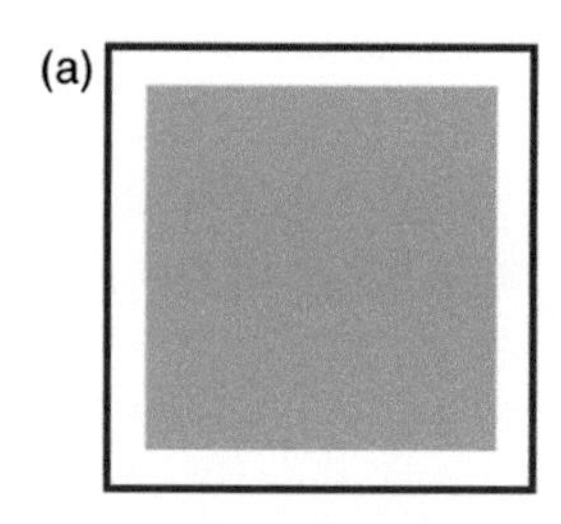

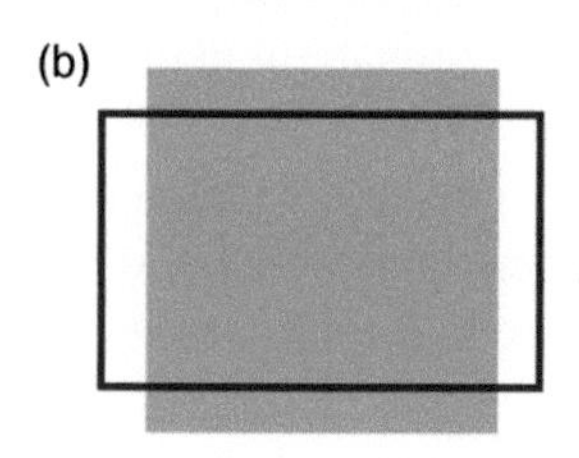

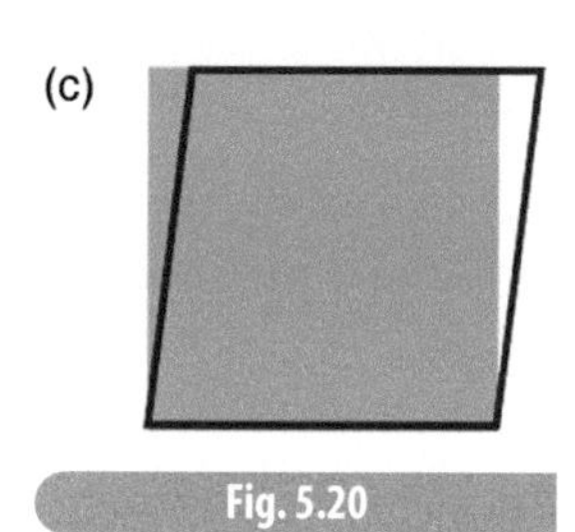

Fig. 5.20

Sample deformation modes for a square: (a) is a pure compression or expansion, (b) is a pure shear, and (c) is a simple shear. The original shape is indicated by the shaded region, and the deformed shape is the wire outline. In the calculation leading to Eq. (5.39), the sides of the shaded square have length s_T.

and leave the proof to the problem set at the end of this chapter. These moduli are linear in stress and are found to describe computer simulations of four-fold spring networks at low temperature ($\beta k_{sp} s_o^2 >> 1$) and high tension rather well (Tessier *et al.*, 2003). Of particular interest is the prediction that μ_s vanishes in the stress-free limit at low temperature. This behavior reflects the fact that the change in the spring energy of the simple shear deformation in Fig. 5.20(c) vanishes at zero tension, to leading order, and hence the plaquette is unstable against simple shear modes to lowest order in the energy (see problem set).

Although the square plaquette approximation successfully describes networks under large tension, its domain of validity is limited. For example, it predicts that the area per vertex should be $A_v = s_o^2$ at $\tau = 0$, according to Eq. (5.36), whereas the measured value from simulations at low temperature extrapolates to $0.62 s_o^2$ at $T = \tau = 0$. Further disagreements are found for the elastic moduli at low stress. The reason that $A_v < s_o^2$ at $\tau = 0$ is the degenerate ground state: the square configuration has the largest area for a given spring length, but the system can explore many other configurations with $0 < A_v < s_o^2$ at no cost in energy. That the network samples a wide array of plaquette shapes can be seen in Fig. 5.21, which displays a configuration from a computer simulation of a periodic network with four-fold connectivity. The temperature of this network is moderate, so the springs vary in length, as does the plaquette area. At low temperature, the springs should be close to s_o in length, and the plaquettes should be parallelograms. If the network samples these parallelogram shapes with equal probability, then the area per plaquette (at one vertex per plaquette) is given by

$$\langle A_v \rangle = s_o^2 \int \sin\theta \, d\theta \, / \int d\theta = (2/\pi) s_o^2, \tag{5.40}$$

where θ is the angle between adjacent sides and covers the range 0 to $\pi/2$. This description predicts that the mean area per vertex is $(2/\pi)s_o^2 = 0.64 \, s_o^2$, in very good agreement with the simulation result of $0.62 \, s_o^2$.

Encouraged by our success in describing the network in terms of parallelogram plaquettes at vanishing stress and temperature, let us extend this approach to non-zero temperature at zero stress. Displayed in Fig. 5.22 is the area per vertex for four- and six-fold networks at zero stress. The areas are normalized to an area per vertex A_o, namely $(2/\pi)s_o^2$ for four-fold and $\sqrt{3}\, s_o^2/2$ for six-fold networks, respectively. The first feature to note in Fig. 5.22 is that the networks behave oppositely at low temperature: four-fold networks expand while six-fold ones contract, indicating that the thermal behavior of bond lengths and angles both must be important.

Is there a simple model that approximately captures the average behavior of the plaquettes in Fig. 5.21? In particular, is there a way of sampling the configurations of one plaquette that yields similar averages as found in the network as a whole? One can imagine many different approximations,

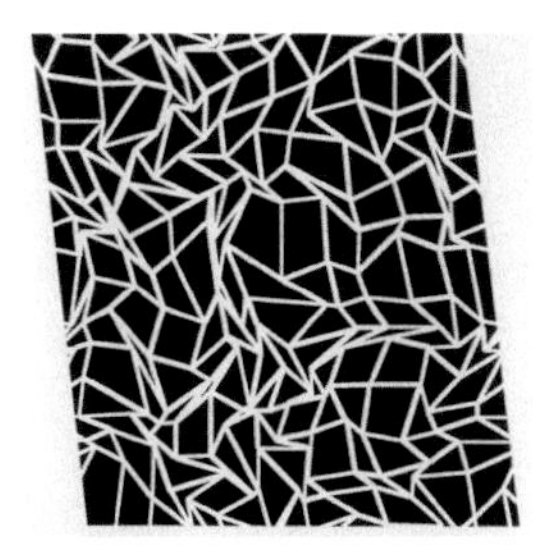

Fig. 5.21

Configuration of a square network of springs taken from a Monte Carlo simulation at zero stress and a temperature of $k_BT = k_{sp}s_o^2/4$. The network is subject to non-orthogonal periodic boundaries whose location is indicated by the black background (from Tessier, Boal and Discher, personal communication).

but an approach that works fairly well is to sample the shapes of parallelograms with one side fixed at its zero-temperature length s_o. That is, one follows the shapes of parallelograms generated by a single vertex as it explores a two-dimensional coordinate space, as illustrated in Fig. 5.23. We refer to this as a mean field model, in the sense that all network plaquettes are assumed to be identical, such that the behavior of the network is the same as that of an individual plaquette. In reality, the shape of a plaquette is not independent of its neighbor, so that this simple model does not include local correlations in shape, nor does it include general quadrilateral shapes which proliferate as the temperature rises. Further, to be closer to the mean field approach, the length of the fixed side should be varied until the value that minimizes the free energy is found at each temperature. We introduce this model not so much because of its accuracy, as to illustrate how to treat a network at $T > 0$.

Our mean field model can be evaluated analytically for four-fold networks, as we now show. The network as a whole has N plaquettes, N vertices, and $2N$ springs, so that each plaquette corresponds to a single vertex and two springs, one of length s and the other s_o. Hence, the potential energy of the plaquette is

$$E = (k_{sp}/2)(s - s_o)^2, \tag{5.41}$$

and the corresponding Boltzmann factor for a single vertex must be

$$\exp(-\beta E) = \exp(-\beta k_{sp}[s - s_o]^2/2) = \exp(-\alpha[\sigma - 1]^2), \tag{5.42}$$

where the dimensionless variables

$$\alpha = \beta k_{sp} s_o^2/2 \tag{5.43a}$$

$$\sigma = s/s_o, \tag{5.43b}$$

have been introduced.

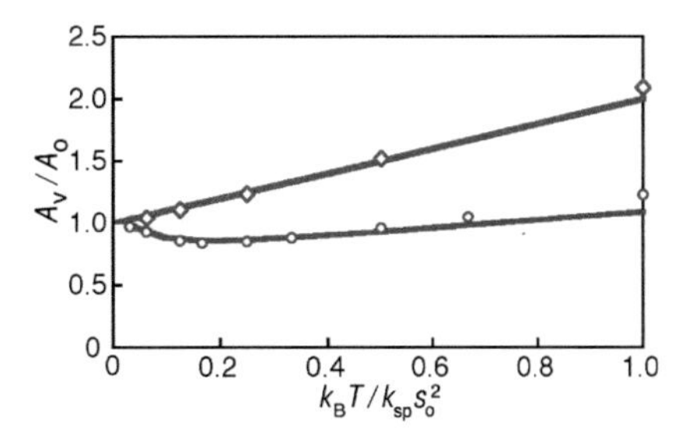

Fig. 5.22

Mean area per vertex for spring networks with four-fold (diamonds) and six-fold (circles) connectivity, normalized to $A_o = (2/\pi)s_o^2$ for four-fold networks and $A_o = \sqrt{3}\, s_o^2/2$ for six-fold networks, respectively. Solid lines are mean field calculations: Eq. (5.52) for four-fold networks or a numerical integration of the corresponding expression for six-fold spring networks (from Tessier *et al.*, 2003).

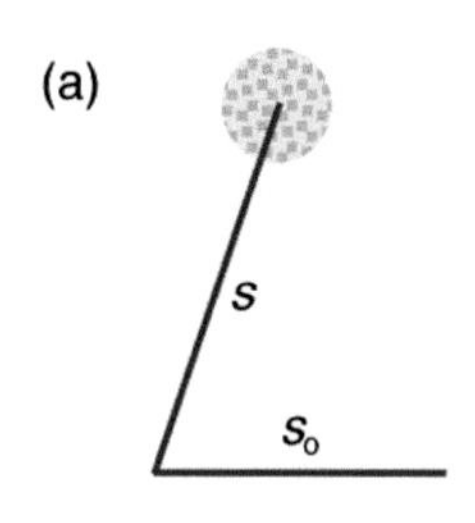

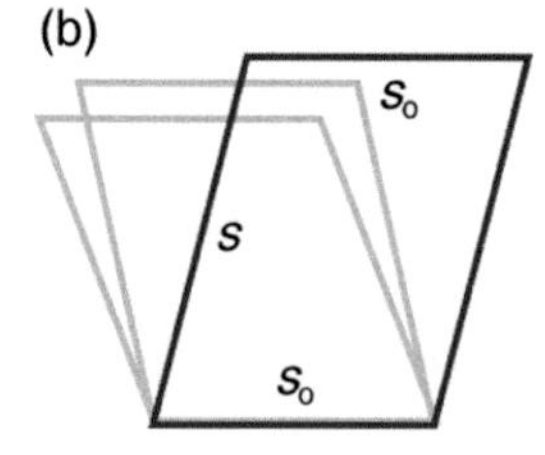

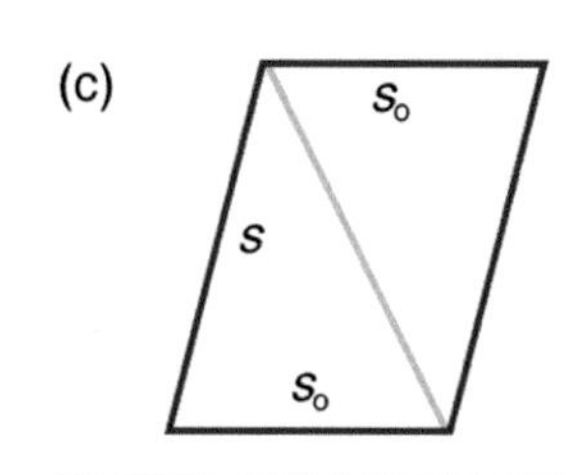

Fig. 5.23

(a) Parallelogram shapes generated by a fixed bond of length s_o and a freely moving vertex, indicated by the fuzzy shading. The plaquettes can represent uniform networks with either four-fold (b) or six-fold (c) connectivity.

The Boltzmann factor can be used to construct a probability distribution for plaquette shapes analogous to Eq. (5.34) for isolated springs. As the vertex moves in a two-dimensional plane, the probability distribution involves two Cartesian directions $dx\, dy$, which becomes $s\, ds\, d\theta$ in polar coordinates. For the four-fold plaquette in Fig. 5.23(b) (but not the six-fold version in part (c)), the energy does not depend upon θ, which can then be integrated separately from s. Thus, the two-dimensional generalization of Eq. (5.34) for the probability distribution in our mean field approximation is

$$\mathcal{P}(\sigma)\sigma d\sigma = \exp(-\alpha[\sigma - 1]^2)\, \sigma d\sigma \,/ \int \exp(-\alpha[\sigma - 1]^2)\, \sigma d\sigma, \qquad (5.44)$$

where s has been rescaled according to Eq. (5.43b). The ensemble average of the area per vertex $\langle A_v \rangle$ in the mean field approximation is proportional to an integral over the side length s through

$$\langle A_v \rangle = s_o \langle s \rangle \langle \sin\theta \rangle = (2/\pi)\, s_o^{\,2} \langle \sigma \rangle, \qquad (5.45)$$

where

$$\langle \sigma \rangle = \int \sigma\, \mathcal{P}(\sigma)\sigma d\sigma = \int \sigma^2 \exp(-\alpha[\sigma - 1]^2)\, d\sigma \,/ \int \exp(-\alpha[\sigma - 1]^2)\, \sigma d\sigma. \qquad (5.46)$$

The range of σ in Eq. (5.46) runs from zero to infinity.

At low temperature, the integrands are concentrated around $\sigma = 1$, and it is convenient to change variables to

$$\varepsilon = \sigma - 1, \qquad (5.47)$$

so that Eq. (5.46) becomes

$$\langle \sigma \rangle = \int \exp(-\alpha\varepsilon^2)\, (\varepsilon + 1)^2 d\varepsilon \,/ \int \exp(-\alpha\varepsilon^2)\, (\varepsilon + 1) d\varepsilon, \qquad (5.48)$$

where ε runs from -1 to $+\infty$. Expanding the polynomials in Eq. (5.48) leads to a number of integrals of the form

$$\int \exp(-\alpha\varepsilon^2)\varepsilon^n d\varepsilon = \alpha^{-(n+1)/2} \int \exp(-\tau^2)\tau^{\,n} d\tau \qquad (5.49)$$

where the integration limits on τ are from $-\sqrt{\alpha}$ to $+\infty$. Eq. (5.43a) shows that α tends to infinity at low temperature; as a consequence, the integrands are approximately symmetric about $\tau = 0$ at low temperature, and integrals with odd n must vanish. Thus, Eq. (5.48) becomes

$$\langle \sigma \rangle = 1 + \{\alpha^{-1} \int \exp(-\tau^2)\, \tau^2 d\tau \,/ \int \exp(-\tau^2) d\tau\}. \qquad (5.50)$$

This expression appears in the mean square displacement of a spring in one dimension (see Appendix C), and can be evaluated using $\int \exp(-\tau^2)\, \tau^2 d\tau = \sqrt{\pi}\,/2$ and $\int \exp(-\tau^2) d\tau = \sqrt{\pi}$, where the integration limits are $-\infty$ and $+\infty$. Hence, we find

$$\langle \sigma \rangle = 1 + 1\,/\,(2\alpha), \qquad (5.51)$$

and, finally, the area per vertex is

$$\langle A_v \rangle = (2/\pi)s_o^2(1 + k_BT/k_{sp}s_o^2). \quad (5.52)$$

This mean field result for the area per vertex is plotted in Fig. 5.22, and the agreement with the simulations of a complete network is very good, even for temperatures as high as $k_BT = k_{sp}s_o^2$. Somewhat more impressive are the predictions of this approach for the six-fold spring network, also shown in Fig. 5.22. The methodology for six-fold networks is the same as that for four-fold, except that Eq. (5.41) for the energy per vertex now contains the energy of the second spring running along a diagonal of the plaquette. Unfortunately, this extra energy depends on the angle of the "moving" vertex with respect to the fixed base. The resulting probability distribution involves both s and θ, but is straightforward to integrate numerically, predicting the interesting behavior that A_v initially *decreases* with temperature, as is indeed observed for computer simulations of the network. The predictions of this approximation lie within a few percent of the simulated networks up to $k_BT = k_{sp}s_o^2/2$.

Our mean field approach assumes that one spring in the plaquette has a fixed length, and that the shape fluctuations of plaquettes in the form of parallelograms mimic those of the network as a whole. Although this model is clearly useful in determining averages of some simple observables, it is less accurate for observables that probe the distribution of spring lengths. For example, the parallelogram approximation to four-fold networks at low temperature predicts that the area compression modulus, which is a measure of the fluctuations in plaquette area, should obey (see problem set)

$$\beta(K_A + \mu_p)s_o^2 = 4\pi / (\pi^2 - 8). \quad (5.53)$$

Equation (5.53) is 15% above what is found from simulations of four-fold networks at low temperatures, where spring lengths are relatively constant even though the plaquette shape exhibits strong fluctuations. (Tessier *et al.*, 2003).

5.5 Membrane-associated networks

Before we describe and interpret measurements of network elasticity in two dimensions, let us quickly review some relevant results from Sections 5.2–5.4. When an object deforms in response to an applied stress, an element of the object moves from its original position $\mathbf{x}$ to a new position $\mathbf{x} + \mathbf{u}$, where $\mathbf{u}$ is a displacement vector whose magnitude and direction vary locally across the object. The nature and severity of the deformation can be characterized by the strain tensor u_{ij}, a unitless quantity proportional to the rate of change of $\mathbf{u}$ with respect to $\mathbf{x}$. To lowest order, the change in the free energy density associated with the deformation is proportional to the squares of various combinations of components of the strain tensor, and each combination can be made to represent a physical deformation mode such as compression or shear. The forces on an object can be represented by a stress tensor σ_{ij}, which, in two dimensions, has units of energy per unit area. For Hooke's Law materials, the stress is proportional to the strain, $\sigma_{ij} = \Sigma_{k,l} C_{ijkl} u_{kl}$, much like $f = -k_{sp}x$ of a one-dimensional spring. Although the number of elastic moduli C_{ijkl} is large in principle, symmetry considerations show that isotropic or six-fold materials in two dimensions can be described by just two moduli (the area compression and shear moduli, K_A and μ, respectively), while two-dimensional materials with four-fold symmetry can be described by three moduli (K_A, μ_p and μ_s, where the two shear moduli correspond to pure and simple shear modes, respectively).

The moduli depend upon both temperature and stress. For example, two-dimensional spring networks under an in-plane tension τ approximately obey Eqs. (5.32) and (5.39)

$$K_A = (\surd 3\, k_{sp}/2) \bullet (1 - \tau / [\surd 3\, k_{sp}]) \quad \mu = (\surd 3\, k_{sp}/4) \bullet (1 + \surd 3\, \tau / k_{sp}), \quad \text{(six-fold symmetry)} \tag{5.54a}$$

and

$$K_A = (k_{sp} - \tau) / 2 \quad \mu_p = (k_{sp}/2) \bullet (1 + \tau/k_{sp}) \quad \mu_s = \tau, \quad \text{(four-fold symmetry)} \tag{5.54b}$$

where $\tau > 0$ ($\tau < 0$) corresponds to tension (compression). These equations were derived under the assumption that the springs have uniform length in the deformed state. The temperature-dependence of the moduli is weaker than the stress-dependence, and varies with the connectivity of the network. For instance, the behavior of the area per network vertex at low temperature and zero stress is opposite for the two spring models investigated here: six-fold networks initially shrink with rising temperature while four-fold networks expand.

5.5.1 Measured network elasticity

The elastic characteristics of many different cell types have been determined, although not all cells are amenable to theoretical analysis because of their complex cytoskeleton or inhomogeneous structure. Here, we concentrate on just a few cells with a simple architecture: red blood cells, auditory outer hair cells and bacteria. One set of techniques for measuring elasticity can be broadly categorized as micromechanical manipulation: a cell or extracted network is subject to a known stress – by means of a micropipette, for instance – and the response of the cell is observed by optical microscopy or another imaging method. The stress–strain curves

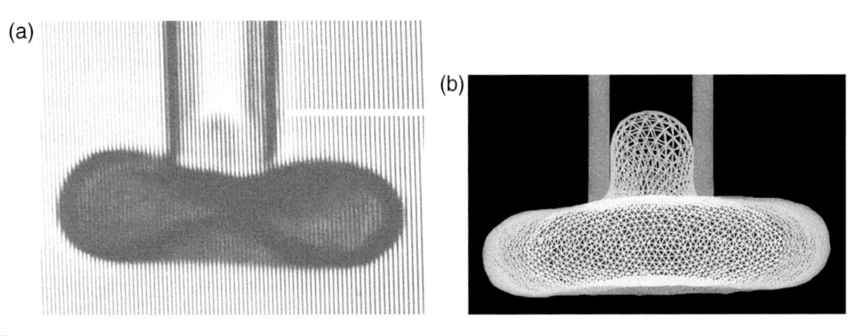

Fig. 5.24 (a) Micropipette aspiration of a red blood cell. The diameter of the cell is about 8 μm, and the inner diameter of the pipette is roughly 1 μm. In line with the horizontal bar, a small segment of the membrane can be seen inside the pipette (from Evans and Waugh, 1980; courtesy of Dr. Evan Evans, University of British Columbia). (b) Computer simulation of an aspiration experiment, using a computational blood cell about 4 μm in diameter (from Discher *et al.*, 1998).

found in these experiments are interpreted using conventional mechanics to yield one or more elastic moduli. A second experimental approach records the thermal fluctuations in a system's shape; as demonstrated for filaments in Chapter 3, the amplitude of these fluctuations is inversely proportional to the deformation resistance of the system.

One of the most thoroughly studied two-dimensional networks in the cell is the membrane-associated cytoskeleton of the human erythrocyte. A classic procedure for investigating the mechanical properties of this cytoskeleton is to apply a suction pressure to a red cell through a micropipette, pulling a segment of its plasma membrane into the pipette, as illustrated in Fig. 5.24 (Evans, 1973a, 1973b; Waugh and Evans, 1979). The figure shows a flaccid human red blood cell, with a small segment of its plasma membrane and cytoskeleton being drawn up into the pipette. The length of the aspirated segment L is related to the applied pressure P and the two-dimensional shear modulus μ through

$$P = (\mu / R_p) [(2L / R_p) - 1) + \ln(2L / R_p)], \tag{5.55}$$

where R_p is the pipette radius (see Section 10.4). Several independent experiments based upon this technique yield a shear modulus of 6–9 × 10^{-6} J/m^2 for human erythrocytes (Waugh and Evans, 1979, and references therein). In fact, the shear moduli of red cells from several vertebrates have been measured by aspiration, and their reported values generally lie not far from 10^{-5} J/m^2 for red cells without a nucleus, and about an order of magnitude higher for nucleated cells (Waugh and Evans, 1976). One experiment with optical tweezers yields a value for μ of 2×10^{-4} J/m^2 (Sleep *et al.*, 1999) while an analysis of the thermal fluctuations of red cell shape places μ below 10^{-5} J/m^2 (Peterson *et al.*, 1992). However, several tweezer-based experiments find $\mu = 2.5 \pm 0.4 \times 10^{-6}$ J/m^2 and an area compression modulus of 4.8 ± 2.7

$\times 10^{-6}$ J/m^2 (Hénon *et al.*, 1999; Lenormand *et al.*, 2001). A similar measurement on the erythrocyte cytoskeleton in isolation (no attached bilayer) give higher values, but with larger uncertainties: $\mu = 5.7 \pm 2.3 \times 10^{-6}$ J/m^2 and $K_A = 9.7 \pm 3.4 \times 10^{-6}$ J/m^2 (Lenormand *et al.*, 2003).

The variation of the red cell cytoskeleton elasticity with temperature has been measured in two experiments. Stokke *et al.* (1985a) observe that the spectrin persistence length decreases with increasing temperature, as expected for flexible filaments. Waugh and Evans (1979) find that the shear modulus decreases from 8×10^{-6} J/m^2 to 5.5×10^{-6} J/m^2 over the temperature range of 5 °C–45 °C. We return to the significance of these results after describing related measurements on other simple cells. We should also point out that the measured value of μ decreases with spectrin density, roughly as expected from Eq. (5.56) below, although there are subtleties associated with network behavior at low density that will be described further in the treatment of percolation in Chapter 6 (Waugh and Agre, 1988). Although we have treated the spectrin network as a two-dimensional quantity, in fact its elements fluctuate out of the membrane plane, giving the network a soft elastic behavior transverse to the plane, an effect studied both theoretically (Boal, 1994) and experimentally (Heinrich *et al.*, 2001). Lastly, we mention that the area compression modulus is close to twice the shear modulus in most experiments (Discher *et al.*, 1994; Lenormand *et al.*, 2001; Lenormand *et al.*, 2003).

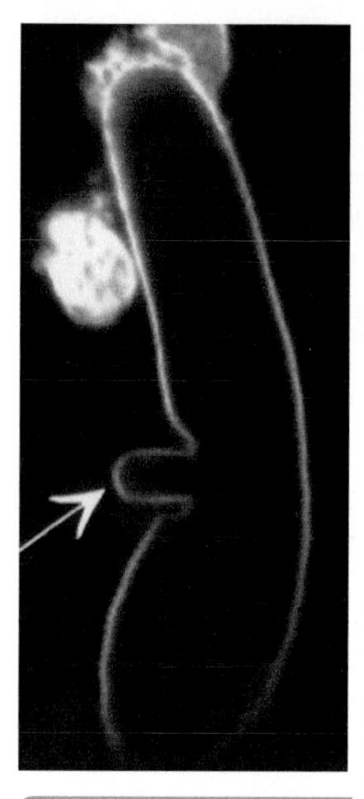

Fig. 5.25

Fluorescence image of an auditory outer hair cell deformed by micropipette aspiration. About 80 μm long and 10 μm wide, the cell is from a guinea pig; the position of the aspirated segment is indicated by the arrow (reprinted with permission from Oghalai *et al.*, 1998; ©1998 by the Society for Neuroscience; courtesy of Dr. William Brownell, Baylor College of Medicine).

The ratio of the area compression modulus to the shear modulus has also been measured for the nuclear envelope, where it is found to be 1.5 ± 0.3 in *HeLa* cells (Rowat *et al.*, 2004) and 2–5 in mouse embryo fibroblasts (Rowat *et al.*, 2006). In addition, the individual elastic moduli have been studied using the aspiration technique, yielding an effective Young's modulus of 2.5×10^{-2} J/m^2 for the nuclear membranes of *Xenopus* oocytes (Dahl *et al.*, 2004); similarly large values are found for the nuclear envelope of TC7 cells (a type of epithelial cell) by Dahl *et al.* (2005). Observed for both swollen and unswollen nuclei, this modulus is three orders of magnitude larger than the moduli of the spectrin network of the human erythrocyte. Given the stiffness of the filaments in the nuclear membrane compared to spectrin, this result is not surprising.

Auditory outer hair cells are relatively large, and can be aspirated in the same way as erythrocytes to determine the elastic properties of the lateral cortex, as displayed in Fig. 5.25. Under the assumption that the cortex is isotropic, its effective shear modulus is found to be $1.5 \pm 0.3 \times 10^{-2}$ J/m^2, which is about 1000 times larger than that of the erythrocyte (Sit *et al.*, 1997). However, the lateral cortex is known to be anisotropic, as seen in Fig. 5.1(b): it is stiffer in the circumferential direction than in the axial direction. On the basis of the response when an outer hair cell is gently bent by an external probe, the circumferential filaments are estimated to have a Young's modulus of 1×10^7 J/m^3 while the axial cross-links have a smaller value of $Y = 3 \times 10^6$ J/m^3 (Tolomeo *et al.*, 1996). These results are significantly lower than Y estimated for actin in Section 3.6, perhaps reflecting different experimental conditions, in part (for a summary of experimental measurements, see Brownell *et al.*, 2001). The surface shear modulus of a particular type of epithelial cell has been observed at 2.5×10^{-4} J/m^2, an order of magnitude higher than the erythrocyte cytoskeleton (Feneberg *et al.*, 2004). Lastly, we mention that a shear modulus has been obtained for fibroblasts on a substrate, where the deformation is generated by applying a magnetic field to microscopic magnetic beads attached to the plasma membrane. The cytoskeleton of a fibroblast is more complex than the cells we have discussed so far, and the measured shear resistance is correspondingly large, $\mu \sim 2\text{–}4 \times 10^{-3}$ J/m^2 (Bausch *et al.*, 1998).

Elastic properties of several thin structural components of bacteria have been measured, although the analysis has been in terms of three-dimensional moduli, rather than the two-dimensional quantities of this chapter. For example, the cell walls of Gram-positive bacteria, containing a thick layer of peptidoglycan, can be extracted and stretched, allowing the determination of a Young's modulus from the resulting stress/strain curve. As applied to *Bacillus subtilis* (Thwaites and Surana, 1991), the technique shows that the Young's modulus of a dry cell wall is $1.3 \pm 0.3 \times 10^{10}$ J/m^3, decreasing by several orders of magnitude to about 3×10^7 J/m^3 when the wall is wet. These values for Y, corresponding to a deformation along the axis of the bacterium, can be compared to the generic $Y \sim 10^9$ J/m^3 of biofilaments (see Section 3.6).

In some cases, individual bacteria are aligned end-to-end like sausages, surrounded by a hollow cylindrical sheath which can be isolated and manipulated as shown in Fig. 5.26 (Xu *et al.*,1996). A Young's modulus can

Fig. 5.26 Scanning tunneling microscope image of a bacterial sheath suspended across a 0.3 μm gap in a Ga–As substrate. The sheath can be deformed by an atomic force microscope to determine an elastic modulus (reprinted with permission from Xu *et al.*, 1996; ©1996 by the American Society for Microbiology).

be extracted from the stress–strain curve obtained when a known force is delivered to the sheath by the tip of an atomic force microscope. For the bacterium *Methanospirillum hungatei*, Y lies between 2×10^{10} and 4×10^{10} J/m^3, a value that is larger than that of typical biopolymers but similar to collagen ($1–2 \times 10^{10}$ J/m^3). The AFM approach has also been applied to the Gram-negative bacterium *Escherichia coli* by Yao *et al.* (1999), yielding an average $Y = 2.5 \times 10^7$ J/m^3 for a hydrated cell wall, where Y changes by a factor of two depending upon direction. The elasticity of the complete cell boundary has been obtained for the Gram-positive bacterial genus *Lactobacillus* (Schär-Zammaretti and Ubbink, 2003). The measurements were performed with AFM, where the slope of the force-indentation curves gave values for an elastic constant of $1–2 \times 10^{-2}$ J/m^2.

5.5.2 Interpretation of measurements

The values of the elastic moduli for many of the two-dimensional networks summarized above are much lower than conventional materials, including plastics. The softness of these networks reflects both the low density and the high flexibility of their filamentous components. The erythrocyte cytoskeleton demonstrates both of these effects, as can be seen in Fig. 5.27. Part (a) of the figure is a drawing of a stretched cytoskeleton, as imaged by an electron microscope, showing once again the approximate six-fold connectivity of the network (the authors of this study find 3% and 8% of the junctions have five- or seven-fold connectivity, respectively; Liu *et al.*, 1987). The distance between junctions in the figure is close to the tetramer contour length of 200 nm, in contrast to the average separation of

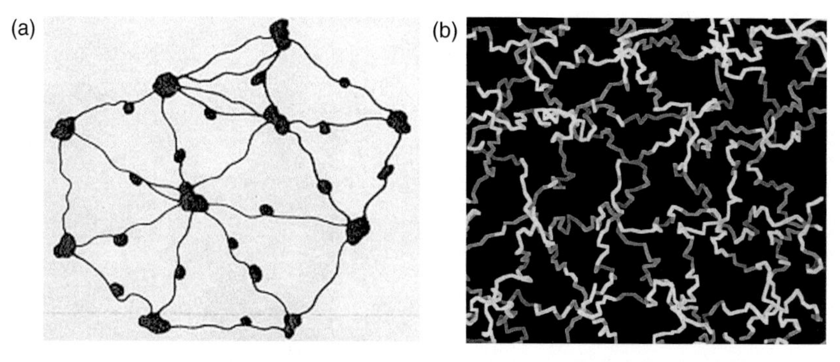

Fig. 5.27 (a) Drawing of a spread erythrocyte cytoskeleton, copied from an electron microscope image; the distance between junctions is close to 200 nm (reprinted with permission from Liu *et al.*, 1987; ©1987 by the Rockefeller University Press). (b) Computer simulation of the cytoskeleton, in which each tetramer is represented by a single polymer chain, linked at six-fold vertices and attached to the bilayer at its mid-point. In this unstretched configuration, elements closer to the viewer are shaded lighter (from Boal, 1994).

75 nm *in vivo*. Unless there are local attractive interactions that tighten up the tetramers when they are in a cell (Ursitti and Wade, 1993; see also McGough and Josephs, 1990), the cytoskeleton should appear something like Fig. 5.27(b), with the tetramers being moderately convoluted and making a large entropic contribution to the elasticity of the network.

The elastic moduli of a network of floppy chains can be estimated from the results we have obtained so far in Chapters 3 and 5. Let's consider six-fold networks as an example. Although the chains are contorted, their junctions, which appear lighter in Fig. 5.27(b), undergo only limited excursions from their equilibrium positions, and behave like the junctions in a spring network at low temperature. Regarded as entropic springs, the chains would have an effective spring constant $k_{sp} = 3k_BT / \langle r_{ee}^2 \rangle$ according to Eq. (3.54), where Eq. (3.33) has been used for the value of the mean square end-to-end distance $\langle r_{ee}^2 \rangle$. Now, the area per vertex A_v of an equilateral plaquette in a six-fold network is $A_v \sim \surd 3 \langle r_{ee}^2 \rangle/2$, so that the spring constant could alternatively be written $k_{sp} \sim 3\surd 3\ k_BT / 2A_v$. Recognizing that the two-dimensional density of chains ρ is given by $\rho = 3/A_v$ (three chains per vertex), the effective spring constant is $k_{sp} = (\surd 3 /2)\rho k_BT$. Lastly, the shear modulus for a six-fold spring network at low temperature is $\mu = \surd 3\ k_{sp}/4$, so that we expect $\mu = (3/8)\rho k_BT$. Now, the implicit assumptions in this derivation, such as the freedom of the chain to move without restrictions imposed by other chains or the bilayer, challenge the accuracy of the factor of 3/8 in this expression. Nevertheless, as obtained on more general grounds in Chapter 6, it is true that μ approximately obeys

$$\mu \sim \rho k_BT. \tag{5.56}$$

The effects on network stiffness caused by unfolding of the spectrin tetramers at large extension has been studied theoretically (Zhu and Asaro, 2008).

Equation (5.56) provides a guide to the entropic contribution to the elasticity of the erythrocyte cytoskeleton. With a spectrin tetramer density of $\rho \sim 800\ \mu m^{-2}$ (see Section 5.1), the cytoskeleton would be expected to have a shear modulus of 3×10^{-6} J/m^2, which is at the low end of the experimental determinations. Computer simulations that include steric interactions of the polymer chains (Boal, 1994) confirm the approximate results, although the predicted shear modulus rises to 1×10^{-5} J/m^2. What about the area compression modulus K_A? If the picture of the erythrocyte cytoskeleton as a low temperature network of entropic or conventional springs is correct, then we would expect $K_A/\mu = 2$ from Eq. (5.32) at zero stress. Experiments using micropipette aspiration find K_A/μ close to 2 as well (Discher *et al.*, 1994). These experiments have been further discussed in terms of effective models of the cytoskeleton by Discher *et al.* (1998).

The temperature-dependence of the network geometry and elasticity provides a test of the importance of entropy in the cytoskeleton. Here, experiment appears to be divided. On the one hand, $\langle r_{ee}^2 \rangle$ of spectrin tetramers falls with rising temperature, and is found to be quantitatively consistent with $\langle r_{ee}^2 \rangle \propto \xi_p$ in Eq. (3.33) and the variation of the persistence length $\xi_p \propto 1/T$ in Eq. (3.21) (Stokke *et al.*, 1985a). On the other hand, if spectrin tetramers do behave like entropic springs, we expect that the elastic moduli should rise roughly linearly with temperature, according to Eq. (5.35). Over the temperature range of 5 °C–45 °C, this change should correspond to an increase of 13%. However, the shear modulus is observed to decrease by about 30% instead (Waugh and Evans, 1979). The reasons for this discrepancy are not yet apparent.

With its composite structure of soft spectrin and somewhat stiffer actin, the cortical lattice of outer hair cells is strongly anisotropic (the flexural rigidity of F-actin is a thousand times that of spectrin). The resistance of this network against simple shear modes should vanish in the absence of angle-dependent interactions at its four-fold junctions. One can obtain a crude estimate of the entropic contribution to the pure shear modulus through the density of polymer chains, namely two chains per 25 × 65 nm rectangular plaquette. For a network of floppy chains, this density corresponds to a shear modulus of 5×10^{-6} J/m^2, using $\mu = \rho k_B T$. In fact, the measured shear modulus is three orders of magnitude larger (Sit *et al.*, 1997; see also Tolomeo *et al.*, 1996), indicating that the shear resistance is of energetic, rather than entropic origin, reflecting the stiffness of the network filaments. Lastly, we note for comparison that a rough estimate of the two-dimensional moduli of the bacterial cell wall may be obtained by multiplying the three-dimensional moduli by the thickness of the wall, as explained further in Chapter 6. With $Y \sim 10^9$ J/m^3 and a thickness of 5 nm, the two-dimensional moduli should be around 5 J/m^2, illustrating the strength of the cell wall compared to the loose networks of the red cell or hair cell. Cell walls are analyzed in more detail in Sections 7.6 and 10.5.

Summary

The soft filaments of a cell can be knitted into a variety of two- or three-dimensional networks, some of which have surprisingly regular connectivity. To within a factor of two larger or smaller, the typical diameter of a network filament is 10 nm, whereas the common spacing between filaments is 50–100 nm. However, some networks, such as the bacterial cell wall, are much denser than this, consisting of thin polymers less than 1 nm wide, connected at intervals of less than 10 nm. Most networks in the cell, with the exception of the cell wall, are moderately flexible and deform in response to ambient forces.

The mathematical description of a deformation involves the dimensionless strain tensor u_{ij}, which is related to the displacement of a given element of an object at position **x** by an amount **u** *via* $u_{ij} = 1/2\ [\partial u_i/\partial x_j + \partial u_j/\partial x_i + \Sigma_k(\partial u_k/\partial x_i)(\partial u_k/\partial x_j)]$, where the subscripts i, j, k refer to the Cartesian coordinate axes. For small deformations, the last term may be safely neglected. To lowest order, the change in free energy density $\Delta\mathcal{F}$ associated with the deformation is quadratic in the strain tensor according to $\Delta\mathcal{F} = \Sigma_{i,j,k,l} C_{ijkl} u_{ij} u_{kl}$, where the elements of the tensor C_{ijkl} are the elastic moduli, the analogs of the scalar spring constant k_{sp} of Hooke's Law for springs. The forces experienced by a surface are described the stress tensor σ_{ij} obtained from $\Delta\mathcal{F}$ through $\sigma_{ij} = \partial\mathcal{F}/\partial u_{ij}$; the stress tensor is treated in greater detail in Chapter 6. The symmetry of the strain tensor, and the internal symmetries of the material, relate different components of its elastic moduli, reducing the number of independent moduli to just a few, in many cases. For instance, the free energy density of two-dimensional materials with six-fold (or full rotational) symmetry is given by $\Delta\mathcal{F} = (K_A/2)\bullet(u_{xx} + u_{yy})^2 + \mu\bullet\{(u_{xx} - u_{yy})^2/2 + 2u_{xy}^2\}$, where K_A and μ are the compression and shear moduli in two dimensions; the corresponding Young's modulus is $Y_{2D} = 4K_A\mu/(K_A+\mu)$. For four-fold materials in two dimensions, $\Delta\mathcal{F} = (K_A/2)\bullet(u_{xx} + u_{yy})^2 + (\mu_p/2)\bullet(u_{xx} - u_{yy})^2 + 2\mu_s u_{xy}^2$, where the two shear moduli, μ_p and μ_s, reflect the resistance against pure shear and simple shear, respectively.

The elastic moduli depend upon both tension τ and temperature T. Considering regular networks of ideal springs (with spring constant k_{sp} and equilibrium length s_o), we find that at low temperature, six-fold networks obey $K_A = (\surd 3\ /2)\bullet(k_{sp} - \tau\ /\surd 3)$ and $\mu = (\surd 3\ /4)\bullet(k_{sp} + \surd 3\ \tau)$, where $\tau > 0$ corresponds to tension. Four-fold networks have a similar form, namely $K_A = (k_{sp} - \tau)/2$, $\mu_p = (k_{sp} + \tau)/2$ and $\mu_s = \tau$, but the relations are valid only at larger values of τ. The low temperature behavior of network geometry at zero stress is intriguing: six-fold networks initially shrink, while four-fold networks expand, as the temperature rises. At modest temperatures, the mean area per vertex A_v of four-fold networks is described by $\langle A_v \rangle = (2/\pi)s_o^2(1 + k_B T\ /\ 2k_{sp}s_o^2)$.

The contribution of energy and entropy to the elastic properties of a network depends, of course, on the flexibility of its filaments. If the filaments are linked together at intervals much less than their persistence length, energy considerations may dominate the elasticity, and the shear resistance is not too far below that of conventional materials. However, if the distance between network junctions is many times the persistence length, the shear modulus should be that expected for a network of floppy chains, namely $\mu \sim \rho k_B T$ where ρ is the chain density in two dimensions.

Problems

Applications

5.1. A six-fold coordinate network of identical springs is expected to display a negative Poisson ratio σ_p over a range of tensions.

(a) Verify the range of tension where $\sigma_p < 0$, based on Eq. (5.33).
(b) Calculate the range of the area per junction vertex where $\sigma_p < 0$. Quote your answer as a ratio to the area per vertex of an unstressed network.
(c) Find the minimum tension needed for $\sigma_p < 0$ in the cytoskeleton of the human red blood cell. Determine the equivalent spring constant from the expression for the shear modulus using $\mu = 5 \times 10^{-6}$ J/m^2 at zero tension.

Notation: each spring has spring constant k_{sp} and equilibrium spring length s_o.

5.2. Six-fold coordinate networks, such as the spectrin cytoskeleton of the human erythrocyte, are predicted to shrink initially as their temperature rises from zero. From Fig. 5.18, determine the (reduced) temperature above which a triangular spring network is predicted to expand. Is this temperature above the (dimensionless) operating temperature of the red cell cytoskeleton, viewed as a triangular network of ideal springs with $\mu = 5 \times 10^{-6}$ J/m^2 and unstressed spring length 75 nm?

5.3. Taking 5×10^{-6} J/m^2 for the shear modulus of the erythrocyte cytoskeleton, predict the areal compression modulus K_A, assuming that the cytoskeleton behaves like a six-fold network of ideal springs (at zero tension and temperature). Calculate the relative change in area $\langle \Delta A^2 \rangle^{1/2} / A_o$ arising from thermal fluctuations at 300 K, where $\Delta A = A - A_o$ and A_o is the unstressed area. Choose two values of A_o:

(i) the area of a triangular plaquette 75 nm to the side,
(ii) 50 μm^2 (roughly one-third the surface area of the human erythrocyte).

[*Hint: use the definition* $(\beta K_A)^{-1} = \langle \Delta A^2 \rangle / \langle A \rangle$; *see Appendix D for the three-dimensional analog.*]

5.4. Evaluate the thermal expansion of a four-fold network like the lateral cortex of the auditory outer hair cell. As a simplification, assume that the network is uniform with an elementary square plaquette of length $s_o = 40$ nm to the side and an areal compression modulus $K_A = 2 \times 10^{-2}$ J/m^2 at $T = 0$ and $\tau = 0$. Treat the network elements as springs, and plot the relative area A_v / s_o^2 from $T = 0$ to 300 K.

5.5. In some networks, long filaments may be cross-linked at regular intervals as shown, where each vertex has three-fold coordination and the magnitude of the bending has been exaggerated. Examples of such networks are given in the following two problems. In one mathematical representation of this situation, the linking parts of the filament are assumed to be rigid so the bending energy of the filament can be written in terms of a bending energy at each vertex of $E_{\mathrm{bend}} = k_{\mathrm{bend}}\theta^2/2$ (where θ is given on the diagram).

(a) Starting from Eq. (3.15), show that the elastic parameter k_{bend} is given in terms of the filament persistence length ξ_{p} by $k_{\mathrm{bend}} = \xi_{\mathrm{p}} k_{\mathrm{B}}T/b$, where b is the separation between successive vertices.

(b) Find k_{bend} for actin with $\xi_{\mathrm{p}} = 15$ μm and microtubules with $\xi_{\mathrm{p}} = 5$ mm. Take the cross-links to be separated by 30 nm. Quote your answers both as a ratio to $k_{\mathrm{B}}T$ at 300 K and in Joules.

5.6. The two-dimensional tension τ experienced by the surface of a sphere under pressure is given by $\tau = RP/2$, where R is the radius of the sphere and P is the pressure difference across its surface. As a specific example, consider a hypothetical bacterium with diameter 1 μm and an internal pressure of 5 atm with respect to its surroundings; the area compression modulus K_{A} of the wall surrounding the bacterium is 5 J/m^2.

(a) Find the tension borne by the surface (quote your answer in J/m^2).

(b) Taking a small section of the surface to be locally flat, find the strain components u_{xx} and u_{yy} induced by this tension. What is the change in surface area (quote your answer in μm^2)?

5.7. Actin can be cross-linked by the proteins α-actinin or fimbrin to form bundles of parallel filaments, whose structures are schematically represented by

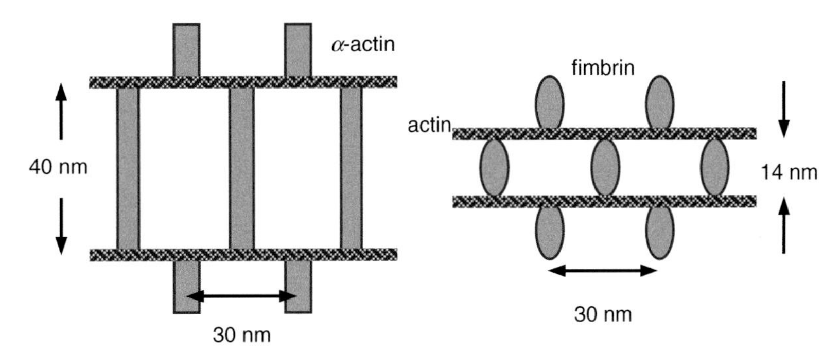

Treating the crosslinks as inextensible, calculate the Young's modulus Y in the direction transverse to the actin filaments for both of these networks. Use the expressions for k_{bend} from Problem 5.5(a) and Y from Problem 5.26(d) without proof. Take the persistence length of actin to be 15 μm at 300 K to set k_{bend}; quote your answers in J/m^2.

5.8. The basal lamina is a mat which, early in its development, is composed largely of the protein laminin. Laminin has the structure of an asymmetric cross roughly 120 nm long and 60 nm across, with arms of varying radius R (taken to be 5 nm here). Consider a network with the rectangular structure (atypical of the basal lamina)

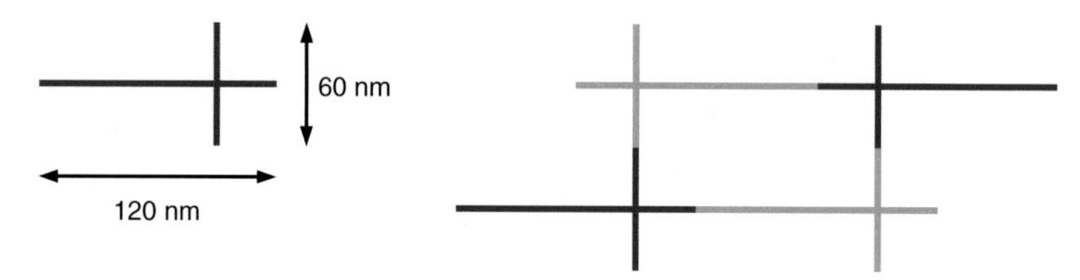

(a) Find the effective spring constant k_{sp} of the network elements using the result from Problem 5.29 that k_{sp} of a uniform rod equals $\pi R^2 Y/L$, where Y is the three-dimensional Young's modulus and L is the length of the filament (assume $Y = 0.5 \times 10^9$ J/m^3).

(b) Calculate the Young's modulus along both axes of this network using the model system in Problem 5.28.

5.9. The cell wall of a Gram-negative bacterium is composed of a thin peptidoglycan network having the approximate two-dimensional structure shown (see Fig. 5.3). The filaments running horizontally in the diagram are sugar rings with a mean separation between successive in-plane crosslinks of ~4 nm; the crosslinks are floppy chains of amino acids having a contour length of about 4 nm, but a much smaller mean end-to-end displacement of 1.3 nm.

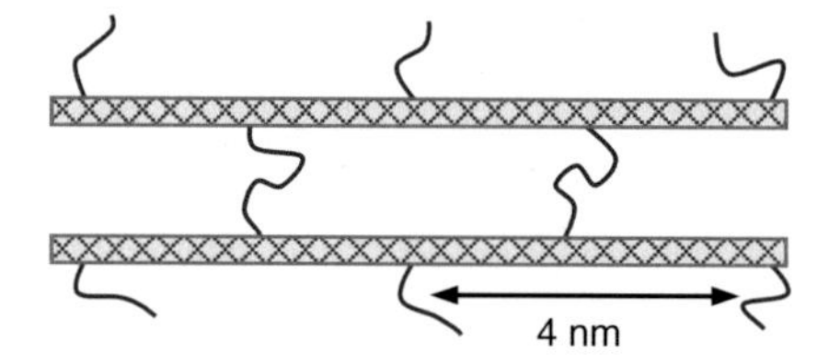

(a) What is the persistence length of the amino-acid chain?

(b) Assuming that the glycan filaments are inextensible rigid rods, calculate the Young's modulus Y transverse to the glycan filaments using results from Problem 5.28.

5.10. The lateral cortex of the auditory outer hair cell in guinea pigs is composed of two different filaments linked to form rectangular plaquettes. The mean dimensions of the network are, with 20% variability, shown in the diagram.

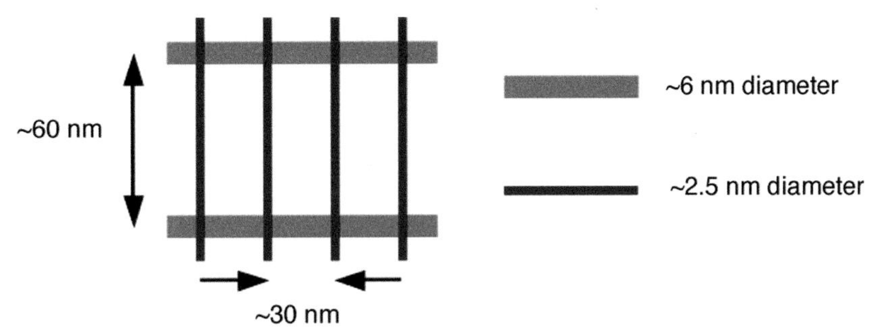

(a) Obtain the effective spring constant of both network elements from $k_{sp} = \pi R^2 Y/L$ (see Problem 5.29 for definitions), assuming $Y = 0.5 \times 10^9$ J/m^3.

(b) Calculate the Young's modulus Y along both axes from the expression in Problem 5.28. For comparison, the observed *shear* modulus of this network is $(1.5 \pm 0.3) \times 10^{-2}$ J/m^2.

5.11. A two-dimensional honeycomb network like that shown in Problem 5.25 is subject to an isotropic tension τ. Take the network geometry to be described by $a = 12$ nm and $b = 15$ nm; take $k_{bend} = 2 \times 10^{-18}$ J for the bending parameter of the longitudinal filaments.

(a) Find the bending angle θ_{min} and the area per vertex of this network for tensions of 0.001, 0.01 and 0.1 J/m^2. Quote the area per vertex in nm^2.

(b) What is the asymptotic value of θ_{min} at large tensions according to the formula in Problem 5.25? Is your answer realistic, and consistent with the domain of applicability with this formula?

You may use the results of Problem 5.25 without proof.

5.12. Materials fail when subject to sufficiently high tension, although the strain at failure depends strongly on the material. As a hypothetical example, consider a sphere of radius $R = 2$ μm bounded by a two-dimensional membrane that fails when its strain ($\Delta A/A$) reaches 4%. In this example, it will be assumed that the tension τ experienced by the membrane is generated by a pressure difference P across it, where τ and P are related by $\tau = RP/2$.

(a) If the area compression modulus K_A of the membrane is 0.2 J/m^2, at what pressure does the membrane fail? Quote your answer in atm.

(b) If the sphere were the size of a child's balloon, say $R = 10$ cm, and all of its remaining characteristics remained the same, at what internal pressure would it fail?

5.13. Suppose that the body of a bacterium (of length 4 μm and diameter 1 μm) is subject to a torque from the flagella at its poles, such that one end rotates by 0.05 μm with respect to the other end, as shown by the dashed line in the diagram.

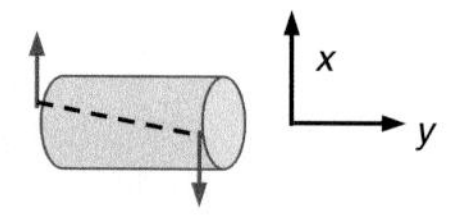

That is, the left-hand end rotates in one sense while the right-hand end rotates in the opposite sense, such that the surface of the bacterium is skewed. As a result, a line initially parallel to the symmetry axis now makes an angle with respect to it, and is displaced 0.05 μm perpendicular to its length.

(a) What type of shear mode does the surface experience?
(b) Assuming that the surface area is unchanged by the deformation, what are the components of the strain tensor within the two-dimensional surface (u_{xx}, u_{yy}, u_{xy}, u_{yx}) for the coordinate system shown?
(c) What is the change in the free energy arising from the deformation, as measured over the entire surface, if $K_A = 2$ J/m^2 and $\mu =$ 1 J/m^2? Quote your answer in J and as a ratio to $k_B T$ at 300 K.

5.14. Suppose that the body of a bacterium (of length 4 μm and diameter 1 μm) is subject to an internal pressure that causes it to elongate by 0.06 μm while its width does not change, as shown in the diagram.

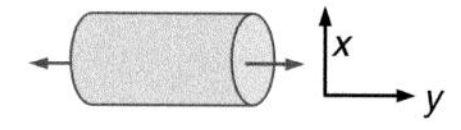

(a) Categorize the deformation of the surface in terms of compression and shear.
(b) What are the components of the strain tensor within the two-dimensional surface (u_{xx}, u_{yy}, u_{xy}, u_{yx}) for the coordinate system shown?
(c) What is the change in the free energy arising from the deformation, as measured over the entire surface, if $K_A = 2$ J/m^2 and $\mu =$ 1 J/m^2? Quote your answer in J and as a ratio to $k_B T$ at 300 K.

5.15. In this chapter it was established that networks of ideal springs confined to a plane may collapse when laterally compressed.

(a) Using the membrane-associated spectrin network of the human erythrocyte as a model, calculate the two-dimensional pressure and the density of network nodes (relative to the unstressed configuration) at the collapse point. In your calculations, assume the network to be six-fold coordinate with spectrin having a mean end-to-end displacement of 75 nm in the network compared to its contour length of 200 nm; obtain its entropic spring constant k_{sp} through Eqs. (5.54).
(b) The collapse transition is predicted to happen for ideal springs that have no spatial volume. How realistic is this approximation for spectrin, given its contour length?

Formal development and extensions

5.16. The area compression modulus K_A is related to the change of area with respect to (two-dimensional) pressure Π by $K_A^{-1} = -(\partial A/\partial \Pi)/A$. For an ideal gas obeying $\Pi A = Nk_BT$, show that $K_A = \Pi$. Note that this can be generalized to $K_V = P$, where K_V is the volume compression modulus and P is the three-dimensional pressure.

5.17. For an isotropic material in two dimensions, the stress is related to the strain by

$$\sigma_{ij} = (K_A - \mu)\, \delta_{ij}\, \mathrm{tr}u + 2\mu u_{ij},$$

which can be obtained by differentiating the free energy density.

(a) Invert this expression to obtain the strain in terms of the stress:

$$u_{ij} = \delta_{ij}\, \mathrm{Tr}\sigma \bullet \{(1/4K_A) - (1/4\mu)\} + \sigma_{ij}/(2\mu).$$

(b) Evaluate u_{xx} and u_{yy} for a uniaxial stress in the x-direction: $\sigma_{xx} \neq 0$, $\sigma_{yy} = 0$.
(c) Using your results from part (b), show that the Poisson ratio $\sigma_p = -u_{yy}/u_{xx}$ is given by $\sigma_p = (K_A - \mu)/(K_A + \mu)$.
(d) The two-dimensional Young's modulus is defined by $Y_{2D} =$ [*stress*]/[*strain*] $= \sigma_{xx}/u_{xx}$. Use your results from part (b) to prove that $Y_{2D} = 4K_A\mu/(K_A+\mu)$.

5.18. Suppose that a two-dimensional isotropic material is stretched uniformly by a fixed scale factor in the x-direction, such that $u_{xx} = \delta$.

(a) Find the corresponding u_{yy} in terms of K_A, μ and δ by minimizing the free energy of the material.
(b) Show that the Poisson ratio $\sigma_p = -u_{yy}/u_{xx}$ is given by $\sigma_p = (K_A - \mu)/(K_A + \mu)$ using your results from part (a).

5.19. In Section 5.3, the area compression modulus for a triangular network of springs under stress was obtained by comparing two expressions for the free energy density, one in terms of spring variables and one in terms of the elastic moduli. As an alternative approach, use the stress-dependent expression for the spring length s_τ to obtain the area compression modulus by differentiation: $K_A^{-1} = (\partial A/\partial \tau)/A$.

5.20. The shear modulus of a spring network with six-fold connectivity is given by $\mu = \sqrt{3}\, k_{sp}/4$, where k_{sp} is the force constant of the springs. To establish this, consider the deformation of a network plaquette whose native shape is that of an equilateral triangle, as shown in the diagram. The sides of the plaquette are the springs; the corners are six-fold vertices, but the four remaining springs connected to them are not show. Under deformation, the top vertex of the triangle is moved by an amount δ in the x-direction.

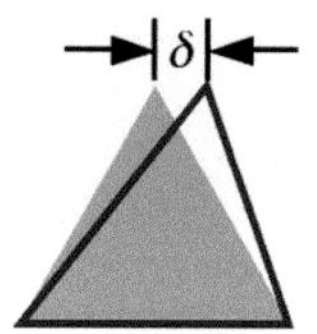

(a) Find the new lengths of the springs forming the left-hand and right-hand sides of the plaquette under the deformation. Define the equilibrium spring length to be s_o; work to lowest order in δ.
(b) Find the energy density associated with the deformation in terms of spring characteristics (k_{sp}, s_o and δ).
(c) Show that the energy density in the strain representation is given by

$$\Delta \mathcal{F} = (2\mu/3)(\delta/s_o)^2.$$

(d) Equate your answers from parts (b) and (c) to obtain the shear modulus.

5.21. Find the tension-dependence of the compression modulus of a spring network with four-fold connectivity. Under tension τ, all springs are stretched from their unstressed length s_o to a stretched length of s_τ. Then, an infinitesimal deformation of the network stretches the springs further from length s_τ to $s_\tau + \delta$ after the deformation.

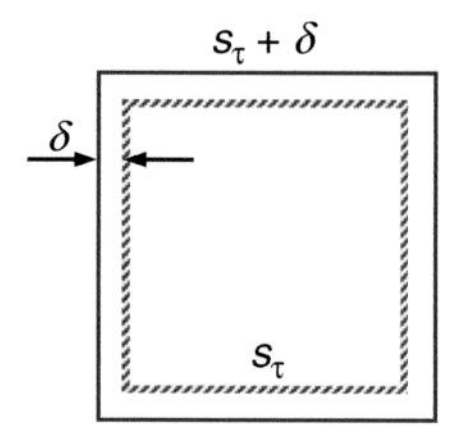

(a) Establish that $s_\tau = s_o / (1 - \tau/k_{sp})$ by minimizing the enthalpy per vertex, which is

$$H_v = k_{sp}(s - s_o)^2 - \tau s^2 \qquad \text{for a square plaquette.}$$

(b) Show that in terms of spring variables k_{sp}, s_τ and δ, the change in H_v/A_v is to order δ^2:

$$\Delta H_v/A_v = (k_{sp} - \tau)\bullet(\delta/s_\tau)^2,$$

where A_v is the area per vertex.

(c) By calculating the strain tensor, show that the change in the free energy density is

$$\Delta\mathcal{F} = 2K_A(\delta/s_\tau)^2,$$

(d) Compare your results from parts (b) and (c), to obtain the compression modulus

$$K_A = (k_{sp} - \tau)/2.$$

5.22. Find the tension-dependence of the elastic modulus for pure shear (μ_p) of a spring network with four-fold connectivity. Under tension τ, all springs are stretched from their unstressed length s_o to a stretched length of s_τ. Then, an infinitesimal deformation of the network changes the springs further from length s_τ to $s_\tau + \delta$ on one side and $s_\tau - \delta$ on the other, as shown.

This is a pure shear mode. In your solution, you may assume without proof that $s_\tau = s_o/(1 - \tau/k_{sp})$, which follows by minimizing the enthalpy per vertex, $H_v = k_{sp}(s - s_o)^2 - \tau s^2$.

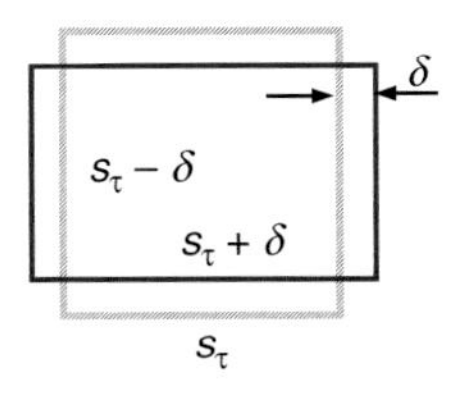

(a) Show that the change in H_v/A_v in terms of spring variables k_{sp}, s_τ and δ is, to order δ^2:

$$\Delta H_v/A_v = (k_{sp} + \tau)\bullet(\delta/s_\tau)^2,$$

where A_v is the area per vertex.

(b) By calculating the strain tensor, show that the free energy density is

$$\Delta\mathcal{F} = 2\mu_p(\delta/s_\tau)^2.$$

(c) Compare your results from parts (a) and (b), to obtain the modulus for pure shear

$$\mu_p = (k_{sp}+\tau)/2.$$

5.23. Find the tension-dependence of the elastic modulus for simple shear (μ_s) of a spring network with four-fold connectivity. Under tension τ, all springs are stretched from their unstressed length s_o to a stretched length of s_τ. Then, an infinitesimal deformation of the network shifts one set of springs by a displacement δ at their ends, as shown.

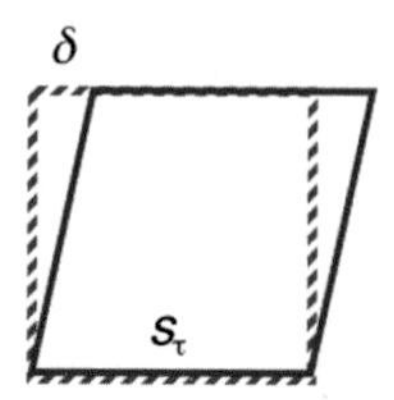

This is a simple shear mode. In your solution, you may assume without proof that $s_\tau = s_o \,/\, (1 - \tau/k_{sp})$ which follows by minimizing the enthalpy per vertex, $H_v = k_{sp}(s - s_o)^2 - \tau s^2$.

(a) Show that the change in H_v/A_v in terms of spring variables k_{sp}, s_τ and δ is, to order δ^2:

$$\Delta H_v/A_v = (\tau/2) \bullet (\delta/s_\tau)^2,$$

where A_v is the area per vertex.

(b) By calculating the strain tensor, show that the free energy density is

$$\mathcal{F} = (\mu_s/2) \bullet (\delta/s_\tau)^2.$$

(c) Compare your results from parts (a) and (b), to obtain the elastic modulus for simple shear

$$\mu_s = \tau.$$

5.24. Consider a network of rigid rods of length s_o joined together at four-fold junctions, around which the rods can rotate freely, so long as they don't intersect. In the mean field limit, all plaquettes of this network have the same shape (as in the drawing) but $u_{xx} = 0$ and there are no fluctuations in the x-direction.

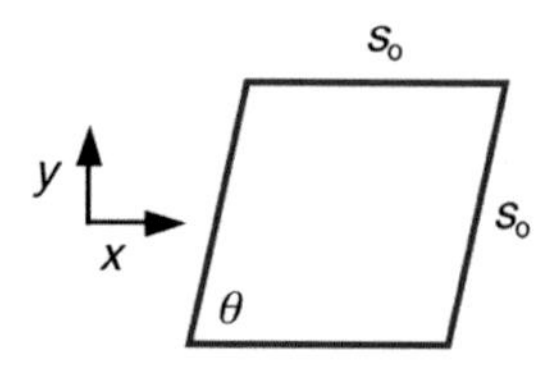

(a) Defining y as the height of the plaquette, determine the expectations $\langle y \rangle$ and $\langle y^2 \rangle$.
(b) Show that the mean area per vertex $\langle A_v \rangle$ is equal to $(2/\pi)s_o^2$.

(c) Because there are no fluctuations in the x-direction, there is no purely compressive mode and the compression modulus K_A and the pure shear modulus μ_p are coupled. The fluctuations in y are related to the sum of K_A and μ_p via

$$\beta(K_A + \mu_p) \langle A_v \rangle = \langle y \rangle^2 / (\langle y^2 \rangle - \langle y \rangle^2),$$

where $\beta = 1/k_BT$. Show that $K_A + \mu_p$ is given by

$$K_A + \mu_p = 4\pi\,(k_BT/s_o^2) / (\pi^2 - 8).$$

5.25. The two-dimensional honeycomb network is an example of a network with three-fold connectivity with unequal bond lengths a and b. Let's assume that the bonds are inextensible, so that the only contribution to the network energy is $k_{bend}\theta^2/2$ at each vertex.

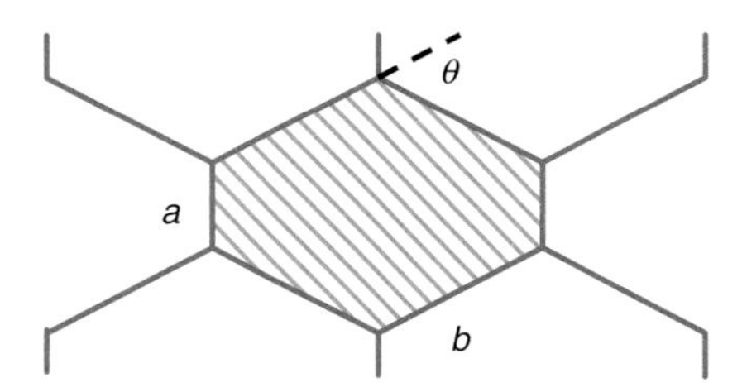

(a) Find the area of the cross-hatched plaquette as a function of θ. How many vertices are there per plaquette (not 6!)? Show that the area per vertex A_v of this network is

$$A_v = ab\cos\theta/2 + (1/2)b^2\sin\theta.$$

(b) Defining E_v as the bending energy per vertex, show that the enthalpy per vertex, $H_v = E_v - \tau A_v$, for this network under tension τ is minimized at an angle

$$\theta_{min} = \tau b^2 / (2k_{bend} + \tau ab/2)$$

if the deformation of the network is small (i.e. small θ).

(c) Find A_v at small tensions from your value of θ_{min}.

5.26. Determine the Young's modulus in the y-direction of the honeycomb network in Problem 5.25 by applying a stress in the vertical direction (σ_{yy}) and finding the resulting strain u_{yy} along the y-axis. (*Note: an alternate approach to this problem is based on enthalpy minimization.*)

(a) Following the notation of Problem 5.25, find the u_{yy} component of the strain tensor as a function of the angle θ working to lowest order in θ.

(b) Apply a force F to each of the vertical bonds, as shown.

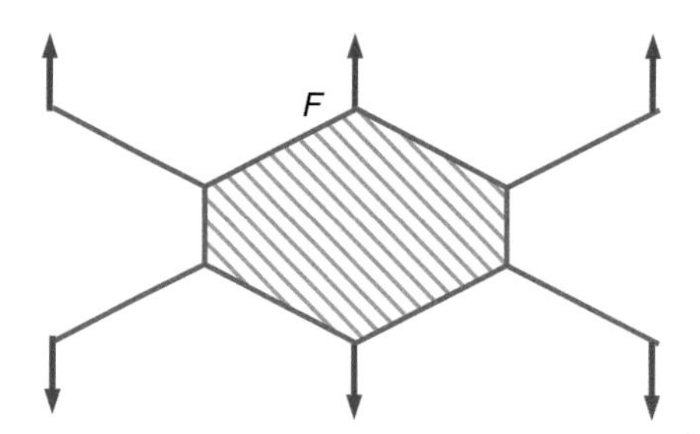

If the bending energy at each vertex is equal to $k_{\text{bend}}\theta^2/2$, show that the deformation angle θ resulting from these forces is governed by

$$F = 4k_{\text{bend}}\theta / b$$

at small θ.

(c) In two dimensions, stress is equal to force divided by length, which for the network here implies $\sigma_{yy} = F/2b$; that is, σ_{yy} is the force in the y-direction divided by the length along the x-axis over which it is spread. From your result in part (b), express σ_{yy} in terms of θ.

(d) From the definition $\sigma_{yy} = Yu_{yy}$, show that the Young's modulus in the y-direction is

$$Y = 4k_{\text{bend}}a/b^3.$$

5.27. The spring network with four-fold connectivity in Section 5.4 can be generalized to a rectangular network of inequivalent springs, as in the diagram here, where L_x, L_y are the unstretched spring lengths in the x and y directions and k_x, k_y are the corresponding spring constants.

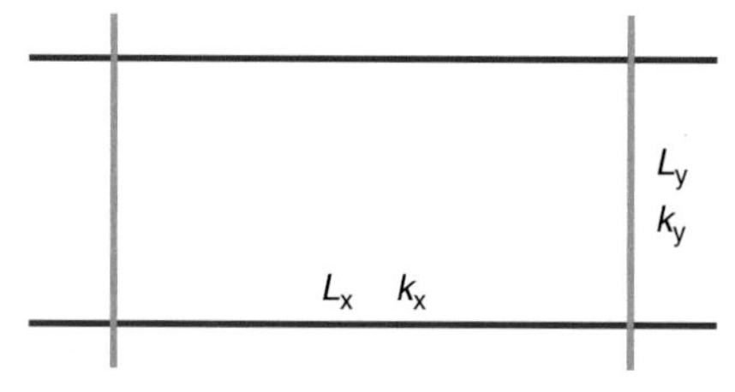

(a) Show that the enthalpy per vertex of this network under a tension τ is given by $H_v = k_x x^2/2 + k_y y^2/2 - \tau(L_x+x)(L_y+y)$, where x and y are the displacements from equilibrium in the x and y directions.

(b) Prove that, once equilibrated at a specific tension, x is given by $x = [\tau L_y k_y + \tau^2 L_x] / [k_x k_y - \tau^2]$; write the corresponding expression for y. Does an isotropic tension produce an isotropic response?

(c) To first order in τ, what is the area per vertex as a function of tension?

5.28. Find the Young's modulus for a deformation along a single axis of the four-fold network in Problem 5.27, where Young's modulus Y is defined via [*stress*] = Y[*strain*].

(a) For the strain, calculate the strain component u_{xx} using the extension along the x-axis calculated in part (b) of Problem 5.27 without proof. Make the approximation that τ is small.
(b) In two dimensions, stress is a force per unit length. Here, the force arises from stretching the springs in the x-direction, and the force is spread across a distance L_y transverse to its direction of application. What is the stress arising from this force?
(c) Combine your results from parts (a) and (b) to obtain $Y_x = k_x L_x / L_y$ for the Young's modulus associated with a uniaxial deformation along the x-axis.
(d) By inspection, write out Y_y, the Young's modulus associated with a uniaxial deformation along the y-axis. Are the two moduli the same?

5.29. A force f is applied uniformly across each end of a cylindrical rod, changing its length by an amount x. Viewing the rod as a spring, show that its effective spring constant is $k_{sp} = \pi R^2 Y / L$, where R and L are the radius and length of the rod, respectively, and Young's modulus Y is defined via [*stress*] = Y[*strain*].

5.30. It was demonstrated in Section 5.2 that the squared displacement $d\ell^2$ between two nearby points separated by the vector (dx_1, dx_2, dx_3) becomes

$$d\ell'^2 = \Sigma_{j,k} (\delta_{jk} + 2u_{jk})\, dx_j\, dx_k,$$

after a deformation characterized by the strain tensor u_{jk}, where δ_{jk} is a 3×3 matrix whose only non-vanishing elements are equal to unity and lie on the diagonal.

(a) At each location, the strain tensor u_{jk} can be diagonalized to give three elements U_α, where $\alpha = 1, 2, 3$. Evaluate $d\ell'^2$ in this representation and show that, as a result

$$dx'_\alpha = (1 + 2U_\alpha)^{1/2}\, dx_\alpha.$$

(b) Apply your result for $d\mathbf{x}$ from part (a) to establish that, for small deformations, the volume element $dV = \Pi_\alpha dx_\alpha$ becomes

$$dV' = dV(1 + \mathrm{tr}u).$$

6 Three-dimensional networks

The filaments of the cell vary tremendously in their bending resistance, having a visual appearance ranging from ropes to threads if viewed in isolation on the length scale of a micron. Collections of these biological filaments have strikingly different structures: a bundle of stiff microtubules may display strong internal alignment, whereas a network of very flexible proteins may resemble the proverbial can of worms. Thus, the elastic behavior of multi-component networks containing both stiff and floppy filaments may include contributions from the energy and the entropy of their constituents. In this chapter, we first review a selection of three-dimensional networks from the cell, and then establish the elastic properties of four different model systems, ranging from entropic springs to rattling rods. In the concluding section, these models are used to interpret, where possible, the measured characteristics of cellular networks.

6.1 Networks of biological rods and ropes

The filaments of the cytoskeleton and extracellular matrix collectively form a variety of chemically homogeneous and heterogeneous structures. Let's begin our discussion of these structures by describing two networks of microtubules found in the cell. The persistence length of microtubules is of the order of millimeters, such that microtubules bend only gently on the scale of microns and will not form contorted networks (see Section 3.6). For example, the microtubules of the schematic cell in Fig. 6.1(a) are not cross-linked, but rather extend like spikes towards the cell boundary, growing and shrinking with time. However, there exist varieties of microtubule-associated proteins (or *MAPs*) that can form links between microtubules, resulting in a bundled structure like Fig. 6.1(b): linking proteins are spaced along the microtubule like the rungs of a ladder, tying two microtubules parallel to one another. Such microtubule bundles are present in the long axon of a nerve cell or in the whip-like flagella extending from some cells. Neither of these configurations resembles the regular meshwork of the erythrocyte cytoskeleton or the nuclear lamina described in Section 5.1. The function of the microtubules in both the radial and parallel structures

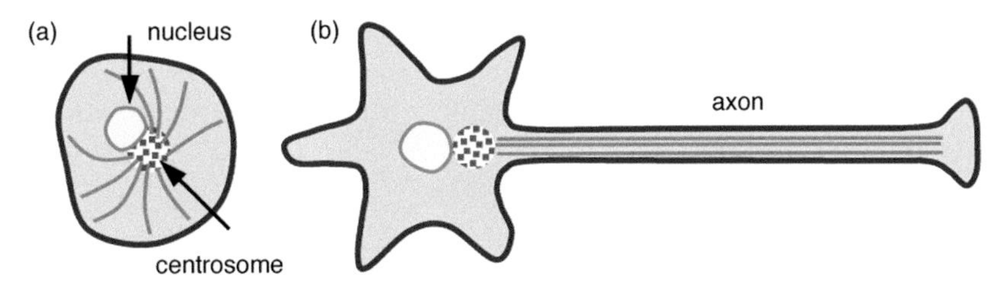

Fig. 6.1 Cartoon of microtubule configurations conventionally found in cells. In part (a), microtubules radiate from the centrosome near the nucleus, while in part (b), they lie parallel to one another along the axon of a nerve cell (not drawn to scale).

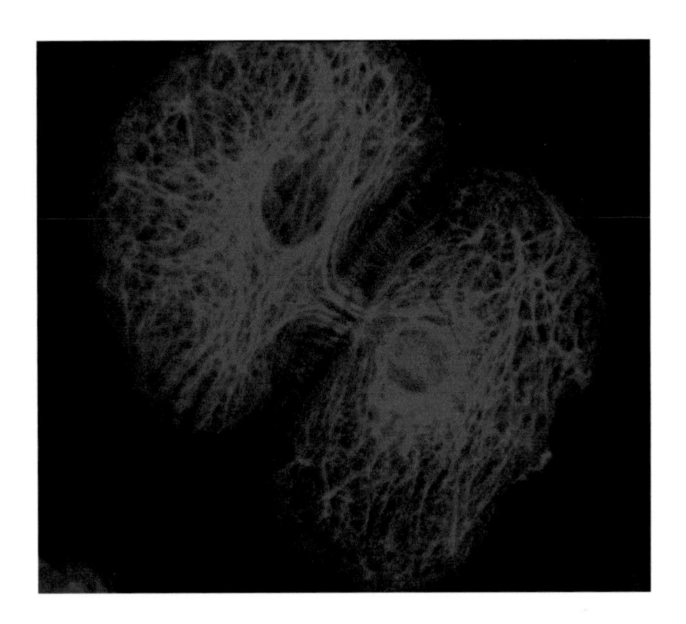

Fig. 6.2 In these newborn mouse keratinocytes, keratin intermediate filaments are seen to form an extended network attached to both the plasma membrane and the surface of the nucleus (courtesy of Dr. Pierre A. Coulombe, Johns Hopkins University School of Medicine).

includes transport: different motor proteins can walk along the tubule, as described further in Chapter 11.

The cell's intermediate filaments form a variety of networks that may run throughout the cell, or may be sheet-like. In some cases, the filaments terminate at a desmosome at the cell boundary, which is attached to a desmosome on an adjoining cell (see Fig. A.2); in these and other ways, extended networks of filaments can bind the cells into collective structures. For example, the immunostained keratin filaments in the cells displayed in Fig. 6.2 extend throughout the cytoplasm and are attached to both the cell nucleus and the plasma membrane. Of the proteins that can form intermediate filaments, several lamins are found in the nuclear laminae of eukaryotic cells, while more than 20 different keratins are present in human epithelial cells alone. The intermediate filament vimentin, discussed in Section 6.6, is observed in a wide range of cells, including fibroblasts and many others.

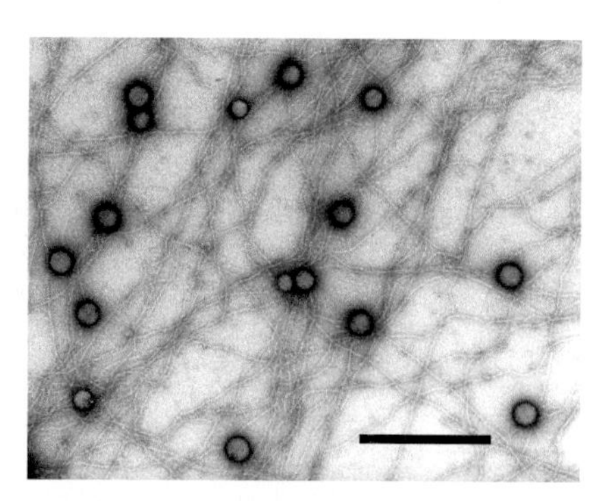

Fig. 6.3 Electron micrograph of a network of actin extracted from rabbit muscle cells and purified before being reconstituted; bar is 500 nm. The dark disks are 100 nm diameter latex spheres introduced in the experimental study of the network (reprinted with permission from Schmidt *et al.*, 1989; ©1989 by the American Chemical Society; courtesy of Dr. Erich Sackmann, Technische Universität München).

With a persistence length of 10–20 μm, actin filaments appear moderately flexible on cellular length scales, as can be seen in the micrograph of a reconstituted actin network in Fig. 6.3 (see Section 6.6 for procedure). These thin filaments may join together as part of a network, although actin filaments, like microtubules, require linking proteins in order to form a network or composite structure. There are a variety of such actin-binding proteins, α-actinin, filamin, and fimbrin being common examples, and they combine with actin to form networks of varying appearance and mechanical properties. Cartoons of composite actin structures are displayed in Fig. 6.4. Fimbrin and α-actinin link actin filaments into parallel bundles, with an average spacing between strands of 14 nm and 40 nm, respectively. The relatively large spacing between filaments linked by α-actinin permits motor proteins such as myosin to slide between adjacent strands. Filamin is a hinged dimer having roughly the shape of a *V*, whose individual arms lie length-wise along distinct actin filaments, resulting in an irregular, gel-like network. Actin also can be linked via spectrin: the auditory outer hair cell is thought to be an actin/spectrin network (see Section 5.1), and the spectrin tetramers of the erythrocyte cytoskeleton are joined, not directly to each other, but to short stubs of actin, each stub having the same average coordination as the network itself.

While actin, intermediate filaments and microtubules are the principal components of the internal networks of the cell, other proteins are present in the two- and three-dimensional networks surrounding the cell, such as the plant cell wall or the extracellular matrix. These networks tend to be composite structures, with a heterogeneous molecular composition and organization, and not many of their components have been isolated and reconstituted into chemically homogeneous networks whose properties can be studied in isolation.

The mechanical description of three-dimensional materials in general requires a larger number of elastic moduli than are needed for

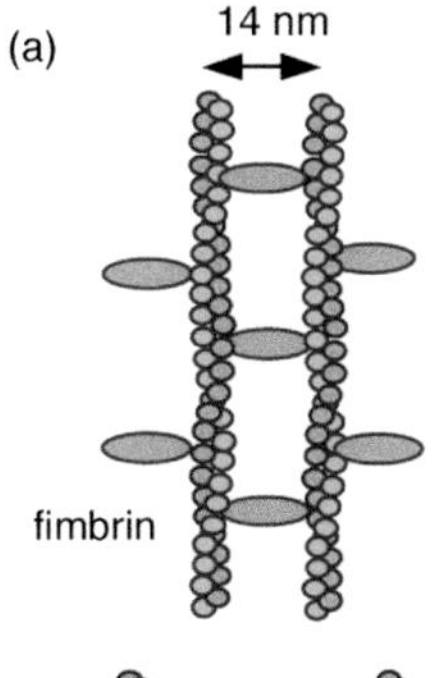

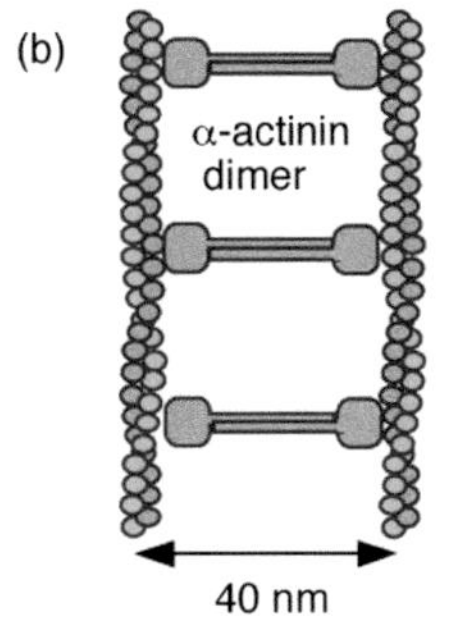

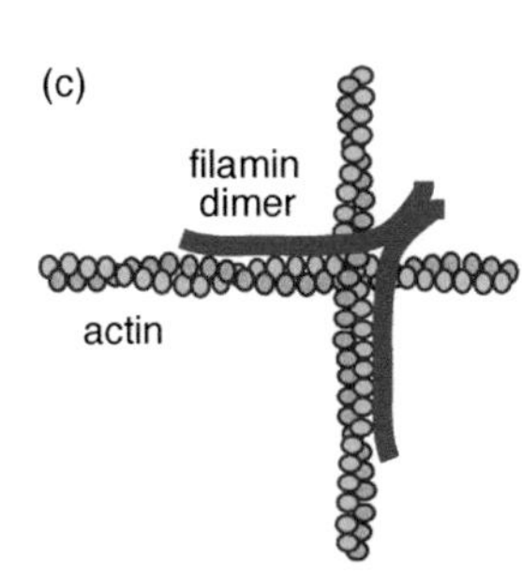

Fig. 6.4

Schematic representation of three structures formed by different actin-binding proteins. (a) Fimbrin has two actin-binding sites and links actin into parallel strands with a separation of 14 nm, while (b) α-actinin, a dimer with one binding site per monomer, results in a looser actin bundle with a separation of 40 nm. (c) In contrast, filamin produces a gel-like network (after Alberts *et al.*, 2008; reproduced with permission; ©2008 by Garland Science/Taylor and Francis LLC).

two-dimensional materials, although materials with high symmetry are simple to describe in any dimension. Both two- and three-dimensional isotropic systems involve only two independent moduli, and a simple cubic material needs three independent moduli in three dimensions. However, the macroscopic elastic parameters of a material may have a slightly more complicated dependence on the characteristics of its microscopic elements in three dimensions than in two. A network of identical springs with spring constant k_{sp} and equilibrium length s_o, for example, has moduli proportional to just k_{sp} in two dimensions but k_{sp}/s_o in three dimensions, with a numerical prefactor that depends upon the network connectivity.

Our treatment of biological networks must recognize that there may or may not be permanent cross-links between its filamentous elements. If the density of filaments and cross-links is sufficiently high, the network possesses both compression and shear resistance, as in Fig. 6.5(a). For instance, in networks composed of floppy chains, the elasticity of the network near its equilibrium configuration is entropic and the moduli are proportional to $\rho k_B T$, where ρ is the density of chains. Of course, if the density of chains is so low that there are few junctions per chain, the network may not be completely connected, and the shear modulus vanishes as in Fig. 6.5(b). There is a well-defined threshold for the average connectivity, referred to as the connectivity percolation threshold, above which network elements are sufficiently linked to their neighbors that a continuous path can be traced along the elements from one side of the network to the other, as in Fig. 6.5(a). At temperatures above zero, the elastic moduli also are believed to be non-vanishing immediately above the connectivity percolation threshold, although this is not true at zero temperature for certain classes of forces.

Suppose that there are no cross-links at all. Clearly, such collections of rods and chains must behave like fluids if they are observed over long time scales: even if the filaments are wrapped around each other forming knots and loops, they can wiggle past each other owing to thermal motion such that the system as a whole ultimately deforms in response to an applied shear. However, on short time scales, the system may respond somewhat elastically, a behavior referred to as viscoelasticity. Thus, the elastic moduli for uncross-linked chains may be time-dependent, with the effective shear modulus decreasing as time increases. The viscoelastic properties of a polymer solution depend upon its density, as illustrated in Fig. 6.6 for stiff filaments. In the dilute regime, the filaments are sufficiently separated in space that they can rotate largely unencumbered and the solution's viscosity is close to that of its solvent. At the opposite extreme, in the concentrated regime the filaments are so dense that they are in frequent contact with each other. In between lies the semi-dilute regime, characteristic of many biopolymer solutions.

The behavior of a polymer solution at moderate to high concentrations depends in part on the stiffness of the polymer. Floppy chains, with an average contour length L_c much longer than the polymer persistence length

ξ_p, form entangled configurations with random orientations of the polymers. Individual stiff rods, with $L_c << \xi_p$, each have a well-defined orientation, which may or may not be aligned with those of its neighbors. Thus, a solution of polymeric rods may be disordered and isotropic at modest concentrations, but ordered at higher density into a liquid crystalline state called the nematic phase, in which individual rods or molecules display long-range orientational order (see Vertogen and de Jeu, 1988).

One way of characterizing the time-dependence of the effective elastic moduli of viscoelastic materials is to apply a strain to the material at a fixed driving frequency, as shown in Fig. 6.7. The relationship between the applied strain u_{xy} and the resulting stress σ_{xy} developed within the system involves two functions of the angular frequency ω of the driving strain, namely $\sigma_{xy} = G'(\omega)u_{xy}(t) + G''(\omega)\bullet(\mathrm{d}u_{xy}/\mathrm{d}t)/\omega$, where $\mathrm{d}u_{xy}/\mathrm{d}t$ is the rate of strain. The functions $G'(\omega)$ and $G''(\omega)$ are called the shear storage and shear loss moduli, respectively, and they are measures of the energy stored (and recovered) or lost during a cycle of the system. At very long times, or, equivalently when ω vanishes, G' becomes the shear modulus and G''/ω becomes the viscosity. Both of these moduli have been measured for a variety of cytoskeletal filaments.

In Section 6.2, we review the static elastic properties of several three-dimensional systems, and relate the elastic constants of a regular spring network to the microscopic characteristics of the springs themselves. In Section 6.3, we develop an approximate expression for the free energy density of a cross-linked network of floppy chains under deformation. Following this, in Section 6.4, we study the mechanical failure of materials and investigate the behavior of networks near the connectivity percolation threshold. The formalism for representing the mechanical properties of viscoelastic materials is presented in Section 6.5. Lastly, our analytical results are summarized and applied to cytoskeletal networks in Section 6.6.

(a)

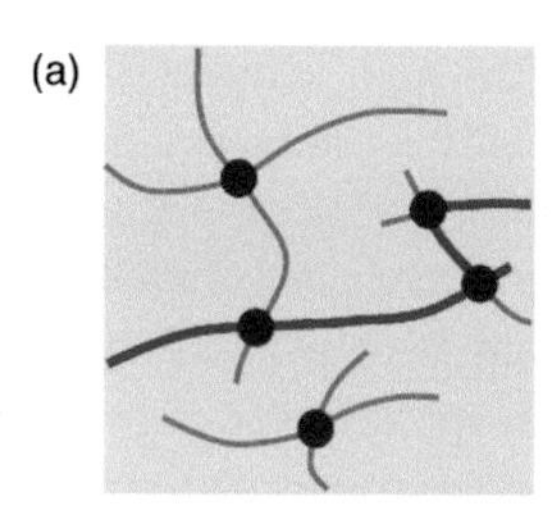

(b)

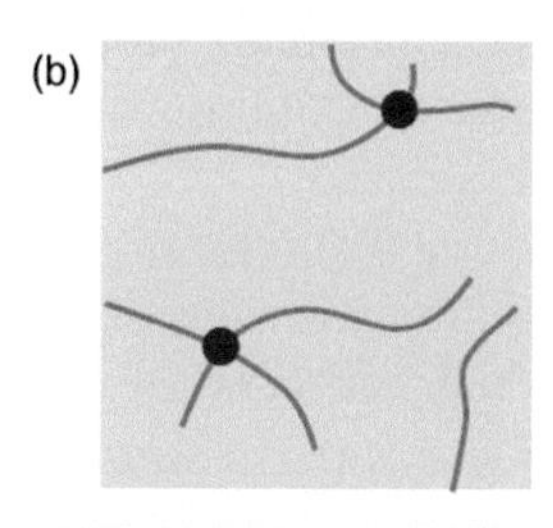

Fig. 6.5

(a) Above the connectivity percolation threshold, a continuous path exists across the network along its elements, as indicated by the thick dark line. (b) Below the percolation threshold, a network is not completely connected and its shear resistance vanishes.

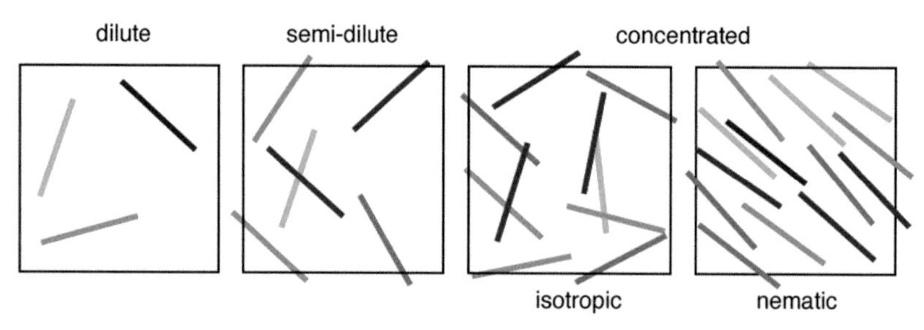

Fig. 6.6

Polymer solutions can be classified according to concentration as dilute, semi-dilute and concentrated. In addition, concentrated solutions of stiff filaments may display a disordered isotropic phase or, at higher density, an ordered nematic phase, in which the filaments possess long-range orientational order.

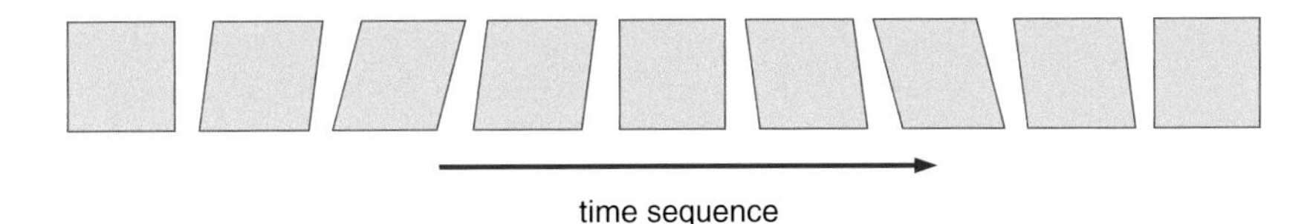

Fig. 6.7 Time evolution of a system being driven by an oscillating shear strain, permitting the frequency-dependence of the strain response of a viscoelastic material to be determined.

6.2 Elasticity in three dimensions

The three-dimensional materials of interest in our treatment of the cell are not so much conventional bulk solids (bone is an obvious exception to this statement) as semiflexible networks having more complex elastic characteristics than networks of floppy chains. In this section, we review the formalism for describing the generic elastic properties of three-dimensional materials, and present the free energy of deformation for both an isotropic material and one with four-fold symmetry. We then discuss the compression modulus of one particular system, namely a spring network with cubic symmetry, as a demonstration of how the microscopic characteristics of a three-dimensional network are reflected in its elastic moduli.

6.2.1 Strain and stress

The starting point for describing the deformation of an object under stress is the displacement vector **u**, which is the (vector) change in the position **x** of an element of the object (see Figs. 5.8 and 5.9). If the object changes shape, then **u** varies locally across it; that is, **u** is a function of the coordinates $x_i = \{x, y, z\}$. The magnitude of the deformation is thus reflected in the rate at which **u** varies from one location to another, $\partial u_i / \partial x_j$. As shown in Section 5.2, these derivatives arise naturally in the expression for the change in separation between two locations under a deformation, and appear in a quantity called the strain tensor u_{ij} in the combination

$$u_{ij} \equiv 1/2\ [\partial u_i / \partial x_j + \partial u_j / \partial x_i + \Sigma_k\ (\partial u_k / \partial x_i)(\partial u_k / \partial x_j)], \qquad (6.1)$$

where the indices i and j refer to the coordinate system. For small deformations, the last term can be ignored, leading to the more familiar form

$$u_{ij} \cong 1/2\ (\partial u_i / \partial x_j + \partial u_j / \partial x_i). \qquad (6.2)$$

Example 5.1 shows how to calculate u_{ij} for a simple deformation. Note that that the trace of the strain tensor is equal to the fractional change in the volume:

$$(\mathrm{d}V' - \mathrm{d}V)/dV = \mathrm{tr}u. \qquad (6.3)$$

For a given force, the magnitude of a deformation depends on the total surface area receiving the force: the deformation is smaller if the force is applied to a large area than if it is applied to a small area. Thus, what is important in determining the strain of the deformation is the force per unit area, or the *stress*. Now, the response of an object to an applied force also depends on the direction of the force with respect to the surface receiving it: if the force is applied perpendicular to the surface, then the deformation will include compression, while if the force is applied parallel to the surface, it will include shear. Thus, stress cannot be just a scalar quantity: it depends on the direction of the force as well as the orientation of the surface that receives it and must be a tensor, just like the strain.

To put this mathematically, we represent an element of the surface by the vector **a** which has a magnitude equal to the surface area a, and a direction parallel to the normal vector to the surface **n**, as in

$$\mathbf{a} = a\,\mathbf{n}. \tag{6.4}$$

The stress tensor σ_{ij} is then related to the components F_i and the area element a_j via

$$F_i = \Sigma_j\, \sigma_{ij}\, a_j, \tag{6.5}$$

where σ_{ij} has units of force per unit area (or energy per unit volume) in three dimensions. In two dimensions, σ_{ij} still has units of energy density, where it is the energy per unit area or the force per unit length. By considering moments of forces (products of the force and the distance from a rotational axis) it can be shown that

$$\sigma_{ij} = \sigma_{ji}, \tag{6.6}$$

i.e., the stress tensor is symmetric (see Landau and Lifshitz, 1986).

Example 6.1. As a simple example of the stress tensor, consider an object under hydrostatic pressure P, in which the force has a constant magnitude throughout the medium. The force **F** experienced by this pressure on a surface is in the opposite direction to the vector **a** describing the surface:

$$F_i = -P\, a_i = -P\, \Sigma_j\, \delta_{ij}\, a_j. \tag{6.7}$$

Comparing Eqs. (6.5) and (6.7), the form of the stress tensor for hydrostatic pressure must be

$$\sigma_{ij} = -P\, \delta_{ij}. \tag{6.8}$$

In general, the diagonal elements of the stress tensor correspond to compressions and the off-diagonal elements to shear. The fact that the off-diagonal elements vanish for hydrostatic pressure indicates that the system is not subject to shear in this example.

6.2.2 Elastic moduli

For ideal springs in one dimension, we know that the restoring force f is proportional to the displacement from equilibrium x: $f = -k_{sp}x$, where k_{sp} is the spring constant. The corresponding relationship for continuous materials is the general form of Hooke's Law and reads $[stress] \propto [strain]$, where stress is the analog of force and strain is the analog of the displacement. In principle, the analog of the spring constant is more complicated, as it must relate one tensor to another tensor, each tensor having components in two directions. Thus, the generalization of the spring constant involves a quantity with four indices, two for u_{ij} two for σ_{kl}. There is a choice of conventions for defining the generalized spring constants: one can write

$$u_{ij} = \Sigma_{k,l}\, S_{ijkl}\sigma_{kl}, \tag{6.9}$$

where the constants S_{ijkl} are called the elastic compliance constants (or elastic constants), or one can choose

$$\sigma_{ij} = \Sigma_{k,l}\, C_{ijkl}u_{kl}, \tag{6.10}$$

where the constants C_{ijkl} are called the elastic stiffness constants (or elastic moduli). In this text, we use the elastic moduli C_{ijkl} which have units of energy density because the strain tensor is dimensionless.

Just like the potential energy of a simple spring is proportional to the square of the displacement from equilibrium, as in $V = k_{sp}x^2/2$, here the free energy density is proportional to the square of the strain tensor, as in

$$\Delta\mathcal{F} = 1/2\ \Sigma_{i,j,k,l}\, C_{ijkl}\, u_{ij}\, u_{kl}. \tag{6.11}$$

With simple springs, rather than start with the force law and obtain the potential energy by integrating the force over the distance through which it acts, one could instead start with the potential energy and differentiate with respect to position. For stresses and strains, this approach becomes

$$\sigma_{ij} = \partial\mathcal{F} / \partial u_{ij}, \tag{6.12}$$

a result proven in standard texts on continuum mechanics (see, for example, Landau and Lifshitz, 1986).

We now describe in some detail the elastic parameters of an isotropic material, which means that its characteristics are independent of direction. An object made from this material may have an arbitrary shape: "isotropic" only refers to the material itself. The first step is to determine the minimum number of elastic moduli needed to describe the material. In Appendix D, we show how to apply symmetry considerations to C_{ijkl} to find relationships among its elements, ultimately reducing the number of independent components to two for an isotropic material. An alternative approach that we use here is to determine the lowest order contributions to an expansion of the free energy density $\mathcal{F}$ in powers of the strain tensor and then to identify the elastic moduli with specific deformation modes. From Eq. (6.11), the smallest deformations of a continuous material have an energy density that is quadratic in the strain tensor. To describe a scalar quantity like the energy, the quadratic combinations of u also must be scalars, of which there are only two possibilities: the squared sum of the diagonal elements, $(\mathrm{tr}u)^2$, and the sum of the elements squared, $\Sigma_{i,j}u_{ij}^2$. Thus, for small strains, the free energy density can be written as

$$\mathcal{F} = \mathcal{F}_o + (\lambda/2)(\mathrm{tr}u)^2 + \mu\, \Sigma_{i,j}u_{ij}^2. \tag{6.13}$$

The constant $\mathcal{F}_o$ represents the free energy density of the unperturbed system. The constants λ and μ are called the Lamé coefficients and have units of energy density, just like C_{ijkl}.

The Lamé coefficients can be related to the elastic moduli by rewriting the strain tensor as the sum of a pure shear and a hydrostatic compression through the simple rearrangement

$$u_{ij} = (u_{ij} - \delta_{ij}\,\mathrm{tr}u/3) + \delta_{ij}\,\mathrm{tr}u/3. \tag{6.14}$$

If the deformation is a pure shear (no change in volume), then only the first term on the right-hand side survives. In contrast, for a hydrostatic compression, the trace of the first term vanishes and the second term measures the change in volume associated with the deformation, according to Eq. (6.3). Thus, the first and second terms of Eq. (6.14) correspond to shear and compression modes, respectively. The factor of 3 appearing in both terms is dimension-dependent and arises from the number of elements in the trace; for two-dimensional systems, 3 is replaced by 2.

The free energy density also can be decomposed into independent deformation modes by substituting Eq. (6.14) into Eq. (6.13):

$$\begin{aligned}\Delta\mathcal{F} &= (\lambda/2)(\mathrm{tr}u)^2 + \mu\, \Sigma_{i,j}\,[(u_{ij} - \delta_{ij}\,\mathrm{tr}u/3) + \delta_{ij}\,\mathrm{tr}u\,/3]^2 \\ &= \mu\, \Sigma_{i,j}\,(u_{ij} - \delta_{ij}\,\mathrm{tr}u/3)^2 + 1/2\,(\lambda + 2\mu/3)\,(\mathrm{tr}u)^2,\end{aligned} \tag{6.15}$$

where the first term is a pure shear and the second is a hydrostatic compression. Just as the elastic parameter k_{sp} multiplies the square of the displacement in the potential energy of a Hookean spring, so too here: the energy density is proportional to the relevant elastic modulus times the square of the corresponding strain. Hence, the shear modulus is simply the Lamé coefficient μ, while the compression modulus K_V must be

$$K_V = \lambda + 2\mu/3. \tag{6.16}$$

Written in terms of K_V and μ, Eq. (6.15) becomes

$$\Delta\mathcal{F} = (K_V/2)\bullet(\mathrm{tr}u)^2 + \mu\, \Sigma_{i,j}\,(u_{ij} - \delta_{ij}\,\mathrm{tr}u/3)^2. \tag{6.17}$$

It is clear from the quadratic form of Eq. (6.15) that both K_V and μ must be positive, or else the system could spontaneously deform by a shear or compression mode. However, the Lamé coefficient λ is not constrained to be positive and a very limited number of systems are known in which λ is negative.

As they are directly related to specific deformation modes, the pair K_V and μ provide a useful parametrization of the two independent elastic moduli expected for isotropic materials. The stress–strain relations can be expressed in terms of K_V and μ by substituting Eq. (6.15) into Eq. (6.12):

$$\sigma_{ij} = \delta_{ij} K_V \operatorname{tr}u + 2\mu(u_{ij} - \delta_{ij} \operatorname{tr}u / 3), \qquad (6.18)$$

or its inverse

$$u_{ij} = \delta_{ij} \operatorname{tr}\sigma / 9K_V + (\sigma_{ij} - \delta_{ij} \operatorname{tr}\sigma / 3) / 2\mu. \qquad (6.19)$$

Under isotropic pressure, for example, $\sigma_{ij} = -P\delta_{ij}$ from Eq. (6.8), so that the trace of Eq. (6.18) yields $P = -K_V \operatorname{tr}u$, where $\operatorname{tr}u$ is the relative change in volume. In two dimensions, this expression becomes $\tau = +K_A(u_{xx} + u_{yy})$ for an isotropic tension τ.

6.2.3 Young's modulus and Poisson's ratio

Another representation for the elastic characteristics of isotropic materials can be found by analyzing the deformation of a uniformly thin rod by a tensile force applied perpendicular to its ends. The coordinate system is chosen such that the rod lies along the z-axis, and the forces are applied along the z-axis as well. With no x or y components to the force, all components of σ_{ij} vanish except σ_{zz}. Under such a simple stress tensor, Eq. (6.19) can be solved easily for the components of the strain tensor:

$$u_{xx} = u_{yy} = \sigma_{zz} (1 / 3K_V - 1 / 2\mu) / 3 \qquad (6.20a)$$

$$u_{zz} = \sigma_{zz} (1 / 3K_V + 1/\mu) / 3. \qquad (6.20b)$$

Note the sign convention: a tensile force corresponds to $\sigma_{zz} > 0$ [opposite sign to compression in Eq. (6.8)] so that $u_{zz} > 0$ in Eq. (6.20b) and the rod stretches in the direction of the applied force. If $K_V \sim 3\mu$, as it is for many materials, both u_{xx} and u_{yy} are less than zero, and the rod shrinks transversely. The Poisson ratio σ_p is a measure of how much a material *contracts* in the x- or y-direction when it is stretched in the z-direction:

$$\sigma_p = -u_{xx}/u_{zz}. \qquad (6.21)$$

Substituting Eq. (6.20) from the analysis of the thin rod gives

$$\sigma_p = (3K_V - 2\mu)/(6K_V + 2\mu). \qquad \text{(three dimensions)} \qquad (6.22)$$

The Young's modulus Y, defined by the stress–strain relation, is

$$u_{zz} = \sigma_{zz} / Y. \qquad (6.23)$$

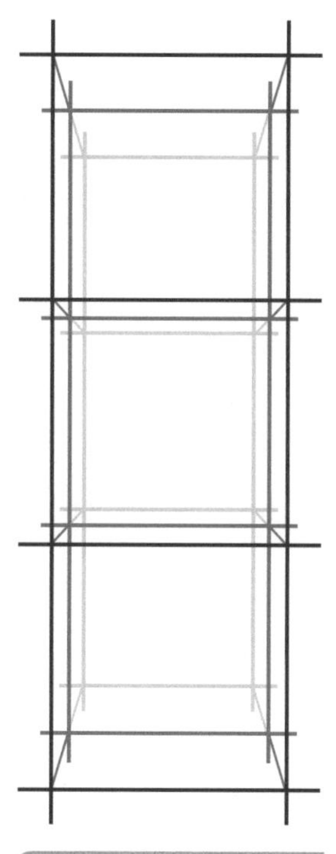

Fig. 6.8

Example of a network with cubic symmetry in three dimensions.

This commonly measured elastic quantity can be expressed in terms of the fundamental elastic constants K_V and μ by substituting Eq. (6.20) for the strain tensor into Eq. (6.23):

$$Y = 9K_V\mu \,/\, (3K_V + \mu). \tag{6.24}$$

Many materials display $\sigma_p = 1/3$, which corresponds to $K_V = (8/3)\mu = Y$. Equations (6.22) and (6.24) permit the combination Y and σ_p to be used as an alternate pair to K_V and μ for describing the elastic characteristics of isotropic materials.

6.2.4 Spring network with cubic symmetry

To see how these moduli are related to the microscopic structure of a material, we turn to our favorite example, a network of identical springs, each having a potential energy $V(s) = (k_{sp}/2) \cdot (s - s_o)^2$, where k_{sp} is the spring constant and s_o is the unstretched spring length. Suppose the springs are linked at six-fold junctions to form a network with cubic symmetry, as in Fig. 6.8. This is a somewhat dangerous example because, as we saw for two-dimensional spring networks with four-fold connectivity in Section 5.4, the ground state of this network is degenerate unless interactions between non-adjacent network nodes are introduced to provide rigidity. However, the simple cubic network is easy to evaluate and yields a qualitatively correct relationship between k_{sp}, s_o, and the elastic moduli. To find the relation between an elastic modulus of the network and its microscopic description, we compare the change in the free energy according to Eq. (6.15) with an equivalent expression involving the spring variables k_{sp} and s_o. Specifically, let's consider the deformation mode illustrated in Fig. 6.9, wherein all springs change in length from s_o to $s_o + \delta$, which is a pure compression. For this mode, the off-diagonal elements of the strain tensor vanish, leaving only the diagonal elements $u_{xx} = u_{yy} = u_{zz} = \delta \,/ s_o$. The change in the free energy density is, according to Eq. (6.15),

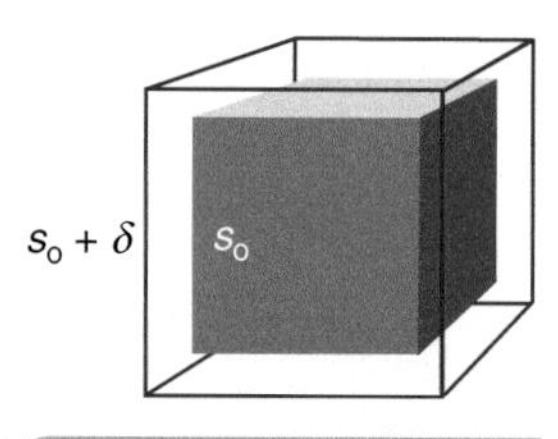

Fig. 6.9

Sample deformation mode of a spring network with cubic symmetry. The springs change in length from s_o to $s_o + \delta$. The original shape of a unit cell of the network is shaded, while the deformed shape is indicated by the wire outline.

$$\Delta\mathcal{F} = (9K_V/2) \cdot (\delta \,/ s_o)^2. \tag{6.25}$$

Now, a cubical unit cell of the network contains a single vertex and three springs, such that the change in the potential energy of this cell upon deformation is $3k_{sp}\delta^2/2$. Dividing the change in potential by the volume of the unit cell s_o^3, yields

$$\Delta\mathcal{F} = (3k_{sp}/2) \cdot (\delta^2/s_o^3). \tag{6.26}$$

The compression modulus can be found by comparing Eqs. (6.25) and (6.26)

$$K_V = k_{sp} \,/\, 3s_o, \qquad \text{(rigid cubic symmetry)} \tag{6.27}$$

showing that the elastic moduli depend upon both the spring constant and the equilibrium spring length in three dimensions. Two other regular spring networks are treated in the end-of-chapter problems, also yielding $K_V \propto k_{sp}/s_o$, with a proportionality constant of about 1/2.

6.3 Entropic networks

We showed in Section 5.5 for a specific two-dimensional network of floppy chains that its elastic moduli are roughly equal to $\rho k_B T$, where T is the temperature and ρ is the chain density. A similar relation holds in three dimensions, with ρ becoming the three-dimensional density of chains. We now derive this result more generally for a network of chains joined at random locations.

6.3.1 Construction of random chain networks

The model investigated here is one developed some time ago for interpreting the properties of vulcanized rubber (Flory, 1953; Treloar, 1975). To construct the model network, flexible chains are packed together with sufficient density that a given chain is close to its neighbors at many locations along its contour length. The network is then given rigidity by welding different chains to one another at points where they are in close proximity, as illustrated in Fig. 6.10. After welding has occurred, there are n chain segments, each a random chain in its own right (having been cut from a larger random chain) with an end-to-end displacement $\mathbf{r}_{ee}$ obeying the conventional Gaussian probability distribution, Eq. (3.42). Despite our freezing the positions of the chains during network construction, the network nodes are not fixed in space at non-zero temperature and the instantaneous end-to-end displacement of a segment may change, even though the contour length is fixed. When the network deforms in response to an external stress, the average value of $\mathbf{r}_{ee}$ changes as well.

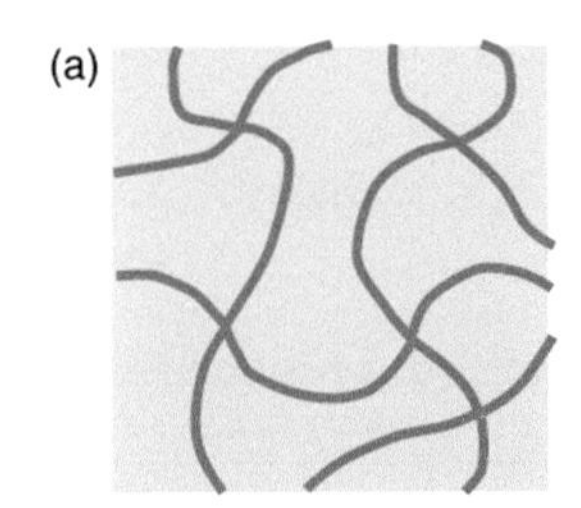

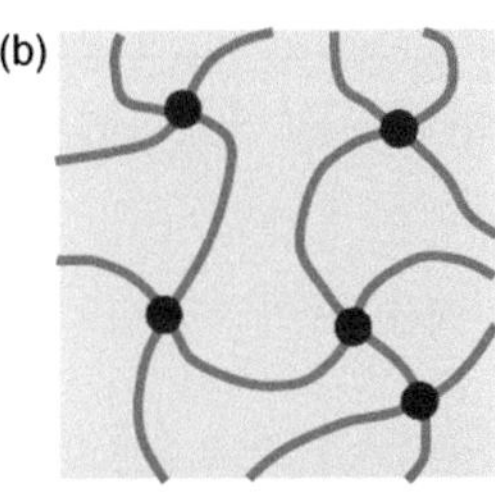

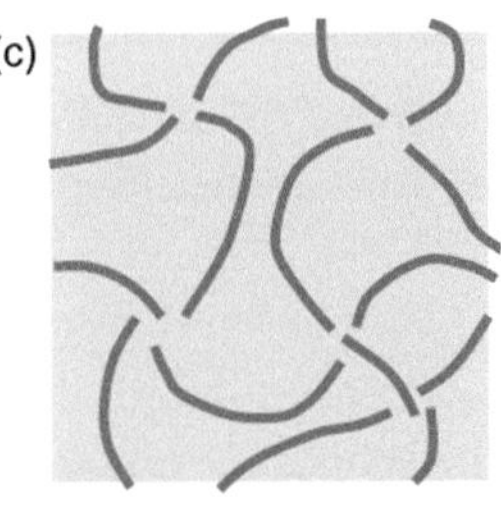

Fig. 6.10

Construction of a random network of random chains. Six long chains in (a) are welded at five points to produce the shear-resistant network in (b) with $n = 16$ segments, shown as isolated units in (c).

6.3.2 Deformation of the network

To describe quantitatively the deformation of the network as a whole, we introduce the scale factors Λ_x, Λ_y, Λ_z, such that a system in the shape of a rectangular prism of sides L_x, L_y, L_z before the deformation becomes $\Lambda_x L_x$, $\Lambda_y L_y$, $\Lambda_z L_z$ after the deformation, as illustrated in Fig. 6.11. Thus, an extension (compression) of the network in a given Cartesian direction corresponds to $\Lambda > 1$ ($\Lambda < 1$). We follow Chapter XI of Flory (1953) to determine how the entropy of the network depends upon the scale factors.

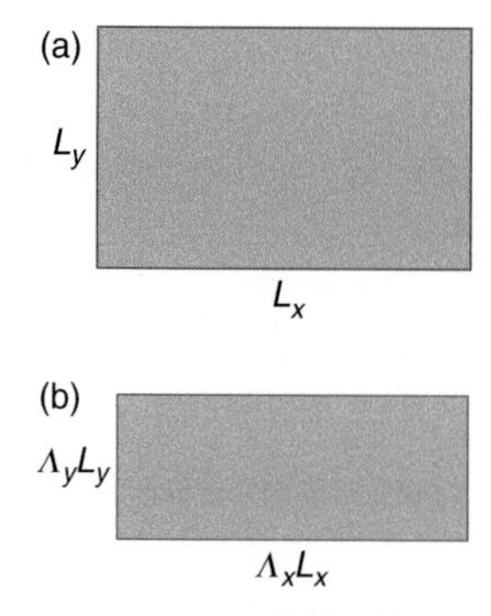

Fig. 6.11

Under deformation, the rectangle of sides L_x and L_y in (a) changes to a different rectangle of sides $\Lambda_x L_x$ and $\Lambda_y L_y$ in (b), where the scale factors satisfy $\Lambda_x > 1$ and $\Lambda_y < 1$ for the deformation shown.

As is apparent in Fig. 6.10(c), the chains possess a distribution of end-to-end displacement vectors that changes with deformation as the chains are stretched, compressed and reoriented. The probability that a given chain i has a particular displacement vector $\mathbf{r}_i = (x_i, y_i, z_i)$ *after* deformation can be found from the probability that it had a displacement vector $(x_i/\Lambda_x, y_i/\Lambda_y, z_i/\Lambda_z)$ in the unstressed state *before* the deformation. To be a little more precise, the probability that the displacement of a chain lies in the range $\mathbf{r}_i$ to $\mathbf{r}_i + \Delta\mathbf{r}$ *after* deformation is given by

$$\mathcal{P}(x_i/\Lambda_x, y_i/\Lambda_y, z_i/\Lambda_z)\,(\Delta x/\Lambda_x)\bullet(\Delta y/\Lambda_y)\bullet(\Delta z/\Lambda_z), \tag{6.28}$$

where the three-dimensional probability density $\mathcal{P}$ is given by Eq. (3.47). The actual number of chains n_i lying in this range of $\mathbf{r}_i$ is equal to the product of this probability and the total number of chains n in the system, resulting in

$$n_i = n\,(2\pi\sigma^2)^{-3/2}\exp\{-[(x_i/\Lambda_x)^2 + (y_i/\Lambda_y)^2 + (z_i/\Lambda_z)^2]\,/\,2\sigma^2\}\;\Delta x\Delta y\Delta z\,/\,\Lambda_x\Lambda_y\Lambda_z, \tag{6.29}$$

after substitution of Eq. (3.47), where $\sigma^2 = Nb^2/3$ for a three-dimensional chain of N equal segments of length b. We will use n_i in determining the change in the free energy during deformation.

Before evaluating the entropy of the network, let us review the method of constructing one of its configurations. To generate a specific configuration such as Fig. 6.10(b), two conditions must be satisfied:

(a) the distribution of n individual segments must have the correct number of chains n_i for each range of $\mathbf{r}_i$, and
(b) each weld site must be within the appropriate distance of a site on a neighboring chain.

The probability of the chains satisfying conditions (a) and (b) are defined as $\mathcal{P}_\mathrm{a}$ and $\mathcal{P}_\mathrm{b}$, respectively; the combined probability for constructing the configuration is $\mathcal{P}_\mathrm{a}\mathcal{P}_\mathrm{b}$.

To evaluate $\mathcal{P}_\mathrm{a}$, we need to know two quantities: a number and a probability. First, the number of chains available in the original configuration to change to the new configuration is n_i of Eq. (6.29). Second, the probability p_i that any one of these n_i chains has the correct $\mathbf{r}_i$ is

$$p_i = \mathcal{P}(x_i, y_i, z_i)\,\Delta x\,\Delta y\,\Delta z, \tag{6.30}$$

so that the probability of all n_i chains in this range have the correct $\mathbf{r}_i$ is $p_i^{\mathrm{n}(i)}$, ignoring permutations. For all n chains in the network, the total probability of the correct position involves the product of the individual terms, or $\Pi_i\, p_i^{\mathrm{n}(i)}$. However, the individual chains can be permuted a total of $n!\,/\,\Pi_i n_i!$ ways, so that the probability of satisfying criterion (a) is

$$\mathcal{P}_\mathrm{a} = (\Pi_i\, p_i^{\mathrm{n}(i)})\bullet(n!\,/\,\Pi_i n_i!) = n!\;\Pi_i(p_i^{\mathrm{n}(i)}/n_i!). \tag{6.31}$$

Let's review the welding process before determining $\mathcal{P}_b$. How many welds are there for n segments? Each weld links four chain segments; however, the number of welds is not $n/4$, but rather $n/2$, since each segment is defined by two welds (we neglect dangling chains; see Chapter XI of Flory, 1953, for a discussion of this approximation). The number of welding sites on the chains must be twice the number of welds, or n, because each weld involves two separate weld sites, usually on different chains. Thus, there are n segments, n welding sites and $n/2$ welds. Now, what is the probability of finding a welding site in the correct position with respect to neighboring chains? Starting with site number 1, the probability that one of the $n - 1$ remaining welding sites lies within a volume δV of site number 1 is just $(n - 1)\delta V / V$, where V is the volume of the entire network. After the first weld, two sites have been removed from the list of potential weld locations. For the second weld, the probability that one of the remaining $(n - 3)$ welding sites lies within a volume δV of site number 3 is then $(n - 3)\delta V / V$. We repeat this argument for all $n/2$ welds, to obtain

$$\begin{aligned}\mathcal{P}_b &= (n - 1)(n - 3)(n - 5)\ldots 1\ (\delta V / V)^{n/2} \\ &\cong (n/2)!\ (2\ \delta V / V)^{n/2}. \end{aligned} \tag{6.32}$$

where we have used the approximation $(n/2 - 1/2)(n/2 - 3/2)\ldots 1/2 \cong (n/2)!$ In this expression, V is the deformed volume $\Lambda_x\Lambda_y\Lambda_z V_o$, where V_o is the original volume of the network.

6.3.3 Entropy change under deformation

The probability for the network to have a particular shape is $\mathcal{P}_a\mathcal{P}_b$, permitting its entropy to be written as $S = k_B\ln\mathcal{P}_a + k_B\ln\mathcal{P}_b$ (see Appendix C). Our next task is explicitly to evaluate $\ln\mathcal{P}_a$ and $\ln\mathcal{P}_b$. The logarithm of Eq. (6.31) can be simplified by using Stirling's approximation for factorials, $\ln n! = n \ln n - n$, such that

$$\begin{aligned}\ln\mathcal{P}_a &= n \ln n + \Sigma_i\, n_i \ln(p_i / n_i) \\ &= \Sigma_i\, n_i \ln(n\, p_i / n_i). \end{aligned} \tag{6.33}$$

The product np_i can be obtained from Eq. (6.30), and the n_i themselves are given by Eq. (6.29). The calculation is performed in the problem set, and leads to

$$\ln\mathcal{P}_a = -(n/2)[\Lambda_x^2 + \Lambda_y^2 + \Lambda_z^2 - 3 - 2\ln(\Lambda_x\Lambda_y\Lambda_z)]. \tag{6.34}$$

Mercifully, it takes considerably less effort to obtain $\ln\mathcal{P}_b$. From Eq. (6.32), we find

$$\ln\mathcal{P}_b = \ln([n/2]!) + (n/2)\ln(2\ \delta V / V_o) - (n/2)\ln(\Lambda_x\Lambda_y\Lambda_z) \tag{6.35}$$

after substituting $V = \Lambda_x\Lambda_y\Lambda_z V_o$. Combining $\ln\mathcal{P}_a$ and $\ln\mathcal{P}_b$, we obtain the entropy S of the deformation

$$S = -(k_B n/2)[\Lambda_x^2 + \Lambda_y^2 + \Lambda_z^2 - 3 - \ln(\Lambda_x\Lambda_y\Lambda_z) - \ln\,(n/2)! - (n/2)\ln(2\,\delta V\,/\,V_o)]. \tag{6.36}$$

The last two terms in this expression are independent of the deformation and will not appear in the entropy difference ΔS with respect to a particular reference state. Taking the reference state to be $\Lambda_x = \Lambda_y = \Lambda_z = 1$, Eq. (6.36) leads to

$$S = -(k_B n/2)[\Lambda_x^2 + \Lambda_y^2 + \Lambda_z^2 - 3 - \ln(\Lambda_x\Lambda_y\Lambda_z)]. \tag{6.37}$$

Because there is no internal energy scale associated with the chains, the free energy ΔF is just $-T\Delta S$, and

$$\Delta F = (k_B Tn/2)[\Lambda_x^2 + \Lambda_y^2 + \Lambda_z^2 - 3 - \ln(\Lambda_x\Lambda_y\Lambda_z)]. \tag{6.38}$$

The appeal of having an analytical expression for ΔF should not cause us to forget the physical and numerical approximations in its derivation. In particular, opinion is divided regarding the presence of the logarithmic term (see Flory, 1976; Deam and Edwards, 1976) and we will work under the constraint of constant volume with $\Lambda_x\Lambda_y\Lambda_z = 1$. Thus

$$\Delta F = (k_B Tn/2) \bullet (\Lambda_x^2 + \Lambda_y^2 + \Lambda_z^2 - 3). \qquad \text{(constant volume)} \tag{6.39}$$

6.3.4 Elastic moduli

Elastic moduli can be obtained from Eq. (6.39) by imposing specific deformations. For illustration, we extract the shear modulus from ΔF by performing a pure shear, with $\Lambda_x = \Lambda = 1/\Lambda_y$ and $\Lambda_z = 1$, yielding

$$\Delta F = (k_B Tn/2) \bullet (\Lambda^2 + 1/\Lambda^2 - 2). \qquad \text{(pure shear)} \tag{6.40}$$

The expression in the brackets can be rewritten as $(\Lambda - 1/\Lambda)^2$, which becomes $4\delta^2$ when $\Lambda = 1 + \delta$ and δ is small. Dividing Eq. (6.40) by the network volume, which has not changed from its undeformed value V_o by shear, the free energy density $\Delta\mathcal{F}$ for the pure shear mode is

$$\Delta\mathcal{F} = 2\delta^2 \rho k_B T, \tag{6.41}$$

where $\rho = n\,/\,V$ is the density of chains. To find the shear modulus, we evaluate $\Delta\mathcal{F}$ for isotropic materials in Eq. (6.17) under pure shear conditions corresponding to $\Lambda = 1 + \delta$, namely $u_{xx} = \delta$, $u_{yy} = -\delta$, $u_{zz} = 0$, to obtain

$$\Delta\mathcal{F} = 2\delta^2\mu. \tag{6.42}$$

A comparison of Eqs. (6.41) and (6.42) gives the long-awaited result

$$\mu = \rho k_B T. \tag{6.43}$$

Although our expression for the free energy of random networks qualitatively describes their behavior under compression and tension, we should not lose sight of the fact that several approximations were used in its derivation so that the expression may not be completely accurate. For example, our approach omits self-avoidance within a chain, repulsive interactions between chains, and the effects of chain entanglement (see references in Everaers and Kremer, 1996). Although computer simulations of random networks (for example, see Plischke and Joos, 1998) confirm the density-dependence of the shear modulus predicted by Eq. (6.43), they have not yet verified the numerical prefactor. Nevertheless, the approach to random networks presented above does capture their most important features at modest chain densities.

6.4 Network percolation and failure

A solid object responds elastically when placed under small tension or compression, in the sense that the object returns to its equilibrium density and original shape once the stress is removed. Under similar conditions, a fluid body returns to its equilibrium density, but its shape may have changed because a fluid cannot resist shear. Under moderate stress, however, even a solid body may deform permanently; for example, a ductile metal may stretch in the direction of a uniaxial tension and consequently assume a different shape at equilibrium density when the tension is no longer present. At higher tensions still, the deformation may become non-uniform and the material fails: a solid may fracture and holes may form in a fluid or a network. The strain at failure varies with composition, ranging from just a few percent for some solids and two-dimensional fluid membranes to many tens of percent in other systems.

Once formed, a crack in a fractured solid may follow a complex pathway, depending on the graininess of the solid and the inhomogeneity of the stress field. The simplest cracks can be classified into three modes as illustrated in Fig. 6.12 for a plate of modest thickness. Mode I is also known as the opening mode, mode II is the forward shear mode and mode III is

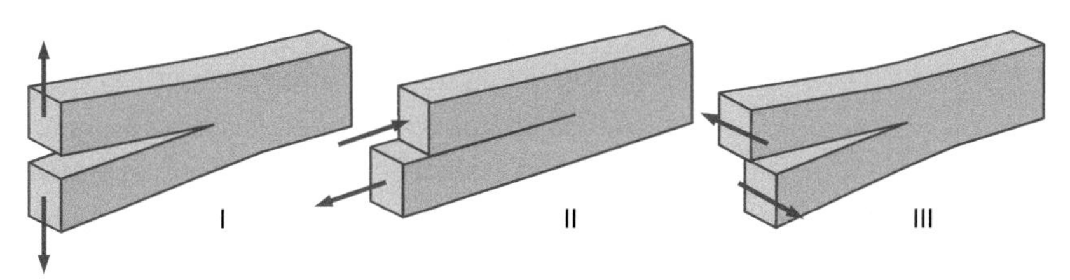

Fig. 6.12 Three simple modes of crack formation in solids: opening mode (I), forward shear mode (II) and antiplane shear mode (III).

antiplane shear mode. The crack in mode II is along the direction of the shear axis whereas it is transverse to the shear axis in mode III.

The mathematical description of the formation and propagation of cracks in solids goes back a century, although it is by no means a closed subject. The starting point for many treatments is the Griffith (1921) solution for the opening mode in brittle solids. A number of assumptions must be made in order to keep the mathematics tractable: among other things, the material adjacent to the crack is taken to be at zero stress and undeformed (in mode I). To within a factor of two (depending on the boundary conditions), the critical stress for fracture σ_c is given by

$$\sigma_c \sim (2Y\gamma_s / \pi a)^{1/2}, \tag{6.44}$$

where Y is Young's modulus and γ_s is the surface energy per unit area of the solid in bulk. The crack is isolated in an infinite medium and has length $2a$. For illustration, the work per unit thickness associated with creating the crack is $4\gamma_s a$ because the crack has an upper and lower surface. For a more complete description of fracture in solids, see Hellan (1984) or Sanford (2003).

The Griffith expression for the formation of a crack in an infinite medium assumes that the medium is truly brittle, like a glass or ceramic. When applied to softer materials that are capable of plastic deformation, the theory grossly underestimates the threshold stress for crack formation by up to two or three orders of magnitude. The interested reader is directed elsewhere (for example, Chapter 7 of Sanford, 2003) for a discussion of ways of modifying Eq. (6.44) to incorporate plasticity.

Commonly, the most important structural elements in cells are soft networks such as the cytoskeleton and two-dimensional fluids such as the plasma membrane. The failure of these components has a somewhat different description than the failure of a solid such as bone. The rupture of membranes such as the lipid bilayer is treated at length in Section 7.5; here, we present a brief overview of such rupture on a microscopic length scale.

The structure of the lipid bilayer is shown in cross section in Fig. 1.6: two monolayers of dual-chain lipid molecules form a sandwich with the polar head group of each molecule facing into the aqueous medium surrounding the bilayer, and the hydrocarbon chains of the lipids lined up in the interior of the membrane. When it is subject to a tensile stress in the bilayer plane, the hydrophilic head groups of the lipid molecules are pulled further from each other, exposing more of the hydrophobic interior. Of course, this reorientation is energetically unfavorable, and the resulting increase in potential energy confers stretching resistance to the membrane. After a few percent strain in the membrane plane, it becomes energetically more favorable for a pore to form in the bilayer rather than for its density to continue falling as it is stretched. The pore permits the aqueous regions on each side of the membrane to fuse, creating a passageway for water-soluble compounds to traverse the bilayer without encountering the hydrocarbon chains in its interior. The in-plane area of the combined membrane + pore system rises because of the presence of the pore, with the result that the density of the lipid molecules relaxes back toward its equilibrium value even though the system remains under tension. The area of the pore rises with increasing tensile stress in the bilayer plane.

The mathematical treatment of pore formation is provided in Section 7.5. The pore should not be regarded in a cookie-cutter sense as a flat segment chopped cleanly out of the bilayer while leaving neighboring lipid molecules untouched. Rather, the lipid molecules reorient themselves around the pore, creating a smoother surface in which hydrophilic head groups of the lipids line the aqueous region within the pore, as well as the external aqueous environment. Of course, there is an energy penalty associated with the lipid rearrangement at the pore, and this can be categorized as a line tension or energy per unit length of the pore boundary, such that the longer the perimeter of the pore, the higher the energy penalty.

The creation of holes in a membrane under stress has certain similarities to the fracture of a solid: the stress has been relieved in a spatially inhomogeneous way after the material has failed. For a solid object, a crack has appeared and the shape is deformed; for a membrane, the total membrane area has been enlarged by the presence of a hole. However, the shape, location and even the number of pores may fluctuate if the membrane is a two-dimensional fluid. In addition, we point out that pores might arise in fluid membranes even in the absence of a tensile stress: at higher temperatures, thermal fluctuations may provide sufficient energy to overcome the activation energy for creating a pore.

We now turn to networks, where the failure mechanism involves the breaking of many individual bonds. In a cell, this can occur under tensile stress, similar to situations that we have just encountered, or it can occur when individual bonds are cleaved by external agents. The failure mechanism can be visualized by returning to the network shown in Fig. 6.10(a). In constructing this network, we assumed that each initial chain is sufficiently long that it is welded at many points to neighboring chains. What happens when the density of welds decreases to the point where there is less than a single weld per chain, on average? Clearly, if the density is so low that many chains are physically separated and sustain no welds, then we may not even have a connected network. This phenomenon has been studied in a variety of contexts, including the gelation of polymers (see Flory, 1953, Chapter IX) and goes under the generic name of percolation (a good introductory text is Stauffer and Aharony, 1992).

Consider the two-dimensional square lattice in Fig. 6.12, in which bonds have been placed between lattice sites in a random fashion. The population of bonds on the lattice as a whole can be described by a parameter p, which is the probability that two nearest-neighbor sites are connected by a bond.

Table 6.1 Percolation threshold for a selection of regular lattices in two and three dimensions

Lattice	z	p_C (bond)	p_C (site)	p^* (bond)
Two dimensions				
honeycomb	3	0.653	0.696	1
square	4	0.500	0.593	1
triangular	6	0.347	0.500	2/3
Three dimensions				
simple cubic	6	0.249	0.312	1
body-centred cubic	8	0.180	0.246	3/4
face-centered cubic	12	0.119	0.198	1/2

Notes. "Bond" and "site" refer to different measures of defects in the lattice, as described in the text; z is the maximum coordination of a site and p^* is the rigidity percolation threshold (for bonds) as calculated from Eq. (6.47). Data from Chapter 2 of Stauffer and Aharony (1992).

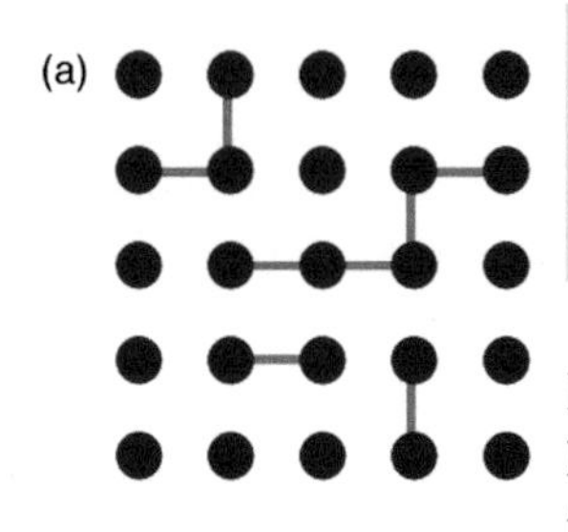

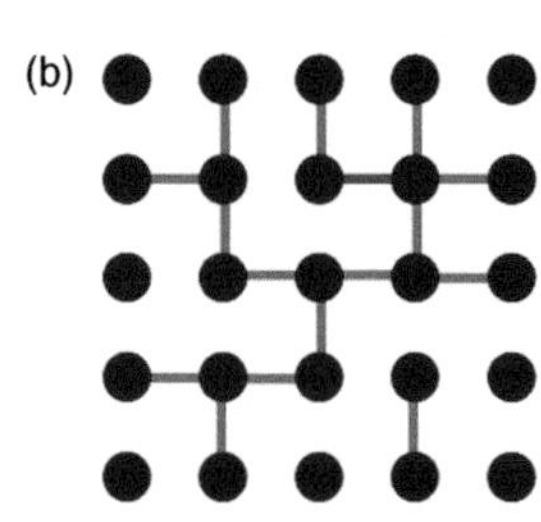

Fig. 6.13

Percolation phenomena on a square lattice in two dimensions. For the purpose of this example only, we ignore bonds along the edge of the lattice, so that the maximum allowable number of bonds is 24. Configuration (a), with bond occupation probability $p = 8/24 = 1/3$, is below the connectivity percolation threshold p_C, while configuration (b), with $p = 16/24 = 2/3$, is above p_C, resulting in at least one path completely traversing the network.

If the lattice has a full complement of bonds, $p = 1$, and each site on the lattice has four-fold connectivity. If no bonds are present, p obviously vanishes. The sample configurations in the figure have values of p intermediate between 0 and 1.

One attribute of a configuration is whether a continuously connected path traverses the lattice. Such a path is absent in configuration (a) of Fig. 6.13 (with $p = 1/3$), but present in configuration (b) (with $p = 2/3$). For infinite systems, the existence of a connecting path across the lattice is a discontinuous function of p and there is a well-defined value of p, called the connectivity percolation threshold p_C, below which there is no connecting path. The connectivity threshold is lattice-dependent, as can be seen from the sample of values in Table 6.1. The two columns in the table labelled *bond* and *site* refer to which element of the lattice is used to parametrize its defects. We have so far discussed bond depletion, where p is the fraction of occupied bonds. However, one can also use site occupation as a parameter, bonds then being added between all occupied nearest neighbor sites. As they should be, the occupation probabilities at percolation for these two measures are similar, but not identical. Note that the percolation threshold decreases with increasing z, the maximum number of bonds that can be connected to each site.

The elastic moduli also depend upon p: for $p < p_C$, the network is sufficiently dilute as to be disconnected, so that both μ and K_A vanish. However, the presence of a connecting path at $p > p_C$ may not guarantee that the network can resist deformation. Even though p is safely above p_C in configuration (b), there is only one connecting path from left to right – which is not a lot of paths to provide shear resistance. In fact, even at $p = 1$, square networks do not resist simple shear, as established in Chapter 5. It is found

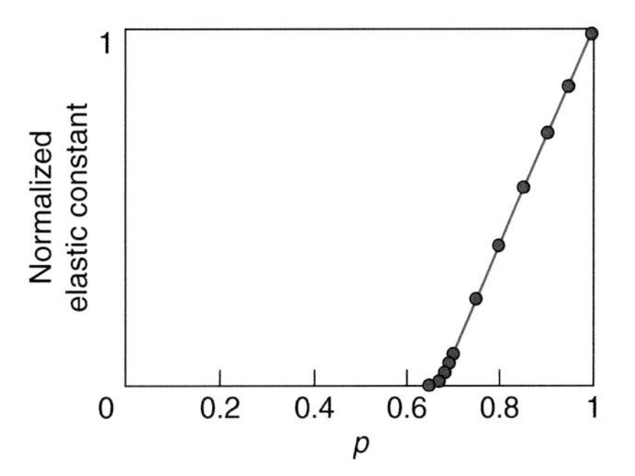

Fig. 6.14 Behavior of an elastic constant for a bond-diluted triangular network. For this network, $p_C = 0.325$, well below the apparent value of p_R (data reprinted with permission from Feng *et al.*, 1985; ©1985 by the American Physical Society).

(Feng and Sen, 1984) that the elastic moduli vanish below a well-defined threshold, referred to as the rigidity percolation threshold p_R. At zero temperature, p_R may be considerably larger than p_C if the forces between network nodes are *central* forces, i.e. they are directed only along the bonds and do not depend upon bond angles (Feng *et al.*, 1985). Once the percolation threshold has been crossed from below, the elastic moduli rise with a power-law dependence on $p - p_R$ towards their values for a fully connected network, as shown in Fig. 6.14. As $p \to 1$, an elastic modulus C is approximately described by

$$C(p) / C(p = 1) \cong (p - p_R) / (1 - p_R) \qquad (p \to 1), \tag{6.45}$$

which vanishes at p_C and is equal to unity at $p = 1$.

As we are reminded by Thorpe (1986), the study of rigidity goes back more than a century to Maxwell (1864), who developed a simple model for the rigidity percolation threshold based on bond-induced constraints on the motion of lattice points. Let's consider a regular lattice with N sites embedded in a d-dimensional space. Each site can be linked to its z nearest neighbors with a probability p. If they are unconstrained, the N sites have Nd variables describing their possible motion. The presence of a bond between sites provides one constraint on the number of ways the sites can move; we will say that the constraint reduces the number of floppy modes of the system. For example, two particles in a plane may move in four ways if unconstrained, but only three ways if they are joined by a rigid bar (two translations in the plane, and one rotation).

How many constraints are present at an occupation p? The number of bonds linked to a given vertex is zp on average; however, the total number of bonds is not Nzp, because each bond is shared by two sites. Thus, the number of bonds or constraints is $Nzp/2$, reducing the number of floppy modes to

$$[\mathit{floppy\ modes}] \cong Nd - Nzp/2. \tag{6.46}$$

This equation is an approximation because some bonds are redundant, i.e. their presence does not reduce the number of floppy modes. The structure becomes rigid when the number of floppy modes vanishes, which occurs at a probability p^* of

$$p^* = 2d / z, \tag{6.47}$$

according to Eq. (6.46). As indicated by the right-hand column in Table 6.1, p^* is always larger than the observed p_C, consistent with the idea that rigidity requires more bonds than does connectivity. Further, Eq. (6.47) is in good agreement with p_R obtained by simulations at zero temperature (Feng and Sen, 1984; Feng *et al.*, 1985).

At finite temperature, however, nodes can depart from their zero-temperature lattice sites and interact entropically, generating a resistance to deformation. Computer studies of defective networks at finite temperature argue that entropic repulsion is enough to restore elasticity for $p > p_C$, such that $p_R = p_C$ even for central-force networks at non-zero temperature (Plischke and Joos, 1998; Farago and Kantor, 2002; Plischke, 2006). The mechanical behavior of random fiber networks with bond-bending forces is explored further by Heussinger and Frey (2006) and Plischke (2007).

6.5 Semiflexible polymer solutions

The networks that we have investigated thus far have permanent connection points linking together different structural elements such as polymer chains. If the number of connection points per chain is high enough, at least some of the elastic moduli are non-vanishing, depending on the nature of the connectivity. Consider now a situation like Fig. 6.15, in which there is a reasonable density of chains, but no permanent cross-links. Subjected to a shear, this network behaves like a fluid if allowed to relax completely, in the sense that the filaments ultimately slide around each other, and the network does not recover its original shape when the stress is released. Raising the filament density causes the network to become ever more entangled, such that its fluid behavior emerges only at ever longer time scales. One can imagine that, at sufficiently high density, the fluid response of the network takes so long that it behaves like an elastic solid on time scales of interest.

Solutions of polymer chains are examples of viscoelastic materials, behaving like elastic solids on short time scales and viscous fluids on long time scales. Their mechanical properties depend upon such characteristics as the polymer contour length and density, as we discuss momentarily. The importance of viscoelasticity to biofilaments in the cell is the subject of current research: chemically pure actin, intermediate filaments and microtubules can be studied in solution at concentrations comparable to

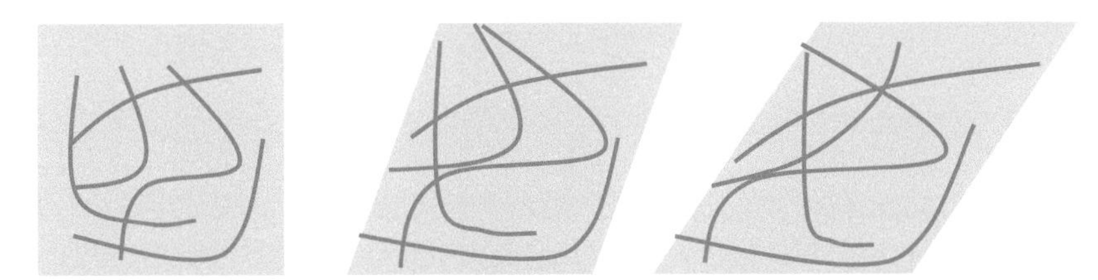

Fig. 6.15 Application of a shear stress (from left to right) to a network of short filaments. If the filaments are not cross-linked the network responds like a fluid, when observed on long enough time scales.

those in the cell; the properties of such solutions are presented in Section 4.5. In this section, we describe universal features of viscoelastic materials, such as frequency-dependent elastic moduli, and then report without proof several predictions of viscoelastic behavior from model systems.

6.5.1 Classification of polymer solutions

The viscoelastic characteristics of a polymer solution depend upon its concentration c and the ratio of the polymer contour length L_c to its persistence length ξ_p. For simplicity, we present the generic features of two polymeric systems: one with very flexible chains ($L_c >> \xi_p$) and a second with moderately stiff rods ($L_c << \xi_p$). Their solutions can be classified as dilute, semi-dilute and concentrated, although the boundaries between these categories are not necessarily sharp. The behavior of a solution is uniform within a given concentration regime, and may change smoothly or abruptly in crossing a transition region. Cartoons of polymer configurations expected in the three concentration regions are drawn in Fig. 6.16; many biological systems of interest fall into the semi-dilute category.

Individual polymers can rotate without obstruction in the dilute regime, which spans a concentration range determined by the effective volume per polymer. For a random coil, the effective volume is about $4\pi R_g^3/3$, where R_g is the radius of gyration, defined as $R_g^2 = N^{-1} \Sigma_i (\mathbf{r}_i - \mathbf{r}_{cm})^2$ for a discrete chain of N segments located at positions $\mathbf{r}_i$ (the center-of-mass position is at $\mathbf{r}_{cm} = N^{-1}\Sigma_i \mathbf{r}_i$). For instance, if the chain is uniformly distributed within a sphere of radius R, the radius of gyration is $(3/5)^{1/2}R$. In contrast, to rotate freely of its neighbors, a stiff rod of length L_c must have an effective volume much larger than that of a random chain with the same L_c: the centers of mass of two hard rods are excluded from a volume of the order L_c^3. In the dilute regime, then, the average volume occupied by a single filament (stiff or floppy) must be greater than the effective volume that we have just estimated. Equivalently, the concentration in the dilute regime must be *less* than a transition number density ρ^* of about

$$\rho^* \sim (4\pi R_g^3/3)^{-1} \text{ (floppy)} \qquad \text{or} \qquad \rho^* \sim L_c^{-3} \text{ (stiff)}. \qquad (6.48)$$

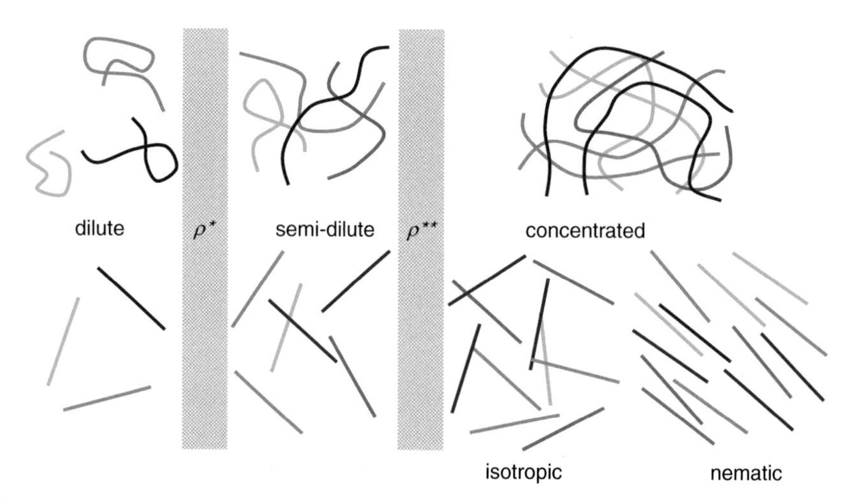

Fig. 6.16 Generic behavior of a solution of polymer chains as a function of chain density, which increases from left to right. Densities at the transitions between regions are indicated by ρ^* and ρ^{**}.

At the other extreme are concentrated solutions, wherein the individual polymeric units strongly overlap and intertwine. For flexible polymers, these systems may have the appearance of a polymer melt; the concentrated regime occurs at number densities above a transition density ρ^{**} of

$$\rho^{**} \sim v_{ex}/b^6 \text{ (floppy)}, \tag{6.49}$$

where v_{ex} is the approximate volume occupied by a chain segment, and b is the average segment length. The proof of this result is somewhat involved; we refer the interested reader to Section 5.1 of Doi and Edwards (1986). The persistence lengths of the filaments discussed in Section 6.6 are all sufficiently long ($\xi_p > 1$ μm) that our primary interest here is in the properties of semiflexible rods, rather than floppy chains.

The transition concentration to the high-density region for rigid rods depends upon their diameter D_f, where the subscript f denotes *filament*. Suppose that one rod, A, lies along the y-axis while another rod, B, is free to move, so long as its axis is parallel to the x-axis, as illustrated in Fig. 6.17. These rods intersect each other if the center of rod B lies within the box shown in the figure. That is, the centers are excluded from a volume equal to $2D_fL_c^2$ if the rods are forbidden from intersecting. Now, angular averages of rod orientations must be performed to obtain an exact expression for the excluded volume, but the result must be of order $D_fL_c^2$, corresponding to a transition number density of

$$\rho^{**} \sim (D_fL_c^2)^{-1} \text{ (stiff)}. \tag{6.50}$$

For densities not too far above ρ^{**}, the rods are randomly oriented, and the system is isotropic. However, rigid rods may order into a so-called

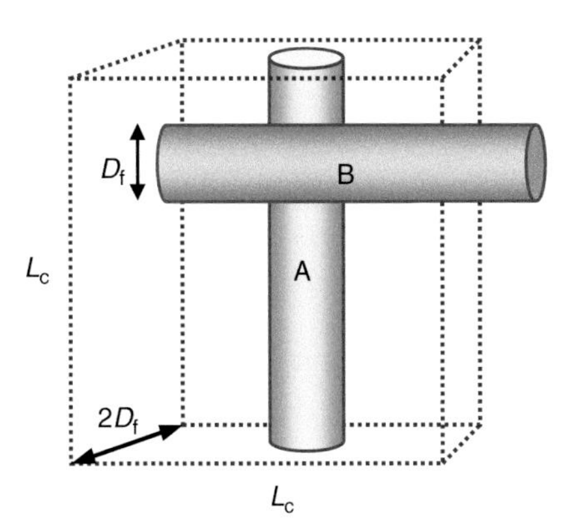

Fig. 6.17 Configurations of two identical rods of length L_c and diameter D_f. Rod A is fixed, while rod B moves with its orientation always horizontal. If the rods cannot overlap, the center of rod B is excluded from the region indicated by the box of dimension $L_c \times L_c \times 2D_f$.

nematic phase at high densities, in which they display long-range orientational order even though their positions are not regularly spaced as they might be on a crystal lattice. The differences between isotropic and nematic phases are illustrated in Fig. 6.16. Whether the nematic phase is achievable depends upon the anisotropy of the interaction between rods. Considering hard rods of diameter D_f and length L_c, Onsager (1949) showed that the nematic phase exists only for L_c/D_f greater than approximately 3 (see also Vertogen and de Jeu, 1988, Chapter 13). Note that the highest number density permitted for hard rods is roughly $(L_c D_f^2)^{-1}$.

6.5.2 Storage and loss moduli

How do we formally describe a material whose response to an imposed stress or strain is time-dependent? Let's consider a fixed-strain measurement in which a shear strain u_{xy} is imposed upon an object, creating a stress σ_{xy}. If the system were an elastic solid, the stress would be proportional to the strain, $\sigma_{xy} = \mu u_{xy}$, with the shear modulus μ fixed for the duration of the measurement. However, for a viscoelastic material like an uncross-linked polymer solution, the stress slowly decays to zero from its initial value as the polymers untangle and the network relaxes. Thus, the effective elastic modulus is time-dependent, and the stress–strain relationship can be written as

$$\sigma_{xy} = \int_{-\infty}^{t} G(t - t')(\mathrm{d}u_{xy} / \mathrm{d}t')\, \mathrm{d}t' \tag{6.51}$$

where $\mathrm{d}u_{xy}/\mathrm{d}t$ is the shear rate and $G(t - t')$ is the relaxation modulus, having the same units as the shear modulus (see Ferry, 1980, for alternate formulations of this equation). Equation (6.51) implies that the response

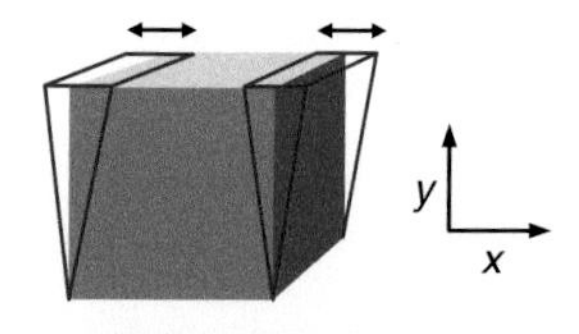

Fig. 6.18

Application of an oscillating shear strain $u_{xy}(t)$ to a three-dimensional system. The undisturbed system is indicated by the shaded region, and the time-dependent oscillation is shown by the wire outline.

of the system is additive in time: small changes in stress result from small changes in strain, which are integrated to yield the total stress.

A variety of experimental configurations can be employed to measure the relaxation modulus. Commonly, the frequency-dependence of the elastic response is measured, rather than its time-dependence. In experiments illustrated in Fig. 6.18, the system is subjected to an oscillating strain of the form

$$u_{xy}(t) = u_{xy}{}^{\mathrm{o}} \sin\omega t, \tag{6.52}$$

where $u_{xy}{}^{\mathrm{o}}$ is the amplitude of the strain, ω is its frequency, and the initial value of the strain has been set to zero for convenience. The corresponding strain rate is simply

$$\mathrm{d}u_{xy} /\mathrm{d}t = \omega\, u_{xy}{}^{\mathrm{o}} \cos\omega t. \tag{6.53}$$

Equations (6.52) and (6.53) can be substituted into the definition of the relaxation modulus to obtain

$$\begin{aligned}\sigma_{xy}(t) &= \int_0^\infty G(\tau)\omega u_{xy}{}^{\circ}\cos[\omega(t-\tau)]\,\mathrm{d}\tau \\ &= u_{xy}{}^{\circ}\left[\omega\int_0^\infty G(\tau)\sin\,\omega\tau d\,\tau\right]\sin\,\omega\tau \\ &\quad + u_{xy}{}^{\circ}\left[\omega\int_0^\infty G(\tau)\cos\,\omega\tau d\,\tau\right]\cos\,\omega\tau,\end{aligned} \tag{6.54}$$

where the integration variable has been changed to $\tau = t - t'$ and we have used the identity $\cos(\alpha - \beta) = \sin\alpha\,\sin\beta + \cos\alpha\,\cos\beta$. The terms in the square brackets are functions of frequency, but their explicit time-dependence has been integrated away. We replace these terms with two new moduli by writing

$$\sigma_{xy} = u_{xy}{}^{\mathrm{o}}G'(\omega)\,\sin\omega t + u_{xy}{}^{\mathrm{o}}G''(\omega)\,\cos\omega t, \tag{6.55}$$

where $G'(\omega)$ is the shear storage modulus and $G''(\omega)$ is the shear loss modulus.

The physical meaning of the moduli can be seen by considering their phase with respect to the applied strain, $u_{xy}(t) = u_{xy}{}^{\mathrm{o}}\sin\omega t$. First, we recast Eq. (6.55) in terms of the applied strain and its rate of change, Eq. (6.53),

$$\sigma_{xy} = G'(\omega)u_{xy}(t) + G''(\omega)\bullet(\mathrm{d}u_{xy}/\mathrm{d}t)/\omega. \tag{6.56}$$

If only the storage modulus is non-vanishing (i.e., $G'' = 0$), the stress is directly proportional to the strain in Eq. (6.56) and G' appears like a frequency-dependent shear modulus μ. That is, G' is a measure of the elastic energy stored in and retrieved from the system. Conversely, if only the loss modulus is non-vanishing ($G' = 0$), the stress is proportional to the

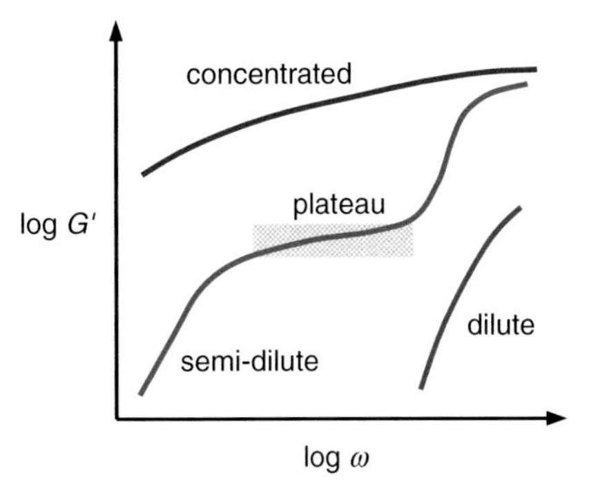

Fig. 6.19 Schematic representation of the storage modulus $G'(\omega)$ as a function of frequency ω for three sample concentrations: dilute, semi-dilute and concentrated. Physical systems may range between the examples shown here.

rate of strain and has the same generic form as Newton's law for a viscous fluid, $\sigma_{xy} = \eta(\mathrm{d}u_{xy}/\mathrm{d}t)$ where η is the viscosity. Thus, G''/ω appears as a frequency-dependent dynamic viscosity and G'' should vanish linearly with ω as $\omega \to 0$.

If the material is truly a fluid, then the storage modulus G' should decrease as the frequency decreases, ultimately vanishing at $\omega = 0$. While the observed behavior of G' is monotonic with frequency, it is not without structure, as the schematic plot in Fig. 6.19 illustrates. Dilute solutions display the smallest storage moduli, as one might expect from their near-fluid behavior. At the other extreme, concentrated polymeric solutions have storage moduli typical of hard plastics, of the order 10^9 J/m^3. Semi-dilute solutions range between these extremes, displaying little resistance to shear at small ω (long times) but a high resistance at large ω (short times). At intermediate frequencies, they may display slowly varying behavior associated with polymer entanglements, referred to as the plateau region.

At each frequency range, different shear relaxation modes are allowed during the oscillation of the applied strain. At short times, only local molecular rearrangements may take place, while over long times, large-scale motion of the polymer as a whole is permitted. Between these time frames, the polymers move on intermediate length scales, which may be comparable to the distances between entanglement points for semi-dilute solutions. By entanglements, we mean loops and even knots that might occur as polymers thread their way around one another. Thus, over a range of frequencies, entanglements behave almost as fixed cross-links, which become unstuck only when viewed on longer time scales. This results in a plateau at intermediate frequency, where the solution behaves like a network of (temporarily) fixed cross-links, resulting in a frequency-independent shear resistance determined by the density of cross-links.

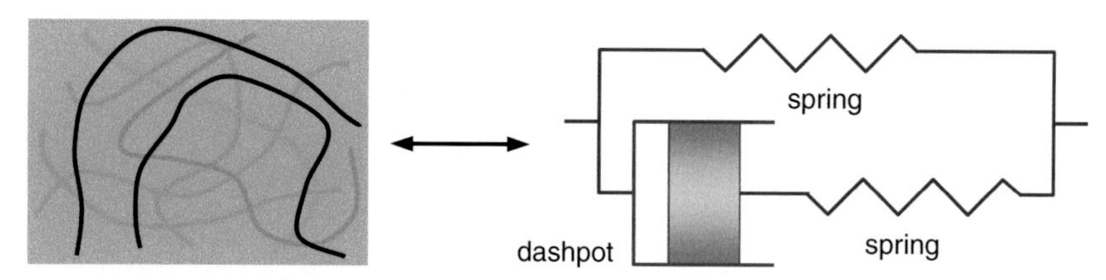

Fig. 6.20 Effective representation of a polymer solution in terms of a simple mechanical model involving springs and viscous dashpots (see Elson, 1988).

6.5.3 Model results for $G'(\omega)$

The formalism that we have developed above employs two frequency-dependent moduli to represent the viscoelasticity of a polymeric solution at fixed concentration. A number of simple mechanical models have been advanced for understanding the measured behavior of $G'(\omega)$ and $G''(\omega)$, as reviewed in Ferry (1980) (see Elson, 1988, for biological applications). For example, the spring-and-dashpot representation shown in Fig. 6.20 permits the ω-dependence of the moduli to be interpreted in terms of effective spring constants and the viscosity of the system as a whole. However, our interest lies in relating the experimental measurements of G' and G'' to the microscopic structure of the material, and this has been the subject of extensive theoretical investigation.

The plateau region of G' at intermediate frequencies tends to cover time scales of importance in the cell. Now, we have shown for networks of floppy chains with permanent cross-links that the shear modulus μ is proportional to the number density ρ (or to the concentration c) of chains: $\mu = \rho k_B T$. The plateau region is a frequency range in which entanglements behave like quasi-permanent cross-links. Thus, if a cellular network consists of floppy chains with a persistence length much shorter than the distance between entanglements, we would expect G' to be proportional to c^1 or ρ^1 (see also Ferry, 1980, Chapter 13). However, for most cellular filaments except spectrin, the persistence length is comparable to, if not much larger than, the diameter of a typical cell, and the filaments appear as gently curving ropes or rods on cellular length scales. Is the concentration-dependence of G' in the plateau region affected by the rigidity of the filament?

Several studies have addressed this question. Decades ago, Kirkwood and Auer (1951) showed that in a particular representation of a rod, the storage modulus obeyed

$$G'(\omega) = (3\rho k_B T/5)\ \omega^2\tau^2 / (1 + \omega^2\tau^2), \tag{6.57}$$

where τ is a relaxation time depending on the rod geometry and solution viscosity (see also Problem 6.29 and Ferry, 1980, Chapter 9). Although this expression does not contain an intermediate plateau, nevertheless G' rises

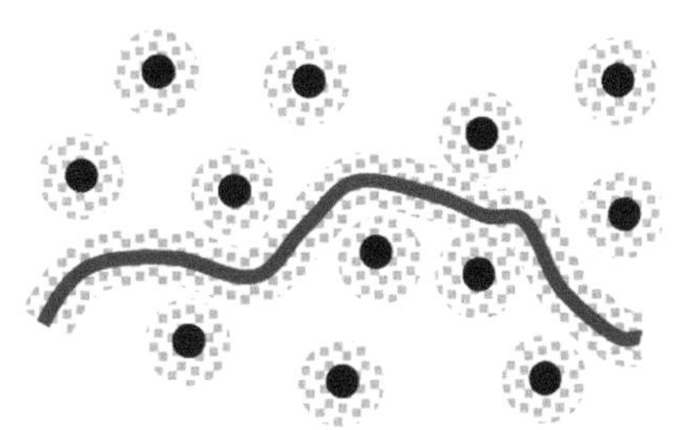

Fig. 6.21 Cartoon of a polymer confined to an effective tube created by its neighbors (indicated by disks). Referred to as the tube model (Doi and Edwards, 1986), this representation has been used to predict the frequency and concentration-dependence of G' (Isambert and Maggs, 1996; Morse, 1998).

with ω to reach an asymptotic value of $(3/5)\rho k_B T$ at high frequency; i.e. in the elastic regime, G' scales like ρ^1. More recently, Isambert and Maggs (1996) and Morse (1998) find that G' is proportional to $\rho^{7/5}$ for moderately flexible filaments whose appearance in solution has the snake-like form of Fig. 6.21 (see also Gittes and MacKintosh, 1998). MacKintosh *et al.* (1995) find that G' in the plateau region grows faster than the square of the density, namely $\rho^{11/5}$ for polymer solutions and $\rho^{5/2}$ for densely cross-linked gels, by considering the effects of tension on rods that are close to straight.

These three calculations emphasize the entropic contribution to network elasticity. In contrast, Satcher and Dewey (1996) examine the elasticity of networks where flexible network elements are joined at right angles to form T-junctions. Deformations of the network bend the elements perpendicular to their length, and the elasticity of the network reflects the energy required to bend the elements. In this approach, the shear modulus is predicted to grow as ρ^2. The concentration-dependence of G' predicted by the above models can be tested experimentally, as we describe in the following section. The moduli of networks of filaments with anisotropic elasticity have been obtained in two dimensions by computer simulation (Wilhelm and Frey, 2003).

6.6 Rheology of cytoskeletal components

In this section, we investigate the rheology of polymer networks and solutions in three dimensions (rheology is the study of the flow and deformation of matter). The systems on which we report are the principal filaments of the cytoskeleton – actin, intermediate filaments, and microtubules. In the case of actin, both pure and cross-linked networks have been investigated. We also provide references to measurements made on the cytoplasm of several cell types, although the findings do not have a complete interpretation owing to the chemical and structural heterogeneity of the cytoplasm. We first review some results from Sections 6.2–6.5 on viscoelasticity, and provide a brief synopsis of several techniques used in rheological studies, before turning our attention to the experimental findings themselves.

6.6.1 Storage and loss moduli

Isotropic materials in three dimensions are characterized by the compression modulus K_V and shear modulus μ. When constructed with permanent connections, networks of floppy polymers obey $\mu \sim \rho k_B T$, where ρ is the chain number density (i.e. the number of chains per unit volume), a single chain being defined as a chain segment running between two junctions. In the absence of permanent cross-links, polymer solutions display a time-dependent elasticity characteristic of viscoelastic materials. When subject to an oscillating strain u_{xy} of angular frequency ω, the stress σ_{xy} developed in the solution obeys $\sigma_{xy}(t) = G'(\omega)u_{xy}(t) + G''(\omega) \bullet (du_{xy}/dt)/\omega$, where $G'(\omega)$ and $G''(\omega)$ are the shear storage and shear loss moduli, respectively. As $\omega \to 0$, G' becomes the shear modulus and G''/ω becomes the viscosity.

The behavior of G' and G'' at intermediate ω depends upon the number density ρ. Solutions of moderately inflexible filaments of contour length L_c and diameter D_f can be categorized according to concentration as dilute, semi-dilute or concentrated, the regions being separated by their transition densities $\rho^* \sim L_c^{-3}$ and $\rho^{**} \sim (D_f L_c^2)^{-1}$. In the dilute regime, G' is close to that of the solvent, while in the concentrated regime, it is similar to the shear modulus of a hard plastic. In semi-dilute solutions, $G'(\omega)$ first increases with ω until reaching a plateau, where polymer entanglements act as transient connection points. Beyond the plateau region, $G'(\omega)$ rises again to reach its asymptotic value in the 10^9 J/m^3 range. The density-dependence of G' at the plateau depends on the nature of the network: floppy chains with permanent connections increase as ρ^1, while semiflexible filaments with no cross-links rise like $\rho^{7/5}$. Solutions of rigid rods do not exhibit a plateau at intermediate frequency, but rather rise smoothly towards their high-frequency limit, where G' is proportional to ρ^1.

6.6.2 Measurements of viscoelasticity

Several methods for measuring the viscoelastic properties of polymer solutions can be applied to biofilaments, although care must be taken to ensure that the technique does not disrupt the fragile biological network being measured. The procedures can be assigned to one of two broad categories involving either the observation of a response to an externally applied stress, or the measurement of thermal fluctuations in a sample. Methods of applying a stress or strain to a system include conventional viscometers; the cone-and-plate device in Fig. 6.22 is but one of many geometries that

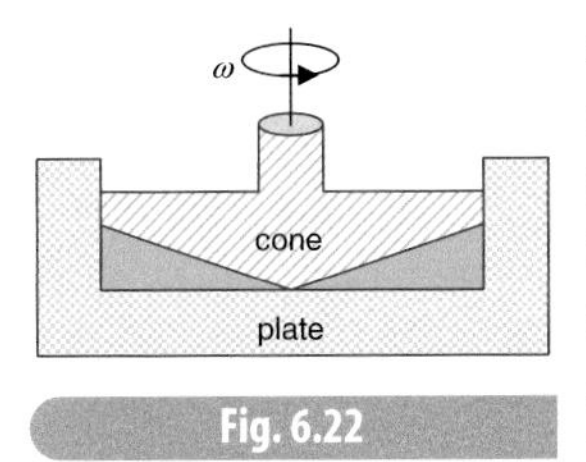

Fig. 6.22

Schematic cross section of a cone-and-plate device for measuring viscoelastic properties. The sample occupies the shaded region between the externally driven cone, and the plate (see Tran-Son-Tay, 1993).

are in use (see Ferry, 1980, Chapter 5). The cone is driven at a frequency ω to impose a strain on the sample, which fills the gap between the cone and the plate. A variation of this method has been developed for observing the shear response of cells in a solution: the cone and plate are counter-rotating and made of optically transparent materials, permitting the deformation of the cells to be observed through a microscope (see review by Tran-Son-Tay, 1993).

Viscometers such as the cone-and-plate device measure the bulk properties of systems and require sample sizes much larger than a typical cell volume. In contrast, deformations of an individual cell and its internal components can be made visible by the injection of magnetic or fluorescent beads into its cytoplasm. Such beads might be simply passive markers aiding the analysis of strain, or may contain magnetic material permitting the bead to be manipulated by a constant or oscillating external magnetic field (Crick and Hughes, 1950; Ziemann *et al.*, 1994, and references therein). Analysis of video recording of the beads through a microscope fixes their position to a precision of better than about 0.05 μm, compared to a bead diameter of about 2–3 μm. Although such beads are not tiny by cell standards, they are certainly small enough to be sensitive to local inhomogeneities of the cytoskeleton.

Earlier, we introduced the relation between thermal fluctuations and elasticity in the context of filaments and two-dimensional networks. The elastic characteristics of the cytoskeleton also can be determined by observing thermal fluctuations in the position of marker beads in the network, thus reducing disturbances to the material caused by the measuring device. Several techniques based on fluctuations have been developed, such as the recording of single bead motion (Schnurr *et al.*, 1997) or measuring interference of scattered light from an ensemble of beads (Palmer *et al.*, 1999). These procedures permit the study of oscillations at higher frequency, into the kilohertz range, than may be possible with mechanical rheometers. Bead diameters of 0.5–5 μm have been used, permitting the study of micron-size domains of a sample.

6.6.3 Chain densities in actin networks

Filament-forming proteins such as actin can be extracted from cells, purified, and then repolymerized into filaments and networks in a controlled fashion. Let us determine the filament density of a reference solution of 1 mg/ml F-actin, lying midway in the range often studied in the laboratory; knowing the chain density for this reference concentration, the chain density at an arbitrary concentration can be obtained through multiplying by their concentration ratio. At 370 monomers per micron of contour length, each monomer having a molecular mass of 42 kDa, a filament of F-actin possesses a mass per unit length λ_p of 16×10^6 Da/μm (see Table 3.2). A solution of 1 mg/ml (equal to 24 μM) corresponds to $c = 6.1 \times 10^{26}$ Da/m^3, leading to a density of filament length $\rho_L = c/\lambda_p = 3.8 \times 10^{19}$ μm/m^3; i.e. a cubic meter of this solution contains 3.8×10^{19} μm of F-actin. The chain density ρ can be determined from ρ_L knowing the average filament length.

6.6.4 Cross-linked actin networks

Our discussion of actin rheology begins with cross-linked networks, whose properties have been studied for several decades with conventional viscometers (see, for example, Maruyama *et al.*, 1974, and Brotschi *et al.*, 1978). At our reference F-actin concentration of 1 mg/ml, the minimum ratio of cross-linking proteins to actin monomers for gel formation is observed to be roughly 1:100 or lower, depending on the cross-linker (Brotschi *et al.*, 1978). To estimate the shear modulus expected for such gels, let us assume a 1:100 ratio of cross-linker to actin monomer. If the cross-linking protein creates a four-fold junction, we then assign a quarter of a cross-linker per segmented end. As a segment has two ends, there is half a cross-linker per segment, or 50 monomers per segment at a ratio of 100 monomers per cross-linker. At 370 F-actin monomers per micron, the mean distance between network nodes must then be 50/370 = 0.14 μm, corresponding to a filament density of 2.7×10^{20} filaments/m^3 if $\rho_L = 3.8 \times 10^{19}$ μm/m^3 at 1 mg/ml. Our estimate of the shear modulus from $\mu \sim \rho k_B T$ for this model network is thus 1.1 J/m^3 with $k_B T = 4.0 \times 10^{-21}$ J at 300 K.

The observed mechanical properties of cross-linked networks depend upon the stability of the cross-link, among other things. Janmey *et al.* (1990) used an actin-binding protein (ABP-280) to provide robust cross-links at concentration ratios to actin monomers of 1:100 or less. The shear storage modulus $G'(\omega)$ was observed to be almost independent of frequency in the range examined, having a value of 16–18 J/m^3 for an actin concentration of 1.6 mg/ml and ABP:actin monomer ratio of 1:133. This value is about an order of magnitude larger than expected from $\mu \sim \rho k_B T$: the actin concentration is 60% higher than our reference concentration, but the number of cross-links per monomer is 30% lower, yielding $\rho k_B T \sim 1.3$ J/m^3. A different study of strongly cross-linked actin (Xu *et al.*, 1998a) used 2% biotinylated actin (i.e. the protein biotin was attached to 2% of the actin monomers), yielding a gel in the presence of avidin, a protein capable of linking with four biotins per avidin. Even the small concentration of avidin used in this study (1:500 actin monomers) raises G' by almost two orders of magnitude compared to that of pure 2% biotinylated actin, illustrating the rigidity imparted by the cross-links. Assuming that all available biotins are linked to an avidin, actin concentrations of 0.65 mg/ml correspond to $\rho k_B T \sim 0.7$ J/m^3, which is more than an order of magnitude smaller than the measured G' in this experiment, namely ~20 J/m^3, independent of ω. The main result of both of these studies is that G' is observed to be largely independent of frequency, like a permanently cross-linked network; further, G' is about an order of magnitude larger than the nominal $\rho k_B T$.

The elastic behavior of actin networks is markedly different when α-actinin (from *Acanthamoeba*) is the cross-linker. At the relatively high ratio of α-actinin to actin monomer of 1:15 in a 1 mg/ml solution, G' is observed to be about 1.4 J/m^3 at low frequency, rising to more than 100 J/m^3 at frequencies exceeding 1 Hz (Sato *et al.*, 1987). At these concentrations, the average distance between cross-links is a scant 20 nm (if all available α-actinin proteins form elastically active four-fold junctions) and the nominal value of $\rho k_B T$ is 7 J/m^3, this time much higher than the observed value. The origin of this discrepancy is believed to lie in the weaker affinity of α-actinin to actin, a conclusion deduced in part from the temperature dependence of $G'(\omega)$ for these networks. Over a broad frequency range, G' drops by about an order of magnitude as the temperature rises from 8 °C to 25 °C, implying that the linkage is susceptible to rupture by thermal energies at 25 °C. In contrast, the elastity of actin networks with biotin/avidin junctions displays no change from 8 °C to 25 °C. At physiological temperatures, then, actin networks cross-linked by α-actinin tend to be stiff at short time scales and soft at long time scales, typical of a viscoelastic material (see also Tempel *et al.*, 1996, for a percolation approach to this network). The elastic properties of cross-linked actin networks also depend on the mean length of the individual polymers (Kasza *et al.*, 2010).

6.6.5 Actin solutions

The behavior of actin networks without permanent cross-links has been the subject of intense study for some years (see Xu *et al.*, 1998b, Hinner *et al.*, 1998, Palmer *et al.*, 1999, Tang *et al.*, 1999, and references therein). Many studies have been made of the semi-dilute regime with concentrations around 0.5–2 mg/ml (approaching the concentration of the actin cortex of some cells) finding that the shear storage modulus is independent of frequency at $10^{-3} < \omega < 10$ rad/s, as shown in Fig. 6.23. The measured values of G' in the low frequency range depend on preparation conditions: at concentrations of 1 mg/ml, fresh actin has G' of the order 0.1–1 J/m^3, with several measurements close to 1 J/m^3 (Hinner *et al.*, 1998; Palmer *et al.*, 1999; Xu *et al.*, 1998b), although G' rises by an order of magnitude as the actin ages. If the actin filaments have a mean length of 2 μm, for example, the corresponding $\rho k_B T$ is just under 0.1 J/m^3. These observations suggest that the filaments are sufficiently entangled to behave like a cross-linked network on time scales long compared with many cellular time scales even though the linkages are weak. However, the behavior of the plateau modulus is more subtle than $\rho k_B T$, as G' increases with chain density like $\rho^{1.2-1.4}$ (Xu *et al.*, 1998b; Hinner *et al.*, 1998; Palmer *et al.*, 1999;

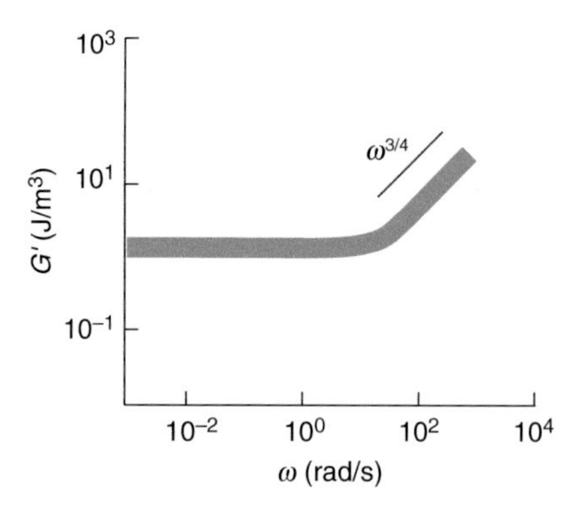

Fig. 6.23 Approximate shear storage modulus $G'(\omega)$ of a 1 mg/ml actin solution, as measured by light scattering (redrawn with permission, after Palmer *et al.*, 1999; ©1999 by the Biophysical Society). At small ω, G' is frequency-independent, but scales like $\omega^{3/4}$ at higher frequency.

see also Gardel *et al.*, 2003). This concentration-dependence agrees with theoretical expectations of semi-dilute solutions of semiflexible filaments which predict a scaling of $\rho^{7/5}$ (Isambert and Maggs, 1996; Morse, 1998). At high frequency, $G'(\omega)$ rises like $\omega^{3/4}$ (Hinner *et al.*, 1998; Palmer *et al.*, 1999; Schnurr *et al.*, 1997), again as expected on theoretical grounds (Isambert and Maggs, 1996; Morse, 1998; Gittes and MacKintosh, 1998).

6.6.6 Actin filaments at high density

Depending on location within the cell, actin concentrations may be as high as 5 mg/ml or larger, corresponding to filament densities of 1.9×10^{20} chains/m^3 for filaments of length 1 μm (see Podolski and Steck, 1990, for actin filament lengths). Such densities lie in the concentrated solution regime, whose lower limit from Eq. (6.49) is $\rho^{**} \sim (D_f L_c^2)^{-1} \sim 1.2 \times 10^{20}$ chains/m^3 if $L_c = 1$ μm and $D_f = 8$ nm. Is the actin concentration in cells sufficiently high to form a nematic phase? Onsager (1949) predicted that the threshold for nematic ordering of hard, rigid rods of uniform length occurs at a density (see also de Gennes and Prost, 1993, Section 2.2)

$$\rho_N = 4.25 / D_f L_c^2. \tag{6.58}$$

This is a factor of four larger than ρ^{**} and has a value of 5.3×10^{20} chains/m^3 (at $L_c = 1$ μm, $D_f = 8$ nm), clearly larger than ρ at actin concentrations of 5 mg/ml (if $L_c = 1$ μm). However, studies of actin with a distribution of filament lengths indicate that the nematic phase may be reached at densities lower than predicted by ρ_N at the mean filament length (Suzuki *et al.*, 1991; Coppin and Leavis, 1992; Furukawa *et al.*, 1993). Figure 6.24 illustrates how Eq. (6.58) exceeds the observed transition density to the nematic phase found by Furukawa *et al.* (1993), although the filament length was not measured directly in the experiment. Käs *et al.* (1996) observe an onset of

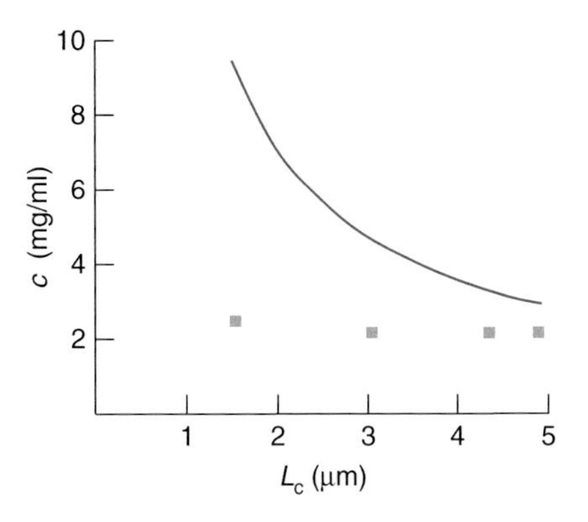

Fig. 6.24 Minimum concentration required for nematic ordering in actin/gelsolin mixtures observed by Furukawa *et al.* (1993) as a function of their estimated actin filament length (reprinted with permission; ©1993 by the American Chemical Society). The solid line is the prediction of the threshold concentration for hard rods from Eq. (6.58) with $D_f = 8$ nm (Onsager, 1949).

nematic ordering at concentrations exceeding 2.5 mg/ml in experiments with a measured distribution of filament lengths; for this experiment, Eq. (6.58) predicts the onset should occur at $\sim 3 \times 10^{18}$ filaments/m^3 or 1.0 mg/ml (using $\langle L_c \rangle = 14$ μm), which is lower than the observed experimental threshold.

6.6.7 Intermediate filament and microtubule solutions

A comparative rheological study has been made of microtubules and vimentin (a common intermediate filament), both of which are more rigid than actin (Janmey *et al.*, 1991). In the experiment, both protein solutions were held at concentrations of 2 mg/ml, which corresponds to a lower density of contour length than actin solutions owing to the higher mass per unit length λ_p of the filament. Repeating the same calculation as performed for actin, the density of contour length ρ_L is 7.5×10^{18} μm/m^3 for microtubules ($\lambda_p = 16 \times 10^7$ Da/μm from Table 2.2) and 3.4×10^{19} μm/m^3 for vimentin (assuming $\lambda_p = 3.5 \times 10^7$ Da/μm, typical of intermediate filaments) at a concentration $c = 2$ mg/ml. The measured shear and storage moduli of both protein solutions show little frequency-dependence, and are in the range 2–3 J/m^3 over an angular frequency range $10^{-2} < \omega < 10^2$ rad/s. Not unexpectedly, these values are about a factor of five lower than that of F-actin networks measured in the same experiment; $\rho_L = 5.7 \times 10^{19}$ μm/m^3 at $c =$ 1.5 mg/ml for actin, which is 1.7 times that of vimentin and 7.6 times that of microtubules at 2 mg/ml. The concentration-dependence of the shear modulus of the microtubule solutions was found to obey $c^{1.3}$, not inconsistent with the behavior expected for rigid or semiflexible filaments, given the uncertainty of the measurement. However, vimentin solutions obey $c^{0.5}$, an unusually low exponent that may arise if the average length of the vimentin filaments is concentration-dependent (Janmey *et al.*, 1991). Just for comparison, the study quotes an average vimentin filament length of $L_c = 3.5$ μm, giving $\rho = \rho_L / L_c \sim 1 \times 10^{19}$ chains/m^3 and $\rho k_B T = 0.04$ J/m^3.

6.6.8 Other measurements of soft networks

The networks discussed in this text are chosen on the basis of their structural simplicity, permitting the microscopic origin of their elastic properties to be investigated. The characteristics of other homogeneous networks, such as collagen (Harley *et al.*, 1977), elastin (Cirulis *et al.*, 2009) and reconstituted spectrin (Stokke *et al.*, 1985b), have also been examined. Achieving a moderately complete picture of single-component networks aids the interpretation of multicomponent networks present in the cell; for example, Griffith and Pollard (1982) have determined the viscosity of actin/microtubule networks. Highly heterogenous systems have been investigated as well, although their interpretation may be necessarily qualitative (for a review of pioneering work in this area, see Elson, 1988). The techniques employed include micropipette aspiration (Hochmuth and Needham, 1990), deformation by atomic force microscope (Radmacher *et al.*, 1996; Pesen and Hoh, 2005; Vadillo-Rodriguez *et al.*, 2009) or the use of internal markers to drive or record deformation of the cytoplasm. For example, magnetic beads (Bausch *et al.*, 1999; Feneberg *et al.*, 2004) or fluorescent markers can be injected into the cell cytoplasm to follow the local strain it develops in response to a stress (Panorchan *et al.*, 2006; Ragsdale *et al.*, 1997; and references therein). Not unexpectedly, such measurements made on the cytoplasm of a particular macrophage show that the shear storage modulus is in the range quoted above for networks of pure filaments such as actin, ranging from 10–10^2 J/m^3. Further, the elastic response may vary by a factor of two within a given cell, and much more between different cells of a given population (Bausch *et al.*, 1999). Somewhat lower values than this have been found for G' and G'' in *Xenopus* egg cytoplasmic extracts (Valentine *et al.*, 2005).

The mechanical properties of the bacterial cell wall are treated in some detail in Sections 3.5 and 7.5. The measured values of a typical elastic modulus of hydrated wall material is a few times 10^7 J/m^3 (see, for example, Yao *et al.*, 1999), which is well below the value of 10^9 J/m^3 often found for dense polymers. However, this is not unexpected given that the molecular structure of peptidoglycan, the material of the wall, has stiff longitudinal elements linked by floppy transverse elements. Systematic studies also have been made of defective networks in cells, particularly the two-dimensional network of the human erythrocyte (Waugh and Agre, 1988). Spectrin-depleted cytoskeletons exhibit a reduced shear resistance, consistent with their rigidity emerging at the connectivity percolation threshold p_C, although their elasticity is not completely described by percolation (see problem set).

A number of biofilaments exhibit strain hardening, where G' increases as the network becomes more strained, as observed in fibrin (Hudson *et al.*, 2010; Kang *et al.*, 2009) and a suite of filaments common to cells (Storm *et al.*, 2005). The phenomenon is explored further in the problem set and has been the focus of extensive computer simulations (for example, Åström *et al.*, 2008; Huisman *et al.*, 2008). Other features beyond those described here may appear in G' and G'' for networks composed of coiled filaments (Courty *et al.*, 2006). Lastly, we describe in Chapter 5 the two-dimensional elastic moduli of the nuclear envelope; an introduction to the viscoelastic behavior of the envelope can be found in Dahl *et al.* (2005).

Summary

We have focused in this chapter on the three principal components of the cytoskeleton – actin, intermediate filaments and microtubules. With persistence lengths greater than or equal to the diameter of a typical cell, these filaments have gentle curvature when viewed in their cellular environment. Solutions of chemically pure filaments of each of these proteins can be studied in the laboratory. In the cell, binding proteins may be present that link the filaments into three-dimensional networks with fixed connection points.

The static elastic properties of three-dimensional materials with high symmetry can be summarized in just a few elastic moduli. For materials with cubic symmetry, three moduli appear in the free energy of deformation (see Appendix D); for isotropic materials, only two moduli (K_V and μ) are present: $\Delta\mathcal{F} = \mu\, \Sigma_{I,j}\, (u_{ij} - \delta_{ij}\, \mathrm{tr}u/3)^2 + K_V\, (\mathrm{tr}u)^2/2$, where u_{ij} is the strain tensor. The moduli are rooted in the microscopic characteristics of the material; for example, at zero temperature the moduli of a network of springs with force constant k_{sp} and equilibrium spring length s_o are proportional to k_{sp}/s_o. The elastic moduli of floppy chain networks with fixed connection points are close to $\rho k_B T$, where ρ is the number of chains per unit volume. However, when the average number of connection points per chain is less than the connectivity percolation threshold, the shear modulus vanishes. Under certain circumstances (e.g. central forces at zero temperature) the shear modulus vanishes below a separate rigidity percolation threshold, at a density higher than the connectivity threshold. For a network on a lattice in d dimensions, the bond occupation probability at the rigidity percolation threshold is not far from $2d/z$, where z is the number of nearest neighbors to a lattice site.

Without permanent cross-links, polymers in solution may behave like fluids over very long time scales. However, on short time scales, their mechanical elements may be sufficiently dense or entangled that they move past each other only very slowly, and exhibit shear resistance. Such viscoelastic materials are characterized by time- (or frequency-) dependent shear moduli: subject to an oscillating strain of angular frequency ω, they develop a stress $\sigma_{xy} = G'(\omega)u_{xy}(t) + G''(\omega) \bullet (du_{xy}/dt)/\omega$, where $G'(\omega)$ and $G''(\omega)$ are the shear storage modulus and shear loss modulus, respectively. In the zero-frequency limit, G' becomes the shear modulus and G''/ω becomes the viscosity. In general, G' increases with frequency, although its specific behavior depends upon concentration. Of interest in the cell is the behavior of a solution of stiff filaments, say with contour length L_c held at a number density ρ. In the dilute regime, $\rho < \rho^* \sim L_c^{-3}$, the polymer solution behaves like a simple fluid with small values of G'. In contrast, at high filament concentrations, $\rho > \rho^{**} \sim (D_f L_c^2)^{-1}$, the solution lies in the concentrated regime and G' approaches 10^9 J/m^3. Nematic ordering may occur at such high concentrations if L_c/D_f exceeds approximately 3. Less uniform behavior is seen at concentrations between ρ^* and ρ^{**}, where G' first rises with frequency, then becomes roughly constant through a plateau region, and finally rises again towards an asymptotic value. The plateau reflects the presence of chain entanglements, and is predicted to scale with concentration like ρ^1 for floppy chains or $\rho^{7/5}$ for semiflexible filaments. Solutions of rigid rods also display a high-frequency regime that increases like ρ^1.

Networks of biofilaments in the cell display most of the theoretical expectations described above; in fact, biofilaments represent an excellent laboratory for testing some concepts from polymer physics because, unlike alkanes, they can be fluorescently labeled and studied individually by microscopy. Filamentous networks with strong permanent cross-links are found to have frequency-independent storage moduli G' that are an order of magnitude larger than $\rho k_B T$, depending on the stiffness of the filament. Semidilute solutions of actin and microtubules individually show that G' in the plateau region scales with concentration like $\rho^{7/5}$, and with frequency like $\omega^{3/4}$ at high frequency, in agreement with theoretical predictions. At high concentations, actin solutions are observed to form a nematic phase, although the concentration at the onset of ordering may be different than expected theoretically for systems of rigid rods of uniform length.

Problems

Applications

6.1. Suppose that a spherical cell of radius 5 μm contains a solution of actin filaments at a concentration of 3 mg/ml.

(a) Calculate the nominal value of $\rho k_B T$ of the filaments if they have a length of 5 μm (where $T = 300$ K).
(b) Taking the volume compression modulus K_V to be $\rho k_B T$, calculate the fluctuations in the volume occupied by the network $\langle \Delta V^2 \rangle^{1/2} / V$ using Eq. (D.35), assuming that the cell membrane permits these fluctuations.

6.2. A simplification of the peptidoglycan network of the bacterial cell wall treats the glycan strands as rigid rods and the string of amino acids as entropic springs, as in the figure in Problem 6.20. There, we show that the compression modulus of this model at zero tension is $K_V = k_{sp}/(8b)$, with symbols defined in the figure.

(a) If the contour length of the peptide chains is 4 nm, but their mean end-to-end displacement is $a = 1.3$ nm, what is their persistence length and effective spring constant k_{sp} (at $T = 300$ K)?
(b) If $b = 1.0$ nm, what is K_V using your results from part (a)?
(c) Compare the result from (b) with $\rho k_B T$ at $T = 300$ K using the bond density (transverse plus longitudinal) calculated in Problem 6.20(a).

6.3. Hereditary spherocytosis is a red blood cell disorder in which a reduced spectrin content of the cytoskeleton is associated with a reduced shear modulus. A sample of data taken from a study by Waugh and Agre (1988) is shown in the table below (all quantities have been normalized to a control sample). Plot these data and compare them with two predictions from percolation theory: $\mu/\mu_0 = (p - p_{perc})/(1 - p_{perc})$, where p_{perc} is either p_C or p^* from Eq. (6.47) for a triangular lattice in two dimensions. Interpreting the measurements in terms of a randomly depleted network, with which value of p_{perc} are the data more consistent?

Spectrin content	Shear modulus	Spectrin content	Shear modulus
0.54	0.62±0.04	0.52	0.62±0.03
0.97	0.82±0.09	0.68	0.73±0.05
0.73	0.77±0.04	0.55	0.61±0.04
0.65	0.61±0.03	0.71	0.73±0.05
0.43	0.62±0.09	0.43	0.63±0.07

6.4. What is the lowest density of nearest-neighbor bonds (i.e. bonds per unit volume) required for rigidity percolation for the three-dimensional lattices in Table 6.1 – simple cubic, body-centered cubic and face-centered cubic? Take the nearest neighbor bonds in each lattice to have the same length b and assume that the rigidity percolation threshold p_R is given by Eq. (6.47).

6.5. In one experiment (Abraham *et al.*, 1999), actin filaments at the leading edge of a migrating cell were found to have concentrations as high as 1600±600 μm of filament per cubic micron of cell volume.

(a) If the filaments were 1 μm in length, in what concentration regime would this system fall (in number of filaments per m^3)? Would you expect the filaments to form a nematic phase?
(b) From the density you obtain in (a), what is the pressure of the filaments, regarding them as non-interacting objects (at $T = 300$ K)? Quote your answer in J/m^3 and in atmospheres.

6.6. The bacterium *E. coli* has the approximate shape of a cylinder capped at each end by hemispheres. Use the image in Fig. 1.1 to obtain a typical length and diameter of this bacterium.

(a) Evaluate the densities ρ^*, ρ^{**} separating the dilute, semi-dilute and concentrated regions of bacterial solutions. Quote your answer in bacteria per cubic meter.
(b) Does the shape of the bacterium satisfy Onsager's condition for the existence of a nematic phase?

6.7. The tobacco mosaic virus (TMV) has the shape of a cylindrical rod 260 nm in length and 18 nm in diameter; its mass per unit length is given in Table 3.2.

(a) Does the shape of the virus satisfy Onsager's condition for the existence of a nematic phase?
(b) Find the transition concentrations separating the four possible phases of this virus in solution. Calculate ρ^*, ρ^{**} and ρ_N in filaments/m^3, and c^*, c^{**} and c_N in mg / ml.

6.8. Under certain conditions, disk-like objects can form an ordered phase in which the disks display long-range orientational order. Characterizing a disk by its diameter D and thickness T, the following conditions are found.

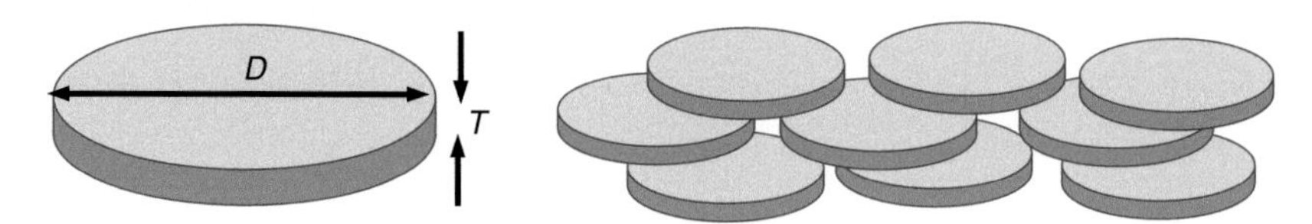

Condition I: the disks may form an ordered phase if $T/D < 0.2$ (with some uncertainty, see Veerman and Frenkel, 1992);

Condition II: an ensemble of very thin disks ($T \ll D$) may form an ordered phase if its number density satisfies $\rho > 4/D^3$ (see Eppenga and Frenkel, 1984).

(a) Look up the approximate diameter of a human red cell and estimate its effective thickness assuming it is a cylinder with volume 140 μm^3. Compare the dimensions with condition I.
(b) Find the approximate concentration of red cells in the blood (see Chapter 23 of Alberts *et al.*, 2008) and compare with condition II.

6.9. A hypothetical bacterium has the shape of a spherocylinder with diameter 1 μm and length 4 μm. To strengthen its plasma membrane, the outer surface of the cell is covered with a cell wall in the form of a four-fold network in a single layer; to a good approximation, all the bonds in the network have the same length of 4 nm. An antibiotic has been developed to attack the bonds in this network, causing it to fail and the bacterium to rupture. How many moles of this antibiotic are needed to kill a single bacterium?

6.10. Find the threshold stress for crack propagation in a lipid bilayer according to the Griffith formula for brittle materials (Problem 6.26). Assume the crack length to be 20 nm and take the material parameters to be $Y = 10^9$ J/m^3 and $\gamma_s = 5 \times 10^{-2}$ J/m^2 (the Young's modulus and surface energy density, respectively). Quote your answer in atmospheres. Is this result quantitatively consistent with your expectations for the strength of lipid bilayers (see the Summary for the strength of membranes and cell walls)?

6.11. Fresh bone has a Young's modulus of about 20 GPa, with some variation.

(a) Use the Griffith expression from Problem 6.26 to estimate the threshold stress for the propagation of a crack of 20 μm, which is a few cell diameters long. Assume without justification that the applicable interfacial tension is 0.04 J/m^2.
(b) Estimate the percent strain on the bone at the threshold stress.
(c) Compare your answer in part (c) with the measured facture threshold and tensile strength for bone of 160 and 110 MPa, respectively. Compared to other situations mentioned in Section 6.4, is the crack formula more accurate in this example?

6.12. The Maxwell element of a viscoelastic system consists of a spring (Young's modulus Y) and dashpot (viscosity η) connected in series. The dependence of the shear storage modulus G' on frequency ω is $Y\omega^2\tau^2 / (1 + \omega^2\tau^2)$, where the time scale τ is η / Y (see Problem 6.29).

(a) In terms of τ, at what value of ω does the storage modulus reach 90% of its asymptotic value at high frequency? Define this as ω_{90}.
(b) Using $\rho k_B T$ as an approximation for the Young's modulus, estimate the time scale τ of a soft network with a node density of

2×10^{19} m^{-3}. Assume the temperature to be 300 K and the viscosity to be that of water.

(c) What is ω_{90} from part (a) for this network?

6.13. Collagen fibrils are biological ropes composed of many strands of the collagen triple helix. Fibrils with diameters of 200–300 nm are observed to yield when the applied stress reaches a few hundred MPa (Shen *et al.*, 2008).

(a) What is the maximum mass that can be hung from a single vertical fibril before it yields? Assume the fibril has a diameter of 200 nm and a yield stress σ^* of 200 MPa.

(b) If the load-bearing component of a tendon has the same yield stress, what is the maximum force the tendon can sustain, assuming it has a diameter of 0.8 cm?

(c) Is your answer in part (b) consistent with the forces you would expect the lower leg to experience?

6.14. Elastic moduli such as K_V and μ are usually quoted at zero strain. However, as we saw for spring networks, the moduli may depend on the strain or stress state of the system. For example, the shear storage modulus G' of a particular vimentin network obtained in an oscillatory experiment exemplified in Fig. 6.18 can be roughly fitted by $G' = (4 + 125\gamma^{5/2})$ Pa. In this expression, γ is the strain amplitude appearing in $[strain] = \gamma \sin\omega t$, where ω is the oscillating frequency, fixed at 10 rad s^{-1} here.

(a) If G' has no frequency dependence, what is the shear modulus μ of the network?

(b) What is the ratio of G' at 50% strain compared to 0% strain. Does the network harden or soften with increasing strain?

(c) At what strain is G' equal to 1 MPa according to this expression? Is such a strain state physically achievable?

(d) For a polymer network with fixed welds at polymer junctions, explain why you would expect the shear resistance to increase as the network itself became sheared; use simple shear as the deformation.

6.15. As introduced in Problem 6.14, some systems display strain hardening at moderate to high strains (although strain softening is also possible). In one set of measurements, the shear storage modulus of four cross-linked networks can be described approximately by

$$G'(\text{actin}) = 90 + 5800\gamma^{3/2},$$

$$G'(\text{fibrin}) = 20 + 240\gamma^{1.7},$$

$$G'(\text{vimentin}) = 4 + 250\gamma^{5/2},$$
$$G'(\text{neurofilaments}) = 3 + 0.032\gamma^{4.5},$$

where γ is the strain amplitude in an oscillating strain field and all values are quoted in Pa. Note that these measurements are for specific concentrations of biofilaments and a driving frequency of 10 rad/s (Storm *et al.*, 2005).

(a) On a single graph, plot each of these functions logarithmically over the strain range $0.01 \leq \gamma \leq 1$; for neurofilaments, use the range $0.01 \leq \gamma \leq 10$.
(b) For each biofilament, find the value of γ at which the constant term in the expression is equal to the strain-dependent term. Call this strain γ^*.
(c) For the three types of filaments, make a log–log plot of γ^* against G'_0, where G'_0 is the value of G' at zero strain. Very crudely assuming that γ^* is proportional to a power of G'_0, what is the exponent of the power law?

Formal development and extensions

6.16. Suppose that we have a three-dimensional isotropic material whose compression resistance is much greater than its shear resistance: $K_V >> \mu$.

(a) Find the strain tensor for the deformation $x \rightarrow \Lambda_- x$, $y \rightarrow \Lambda_- y$, $z \rightarrow \Lambda_+ z$ in terms of δ, where $\Lambda_+ = 1 + \delta$ and there is no volume change.
(b) Predict the Poisson ratio from your results in (a) using the definition $\sigma_p = -u_{xx}/u_{zz}$.
(c) Compare your prediction from (b) with Eq. (6.21) in the appropriate limit.

6.17. Consider a network of springs connected in the shape of a cube, as in Fig. 6.8, such that the springs are forced to lie at right angles to one another. Each spring has the same unstretched length s_o and force constant k_{sp}.

(a) By minimizing the enthalpy per vertex H_v of the network under pressure ($H = E + PV$), find the spring length s_p at pressure P, and show that $s_p = s_o(1 - Ps_o/k_{sp})$ at small P (where $P > 0$ is compression).
(b) Using $V = s_p^3$ of a single cube, determine the compression modulus K_V at $P = 0$ from your results in part (a).

6.18. A spring network with the symmetry of a body-centered cubic lattice has the structure shown in the diagram.

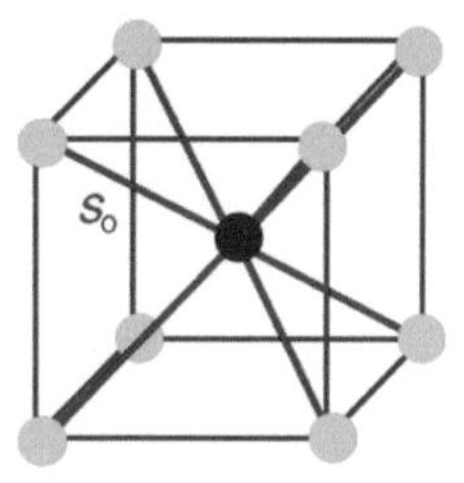

Each vertex is connected to its eight nearest neighbors by springs of unstretched length s_o and spring constant k_{sp}.

(a) Find the volume per vertex in terms of s_o.
(b) Find the strain tensor and the change in energy density if all springs are stretched from s_o to $s_o + \delta$.
(c) Show that the compression modulus is given by $K_V = (1/\sqrt{3}) \bullet (k_{sp}/s_o)$.

6.19. A spring network with the symmetry of a face-centered cubic lattice has the structure shown in the diagram.

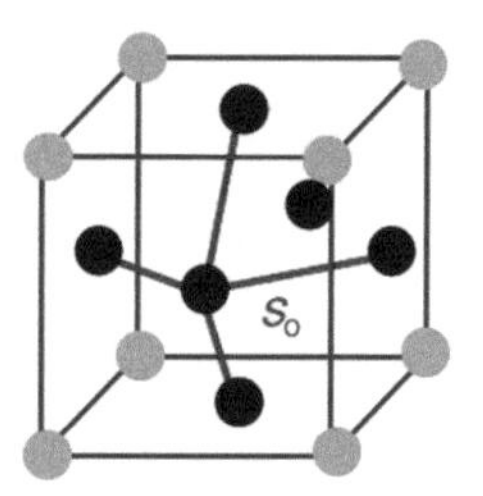

Each vertex is connected to its twelve nearest neighbors by springs of unstretched length s_o and spring constant k_{sp}.

(a) Find the volume per vertex in terms of s_o.
(b) Find the strain tensor and the change in energy density if all springs are stretched from s_o to $s_o + \delta$.
(c) Show that the compression modulus is given by $K_V = (2\sqrt{2}/3) \bullet (k_{sp}/s_o)$.

6.20. The box in the diagram represents a section of an extended network that we will use as a simplified model of peptidoglycan, the material of the bacterial cell wall. The multisegmented rods with monomer length b represent the glycan strands, while the five short bonds of length a perpendicular to the glycan are strings of amino acids. All bonds (indicated by the thick lines) and vertices (indicated by the disks) are shared with neighboring sections of the wall (we use the word "section" to represent a specific volume of the cell wall with a specific structure; here, the volume of a section is $4a^2b$).

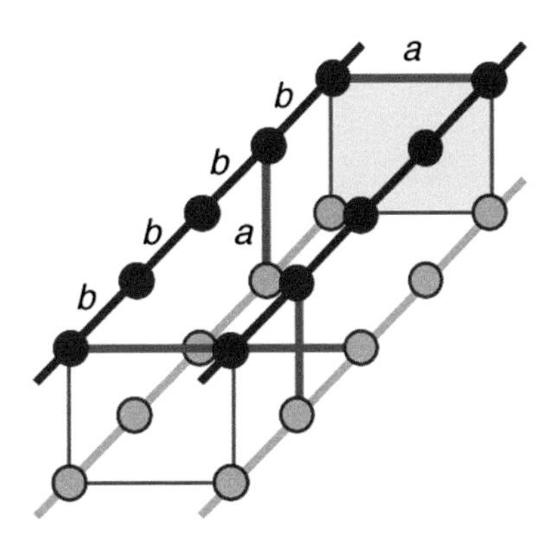

(a) How many vertices and bonds (both transverse and longitudinal) are there in this section? Show that the volume per vertex is a^2b and the volume per bond is $2a_2b/3$ (*If you haven't done this type of problem before, determine what "fraction" of a bond or vertex belongs to each section.*)

(b) Show that the enthalpy per vertex H_V for this network under tension τ is

$$H_V = (k_{sp}/4) \bullet (a - a_o)^2 - \tau a^2 b,$$

where k_{sp} is the effective spring constant of the peptide chain and a_o is its unstressed (but not contour) length. Assume that the glycan strand (b) is a rigid rod, but a can vary about a_o.

(c) Show that H_V is minimized by the bond length

$$a_\tau = a_o / (1 - 4\tau b/k_{sp}).$$

(d) Find the volume per vertex, and show that the volume compression modulus is

$$K_V = (k_{sp}/ 8b) \bullet (1 - 4\tau b/k_{sp}).$$

6.21. A random polymer network is subject to a strain in the z direction characterized by scaling factor $\Lambda_z = \Lambda$; the scaling factors in the x and y directions are equal and assume values to conserve the volume of the network.

(a) How does this deformation differ from the pure shear mode of Eq. (6.40)?

(b) Using Eq. (6.39) as a model free energy, show that the free energy density $\Delta\mathcal{F}$ of this deformation is given by

$$\Delta\mathcal{F} = (k_B T\rho/2) \bullet (\Lambda^2 + 2/\Lambda - 3).$$

(c) To order δ^2, what is $\Delta\mathcal{F}$ for small deformations when $\Lambda = 1 + \delta$?
(d) Evaluate the strain tensor and determine $\Delta\mathcal{F}$ to the same level of approximation as part (c).
(e) Combining your results from parts (c) and (d), what is the shear modulus μ? Does your result agree with the expression for μ obtained from a pure shear mode?

6.22. Consider a three-dimensional cage formed from 12 inextensible rigid rods of length s_o. If the rods join at right angles, the cage is a cube of volume s_o^3, as in Fig. 6.9.

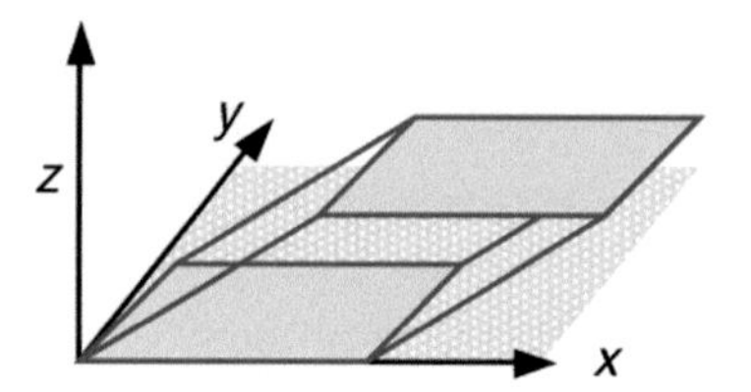

Find the average volume of the cage if the rods freely pivot at the junctions. (You may find it calculationally helpful to fix one rod to lie along the x-axis and a second rod to lie in the xy-plane.)

6.23. Determine the mean value of the squared volume $\langle V^2 \rangle$ for the freely jointed cage in Problem 6.22, and extract the volume fluctuations $(\langle V^2 \rangle - \langle V \rangle^2) / \langle V \rangle$.

6.24. Complete the mathematical proof leading from Eq. (6.33) to (6.34). The steps are as follows:

- Show that $\ln(np_i/n_i) = \ln(\Lambda_x\Lambda_y\Lambda_z) + (1/2\sigma^2) \bullet [x_i^2(1/\Lambda_x^2 - 1) + y_i^2(1/\Lambda_y^2 - 1) + z_i^2(1/\Lambda_z^2 - 1)]$.
- Substitute n_i and p_i into this expression and show that $\ln\mathcal{P}_a = n[K/2\sigma^2 + \ln(\Lambda_x\Lambda_y\Lambda_z)]$, where K is the integral

$$K = (\Lambda_x\Lambda_y\Lambda_z)^{-1} (2\pi\sigma^2)^{-3/2} \iiint \exp\{-[(x/\Lambda_x)^2 + (y/\Lambda_y)^2 + (z/\Lambda_z)^2]/2\sigma^2\} \bullet$$
$$\bullet\ [x^2(\Lambda_x^{-2}-1) + y^2(\Lambda_y^{-2}-1) + z^2(\Lambda_z^{-2}-1)]\, \mathrm{d}x\mathrm{d}y\mathrm{d}z.$$

In establishing this, you will have converted summations of the form $\Sigma_i\, f(x_i)\Delta x$ into integrals, where the integration limits run from $-\infty$ to $+\infty$.

- Break K into a sum over three terms in x, y, z and prove:

$$\Lambda_x^{-1}(\Lambda_x^{-2}-1) \bullet (2\pi\sigma^2)^{-1/2} \int \exp[-(x^2/2\sigma^2)]\, x^2\, \mathrm{d}x = 2\sigma^2(1 - \Lambda_x^2)/2.$$

- Combine these results to obtain Eq. (6.34):

$$\ln\mathcal{P}_a = -(n/2)[\Lambda_x^2 + \Lambda_y^2 + \Lambda_z^2 - 3 - 2\ln(\Lambda_x\Lambda_y\Lambda_z)].$$

6.25. The ratio of the flexural rigidity κ_f to torsional rigidity κ_{tor} in Chapter 4 is proportional to the ratio of the Young's modulus Y to the shear modulus μ. (a) Find Y/μ over the range $1/4 \le \sigma_p \le 1/3$, where σ_p is the Poisson ratio. (b) Find the corresponding range of κ_f/κ_{tor} for a solid cylinder.

6.26. A simple model for the formation of crack in a brittle medium takes into account only the idealized deformation energy of the medium and the work for creating new surface. Consider a crack of length $2a$ in an infinite medium of unit thickness. Assume without proof that the decrease in potential energy of the medium upon the creation of the crack is given by $\Delta U = -\pi a^2\sigma^2/Y$, where σ is the applied stress and Y is the Young's modulus.

(a) If the energy per unit area for creating new surface at the crack is γ_s, what is the total energy ΔE of the system (per unit thickness)?
(b) For a crack to propagate, the minimum rate of energy change with crack length must obey $d(\Delta E)/da = 0$. Explain.
(c) Show that the minimum stress for crack formation is then given by the Griffith expression $\sigma_c = (2Y\gamma_s / \pi a)^{1/2}$.

6.27. The *Maxwell* element for modeling a viscoelastic system treats it as a spring and a viscous dashpot connected in *series* as shown in the diagram. The arrows indicate where the stress is applied.

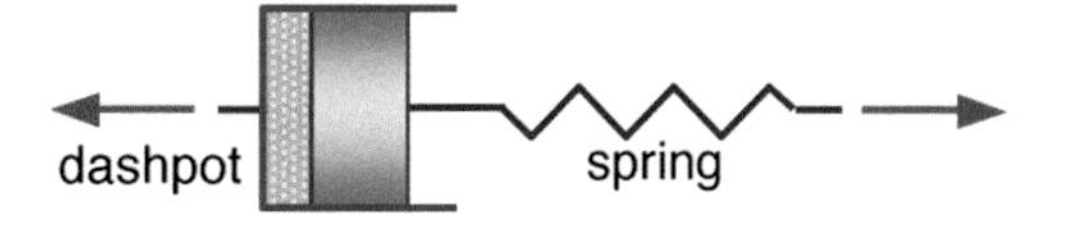

(a) The stress σ is the same in both components. What is the behavior of the strain u along the axis?
(b) Show that the time rate of change of u is governed by

$$du/dt = (1/Y)(d\sigma/dt) + (\sigma/\eta),$$

where Y and η are the Young's modulus of the spring and the viscosity of the dashpot, respectively.
(c) If the stress is constant at σ_o, show that $u(t)$ increases linearly with time.
(d) If the strain is instantly changed to u_o at time $t = 0$ and remains unchanged at that value, show that the stress decreases exponentially with a time constant of $\tau = \eta/Y$.

6.28. The *Voigt* element for modeling a viscoelastic system treats it as a spring and a viscous dashpot connected in *parallel* as shown in the diagram. The arrows indicate where the stress is applied.

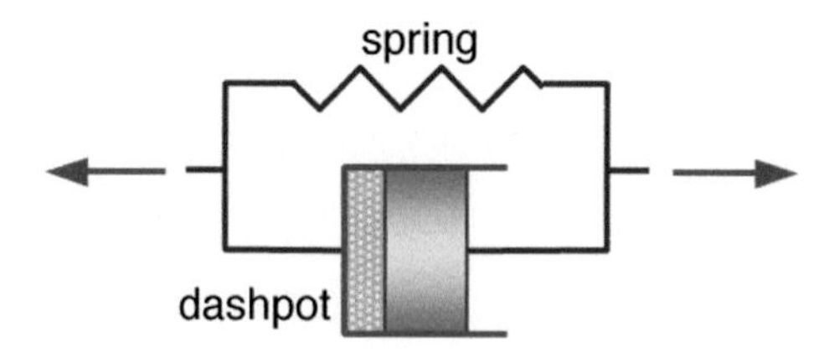

(a) The strain u along the axis is the same in both components. What is the behavior of the stress σ?

(b) Show that the time rate of change of u is governed by

$$\sigma = Yu + \eta \, \mathrm{d}u/\mathrm{d}t,$$

where Y and η are the Young's modulus of the spring and the viscosity of the dashpot, respectively.

(c) If the strain is instantly changed to u_o at time $t = 0$ and remains unchanged at that value, what is the behavior of the stress?

(d) If the stress is fixed at σ_o, show that $u(t)$ increases exponentially to σ_o/Y with a time constant of $\tau = \eta/Y$. Hint: evaluate the integral $\int_0^t \mathrm{d}\{u(t) \cdot \exp(t/\tau)\}$; assume $u(0) = 0$.

6.29. Consider an oscillating stress (of angular frequency ω) having the complex form $\sigma(t) = \sigma_o \exp(i\omega t)$ applied to the Maxwell element discussed in Problem 6.27. The mechanical response is governed by the equation $\mathrm{d}u/\mathrm{d}t = (1/Y)(\mathrm{d}\sigma/\mathrm{d}t) + (\sigma/\eta)$, where the symbols are defined in Problem 6.27. Define an elastic parameter C via the ratio of the stress difference to the strain difference at two times t_2 and t_1; that is $C \equiv \Delta\sigma / \Delta u$, where $\Delta\sigma = \sigma(t_2) - \sigma(t_1)$ and $\Delta u = u(t_2) - u(t_1)$. Show that the real and imaginary parts of C are given by $\mathrm{Re}C = Y\omega^2\tau^2 / (1 + \omega^2\tau^2)$ and $\mathrm{Im}C = Y\omega\tau / (1 + \omega^2\tau^2)$, where the decay time τ is defined as $\tau = \eta/Y$.

6.30. A longitudinal compressive force F is applied to the ends of a cylinder of length L and radius R, as in the diagram.

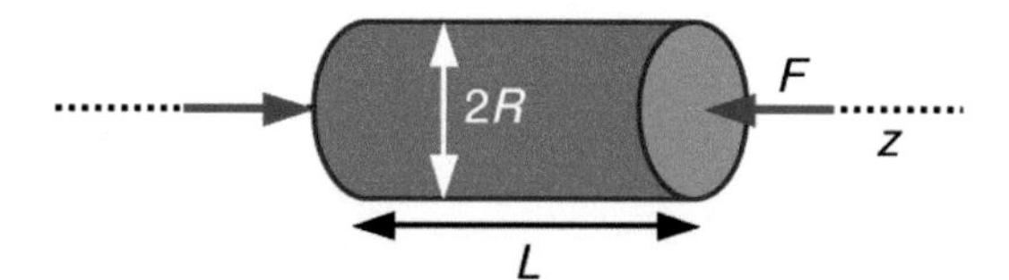

If F is less than $F_{\mathrm{buckle}} = \pi^2 Y\mathcal{I}/L^2$ (see Section 3.5), the rod has a non-zero strain in the z-direction, where the axis of symmetry lies along the z-axis

(a) Find the stress in terms of R, L and Y just at the buckling point $F = F_{\mathrm{buckle}}$.

(b) Find the corresponding u_{zz}.

PART II

MEMBRANES

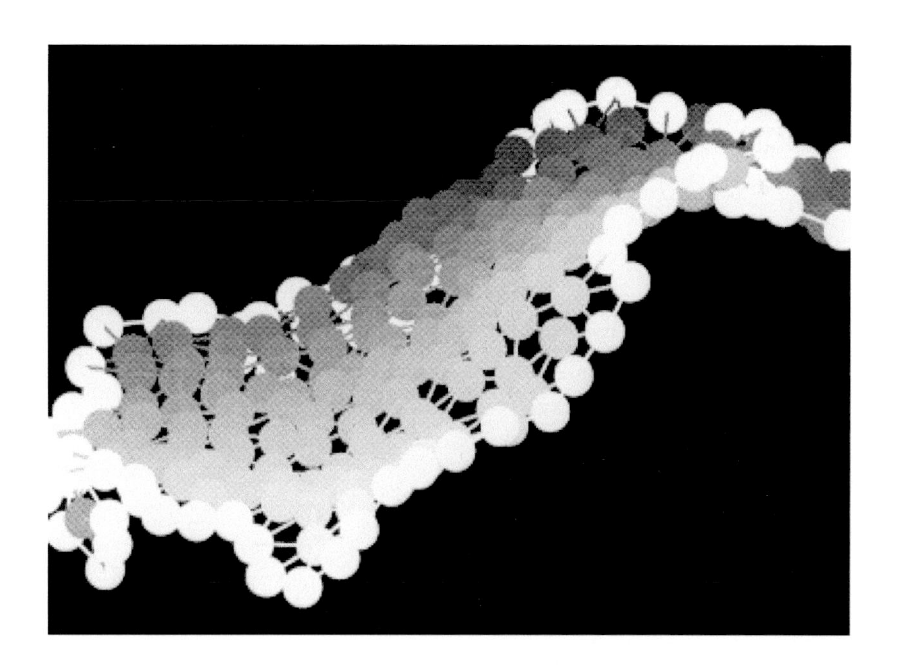

7 Biomembranes

The two broad categories of structural components of the cell are filaments, the focus of Part I of this text, and sheets, which we treat in Part II. In principle, two-dimensional sheets may display more complex mechanical behavior than one-dimensional filaments, including resistance to both out-of-plane bending and in-plane shear. As examples, the plasma membrane is a two-dimensional fluid having no resistance to in-plane shear, whereas the cell wall possesses shear rigidity as a result of its fixed internal cross-links. In this first chapter of Part II, we introduce the chemical composition of the membranes, walls and lamina that are of mechanical importance to the cell and its environment. Our theoretical framework is continuum mechanics, within which we describe the elasticity and failure of membranes, with an emphasis on experiment. However, the softness of biomembranes means that their thermally induced undulations are important, as will be explored in Chapters 8 and 9, both of which are more mathematical than Chapter 7. More extensive reviews of biomembranes than those provided here can be found in Evans and Skalak (1980), Cevc and Marsh (1987) and Sackmann (1990).

7.1 Membranes, walls and lamina

The design principles of Chapter 1 argue that the cell's membranes should be very thin, perhaps even just a few molecules in thickness, if their sole purpose is to isolate the cell's contents. The membrane need not contribute to the cell's mechanical strength, as this attribute can be provided by the cytoskeleton or cell wall. Are there materials in nature that meet these design specifications? Evidence for the existence of monolayers just a single molecule thick goes back to Lord Rayleigh, in an experiment that would be environmentally frowned upon today. He poured a known volume of oil onto a calm lake and observed, by reflection, the area covered by the oil. Dividing the volume of the oil by the area over which it spread gave him the thickness of the layer – a good estimate of the size of a single oil molecule.

The oil/water interface in Rayleigh's experiment is flat because the fluids are immiscible and have different densities. In contrast, interfaces with more

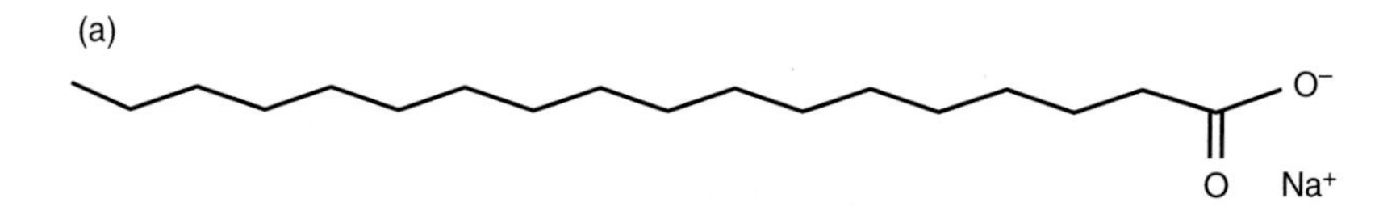

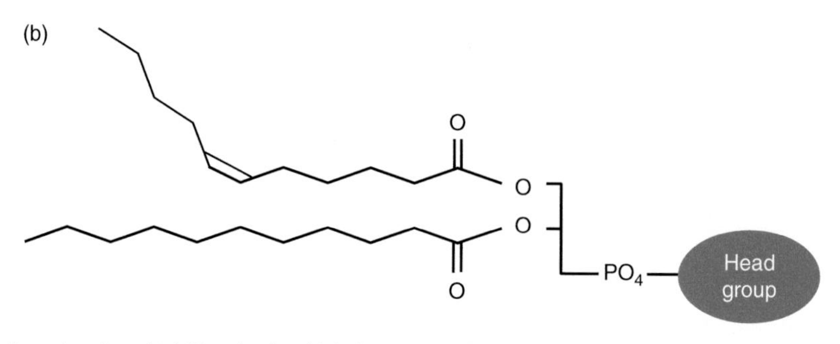

Fig. 7.1 Examples of amphiphilic molecules. (a) Sodium stearate has a single hydrocarbon chain terminating at a carboxyl group. (b) Phospholipids of the cell's membranes generally have a pair of inequivalent hydrocarbon chains.

diverse geometries arise from a class of compounds called surfactants (surface active agents), possessing attractive interactions with a range of fluids whose composition may be so dissimilar as to make them immiscible. For example, sodium stearate (common soap) has a long alkane chain terminating at a carboxylic acid group, as illustrated in Fig. 7.1(a). The hydrocarbon chain is non-polar, and is said to be water-avoiding or hydrophobic, while the (polar) carboxyl group is water-loving or hydrophilic; this dual nature makes the molecule amphiphilic. As a detergent, the hydrophobic end of this amphiphile embeds itself in an oily droplet, leaving the surface of the droplet decorated with polar carboxyl groups that are attractive to water. Figure 7.2(a) illustrates how a continuous molecular monolayer of amphiphiles can form between oil and water, with the hydrophobic component of the surfactant penetrating the oil phase and the hydrophilic component facing the aqueous phase. Such a structure is close to what we are seeking for the cell boundary, except that we want to have water on both sides of the boundary, not water on one side and oil on the other. By cutting and pasting two copies of Fig. 7.2(a) and omitting the oil, we can construct a bilayer with two leaflets back-to-back, as in Fig. 7.2(b). Are there surfactants capable of preferentially assembling into such bilayers?

A class of molecules broadly referred to as liquid crystals form a variety of condensed phases with properties in between those of solids and isotropic fluids (see Vertogen and de Jeu, 1988). These intermediate phases, called mesophases or mesomorphic phases, are characterized by partial ordering of the molecules over long distances; for instance, the nematic phase

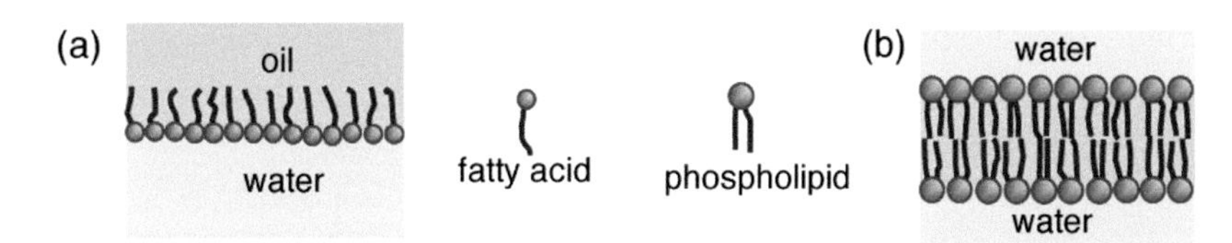

Fig. 7.2 Examples of extended structures formed by amphiphilic molecules: (a) a *monolayer* of soap molecules at the interface between oil and water, (b) a *bilayer* of dual-chain lipid molecules separating two aqueous domains.

of rod-like molecules described in Chapter 6 has long-range orientational order without long-range positional order. Molecules capable of forming mesophases are referred to as mesogens, and many amphiphiles of importance in cells are members of this family. Depending on temperature, the amphiphiles present in a bilayer may form a two-dimensional mesophase, although the membranes of most cells operate at sufficiently high temperature, or are sufficiently heterogeneous chemically, that they are isotropic fluids. More detailed treatments of the formation and characterization of bilayer phases can be found in the reviews by Bloom *et al.* (1991) or Nagle and Tristram-Nagle (2000). The phases accessible to binary or ternary systems containing lipids is truly impressive, but space limitations restrict our treatment to only a few of these phases, primarily the lamellar phase.

Single-chain fatty acids such as sodium stearate assemble into bilayers only at high concentrations, greater than 50% soap by weight, for instance. This is not what is needed for a cell: the membrane should form at low amphiphile concentration if its materials are to be used efficiently. Consequently, the cell's bilayers are based upon particular varieties of phospholipid molecules having two hydrocarbon chains. Lipids themselves are a broad group of organic compounds that are soluble in organic solvents, but not in water; the generic phospholipid in Fig. 7.1(b) comprises two fatty acids linked to a glycerol, which in turn is linked to a polar head group via a phosphate PO_4. The polar head groups of the cell's phospholipids are selected from several organic compounds, none of which is particularly large or complex, as can be seen in Fig. 7.3. In many phospholipids found in biomembranes, one of the fatty acids contains only C–C single bonds while the other contains a C–C double bond. A chain may twist around a C–C single bond, leading to a variety of conformations for alkane chains in isolation, as described in Chapter 3. Yet when present in a layered structure, the saturated hydrocarbon chain (single bonds only) of a phospholipid tends to straighten out through steric interactions with its neighbors. In contrast, a double bond generates a permanent kink in the hydrocarbon chain, making the dense packing of phospholipids more difficult and reducing the in-plane viscosity of the bilayer. Other lipids such

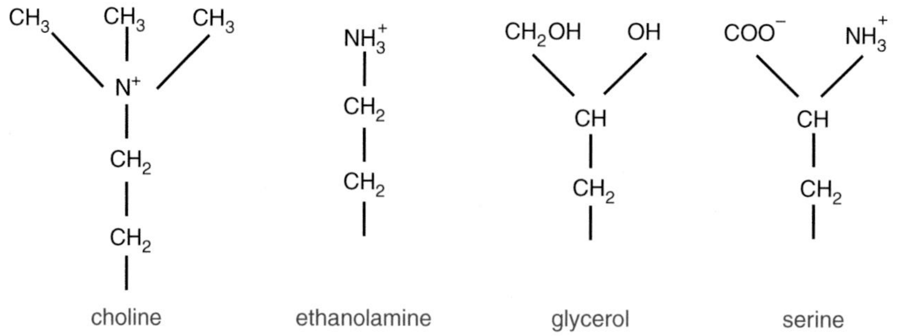

CH_3 CH_3 CH_3 N^+ CH_2 CH_2

NH_3^+ CH_2 CH_2

CH_2OH OH CH CH_2

COO^- NH_3^+ CH CH_2

choline ethanolamine glycerol serine

Fig. 7.3 Examples of polar head groups commonly found on phospholipid molecules in cellular membranes.

as cholesterol may fill in the spatial gaps formed by kinks (see Appendix B for a longer discussion of lipid geometry).

Biomembranes are not composed of just one type of lipid, although it is possible to produce and study pure lipid bilayers in the laboratory. Rather, a selection of lipids possessing a range of hydrocarbon chain lengths and polar head groups are found in cell membranes, as illustrated in Table 7.1. The range of chain lengths for membrane lipids is centered around 15–18 carbon atoms, reflecting the operating temperature and lipid concentration of the cell. If the chains are rather short, the lipids won't form bilayers at low concentrations, while if the chains are overly long, the bilayer is too viscous and lateral diffusion of molecules within the bilayer is restricted. At ~0.1 nm per CH_2 group on the chain, each hydrocarbon chain has a length close to 2 nm in a single layer, for a total bilayer thickness of 4–5 nm. The mean cross-sectional area of a single chain is about 0.20 nm^2, while the average surface area of the bilayer occupied by most membrane lipids is 0.4–0.7 nm^2 (see Nagle and Tristram-Nagle, 2000). Furthermore, membranes contain more than just phospholipids. For example, rigid cholesterol molecules also may be present, ranging from 0% to 17% by total lipid weight in the plasma membranes of *E. coli* and the red blood cell, respectively. Important for the functioning of the cell, proteins are embedded in membranes to selectively permit and control the passage of material across the bilayer. Such membrane proteins are much larger than lipid molecules, and they enhance the thickness of the membrane.

The bilayer is an attractive arrangement for the binary surfactant/water mixture because the hydrophobic chains of the lipid are not exposed to water, while the hydrophilic head groups are solvated. However, this is not the only way of organizing the hydrophobic chains so that they avoid an aqueous environment: a sphere or cylinder, whose surface is lined by polar groups and whose interior is filled with hydrocarbons, also does the job. Which of these phases is most favored for a given amphiphile? Consider first single-chain fatty acids. At very low concentrations, the molecules disperse in water, as illustrated in Fig. 7.4(a). Yet, owing to their hydrophobic

Table 7.1 Chain lengths of fatty acids in rat liver membranes, where n_c is the number of carbon atoms in the hydrocarbon chain, including the COO group

	Percent by weight					
Membrane	$n_c = 14$	15	16	17	18	20
mitochondrial (outer)	<1	27	25	14	14	16
mitochondrial (inner)	<1	27	22	16	16	19
plasma membrane	1	37	31	6	13	11
Golgi apparatus	1	35	23	9	18	15

Note. The distributions are a percent of the total fatty acid content by weight; not all values of n_c are listed and $n_c = 20$ includes only arachidonic acid. Note that the fatty acids are generally bound as phospholipids and other molecules (from Gennis, 1989).

(a)

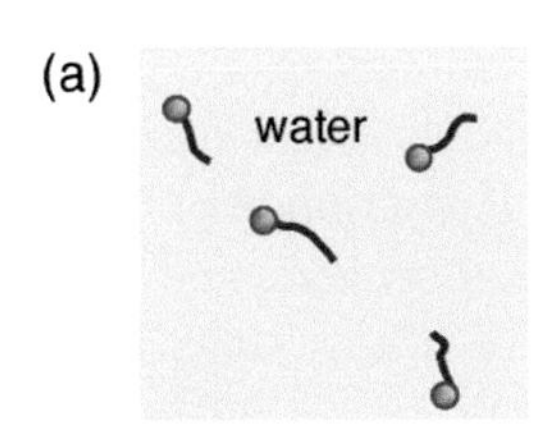

(b)

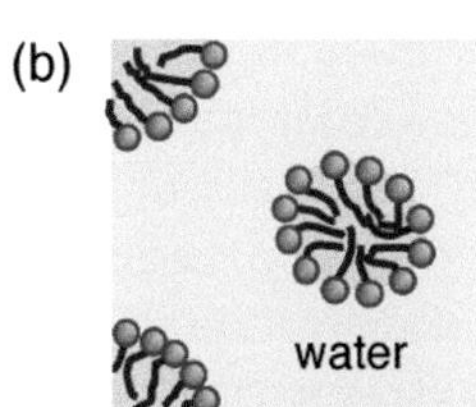

(c)

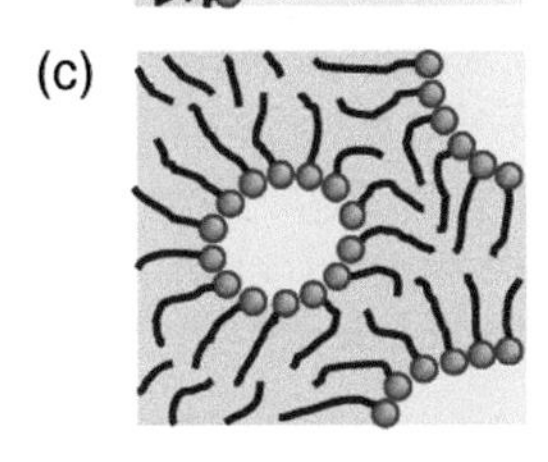

Fig. 7.4

Phases expected for fatty acids in water, where the linear hydrocarbon chain of the acid is not too short. (a) Homogeneous solution at low amphiphile concentration, (b) micelle formation at intermediate concentration, and (c) inverted micelles at high concentration. The roughly circular structures in (b) and (c) are a few nanometers in radius for fatty acids with 10–20 carbon atoms.

chains, the amphiphiles rapidly reach the saturation point when their concentration is raised, typically in the 10^{-3} molar range (or less for long acyl chains). At higher concentrations, the amphiphiles assemble into collective structures such as the micelles shown in Fig. 7.4(b). Typically formed by fatty acids, the surface of the micelle is covered by polar groups while the hydrocarbon tails gather in the interior. The micelle may have the shape of a spherical droplet, or a long cylinder like a sausage, having dimensions of nanometers (transversely) to microns (longitudinally). The onset of micelle formation occurs at a threshold called the critical micelle concentration (CMC); raising the amphiphile concentration above the CMC increases the number of micelles, but only weakly affects their average size. At high amphiphile concentrations, micelles fill the volume of the system or form the inverted micelle phase of Fig. 7.4(c).

The phase favored by a particular amphiphile partly reflects its molecular shape. Consider the three configurations displayed in Fig. 7.5. The spatial region occupied by a molecule in a spherical or cylindrical micelle could be imagined to look like an ice-cream cone or a wedge-shaped slice of pizza, as in Fig. 7.5(a). The interior of the micelle is cramped, and only an amphiphile with either a single hydrocarbon chain, or perhaps two rather short chains, would make a comfortable fit. In contrast, the interior of a bilayer is somewhat roomier, and is favored by dual-chain lipids whose head group and chains have a similar cross-sectional area, like the cylinder shown in Fig. 7.5(b). However, if the head group of a dual-chain lipid is too small (say, ethanolamine, which is the smallest of the four common species displayed in Fig. 7.3), the bilayer is less favorable than the inverted micelle phase of Fig. 7.5(c). We see from these examples that the preferred geometry is affected by the ratio of head group area to the cross-sectional area of the hydrocarbon region.

(a)

(b)

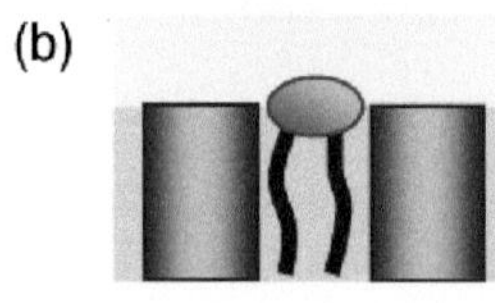

(c)

Fig. 7.5

Interface shape is related to molecular geometry. (a) Single-chain fatty acids tend to form micelles, while (b) dual-chain phospholipids with moderate size head groups prefer bilayers. (c) Phospholipids with small head groups may form inverted micelles. The mean molecular shape is indicated by the cones and cylinders drawn beside the amphiphile.

A flat bilayer has a mean separation between lipid molecules that minimizes their free energy. When subjected to a deformation such as the bending mode displayed in Fig. 7.6, the mean spacing between head groups in one layer is less than optimal, while the spacing in the other layer is greater than optimal. Thus, a bending mode usually requires an input of energy determined by the bending resistance of the bilayer. Depending upon its structure and environment, the bending resistance of a sheet increases like the square or cube of its thickness, so that very thin sheets are highly flexible. We are all familiar with the unruly behavior of thin sheets of plastic wrap or aluminum foil which, in spite of the rigidity of their parent material, twist and fold in the most exasperating manner. The same is true of bilayers, which, being much thinner than household materials, are very floppy and may exhibit thermal undulations on cellular length scales. In this chapter, we introduce a very simple mathematical description of bending elasticity, exploring its zero-temperature implications here, and finite temperature effects in Chapters 8 and 9. The role of bilayer elasticity in determining cell shape is investigated in Chapter 10.

Membranes also may be under tension or compression in a cell. For instance, a plant cell wall may bear a mechanical load from the bending of the plant, or a cell in the circulatory system may be subject to shear stress. Internal stresses may arise from an elevated osmotic pressure, or from contractile forces generated by interacting filaments within the cell. Small stresses can be accommodated by small alterations in molecular configurations within the membrane, perhaps including a change in the mean separation between head groups, the magnitude of the change reflecting the membrane's compression modulus. At sufficiently high stress, the membrane may tear or rupture, the amphiphiles rearranging themselves to reduce exposure of their hydrocarbon regions to water, as illustrated in Fig. 7.7. However, the rearrangement of the lipids at the pore boundary results in configurations that are not as energetically favorable as the bilayer itself, such that there is an energy penalty for the formation of holes; this can be characterized by a line tension or edge tension, which is an energy per unit length along the boundary of the hole. The effective edge tension is temperature-dependent and vanishes at sufficiently high temperature, where the membrane is unstable against hole formation even in the absence of mechanical stress.

In this section, the boundaries of cells or organelles are viewed as effectively uniform structures as far as their mechanical properties are concerned. Nevertheless, we noted that the cell's membranes themselves contain many different molecular species: there are a variety of lipids as well as embedded proteins. Further, the membrane-associated cytoskeleton provides the boundary with shear resistance and alters its mechanical behavior. For example, the suite of ground-state configurations in human

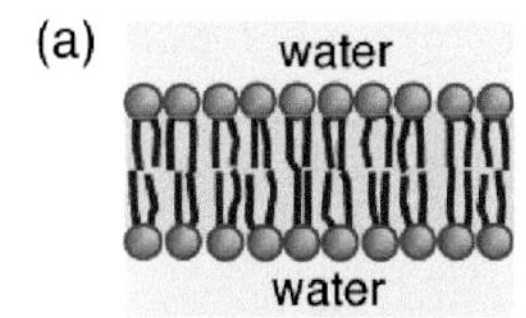

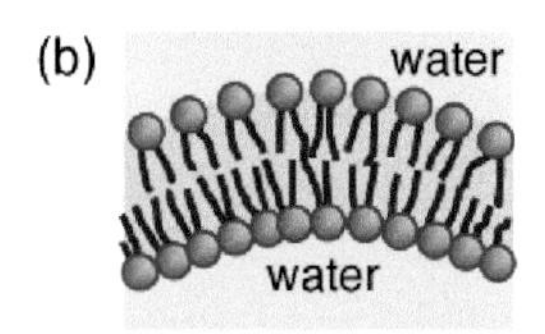

Fig. 7.6

Bending a symmetric bilayer from a planar configuration (a) to a curved shape (b) involves stretching (top) and compressing (bottom) the individual leaflets.

erythrocytes reflects the interplay between the fluid bilayer and the shear-resistant spectrin cytoskeleton (Lim *et al.*, 2008). In addition, the properties of the boundary's structural components influence each other, an example being the lateral diffusion of proteins within the bilayer. Small molecules diffuse within the plasma membrane at rates similar to lipids, with appropriate correction factors. However, large objects such as proteins may be stuck inside a corral created by the associated cytoskeleton (Sheetz, 1983), reducing their lateral diffusion coefficient by an anomalously large amount (Kusumi and Sako, 1996; Saxton and Jacobson, 1997; Tang and Edidin, 2003; reviewed further in Kusumi *et al.*, 2010). The effect on diffusion arising from fluctuations in the cytoskeleton–membrane spacing has been investigated by computer simulation (among many others, Boal and Boey, 1995; Brown, 2003).

As the plasma membrane is examined ever more closely, it emerges that it may not just be compositionally inhomogeneous, but also spatially inhomogeneous because of the presence of lipid rafts (Simons and Ikonen, 1997). In synthetic membranes, the possibility that a multi-component system may separate into distinct compositional domains has been studied most intensely in ternary systems containing, for example, sphingomyelin or cholesterol (Dietrich *et al.*, 2001; Veatch and Keller, 2003; summarized in Veatch *et al.*, 2008). The difference between the two types of domains commonly involves a different degree of ordering among the lipid hydrocarbons, although even more distinct phases may be present, for example, the ripple phase (Leidy *et al.*, 2002). The fact that the two phases have distinct mechanical and structural properties opens up the possibility that they can preferentially incorporate proteins, and act as protein-sorting centers (McIntosh *et al.*, 2003). The application of the phase boundaries obtained in synthetic systems to the plasma membrane must be done with some caution, however, as the phase diagram has been shown to be sensitive to the probe species used in the measurements (Veatch *et al.*, 2007) and may also be sensitive to the presence of small amounts of cross-linked lipids (Putzel and Schick, 2009).

The lipid membranes described in this chapter are two-dimensional fluids: they cannot resist a shear stress and their molecular components

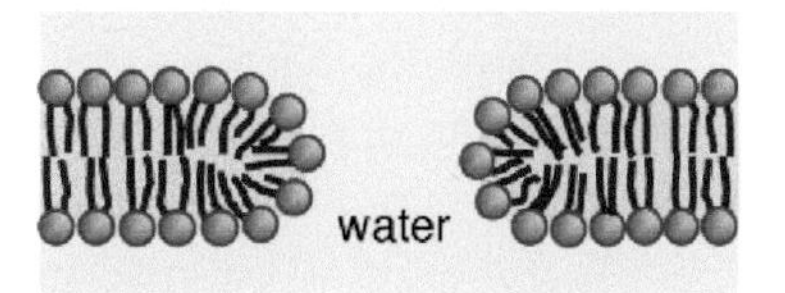

Fig. 7.7

The formation of a hole in a fluid bilayer forces amphiphilic molecules to rearrange themselves to protect the hydrocarbon interior of the bilayer from exposure to water.

diffuse within the membrane plane. However, there are other examples of sheet-like structures in the cell that do resist shear. Some of these, like the membrane-associated cytoskeleton of the erythrocyte, are molecularly thin, while others, such as the peptidoglycan network in Gram-negative cells, have just a few molecular layers. Much thicker than the wall around a small bacterium, sheath-like structures may provide organization to strings of cells in a filament, and lamina may supply cohesion to large, extended sheets of cells. Even such thick structures can be described using two-dimensional elastic parameters such as the bending resistance or area compression modulus, so long as the thickness of the structure is much less than the dimensions of the cell as a whole. For example, the cell wall of a Gram-positive bacterium may have a thickness of 25 nm (see Section 5.1) and the plant cell wall (see Section 1.2) may be 200 nm across, yet both of these can be analyzed within the framework of two-dimensional mechanics if desired.

We cover a variety of membrane characteristics in this chapter, although the formalism invoked is neither extensive nor challenging. Rather than segregate the formal development of a topic from its application to cell mechanics, as we do in many chapters of this text, here we treat theory and experiment together within each section. We present a mathematically simple model for self-assembly in Section 7.2 that avoids the use of chemical potentials, at the loss of some predictive power available in a completely rigorous treatment. This section also demonstrates the relationship between aggregate structure and molecular shape. Of importance in cell function is not only the self-assembly of membranes, but also their behavior under mechanical stress, and Sections 7.3 and 7.4 begin a discussion of the compression and bending resistance (to be continued throughout Chapters 8 and 10) of bilayers. Section 7.5 examines the energetics of hole formation and its relation to membrane topology and structural failure under stress. Lastly, the properties of shear-resistant structures such as the cell wall and basal lamina are presented in Section 7.6.

7.2 Self-assembly of amphiphiles

Amphiphiles, such as the phospholipids of cell membranes, can form aggregates in aqueous solution if their concentration is above a threshold value commonly called the critical micelle concentration (CMC, although the threshold is more accurately the critical aggregation threshold). The driving force for aggregation is the aversion of the amphiphile's hydrophobic regions to exposure to water: the interior of a micelle provides a safe haven for hydrocarbon chains and reduces their contact energy with the

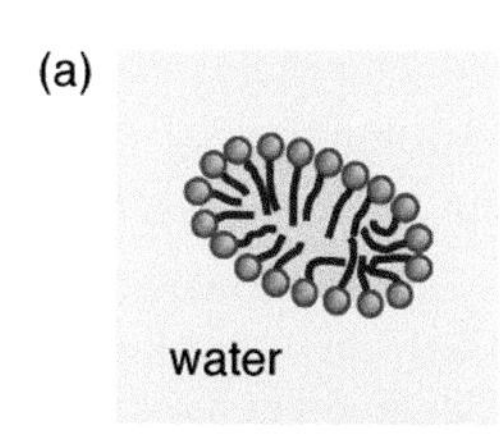

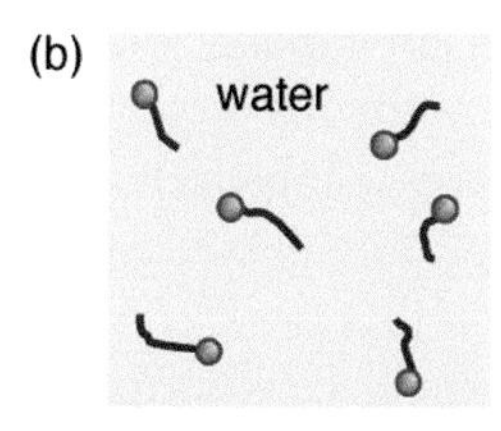

Fig. 7.8

A simple two-phase model for the formation of aggregates in a water/amphiphile system. In (a), the amphiphiles are condensed into a single blob, while in (b) they are dispersed throughout the aqueous phase.

aqueous environment. However, the formation of a cluster like a micelle lowers the number of objects in a system, and hence reduces the overall entropy. Thus, there is a competition between energy, which favors the formation of clusters, and entropy, which favors the distribution of molecules throughout the solution.

Although most treatments of aggregation use the language of chemical potentials (for example, Chapters 2 7 of Tanford, 1980, and Chapter 16 of Israelachvili, 1991), we present here a description based upon the free energies of two idealized phases, invoking just the thermodynamic relations developed in Appendix C. As our reference state, we consider a single blob of amphiphiles in a bath of polar water molecules, as in Fig. 7.8(a). We compare this condensed phase with the solution in Fig. 7.8(b), which we view as a gas of amphiphiles in an aqueous medium. Clearly this picture is an oversimplification: there is a distribution of cluster sizes in any physical system and individual amphiphiles are present in solution even once an aggregate has formed.

We define the reference state in Fig. 7.8(a) to have zero free energy: the entropy of the blob is neglected, and the potential energy between molecules within the blob is defined to be zero. With these definitions, each amphiphile pays an energy penalty $E_{\rm bind}$ to escape from the cluster; we approximate $E_{\rm bind}$ as the energy required to create the new water/hydrocarbon interface once its hydrophobic region is exposed to water in the solution phase. This interfacial energy is crudely equal to the water/hydrocarbon surface tension γ multiplied by the effective area of the hydrocarbon chains. We take the hydrocarbon region of the amphiphile to have the geometry of a cylinder of radius $R_{\rm hc}$ and length $n_{\rm c}\ell_{\rm cc}$, where $n_{\rm c}$ is the number of carbon atoms along the hydrocarbon chain and $\ell_{\rm cc} = 0.126$ nm is the average C–C bond length projected on the chain (see problem set and Chapter 16 of Israelachvili, 1991). Hence, the area of the hydrophobic region is approximately $2\pi n_{\rm c} R_{\rm hc}\ell_{\rm cc}$, yielding a removal energy per molecule of

$$E_{\rm bind} = 2\pi n_{\rm c} R_{\rm hc}\ell_{\rm cc}\gamma, \tag{7.1}$$

where we have omitted the end cap area of the cylinder.

In entering the aqueous phase, an amphiphilic molecule pays an energy penalty, but acquires a much larger configuration space to explore, hence increasing the entropy of the system. Let us assume that the solution phase is sufficiently dilute that dissolved amphiphiles behave like an ideal gas. As established in Appendix C (see also Section 9.10 of Reif, 1965), the entropy per molecule $S_{\rm gas}$ of an ideal gas at number density ρ is

$$S_{\rm gas} = k_{\rm B}\{5/2 - \ln(\rho \bullet [h \,/\, \{2\pi m k_{\rm B}T\}^{1/2}]^3)\}, \tag{7.2}$$

where m is the mass of each molecule and h is Planck's constant. The length scale $h/\{2\pi m k_{\rm B}T\}^{1/2}$ is provided by the translational motion of the molecule at $T > 0$.

Given the expressions for E_{bind} and S_{gas}, in our model system, the free energy per molecule F_{sol} in the solution phase at fixed volume is

$$F_{sol} \sim E_{bind} - TS_{gas}, \tag{7.3}$$

where we assume that the entropy of bulk water is largely unchanged by the presence of amphiphiles in solution, and that γ includes changes in entropy arising from the ordering of water molecules near an amphiphile. Whether the energy or entropy of the amphiphiles dominates the free energy F_{sol} depends upon the density ρ. At low densities, the entropy per particle dominates and the solution phase is favored, while at high amphiphile density, E_{bind} favors the condensed phase. With our definition that the free energy of the condensed (reference) state is zero, the cross-over between phases occurs at $F_{sol} = 0$, or $E_{bind} = TS_{gas}$, corresponding to an aggregation density ρ_{agg} of

$$\rho_{agg} \bullet [h \,/\, \{2\pi m k_B T\}^{1/2}]^3 = \exp(5/2 - E_{bind}/k_B T), \tag{7.4}$$

from Eqs. (7.1) and (7.2). This expression confirms our intuition that the threshold for aggregation *decreases* as the binding energy of an amphiphile *increases*.

To evaluate the accuracy of Eq. (7.4), we calculate ρ_{agg} for two generic amphiphiles – single-chain and double-chain phospholipids. Both amphiphiles are taken to have 10 carbon atoms per hydrocarbon chain, corresponding to a length $n_c \ell_{cc}$ of 1.26 nm. With molecular masses in the range of 400 Da (single) and 570 Da (double), the phospholipids have length scales $h \,/\, \{2\pi m k_B T\}^{1/2}$ of 5.1×10^{-12} m (single) and 4.3×10^{-12} m (double) at $T = 300$ K. To estimate E_{bind}, we use a surface tension of $\gamma = 0.05$ J/m^2 (this is γ for short alkanes and water; Weast, 1970), although γ may be as low as 0.02 J/m^2 for some lipids (Parsegian, 1966). The effective hydrocarbon radius of a single chain is 0.2 nm; the effective radius of a double-chain lipid is less than twice that of a single chain, and we take $\sqrt{2} \bullet 0.2 = 0.3$ nm as a reasonable approximation. These values for γ, n_c and R_{hc} give $E_{bind}/\, k_B T = 20$ (single) and 30 (double), such that Eq. (7.4) yields

$$\rho_{agg}\,\text{(single)} \sim 2 \times 10^{26} \,/\, \text{m}^3 \sim 0.3 \text{ molar} \tag{7.5}$$

$$\rho_{agg}\,\text{(double)} \sim 1.4 \times 10^{22} \,/\, \text{m}^3 \sim 2 \times 10^{-5} \text{ molar.}$$

We see immediately that the aggregation threshold for double-chain lipids is much less than that of single-chain lipids, and is low in an absolute sense; we return to the importance of this finding to properties of the cell at the end of this section. How does Eq. (7.5) compare to experiment? The observed values of the CMC of single- and double-chain lipids depend strongly on the head group and lie between 10^{-3} to a few times 10^{-2} molar for single-chain lipids and 10^{-5} to 10^{-3} molar for dual-chain lipids, all with 10 carbon atoms per hydrocarbon chain. The values calculated in Eq. (7.5) are not in strong disagreement with the range observed experimentally.

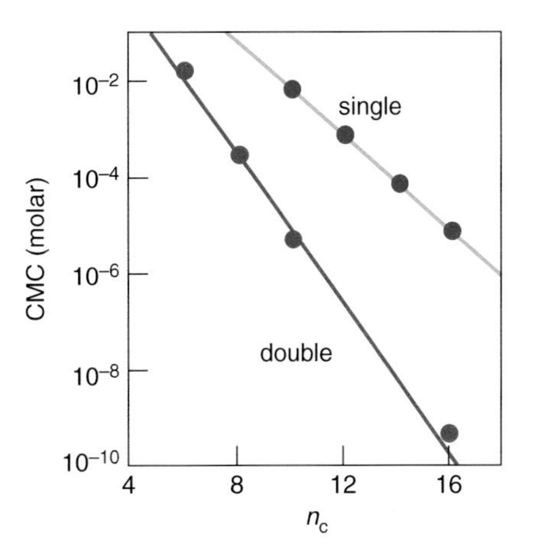

Fig. 7.9 Semilogarithmic plot of critical micelle concentration (molar) as a function of the number of carbon atoms n_c in an individual chain. The single-chain lipids are lyso-phosphatidylcholines and the double-chain ones are di-acyl phosphatidylcholines. The straight lines are the fitted functions 690 exp(−1.15n_c) molar and 570 exp(−1.8n_c) molar for the single- and double-chain lipids, respectively. Data compilations are from Marsh (1990) and Chapter 16 of Israelachvili (1991).

A better test of the physics underlying Eq. (7.4) can be made by examining the chain length dependence of the CMC for a given type of lipid. Examples of both single- and double-chain lipids with the chemical composition R_nPC and R_nR_nPC are displayed in Fig. 7.9, where R_n is a linear hydrocarbon with n_c carbon atoms and PC refers to the phosphatidylcholine head group. The first thing to note about the figure is that the single-chain lipids have uniformly higher CMCs than double-chain lipids for the same number of carbon atoms per chain. This is as expected, because the symmetric dual-chain lipids in Fig. 7.9 have twice as many carbon atoms as their single-chain counterparts, resulting in a larger hydrophobic area and hence a greater tendency to form aggregates at low concentration. The second feature is the exponential dependence of the CMC on the number of carbon atoms, namely 690 exp(−1.15n_c) molar and 570 exp(−1.8n_c) molar for the specific R_nPC and R_nR_nPC sequences in the figure. For each additional CH_2 group added to one of the chains, the CMC drops approximately by factors of 3 and 2.5 for the single- and dual-chain phospholipids in the figure, respectively. This exponential behavior of the CMC can be seen in Eq. (7.4) when it is written as

$$\rho_{\mathrm{agg}} = [\{2\pi m k_{\mathrm{B}} T\}^{1/2} / h]^3 \, e^{5/2} \exp(-\,2\pi n_{\mathrm{c}} R_{\mathrm{hc}} \ell_{\mathrm{cc}} \gamma / k_{\mathrm{B}} T), \qquad (7.6)$$

after substituting Eq. (7.1). The exponential factors 1.15 and 1.8 from the fits to the data in Fig. 5.9 can be compared directly to $2\pi R_{hc}\ell_{cc}\gamma/k_BT$ in Eq. (7.6). Taking the values for R_{hc}, ℓ_{cc} and γ used in the calculation of Eq. (7.5), the exponential slopes of Fig. 7.9 are predicted to be 2.0 and 3.0 respectively, about a factor of two larger than observed in the fit. Considering the approximations involved in obtaining Eq. (7.6), and the assumed values of the parameters, this prediction is not unimpressive. In particular, if the surface tension at the molecular scale is closer to 0.030 J/m^2, the predicted exponents are close to the measured values ($\gamma = 0.02$ J/m^2 has been obtained in an analysis of phase transitions in water + fatty-acid salt systems by Parsegian, 1966).

A more rigorous treatment of aggregation, which yields a result similar to Eq. (7.6), is performed in Chapter 16 of Israelachvili (1991), where the two-phase assumption (solution or aggregate) is dropped and a distribution of aggregate sizes is permitted. This approach allows one to calculate the average cluster size as a function of amphiphile concentration, and to show how the free amphiphile concentration in solution approaches its asymptotic value near the CMC. Readers familiar with the concepts of chemical equilibrium will not find this approach difficult. The phase behavior of multicomponent systems, for example with more than one type of amphiphile, is more complex than the two-component system explored here; introductions to this topic can be found in Mouritsen (1984) and Gompper and Schick (1994).

Our model for aggregation considers only two states – a single cluster vs. an ideal solution – without regard to the shape of the cluster. However, we know experimentally that aggregates can have different geometries, including spherical or cylindrical micelles and bilayers. Is there a way of understanding which of these geometries is favored by a given amphiphile without having to calculate the free energy of each possible phase? Israelachvili *et al.* (1976) demonstrated the relationship between amphiphile shape and aggregate geometry by comparing the mean shape of the amphiphile with the optimal molecular packing arrangement in a cluster.

Using known carbon–carbon bond lengths and angles, a single saturated hydrocarbon chain with n_c carbon atoms has a volume of $v_{hc} = 27.4 + 26.9n_c \times 10^{-3}$ nm^3 (consistent with data from Lewis and Engelman, 1983) and a maximum chain length of $\ell_{cc} = 0.154 + 0.126n_c$ nm (see Tanford, 1980). Note that the actual thickness of a single bilayer leaflet will be less than the length of a fully extended chain because of the presence of kinks (Lewis and Engelman, 1983; Bloom *et al.*, 1991, review bond orientation; Nagle and Tristram-Nagle, 2000, review lipid sizes). As n_c increases, the constant terms in the expressions for v_{hc} and ℓ_{hc} become less important, and the effective cross-sectional area of the chain becomes $a_{hc} = v_{hc}/\ell_{hc} = 0.21$ nm^2. In a micelle or bilayer, the average surface area a_o occupied by the head group of an amphiphile is greater than a_{hc} of a single chain, being near 0.5 nm^2 for many phospholipids. The optimal packing arrangement of molecules in a cluster reflects the difference between a_o and the average cross-sectional area of the hydrocarbon chain or chains, taking into account any kinks or coils in the chains. We now examine the molecular packing in several aggregate shapes.

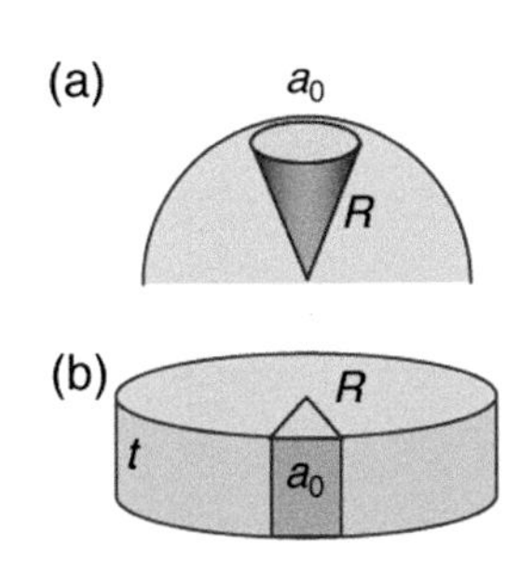

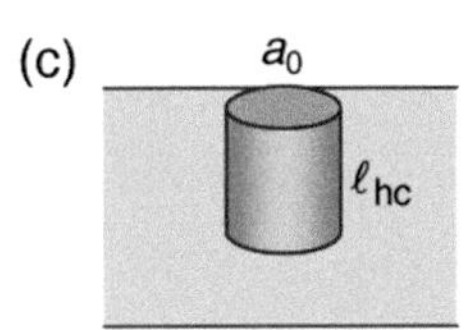

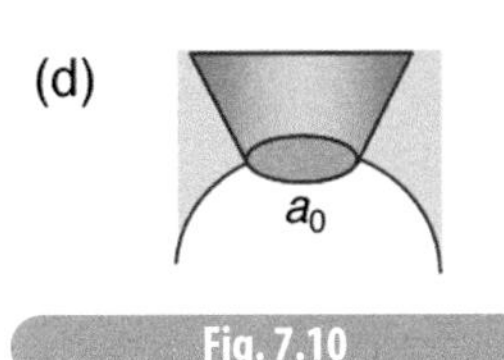

Fig. 7.10

Packing constraints experienced by a typical amphiphile in four aggregates: (a) spherical micelle, (b) cylindrical micelle, (c) bilayer, and (d) inverted micelle. The cross-sectional area of a head group is denoted by a_o, while the radius and thickness of the molecular shape are R and t, as applicable.

7.2.1 Spherical micelles

As shown in Fig. 7.10(a), a spherical micelle of radius R has a surface area $4\pi R^2$ and volume $4\pi R^3/3$. Knowing the average surface area a_o occupied by a molecule, as well as the volume of its hydrocarbon chain(s) v_{hc}, one can deduce the number of molecules in the micelle as either $4\pi R^2/a_o$ (i.e. the total area divided by the area per molecule) or $(4\pi R^3/3)/v_{hc}$ (i.e. the total volume divided by the volume per molecule). For both of these expressions to give the same number of molecules, R must satisfy

$$R = 3v_{hc}/a_o. \tag{7.7}$$

Now, the molecules of a micelle are arranged so that their polar head groups reside on the exterior of the cluster, while the chains reside inside. The interior of the cluster cannot contain a void, so the distance from the centre to the surface cannot exceed the length of one amphiphile, as shown in Fig. 7.4(b). Thus, the radius of the micelle must be less than or equal to the projected length of the hydrocarbon chain ℓ_{hc} (shown in Fig. 7.1 for a saturated chain; note that a saturated chain would have to be placed under considerable tension to produce a completely linear chain from the zig-zag configuration displayed). Since $R \leq \ell_{hc}$, Eq. (7.7) becomes

$$v_{hc}/a_o\ell_{hc} \leq 1/3. \qquad \text{spherical micelles} \tag{7.8}$$

According to this expression, spherical micelles are favored by large values of a_o compared to v_{hc}/ℓ_{hc}, corresponding to the conical shape shown in Fig. 7.10(a). The combination $v_{hc}/a_o\ell_{hc}$ is called the shape factor.

7.2.2 Cylindrical micelles

As the volume of the hydrocarbon region of an amphiphile increases for a given head group, perhaps because the chains are longer or because there are two chains instead of one, the micelle must distort from its spherical shape to a prolate ellipsoid (cigar-shaped). The length of the ellipsoid can increase to accommodate a larger internal volume until the micelle becomes truly cylindrical. The maximum value of the shape factor for ellipsoidal and cylindrical micelles can be obtained by considering a section of a uniform cylinder of thickness t as shown in Fig. 7.10(b), where the axis of cylindrical symmetry is vertical. The surface area of this section around the ring is $2\pi Rt$, and the volume is $\pi R^2 t$. Repeating the same argument that led to Eq. (7.7), we determine that the number of molecules in the cylindrical section is either $2\pi Rt/a_o$ (i.e. the total area divided by the area per molecule) or $\pi R^2 t/v_{hc}$ (i.e. the total volume divided by the volume per molecule). Equating these two expressions for the number of molecules, the radius must obey

$$R = 2v_{hc}/a_o. \tag{7.9}$$

Substituting $R \leq \ell_{hc}$, as before, leads to

$$1/3 < v_{hc}/a_o\ell_{hc} \leq 1/2, \qquad \text{cylindrical micelles} \tag{7.10}$$

for the range of shape factor favoring cylindrical micelles.

7.2.3 Bilayers

At even larger values of the internal volume than permitted by Eq. (7.10), cylindrical micelles distort towards bilayers by spreading so that they become elliptical in cross section rather than circular. The packing geometry of the bilayer is best satisfied by cylindrical molecules, whose hydrocarbon volume is given by $v_{hc} = a_o\ell_{hc}$, as shown in Fig. 7.10(c). Hence, the "ideal" bilayer satisfies $v_{hc}/a_o\ell_{hc} = 1$, and the range of the shape factor favored for bilayers is

$$1/2 < v_{hc}/a_o\ell_{hc} \leq 1. \qquad \text{bilayers} \tag{7.11}$$

Given that the typical value of a_o for phospholipids is 0.5 nm^2, about double the average cross-sectional area of a single hydrocarbon chain, it is no surprise that dual-chain phospholipids preferentially form bilayers compared with single-chain lipids.

7.2.4 Inverted micelles

Lastly, we consider what happens when the hydrocarbon volume v_{hc} is larger than the product of the head group area a_o with the chain length ℓ_{hc}, such that the shape factor $v_{hc}/a_o\ell_{hc}$ exceeds unity. Here, the head group is relatively small, and lies near the apex of a truncated cone as shown in Fig. 7.10(d); this situation is the inverse of the molecular shape favored by micelles. Such molecules form inverted micelles, with the head groups on the "inside" of the micelle, and the hydrocarbon regions radiating away from the aqueous core. Thus, the range of shape factor favoring inverted micelles is

$$v_{hc}/a_o\ell_{hc} > 1. \qquad \text{inverted micelles} \tag{7.12}$$

7.2.5 Amphiphiles in the cell

From Eqs. (7.8) to (7.12) we have established a correspondence between the value of the shape factor $v_{hc}/a_o\ell_{hc}$ and aggregate geometry preferred by an amphiphile:

0 to 1/3	spherical micelles
1/3 to 1/2	cylindrical micelles

1/2 to 1	bilayers
greater than 1	inverted micelles.

What does this imply for aggregates in the cell? Let's consider amphiphiles having sufficiently long hydrocarbon chains that their length and volume are linear functions with zero intercept, with $\ell_{hc} = 0.126 n_c$ nm for chains in a linear zig-zag configuration (less for chains in a bilayer; see Lewis and Engelman, 1983) and $v_{hc} = 26.9 n_c \times 10^{-3}$ nm^3, where n_c is the number of carbon atoms in a single chain. These values imply a cross-sectional area of 0.21 nm^2 for a single chain, consistent with observations from X-ray scattering on single layers of single-chain surfactants. The shape factor is then $0.21/a_o$ for a single-chain amphiphile, and $0.42/a_o$ for a double-chain, where a_o is quoted in nm^2. Head groups in the cell commonly have cross-sectional areas near 0.50 nm^2, ranging from ~0.40 nm^2 for phosphatidylethanolamines to 0.5–0.7 nm^2 for phosphatidylcholines, depending upon their liquid crystalline phase (Huang and Mason, 1978; Lewis and Engelman, 1983; Nagle and Tristram-Nagle, 2000). From these estimates of ℓ_{hc}, a_o, and v_{hc}, the shape factor should be roughly 0.4 for single-chain lipids, and 0.8 for double-chain lipids. Thus, molecular packing arguments predict that single-chain lipids in the cell are most likely to be found in spherical or distorted micelles, and double-chain lipids probably form bilayers.

The fact that dual-chain lipids form bilayers has another implication for cells: we have established that such lipids also have a very low CMC, less than 10^{-5} molar for many lipids of interest. This characteristic is consistent with our design principles for the cell: a low CMC means that the concentration of lipids need not be very high before a bilayer will condense from solution, and that the concentration of lipids remaining in solution once the bilayer forms is fairly low. In other words, the building materials of the cell, once produced by metabolic or other means, are used very efficiently to build membranes. Lastly, although we have emphasized the influence of molecular geometry in selecting the condensed states favored by a particular lipid, other factors may play a role as well. For example, egg lecithin lipids in a salt solution form spherical vesicles at elevated salt concentrations, which evolve into worm-like micelles and then bilayer vesicles as the salt concentration is reduced (Egelhaaf and Schurtenberger, 1999).

7.3 Bilayer compression resistance

The molecular shape of an amphiphile influences the preferred geometry of its mesoscopic aggregate, as described in Section 7.2. The conformations of amphiphiles in an aggregate can be investigated by experimental techniques such as NMR (Marcelja, 1974; for a review, see Bloom *et al.*,

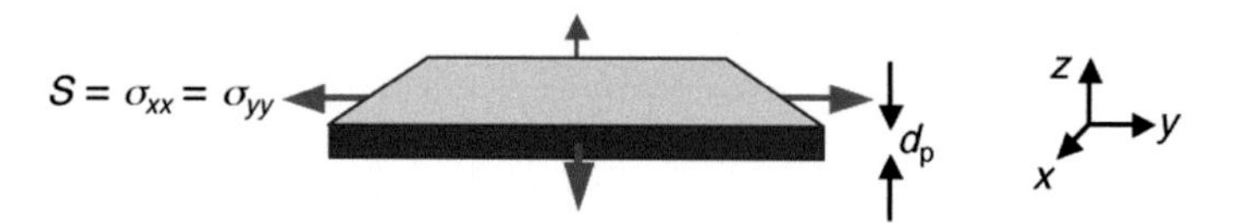

Fig. 7.11 Application of an in-plane tension $\tau = Sd_p$ to a uniform plate of thickness d_p.

1991) and by computer simulations (for example, Pink *et al.*, 1980; van der Ploeg and Berendsen, 1983; Egberts and Berendsen, 1988; Cardini *et al.*, 1988; Heller *et al.*, 1993; for a review, see Pastor, 1994), from which the elastic behavior of membranes can be obtained (see Goetz and Lipowsky, 1998; Goetz *et al.*, 1999, and references therein). In this and the following section, we explore how the elasticity of a bilayer reflects its molecular constituents. Here, we present two models for the area compression resistance of a bilayer and evaluate them experimentally. We do not treat the thermal expansion of bilayers, but refer to Evans and Waugh (1977b), Evans and Needham (1987) and Marsh (1990) for further reading.

As a first and perhaps naïve approach, we view the bilayer as a homogeneous rigid sheet, much like a thin metallic plate in air. To find its area compression modulus K_A, consider what happens when a tensile stress is applied to the square plate of thickness d_p shown in Fig. 7.11. In this situation, the stress tensor has two non-vanishing components, $\sigma_{xx} = \sigma_{yy} = S$, which generate the in-plane elements of the strain tensor

$$u_{xx} = u_{yy} = S\,(2\,/\,9K_V + 1\,/\,6\mu), \tag{7.13}$$

where K_V and μ are the volume compression and shear moduli in three dimensions, respectively [see Eqs. (6.17) – (6.19)]. Now, the description of a deformation in two dimensions involves a two-dimensionsional tension τ, which is a force per unit length, rather than the force per unit area in three dimensions. This tension is related to the three-dimensional stress σ through $\tau = Sd_p$ which gives rise to a relative change in area $u_{xx} + u_{yy}$ according to $\tau = K_A(u_{xx}+u_{yy})$. Substituting Eq. (7.13), we find

$$K_A = d_p K_V\,/\,(4/9 + K_V/\,3\mu). \qquad \text{uniform rigid plate} \tag{7.14}$$

For many materials, $K_V \sim 3\mu$, and the denominator is of order unity. The important result from this equation is that, because K_V and μ are bulk properties independent of the object's shape, then K_A should increase linearly with plate thickness.

Useful as this approach might be, it ignores an important feature of amphiphilic condensates that their constituents rearrange easily in response to a stress. Consider the interface formed by water and a monolayer of

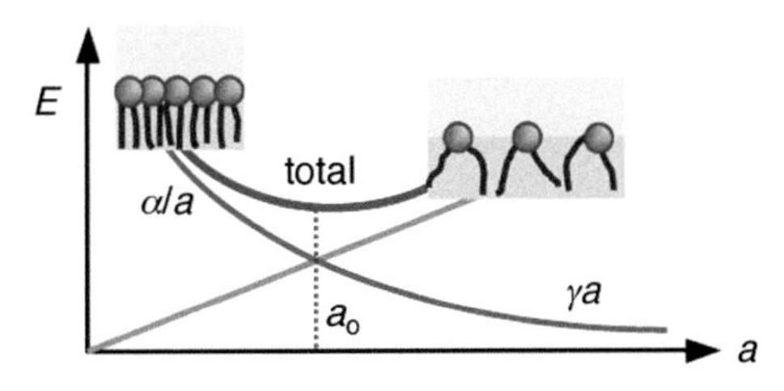

Fig. 7.12 Dependence of the energy per molecule E at a water + amphiphile interface as a function of the mean interface area a occupied by a surfactant molecule. The equilibrium value of a, defined as a_o, occurs where dE/da vanishes (after Chapter 17 of Israelachvili, 1991).

amphiphiles, as in the insets of Fig. 7.12. The mean interfacial area occupied by an amphiphile reflects a competition between the steric repulsion among the molecules and the surface tension of the interface. At low density, as on the right-hand side of Fig. 7.12, the energy penalty per molecule for exposing the amphiphile's hydrocarbon chains is equal to γa, where a is the mean interface area occupied by an amphiphile and γ is the surface tension of the water + amphiphile interface. At high density, as on the left-hand side of Fig. 7.12, the molecules are crowded together, and their repulsive energy rises with density as a power of $1/a$. The simplest dependence of the interface energy per molecule on a is then $E = \alpha / a + \gamma a$, although higher-order terms in a^{-1} become more important as a decreases (Chapter 17 of Israelachvili, 1991). As evident in Fig. 7.12, E has a minimum at the mean value of a at equilibrium, namely a_o. Setting $dE/da = 0$ gives $a_o = \sqrt{(\alpha/\gamma)}$, permitting E to be written as

$$E = 2\gamma a_o + (\gamma / a) \bullet (a - a_o)^2. \tag{7.15}$$

Only the second term changes as a departs from a_o.

Near equilibrium, the energy per amphiphile changes by $(\gamma / a_o) \bullet (a - a_o)^2$ relative to E at $a = a_o$, giving an elastic energy density of $\gamma[(a - a_o)/a_o]^2$ when divided by the area per molecule a_o. Now, this energy density also can be written as $(K_A/2) \bullet (u_{xx} + u_{yy})^2$, where $u_{xx} + u_{yy}$ equals the relative change in area, $(a - a_o)/a_o$. Comparing these expressions for the energy at small deformation, we find that K_A is 2γ for a monolayer, and 4γ for a bilayer:

$$K_A = 2\gamma \qquad \text{monolayer}$$
$$K_A = 4\gamma, \qquad \text{bilayer} \tag{7.16}$$

where the difference between the monolayer and bilayer arises because there are two interfaces to the latter. A polymer brush model for the bilayer gives the closely related expression for $K_A = 6\gamma$ (see Rawicz *et al.*, 2000). Experimentally, the surface tension of the water + amphiphile interface is in the range 0.02–0.05 J/m^2 (see discussion in Section 7.2), so Eq. (7.16) predicts that K_A of a lipid bilayer should lie in the range 0.08–0.2 J/m^2, and should be independent of bilayer thickness.

7.3.1 Experimental measurements of K_A and K_V

The area and volume compression moduli have been measured for a number of lipid bilayers using variations of a technique in which membrane strain is recorded as a function of applied stress. In most cases, the membrane is a component of a cell or a pure bilayer vesicle, with linear dimension of the order of microns, and the deformation is observed microscopically; for example, the membrane may be placed under stress through aspiration by a micropipette with a diameter of a few microns (Evans and Waugh, 1977a; Kwok and Evans, 1981; Evans and Needham, 1987; Needham and Nunn, 1990; Evans and Rawicz, 1990; Zhelev, 1998). The stress-induced change in membrane area is extracted from microscope images, and K_A is found from a fit to the stress–strain curve. Images and schematic representations of aspirated cells appear in Sections 1.4, 5.5 and 7.5. In a rather different approach first suggested by Parsegian *et al.* (1979), a stack of bilayers is placed under osmotic pressure causing the interface area per lipid to decrease, rather than increase as it does in the aspiration technique. Another version of this technique (Koenig *et al.*, 1997) uses X-ray scattering and the ordering of hydrocarbon chains to determine independently the average interface area per lipid; parenthetically, interface areas per lipid of 0.59, 0.61 and 0.69 nm^2 are found for diMPC, SOPC and SDPC, respectively (definitions follow Table 7.2).

Apparent values of K_A from the slope of the stress–strain curve are displayed in Table 7.2 for a number of pure lipid bilayers; many of the measurements can be seen to lie in the range of 0.1–0.2 J/m^2, as expected from the surface tension argument of Eq. (7.16). However, there are two contributions to the apparent area compression resistance for soft membranes, one arising from a change in intermolecular separation (K_A) and one arising from membrane undulations. At low tension, the membrane may fluctuate transversely such that its average in-plane area is less than its true area, an effect with the same entropic roots as the reduction in the end-to-end displacement of a floppy chain compared to its contour length (see Chapter 2). Thus, when the membrane is placed under *small* tensions, its in-plane area increases because out-of-plane undulations are suppressed, not because the interfacial area increases (Helfrich and Servuss, 1984; Evans and Rawicz, 1990; Marsh, 1997). At larger tensions, once the undulations have been ironed out, the membrane area does increase, as governed by K_A (shown analytically in Chapter 8). When bending motion is accounted for, K_A of many lipids in Table 7.1 rises to about 0.24 J/m^2 even though the bilayer thickness varies by up to 25% as measured by X-ray scattering. These data favor the surface tension approach of Eq. (7.16), rather than

Table 7.2 Selected measurements of the apparent area compression modulus K_A of lipid bilayers and cell membranes

Membrane	T (C)	Apparent K_A (J/m^2)	Reference
diAPC	15	0.057±0.014	Needham and Nunn, 1990
	18	0.135±0.020	Evans and Rawicz, 1990
	21	0.183±0.008	Rawicz *et al.*, 2000
diGDG	23	0.160±0.007	Evans and Rawicz, 1990
diMPC	21	0.150±0.014	Rawicz *et al.*, 2000
	29	0.145±0.010	Evans and Rawicz, 1990
	30	0.14	Koenig *et al.*, 1997
diOPC	21	0.237±0.016	Rawicz *et al.*, 2000
SDPC	30	0.12	Koenig *et al.*, 1997
SOPC	15	0.19±0.02	Needham and Nunn, 1990
	30	0.22	Koenig *et al.*, 1997
egg PC		0.14	Kwok and Evans, 1981
		0.17	Zhelev, 1998
red cell plasma membrane		0.45	Evans and Waugh, 1977 a, b

Notes. For a summary of measurements up to 1990, see Table II.10.2 of Marsh (1990). Correcting for bending motion, the extracted value of K_A rises to about 0.24 J/m^2 for many phospholipids with 13–22 carbons atoms per hydrocarbon (Rawicz *et al.*, 2000).

Abbreviations: diAPC = diarachidonyl-phosphatidylcholine; diGDG = digalactosyl-diacylglycerol; diMPC = dimyristoyl-phosphatidylcholine; diOPC = dioleoyl-phosphatidylcholine; SDPC = 1-stearoyl-2-docosahexaenoyl-phosphatidylcholine; SOPC = 1-stearoyl-2-oleoyl-phosphatidylcholine; egg PC = egg phosphatidylcholine.

the rigid plate model of Eq. (7.14); the former model predicts K_A is independent of membrane thickness d_p, while the latter predicts K_A increases linearly with d_p.

More chemically heterogeneous than a pure lipid bilayer is the plasma membrane of the red blood cell, which contains membrane-embedded proteins as well as ~40 mol % cholesterol. As can be seen from Table 7.2, this membrane has a significantly higher compression modulus than the pure lipid bilayers. In general, it is found that the presence of cholesterol increases the membrane's resistance both to compression and to bending (for example, Henriksen *et al.*, 2006; see Section 7.4), although the change in K_A as a function of the cholesterol fraction is not at all linear, as illustrated in Fig. 7.13 for an SOPC bilayer (Needham and Nunn, 1990). There is a gentle linear rise at low cholesterol concentrations, followed by

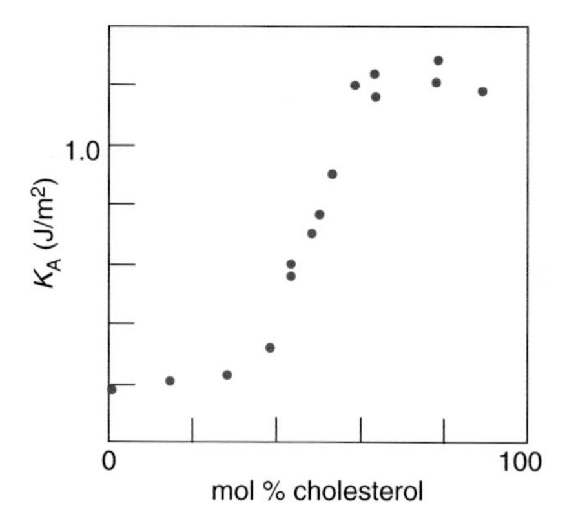

Fig. 7.13 Measured values of the area compression modulus K_A for SOPC bilayers containing cholesterol, as a function of cholesterol concentration expressed in mol % (redrawn with permission; after Needham and Nunn, 1990; ©1990 by the Biophysical Society).

a dramatic jump as the concentration passes through 50 mol %, reaching a plateau value that is about a factor of six larger than K_A of pure SOPC. Similar behavior is observed for mixtures of diMPC and cholesterol (Evans and Needham, 1987). In contrast, the presence of salicylate tends to lower K_A with increasing concentration (Zhou and Raphael, 2005) as do short-chain alcohols. Further, at fixed alcohol concentration, the lowering of K_A becomes more pronounced as the hydrocarbon chain length of the alcohol rises: the effectiveness in lowering K_A follows the order methanol < ethanol < propanol < butanol (Ly and Longo, 2004). The presence of proteins also alters the compression modulus (McIntosh and Simon, 2007).

Lastly, we note that the volume compression modulus K_V has been measured for some phospholipids commonly used in the study of bilayers (for a summary, see Table II.10.1 of Marsh, 1990). For example, the compression modulus of diPPC (dipalmitoyl-phosphatidylcholine) is observed to be 2–3 × 10^9 J/m^3 at physiologically relevant temperatures (Mitaku *et al.*, 1978). These values of the bulk modulus are in the same range as everyday liquids such as water ($K_V = 1.9 \times 10^9$ J/m^3) and ethyl alcohol ($K_V = 1.1 \times 10^9$ J/m^3) at 0 °C and atmospheric pressure.

7.3.2 Benchmark estimates of K_A

Simple models exist for estimating the area compression modulus, although they are less accurate than the surface tension argument leading to Eq. (7.16). For example, K_A of a two-dimensional ideal gas is equal to the pressure of the gas, which in turn equals $\rho k_B T$, where ρ is the molecular number density. Viewing the interface as a two-dimensional gas, we would assign $\rho = a_o^{-1}$ for a monolayer or $2a_o^{-1}$ for a bilayer. Taking a typical value of a_o

to be 0.5 nm^2 for phospholipids gives $\rho k_B T = 0.016$ J/m^2 for a bilayer, about an order of magnitude low compared with experiment. Alternatively, the thin plate model of Eq. (7.14) predicts $K_A \sim K_V d_p$, where d_p is the thickness of the plate. A typical value of K_V is 2–3 × 10^9 J/m^3 and a typical bilayer thickness is 4 nm, yielding ~10 J/m^3 for $K_V d_p$. This estimate is more than an order of magnitude larger than the measured values of K_A, indicating that the sliding of amphiphiles reduces the bilayer's compression resistance. In summary, the surface tension approach best captures both the bilayer thickness dependence and the absolute magnitude of K_A.

7.3.3 Rupture tension

As the bilayer is stretched, it becomes thinner and its hydrophobic core is increasingly exposed to water. At what point is the bilayer stretched so thin that it ruptures, releasing tension through the formation of a hole that permits the bilayer to return to near its optimal density? The answer to this question is time-dependent: it depends on how long you wait to see the membrane fail. On laboratory time scales, a membrane fails once its area is stretched just a few percent, often 2%–5%, beyond its equilibrium value (see Ly and Longo, 2004, and references therein). That is, applying a small tension to the bilayer causes it to stretch, but applying a sufficiently large tension causes it to fail once its area has increased by a few percent; rupture strains in the few percent range have also been found in computer simulations (Fournier and Joos, 2003). The corresponding tension at which failure occurs on laboratory time scales, referred to as the lysis tension, is of the order 0.01 J/m^2, with significant variation (Evans *et al.*, 2003). In general, the lysis tension increases with K_A of the membrane, and its magnitude can be estimated by noting that the tension is equal to the product of K_A and the critical strain at lysis, typically a few percent. It's possible to probe bilayer failure at a molecular level through the use of AFM (Künneke *et al.*, 2004). We discuss membrane failure again in Section 7.5; the reader is referred to Needham and Nunn (1990) for further discussion.

7.4 Bilayer bending resistance

Displacing the amphiphiles of a bilayer from their mean or equilibrium positions and configurations requires an input of energy. One example of this resistance to deformation is encountered in the compression or stretch of a bilayer, as described in Section 7.3, where compression is opposed by steric interactions between amphiphiles and stretch is opposed by the water + amphiphile surface tension. Another example is the bending of a bilayer, which involves both a stretching of one leaflet, and a compression of the other, as seen in Fig. 7.6(b). In this section, we introduce the energetics of bending deformations, and both summarize and interpret experimental measurements of bending elastic moduli. We defer to Chapter 8 a discussion of membrane structure and its influence on elasticity and cell shape.

7.4.1 Radius of curvature

Consider the two uniformly curved surfaces in Fig. 7.14: a spherical shell and a hollow cylinder, both of radius R. For the time being, we define the curvature C of a surface observed along a particular direction as the rate of change of a unit vector tangent to the surface, similar to our treatment of polymers in Section 3.2. For the sphere in Fig. 7.14(a), for instance, the rate of change of the unit tangent vector along a ring of radius R is equal to $1/R$; hence $C = 1/R$, where R must also be the radius of curvature. The sphere is a special case; in general, two curvature parameters are needed to describe a surface, as can be demonstrated from the cylinder in Fig. 7.14(b). Around the circumference of the cylinder, the tangent vectors change at the same rate as they do along an equator of the sphere, so the curvature is $1/R$. Yet parallel to the symmetry axis, the tangent vectors are parallel and the curvature vanishes, or alternatively, the radius of curvature is infinite. Thus, the cylinder is described by two inequivalent curvatures: $C_1 \equiv 1/R_1 = 1/R$ and $C_2 \equiv 1/R_2 = 0$. The curvature formalism for surfaces of arbitrary shape is developed in Chapter 8; in this section, we consider only simple shapes whose curvature is obvious from inspection.

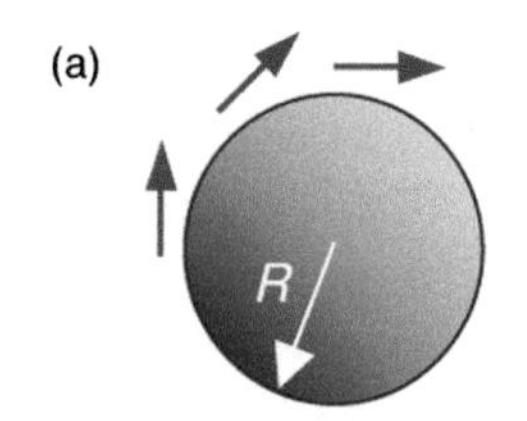

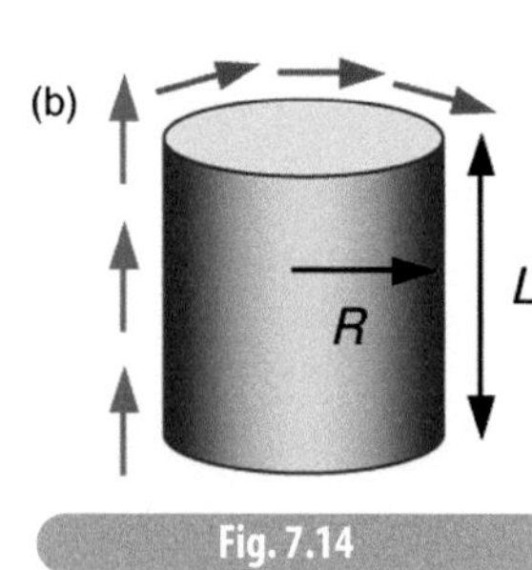

Fig. 7.14

Uniformly curved shells: (a) is a spherical shell of radius R, while (b) is a cylindrical shell of radius R and length L. Sets of vectors tangent to the surface are displayed for both configurations: red tangents define a line with curvature $1/R$; blue tangents define a line with zero curvature.

For a given molecular composition, the energy per unit surface area to bend a bilayer increases with the curvature. The simplest form for this energy density involves the squared mean curvature $(C_1/2 + C_2/2)^2 = (1/R_1 + 1/R_2)^2/4$ and the Gaussian curvature $C_1C_2 = 1/R_1R_2$, where R_1 and R_2 are the two radii of curvature (Canham, 1970; Helfrich, 1973; Evans, 1974). Thus, we write the energy density $\mathcal{F}$ as

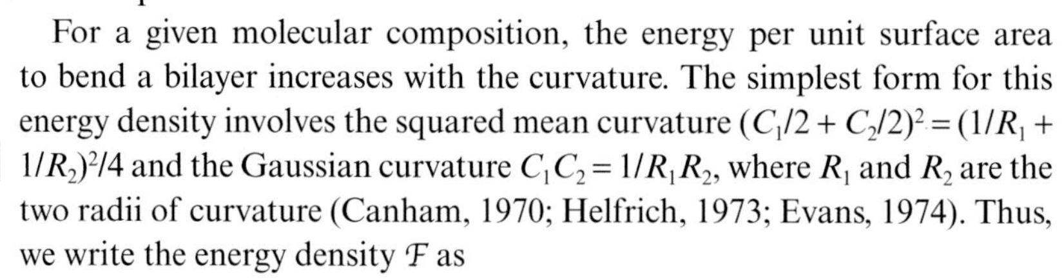

$$\mathcal{F} = (\kappa_b/2) \cdot (1/R_1 + 1/R_2)^2 + \kappa_G/(R_1R_2), \tag{7.17}$$

where the material-specific parameters κ_b and κ_G are the bending rigidity and the Gaussian bending rigidity, respectively; both parameters have units of energy. The Gaussian term has the interesting property that its integral is invariant under shape deformation for a given topology; e.g., if a spherical shell is deformed into an ellipsoid, $\int (R_1R_2)^{-1} dA$ doesn't change. As will be explained in Chapters 8 and 10, Eq. (7.17) can be modified to include bilayers whose native curvature is not zero. For the two shells in Fig. 7.14, the bending energy compared to a flat surface is

$$E = 4\pi(2\kappa_b + \kappa_G) \qquad \text{sphere} \tag{7.18a}$$

$$E = \pi\kappa_b L/R, \qquad \text{cylinder} \tag{7.18b}$$

according to Eq. (7.17). Note that the bending energy of a spherical shell is independent of its radius. Extensive measurements have been reported for κ_b, as we now summarize, but rather few for κ_G; many experiments and analyses find κ_G is negative (Section 4.5.9 of Petrov, 1999; Claessens *et al.*, 2004; Siegel and Kozlov, 2004).

7.4.2 Experimental measurements of κ_b

As will be shown later, the bending resistance of a thin sheet is about a factor of ten less than the product of its area compression modulus K_A with the *square* of its thickness. With $K_A \sim 0.2$ J/m^2 from Table 7.2, the very thinness of bilayers at 4×10^{-9} m means that their bending rigidity must be of the order 10^{-19} J, or some tens of k_BT. As a consequence, an isolated membrane readily undulates at room temperature unless its geometry is constrained by some means such as osmotic swelling (see images in Hategan *et al.*, 2003). Analysis of a bilayer's thermal undulations is the heart of many measurements of the bending modulus, as pioneered by Brochard and Lennon (1975) and Servuss *et al.* (1976). A commonly employed technique examines the shape fluctuations of synthetic vesicles (pure bilayer membranes: Schneider *et al.*, 1984; Faucon *et al.*, 1989; Duwe *et al.*, 1990; Méléard *et al.*, 1997) or flat bilayer sheets (Mutz and Helfrich, 1990). Many of these experiments require the development of extensive theoretical machinery to analyse the amplitudes of the shape fluctuations and obtain a bending modulus. An alternate approach is based on the observation that the effective in-plane area of a membrane is reduced by thermal fluctuations, as described in Section 7.3, such that a tension applied to a membrane irons out these fluctuations first before increasing the true area of the bilayer. The apparent area compression modulus $K_{A,\,app}$ is given by (Evans and Rawicz, 1990; see Section 8.5)

$$K_{A,\,app} = K_A \,/\, [1 + K_A k_B T\,/(8\pi\kappa_b\tau)], \tag{7.19}$$

where τ is the applied tension. Equation (7.19) becomes $K_{A,\,app} \sim K_A$ at low temperature or large tension, but is $K_{A,\,app} \sim 8\pi\kappa_b\tau \,/\, k_BT$ at small tension. Thus, one can extract κ_b from the behavior of $K_{A,\,app}$ obtained from a stress–strain curve at small stress.

A selection of bending moduli obtained experimentally is displayed in Table 7.3, where it is seen that many of the measurements cluster in the

Table 7.3 Bending rigidity κ_b for selected lipid bilayers and cell membranes. Most measurements are made at 20–30 °C; definitions follow Table 7.2

Membrane	κ_b ($\times 10^{-19}$ J)	(k_BT)	Reference
diAPC	0.44±0.05	11	Evans and Rawicz, 1990
diGDG	0.44±0.03	11	Evans and Rawicz, 1990
	0.15–0.4		Duwe *et al.*, 1990
	0.2±0.07	5	Mutz and Helfrich, 1990
diMPC	0.56±0.06	14	Evans and Rawicz, 1990
	1.15±0.15	29	Duwe *et al.*, 1990
	0.46±0.15	11	Zilker *et al.*, 1992
	1.30±0.08	33	Méléard *et al.*, 1997
diMPE	0.7±0.1	18	Mutz and Helfrich, 1990
diOPC	0.85±0.10	21	Rawicz *et al.*, 2000
SOPC	0.90±0.06	23	Evans and Rawicz, 1990
egg PC	1–2		Schneider *et al.*, 1984
	0.4–0.5	11	Faucon *et al.*, 1989
	1.15±0.15	29	Duwe *et al.*, 1990
	0.8	20	Mutz and Helfrich, 1990
	0.5	12	Cuvelier *et al.*, 2005
red blood cell	0.13–0.3	3–8	Brochard and Lennon, 1975
plasma	1.3	32	Evans, 1983
membrane	0.3–0.7	8–18	Duwe *et al.*, 1990
	1.4–4.3	35–108	Peterson *et al.*, 1992
	0.2±0.05	5	Zilker *et al.*, 1992
	2.0	50	Hwang and Waugh, 1997
	2.07±0.32	50	Scheffer *et al.*, 2001
	~9	~225	Evans *et al.*, 2008

range $\kappa_b \sim 10$–$20\ k_BT$. The table also includes the red cell plasma membrane, whose bending modulus is similar to, but somewhat larger than, pure bilayers. However, this similarity may not hold for membranes in general: one study of *Dictyostelium discoideum* cells found bending moduli in the hundreds of k_BT range, more than an order of magnitude larger than that of pure bilayers (Simson *et al.*, 1998). Similar to the behavior of the modulus K_A, the bending modulus κ_b is found to rise with cholesterol concentration. For example, both Duwe *et al.* (1990) and Méléard *et al.* (1997) find that κ_b of diMPC increases by a factor of three as the cholesterol concentration is raised from 0 to 30 mol%. However, just as the presence of short-chain alcohols lowers K_A, they also lower κ_b: the longer the hydrocarbon chain of the alcohol, the stronger the effect, in the order methanol < ethanol < propanol < butanol (Ly and Longo, 2004). The inclusion of salicylate also lowers both K_A and κ_b (Zhou and Raphael, 2005); the effects of proteins on membrane elasticity are reviewed in McIntosh and Simon (2007). Lastly, we mention that the temperature dependence of κ_b has also been reported (Lee *et al.*, 2001), as has the dependence of κ_b and κ_G on salt concentration (Claessens *et al.*, 2004).

7.4.3 Interpretation of κ_b

Several models for the molecular arrangements in a bilayer permit the prediction of how κ_b depends on the bilayer thickness d_{bl}. As displayed in Fig. 7.6, the bending of a bilayer, or any plate for that matter, involves a compression of one surface and an extension of the other, the magnitude of the strain being governed by the area compression modulus K_A. We show in the problem set that the bending rigidity of a uniform plate obeys $\kappa_b = K_A d_{bl}^2/12$, which drops to $\kappa_b = K_A d_{bl}^2/48$ if the plate is sliced into two leaflets free to slide past each other. A polymer brush model for the bilayer predicts the intermediate result $\kappa_b = K_A d_{bl}^2/24$ (Rawicz *et al.*, 2000). These models can be summarized by

$$\kappa_b = K_A d_{bl}^2/\alpha, \tag{7.20}$$

where the numerical constant α = 12, 24 or 48 (see Section 17.9 of Israelachvili, 1991, for a slightly different functional dependence). If K_A is independent of d_{bl} for the bilayers of interest, as indicated in Section 7.3,

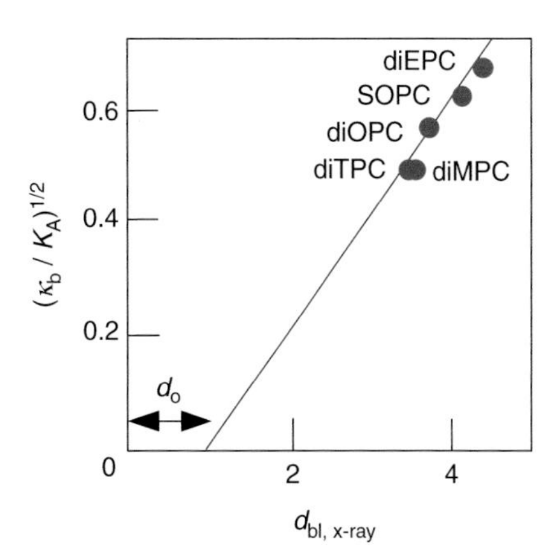

Fig. 7.15 Ratio of bilayer bending rigidity κ_b to area compression modulus K_A as a function of bilayer thickness $d_{bl,\ X\text{-ray}}$ as determined by X-ray scattering; the straight line is Eq. (7.20) with $\alpha = 24$ and offset $d_o = 1$ nm (after Rawicz *et al.*, 2000; ©2000 by the Biophysical Society). Abbreviations: diEPC = dierucoyl-PC, SOPC = stearoyl-oleoyl-PC, diOPC = dioleoyl-PC, diMPC = dimyristoyl-PC, and diTPC = ditridecanoyl-PC, where PC = phosphatidylcholine.

then Eq. (7.20) predicts $\kappa_b \propto d_{bl}^2$; a rigid plate, however, obeys $K_A \propto d_{bl}$, which leads to $\kappa_b \propto d_{bl}^3$.

The bilayer thickness dependence of κ_b has been tested by Rawicz *et al.* (2000) as displayed in Fig. 7.15. For a selection of lipids, the bilayer thickness $d_{bl,\, X\text{-ray}}$ was determined by X-ray scattering, although it is possible that there is an offset d_o between $d_{bl,\, x\text{-ray}}$ and the mechanically relevant d_{bl}. The straight line drawn through the data has $d_o = 1$ nm, and $\alpha = 24$ in Eq. (7.20); the data favor the d_{bl}^2 dependence of κ_b of Eq. (7.20) as well as the value of α from the polymer brush model. Just as the constancy of K_A argues against the rigid-plate model, so too is there little experimental support for the rigid-plate prediction $\kappa_b \propto d_{bl}^3$.

7.5 Edge energy

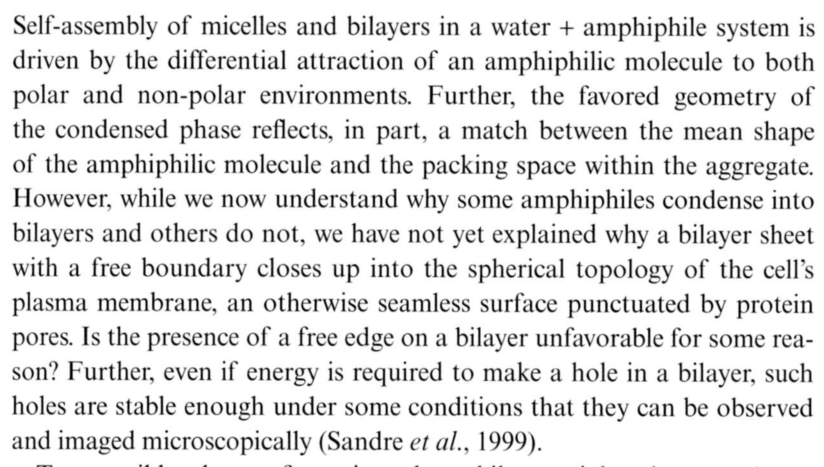

Self-assembly of micelles and bilayers in a water + amphiphile system is driven by the differential attraction of an amphiphilic molecule to both polar and non-polar environments. Further, the favored geometry of the condensed phase reflects, in part, a match between the mean shape of the amphiphilic molecule and the packing space within the aggregate. However, while we now understand why some amphiphiles condense into bilayers and others do not, we have not yet explained why a bilayer sheet with a free boundary closes up into the spherical topology of the cell's plasma membrane, an otherwise seamless surface punctuated by protein pores. Is the presence of a free edge on a bilayer unfavorable for some reason? Further, even if energy is required to make a hole in a bilayer, such holes are stable enough under some conditions that they can be observed and imaged microscopically (Sandre *et al.*, 1999).

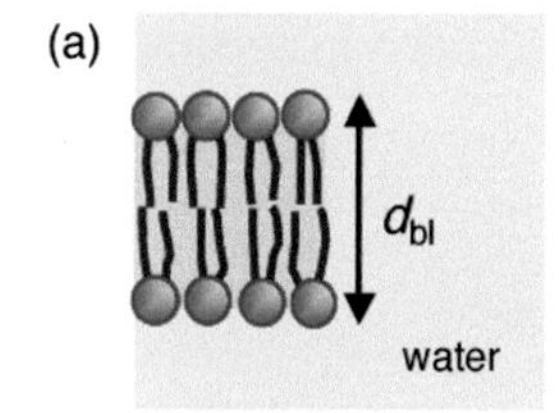

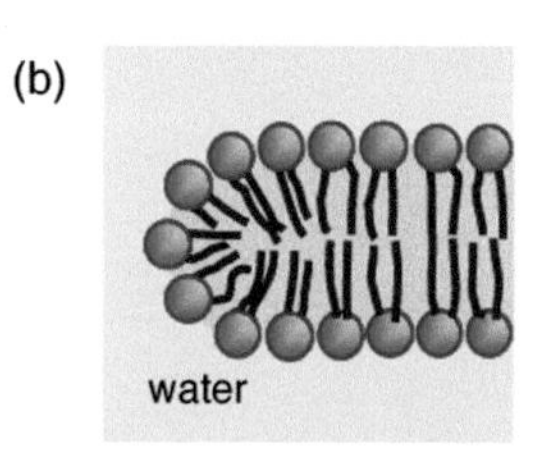

Fig. 7.16

Possible configurations that a bilayer may adopt at a free edge. In (a), the hydrocarbon chains are exposed to water, while in (b), the molecular packing is not matched to the shape of a cylindrical lipid.

Two possible edge configurations that a bilayer might adopt are shown in Fig. 7.16. Configuration (a) is energetically costly because the hydrocarbon core is exposed to water at the boundary (Deryagin and Gutop, 1962; Helfrich, 1974a; Litster, 1975), in contrast to configuration (b) which has the hydrocarbons hidden from water, at the price of introducing a curved surface. Computer simulations strongly favor the toroidal hole configuration (b) (for example, Jiang *et al.*, 2004; Leontiadou *et al.*, 2004; see also Allende *et al.*, 2005). Now, we argued in Section 7.2 that bilayer-forming molecules tend to have a cylindrical shape, implying that the curved surface in Fig. 7.16(b) is not their favorite packing geometry. Thus, both configurations in Fig. 7.16 suffer an energy penalty for creating a free edge, and we parametrize this penalty by an edge tension λ, an energy per unit length, analogous to the surface tension γ.

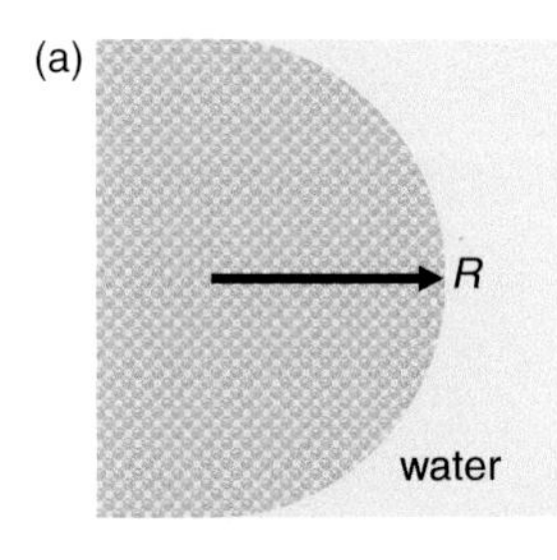

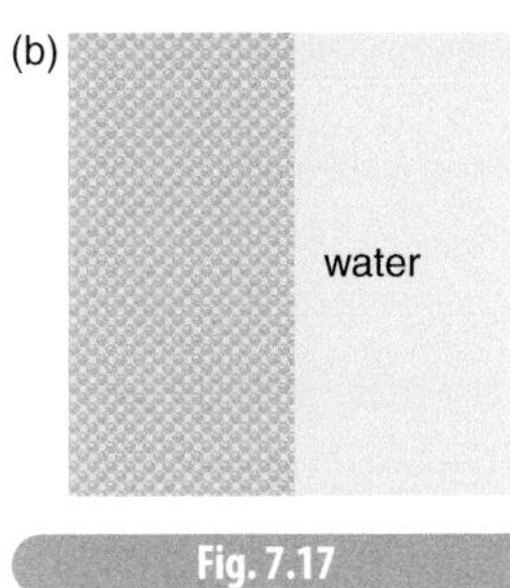

Fig. 7.17

Face view of two bilayer-water boundaries with different curvature, where the small disks represent the lipids (Fig. 7.16 is a cross section or side view). The curvature in the bilayer plane is $1/R$ In panel (a) and zero in panel (b).

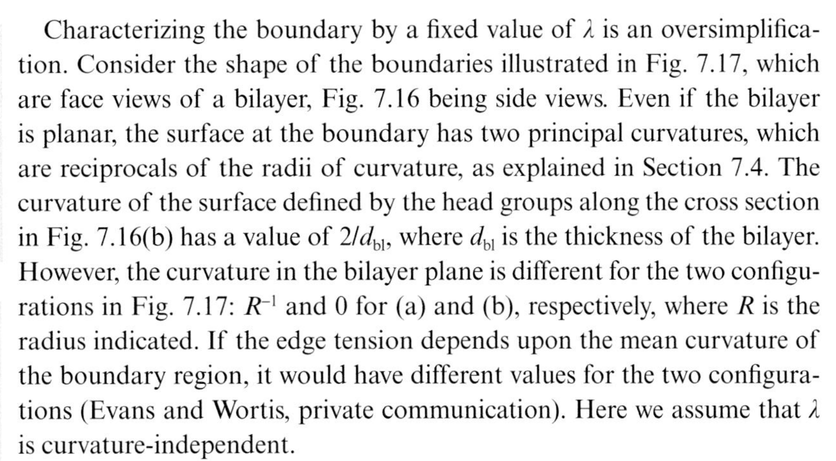

Characterizing the boundary by a fixed value of λ is an oversimplification. Consider the shape of the boundaries illustrated in Fig. 7.17, which are face views of a bilayer, Fig. 7.16 being side views. Even if the bilayer is planar, the surface at the boundary has two principal curvatures, which are reciprocals of the radii of curvature, as explained in Section 7.4. The curvature of the surface defined by the head groups along the cross section in Fig. 7.16(b) has a value of $2/d_{bl}$, where d_{bl} is the thickness of the bilayer. However, the curvature in the bilayer plane is different for the two configurations in Fig. 7.17: R^{-1} and 0 for (a) and (b), respectively, where R is the radius indicated. If the edge tension depends upon the mean curvature of the boundary region, it would have different values for the two configurations (Evans and Wortis, private communication). Here we assume that λ is curvature-independent.

In this section, we explore the implications of edge tension for several aspects of cell stability. First, we examine the competition between bending resistance and edge tension that compels a large, flat membrane to close into a sphere. Next, we calculate the energetics of hole formation in a planar membrane under tension. The analytical treatment of these topics is restricted to zero temperature, and we show by computer simulation what effects membrane and boundary fluctuations have at finite temperature. In closing Section 7.5, the measured values of the edge tension are summarized and interpreted. Although we focus on the edge tension of lipid–water systems, the concept of edge tension applies more generally to membranes where there is a boundary between phases (see Brewster and Safran, 2010, as well as end-of-chapter problems).

7.5.1 Vesicle formation: edge energy vs. bending energy

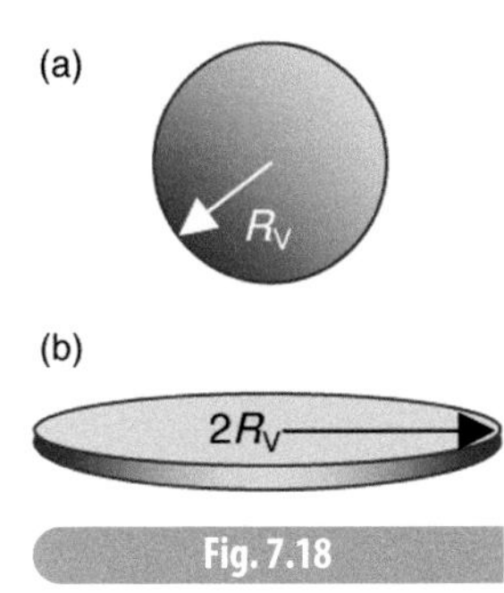

Fig. 7.18

Two idealized membrane shapes: (a) a spherical shell of radius R_V and (b) a flat disk with a circular boundary. The radius of the disk must be $2R_V$ if the disk and sphere have the same area.

To illustrate the energetics driving the closure of fluid membranes into vesicles and cells, we consider two sample shapes shown in Fig. 7.18: a closed spherical shell of radius R_V and a flat disk with a circular boundary (Helfrich, 1974a; Fromhertz, 1983). If the sphere and disk have the same area, the radius of the disk must be $2R_V$. Describing the bending energy of the membrane by the simple curvature model presented in Section 7.4, the energy E_{sphere} required to bend a flat membrane into the shape of a sphere is independent of the sphere radius, and is given by

$$E_{sphere} = 4\pi(2\kappa_b + \kappa_G), \tag{7.21}$$

where κ_b and κ_G are the bending rigidities. This energy may be compared to that of a flat disk,

$$E_{disk} = 4\pi R_V \lambda, \tag{7.22}$$

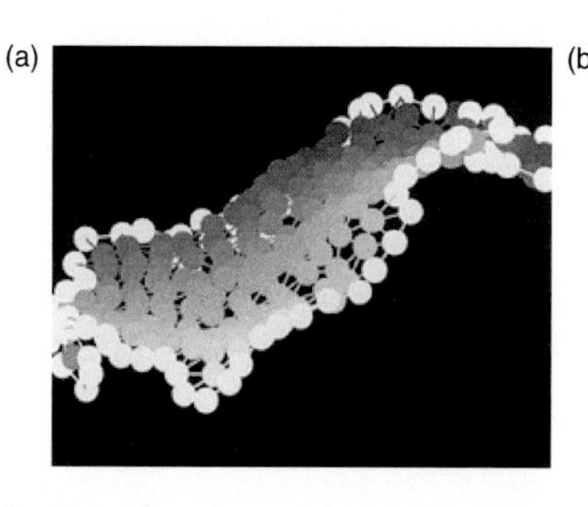

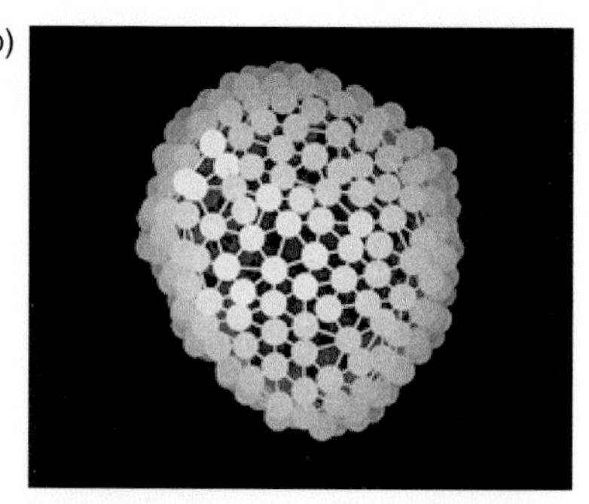

Fig. 7.19 Simulation of membranes with open (a) and closed (b) topologies. The membrane boundary is indicated by the white spheres (from Boal and Rao, 1992b).

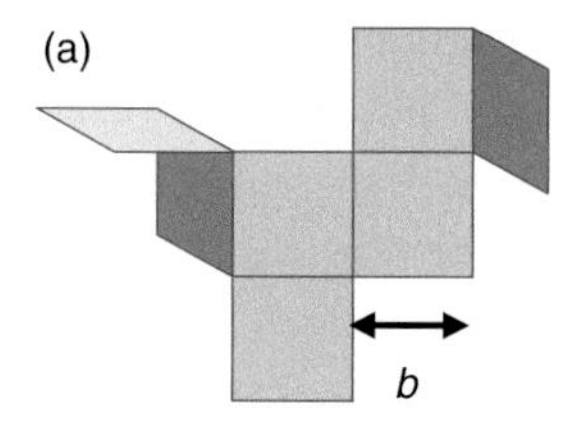

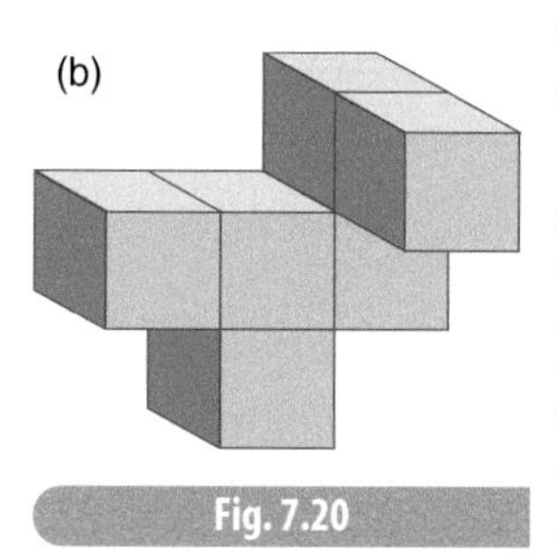

Fig. 7.20 Sample configurations of an open (a) and closed (b) self-avoiding surface on a cubic lattice. Each plaquette has length b to the side.

to which the only contribution is from the edge energy along the perimeter $4\pi R_V$. At zero temperature, the shape adopted by the membrane is the one with the lowest energy. If R_V is small, the disk is favored. However, the disk energy increases with perimeter, and the sphere becomes the preferred shape for radii above

$$R_V{}^* = (2\kappa_b + \kappa_G) / \lambda, \tag{7.23}$$

although we note that there may be an energy barrier between the disk and sphere configurations (see end-of-chapter problems).

At non-zero temperature, the curvatures of the surface and the boundary fluctuate locally. Two snapshots from a computer simulation of a membrane with bending resistance at $T > 0$ with a bending resistance similar to Eq. (7.21) are displayed in Fig. 7.19, and the undulations of the membrane are clearly visible. Because of their entropy, these shape fluctuations favor the "magic carpet" configuration of Fig. 7.19(a) over the handbag in Fig. 7.19(b); hence, λ must be larger to seal the free boundary at $T > 0$. Simulations show that the sheet closes only if λ exceeds a threshold value of

$$\lambda^* = 1.36\, k_B T / b, \qquad \text{(three dimensions)} \tag{7.24}$$

where b is a length scale from the simulation (Boal and Rao, 1992b). For the simulation of Fig. 7.19, b is set by the mean separation between vertices.

The minimum edge tension needed to close the boundary can be calculated within a lattice model where the surface is represented by a connected set of identical square plaquettes. Two sample configurations of a membrane on a lattice are displayed in Fig. 7.20: configuration (a) has a free boundary while configuration (b) is closed. For a fixed number of plaquettes N, Glaus (1988) has explicitly evaluated by computer the number of simply connected configurations C_N (where "connected" means no internal voids or gaps) for both ensembles, finding

$$C_N \cong 12.8^N \bullet N^{-1.48} \qquad \text{open}$$

$$C_N \cong 1.73^N \bullet N^{-1.51}. \qquad \text{closed} \tag{7.25}$$

The mean perimeter Γ of the open configurations is close to $2Nb$, where b is the bond length, so that the free energy, $F = \Gamma\lambda - TS$, of the two ensembles is

$$F \cong 2N\lambda b - k_B T \ln C_N \cong 2N\lambda b - k_B TN \ln(12.8) \qquad \text{open} \tag{7.26a}$$

$$F \cong -k_B TN \ln(1.73). \qquad \text{closed} \tag{7.26b}$$

Equations (7.26a) and (7.26b) are equal at the transition point, yielding $\lambda^* b = -0.5 k_B T \bullet \ln(1.73 / 12.8)$ or

$$\lambda^* = 1.0\, k_B T / b, \tag{7.27}$$

which is near the value observed in the fluid membrane simulation. We describe below how to relate b to a physical length scale of the membrane, and hence extract a physical bound on the edge tension.

7.5.2 Membrane rupture: edge energy vs. tension

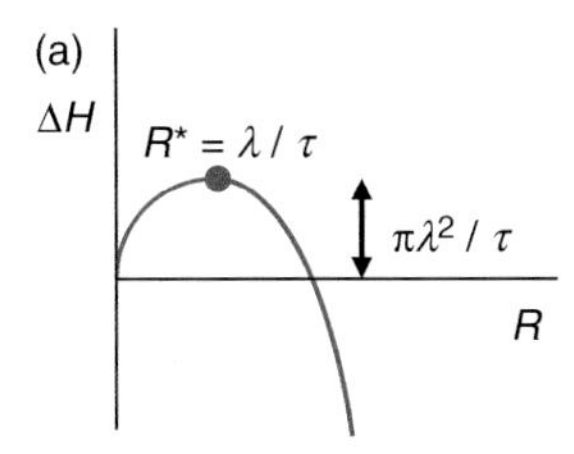

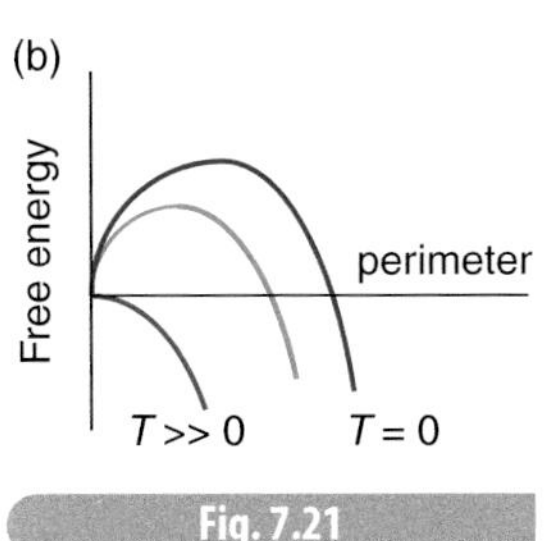

Fig. 7.21

(a) Enthalpy difference ΔH of a membrane + hole system compared to an unbroken membrane as a function of hole radius R at zero temperature. (b) Reduction in the free energy of a membrane + hole as the temperature increases.

Subjected to a small tensile stress compared to the area compression modulus, a bilayer will expand and sustain a strain of a few percent, as described in Section 7.3. However, if the stress is too large, the membrane will rupture. A simple model for membrane failure employs the creation of a circular hole in an incompressible fluid sheet (Deryagin and Gutop, 1962; Litster, 1975; for more general configurations, see Netz and Schick, 1996; for a comparison with solid membranes, see Zhou and Joós, 1997). At zero temperature, the system acts to minimize its enthalpy H,

$$H = E - \tau A, \tag{7.28}$$

where τ is the two-dimensional tension ($\tau > 0$ is tension, $\tau < 0$ is compression). The energy of a circular hole in the sheet is $E = 2\pi R\lambda$, and the area difference of the sheet + hole system with respect to the intact sheet is just πR^2, where R is the radius of the hole. Hence, the difference in enthalpy ΔH of the membrane + hole system compared to the unbroken membrane is

$$\Delta H = 2\pi R\lambda - \tau \pi R^2. \tag{7.29}$$

For small holes, the term linear in R dominates Eq. (7.29) and ΔH grows with R for $\lambda > 0$. However, the term quadratic in R dominates for large holes, and ΔH becomes increasingly negative for membranes under tension, as shown in Fig. 7.21. The maximum value of ΔH occurs at a hole radius of

$$R^* = \lambda / \tau. \tag{7.30}$$

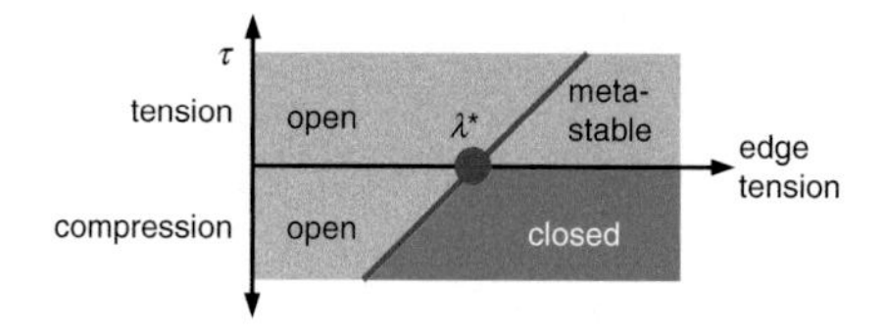

Fig. 7.22 Schematic phase diagram illustrating the stability of a two-dimensional membrane + hole system. Holes may arise from thermal fluctuations anywhere in the parameter space of the diagram, but are thermodynamically favored only in the "open" and "metastable" regions (after Shillcock and Boal, 1996).

The physical meaning of R^* is that, at zero temperature, holes with $R < R^*$ shrink, while those with $R > R^*$ expand without bound.

For a membrane under tension, Fig. 7.21(b) tells us that small holes are metastable in the sense that they do not represent configurations with the minimum free energy. Rather, there is an energy barrier of height $\pi\lambda^2/\tau$ (at zero temperature) which a hole must cross before it can expand without bound, as favored thermodynamically. At finite temperature, energy exchange with its environment permits a hole to form and cross the energy barrier (for discussion of the dynamics of hole formation, see Deryagin and Gutop, 1962; Wolfe *et al.*, 1985; Wilhelm *et al.*, 1993). Further, the entropy of a membrane + hole system at $T > 0$ lowers its free energy as shown in Fig. 7.21(b). Simulations not only confirm that holes form more easily as the temperature rises (or the edge tension decreases), they also show that, even at $\tau = 0$, there is a minimum λ required to keep the membrane from fragmenting (Shillcock and Boal, 1996). For planar membranes in two dimensions, this minimum edge tension λ^* is

$$\lambda^* = 1.66\, k_B T / b, \qquad \text{two dimensions} \qquad (7.31)$$

where b is the mean separation between network points in the simulation (the points define the discrete membrane of the simulation; they appear as spheres in Fig. 7.19). This threshold λ is somewhat larger than what is obtained in three dimensions, Eq. (7.24). A schematic phase diagram found in the simulation of membrane rupture in two dimensions is shown in Fig. 7.22. Membranes are unstable against hole formation, even under compression, in the regions labelled "open", and have a finite lifetime in the "metastable" region. The threshold λ^* is clearly visible on the $\tau = 0$ axis. Systems with multiple holes have been investigated by simulation as well, although a consensus on the effects of multiple holes has not yet emerged (Nielsen, 1999; Shillcock and Seifert, 1998a).

7.5.3 Measured edge tensions

Physical values of the edge tension have been obtained from several different techniques, which have in common the measurement of a stress and, directly

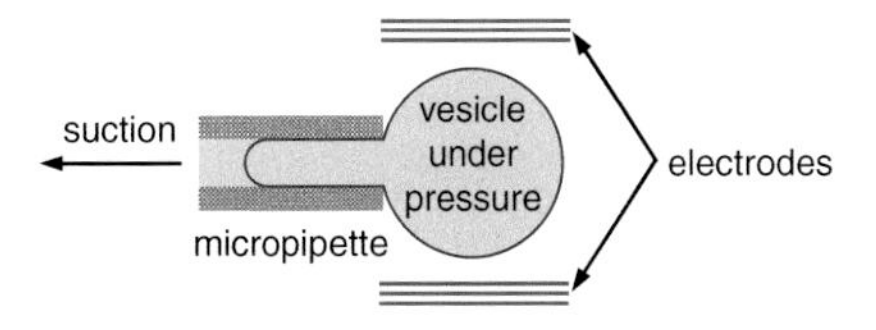

Fig. 7.23 In the experiments of Zhelev and Needham (1993), a pure bilayer vesicle is placed under pressure by aspirating it with a micropipette. Applying an electric field across the vesicle thins the bilayer, creating a hole.

or indirectly, a hole size. One approach uses the competition between bending resistance and edge tension in the formation of vesicles, as described earlier in this section. Fromhertz *et al.* (1986) measured the maximum size of disk-like membranes containing egg lecithin and taurochenodesoxycholate. The mixture was sonicated to destroy large-scale aggregates, then stained and imaged by electron microscopy. The images showed uniform disk-like membranes arranged in stacks, like columns of coins, with average radii of ~20 nm, and a distribution of small vesicles corresponding to disk radii of 30 nm. From these and related observations, the authors conclude that the edge tension of pure lecithin bilayers is 4×10^{-11} J/m, based upon an analysis similar to Eq. (7.23) to relate λ to the maximum disk size.

A second approach places a bilayer in a vesicle under large enough tension to cause pore formation. The earliest version of this technique used osmotic pressure to swell a small vesicle (Taupin *et al.*, 1975), while a more recent refinement uses micromechanical manipulation of a vesicle (Zhelev and Needham, 1993). As illustrated in Fig. 7.23, a giant bilayer vesicle (25–50 μm in diameter) can be aspirated into a micropipette, and a hole created in the membrane by applying an electric field (see Chernomordik and Chizmadzhev, 1989). In some cases, the hole is long-lived and its radius R can be determined by measuring the rate of fluid loss through the opening. The tension τ experienced by the membrane can be determined from the known aspiration pressure. Balancing the force from the edge tension against that of the tensile stress (where the force $F = -\mathrm{d}V / \mathrm{d}x$ is obtained from the derivative of Eq. (7.29)) implies $\lambda = R / \tau$. The experiment yields $\lambda = 0.9 \times 10^{-11}$ J/m for pure stearoyl-oleoyl phosphatidylcholine (SOPC) and 3×10^{-11} J/m for SOPC with 50 mol % cholesterol. These results are similar to the osmotic swelling experiments of Taupin *et al.* (1975), who find $\lambda \sim 0.7 \times 10^{-11}$ J/m for dipalmitoyl phosphatidylcholine (diPPC). The conditions under which nominally unstable holes can have unexpectedly long lifetimes are discussed in Moroz and Nelson (1997a) who extend the analysis of Zhelev and Needham (1993).

How do these measured values of λ compare with the bound $\lambda^* > k_B T / b$ found in simulations of rupture at $T > 0$? The length scale b of the simulations is an effective length scale associated with the bending of the boundary, similar to the persistence length of polymer chains in Chapter 3, and is not necessarily a molecular dimension. The persistence length of the boundary is a measure of the distance over which the direction of the membrane boundary is decorrelated and must be at least equal to the membrane thickness. If the membrane has the cross section of Fig. 7.16(b), the length scale of the curved region in the bilayer plane must be greater than or equal to the inverse curvature perpendicular to the plane, namely $d_{bl}/2$. Taking $b \sim d_{bl} \sim 4 \times 10^{-9}$ m, the simulations predict that the edge tension must exceed $\lambda^* \sim 10^{-12}$ J/m at ambient temperatures for membrane stability ($k_BT = 4 \times 10^{-21}$ J at $T = 300$ K). The measured values of λ for bilayers in the cell exceed this estimate of λ^* by an order of magnitude, easily sufficient to make the bilayer resistant against rupture under moderate tensions, but not sufficient to enable it to operate at an elevated osmotic pressure. To withstand such pressures, a cell must augment its plasma membrane with a cell wall.

Yet another technique is based on direct imaging of a hole in a membrane, from which the line tension can be obtained by measuring the growth rate of the hole (Sandre *et al.*, 1999). This approach yields $\lambda \cong 1.5 \times 10^{-11}$ J/m for DOPC bilayers, a value that is reduced when detergent is added to the system and is absorbed at the pore boundary (Puech *et al.*, 2003). As with SOPC bilayers, the line tension is found to increase with the addition of cholesterol (Karatekin *et al.*, 2003).

The picture of the membrane boundary in Fig. 7.16(b) is a region where a leaflet of the bilayer is curved, with the outer radius of curvature of approximately half the bilayer thickness. This picture implies that the line tension should be proportional to the membrane bending resistance. This prediction has been tested experimentally, and the linear proportionality $\lambda \propto \kappa_b$ has been confirmed for lipids with chain lengths between 13 and 22 carbon atoms (Rawicz *et al*, 2000; see also Olbrich *et al.*, 2000, and Evans *et al.*, 2003).

7.6 Cell walls and sheaths

Bacteria are among the most mechanically simple of cells, although they are now known to possess a number of subtle structures that influence the cell division cycle (see Section 10.5). Detailed images of several types of bacteria are displayed in Figs. 1.1, 1.6 and 1.7: the interiors of the cells are devoid of organelles, yet the cell boundary may be complex. In this section, we study the mechanical properties of the cell boundary, of which there are three common variants. The smallest bacteria are bounded simply by a plasma membrane without a cell wall, because the mechanical necessity for a strong wall to withstand a cell's internal pressure is less important for small cells. However, most bacteria come in two designs: Gram-negative

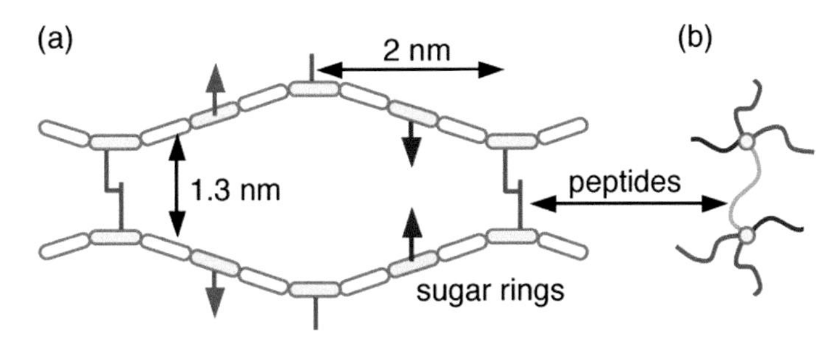

Fig. 7.24 (a) Face view of a section of a peptidoglycan network: polymeric ropes of sugar rings run in one direction and peptide strings form transverse links. (b) As seen along the glycan axis, peptides can link glycan ropes in several directions. Bending of the glycan chains have been exaggerated for effect.

strains have a thin wall sandwiched between the plasma membrane and an outer membrane; Gram-positive strains have a thick wall exterior to the plasma membrane, but do not possess a secondary membrane. As their outermost shell, many bacteria possess a highly regular structure called the S-layer adjacent to the outer membrane (for Gram-negative bacteria) or the cell wall material (Gram-positive bacteria).

The wall itself is composed of layers of peptidoglycan, an inhomogeneous network in which stiff polysaccharide chains are cross-linked by floppy polypeptides, as illustrated in Fig. 7.24. The glycan ropes have one cross-link available on each of the 1.0 nm long disaccharide units, although not all links are attached to neighboring glycan strands. For the network connectivity to be uniform in three dimensions, the peptides would have to radiate like a helix around the glycan axis, as in Fig. 7.24(b), such that there are four disaccharides, or 4 nm, between peptides linking the same pair of adjacent glycans (Koch, 2006). The peptide strings themselves have a contour length of 4.3 nm (Braun *et al.*, 1973) but an end-to-end displacement of just 1.3 nm (Burge *et al.*, 1977), a reduction that one would expect for floppy chains. The cell wall of a Gram-negative bacterium is relatively thin, and a thickness of 3–8 nm would correspond to 2–6 glycan strands, given an interchain separation of 1.3 nm. In contrast, Gram-positive cell walls are much thicker at 25 nm, corresponding to 20 or so glycans. It's no surprise, then, that Gram-positive bacteria can support more than 20 atmospheres pressure, compared with 3–5 atmospheres for Gram negative bacteria (as low as $0.85–1.5 \times 10^5$ J/m^3 in some cases, see Arnoldi *et al.*, 2000). Clearly, there must be unconnected peptides in Gram-negative walls owing to their small thickness, and this is observed for Gram-positive walls as well (Zipperle *et al.*, 1984).

Measurements of the elastic moduli of the bacterial cell wall have concentrated on obtaining direction-dependent Young's moduli Y. In one approach, a sample of highly aligned walls from the Gram-positive bacterium *Bacillus subtilis* was found to have $Y = 1.3 \pm 0.3 \times 10^{10}$ J/m^3 (dry) and 3×10^7 J/m^3 (hydrated) from an analysis of stress–strain curves (Thwaites and Surana, 1991). The wall material in these experiments was oriented with the peptide strings largely parallel to the axis of applied tension, so $Y \sim 3 \times 10^7$ J/m^3 reflects the "soft" direction of peptidoglycan. Another approach using an atomic force microscope to deform peptidoglycan from the Gram-negative bacterium *E. coli* gave a similar value for the average Young's modulus of 2.5×10^7 J/m^3 for hydrated material, which varied by a factor of two according to direction (Yao *et al.*, 1999). Dried samples of the *E. coli* wall were more rigid at 3–4 $\times 10^8$ J/m^3, although not as stiff as *B. subtilis*.

The lateral stress within the membrane plane depends on its curvature and the pressure difference across it. For instance, the surface stress within the thin rubber of a spherical balloon is the same in any direction (in the plane of the rubber) and it increases with the radius of the balloon for a given pressure difference. However, for non-spherical shapes, the stress depends on direction, and we will show in Chapter 8 that for the cylindrical section of a spherocylinder, the stress is larger in a direction around the equator than it is along the axis of the cylinder. Thus, the cell wall of a bacterium like *E. coli*, which has a shape like a spherocylinder, must support a larger stress around its equator than along its axis. Indeed, the structure of the cell wall has been designed to accommodate this: Young's moduli for *E. coli* walls are softer along the cylindrical axis. Other studies of *E. coli* reveal that the (softer) peptide strands tend to be oriented along the cylindrical axis (Verwer *et al.*, 1978), while the (stiffer) glycans lie in the hoop direction. This design strategy has been employed elsewhere in the bacterial world in the construction of tubular sheaths within which some bacteria elongate and divide. For example, the sheath of the bacterium *Methanospirillum hungatei* has been found to contain reinforcement hoops around its equator (Southam *et al.*, 1993), although one should note that this sheath material is much stiffer than peptidoglycan (Xu *et al.*, 1996).

Let's now try to understand the elastic behavior of peptidoglycan at a microscopic level, beginning with the properties of the individual chains. The peptide appears to have the classic properties of an entropic spring, given that its end-to-end length in the network (r_{ee} = 1.3 nm) is much less than its contour length (L_c = 4.2 nm). We analyze the peptide strand using two relationships valid for single chains, recognizing that a chain may have different behavior in isolation than in a network. The persistence length ξ_p of the peptide chain can be obtained from $\langle r_{ee}^2 \rangle = 2\xi_p L_c$, yielding $\xi_{p,pep}$ = 0.2 nm for peptides; this value is somewhat lower than alkanes, perhaps indicating the presence of other attractive interactions within the chain. The effective spring constant for an ideal chain in three dimensions is $k_{sp} = 3k_BT/\langle r_{ee}^2 \rangle$, which gives $k_{sp,pep} = 7.1 \times 10^{-3}$ J/m^2 for peptides. In contrast, the glycan chains are much stiffer: $\xi_p \sim 10$ nm for a typical unbranched

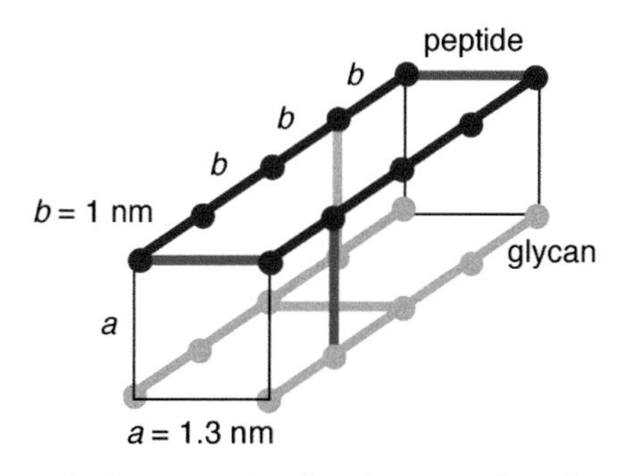

Fig. 7.25 Section of an idealized peptidoglycan network in three dimensions; heavy lines represent peptide (red) or glycan (magenta) chains and disks designate the junctions between chains. The segment lengths are $b = 1$ nm for dissacharides and $a = 1.3$ nm for peptides.

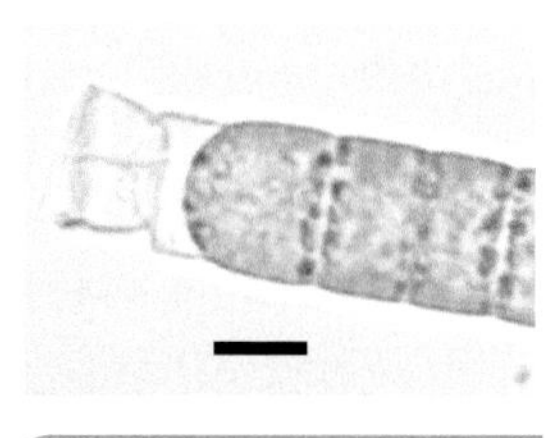

Fig. 7.26 Section of a filamentous cyanobacteria (here, a strain of *Oscillatoria*) showing the remnant of an envelope at its torn end. Scale bar is 5 μm.

polysaccharide (Stokke and Brant, 1990; Cros *et al.*, 1996). In other words, the filaments in the longitudinal direction (polypeptides) are much softer than the filaments around the equator (glycans).

Suppose we assume for simplicity that all filaments in peptidoglycan are sufficiently flexible that the material behaves like the networks of floppy chains studied in Chapters 5 and 6. There, we showed that the shear modulus μ was close to $\rho k_B T$ for both two- and three-dimensional polymer networks, where ρ is the (2D or 3D) density of chains. To apply this benchmark to peptidoglycan, we evaluate the chain density using the peptide connectivity in the three-dimensional model of Fig. 7.25, where the network "bonds" are drawn as heavy lines and the glycan–peptide junctions are shown as disks. Four glycan chains, with disaccharide length $b = 1$ nm, are displayed in the figure, linked by five peptide strings of length $a = 1.3$ nm. It is shown in Problem 7.30 that the bond density is $\rho = 3 / 2a^2b$, such that the benchmark modulus is $\mu = \rho k_B T = 3.6 \times 10^6$ J/m^3. For many materials, $Y = K_V = (8/3)\mu$ so we expect Y to be about 1×10^7 J/m^3 in this representation, which is surprisingly close to $Y = 2\text{–}3 \times 10^7$ J/m^3 observed experimentally. The accuracy of this model can be improved by recognizing that the glycan strands do not extend unbroken around the equator of the bacterial cylinder; instead, the strands are rather short with a mean length of 5–10 disaccharide units (or 5–10 nm; Höltje *et al.*, 1998). Fragmentation of the glycan reduces the anisotropy of the Young's modulus, and helps explain why a floppy network provides a good approximation for peptidoglycan. The analysis of peptidoglycan is taken further by Boulbitch *et al.* (2000).

There are many examples of sheet-like structures beyond the cell proper. For example, Fig. 7.26 displays the remnant of a sheath-like structure at the end of a torn filamentous cyanobacterium, in this case a strain of *Oscillatoria* with a diameter of about 8 μm. Generally speaking, the quasi-two-dimensional networks outside of the cell tend to be thicker than the bacterial cell wall; here, we describe two such systems, the basal lamina and the plant cell wall.

7.6.1 Basal lamina

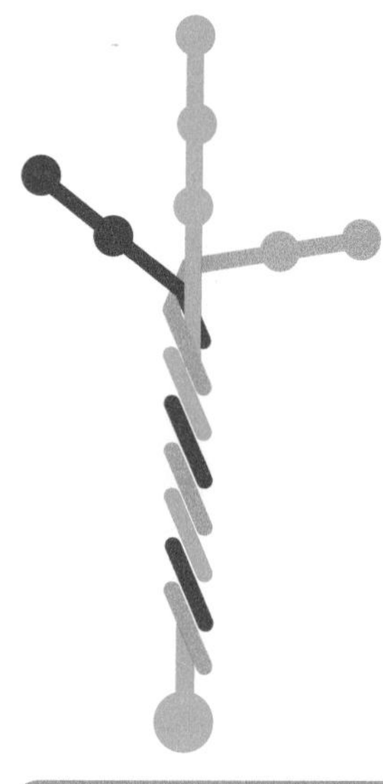

Fig. 7.27

The protein laminin is composed of three glycoprotein chains partly wound into a triple helix. The contour length along the main axis of laminin is about 120 nm, with the arms extending about 60 nm across; both measurements are approximate.

The construction of a multicellular organism such as ourselves involves more than just cells in contact with other cells. There are regions where the cells are embedded in connective tissue, such as tendon, bone, etc. The deformation resistance of connective tissue spans several orders of magnitude: human cartilage and tendon have Young's moduli of 0.24×10^9 and 0.6×10^9 J/m^3 compared to fresh bone with $Y = 21 \times 10^9$ J/m^3. Connective tissue may contain a variety of molecular components, with collagen and elastin being among the more common proteins present. The mechanical interface between cells and connective tissue is supplied by the basal lamina. In vertebrates, for instance, epithelial tissue such as the intestinal lining consists of a two-dimensional layer of close-packed cells, with the basal lamina providing a tough mat for organization and stability (see Figs. A.1 and A.2 in Appendix A). Similarly, individual muscle cells may be embedded in connective tissue, with the basal lamina once again acting as the mechanical interface. The thicknesses of basal laminae are commonly in the 40–120 nm range, which is more than a hundred times smaller than the 10–20 μm width of a typical cell to which it is attached. In other words, the basal lamina is effectively a two-dimensional sheet as measured by the size of cells; also referred to as the basement membrane, it is classified as a type of extracellular matrix.

The basal lamina has many different components: there are two principle structural components (laminin and type IV collagen) as well as several types of proteins that anchor the lamina to the plasma membrane of its associated cells. The composition not only varies from tissue to tissue, but also evolves with time as the lamina is built. The glycoprotein laminin, briefly introduced in Problem 5.8, is actually three different proteins (labeled α, β and γ) wound together into the shape of a cross. This is shown in Fig. 7.27: the three short arms of the cross are elements of longer proteins, the main lengths of which are wound into a triple helix. There are several variants of each of the α, β and γ constitutive proteins, with the result that laminin has many isoforms; we quote a representative length as 120 nm along the main axis, and a total width of 60 nm for the two arms that cross it. Laminin has binding sites at the end of each arm, permitting it to bind to other laminins as well as to the anchoring proteins (integrins) in the plasma membrane of the neighboring cell. The resulting laminin network is two-dimensional in the sense that it contains just a single layer of laminin; however, the connectivity within the network plane is irregular.

Early in the construction of the basal lamina, laminin is its primary component and provides its organization. However, as the lamina matures, a second scaffolding is added, composed of type IV collagen. This type of collagen has a much looser structure than the tight fibrils made from type I collagen described in Section 4.1; rather, the three collagen chains are more informally wrapped about each other. The resulting two-dimensional network has more uniform connectivity than the laminin network, with most junctions among the collagen filaments being 4- to 6-fold coordinate.

Although measurements of the elastic properties of basal laminae go back many years, we mention one determination of the storage and loss moduli G' and G'' from the viscoelastic framework of Section 6.5. The experiment (Tidball, 1986) was performed at an oscillation frequency of 1 Hz on skeletal muscle cells of frogs, yielding a storage modulus G' of 1.5×10^8 J/m^3 and a loss modulus G'' of 0.68×10^8 J/m^3. Not surprisingly, these values are orders of magnitude higher than many of the low-density networks described in Section 6.6.

7.6.2 Plant cell wall

Like the basal lamina, the plant cell wall is a type of extracellular matrix. However, with a width of one or two hundred nanometers, it is generally thicker than either the basal lamina or the bacterial cell wall. Similar to the bacterial cell wall, one of the roles of the plant cell wall is to provide tensile strength against the internal pressure (or turgor pressure) of the cell, which may reach an impressive ten atmospheres. In addition, the plant cell wall provides support and organization to the cells of the plant on the large scale through its bending, shear and compression resistance.

Cell growth under pressure, without membrane failure, is just as much a problem for plant cells as it is for bacteria, and this is reflected in the composition of the wall. In the early stages of cell growth, what is called the *primary* cell wall is laid down, consisting of cellulose microfibrils, separated from one another by 20–40 nm and interwoven with the polymer pectin. Cross-linking glycans provide the mechanical links between the microfibrils. As the cells grow and mature, lignin is added to the matrix to create a *secondary* wall, which may be much thicker and stiffer than the primary wall. The cellulose microfibrils are laid down in planes with orthogonal orientation to provide tensile strength in different directions as shown in Fig. 4.5, where the pectin network has been omitted for clarity.

Unlike cellular filaments such as spectrin, actin and microtubules, which are proteins, the cellulose and pectin filaments of the cell wall are long unbranched polysaccharides. In cellulose, individual polysaccharide molecules contain between several hundred and many thousand residues of D-glucose, depending on the plant type, with wood generally having shorter chains than cotton. The fundamental disaccharide is shown in Fig. 7.28: hydrogen

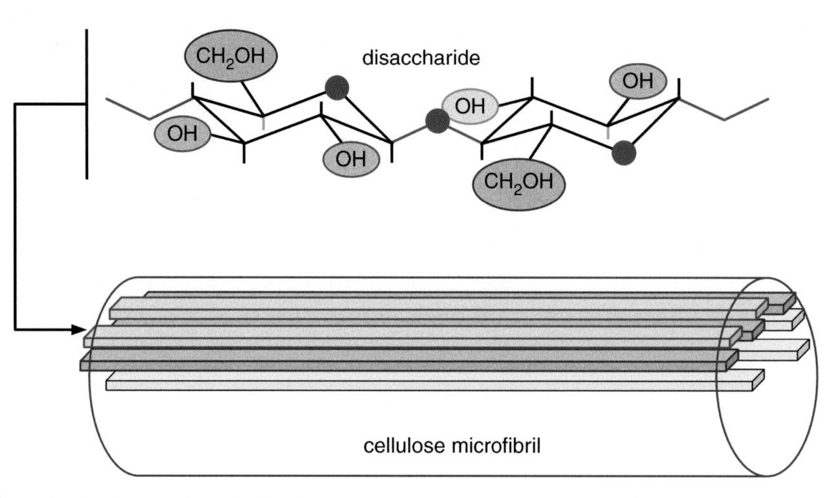

Fig. 7.28 Cellulose is a linear polysaccharide whose fundamental disaccharide is shown. Hydrogen-bonding within individual chains gives them a planar structure, and enables them to associate into microfibrils. Commonly, microfibrils contain 40–90 individual chains, depending on the plant species.

atoms are omitted from the drawing, and each ring of the unit is identical, but rotated 180° with respect to its neighbor on the chain. The $-CH_2OH$ sidegroups to the ring support extensive hydrogen bonding along the chain (making it planar) and between adjacent chains. As a result, many individual chains may bundle up to form a microfibril, the diameter of which varies according to plant species from 2 nm in quince to more than 20 nm in some green algae (reviewed in Brett, 2000). In many higher plants, the microfibril diameter is 5–10 nm, corresponding to 60–90 chains in cross-section.

Given the wide variety of plants and the hydration state of their components, it is no surprise that measured values of their elastic moduli also span a large range. When dried as wood, the Young's modulus is $\sim 1\text{–}4 \times 10^{10}$ J/m^3, with significant dependence on species, hydration state and strain angle with respect to the cell wall (for an introduction, see Cave, 1968).

Summary

Membranes in the cell are based upon a lipid bilayer structure with thickness d_{bl} of about 4 nm and vanishing in-plane shear resistance, like a two-dimensional fluid. Belonging to a class of molecules called amphiphiles, membrane lipids have a relatively small electrically polar region called the head group, to which are indirectly attached one or more non-polar hydrocarbon chains. In aqueous solution, amphiphiles condense into aggregates if their concentration exceeds the so-called critical micelle concentration

(CMC). In some models of micelle formation, the CMC is proportional to the product of a volume term $[\{2\pi mk_BT\}^{1/2}/h]^3$ and a probability factor $\exp(-E_{bind}/k_BT)$, where m and E_{bind} are the molecular mass and the energy required to remove the amphiphile from an aggregate, respectively (h and k_B are the usual Planck and Boltzmann constants). The more energy required to extract an amphiphile, the smaller the CMC, which, for the types of lipid found in membranes, may be as low as 10^{-5} molar (single hydrocarbon chain) and 10^{-10} molar (two chains). The aggregate geometry preferred by a lipid depends upon its molecular geometry as characterized by a shape factor $v_{hc}/a_o\ell_{hc}$, where a_o is the interface area occupied by the polar head group and v_{hc} and ℓ_{hc} are the volume and length of the hydrocarbon region, respectively. Geometrical packing arrangements argue that $v_{hc}/a_o\ell_{hc} \leq 1/3$ favors spherical micelles, while $1/3 < v_{hc}/a_o\ell_{hc} \leq 1/2$ favors cylindrical micelles. Bilayers are preferred in the range $1/2 < v_{hc}/a_o\ell_{hc} \leq 1$, and inverted micelles are the choice if $v_{hc}/a_o\ell_{hc}$ exceeds 1.

With their symmetry axis normal to the bilayer interface, amphiphiles occupy an interfacial area per molecule a_o that represents a competition between steric repulsion (at high density) and water–amphiphile surface tension γ (at low density). Viewed as an isotropic plate, a membrane is expected to obey $K_A \sim K_V d_{bl}$, where K_A and K_V are the area and volume compression moduli, respectively, and d_{bl} is the bilayer thickness. This prediction is not supported by experiment, however, and K_A is relatively constant at 0.2 J/m^2 for many bilayers, a behavior consistent with $K_A \sim (4\text{–}6)\gamma$ from models based on the surface tension (γ) of bilayers. Bending deformations of the membrane alter the area per molecule at an energy cost characterized by two material parameters: the bending rigidity κ_b and the Gaussian rigidity κ_G. The simplest energy density for bending has the form $(\kappa_b/2)\bullet(1/R_1 + 1/R_2)^2 + \kappa_G/R_1R_2$, where the local shape of the membrane is described by the two radii of curvature R_1 and R_2. Typical values found experimentally for lipid bilayers are $\kappa_b \sim 0.2\text{–}1.0$ J $\sim 5\text{–}25\ k_BT$. Several models for bending deformations lead to the general expectation $\kappa_b = K_A d_{bl}^2/\alpha$ where α may be 12, 24 or 48 according to the model. Current experiment supports $\kappa_b \propto d_{bl}^2$ (not d_{bl}^3 of rigid plates) and favors $\alpha = 24$ of a polymer brush model.

Under moderate tension, membrane strain is governed by the area compression modulus K_A. However, if the strain exceeds 2%–4%, most membranes rupture, at a corresponding stress usually less than $K_A/20 \sim 10^{-2}$ J/m^2. Rupture is accompanied by the opening of a hole in the bilayer which, at its simplest level, can be characterized by an edge tension λ along its boundary. The bilayer boundary probably involves the formation of a smooth interface with the headgroups arranged in cross section like the letter U, leading to the crude estimate that $\lambda \sim \kappa_b/d_{bl}$, or $\lambda \sim 2 \times 10^{-11}$ J/m; in fact, many measurements of λ cluster around $1\text{–}4 \times 10^{-11}$ J/m. The edge energy is one of the forces driving an open membrane with a free boundary to close up into a sphere if its perimeter exceeds $(2\kappa_b + \kappa_G)/\lambda$, which is in the range of a few tens of nanometers for typical values of λ and κ_b. The simplest free energy density $\mathcal{F}$ describing the formation of a circular hole in a planar membrane under a tension τ at zero temperature is $\mathcal{F} = 2\pi R\lambda - \pi R^2\tau$, which predicts that the thermodynamically favored state for $\tau > 0$ (tension) is a hole, although to reach this state, the system faces an energy barrier of $\pi\lambda^2/\tau$, often 10–100 k_BT or more. However, at finite temperature, the barrier is reduced, and vanishes for λ less than a critical value $\lambda^* \sim k_BT/b$, where b is a length scale characterizing the curvature of the boundary. As b is greater than d_{bl}, then the threshold λ^* must be less than 10^{-12} J/m, which is an order of magnitude below the physical values observed for the bilayers in a cell.

Problems

Applications

7.1. (a) Find the total length of a saturated hydrocarbon chain as a function of the number of carbon atoms $n_c \geq 3$, assuming that all bond angles are 109.5° and the bond lengths and radii are:

$$\text{C–C} = 0.154 \text{ nm} \qquad \text{C–H} = 0.107 \text{ nm}.$$

Take the chain to have a linear zig-zag configuration.

(b) Evaluate this length for $n_c = 14$ and 20, the range of importance for membrane lipids.

7.2. Evaluate S_{gas}/k_B per molecule for an ideal gas of hydrogen molecules at STP ($T = 273$ K and $P = 1$ atmosphere). Compare this contribution to the free energy (i.e. TS_{gas}/k_BT) with PV/k_BT per molecule.

7.3. The CMCs of several single-chain lipids are:

lauroyl PC	7.0×10^{-4} molar
myristoyl PC	7.0×10^{-5} molar
palmitoyl PC	7.0×10^{-6} molar.

After determining the number of carbons for each hydrocarbon chain, use Eq. (7.6) to find the nominal surface tension γ of the hydrocarbon region at $T = 300$ K (see Table B.1). For simplicity, take the lipid mass to be 500 Da throughout.

7.4. Find the range of the head group area a_o expected for bilayer formation for lipids with 1, 2, or 3 chains. Assume that $v_{hc} = 0.2\ \ell_{hc}$ for a single hydrocarbon chain, where v_{hc} and ℓ_{hc} are quoted in cubic nanometers and nanometers, respectively. Do the results of Koenig *et al.* (1997) agree with your predictions?

7.5. Suppose that the plasma membrane of a bacterium were constructed from an initially flat fluid membrane. How much energy (in units of k_BT) is required to bend this membrane into the shape of a spherocylinder 3 μm long and 1 μm in diameter? Assume $\kappa_b = \kappa_G = 10\ k_BT$.

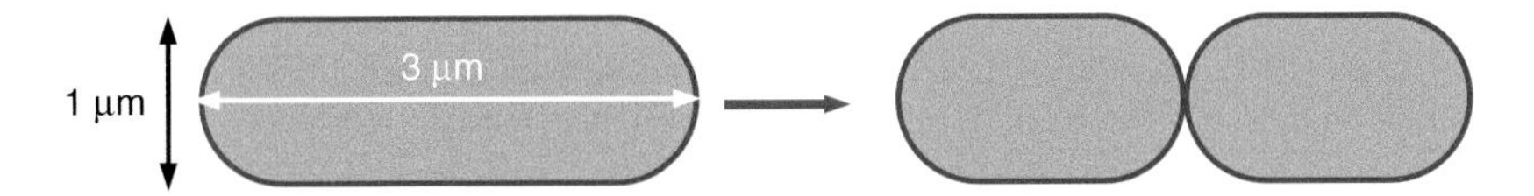

Now let this cell divide, keeping the volume and diameter the same. What is the total bending energy of the daughter cells?

7.6. Estimate the bending rigidity of the red blood cell cytoskeleton using $\kappa_b = K_A t^2/12$. Take the thickness t of the cytoskeleton to be 25 nm. Obtain K_A from the shear resistance $\mu = 10^{-5}$ J/m^2 by modeling the cytoskeleton as a triangular network of springs at zero temperature and zero stress (see Section 5.3). Quote your answer in k_BT (at T = 300 K). Compare this value of κ_b with that of a typical lipid bilayer such as diMPC in Table 7.3 to determine the relative importance of the cytoskeleton to the bending resistance of the red cell.

7.7. Find the edge tension λ for the two bilayer configurations in Fig. 7.16 as a function of the bilayer thickness d_{bl}.

(a) In Fig. 7.16(a), assume the hydrocarbons form a planar surface with $\gamma = 0.050$ J/m^2 as the hydrocarbon–water surface tension.

(b) In Fig. 7.16(b), assign a curvature resistance $\kappa_b/2$ to the curved monolayer at the bilayer edge to find the bending energy density of a half-cylinder with radii of curvature $R_1 = d_{bl}/2$ and $R_2 = \infty$. Determine λ from this expression.

(c) If the bilayer thickness is approximately $d_{bl} = 0.26n_c$ nm, calculate λ from parts (a) and (b) for $n_c = 10$ and 20. Use $\gamma = 0.050$ J/m^2 in part (a) and $\kappa_b = 10\ k_BT$ (at $T = 300$ K) in part (b). Which picture of the bilayer edge has the lower λ?

7.8. The disk and sphere configurations of a fluid membrane have the same energy at the special disk radius $2(2\kappa_b + \kappa_G)/\lambda$, but there is an energy barrier between them of $\Delta E = \pi(2\kappa_b + \kappa_G)$ (see Problem 7.21). The probability of a system to hop over an energy barrier is proportional to $Q \equiv \exp(-\Delta E/k_BT)$, among other terms. Evaluate Q for $\kappa_b = \kappa_G = 10k_BT$, typical of a lipid bilayer. What does your result tell you about the likelihood of a disk-like bilayer closing up into a spherical cell or vesicle at this special radius and temperature?

7.9. A planar membrane (with edge tension λ) subject to lateral tension τ experiences an energy barrier ΔH against hole formation.

(a) Find an algebraic expression for ΔH and evaluate it for $\lambda = 2 \times 10^{-11}$ J/m and $\tau = 10^{-3}$ J/m^2; quote your answer in units of k_BT.

(b) The probability that a hole could "hop" over the barrier at $T >$ 0 is proportional to $Q = \exp(-\Delta H / k_B T)$, among other factors. Calculate Q for your value of ΔH from part (a).
(c) At what value of τ is the barrier height equal to $k_B T$ if $\lambda = 2 \times 10^{-11}$ J/m?

7.10. A particular bilayer contains two types of lipids with very different chain lengths, one with $n_c = 12$ and the other with $n_c = 20$, where n_c is the number of carbons in each chain.

(a) If the lipids were to phase separate into two regions as in the grossly oversimplified picture shown, what would be the edge tension λ along the phase boundary? Assume that each carbon atom adds 0.126 nm to the chain length, and that the water–hydrocarbon interfacial tension is $\gamma = 0.05$ J/m^2.
(b) Compare your answer to a commonly quoted value for the bilayer/water edge tension.
(c) Suppose there are one hundred circular domains where the phases have separated, each of radius 5 nm. What energy would be released if all these domains coalesced into one circular region with the equivalent area? Quote your answer in terms of $k_B T$ at $T = 300$ K.

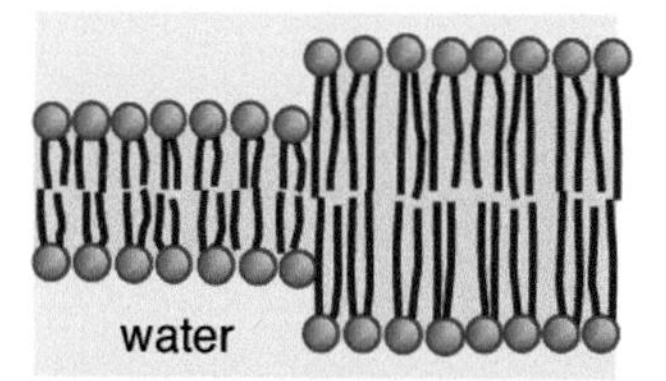

7.11. Thermophiles are a type of bacterium that can survive in high temperature environments, such as hot springs. Their plasma membranes contain an unusual type of dual-chain amphiphile in which the hydrocarbon chains are twice as long as conventional lipids and two such chains are attached to each other at both ends via polar head groups. These amphiphiles span the membrane, having one head group on each side of the membrane as shown in the figure.

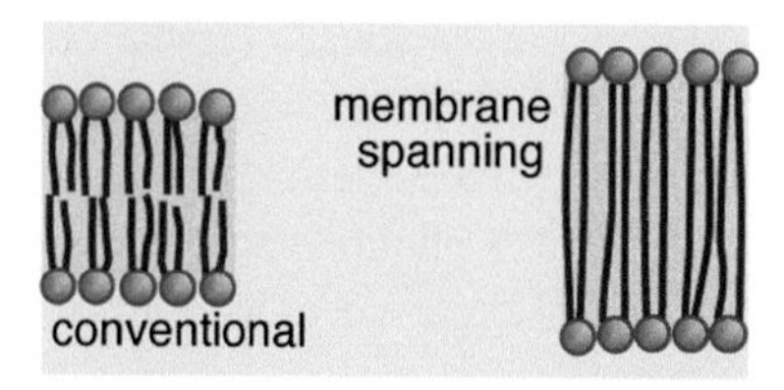

Consider a membrane composed of a single species of such amphiphiles, with 40 carbon atoms along each chain, each contributing 0.126 nm to the chain length.

(a) How thick is the hydrocarbon core of this membrane?
(b) Assuming a water–hydrocarbon surface tension $\gamma = 0.05$ J/m^2, estimate the area compression modulus K_A of the membrane.
(c) Estimate the bending modulus κ_b for the membrane using your result from part (b). If the membrane were a bilayer with the same K_A, by what factor would κ_b be changed?
(d) Qualitatively compare the line tension expected for a bilayer and for a membrane composed of membrane-spanning lipids; in particular, discuss the arrangement of the lipids at the boundary of a hole. Which type of membrane would be easier to rupture, assuming they have same thickness?

7.12. As described in Problem 7.11, the plasma membranes of high-temperature bacteria contain dual-chain amphiphiles which span the membrane such that both of their two polar head groups reside in an aqueous phase, but on opposite sides of the membrane. As an example, suppose that each of the two hydrocarbon chains has n_c = 40 carbon atoms.

(a) Find the energy E_{bind} to remove one of these molecules from a membrane under the same approximations as in Section 7.2. Quote your answer in terms of k_BT at $T = 80$ °C.
(b) Estimate the critical aggregation threshold ρ_{agg} as in Section 7.2. Assume a molecular mass of 1300 Da and quote your answer as a molar concentration.
(c) Compare your answer in part (b) with ρ_{agg} for dual chain lipids in Fig. 7.9.

Take the radius of the amphiphile to be the same as a conventional dual chain lipid, and use $\gamma = 0.05$ J/m^2 for the hydrocarbon/water interfacial tension.

7.13. In Problem 7.25 we examine the strain experienced by a cell wall arising from its growth. To apply the analysis to a physical system, consider a Gram-negative bacterium of radius 0.5 μm, surrounded by a wall having thickness 25 nm and area compression modulus $K_A = 1$ J/m^2.

(a) Assuming that the inner surface of the wall has zero strain, what is the strain (in percent) on the outer surface? Do you think that this strain is large enough to cause the material to fail?
(b) What is the root mean square strain $\langle u_{xx}{}^2 \rangle^{1/2}$ averaged over the wall?

(c) Assuming that your results for $\langle u_{xx}^2 \rangle$ in part (b) apply to the whole cell, what is the total energy associated with the strain in the cell wall. Treat the cell as a spherocylinder of total length 3 μm.

You may use results from Problem 7.25 without proof.

7.14. The peptidoglycan network in a Gram-negative cell wall has many disconnected peptide links so that a two-dimensional representation for the network may be approximately valid. Let's examine the planar rectangular network shown in the figure, with two inequivalent springs of force constant k_x and k_y, with rest lengths L_x and L_y.

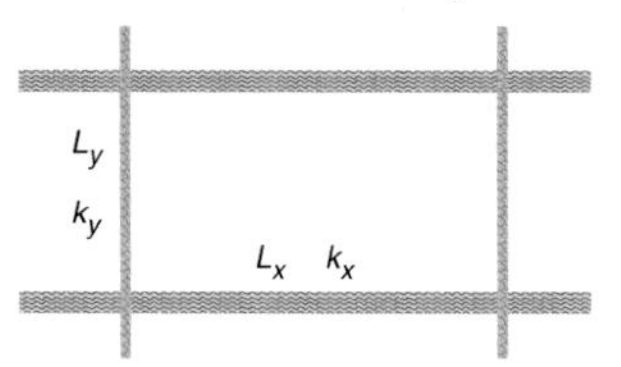

(a) Assuming $k_y = 7 \times 10^{-3}$ J/m^2, calculate the two-dimensional Young's modulus in the y-direction, Y_{2y}. Compare your result against the benchmark $\rho k_B T$ using an in-plane area of 2.5 nm^2 per disaccharide unit (Wientjes *et al.*, 1991), and 1.5 effective chains per unit. Use without proof the results from Problem 5.28; assume $T = 300$ K.

(b) Find the corresponding three-dimensional Young's modulus $Y_{3y} \sim Y_{2y}/d$ for a thin plate of thickness d. Compare your result to the measured values of Y quoted in the text, and comment on any discrepancies. Use the interchain spacing in Fig. 7.25 for d.

7.15. The persistence length of individual chains of cellulose often lies in the 5–10 nm range, depending on conditions (Muroga *et al.*, 1987). Based on this, find the effective Young's modulus of a single chain, and compare your answer to the measured Young's modulus of wood. You may estimate the effective radius of the polysaccharide chain from the observation that very thin microfibrils of cellulose, with diameters of 2 nm, contain 10–15 cellulose chains (Brett, 2000).

Formal development and extensions

7.16. Consider a thin solid sheet of thickness t made from an isotropic material having a Young's modulus Y and a Poisson ratio σ_p. When subjected to a pressure P along its

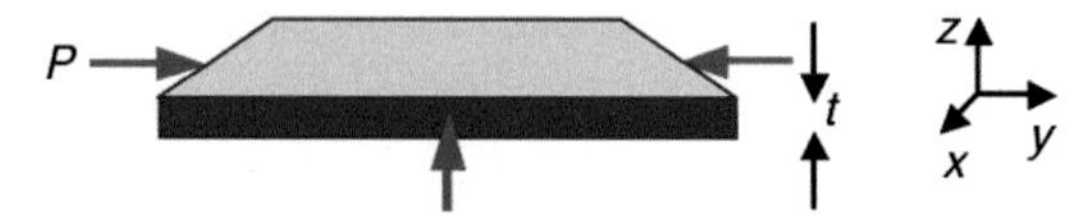

edges, it deforms according to a strain tensor whose only non-vanishing components are (Section 5 of Landau and Lifshitz, 1986):

$$u_{xx} = u_{yy} = -(P/Y)(1 - \sigma_p) \qquad u_{zz} = +2P\sigma_p/Y.$$

Prove that the three-dimensional energy density of this deformation is given by $\Delta\mathcal{F}_{3D} = P^2(1 - \sigma_p)/Y$. You will need the relations $\mu = Y/2(1 + \sigma_p)$ and $K_V = Y/3(1 - 2\sigma_p)$.

7.17. Treat the plate in Problem 7.16 as a two-dimensional system with strain elements $u_{xx} = u_{yy}$ as given, and $u_{xy} = 0$.

(a) Show that the two-dimensional energy density of this deformation is $\Delta\mathcal{F}_{2D} = 2K_A P^2(1 - \sigma_p)^2/Y^2$, where K_A is the area compression modulus.

(b) Compare your results from (a) with $\Delta\mathcal{F}_{2D}$ from Problem 7.16 to establish

$$K_A = K_V t \bullet \alpha,$$

where K_V is the volume compression modulus and

$$\alpha = 3(1 - 2\sigma_p) / 2(1 - \sigma_p).$$

(c) Evaluate α for $\sigma_p = 1/3$, which is a common value for many solids. At what value of σ_p does $\alpha = 1$?

7.18. Obtain a relationship between the bending rigidity κ_b and the two-dimensional area compression modulus K_A by considering the gentle bending of a sheet into a section of a cylinder with principal curvatures $1/R$ and 0, shown in cross-section in the diagram. Take the surface of zero-strain (the neutral surface) to run through the middle of the sheet, and assume that the thickness is unchanged by the deformation.

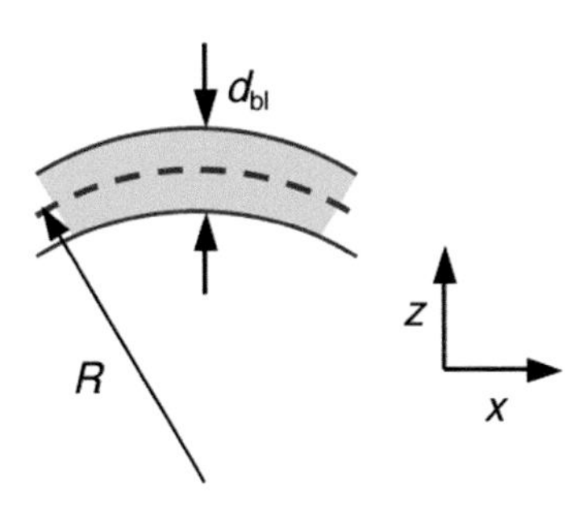

(a) Show that the energy density of the cylindrical deformation is $\Delta\mathcal{F} = \kappa_b/2R^2$.

(b) The strain u_{xx} varies linearly with z from 0 at the neutral surface to $\pm u_m$ at the boundary. Perform a simple average over the

z-coordinate to find the average value of u_{xx}^2 across the membrane, and find the corresponding energy density if $\Delta\mathcal{F} \sim \langle u_{xx}^2 \rangle K_A/2$.
(c) For small deformations, demonstrate that $u_m = d/2R$.
(d) Combine your results from (a) to (c) to obtain $\kappa_b = K_A\, d^2/12$.

This problem is treated more thoroughly in Section 11 of Landau and Lifshitz (1986).

7.19. Use the result $\kappa_b = K_A d_{bl}^2/12$ from Problem 7.18 for a single uniform sheet to show that $\kappa_b = K_A d_{bl}^2/48$ if the sheet is sliced in two along the neutral surface, with each slice permitted to slide smoothly past the other.

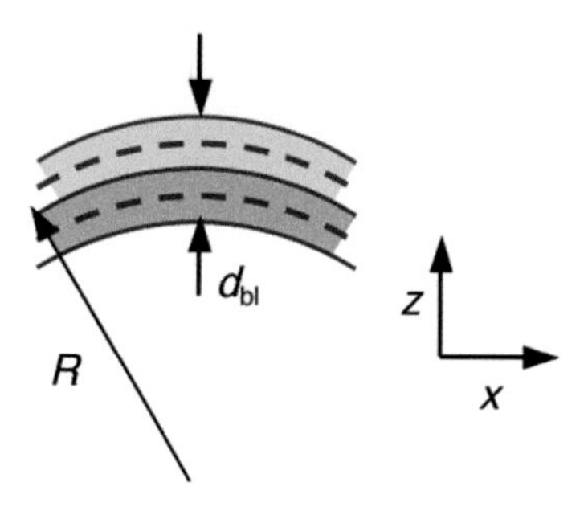

7.20. Consider an open membrane with the shape of a bowl of constant radius of curvature R.

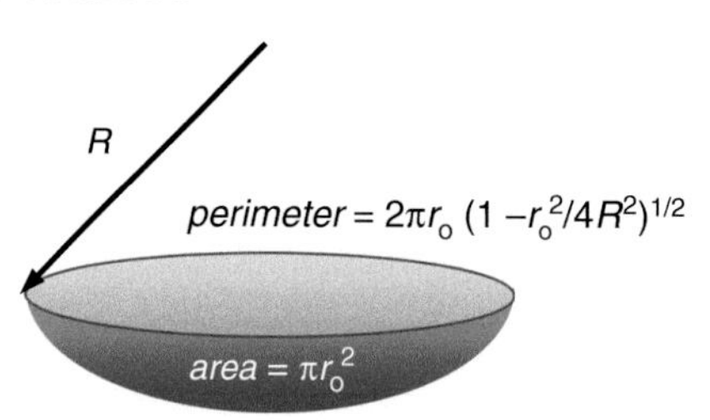

(a) If the radius of the boundary is r, show that the area of the membrane is

$$A = 2\pi R^2\,[1 \pm (1 - r^2/R^2)^{1/2}].$$

(b) To what shapes do the +/– signs in part (a) correspond?
(c) Defining the area of the surface to be πr_o^2, show that the perimeter of the bowl is equal to $2\pi r_o(1 - r_o^2/4R^2)^{1/2}$.

7.21. The geometry of an open membrane with the shape of a bowl of constant radius of curvature R was established in Problem 7.20, for a membrane with fixed area πr_o^2.

(a) Show that the contributions of the edge energy and bending energy give a total energy of

$$E = 2\pi\lambda r_o(1 - r_o^2C^2/4)^{1/2} + (2\kappa_b + \kappa_G)(\pi r_o^2C^2),$$

where the curvature C is equal to $1/R$.

(b) The disk and sphere have the same energy at $r_o = 2(2\kappa_b + \kappa_G)/\lambda$. Plot the reduced energy E/E_{sphere} at this value of r_o from $C = 0$ (disk) to $C = 2/r_o$ (sphere), where $E_{sphere} = 4\pi(2\kappa_b + \kappa_G)$.

(c) At what value of C is E/E_{sphere} in (b) a maximum? Establish that $E_{max}/E_{sphere} = 5/4$ at this value. What does the fact that $E_{max} > E_{sphere}$ and $E_{max} > E_{disk}$ imply about the stability of the sphere and disk configurations?

7.22. Show that the energy barrier between open and closed configurations for the idealized membrane in Problems 7.20 and 7.21 disappears when the radius of the flat disk r_o exceeds $4(2\kappa_b + \kappa_G)/\lambda$. Are flat disks with radii above this value stable or unstable compared to a closed sphere? You may use the expression for the energy in Problem 7.21(a) without proof.

7.23. A closed shell has the shape of a pancake with rounded edges as show in cross section in the diagram, where the vertical dashed line is an axis of rotational symmetry.

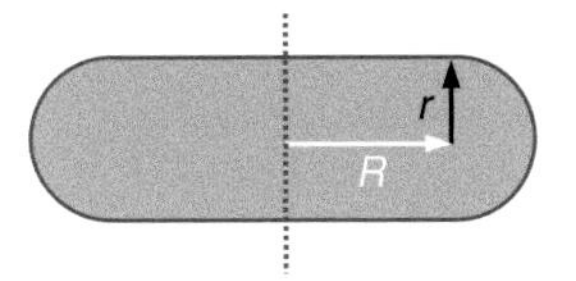

(a) Prove that the area of the curved region is

$$2\pi r^2(\pi\rho + 2),$$

where $\rho = R/r$.

(b) Find the energy density at the equator using Eq. (7.17).

(c) If your result from (b) applied to all curved surfaces (is this an underestimate or overestimate?) show that the corresponding bending energy of the shell would be

$$2\pi\,(\pi\rho + 2) \bullet [(\kappa_b/2) \bullet (\rho + 2)^2 + \kappa_G(\rho + 1)]\,/\,(\rho + 1)^2.$$

(The energy density is treated correctly in Chapter 10.)

7.24. An initially flat plate is gently bent in one direction to form a section of a cylindrical surface, with curvature $1/R$ in one direction and 0 in the other. In the diagram, the inner surface is compressed and the outer one is stretched compared to the unstrained ("neutral") surface indicated by the dashed line. The energy per molecule associated with the interfacial tension γ is $E = \gamma(a - a_o)^2/a$ for small deformations, where a is the area per molecule (with unstressed value a_o).

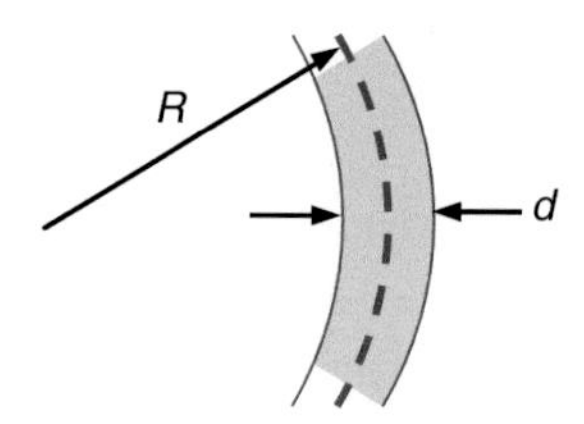

(a) Show that the energy density of the deformation is $\mathcal{E} = 2\gamma (d / 2R)^2$, where d is the thickness of the plate ($d << R$).
(b) If the bending energy were due entirely to interfacial tension, show that $\kappa_b = \gamma d^2$, where κ_b is the usual bending resistance appearing in Eq. (7.17).

7.25. Gram-positive bacteria are surrounded by a thick wall of peptidoglycan, which grows through the addition of new material to the inner surface of the wall. Suppose that the radius R of the cell does not change throughout the growth process of the wall, and that, once formed, wall material moves outwards due to the addition of new wall at the inner surface. The wall has thickness d, with $d << R$; the inner surface lies at $z = 0$.

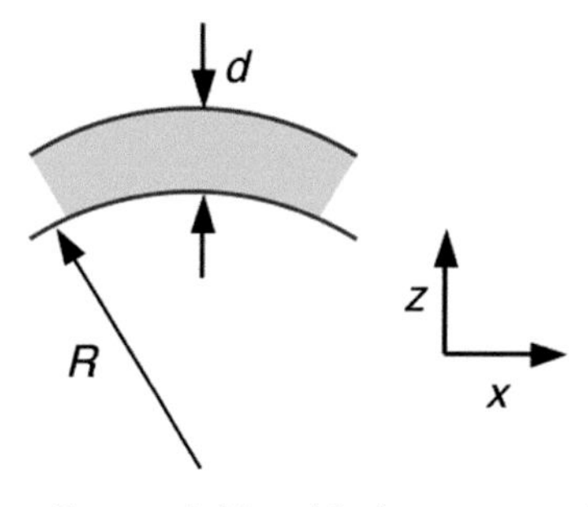

(a) If the new wall material is added at zero strain, what is the lateral strain u_{xx} at the outer surface of the wall?
(b) Perform a simple average along the z-axis to find the mean square strain $\langle u_{xx}^2 \rangle$.
(c) If the deformation energy density is $\Delta\mathcal{F} \sim \langle u_{xx}^2 \rangle K_A/2$, show that $\Delta\mathcal{F} = (d/R)^2 K_A/6$.

7.26. In Problems 7.21 and 7.22 we determine the location of an energy barrier by equating to zero its derivative with respect to a parameter such as the membrane curvature C. But the condition $dE/dC = 0$ by itself does not insure that the energy is at a maximum.

(a) What behavior of the energy as a function of C is permitted by $dE / dC = 0$? What additional constraint must be imposed to make the energy a local maximum?
(b) Confirm that your additional constraint from part (b) is satisfied by the energy function in Problems 7.21 and 7.22.

7.27. Transmembrane proteins are folded such that their hydrophobic residues align with the hydrocarbon core of the lipids in the surrounding bilayer. Calculate the energy required to remove such a protein from its membrane assuming the following parameters: the protein has a diameter of 16 nm, the membrane has a thickness of 4 nm and the interfacial tension γ between the hydrophobic region of the protein and water is 0.050 J/m^2. Compare this with a dual-chain lipid with $n_c = 16$. Quote both of your answers in k_BT at $T = 300$ K. How difficult do you think it would be to remove the protein compared to a lipid?

7.28. The diffusion of proteins within a membrane is reminiscent of a two-dimensional gas. Starting with the partition function of an ideal gas in two dimensions,

$$Z = (N!)^{-1} A^N h^{-2N} (2\pi m k_B T)^N,$$

show that corresponding entropy is

$$S = k_B N[\ln(A/N) + \ln T + \ln(2\pi m k_B /h^2) + 2].$$

In these expressions, A is the area of the system and N is the number of particles of mass m; h, k_B and T have their usual meaning. Please see the end of Appendix C for an outline of the proof in three dimensions.

7.29. It is stated in the text that the surface integral over the Gaussian curvature is an invariant for a given topology; for example, all shapes that can be obtained by deforming a spherical shell have the same form of $\int (R_1 R_2)^{-1} dA$ for principal curvatures R_1 and R_2. This implies that for shells described by the free energy density

$$\Delta\mathcal{F} = (\kappa_b/2)(1/R_1 + 1/R_2)^2 + \kappa_G/R_1 R_2$$

the Gaussian contribution should be constant. Confirm this using $\Delta\mathcal{F}$ to explicitly calculate the bending energy of a spherocylinder of radius R and overall length $L + 2R$, and comparing your result to the bending energy of a sphere. Calculate E(sphere) first, then E(spherocylinder).

7.30. A model for the bonding of peptidoglycan in a three-dimensional cell wall is shown in Fig. 7.25.

(a) Find an algebraic expression for the bond density in terms of bond lengths a and b.

(b) Find the shear modulus μ using Eq. (6.43) at $T = 300$ K for the quoted a and b.

(c) Use Eqs. (6.22) and (6.24) to establish $Y = (8/3)\mu$ for materials whose Poisson ratio is 1/3. Find Y for your value of μ in part (b), and compare it with the measured values of Y for peptidoglycan.

8 Membrane undulations

Membranes of the cell are characterized by several elastic parameters, such as the area compression modulus, that reflect the membrane's quasi-two-dimensional structure. As described in Chapter 7, these parameters have small values for a lipid bilayer just 4–5 nm thick, yet they properly describe the energetics of membrane deformation at zero temperature where thermal fluctuations in shape are unimportant. But what happens at finite temperature? In the discussion of polymers and networks in Part I, we saw that the entropic contribution to the elasticity of very flexible filaments is significant at ambient temperatures, owing to the large configuration space available to these filaments. Do we expect similar behavior for flexible sheets? In this chapter, we develop a mathematical description of surfaces, and explore the characteristics of membrane undulations. Membranes are treated in isolation here, and in interaction with other surfaces in Chapter 9. A more extensive treatment of membrane defects and fluctuations can be found in Leibler (1989) and Chaikin and Lubensky (1995).

8.1 Thermal fluctuations in membrane shape

The bending rigidity κ_b of a phospholipid bilayer lies close to 10^{-19} J, or 10–20 k_BT at ambient temperatures, as summarized in Table 7.3. What does such a small value of κ_b imply about the undulations of a membrane with the dimensions of a cell? For illustration, we calculate the change in energy of the flat, disk-shaped membrane in Fig. 8.1(a) as it is deformed into the surface of constant curvature in Fig. 8.1(b). Configuration (b) has radius of curvature R and energy density $2\kappa_b/R^2$, where the Gaussian rigidity is neglected (see Sections 7.4 and 8.2). Taking the disk diameter as 2 μm and $\kappa_b \sim 10\ k_BT$, the deformation energy is $20\pi k_BT/R^2$, where R must be quoted in microns. What is the typical value of R when the disk flexes at non-zero temperature? The thermal energy scale is set by k_BT and the disk energy rises to this value (from zero for a flat disk with $R = \infty$) when R declines to ~8 μm. This radius of curvature corresponds to $\theta \sim 15°$ in Fig. 8.1(b), meaning that undulations may have measurable effects in common cells.

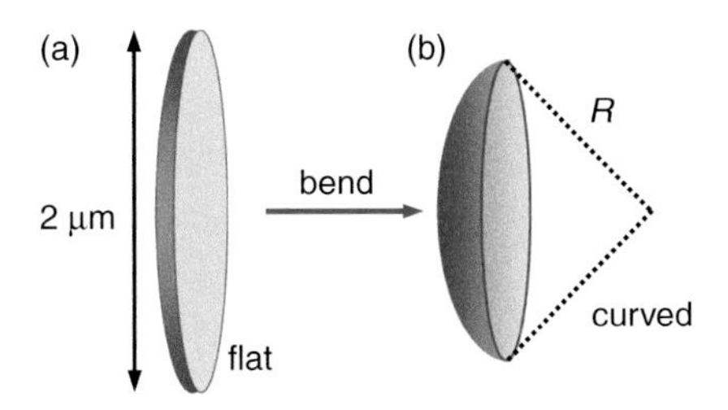

Fig. 8.1 Bending of a disk with diameter 2 μm (a) into a surface with the same area and a constant radius of curvature R (b).

As introduced in Chapter 3, thermal fluctuations of flexible filaments are characterized by a temperature-dependent length scale called the persistence length ξ_p, equal to $\kappa_f / k_B T$, where κ_f is the flexural rigidity. Can membrane undulations be characterized by a persistence length in the same way as a filament? Inspection of Fig. 8.2(a) reminds us that the persistence length of a filament is a measure of the contour distance over which tangent vectors become decorrelated, as indicated by the directions of the arrows. For membranes, there is more than one tangent vector to a given surface element, and a better indicator of the surface orientation is provided by the normal vector. As shown in Fig. 8.2(b), normal vectors wobble and sway across a wavy surface, and their relative orientation over large length scales also can be described by a persistence length. However, for sheets, ξ_p depends exponentially on the bending rigidity as $\xi_p \sim b \exp(-4\pi\kappa_b/3k_B T)$, where b is an elementary length scale of the membrane. For example, if b is of the order of nanometers, and κ_b is $10k_B T$, then ξ_p is of the order of kilometers, implying that the undulations of cellular membranes are smooth. The exponential dependence of ξ_p on κ_b in membranes, vs. a linear dependence on κ_f in polymers, is not unexpected: a change in direction of a filament involves just a few neighboring segments, whereas a change in direction of a surface affects many adjacent elements in the membrane plane.

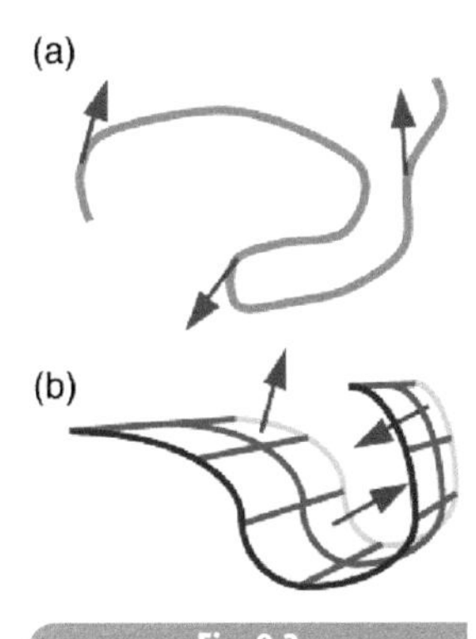

Fig. 8.2 Just as the persistence length of a polymer characterizes the change in orientation of its tangent vectors (a), so too the persistence length of a membrane characterizes the change in orientation of the surface normal vectors (b).

Flexible polymers exhibit a universal scaling behavior with contour length; for ideal polymers, the mean linear dimension (e.g., the end-to-end displacement) scales like $N^{1/2}$, where N is the number of segments on the chain (see Section 2.3). Is there analogous scaling behavior for membranes? The answer is a subtle "yes", with the scaling exponent depending on the self-avoidance and connectivity properties of the membrane. As exemplified in Fig. 8.3(a), most bilayers in the cell at physiologically relevant temperatures are fluid in the sense that neighboring lipid molecules are not covalently bound, and slide past each other when the membrane is subject to an in-plane shear. These fluid membranes can be contrasted with what we will refer to as polymerized membranes, such as the MoS_2 monolayer illustrated in Fig. 8.3(b), which acquires in-plane shear resistance from ionic or

(a)

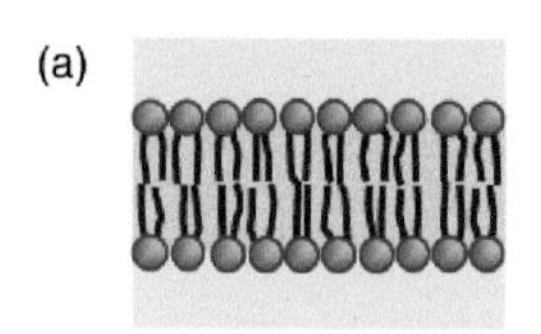

(b)

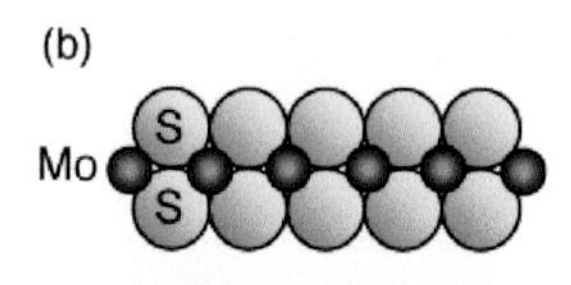

Fig. 8.3

Schematic cross section through two membranes: (a) a fluid bilayer with no shear resistance, (b) a polymerized sheet that resists in-plane shear (e.g., MoS_2). The thickness of the membrane is nanometers in (a) and ångstroms in (b).

covalent bonds between neighboring atoms or molecules. Although either type of membrane may be viewed as a two-dimensional generalization of a one-dimensional polymer chain, in fact fluid and polymerized membranes each have distinctive geometrical and elastic properties.

As evidenced in many simulation studies, polymerized membranes are commonly flat over long length scales, meaning that transverse undulations decrease relative to the longitudinal dimension of the membrane (e.g., the square root of the area). In contrast, a fluid membrane behaves like a branched polymer at distances long compared to the membrane persistence length, developing wavy arms and protrusions. Such structures cannot arise from an initially flat polymerized membrane, in the same way as a piece of paper cannot wrap an apple without developing creases and folds. Two sample configurations of polymerized and fluid membranes are displayed in Figs. 8.9 and 8.11; we caution that the complete geometrical description of membranes is more complicated than just these two phases (flat and branched polymer) as will be established later.

This chapter is somewhat more mathematical than others in this book, even though the formal proofs are kept to a minimum. In Section 8.2, some aspects of the mathematical definition and properties of surfaces are presented, including a simplified representation of a gently undulating surface. This formalism is applied immediately in Section 8.3 to obtain the persistence length of a fluid membrane. The mathematical machinery required to predict the geometrical scaling of membranes is too intricate for inclusion here, and so, in Section 8.4, we sketch out some simple expectations for scaling, which are then confirmed by computer simulation. Lastly, several experimental analyses of membrane undulations are reported in Section 8.5; the values of the elastic moduli deduced from these experiments can be found in Section 7.4.

8.2 Mathematics of curvature

For a moment, let's return to the properties of a one-dimensional curve embedded in two dimensions. To give the curve a physical meaning, suppose that it is the trajectory of a ball thrown from one person to another, executing an arc as it travels. Measuring the height h of the ball as a function of the distance x along the ground allows us to construct the function $z = h(x)$ of the trajectory. Alternatively, if we know the time evolution of the trajectory, we can describe the motion by two parametric functions $x(t)$ and $h(t)$, where t is the time. The utility of each of these descriptions varies with the application. As introduced in Section 3.2, the parametric approach based on the arc length s provides simple expressions for geometrical quantities

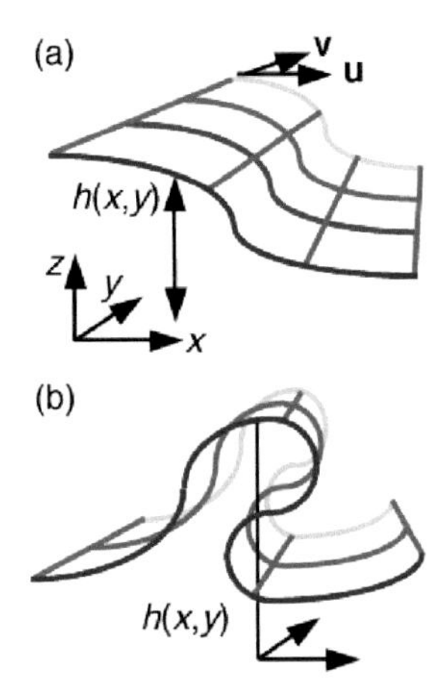

Fig. 8.4

(a) A surface can be defined by a function $h(x,y)$ of a set of external Cartesian coordinates x,y or a set u,v resident on the surface itself. (b) The function $h(x,y)$ is not single-valued if overhangs are allowed.

such as the unit tangent vector $\mathbf{t} = \partial\mathbf{r}/\partial s$ and the curvature $C = \mathbf{n} \cdot (\partial\mathbf{t}/\partial s) = \mathbf{n} \cdot (\partial^2\mathbf{r}/\partial s^2)$, where $\mathbf{n}$ is the normal to the curve at position $\mathbf{r}$.

The analytical description of a surface involves a choice of coordinates and representation (the parametric representation, for example). One coordinate convention uses a set of basis vectors embedded in the surface itself, much like the arc length along a curve, as indicated by the pair $\mathbf{u},\mathbf{v}$ in Fig. 8.4(a). However, a common approach in membrane studies employs Cartesian coordinates to write a point $\mathbf{r}$ on the surface as

$$\mathbf{r} = [x, y, h(x,y)], \tag{8.1}$$

where h is the "height" away from the xy plane, as illustrated in Fig. 8.4(a). Unfortunately, $h(x,y)$ may not be single-valued for general surfaces: for instance, $h(x,y)$ has three values in the overhang region of Fig. 8.4(b). In this text, we adopt the Monge representation in which $h(x,y)$ of Eq. (8.1) is constrained to be single-valued; i.e. overhangs are forbidden. This formalism allows us to establish results applicable at long length scales, so long as the surfaces are not too large compared with their persistence length.

Our first task is to express tangent vectors and normals within the Monge representation. Suppose that we take a slice of a surface along the x-direction (i.e. at constant y). We can construct a tangent vector in this direction by making a unit step in the x-direction, and a step $\partial h/\partial x \equiv h_x$ in the z-direction, as shown in Fig. 8.5(a). Repeating this procedure for the y-direction gives the pair of tangent vectors

$$\partial_x\mathbf{r} = (1, 0, h_x) = (1, 0, \partial_x h) \tag{8.2a}$$

$$\partial_y\mathbf{r} = (0, 1, h_y) = (0, 1, \partial_y h), \tag{8.2b}$$

where $\partial_x \equiv \partial/\partial x$. The vectors $\partial_x\mathbf{r}$ and $\partial_y\mathbf{r}$ are *not* unit vectors, and are *not* generally orthogonal; however, they do define the plane tangent to the surface at point $[x, y, h(x,y)]$, and can generate the unit normal vector $\mathbf{n}$ to the surface via the cross product

$$\mathbf{n} \equiv (\partial_x\mathbf{r}) \times (\partial_y\mathbf{r}) \,/\, |(\partial_x\mathbf{r}) \times (\partial_y\mathbf{r})| = (-h_x, -h_y, 1) \,/\, (1 + h_x^{\,2} + h_y^{\,2})^{1/2}. \tag{8.3}$$

As shown in Fig. 8.5(b), a segment of length $\mathrm{d}x$ along the x-axis corresponds to a vector $(\partial_x\mathbf{r})\mathrm{d}x$ along the surface. The cross product of the vectors $(\partial_x\mathbf{r})\mathrm{d}x$ and $(\partial_y\mathbf{r})\mathrm{d}y$ gives the area element $\mathrm{d}A$ on the surface corresponding to $\mathrm{d}x\,\mathrm{d}y$ in the coordinate plane:

$$\mathrm{d}A = |(\partial_x\mathbf{r}) \times (\partial_y\mathbf{r})|\, \mathrm{d}x\, \mathrm{d}y = (1 + h_x^{\,2} + h_y^{\,2})^{1/2}\, \mathrm{d}x\, \mathrm{d}y, \tag{8.4}$$

according to Eq. (8.2). The quantity $(1 + h_x^{\,2} + h_y^{\,2})$ is called the metric g of the surface

$$g \equiv 1 + h_x^{\,2} + h_y^{\,2} = 1 + (\partial_x h)^2 + (\partial_y h)^2, \tag{8.5}$$

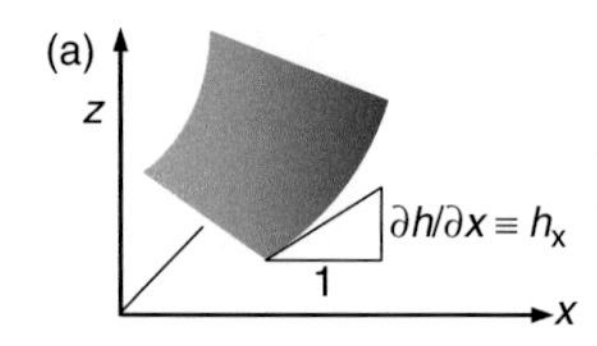

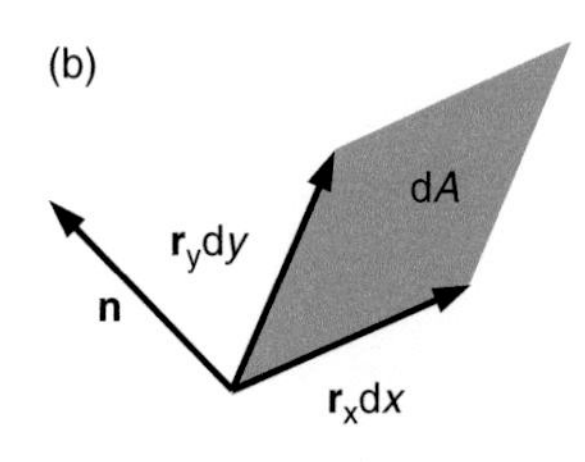

Fig. 8.5

(a) For a line tangent to the surface and lying in the xz-plane, a step $\Delta x = 1$ corresponds to $\Delta z = \partial h/\partial x$, resulting in a tangent vector $\mathbf{r}_x = (1, 0, \partial h/\partial x)$. (b) An area element $dx\, dy$ on the xy-plane corresponds to an element $dA = |\mathbf{r}_x \times \mathbf{r}_y|\, dx\, dy$ on the surface. The unit normal to the surface is $\mathbf{n} = \mathbf{r}_x \times \mathbf{r}_y / |\mathbf{r}_x \times \mathbf{r}_y|$.

such that $\mathrm{d}A = \sqrt{g}\, \mathrm{d}x\, \mathrm{d}y$. A more thorough description of the metric and parallel transport on curved surfaces can be found in texts such as Goetz (1970) or Kreyszig (1959); we return to the metric momentarily.

The curvature of a surface can be obtained from the same definition as that for a line, namely $C = \mathbf{n} \bullet (\partial^2\mathbf{r}/\partial s^2)$, where s is the arc length. However, the curvature of a surface is not unique, and a direction must be specified for the path on a surface to which the curvature applies. For example, the curvature of a cylindrical shell of radius R is zero in a direction parallel to the cylindrical axis, and $1/R$ along a circle perpendicular to the axis. One could choose other directions along the surface in this example as well, each having a different curvature. The extremal values of the curvature as a function of direction are called the principal curvatures, denoted here by C_1 and C_2; the combinations $(C_1 + C_2)/2$ and $C_1 \bullet C_2$ are the mean and Gaussian curvatures, respectively. Unfortunately, it takes some mathematical effort to obtain expressions for the curvatures; here, we forgo some elegant results from differential geometry and work strictly within the Monge representation.

To determine the curvature $C = \mathbf{n} \bullet (\partial_s^2\mathbf{r})$, we need Eq. (8.3) for $\mathbf{n}$, as well as the derivatives $\mathbf{r}' = \partial_s\mathbf{r}$ and $\mathbf{r}'' = \partial_s^2\mathbf{r}$:

$$\mathbf{r}' = (\partial_x\mathbf{r})x' + (\partial_y\mathbf{r})y', \tag{8.6}$$

$$\mathbf{r}'' = (\partial_x^2\mathbf{r})(x')^2 + (\partial_y^2\mathbf{r})(y')^2 + 2(\partial_x\partial_y\mathbf{r})x'y' + (\partial_x\mathbf{r})x'' + (\partial_y\mathbf{r})y'', \tag{8.7}$$

where $x' \equiv \partial_s x$, $x'' \equiv \partial_s^2 x$, etc. The last two terms in Eq. (8.7) do not contribute to $\mathbf{n} \bullet (\partial_s^2\mathbf{r})$, owing to the orthogonality of $\mathbf{n}$ with $\partial_x\mathbf{r}$ and $\partial_y\mathbf{r}$, leaving

$$C = \mathbf{n} \bullet \mathbf{r}'' = b_{xx}(x')^2 + b_{yy}(y')^2 + 2b_{xy}x'y'. \tag{8.8}$$

The scalar coefficients $b_{\alpha\beta}$ are

$$b_{\alpha\beta} \equiv \mathbf{n} \bullet (\partial_\alpha\partial_\beta\mathbf{r}) \qquad (\alpha, \beta = x, y). \tag{8.9}$$

In the literature, the coefficients b_{uu} are sometimes replaced by $b_{uu} = L$, $b_{uv} = M$, $b_{vv} = N$ in notation introduced by Gauss (see Chapter IV of Kreyszig, 1959), where $\mathbf{u}$ and $\mathbf{v}$ are a general coordinate set, as in Fig. 8.4(a). Differentiating the orthogonality condition $\mathbf{n} \bullet (\partial_\alpha\mathbf{r}) = 0$ leads to the relation $\partial_\beta[\mathbf{n} \bullet (\partial_\alpha\mathbf{r})] = \mathbf{n} \bullet (\partial_\alpha\partial_\beta\mathbf{r}) + (\partial_\beta\mathbf{n}) \bullet (\partial_\alpha\mathbf{r}) = 0$, or, from the definition (8.9),

$$b_{\alpha\beta} = -(\partial_\alpha\mathbf{r}) \bullet (\partial_\beta\mathbf{n}) \qquad (\alpha, \beta = x, y). \tag{8.10}$$

To summarize, Eqs. (8.8) and (8.10) provide a formal way of obtaining the curvature from a knowledge of $\mathbf{r}$.

The direction-dependence of the curvature is implicit in Eq. (8.8) through $\mathbf{r}''$, the second derivative of the position with respect to arc length along a particular direction. The maximum and minimum values of $\mathbf{n} \bullet \mathbf{r}''$ are the principal curvatures, and can be obtained from Eq. (8.8) using the method of Lagrange multipliers (see Section 1.5 of Safran, 1994) or other techniques (see David, 1989, or Chapter IV of Kreyszig, 1959). The mathematics is something of a digression, so we simply quote the results for the mean and Gaussian curvatures:

$$(C_1+C_2)/2 = (g_{xx}b_{yy} + g_{yy}b_{xx} - 2g_{xy}b_{xy}) \,/\, 2g \tag{8.11}$$

$$C_1C_2 = (b_{xx}b_{yy} - b_{xy}^2) \,/\, g, \tag{8.12}$$

where g is the metric familiar from Eq. (8.5). The metric is the determinant of the metric tensor $g_{\alpha\beta}$, whose components here are

$$g_{\alpha\beta} = (\partial_\alpha \mathbf{r}) \bullet (\partial_\beta \mathbf{r}) \qquad (\alpha, \beta = x, y). \tag{8.13}$$

It is straightforward to verify Eq. (8.5) by substituting Eq. (8.2) into (8.13). In the general coordinate set $(\mathbf{u},\mathbf{v})$ the element of arc length squared is written in terms of the metric tensor as

$$\mathrm{d}s^2 = g_{uu}(\mathrm{d}u)^2 + g_{vv}(\mathrm{d}v)^2 + 2g_{uv}\mathrm{d}u\,\mathrm{d}v. \tag{8.14}$$

The tensors $g_{\alpha\beta}$ and $b_{\alpha\beta}$ are referred to as the first and second fundamental forms, respectively, in some texts on differential geometry.

We now determine the mean and Gaussian curvatures in the Monge representation. First, we evaluate $\partial_x \mathbf{n}$ using Eq. (8.3), finding

$$\partial_x \mathbf{n} = -\ \{([1+h_y^2]h_{xx} - h_x h_y h_{xy}),\ ([1+h_x^2]h_{xy} - h_x h_y h_{xx}),\ (h_x h_{xx} + h_y h_{xy})\} \,/\, (1 + h_x^2 + h_y^2)^{3/2}, \tag{8.15}$$

$$\partial_y \mathbf{n} = -\ \{([1+h_y^2]h_{xy} - h_x h_y h_{yy}),\ ([1+h_x^2]h_{yy} - h_x h_y h_{xy}),\ (h_x h_{xy} + h_y h_{yy})\} \,/\, (1 + h_x^2 + h_y^2)^{3/2}.$$

These somewhat intimidating relations can be combined with Eqs. (8.2) and (8.10) to yield the simple

$$b_{\alpha\beta} = h_{\alpha\beta} \,/\, (1 + h_x^2 + h_y^2)^{1/2} \qquad (\alpha, \beta = x, y). \tag{8.16}$$

Lastly, Eqs. (8.13) and (8.16) can be substituted into (8.11) and (8.12) to find the mean and Gaussian curvatures:

$$(C_1+C_2)/2 = \{(1+h_x^2)h_{yy} + (1+h_y^2)h_{xx} - 2h_x h_y h_{xy}\} \,/\, 2(1 + h_x^2 + h_y^2)^{3/2} \tag{8.17}$$

$$C_1C_2 = (h_{xx}h_{yy} - h_{xy}^2) \,/\, (1 + h_x^2 + h_y^2)^2. \tag{8.18}$$

For many situations of interest, $h(x,y)$ is a slowly varying function corresponding to gentle undulations, such that the curvatures can be approximated by

$$(C_1 + C_2)/2 \cong (h_{xx} + h_{yy})/2 \tag{8.19}$$

$$C_1C_2 \cong h_{xx}h_{yy} - h_{xy}^2, \tag{8.20}$$

if we retain just the leading-order terms from Eqs. (8.17) and (8.18).

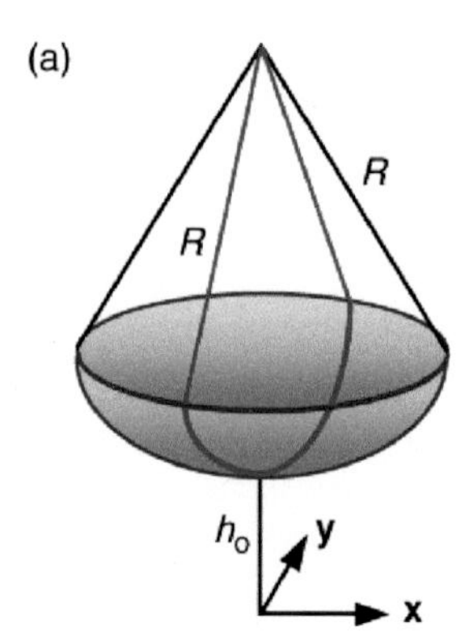

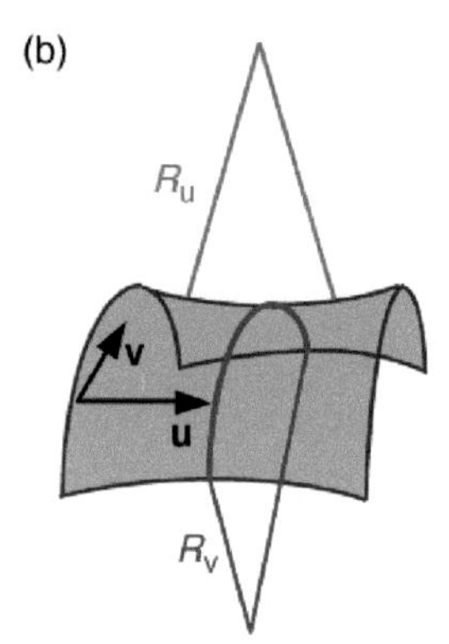

Fig. 8.6

The derivatives h_x and h_y vanish at the minimum or inflection points in these two shapes. Part (a) is a section of a sphere where the principal curvatures have the same sign; (b) is a saddle region where the curvatures have opposite signs: $R_u \bullet R_v < 0$.

To confirm the geometrical meaning of Eqs. (8.19) and (8.20) consider the shape of a surface in the region of a minimum or inflection point, as illustrated in Fig. 8.6. For notational simplicity, place the minimum directly over the coordinate origin at $x = y = 0$. A Taylor series expansion describes the height of the surface for small regions near the minimum, namely

$$h(x, y) = h_o + h_x x + h_y y + h_{xx} x^2/2 + h_{yy} y^2/2 + h_{xy} xy + \ldots, \tag{8.21}$$

where the height at the origin is h_o. For surfaces like Fig. 8.6(a), the derivatives h_x and h_y vanish at the local minimum in height, showing that the surface is quadratic in x and y with coefficients proportional to the local curvature. Equations (8.19) and (8.20) also establish that principal curvatures need not have the same sign. If Fig. 8.6(a) is a section of a sphere, for instance, the principal curvatures have the same sign and are both equal to $1/R$ in magnitude, where R is the radius of the sphere. In contrast, the curvatures at the saddle point in Fig. 8.6(b) have opposite sign: the normals are canted towards each other in the **u**-direction, but away from each other in the **v**-direction; that is $R_u \bullet R_v < 0$, where R_u and R_v are the radii of curvature in the **u** and **v** directions, respectively.

The simplest form for the energy density of bending deformations involves the mean and Gaussian curvatures, usually written in the combination (Canham, 1970; Helfrich, 1973; Evans, 1974)

$$\mathcal{F} = (\kappa_b/2)(C_1 + C_2 - C_0)^2 + \kappa_G C_1 C_2, \tag{8.22}$$

where κ_b and κ_G are the bending rigidity (or bending modulus) and Gaussian bending rigidity (or saddle-splay modulus) respectively. Introducing the constant term C_0, known as the spontaneous curvature, permits Eq. (8.22) to describe bilayers that are curved in their equilibrium state because their two (monolayer) leaflets are compositionally inequivalent. In most of our discussions, C_0 will be set equal to zero. The radii of curvature corresponding to C_0^{-1} can be as small as a few nanometers for phospholipid monolayers and bilayers, depending on their composition (see, for example, Szule *et al.*, 2002 and Fuller *et al.*, 2003). Intrinsic curvature may also arise in bilayers with an asymmetric charge distribution (Ha, 2003).

Now, it is shown in the end-of-chapter problems that Eq. (8.22) has a particularly appealing form when $\kappa_b = -\kappa_G$ (at $C_0 = 0$), namely

$$\mathcal{F} = (\kappa_b/2)[(\partial_x \mathbf{n})^2 + (\partial_y \mathbf{n})^2], \tag{8.23}$$

in the limit of gentle undulations (Kantor, 2004; Nelson, 2004). This form of the energy density is a natural generalization of the bending resistance of a flexible filament, introduced in Section 3.2. Whether physical bilayers are described by $\kappa_b = -\kappa_G$ is not resolved; the existing measurements are consistent with the rather broad range $0 > \kappa_G/\kappa_b > -2$ (see Section 4.5.9 of Petrov, 1999; Siegel and Kozlov, 2004). Fortunately, the integral over the surface of the Gaussian curvature does not change with the shape of the surface as long as the topology of the surface is fixed, courtesy of the Gauss–Bonnet theorem, and so our ignorance of κ_G does not impede our describing shape changes of cells with fixed topology.

8.3 Membrane bending and persistence length

At zero temperature, a membrane minimizes its bending energy by adopting a shape that is flat or uniformly curved, unless factors like compositional inhomogeneity force it to do otherwise. However, at finite temperature, the membrane exchanges energy with its environment, allowing it to explore a configuration space that includes energetically expensive bending deformations. The presence of wrinkles and ripples cause the orientation of the surface to change with position, as illustrated in Fig. 8.7. Mathematically, we represent the local orientation of a surface at position $\mathbf{r}$ by a normal vector $\mathbf{n}(\mathbf{r})$. For the flat configuration in Fig. 8.7(a), the normals are parallel such that the scalar product $\mathbf{n}(\mathbf{r}_1) \bullet \mathbf{n}(\mathbf{r}_2)$ is independent of positions $\mathbf{r}_1$ and $\mathbf{r}_2$. For the wavy configuration in Fig. 8.7(b), the normals change with position and $\mathbf{n}(\mathbf{r}_1) \bullet \mathbf{n}(\mathbf{r}_2)$ follows suit.

The scalar product of the normals provides a measure of the surface fluctuations. For nearby positions on the surface, $\mathbf{n}(\mathbf{r}_1)$ and $\mathbf{n}(\mathbf{r}_2)$ are roughly parallel and $\mathbf{n}(\mathbf{r}_1) \bullet \mathbf{n}(\mathbf{r}_2)$ should not be too much less than unity. In contrast, if the positions are far apart, $\mathbf{n}(\mathbf{r}_1)$ and $\mathbf{n}(\mathbf{r}_2)$ may point in such different directions that $\mathbf{n}(\mathbf{r}_1) \bullet \mathbf{n}(\mathbf{r}_2)$ may not even be positive. Thus, the average of the normal–normal product, $\langle \mathbf{n}(\mathbf{r}_1) \bullet \mathbf{n}(\mathbf{r}_2) \rangle$, taken over the entire surface, should decrease from a value near unity at small separations to near zero at large separations. The spatial decorrelation of the normals is characterized by the persistence length ξ_p through

$$\langle \mathbf{n}(\mathbf{r}_1) \bullet \mathbf{n}(\mathbf{r}_2) \rangle = \exp(-\Delta r / \xi_p), \tag{8.24}$$

where $\Delta \mathbf{r} \equiv \mathbf{r}_1 - \mathbf{r}_2$. This formalism is the same as the persistence length of a polymer, as described in Chapter 3, except that tangent vectors $\mathbf{t}$ are used for polymers whereas normals $\mathbf{n}$ are used for surfaces.

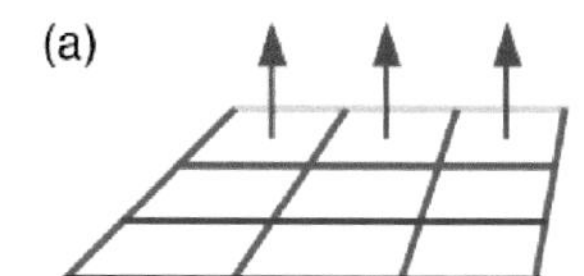

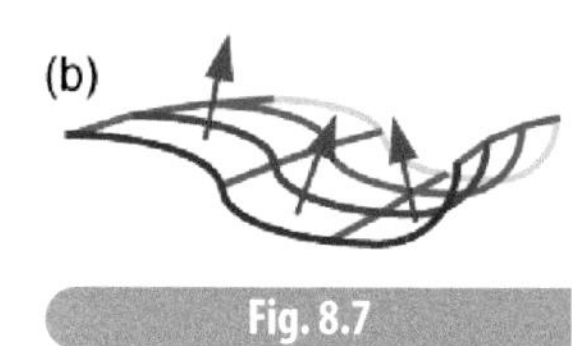

Fig. 8.7

The orientation of a surface normal is independent of position on a planar surface (a) but changes with position on a rough surface (b).

The persistence length of a membrane depends upon its bending resistence κ_b, just as ξ_p of a polymer depends upon the filament's flexural rigidity. Obtaining a mathematical formula for ξ_p for a surface is rather involved; the steps we follow are:

- transform $\langle \mathbf{n}(\mathbf{r}_1) \bullet \mathbf{n}(\mathbf{r}_2) \rangle$ to the Fourier representation $h(\mathbf{q})$ of the height function,
- find $h(\mathbf{q})$ in terms of κ_b and temperature,
- solve for ξ_p.

The proof is presented in detail because we draw intermediate results from it when discussing membrane undulations elsewhere in the text.

8.3.1 Fourier components of $\langle \mathbf{n}(\mathbf{r}_1) \bullet \mathbf{n}(\mathbf{r}_2) \rangle$

As displayed in Eq. (8.3), the surface normal is $(-h_x, -h_y, 1) \bullet (1+h_x^2+h_y^2)^{-1/2}$ in the Monge representation. Using the binomial approximation $(1+x)^\alpha \sim 1 + \alpha x$ at small x permits $\mathbf{n}$ to be expressed to order h^2 as

$$\mathbf{n}(\mathbf{r}) = (-h_x, -h_y, 1 - [h_x^2+h_y^2]/2). \tag{8.25}$$

Accordingly, the product $\mathbf{n}(\mathbf{r}_1) \bullet \mathbf{n}(\mathbf{r}_2)$ must be

$$\begin{aligned}\mathbf{n}(\mathbf{r}_1) \bullet \mathbf{n}(\mathbf{r}_2) &= h_x(\mathbf{X}_1)h_x(\mathbf{X}_2) + h_y(\mathbf{X}_1)h_y(\mathbf{X}_2) \\ &\quad + 1 - [h_x(\mathbf{X}_1)^2+h_y(\mathbf{X}_1)^2]/2 - [h_x(\mathbf{X}_2)^2+h_y(\mathbf{X}_2)^2]/2 + O(h^3) \\ &= 1 - \{\partial_x[h(\mathbf{X}_1)-h(\mathbf{X}_2)]\}^2/2 - \{\partial_y[h(\mathbf{X}_1)-h(\mathbf{X}_2)]\}^2/2 + O(h^3),\end{aligned} \tag{8.26}$$

also to order h^2, where $\mathbf{X}$ is a two-dimensional vector comprising the xy coordinates of the position $\mathbf{r}$ and $\partial_x = \partial/\partial x$. It is convenient to express $\mathbf{n} \bullet \mathbf{n}$ in terms of the Fourier representation of the height function. For readers who have not seen the Fourier representation before, the idea is to expand an arbitrary function $f(x)$ in terms of a series of standard functions such as trigonometric functions. For example, the linear function $f(x) = x$ on the domain $-1 < x < 1$ can be written as a sum of sine functions multiplied by coefficients b_j:

$$f(x) = \Sigma_j\, b_j \sin(j\pi x) \qquad b_j = 2(-1)^{j+1} / j\pi, \tag{8.27}$$

where the sum runs from $j = 1$ to infinity (see Chapter 1 of Powers, 1999). For most numerical problems, the series is truncated at the accuracy appropriate to the situation. Functions other than sines can be used in Fourier representations, although something like $\cos(j\pi x)$ obviously won't work in Eq. (8.27) because our function $f(x)$ vanishes at $x = 0$ whereas $\cos(j\pi x)$ does not.

The height function $h(\mathbf{X})$ can be written in terms of a set of complex functions $\exp(\mathrm{i}\mathbf{q} \bullet \mathbf{X}) = \cos(\mathbf{q} \bullet \mathbf{X}) + \mathrm{i} \sin(\mathbf{q} \bullet \mathbf{X})$ as

$$h(\mathbf{X}) = (A / 4\pi^2) \int \mathrm{d}\mathbf{q} \exp(\mathrm{i}\mathbf{q} \bullet \mathbf{X}) h(\mathbf{q}), \tag{8.28}$$

where $\mathbf{q}$ is a wavevector with units of inverse length and A is the membrane area, a normalization constant. The function $h(\mathbf{q})$ is the continuous analog of the discrete coefficients b_j in Eq. (8.27). Both $\mathbf{X}$ and $\mathbf{q}$ are two-dimensional quantities, as must be $\mathrm{d}\mathbf{q}$ in the integral. We first evaluate $\partial_x[h(\mathbf{X}_1)-h(\mathbf{X}_2)]$ and its square in the Fourier representation before attempting to transform Eq. (8.26):

$$\partial_x[h(\mathbf{X}_1)\text{-}h(\mathbf{X}_2)] = (A / 4\pi^2)\, \partial_x \{\int d\mathbf{q} \exp(i\mathbf{q}\bullet\mathbf{X}_1)h(\mathbf{q}) - \int d\mathbf{q}' \exp(i\mathbf{q}'\bullet\mathbf{X}_2)h(\mathbf{q}')\}$$
$$= (A / 4\pi^2) \int iq_x\, d\mathbf{q}\, [\exp(i\mathbf{q}\bullet\mathbf{X}_1) - \exp(i\mathbf{q}\bullet\mathbf{X}_2)]\, h(\mathbf{q}). \quad (8.29)$$

We have permitted the functions $h(\mathbf{q})$ in Eq. (8.28) to be complex, so we must take the complex square of Eq. (8.29) to ensure that Eq. (8.26) is real:

$$|\, \partial_x[h(\mathbf{X}_1)\text{-}h(\mathbf{X}_2)]\, |^2 = (A / 4\pi^2)^2 \int q_x q'_x\, d\mathbf{q}\, d\mathbf{q}'\, [\exp(i\mathbf{q}\bullet\mathbf{X}_1) - \exp(i\mathbf{q}\bullet\mathbf{X}_2)]$$
$$\bullet\, [\exp(-i\mathbf{q}'\bullet\mathbf{X}_1) - \exp(-i\mathbf{q}'\bullet\mathbf{X}_2)]\, h(\mathbf{q})h^*(\mathbf{q}'), \quad (8.30)$$

where $h^*(\mathbf{q})$ is the complex conjugate of $h(\mathbf{q})$.

Now, the expectation $\langle \mathbf{n}(\mathbf{r}_1)\bullet\mathbf{n}(\mathbf{r}_2)\rangle$ in Eq. (8.24) involves both an average over thermally allowed configurations, and an average over the positions $\mathbf{r}_1$ and $\mathbf{r}_2$ within each configuration, subject to the constraint $\mathbf{r}_1 - \mathbf{r}_2 = \Delta\mathbf{r}$. The positional average of Eq. (8.30) can be accomplished by integrating with $A^{-1} \int d\mathbf{X}_1\, d\mathbf{X}_2\, \delta(\mathbf{X} - [\mathbf{X}_1 - \mathbf{X}_2])$, leading to

$$\langle|\partial_x[h(\mathbf{X}_1)-h(\mathbf{X}_2)]|^2\rangle_{\text{pos}} = [A /(4\pi^2)^2] \int d\mathbf{X}_1\, d\mathbf{X}_2\, \delta(\mathbf{X} - [\mathbf{X}_1-\mathbf{X}_2])\, \bullet$$
$$\int q_x q'_x\, d\mathbf{q}\, d\mathbf{q}'\, [\exp(i\mathbf{q}\bullet\mathbf{X}_1) - \exp(i\mathbf{q}\bullet\mathbf{X}_2)]$$
$$\bullet\, [\exp(-i\mathbf{q}'\bullet\mathbf{X}_1) - \exp(-i\mathbf{q}'\bullet\mathbf{X}_2)]\, h(\mathbf{q})h^*(\mathbf{q}'). \quad (8.31)$$

Integrating the delta function $\delta(\mathbf{X} - [\mathbf{X}_1-\mathbf{X}_2])$ over $\mathbf{X}_2$ has the effect of replacing $\mathbf{X}_2$ by $\mathbf{X}_1 - \mathbf{X}$; collecting terms in $\mathbf{X}_1$ yields

$$\langle|\partial_x[h(\mathbf{X}_1)-h(\mathbf{X}_2)]|^2\rangle_{\text{pos}} = [A /(4\pi^2)^2] \int d\mathbf{X}_1 \int q_x q'_x\, d\mathbf{q}\, d\mathbf{q}' \exp(i[\mathbf{q}-\mathbf{q}']\bullet\mathbf{X}_1)\bullet$$
$$[1 - \exp(-i\mathbf{q}\bullet\mathbf{X})][1 - \exp(i\mathbf{q}'\bullet\mathbf{X})]\, h(\mathbf{q})h^*(\mathbf{q}'). \quad (8.32)$$

The integral over $\mathbf{X}_1$ involves just the exponential $\exp(i(\mathbf{q}-\mathbf{q}')\bullet\mathbf{X}_1)$, and can be replaced by the two-dimensional generalization of $(2\pi)^{-1} \int \exp(iqx)\, dx = \delta(q)$ to give

$$\langle|\partial_x[h(\mathbf{X}_1)-h(\mathbf{X}_2)]|^2\rangle_{\text{pos}} = (A / 4\pi^2) \int q_x q'_x\, d\mathbf{q}\, d\mathbf{q}'\, \delta(\mathbf{q}-\mathbf{q}')$$
$$\bullet[1 - \exp(-i\mathbf{q}\bullet\mathbf{X})][1 - \exp(i\mathbf{q}'\bullet\mathbf{X})]\, h(\mathbf{q})h^*(\mathbf{q}'), \quad (8.33)$$

or, after a trivial integration over $\mathbf{q}'$

$$\langle|\partial_x[h(\mathbf{X}_1)-h(\mathbf{X}_2)]|^2\rangle_{\text{pos}} = (A / 4\pi^2) \int q_x^2\, d\mathbf{q}\, [2\text{–}2\cos(\mathbf{q}\bullet\mathbf{X})]\, h(\mathbf{q})h^*(\mathbf{q}), \quad (8.34)$$

where we have used the trigonometric identity $[\exp(ix)-1]\bullet[\exp(-ix)-1] = 2\text{–}2\cos(x)$. Eq. (8.34) and the corresponding expression for $\langle|\partial_x[h(\mathbf{X}_1)-h(\mathbf{X}_2)]|^2\rangle_{\text{pos}}$ give the full thermal average of Eq. (8.26):

$$\langle \mathbf{n}(\Delta\mathbf{r})\bullet\mathbf{n}(0)\rangle = 1 - (A / 4\pi^2) \int q^2\, d\mathbf{q}\, [1-\cos(\mathbf{q}\bullet\mathbf{X})]\, \langle h(\mathbf{q})h^*(\mathbf{q})\rangle, \quad (8.35)$$

which can be evaluated knowing the thermal average of the Fourier components.

8.3.2 Fourier components and the equipartition theorem

Let's consider a membrane under tension such that a given configuration has an energy E governed by the applied tension τ and bending rigidity κ_b,

$$E = \tau \int dA + (\kappa_b/2) \int (C_1 + C_2)^2 dA, \tag{8.36}$$

where we assume that the topology of the surface is fixed so that the Gaussian curvature term is simply a constant. The tension opposes the creation of new contour area compared to the projected area on the xy plane. In the Monge representation, the surface area element dA is $(1+h_x^2+h_y^2)^{1/2}d\mathbf{X}$ from Eq. (8.4), and the sum of the curvatures is $C_1+C_2 = h_{xx}+h_{yy}$ from Eq. (8.19). Expanding the square root of the metric, and keeping only terms of order h^2, the energy of a configuration can be written as

$$E = (1/2) \int d\mathbf{X} \; \{\tau(h_x^2+h_y^2) + \kappa_b(h_{xx}+h_{yy})^2\} \tag{8.37}$$

where an overall additive constant of the form $\tau \int d\mathbf{X}$ has been neglected.

As before, we use the Fourier representation, Eq. (8.28), to write the surface tension term as

$$h_x(\mathbf{X})^2+h_y(\mathbf{X})^2 = [A/(4\pi^2)]^2 \int [q_x q'_x + q_y q'_y] d\mathbf{q}\, d\mathbf{q}' \exp(i[\mathbf{q}-\mathbf{q}']\bullet\mathbf{X}) h(\mathbf{q}) h^*(\mathbf{q}'), \tag{8.38}$$

where the left-hand side implicitly involves a complex square. This expression is integrated over $d\mathbf{X}$ in Eq. (8.37), allowing us to replace $(2\pi)^{-2}\int d\mathbf{X} \exp(i[\mathbf{q}-\mathbf{q}']\bullet\mathbf{X})$ with the delta function $\delta(\mathbf{q}-\mathbf{q}')$. Then integrating over $\mathbf{q}'$ gives the simple

$$\int d\mathbf{X}\, [h_x(\mathbf{X})^2 + h_y(\mathbf{X})^2] = (A^2 / 4\pi^2) \int q^2 d\mathbf{q}\; h(\mathbf{q}) h^*(\mathbf{q}). \tag{8.39}$$

These steps can be repeated for $(h_{xx} + h_{yy})^2$ to give

$$\int d\mathbf{X}\, [h_{xx}(\mathbf{X}) + h_{yy}(\mathbf{X})]^2 = (A^2 / 4\pi^2) \int q^4 d\mathbf{q}\; h(\mathbf{q}) h^*(\mathbf{q}). \tag{8.40}$$

Substituting Eqs. (8.39) and (8.40) into (8.37) yields

$$E = (1/2)\bullet(A^2 / 4\pi^2) \int d\mathbf{q}\; (\tau q^2 + \kappa_b q^4) h(\mathbf{q}) h^*(\mathbf{q}), \tag{8.41}$$

up to an overall additive constant, as before.

Our next task is to find the thermal expectation $\langle h(\mathbf{q}) h^*(\mathbf{q}) \rangle$ given an energy of the form Eq. (8.41). We proceed by recalling the behavior of a one-dimensional harmonic oscillator, as described in Appendix C. This system is governed by the energy $E(x) = k_{sp}x^2/2$, and its thermal average is

$$\langle E \rangle = k_{sp}\langle x^2 \rangle/2 = (k_{sp}/2) \int x^2 \exp(-E/k_BT)dx \,/ \int \exp(-E/k_BT)dx. \tag{8.42}$$

Integrating this expression gives $\langle x^2 \rangle = k_BT/k_{sp}$, and, consequently, $\langle E \rangle = k_BT/2$. This latter result arises from the equipartition of energy: each oscillation mode has an average energy of $k_BT/2$. Our surface energy Eq. (8.41) can be looked upon as a generalization of the harmonic oscillator. The factor $\int d\mathbf{q}$ is a sum over oscillator modes, with one mode per $(2\pi)^2/A$ in $\mathbf{q}$-space, each with an energy $(A/2)(\tau q^2+\kappa_b q^4)h(\mathbf{q})h^*(\mathbf{q})$. Equating the average energy of an individual mode with $k_B T/2$ gives

$$\langle h(\mathbf{q})h^*(\mathbf{q})\rangle = k_B T / A \cdot (\tau q^2 + \kappa_b q^4). \quad (8.43)$$

which demonstrates that $\langle |h(q)|^2 \rangle$ increases with temperature.

8.3.3 Persistence length

We can now obtain the persistence length by substituting Eq. (8.43) into Eq. (8.35). The resulting integral is

$$\langle \mathbf{n}(\Delta\mathbf{r}) \cdot \mathbf{n}(0) \rangle = 1 - (k_B T / 4\pi^2) \int q^2 \, d\mathbf{q} \, [1 - \cos(\mathbf{q} \cdot \mathbf{X})] / (\tau q^2 + \kappa_b q^4), \quad (8.44)$$

where the area factors A have canceled. We evaluate this in the tensionless limit $\tau = 0$,

$$\langle \mathbf{n}(\Delta\mathbf{r}) \cdot \mathbf{n}(0) \rangle = 1 - (k_B T / 4\pi^2 \kappa_b) \int d\mathbf{q} \, [1 - \cos(\mathbf{q} \cdot \mathbf{X})] / q^2. \quad (8.45)$$

This is a two-dimensional integral with $d\mathbf{q} = q \, dq \, d\theta$, where θ is the angle between $\mathbf{q}$ and $\mathbf{X}$, or $\mathbf{q} \cdot \mathbf{X} = q|\mathbf{X}|\cos\theta$. Defining the θ part of the integral to be $I_\theta(z)$, we have

$$\langle \mathbf{n}(\Delta\mathbf{r}) \cdot \mathbf{n}(0) \rangle = 1 - (k_B T / 2\pi\kappa_b) \int I_\theta(z) \, dq / q, \quad (8.46)$$

with $z = q|\mathbf{X}|$ and

$$I_\theta(z) = 1 - (2\pi)^{-1} \int d\theta \cos(z \cos\theta), \quad (8.47)$$

where the θ-integral runs from 0 to 2π. This integral appears frequently in physical problems, and is one of a class of functions called Bessel functions; this particular one is defined as $J_0(z) \equiv (2\pi)^{-1} \int d\theta \cos(z \cos\theta)$. Bessel functions do not have an analytical form, but they do have approximate representations (see Chapter 9 of Abramowitz and Stegun, 1970) which are used to construct the graph of $I_\theta(z)$ in Fig. 8.8. One sees that $I_\theta(z)$ rises from 0 at $z = 0$, passes by unity at $z \sim 2$–3 after which it oscillates ever closer to 1 as z increases. Hence,

- at large z, $I_\theta(z) \sim 1$ and can be ignored in Eq. (8.46); at large q, the integral is truncated by the intermolecular spacing b;
- at small z, $I_\theta(z) \sim 0$, and the integral is truncated at $q \sim 2$–$3 / \Delta r$.

With q ranging from approximately $\pi/\Delta r$ to π/b, Eq. (8.46) becomes

$$\begin{aligned}\langle \mathbf{n}(\Delta\mathbf{r}) \cdot \mathbf{n}(0) \rangle &\sim 1 - (k_B T / 2\pi\kappa_b) \int_{\pi/\Delta r}^{\pi/b} dq / q \\ &\sim 1 - (k_B T / 2\pi\kappa_b) \ln(\Delta r / b). \quad (8.48)\end{aligned}$$

How do we relate this to the persistence length? The correlation function approaches zero (or $1/e$, strictly speaking) at distances close to the

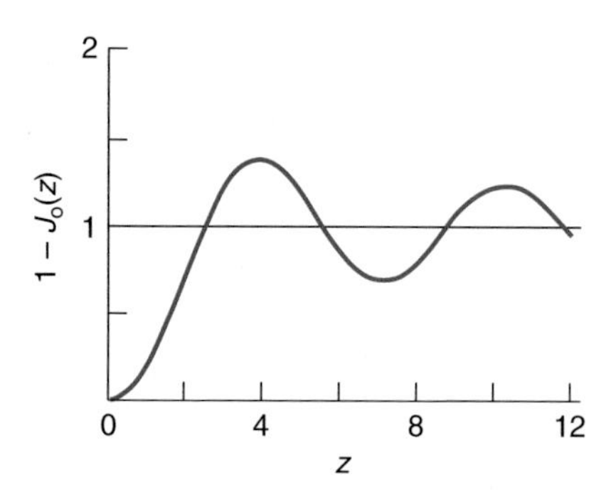

Fig. 8.8 Behavior of the integral $I_\theta(z) = 1 - J_0(z)$ as a function of the argument z, where J_0 is a Bessel function.

persistence length ξ_p. The condition $\langle \mathbf{n}(\Delta\mathbf{r}) \bullet \mathbf{n}(0) \rangle \sim 0$ in Eq. (8.48) is satisfied by

$$\xi_p \sim b \exp(2\pi\kappa_b / k_B T), \qquad \text{(de Gennes and Taupin)} \qquad (8.49)$$

a result first obtained by de Gennes and Taupin (1982). A more detailed treatment by Peliti and Leibler (1985), based on a formalism called the renormalization group, leads to a similar expression with a numerical factor in the exponent which is 2/3 of that in Eq. (8.49),

$$\xi_p \sim b \exp(4\pi\kappa_b / 3k_B T). \qquad \text{(Peliti and Leibler)} \qquad (8.50)$$

Computer simulations (Gompper and Kroll, 1995) are consistent with Eq. (8.50). The exponential dependence of ξ_p on the bending resistance of a surface should be contrasted with the linear dependence of ξ_p on the bending resistance of a polymer.

As demonstrated by Eq. (8.50), the persistence length decreases as the temperature rises, meaning that the undulations have shorter wavelengths. Put another way, it becomes easier to bend a membrane as the temperature increases, an observation that can be quantified by an effective bending rigidity

$$\kappa(\ell) = \kappa_b - (3k_B T / 4\pi) \ln(\ell / b), \qquad (8.51)$$

that depends on the length scale ℓ of the undulations, as introduced by Peliti and Leibler (1985; see also Helfrich, 1986; however, the opposite sign is reported in Helfrich, 1998). For $\ell > b$, the logarithm in Eq. (8.51) is positive and $\kappa(\ell)$ decreases from its zero-temperature value κ_b. This relation also shows that the surface becomes softer when viewed at longer wavelengths. What does Eq. (8.50) tell us about the persistence lengths of lipid bilayers? Numerically, if we set $b \sim 1$ nm, and $\kappa_b \sim 10k_BT$, we expect $\xi_p \sim 10^6$ km. Dramatic as this result is, it does not mean that the membranes of a cell are planar, only that they undulate smoothly on cellular length scales.

The effect of thermal fluctuations on closed membrane shapes is derived by Morse and Milner (1995).

8.4 Scaling of polymers and membranes

If they are sufficiently large, self-avoiding fluid membranes appear like the branched polymers described in Section 3.3, in that they form dangling arms, such as illustrated in Fig. 8.9. In contrast, a fluid sheet without self-avoidance adopts exotic configurations in which the floppy arms can intersect each other, creating a topologically daunting labyrinth of tubes, pockets and chambers. When taken as an ensemble, do membrane shapes have universal characteristics, like the scaling behavior displayed by ideal polymers, branched polymers, etc.? We know that polymer shapes obey a scaling law that depends upon the polymer's interaction and structure: for instance, the linear size of an ideal unbranched polymer scales like $(L_c/\xi_p)^{1/2}$, where L_c is the contour length and ξ_p the persistence length. In analogy, how does the linear size of a membrane scale with its area or, in the case of enclosed surfaces, its volume? Not unexpectedly, distinct classes of membranes possess different scaling behavior.

8.4.1 Membranes without self-avoidance

We can use the squared radius of gyration R_g^2 to represent the linear size of a square membrane of contour area L^2; for instance, a flat membrane obeys $R_g^2 \sim L^2$. Subject to thermal fluctuations, the size of a fluid membrane

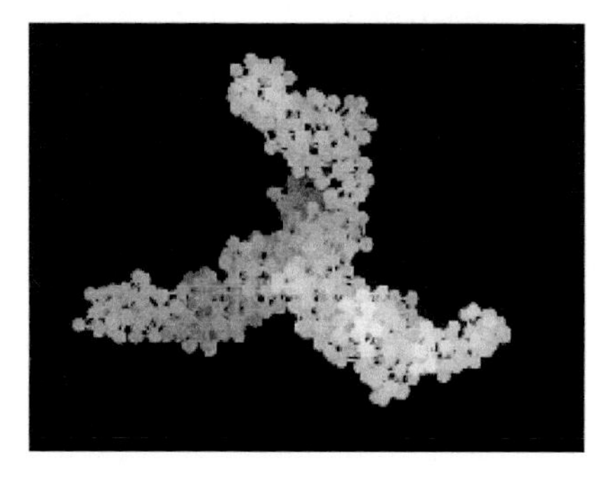

Fig. 8.9 Computer simulation of a fluid membrane with no bending resistance other than self-avoidance; beads and tethers represent the surface. This membrane has the topology of a sphere (no holes or handles) with a surface area that hovers close to a fixed mean and an enclosed volume that fluctuates freely (Boal and Rao, 1992a).

without self-avoidance (called a *phantom* membrane) grows very slowly with contour area, obeying (see Nelson, 2004, for a review)

$$R_g^{\,2} \propto \ln L, \tag{8.52}$$

when the reduced bending resistance $\kappa_b / k_B T$ is small. This result can be proven using a Fourier decomposition of the surface (reproduced in Section 8.5 of Plischke and Bergersen, 1994) and is easy to verify by computer simulation (Kantor *et al.*, 1986, 1987). We conclude that the geometry of membranes can be characterized by observables such as R_g, although the power-law exponents, and even the functional form of the scaling, may be very different from polymers.

8.4.2 Flory scaling of polymers and membranes

Self-avoidance tends to increase the linear size of chains and sheets. The effects are strongest when the embedding dimension is low; for instance, because it cannot double-back on itself, a self-avoiding chain in one dimension looks like a rod and obviously obeys $R_g \sim L_c^{\,1}$, not the scaling $R_g \sim L_c^{\,1/2}$ of ideal chains. We summarized the scaling behavior of polymers in Table 3.1, but at that time only derived the scaling exponents of the simplest systems. In 1953, Flory introduced a generic model for self-avoiding polymers that used energy and entropy arguments to successfully predict their scaling behavior (see Chapter X of Flory, 1953). Here, we obtain the Flory scaling exponent for polymers, then quote its extension to membranes. The proof has two components:

- determining the dependence of the free energy on the chain's linear size r and number of segments N,
- finding the value of r that minimizes the free energy as a function of N.

Following the presentation of de Gennes (1979; see also McKenzie, 1976), the behavior of the free energy is evaluated in two regimes.

(i) Short distances

Steric repulsion between segments causes a chain to swell compared to an ideal chain. The repulsive energy experienced by one segment through its interaction with others is proportional to the product of the concentration of segments and the volume v_{ex} within which they interact. Approximating the segment concentration by N / r^d for a chain in a d-dimensional space, the total repulsive energy experienced by all N segments is proportional to

$$[\textit{number of segments}] \bullet [\textit{interaction volume}] \bullet [\textit{concentration of segments}] = v_{ex} N^2 / r^d.$$

If v_{ex} represents a hard-core steric interaction which defines a length scale $v_{ex}^{1/3}$ but not an energy scale, then the energy scale of the interaction must be k_BT. Thus, the steric contribution to the free energy at short distances behaves like

$$F_{short} = k_BT \, v_{ex} \, N^2/r^d, \tag{8.53}$$

where all constants, including factors of two from double-counting the interacting pairs, have been absorbed into v_{ex}.

(ii) Long distances

As an ideal chain is stretched, the probability of finding a given end-to-end distance r decreases exponentially as $\exp(-dr^2/2Nb^2)$, where b is the elementary segment length (see Section 3.4). Recalling that the entropy S is proportional to the logarithm of the probability, then to within a constant

$$S/k_B = -\, dr^2/2Nb^2. \tag{8.54}$$

Since the free energy is related to the entropy through $F = E - TS$, the entropic contribution to the free energy at long distances is

$$F_{long} = +k_BTdr^2/2Nb^2. \tag{8.55}$$

Combining Eqs. (8.53) and (8.55) and discarding overall normalization constants, the free energy of the self-avoiding chain can be approximated by

$$F \cong k_BTv_{ex}N^2/r^d + k_BTdr^2/2Nb^2, \tag{8.56}$$

showing that there are energy penalties for pushing the chain elements close together (small r) or for stretching the chains out (large r). The value of r that minimizes F can be found by equating to zero the derivative of Eq. (8.56) with respect to r, holding other quantities fixed. This minimization shows that r scales like

$$r \sim N^{3/(2+d)}. \tag{8.57}$$

Equation (8.57) applies to any measure of the linear dimension of the system as a whole, such as the end-to-end distance or the root mean square radius. The exponent on the right-hand side of Eq. (8.57) is called the Flory exponent ν_{Fl}

$$\nu_{Fl} = 3 \,/\, (2+d). \tag{8.58}$$

As a trivial check, Eq. (8.58) predicts $\nu_{Fl} = 1$ for $d = 1$, just as we found for the one-dimensional self-avoiding chain.

For two and three dimensions, the Flory exponent predicts that the end-to-end displacement $\langle r_{ee}^{\,2}\rangle^{1/2}$ scales like $N^{3/4}$ and $N^{3/5}$ respectively. The prediction for $d = 2$ has been obtained by analytic means (Nienhuis, 1982), while the $d = 3$ prediction has been verified numerically (Li *et al.*, 1995, obtain $\nu = 0.5877 \pm 0.0006$). For $d = 4$, the Flory exponent is the same as the ideal exponent, indicating that chain self-avoidance is irrelevant in more than three dimensions. In other words, self-avoidance causes chains to swell in $d = 1$, 2 and 3 dimensions.

A generalized Flory argument for the scaling behavior of a D-dimensional surface embedded in d-dimensional space reads (see de Gennes, 1979, and Kantor, 1989)

$$\nu_{\text{Fl}} = (2+D)/(2+d) \tag{8.59}$$

for the scaling exponent ν in

$$R_{\text{g}}^{2} \propto L^{2\nu}, \tag{8.60}$$

where L is a contour distance along the surface. For polymers, $D = 1$ and Eq. (8.59) reverts to (8.58). For membranes, $D = 2$ and Eq. (8.59) predicts $\nu = 1$ for two dimensions (as expected for a membrane confined to a plane) and $\nu = 4/5$ for three dimensions, a prediction which is not obeyed, as we now explain.

8.4.3 Fluid and polymerized membranes

Membranes are categorized broadly as polymerized or fluid according to whether or not they can resist in-plane shear, although variations of this categorization would include mixed fluid/polymerized systems, sheets with holes, and networks with special types of connectivity. Shear resistance affects the scaling properties of self-avoiding membranes because the smooth arms and other structures illustrated for fluid membranes in Fig. 8.9 are not accessible to a polymerized membrane that is initially flat. In addition to shear, membranes also may resist out-of-plane bending, as quantified by the bending rigidities κ_{b} and κ_{G}. For polymers, the flexural rigidity κ_{f} only sets the length scale beyond which the asymptotic scaling behavior appears; the specific value of κ_{f} does not change the scaling exponent. What about membranes? Does κ_{b} just set the length scale ξ_{p} but not influence the scaling law?

Extensive computer simulations have been performed on self-avoiding fluid membranes and random surfaces (thoroughly reviewed in Gompper and Kroll, 2004). Some studies explicitly count configurations on a lattice (Glaus, 1988), while others may use an algorithm for fluid sheets developed by Baumgärtner and Ho (1990), as illustrated in Fig. 8.10. Most of the simulation work has been performed on closed fluid sheets – computational vesicles, if you will – to avoid interpretational ambiguities arising from edge fluctuations. Analyzing the radius of gyration of a fluid vesicle according to Eq. (8.60) yields $\nu = 0.8$ for systems of modest size, in agreement with the Flory prediction for fluid membranes (Baumgärtner and

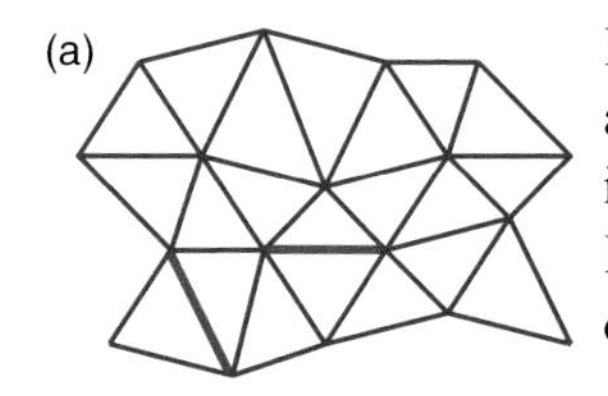

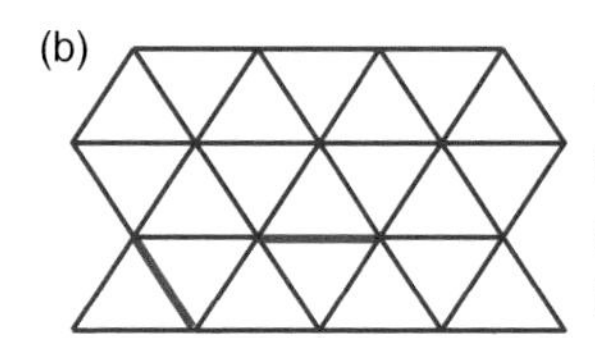

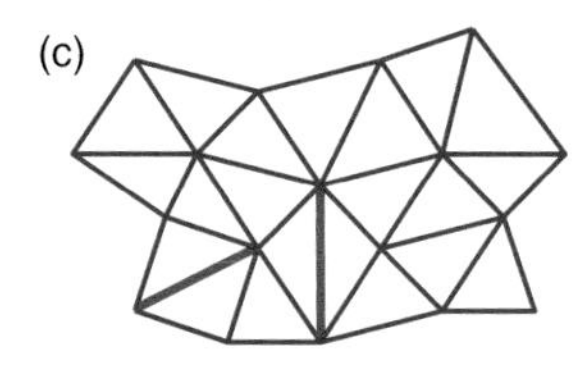

Fig. 8.10

Algorithms for polymerized and fluid surfaces, based on an initial state (b). For polymerized membranes (a), the connectivity of surface elements is fixed, although the elements themselves can move. For fluid membranes (c), the connecting bonds can "flip" between vertices, permitting the bonds to diffuse. The two bonds indicated by heavy lines do not change their connectivity in (a), but have reconnected in (c), creating junctions with 5- and 7-fold connectivity.

Ho, 1990). However, more extensive analyses, which include larger systems and examine the appearance of the vesicle as a function of bending rigidity, favor branched polymer scaling (Kroll and Gompper, 1992; Boal and Rao, 1992a). The exponents of branched polymer scaling are dimension-dependent; for an enclosed shape in three dimensions, the scaling laws are

$$R_g^2 \propto A^1 \text{ and } \langle V \rangle \propto A^1, \text{ (branched polymer)} \qquad (8.61)$$

where A is the surface area and V is the enclosed volume. Analytical treatment (Gross, 1984) and lattice-based simulations (Glaus, 1988) of random surfaces also favor branched polymer scaling for fluid membranes. The branched appearance of a fluid membrane is very clear in the simulation of Fig. 8.9; at large length scales, the membrane gains entropy by forming arms and fingers even if there is a modest cost in energy.

Both phantom and self-avoiding polymerized membranes display different behavior than fluid membranes at long distances. Computer simulations show that phantom polymerized membranes are convoluted at low rigidity κ_b/k_BT as in Eq. (8.52) but become asymptotically flat as $\kappa_b/k_BT >> 0$ (Kantor *et al.*, 1986; Kantor and Nelson, 1987). In the flat phase, $R_g^2 \propto L^2$ (where L is the linear dimension of the membrane) whereas in the crumpled phase, $R_g^2 \propto \ln L$ (see Nelson, 2004). Further, the simulations exhibit a phase transition separating the crumpled phase at small κ_b/k_BT from the flat phase at large κ_b/k_BT, as expected analytically (Peliti and Leibler, 1985; Nelson and Peliti, 1987; Guitter *et al.*, 1989). This behavior is different from phantom (or ideal) chains, which always obey $R_g^2 \propto L_c$ at any flexural rigidity as long as the viewing scale is much larger than the persistence length.

What happens when self-avoidance is included: does the sharp transition between flat and crumpled phases persist? Field-theoretic calculations favor the existence of the phase transition even in the presence of long-range self-avoidance (reviewed in Nelson, 2004). The earliest simulations to address this question supported the idea of a phase transition; however, most simulations find that polymerized membranes are flat, but rough, even in the absence of bending rigidity (Plischke and Boal, 1988, Lipowsky and Girardet, 1990, Gompper and Kroll, 1991, and many others; see also Baumgärtner, 1991, Baillie and Johnston, 1993, and Guitter and Palmeri, 1992). Because of its in-plane shear rigidity, a polymerized membrane cannot form entropy-generating arms, as in Fig. 8.9, without introducing entropy- and energy-consuming folds and creases. As a result, a polymerized membrane at finite temperature looks like Fig. 8.11, with only gentle undulations at long length scales. Choosing a reference plane parallel to the membrane plane, the height fluctuations $\langle h^2 \rangle$ of a polymerized membrane in the Monge representation scale with the membrane area A like

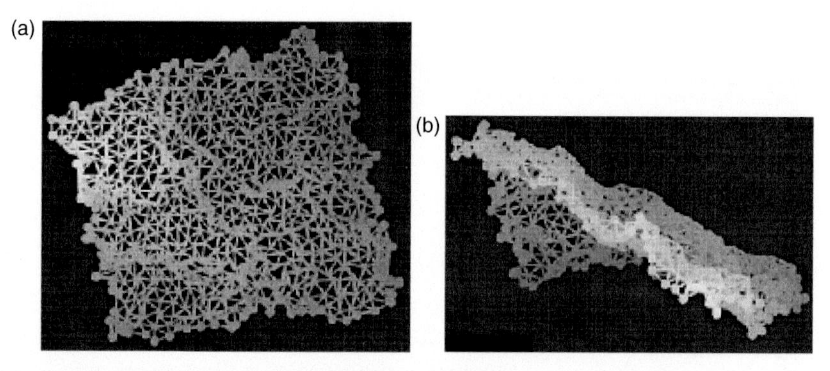

Fig. 8.11 Top (a) and side (b) view of a sample configuration taken from a simulation of a polymerized membrane. The ratio of the membrane's transverse thickness to its overall length declines with increasing membrane size (Boal and Plischke, unpublished).

$$\langle h^2 \rangle \sim A^{\zeta}, \tag{8.62}$$

where the roughness exponent ζ is found analytically to be 0.59 (Le Doussal and Radzihovsky, 1992), in the range anticipated by early simulation studies (Plischke and Boal, 1988; Lipowsky and Girardet, 1990; Gompper and Kroll, 1991). Further numerical work gave $\zeta = 0.58 \pm 0.02$ (Petsche and Grest, 1993) and 0.59 ± 0.02 (Zhang *et al.*, 1996). Allowing a membrane to change topology through the creation of voids or handles does not change the scaling behavior of the systems studied thus far (Levinson, 1991; Plischke and Fourcade, 1991; Jeppesen and Ipsen, 1993). Lastly, we mention that other membrane phases have been predicted (Nelson and Halperin, 1979) and seen in simulation studies (Gompper and Kroll, 1997).

The elastic properties of flat polymerized membranes at finite temperature have been investigated for at least a decade (Aronovitz and Lubensky, 1988; Nelson and Peliti, 1987). One of the interesting features emerging from such studies is the prediction that the Poisson ratio of polymerized membranes is $-1/3$ universally (Le Doussal and Radzihovsky, 1992). As we discussed in our treatment of networks in Chapter 5, a negative Poisson ratio means that the membrane expands transversely when stretched longitudinally. One can imagine how this behavior arises by considering the configuration in Fig. 8.11, or experimenting with a crumpled sheet of aluminum foil: as the folds are pulled out by applying a stress in one direction, the sheet unfolds in the transverse direction as well (D. Nelson, private communication). Simulation studies have confirmed the negative Poisson ratio (Falcioni *et al.*, 1997, and reviewed in Bowick, 2004; see also Zhang *et al.*, 1996). As one might expect, membranes with anisotropic construction possess more complex mechanical properties than the isotropic membranes considered here. Among the conjectured features is a possible tubular phase in which the membrane crumples in one direction while remaining flat in an orthogonal direction (Radzihovsky and Toner, 1995; Bowick *et al.*, 1997; reviewed in Bowick, 2004, and Radzihovsky, 2004).

8.4.4 Effects of pressure and attraction

Beyond bending resistance, polymers and membranes may experience other interactions, like long-range attraction and repulsion, and may be subject to external stresses. Further, a pressure difference may occur across the boundary of closed shapes like rings in two dimensions and vesicles in three dimensions. These factors span a potentially large parameter space; here, we consider only a slice of that space, namely pressure and attraction, to illustrate qualitatively the general scaling behavior of a system. We include both rings and fluid vesicles, the former because of its possible relevance to DNA (Baumgärtner, 1982).

The various scaling laws applicable to closed systems in two and three dimensions are summarized in Table 8.1, showing that the scaling exponents are unique to each system as long as two independent geometrical characteristics are taken into account. These families of configurations are illustrated pictorially in Fig. 8.12, which displays the phase behavior in the pressure/attraction plane at zero bending resistance (Boal, 1991). By *attraction*, we mean short-range attractive interactions among all elements of a chain. Two trends are apparent in the figure:

- the enclosed area increases as the pressure outside the ring drops below the pressure inside, at fixed attraction,
- the linear size of the configuration decreases with increasing attraction at fixed pressure.

Under compression, for example, the gangly arms of a branched polymer stick to each other at high attraction, thus reducing the linear size of the ring. Similar dependence on pressure is found for rings subject to bending rigidity without attraction (Leibler *et al.*, 1987). Note that a finite-size system may not adopt the phase displayed in the figure, particularly for systems under pressure (Maggs *et al.*, 1990).

Studies of closed rings have been extended to both fluid (see Gompper and Kroll, 1995, and references therein) and polymerized (Komura and Baumgärtner, 1991) membranes in three dimensions. Subject to a pressure difference at $\kappa_b = 0$, fluid membranes exhibit a size-dependent transition between the branched polymer phase at zero pressure difference to the inflated phase when the inside pressure greatly exceeds the outside pressure. Inclusion of attractive interactions among elements of a polymerized membrane also leads to a collapsed or dense phase, although the nature of the transition is not as simple as it is for rings (Abraham and Kardar, 1991; Liu and Plischke, 1992).

Table 8.1 Scaling behavior of closed shapes in two and three dimensions (rings of contour length L_c and fluid bags of area A, respectively)

Configuration	Scaling law		Reference
Two dimensions	$\langle R_g^2 \rangle \propto L_c^{2\nu}$	$\langle A \rangle \propto L_c^{2\eta}$	
inflated	$\nu = 1$	$\eta = 1$	
self-avoiding walk	$\nu = 3/4$	$\eta = 3/4$	Duplantier, 1990, Nienhuis, 1982
branched polymer	$\nu = 0.64$	$\eta = 1/2$	Derrida and Stauffer, 1985
dense	$\nu = 1/2$	$\eta = 1/2$	
Three dimensions	$\langle R_g^2 \rangle \propto A^{\nu}$	$\langle V \rangle \propto A^{\eta}$	
inflated	$\nu = 1$	$\eta = 3/2$	
Flory-type argument	$\nu = 4/5$		
branched polymer	$\nu = 1$	$\eta = 1$	
dense	$\nu = 2/3$	$\eta = 1$	

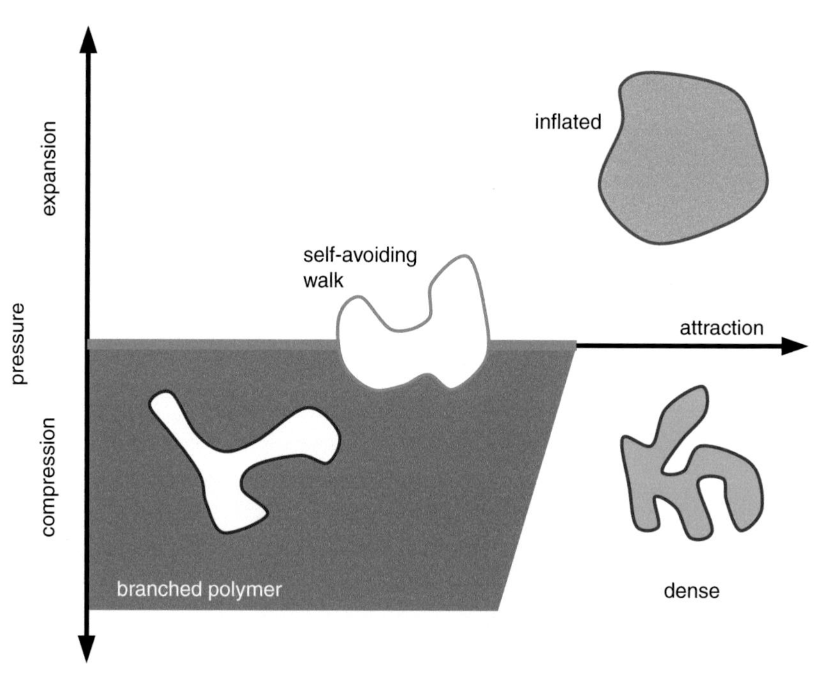

Fig. 8.12 Phase diagram for two-dimensional rings subject to pressure and attraction but very small bending rigidity. There are considerable finite-size scaling effects near the phase boundaries (from Boal, 1991).

8.5 Measurement of membrane undulations

Where not suppressed by the presence of a cell wall or other constraints, the thermal undulations of biomembranes may be large enough to have observable consequences in a cell. For example, flickering can be seen on the plasma membrane of the mature human red blood cell, which is supported only by a membrane-associated cytoskeleton. Undulations are even more pronounced in pure lipid bilayers, as illustrated by the oscillations of a bilayer vesicle in Fig. 1.13. Before we interpret such experimental observations, let us briefly recap some formal results for fluid membranes from Sections 8.2–8.4.

In the Monge representation of a surface, a position **r** has Cartesian coordinates $[x, y, h(x,y)]$, the displacement from the xy-plane [or height function $h(x,y)$] being single-valued if the surface has no overhangs (unlike a cresting wave). An area element $\mathrm{d}A$ on the surface is related to its projected area $\mathrm{d}x\,\mathrm{d}y$ on the xy-plane by the metric g via $\mathrm{d}A = \sqrt{g}\,\mathrm{d}x\,\mathrm{d}y$, where $g = 1 + h_x^2 + h_y^2$ and $h_x = \partial h/\partial x$. The curvature C characterizes the rate at which the orientation of the surface, as measured by the normal **n**, changes with arc length s, according to $\mathbf{n}\bullet\partial^2\mathbf{r} / \partial s^2$ in a given direction. The largest and smallest values of the curvature at a point are called the principal curvatures C_1 and C_2. For gently undulating surfaces, the mean and Gaussian curvatures, which are combinations of C_1 and C_2, are given by $(C_1 + C_2)/2 = (h_{xx}+h_{yy})/2$ and $C_1C_2 = h_{xx}h_{yy} - h_{xy}^2$, respectively, where $h_{xx} = \partial^2h/\partial x^2$, etc. These combinations enter the simplest form for the bending energy density as $\mathcal{F} = (\kappa_b/2)(C_1 + C_2)^2 + \kappa_G C_1 C_2$, where κ_b and κ_G are material constants called the bending rigidity and Gaussian bending rigidity, respectively. The mean direction of the surface normal **n** changes with separation $\Delta\mathbf{r} = \mathbf{r}_1 - \mathbf{r}_2$ between positions $\mathbf{r}_1$ and $\mathbf{r}_2$ as $\langle \mathbf{n}(\Delta\mathbf{r})\bullet\mathbf{n}(0)\rangle = \exp(-\Delta r/\xi_p)$, where ξ_p is the persistence length. For fluid membranes, $\xi_p = b\exp(4\pi\kappa_b/3k_BT)$, demonstrating that the persistence length decreases with increasing temperature as surface undulations become more pronounced.

(a) zero tension

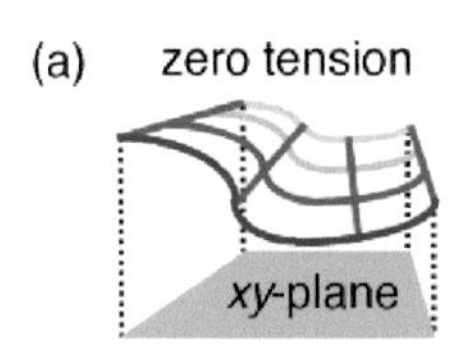

(b) small tension

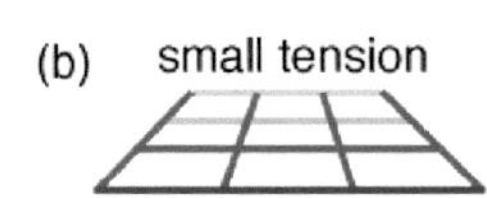

(c) modest tension

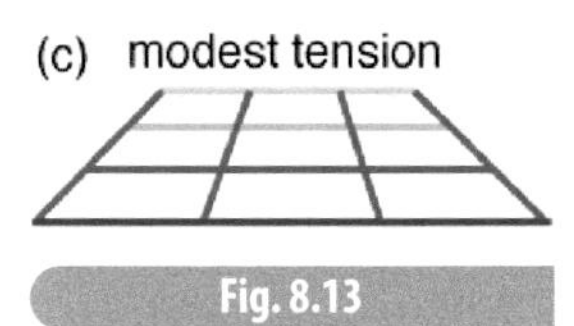

Fig. 8.13

(a) At zero tension, a membrane undergoes thermal fluctuations, reducing its projected area. (b) Applying a small tensile stress suppresses these undulations. (c) At larger tension, the intermolecular separation increases according to K_A.

One manifestation of thermally induced undulations is the response of a membrane to an applied tension τ. As illustrated in Fig. 8.13, the application of a small tension irons out the wrinkles in a bilayer without significantly stretching it. As the applied tension is raised further, the membrane area expands according to the area compression modulus K_A. For the model membrane presented in Section 8.3, the energy E of a particular configuration is described by

$$E = \tau \int \mathrm{d}A + (\kappa_b/2)\bullet\int(C_1+C_2)^2\mathrm{d}A, \qquad (8.63)$$

where $\tau > 0$ corresponds to tensile stress and where the Gaussian curvature term has been omitted (Helfrich, 1975; see also Helfrich and Servuss, 1984). The first term expresses how the tension opposes growth in the contour area $\int dA$ compared to the area in the xy-plane.

The contour area is $\int dA = \int \sqrt{g}\, d\mathbf{X}$, where the two-dimensional integral $d\mathbf{X} = dx\, dy$ runs over the xy-plane and g is the metric. In the Monge representation, $\sqrt{g}$ is equal to $(1 + h_x^2 + h_y^2)^{1/2}$, which is approximately $1 + (h_x^2 + h_y^2)/2$ for gentle undulations corresponding to small derivatives $h_x = \partial h/\partial x$. Thus, the contour area A of a given configuration is

$$A \cong \int d\mathbf{X} + (1/2) \int (h_x^2 + h_y^2)\, d\mathbf{X}. \tag{8.64}$$

In other words, $(1/2)\int(h_x^2+h_y^2)d\mathbf{X}$ is the amount by which the projected area $\int d\mathbf{X}$ is reduced below its contour area A by thermal fluctuations; the reduction must be obtained by averaging over all configurations with the appropriate Boltzmann weight. From Eq. (8.39), the thermal average of $\int(h_x^2+h_y^2)d\mathbf{X}$ is

$$\langle \int (h_x^2+h_y^2)d\mathbf{X} \rangle = (A^2 / 4\pi^2) \int q^2 d\mathbf{q}\, \langle h(\mathbf{q})h^*(\mathbf{q}) \rangle, \tag{8.65}$$

where $\langle \cdots \rangle$ indicates an ensemble average and $\mathbf{q}$ is a two-dimensional wavevector. As established in Eq. (8.43) for a configurational energy of the form in Eq. (8.63), the Fourier transform $h(\mathbf{q})$ of the height function obeys

$$\langle h(\mathbf{q})h^*(\mathbf{q}) \rangle = k_B T / (\tau q^2 + \kappa_b q^4) \bullet A, \tag{8.66}$$

where $h(\mathbf{X}) = (A / 4\pi^2)\int d\mathbf{q}\, \exp(i\mathbf{q} \bullet \mathbf{X}) h(\mathbf{q})$.

Combining Eqs. (8.64) to (8.66), the reduction A_{red} of the in-plane area with respect to the contour area becomes

$$A_{red}(\tau) = (1/2) \langle \int (h_x^2+h_y^2)d\mathbf{X} \rangle = A(k_B T / 8\pi^2) \int d\mathbf{q} / (\tau + \kappa_b q^2), \tag{8.67}$$

or, after making the replacement $d\mathbf{q} = (1/2)\, dq^2\, d\theta$ and integrating $d\theta$ over 2π,

$$A_{red}(\tau)/A = (k_B T / 8\pi) \int dq^2 / (\tau + \kappa_b q^2), \tag{8.68}$$

As discussed in Section 8.3, the domain of the wavevector q is set by the largest and smallest length scales of the system, and ranges from a low of $\pi/A^{1/2}$, as dictated by the size of the membrane, to π/b, as dictated by the molecular spacing b. Changing integration variables to $z = q^2 + \tau/\kappa_b$ gives

$$A_{red}(\tau)/A = (k_B T / 8\pi\kappa_b) \int dz / z, \tag{8.69}$$

where the integral runs from $\pi^2/A + \tau/\kappa_b$ to $\pi^2/b^2 + \tau/\kappa_b$, yielding a logarithm

$$A_{red}(\tau)/A = (k_B T / 8\pi\kappa_b) \ln([\pi^2/b^2 + \tau/\kappa_b]/[\pi^2/A + \tau/\kappa_b]). \tag{8.70}$$

This expression, obtained by Helfrich (1975), differs somewhat from that found by Milner and Safran (1987) for quasi-spherical vesicles, although both approaches yield the same effective compression modulus. Equation (8.70) predicts that $A_{red}(\tau)/A$ vanishes in two limits: $k_BT/\kappa_b \to 0$ or $\tau/\kappa_b \to \infty$; that is, ripples are suppressed at low temperature or high tension, and the projected area becomes the same as the contour area.

We have shown that the projected area of a membrane is less than its contour area because of thermal fluctuations. The change in contour area arising from increased separation between molecules at higher tensions is governed by the (zero temperature) area compression modulus K_A, which can be superimposed on the area change from fluctuations (Helfrich and Servuss, 1984; see also Helfrich, 1975). To determine the effective area compression modulus, Evans and Rawicz (1990) evaluate the area change with respect to the tensionless reference state, which has the largest area reduction $A_{red}(0)/A = (k_BT / 8\pi\kappa_b) \ln(A/b^2)$. Then, applying a tension increases the projected area by an amount $\Delta A/A$ given by

$$\begin{aligned}\Delta A/A &= [A_{red}(0) - A_{red}(\tau)]/A \\ &= (k_BT / 8\pi\kappa_b) \ln([1 + \tau A/\pi^2\kappa_b]/[1 + \tau b^2/\pi^2\kappa_b]). \qquad (8.71)\end{aligned}$$

Given that the membrane area A is very much larger than the squared distance between molecules, Eq. (8.71) is well approximated by

$$\Delta A/A = (k_BT / 8\pi\kappa_b) \ln(1 + \tau A/\pi^2\kappa_b). \qquad (8.72)$$

Adding τ/K_A to the change in the relative area in the low-tension regime, Eq. (8.72), gives an overall area change of

$$\Delta A/A = (k_BT / 8\pi\kappa_b) \ln(1 + \tau A/\pi^2\kappa_b) + \tau/K_A. \qquad (8.73)$$

The relationship between κ_b and K_A for a membrane with a fluctuating area has been studied by Farago and Pincus (2003).

Equation (8.73) permits determination of both κ_b and K_A from a single experiment by recording the membrane area change as a function of tension, as has been done via micropipette aspiration (Evans and Rawicz, 1990; Rawicz *et al.*, 2000). The equation simplifies in two obvious limits.

Low tension Undulations dominate and the logarithmic term is the most important. Using typical values of $\tau \sim 10^{-4}$ J/m^2, $A \sim 10^{-12}$ m^2, and $\kappa_b \sim 10^{-19}$ J confirms that $\tau A/\pi^2\kappa_b >> 1$ for most experimental situations of interest, such that

$$\Delta A/A = (k_BT / 8\pi\kappa_b) \ln(\tau A/\pi^2\kappa_b). \quad \text{(low tension)} \qquad (8.74)$$

Accordingly, a plot of $\Delta A/A$ vs. $\ln(\tau)$ should be linear with a slope of $k_BT / 8\pi\kappa_b$ at low tension, as illustrated in Fig. 8.14(a).

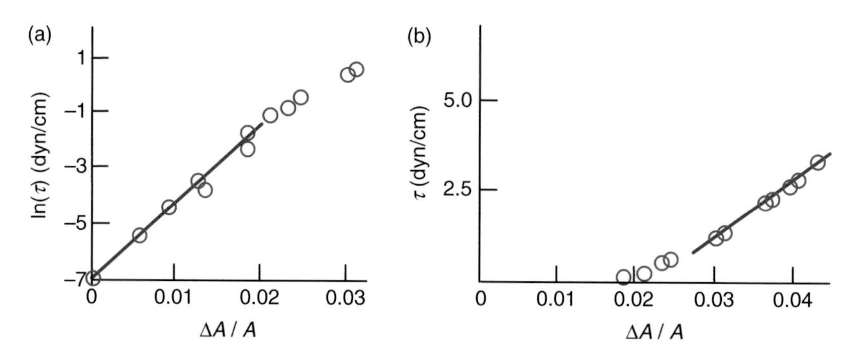

Fig. 8.14 The area change of a lipid bilayer as a function of tension applied to a pure lipid vesicle by aspiration. (a) The measured change in relative area $\Delta A / A$ is proportional to $\ln(\tau)$ at small tension. (b) At higher tension, the area is linear in tension. Data from the application of ascending and descending tension to diGDG bilayers are combined here (redrawn with permission from Evans and Rawicz, 1990; ©1990 by the American Physical Society).

High tension Expansion of the contour area dominates this tension regime, and

$$\Delta A / A = \tau / K_A, \qquad \text{(high tension)} \tag{8.75}$$

showing that the area increases linearly with tension, as seen in Fig. 8.14(b).

The data in Fig. 8.14 demonstrate that the apparent compression modulus $K_{A,app}{}^{-1} = A^{-1}dA/d\tau$ varies with the applied tension; taking the derivative of Eq. (8.73) quickly leads to

$$K_{A,app} = K_A / [1 + K_A k_B T / (8\pi\kappa_b\tau)], \tag{8.76}$$

(Evans and Rawicz, 1990). The crossover between the two regimes is smooth and occurs over a range in tension estimated in the problem set.

What is the amplitude of the thermal undulations? A few mathematical manipulations are needed to answer this question; readers not wishing to make this detour may skip to Eq. (8.82). Using the Fourier transform of $h(\mathbf{X})$ from Eq. (8.28), we have

$$h(\mathbf{X})^2 = (A / 4\pi^2)^2 \int d\mathbf{q}\, d\mathbf{q}' \exp(i\mathbf{q} \cdot \mathbf{X}) \exp(-i\mathbf{q}' \cdot \mathbf{X})\, h(\mathbf{q}) h^*(\mathbf{q}'), \tag{8.77}$$

over which we perform a positional average by integrating with $A^{-1} \int d\mathbf{X}$ to obtain

$$\langle h(\mathbf{X})^2 \rangle_{pos} = (A / 4\pi^2)^2 A^{-1} \int d\mathbf{X} \int d\mathbf{q}\, d\mathbf{q}' \exp(i(\mathbf{q} - \mathbf{q}') \cdot \mathbf{X})\, h(\mathbf{q}) h^*(\mathbf{q}'). \tag{8.78}$$

Grouping the $\mathbf{X}$-dependent terms together and integrating just yields a two-dimensional delta function from $\int d\mathbf{X} \exp(i[\mathbf{q} - \mathbf{q}'] \cdot \mathbf{X}) = (4\pi^2)\delta(\mathbf{q} - \mathbf{q}')$, which can be removed immediately by performing the $d\mathbf{q}'$ integral. Thus, the full ensemble average is

$$\langle h^2 \rangle = (A / 4\pi^2) \int d\mathbf{q}\, d\mathbf{q}'\, \delta(\mathbf{q}-\mathbf{q}') \langle h(\mathbf{q}) h^*(\mathbf{q}') \rangle$$
$$= (A / 4\pi^2) \int d\mathbf{q} \langle h(\mathbf{q}) h^*(\mathbf{q}) \rangle. \quad (8.79)$$

Equation (8.66) provides us with an expression for $\langle h(\mathbf{q})h^*(\mathbf{q})\rangle$ as a function of tension τ and bending rigidity. We set $\tau = 0$ here (the general case is solved in the problem set) and write

$$\langle h^2 \rangle = (A / 4\pi^2) \int d\mathbf{q}\; k_B T / (\kappa_b q^4 A). \quad (8.80)$$

The angular dependence of $d\mathbf{q}$ can be removed immediately, leaving $2\pi q$ dq, and

$$\langle h^2 \rangle = (k_B T / 2\pi\kappa_b) \int dq / q^3. \quad (8.81)$$

As discussed with Eq. (8.69), the q-integral runs from $\pi/A^{1/2}$ to π/b, to give

$$\langle h^2 \rangle = (k_B T / 4\pi\kappa_b)\, (A/\pi^2 - b^2/\pi^2)$$
$$\cong k_B T A / 4\pi^3 \kappa_b, \quad (8.82)$$

where the second equation follows because $A >> b^2$ (Helfrich and Servuss, 1984). As expected, the amplitude of the undulations grows with temperature as $k_B T/\kappa_b$; the amplitude is also proportional to the square root of the contour area as $\langle h^2 \rangle^{1/2} \sim A^{1/2}$.

In addition to the compression and bending resistance of fluid membranes, polymerized membranes possess non-vanishing shear resistance which modifies the q-dependence of the height function $\langle h(\mathbf{q})h^*(\mathbf{q})\rangle$. Simulating a polymerized membrane under pressure against a hard wall, Lipowsky and Girardet (1990) found that their computational membrane was described by

$$\langle h(\mathbf{q}) h^*(\mathbf{q}) \rangle \cong k_B T / [1.3(k_B T Y)^{1/2} q^3 + \kappa_b q^4] \cdot A. \quad (8.83)$$

The two-dimensional Young's modulus Y of the membrane has units of energy per unit area, and is related to the two-dimensional K_A and μ as described in Chapter 5:

$$Y = 4K_A\mu/(K_A+\mu). \quad (8.84)$$

The apparent elastic behavior of the membrane depends on the magnitude of q: at short wavelengths, q is large and the membrane is fluid-like, while at long wavelengths, q is small and the membrane is solid-like. The crossover between these regimes occurs broadly near $q = 1.3(k_B T Y)^{1/2}/\kappa_b$, corresponding to a distance scale

$$L^* = 2\pi\kappa_b/(k_B T Y)^{1/2}, \quad (8.85)$$

whose magnitude is estimated for the red cell in the problem set.

How do we expect membrane undulations to behave in the solid regime? Neglecting the bending resistance in Eq. (8.83), and substituting the result into Eq. (8.79), we have

$$\langle h^2 \rangle = \{k_B T \,/\, [1.3 \cdot 4\pi^2 (k_B T Y)^{1/2}]\} \int d\mathbf{q} \,/ q^3. \tag{8.86}$$

Performing the usual angular integration, $d\mathbf{q}$ can be replaced by $2\pi q\ dq$ to give

$$\langle h^2 \rangle = \{(k_B T / Y)^{1/2} / (1.3 \cdot 2\pi)\} \int dq \,/ q^2, \tag{8.87}$$

where q runs from $\pi / A^{1/2}$ to π / b. As usual, $A^{1/2} >> b$, which simplifies the result of the integral to read

$$\langle h^2 \rangle = (k_B T / Y)^{1/2} A^{1/2} \,/\, 2.6. \tag{8.88}$$

Clearly, the undulations of a solid membrane grow more slowly with A than do those of a fluid membrane ($\langle h^2 \rangle \sim A^1$ from Eq. (8.82) for fluids). Comparing with the scaling form Eq. (8.62), namely $\langle h^2 \rangle \sim A^\zeta$, we see that Eq. (8.88) is consistent with a roughness exponent $\zeta = 1/2$. We note that this value is only slightly less than $\zeta = 0.59$ found analytically (Le Doussal and Radzihovsky, 1992) or the range found in other simulations (Plischke and Boal, 1988; Abraham and Nelson, 1990; Gompper and Kroll, 1991), confirming that polymerized membranes are flat, but rough.

The undulations of the combined membrane plus cytoskeleton of the human red blood cell have been interpreted using Eq. (8.83) (Zilker *et al.*, 1992; Peterson *et al.*, 1992). In the experiment, a red cell is attached to a flat substrate at just enough locations on its plasma membrane to prevent lateral motion without suppressing its undulations. Thermal oscillations of the attached membrane are then observed microscopically by interference, much like Newton's rings. As shown in the problem set, the crossover length L^* for the erythrocyte plasma membrane is of the order half a micron, approximately the cell's thickness, but much less than its diameter. The amplitude of the undulations $\langle h^2 \rangle$ was measured over the wavevector range $0.3 \leq q \leq 4\ \mu m^{-1}$ (or $1.6 \leq 2\pi / q \leq 20\ \mu m$) and was found to obey q^{-4} over much of that range (Zilker *et al.*, 1992). In other words, the undulations of the composite membrane are dominated by bending resistance in this range of q. Although we have presented the analysis of membrane fluctuations in the context of a flat reference state, the theory of membrane fluctuations has been extended to whole cells with multicomponent boundaries (Gov *et al.*, 2003; Gov *et al.*, 2005; Marcelli *et al.*, 2005); we return to these systems in Chapter 10.

The scaling properties of the red cell cytoskeleton have been measured directly by removing the plasma membrane and determining the structure factor of the network by X-ray or light scattering. In a scattering experiment, the intensity of the light scattered into a given angle is proportional to the structure factor $S(\mathbf{q})$ (see Chapter 2 of Kittel, 1966), where $\mathbf{q}$ is the

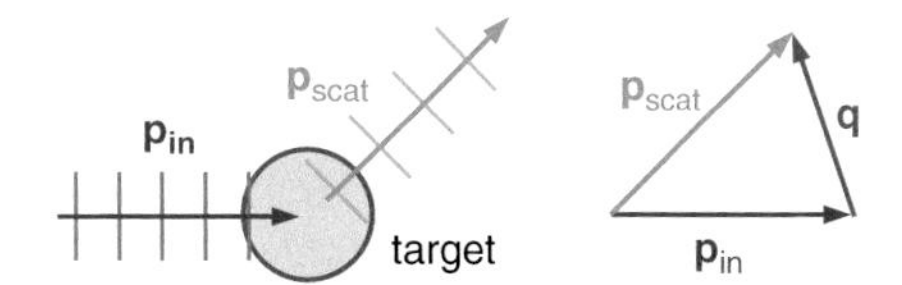

Fig. 8.15 In a light-scattering experiment, an incident photon of momentum $\mathbf{p}_{in}$ may change to momentum $\mathbf{p}_{scat}$ through its interaction with a target.

difference between the momenta of the incident and scattered photons, $\mathbf{p}_{in}$ and $\mathbf{p}_{scat}$, respectively (see Fig. 8.15)

$$\mathbf{q} = \mathbf{p}_{scat} - \mathbf{p}_{in}. \tag{8.89}$$

The experimentally measured structure factor is an average over directions of $\mathbf{q}$ because the cytoskeletons are not constrained to a particular orientation. For a theoretical membrane or network with N vertices, such as the computational membrane displayed in Fig. 8.11, the direction-averaged structure factor has the form (see discussion in Kantor *et al.*, 1986)

$$S(q) = N^{-2} \langle \Sigma_{m,n} \exp[i\mathbf{q} \bullet (\mathbf{r}_m - \mathbf{r}_n)] \rangle, \tag{8.90}$$

where m and n are labels attached to the vertices, permitting $S(q)$ to be predicted for model membranes. The observed q-dependence of the averaged structure factor obeys $q^{-2.35}$, which, when compared to the $q^{-3+\zeta}$ expected for gentle undulations of a polymerized membrane at intermediate q (Goulian *et al.*, 1992), leads to a measured value for ζ of 0.65 ± 0.10 (Schmidt *et al.*, 1993). As indicated in the discussion following Eq. (8.88), this value is in the range expected theoretically.

Summary

In the Monge representation, the position $\mathbf{r}$ on a surface has Cartesian coordinates $[x, y, h(x,y)]$, where h is the displacement of the surface from the xy-plane. An area element dA on the surface is related to its projected area $dx\, dy$ on the xy-plane by the metric g in $dA = \sqrt{g}\, dx\, dy$, where $g = 1 + h_x^2 + h_y^2$ and $h_x = \partial h/\partial x$. The local geometry of a surface or interface can be described by two principal curvatures C_1 and C_2, representing the extremal values of $\mathbf{n} \bullet (\partial^2\mathbf{r}/\partial s^2)$, where $\mathbf{n}$ is the normal to the surface at position $\mathbf{r}$ and s is the arc length along the surface in the direction of interest. For gentle undulations, the combinations $(C_1+C_2)/2$ and C_1C_2, defined as the mean and Gaussian curvatures, are given in terms of the height functions by $(C_1+C_2)/2 = (h_{xx}+h_{yy})/2$ and $C_1C_2 = h_{xx}h_{yy} - h_{xy}^2$, where $h_{xx} = \partial^2 h/\partial x^2$, etc. These combinations enter the simplest form of the bending energy density as $\mathcal{F} = (\kappa_b/2)(C_1+C_2)^2 + \kappa_G C_1 C_2$, where κ_b and κ_G are the bending rigidity and Gaussian bending rigidity, respectively, and carry units of energy. At finite temperature, a surface may experience thermal undulations, causing its normals to vary in direction from one location to another. The mean orientation of the normals at two positions separated by a distance Δr decays exponentially as $\langle \mathbf{n}(\Delta \mathbf{r}) \cdot \mathbf{n}(0) \rangle = \exp(-\Delta r/\xi_p)$, where $\xi_p \sim b \exp(4\pi\kappa_b/3k_B T)$ is the persistence length, b being the elementary length scale of the surface.

It is often convenient to measure the wavelength-dependence of the height function by utilizing the Fourier transformation $h(\mathbf{X}) = (A/4\pi^2) \int d\mathbf{q} \exp(i\mathbf{q} \cdot \mathbf{X}) h(\mathbf{q})$, where $\mathbf{q}$ is a two-dimensional wavevector. Averaged over an ensemble of configurations at temperature T, the Fourier components $h(\mathbf{q})$ of a fluid membrane subject to a surface tension τ and bending rigidity κ_b are given by

$$\langle h(\mathbf{q}) h^*(\mathbf{q}) \rangle = k_B T / (\tau q^2 + \kappa_b q^4) \cdot A, \qquad \text{(fluid)}$$

where h^* is the complex conjugate of h. In simulation studies, terms of order q^3 must appear in the denominator to describe the undulations of a computational polymerized membrane.

The presence of undulations and ripples reduces the in-plane area of a fluid membrane. As tension is applied, the relative area $\Delta A/A$ increases

- according to $(k_B T / 8\pi\kappa_b) \ln(1 + \tau A/\pi^2\kappa_b)$ as the ripples are reduced at low tension,
- according to τ/K_A as the intermolecular separation increases at high tension.

From this behavior, the apparent area compression modulus $K_{A,app}$ is found to be $K_{A,app} = K_A / [1 + K_A k_B T/(8\pi\kappa_b \tau)]$. Polymerized membranes also have unusual elastic properties; in particular, their Poisson ratio is predicted to be $-1/3$, meaning that, in the membrane plane, they expand transversely when stretched longitudinally.

Although membranes do not adopt a unique configuration at $T > 0$, their mean shape can be characterized by sets of scaling exponents. Considering first self-avoiding membranes not subject to internal interactions or external forces, it is found that fluid membranes behave like branched polymers, while polymerized membranes are rough, but flat. As applied to a closed shape like a fluid vesicle, branched polymer scaling means that the enclosed volume scales like the surface area, $V \sim A^1$; in other words, the vesicle reduces its volume so as to form high-entropy arms and fingers. In contrast, a flat polymerized membrane has an in-plane linear size that grows like the square root of the membrane area, $R^2 \sim A^1$, while its transverse fluctuations grow more slowly like $\langle h^2 \rangle \sim A^{0.59}$. With the inclusion of attractive interactions, both fluid and polymerized membranes collapse to dense configurations. When subject to a tensile stress, the ripples on the surface of a fluid vesicle are suppressed as the vesicle inflates, ultimately scaling like $V \sim A^{3/2}$, although the vesicle attains its asymptotic scaling behavior in a size-dependent way. Polymer rings confined to a plane display similar, but not completely identical, scaling behavior to fluid vesicles.

Most of the predicted elastic behavior of fluid and polymerized membranes has been verified experimentally and used in the analysis of data, some of which are presented in Chapter 7. The behavior $\langle |h(\mathbf{q})|^2 \rangle \sim q^{-4}$ of fluid membranes has been confirmed for bilayers, and provides the foundation for an analysis that obtains both κ_b and K_A from the in-plane membrane area as a function of applied tension. The transverse fluctuations $\langle h^2 \rangle \sim A^{0.59}$ predicted for polymerized membranes also have been observed in two-dimensional cytoskeletons isolated from the red cell.

Problems

Applications

8.1. Consider a dimple on a membrane described by the Gaussian height function $h(x,y) = h_o \exp(-[x^2 + y^2]/2w^2)$. The approximate deformation energy of this shape is determined to be $E = \pi\kappa_b(h_o/w)^2$ in Problem 8.15(d).

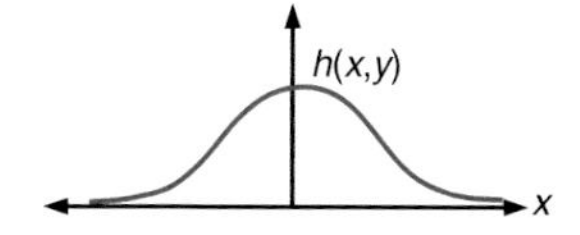

(a) If $w = 1$ μm and $\kappa_b = 10k_BT$, at what value of h_o does the deformation energy of the surface equal k_BT?

(b) Under certain conditions, dimples appear on the surface of a red blood cell, although their occurrence is driven by more factors than bending, and their shape may not be Gaussian. What is the bending contribution to the deformation energy if $w = 1$ μm, $h_o = 0.05$ μm, and $\kappa_b = 30k_BT$? Quote your answer in units of k_BT.

8.2. We establish in Problem 8.16 that the deformation energy of a membrane undulation of the form $h(x,y) = h_o \bullet \cos(2\pi x/\lambda) \bullet \cos(2\pi y/\lambda)$ on a $\lambda \times \lambda$ square patch is approximately $E = 2\pi^2\kappa_b\,(2\pi h_o/\lambda)^2$ where κ_b is the bending rigidity.

(a) What is the value of h_o if $E = k_BT$, assuming $\lambda = 1$ μm and $\kappa_b = 20k_BT$? Note that the peak-to-trough height difference is $2h_o$.

(b) Plot E/k_BT as a function of h_o (in μm) up to 0.2 μm to gauge the magnitude of thermally allowed fluctuations on this membrane patch (use λ and κ_b from part (a)).

8.3. A long persistence length does not suppress all undulations on the cellular length scale. Suppose we observe that the root mean square angle between two normal vectors, $\langle \theta^2 \rangle^{1/2}$, is 1/10 of a radian, or about 6 degrees, when the normals are separated by 10 μm. Using the small-angle expansion $\cos\theta \sim 1 - \theta^2/2$, find the corresponding ξ_p and κ_b/k_BT of the membrane (assume $b = 1$ nm for the molecular length scale). Is this value of κ_b/k_BT extraordinarily different from the values observed for cellular bilayers?

8.4. Suppose that a hypothetical bilayer could be made from short-chain lipids, giving a bilayer thickness of 3 nm, instead of the conventional 4 nm. By what factor would κ_b change according to Eq. (7.20)? By what factor would the persistence length change if $\kappa_b/k_BT = 10$ for the 4 nm bilayer?

8.5. Most pure lipid bilayers of relevance to the cell have a bending modulus κ_b in the range 10–30 k_BT. Plot the persistence length and $\langle h^2 \rangle^{1/2}$ for a thin circular disk of radius 4 μm having this range of κ_b. Use $b \sim 1$ nm for the molecular length scale.

8.6. Plot the difference between a bilayer's contour area and its projected area at zero tension for $\kappa_b = 10k_BT$. Consider the area range 1–100 μm^2 and assume the intermolecular separation is $b = 1$ nm.

8.7. Find the approximate tension at the crossover between the logarithmic and linear behavior of membrane area (defined by $K_{A,app}/K_A = 1/2$). What is the magnitude of this tension if $\kappa_b = 10k_BT$ and $K_A = 0.15$ J/m^2. How does your answer compare with the rupture tension if the membrane in question fails when its area is 3% higher than its unstressed value?

8.8. Plot the amplitude of thermal undulations $\langle h^2 \rangle^{1/2}$ as a function of tension for a bilayer with $\kappa_b = 10k_BT$. Take the membrane area to be 10 μm^2 and assume the intermolecular separation is $b = 1$ nm. Cover the tension range $10^{-4} \leq \tau \leq 10^{-2}$ J/m^2. Use the expression from Problem 8.18 and quote $\langle h^2 \rangle^{1/2}$ in μm.

8.9. (a) Find the length scale for the crossover between fluid and solid behavior for a polymerized network with $\kappa_b = 0.25\ k_BT$ and $K_A = 2\mu = 2 \times 10^{-5}$ J/m^2, representing a two-dimensional cytoskeleton in isolation.

(b) Repeat this calculation for a combined cytoskeleton + bilayer system, with $\kappa_b = 10\ k_BT$, $K_A = 0.2$ J/m^2 (from a bilayer) and μ as in part (a).

For both (a) and (b), determine Y from the two-dimensional relation $Y_{2D} = 4K_A\mu/(K_A+\mu)$ [from Chapter 5 and Eq. (8.84)].

8.10. In an imaginary experiment, the thermal fluctuations in height $\langle h^2 \rangle$ of a patch of material are measured for sheets of three different materials, each having an area $A = 10\ \mu m^2$. Find $\langle h^2 \rangle^{1/2}$ for:

(a) a fluid bilayer with bending modulus $\kappa_b = 15\ k_B T$,
(b) a solid sheet of bacterial cell wall with two-dimensional Young's modulus $Y_{2D} = 1.0\ J/m^2$,
(c) a solid sheet of plant cell wall with two-dimensional Young's modulus $Y_{2D} = 100\ J/m^2$.

Use the appropriate idealized relationship between $\langle h^2 \rangle$ and A presented in Section 8.5; take $k_B T = 4 \times 10^{-21}$ J and quote $\langle h^2 \rangle^{1/2}$ in μm.

Formal development and extensions

8.11. Starting from Eq. (8.3) for the normal vector **n** to a surface in the Monge representation, establish Eq. (8.15) for $\partial_\alpha \mathbf{n}$ component by component.

8.12. Starting with the expressions for $\partial_\alpha \mathbf{r}$ and $\partial_\beta \mathbf{n}$ in the Monge representation [Eqs. (8.2) and (8.15), respectively], establish that $b_{\alpha\beta} = h_{\alpha\beta} / (1 + h_x^2 + h_y^2)^{1/2}$ component by component.

8.13. For bending deformations, show that the free energy density $\mathcal{F} = (\kappa_b/2)(C_1+C_2)^2 + \kappa_G C_1 C_2$ (at zero spontaneous curvature) has the same form as $\mathcal{F} = (\kappa_b/2)[(\partial_x \mathbf{n})^2 + (\partial_x \mathbf{n})^2]$ when $\kappa_b = -\kappa_G$. Use the expressions for the mean and Gaussian curvatures and $\partial_x \mathbf{n}$ in the Monge representation, keeping only the leading order terms in $h_{\alpha\beta}$.

8.14. A hemispherical surface is governed by the equation $x^2 + y^2 + z^2 - R^2 = 0$ with $z > 0$.

(a) Find $h(x,y)$ and $\mathbf{n}(x,y)$ for $x^2 + y^2 < R^2$ using Eq. (8.3).
(b) Find the mean and Gaussian curvatures from Eqs. (8.11) and (8.12) by explicitly determining $b_{\alpha\beta}$ and $g_{\alpha\beta}$.

Do your results for the curvatures and surface normals agree with your intuition for the properties of a sphere?

8.15. Suppose that a dimple on a membrane is described by the Gaussian form

$$h(x,y) = h_o \exp(-[x^2 + y^2] / 2w^2).$$

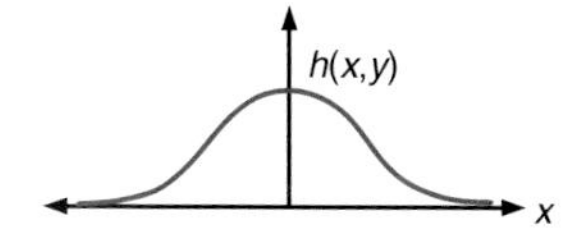

(a) What aspects of the shape do h_o and w represent?
(b) What are the mean and Gaussian curvatures as a function of (x,y) if h_o is small?

(c) Use your results from (b) to determine the explicit form of the energy density of the surface based on Eq. (8.22) with $C_o = 0$.

(d) Assuming that h_o is sufficiently small that the metric is close to unity across the surface, integrate your result from (c) to establish that the energy of the dimple is

$$E = \pi\kappa_b (h_o/w)^2,$$

where κ_b is the usual bending rigidity.

8.16. Consider an "egg-carton" surface governed by the equation

$$h(x,y) = h_o \bullet \cos(2\pi x/\lambda) \bullet \cos(2\pi y/\lambda),$$

where λ is the wavelength of the undulations. For this problem, consider only the xy region $-\lambda/2 \leq x,y \leq \lambda/2$.

(a) Find the normal $\mathbf{n}$ as a function of (x,y) to the leading order in h_o.

(b) Find the mean and Gaussian curvatures if h_o is small. At what points do these curvatures have their largest magnitudes?

(c) Show that the bending energy of a $\lambda \times \lambda$ square of this surface is

$$E = 2\pi^2\kappa_b\,(2\pi h_o/\lambda)^2$$

if the energy density is described by Eq. (8.22) and we take $g = 1$.

8.17. The mean square height of a membrane of area A is found to obey $\langle h^2 \rangle \sim A^{0.6}$. Assuming the Fourier representation has the form $\langle |h(\mathbf{q})|^2 \rangle = \alpha k_B T / q^n A$ (where α is a constant) similar to Eq. (8.66), what value of n is consistent with this scaling behavior?

8.18. Derive the general expression for the amplitude of thermal undulations of a fluid membrane by substituting Eq. (8.66) into Eq. (8.79). Show that when the area $A >> b^2$ (b is the molecular length scale)

$$\langle h^2 \rangle = (k_B T / 4\pi\tau)\ln(1 + \tau A / \pi^2\kappa_b)$$

(Helfrich and Servuss, 1984).

Confirm that this expression reduces to Eq. (8.82) in the limit of small tensions.

8.19. An idealized pancake-shaped cell is shown in cross section in the diagram. Lying in the xy-plane indicated by the dashed line, it has a radius $R + t$ and thickness $2t$, with uniform rounded edges.

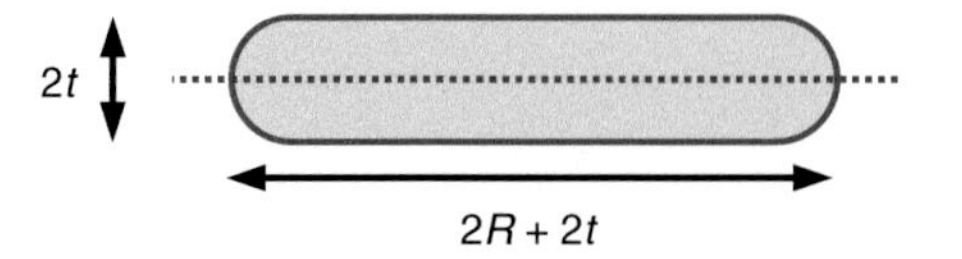

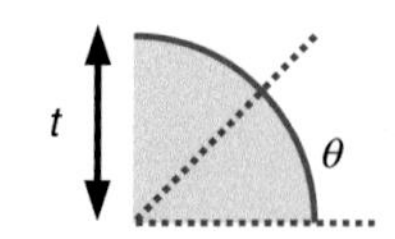

(a) Find the equation for the height $h(x, y)$ of the cell boundary with respect to the xy-plane.
(b) Find expressions for h_x and h_y in Cartesian coordinates.
(c) The expressions for h_x and h_y can be simplified considerably using angular coordinates, such as the conventional polar coordinate $y/x = \tan\phi$ in the xy-plane, and $h/t = \sin\theta$ as shown in the diagram. Rewrite h_x and h_y in terms of these angles and then find the metric g.

8.20. The surface of an elliptical cylinder with its primary axis lying along the y-axis, is governed by the equation $(x/a)^2+(z/b)^2 = 1$, independent of y.

(a) Find $h(x,y)$ of the surface with respect to the xy-plane, as well as its first and second derivatives with respect to x and y.
(b) Find the mean and Gaussian curvatures from Eqs. (8.17) and (8.18).
(c) Setting $a = b = R$, what is $C_1 + C_2$ from part (b)? Does your result agree with the value of $C_1 + C_2$ obtained by simple inspection of the curvature of a uniform cylinder?

9 Intermembrane and electrostatic forces

The variety of single cells capable of living independently is truly impressive, from small featureless mycoplasmas just half a micron across, to elegant protists a hundred times that size, perhaps outfitted with tentacles and even a mouth. Organisms such as ourselves, however, are multicellular, emphatically so at 10^{14} cells per human. How do our cells interact, and adhere when appropriate? In this chapter, we investigate intermembrane forces at large and small separations, beginning with a survey of the different features of membrane interactions in Section 9.1. The Poisson–Boltzmann equation, which we derive in Section 9.2, provides our primary framework for treating membrane interactions at non-zero temperature. This approach is applied to the organization of ions near a single charged plate and to the electrostatic pressure between charged walls in Sections 9.2 and 9.3, respectively. Two other contributors to membrane interactions also are treated in some mathematical detail here: the van der Waals attraction between rigid sheets (Section 9.3) and the steric repulsion experienced by undulating membranes (Section 9.4). Lastly, the adhesion of a membrane to a substrate, and its effect on cell shape, is described in Section 9.5. Our presentation of membrane forces is necessarily limited; the reader is referred to Israelachvili (1991) or Safran (1994) for more extensive treatments.

9.1 Interactions between membranes

Let's briefly review the molecular composition of a conventional plasma membrane. Fundamental to a biomembrane is the lipid bilayer, a pair of two-dimensional fluid leaflets described at length in Chapter 7. The lipid composition of the leaflets is inequivalent: in the human red blood cell, one of the best-characterized examples, the phospholipids in the outer leaflet are predominantly cholines, while lipids containing ethanolamine or serine dominate the inner leaflet facing the cytosol (see Appendix B for head-group structures). Cholesterol also may be abundant in the plasma membrane. The head-group of phosphatidylserine, but not the others, is negatively charged, providing a surface charge density ranging up to 0.3 C/m^2 if every lipid carries a single charge.

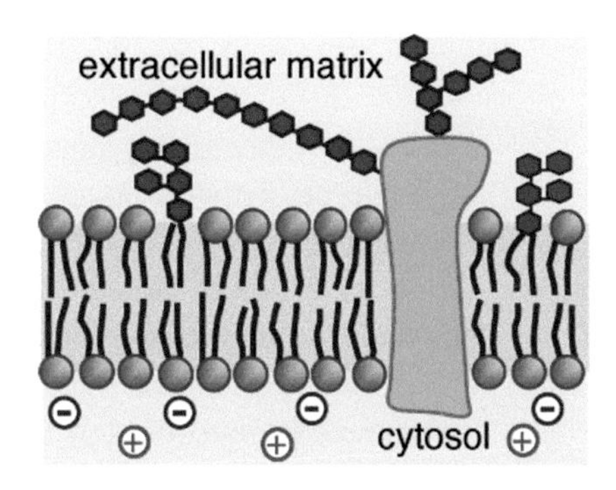

Fig. 9.1 Schematic representation of a plasma membrane. The large blob is an embedded protein; attached to the protein and to two lipids are linear and branched polysaccharides, forming part of the extracellular matrix. Charged lipids, such as phosphatidylserine, and their counterions are indicated by the circles in the cytosol.

(a)

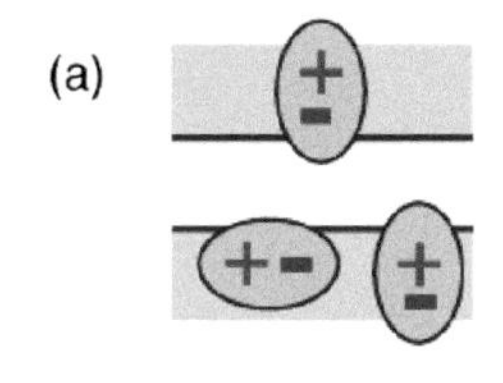

(b)

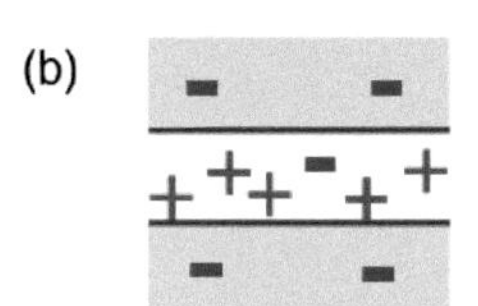

(c)

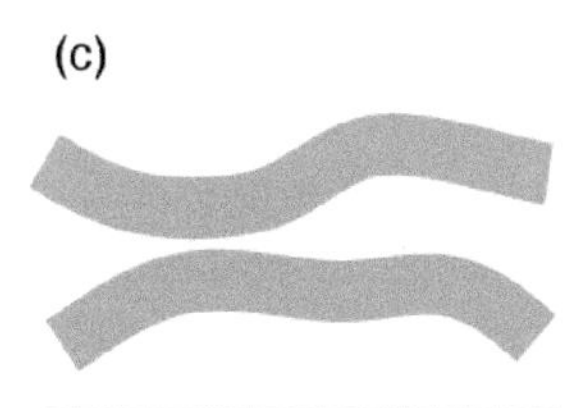

Fig. 9.2 Contributors to the force between membranes include: (a) van der Waals attraction between electric dipoles, (b) electrostatic repulsion between charged bilayers, and (c) entropic repulsion between undulating membranes.

Sometimes embedded in a bilayer, sometimes lying adjacent to it and attached by other means, is a variety of proteins, as displayed in Fig. 9.1. Although proteins make up 50% of a membrane's mass, they are outnumbered by lipids fifty-to-one. Lipids and membrane-bound proteins both are important functional elements of the cell, providing leakage resistance and controlling access, among their other tasks. Not surprisingly, the cell protects these valuable components with the glycocalyx, a sugary coat made from linear and branched polysaccharides attached to lipids (glycolipids) or proteins (glycoproteins and proteoglycans); this coating may extend several nanometers away from the head-group plane of the bilayers. Lastly, sandwiching the membrane and permeating the glycocalyx are aqueous solutions of mobile counterions (for the charged lipids), ions from dissolved salts, and sundry other chemical building blocks.

What is the physical origin of the interaction between membranes? A sample of the forces to be described in this chapter is illustrated in Fig. 9.2. At separations of the order 5 nm or more, contributions are made by the van der Waals forces of molecular polarizability, and by electrostatic forces from charged surfaces. At shorter distances, steric conflict between the extracellular matrix of each membrane adds to the intermembrane force, which is further augmented by the effort needed to expel the solution adjacent to the bilayers. In addition, the membrane itself experiences gentle thermal undulations, which provide entropic resistance against membrane adhesion even at distances of several nanometers. What is the strength of each contribution, and how does it vary with distance? We now provide a synopsis of several forces, a selection of which is described more thoroughly in Sections 9.2–9.4.

9.1.1 van der Waals forces

Even if it is electrically neutral overall, the charge separations within a molecule, say between its electrons and nucleus, provide the basis for an intermolecular force. Collectively called van der Waals forces, the interaction energy between permanent and instantaneous electric dipole moments decreases with distance as $1/r^6$, much faster than the $1/r$ behavior of Coulomb's potential for point charges. For two rigid slabs, an attractive pressure proportional to D_s^{-3} arises from the van der Waals interaction when the thickness of the slabs is appreciably larger than the distance D_s between them. If the contributions from each molecular pair add independently, this pressure is $(\pi^2 C_{vdw}\rho^2) / 6\pi D_s^3$ where C_{vdw} is a constant and ρ is the number of molecules per unit volume in the material. The combination $\pi^2 C_{vdw}\rho^2$ is called the Hamaker constant, with a value of about 2–5×10^{-20} J ~ 5–10 k_BT for many types of material separated by a vacuum. At large distances compared to the thickness of two interacting sheets or films, the van der Waals pressure decreases more rapidly as D_s^{-5}.

9.1.2 Electrostatic forces

According to Coulomb's law, two identical point charges of magnitude q experience a repulsive potential energy of $q^2/4\pi\varepsilon r$, where ε is the permittivity of the medium surrounding the charges. The permittivity of vacuum (ε_o) is 8.85×10^{-12} C^2 / N•m^2, while the permittivities of water and various hydrocarbons are 80 ε_o and 2–3 ε_o, respectively. The interaction between charged membranes at ambient temperatures is considerably more complex than point charges separated by vacuum. First, charged counterions must be present in solution to compensate for the membrane's charge. Being mobile, the counterions may stay close to the membrane at low temperature because of coulombic forces, but may range far and wide at high temperature, enticed by entropy to explore their surroundings. The Poisson–Boltzmann equation provides a framework for describing such charged systems at non-zero temperature. Applied to charged plates separated by a distance D_s, the solution to this equation yields a repulsive pressure of $\pi k_BT/2\ell_B D_s^2$, the Bjerrum length ℓ_B reflecting the properties of the medium through $\ell_B \equiv q^2/4\pi\varepsilon k_BT$ for counterions of charge q. However, this calculated pressure will be strongly suppressed (or *screened*) by the presence of dissolved salts over a distance scale given by the Debye length $\ell_D \equiv (8\pi\ell_B\rho_s)^{-1/2}$, where ρ_s is the molecular number density of the salt (this expression is valid for one species of monovalent salt). Depending on the magnitude of the Debye length, the combined electrostatic and van der Waals force may be attractive at long *and* short distances because of the van der Waals contribution, with electrostatic repulsion emerging only at intermediate separations.

9.1.3 Steric interactions

The structure and movement of the plasma membrane and its polymer coat, the extracellular matrix, provide an often repulsive interaction between cells. Seen in Fig. 9.1, the exterior of many cells is covered by a layer of polymers that can be described as twigs, mushrooms or brushes, depending upon their stiffness and mean separation. The construction of this polymer labyrinth is sufficiently complex that a transparent analytical description of its elastic properties is difficult to achieve; however, the entropic repulsion of the polymers can be felt over distances of many nanometers, obviously dependent on the thickness of the coat. Equally important, the plasma membrane itself experiences thermal undulations that generate a repulsive pressure. This pressure is entropic, and has the form $P = 2c_{fl}\ (k_BT)^2/\kappa_bD_s^3$, where κ_b is the bending rigidity (10–25 k_BT for common bilayers) and c_{fl} is a numerical constant equal to about 0.1. At large values of D_s this pressure may exceed both the van der Waals attraction of thin films and the electrostatic repulsion of screened charges.

9.1.4 Solvation forces

The above forces are most important at distances of several nanometers or more, with the dominant force depending strongly on conditions. However, at distance scales of less than a few nanometers, the solvent becomes important because:

- solvent must be removed from the space between membranes for them to touch,
- solvent molecules near a flat surface may be organized into an irregular structure.

Owing to molecular organization, the solvation force between rigid plates may be oscillatory, with a wavelength comparable to the molecular diameter. Where present, such oscillations may be superimposed on an overall trend that is repulsive or attractive, depending upon the nature of the solvent and boundary surfaces. In the restricted case where the interaction between plate and solvent is ignored, the solvation pressure is attractive as $D_s \to 0$, equal to k_BT times the concentration of the solvent at the plate as $D_s \to \infty$ (see Section 13.3 of Israelachvili, 1991); for an aqueous medium, this calculated pressure is about a thousand atmospheres. However, if there is an attractive interaction between plate and fluid, the liquid density at the plate is higher than in the bulk, and the force is generally repulsive. Owing to the aqueous medium, the solvation force between lipid bilayers may be referred to as a hydration force (see Rand and Parsegian, 1989, for a review). Given its flexible head-group structure, the bilayer is rough on length scales of tenths of nanometers, potentially washing out the ordered

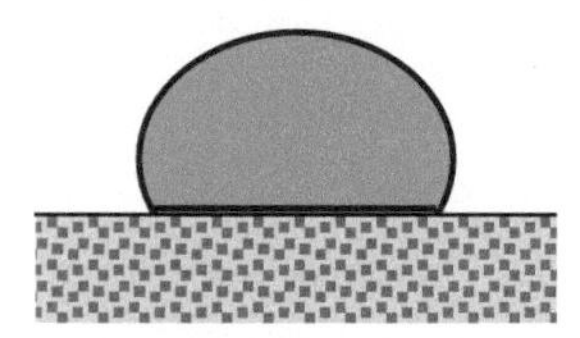

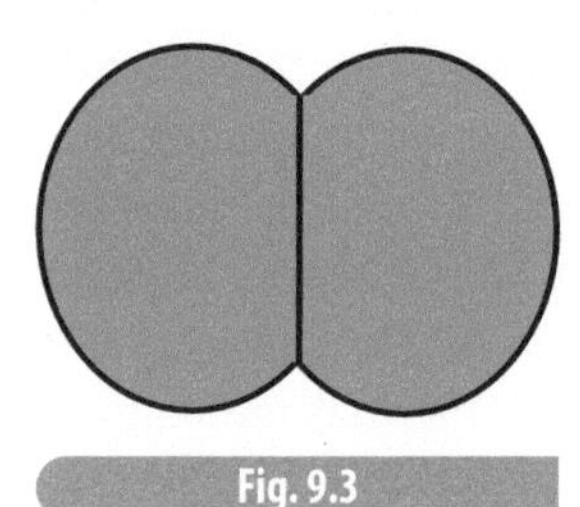

Fig. 9.3

Adhesion energies are large enough to influence the shape of a cell attached to a substrate or of two cells in close contact.

solvent layer that may be present near atomically smooth surfaces. As a consequence, the oscillating behavior of the force is suppressed and it tends to be monotonically repulsive.

9.1.5 Adhesion

At very small separations, membranes may adhere through direct molecular contact, with binding energies that are significant, but not overwhelming. For example, the protein avidin and the vitamin biotin display one of the strongest association energies at 35 k_BT per bond, which is still less than a typical covalent bond energy of 140 k_BT (e.g., C–C or C–O). The difference between the surface energy density of two membranes in contact, or separated by an intervening medium like water, is termed the adhesion energy density. Taken together with the other interactions described above, the adhesion energy density between membranes is about 10^{-5} J/m^2; for instance, the adhesion energy of two membranes with a contact area of 10 μm^2 is 10^{-16} J. What effect does adhesion have on cell shape? We recall from Section 7.4 that the bending energy of a spherical shell is $8\pi\kappa_b + 4\pi\kappa_G$; for many bilayers, the bending rigidity κ_b is around 10 k_BT, so that the bending energy is in the range 10^{-18} J, ignoring the contribution of the Gaussian rigidity (which is shape-invariant for a given topology). Thus, the adhesion energy is sufficient to overcome the change in bending energy arising from shape changes such as those shown in Fig. 9.3.

Let us now formally determine the van der Waals, electrostatic and undulation forces between membranes in Sections 9.2–9.4. We return to membrane adhesion in Section 9.5.

9.2 Charged plate in an electrolyte

The phosphate group of a phospholipid carries a negative charge which may be balanced by the positive charge borne by many lipid-head groups. An exception is the serine group, which is electrically neutral, leaving the phospholipid negative overall (see Appendix B). Thus, a bilayer made from the common phospholipids may have a negative surface charge density. For an area per head-group of 0.5 nm^2 in the bilayer plane, this charge density could be as high as 0.3 C/m^2 if each lipid carries a single electron charge (1.6×10^{-19} C). Overall, the membrane and its environment is electrically neutral, so that positive counterions must be present in the media adjacent to the membrane. Although lipids may diffuse laterally within the bilayer, their charges can be regarded as fixed in the sense that they are confined to a plane. In contrast, the counterions are mobile and may hug the bilayer or diffuse away from it, depending on conditions.

Our task in this chapter is to develop a formalism for calculating the distribution of mobile ions surrounding a charged object with fixed geometry. The formalism must recognize the effects of entropy, which encourages mobile ionic species to explore the physical space available to them, a tendency that increases with temperature. We adopt a mean field approach which ignores local fluctuations in charge, leading to the Poisson–Boltzmann equation. As a first application, we obtain the ion distribution generated on one side of a rigid charged plate, which we interpret as a model system for how mobile ions arrange themselves near a charged bilayer. In Section 9.3, we tackle the more general problem of the forces between two charged plates when there are counterions and perhaps other species present in the environment.

9.2.1 Poisson–Boltzmann equation

A widely applicable approach to calculating charge distributions at non-zero temperature is based on a mean-field approximation which ignores local variations in the charge density. We start with Gauss' law from introductory physics, which, in its integral form, reads

$$\int \mathbf{E} \bullet \mathrm{d}\mathbf{A} = (1/\varepsilon) \int \rho_{\mathrm{ch}} \, \mathrm{d}V, \tag{9.1}$$

where ε is the permittivity of the medium (in vacuum, $\varepsilon_o = 8.85 \times 10^{-12}$ C^2/Nm^2). The integrals are performed over a closed surface with area $\int \mathrm{d}A$ and enclosed volume $\int \mathrm{d}V$; as in Fig. 8.5, $\mathrm{d}\mathbf{A}$ is a vector representing an infinitesimal area element where the orientation of $\mathrm{d}\mathbf{A}$ is locally normal to the surface. On the right-hand side, the integral over the charge density ρ_{ch} yields the net charge enclosed by the surface; ρ_{ch} is not a number density, but rather is the charge per unit volume (in C/m^3 in MKSA units). For example, if the mathematical surface is a spherical shell of radius R surrounding a point charge Q, Eq. (9.1) gives the familiar $E = Q / 4\pi\varepsilon R^2$. The divergence theorem allows the left-hand side to be written as $\int \nabla \bullet \mathbf{E} \, \mathrm{d}V$, where $\nabla \bullet \mathbf{E} = \partial_x E_x + \partial_y E_y + \partial_z E_z$. With this substitution, both integrals in Eq. (9.1) are of the form $\int \ldots \mathrm{d}V$, so their integrands must be equal,

$$\nabla \bullet \mathbf{E} = \rho_{\mathrm{ch}}/\varepsilon, \tag{9.2}$$

because the equation is valid for any arbitrary closed surface. This is the differential form of Gauss' law.

Now, just as force $\mathbf{F}$ and potential energy $V(\mathbf{r})$ are related by $\mathbf{F} = -\nabla V$ for a conservative force, so too electric field $\mathbf{E}$ and potential ψ are related by $\mathbf{E} = -\nabla\psi$, since $\mathbf{E}$ and ψ are just $\mathbf{F}$ and V divided by charge. Replacing $\mathbf{E}$ by the negative gradient of ψ permits Eq. (9.2) to be recast in a form called Poisson's equation

$$\nabla^2\psi = -\rho_{\mathrm{ch}}/\varepsilon, \qquad \text{Poisson's equation} \tag{9.3}$$

which relates the charge density to the electric potential (for applications, see Chapter 4 of Lorrain *et al.*, 1988). As we stated above, the charge distribution reflects a competition between energy and entropy, with the entropic contribution becoming more important with rising temperature. For instance, positive ions cluster near a negatively charged bilayer at low temperature, but roam further afield at high temperature. In our treatment of the harmonic oscillator in Appendix C, we show that the relative probability of the oscillator having a displacement x is proportional to the Boltzmann factor $\exp(-V(x)/k_B T)$, where $V(x)$ is the potential energy. The same reasoning applies here: the density of counterions at a given position is proportional to the Boltzmann factor with a potential energy evaluated at that position. We assume for now that only one species of mobile ion is present, each ion with charge q experiencing a potential energy $V(\mathbf{r}) = q\psi(\mathbf{r})$. Defining ρ_o to be the number density of ions at the reference point $\psi = 0$, the Boltzmann expression for the density profile is

$$\rho(\mathbf{r}) = \rho_o \exp(-q\psi(\mathbf{r})/k_B T). \tag{9.4}$$

Here, the electrostatic potential ψ represents an average over local fluctuations in the ion's environment. This form for $\rho(\mathbf{r})$ can be substituted into the Poisson expression of Eq. (9.3) using $\rho_{ch}(\mathbf{r}) = q\rho(\mathbf{r})$ to give the Poisson–Boltzmann equation, which is a differential equation in ψ

$$\nabla^2\psi = -(q\rho_o/\varepsilon)\exp(-q\psi(\mathbf{r})/k_B T). \quad \text{Poisson–Boltzmann equation} \tag{9.5}$$

Equation (9.5) can easily be generalized to include more than one species of charge (e.g., Chapter 12 of Israelachvili, 1991).

9.2.2 Charged plate with one counterion species

In this section and Section 9.3, the Poisson–Boltzmann equation is applied to several situations involving very large, but thin, charged plates surrounded by a medium of permittivity ε. Let's briefly review the electrostatics of a single plate with charge per unit area σ_s in the absence of counterions, using Eqs. (9.1) or (9.2). Because the plate is large, the electric field is uniform across its face, except near the plate boundary. The contributions to $\mathbf{E}$ from each element of charge on the plate can be resolved into components parallel and perpendicular to the plane of the plate. By symmetry, the parallel components from all regions of the plate cancel, leaving the electric field pointing in a direction normal to the plate, the same as the area element $d\mathbf{A}$. To evaluate the integrals in Eq. (9.1) for this situation, we choose a mathematical surface like a pancake or tabletop: two identical flat sheets lie parallel to the plate and are joined along their perimeter by a surface perpendicular to the plate. The area of this mathematical pancake is much smaller than the plate itself, and it does not approach the plate boundary. The plate slices through the middle of the pancake, parallel to

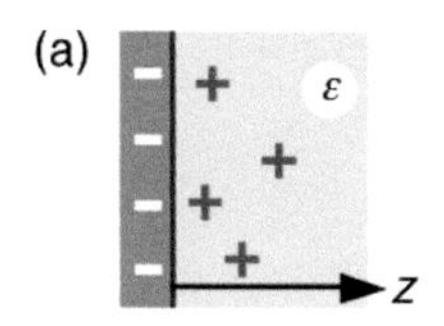

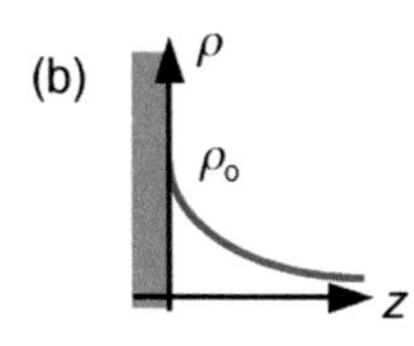

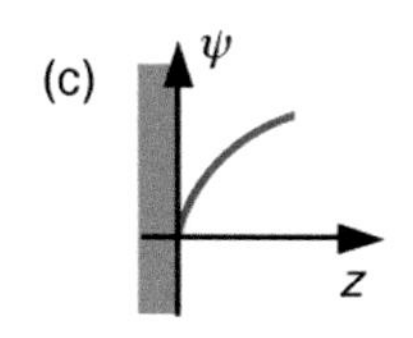

Fig. 9.4

(a) Positively charged counterions swarm near a large negatively charged plate. (b) The density ρ of positive charges decreases as a function of distance from the plate z, having a value ρ_o where the potential ψ vanishes. (c) For a negative plate, the electrostatic potential ψ rises with z from its value at the plate boundary.

its two faces, such that all sections of a face are the same distance from the plate and experience the same electric field E (ignoring direction). Thus, each face of the pancake contributes EA to the integral on the left-hand side of Eq. (9.1), but no contribution arises from the rim of the pancake where **E** is perpendicular to d**A**. The left-hand side of Eq. (9.1) is then $2EA$, and the charge enclosed on the right-hand side is $\sigma_s A$. Solving the equality, the magnitude of the electric field must be $E = |\sigma_s|/2\varepsilon$, independent of the distance from the plate. Since the electric field is the (negative) derivative of the potential, $\mathbf{E} = -\nabla\psi$, a constant value for E implies that the magnitude of ψ must grow linearly with z as $\sigma_s z/2\varepsilon$. As usual, the location where the potential vanishes can be chosen arbitrarily.

Now, let's add counterions to one side of the plate as in Fig. 9.4(a), taking the plate to be negative and the counterions positive. At low temperature, the counterions cluster near the plate, their density ρ falling with distance z from the plate as in Fig. 9.4(b). As the temperature increases, the tail of the charge distribution extends further into the medium on the right. We choose the origin of the z-axis to lie at the plate boundary, where we fix $\psi = 0$. If the field is independent of direction parallel to the plate, the only non-vanishing part of $\nabla^2\psi$ is $\mathrm{d}^2\psi / \mathrm{d}z^2$ and Eq. (9.5) reads

$$\mathrm{d}^2\psi/\mathrm{d}z^2 = -(q\rho_o/\varepsilon)\exp(-q\psi / k_B T). \tag{9.6}$$

It is useful to incorporate the constants $q/k_B T$ into the potential to render it unitless through the definition

$$\Psi(z) \equiv q\psi(z)/k_B T, \tag{9.7}$$

such that Eq. (9.6) becomes

$$\mathrm{d}^2\Psi/\mathrm{d}z^2 = -\rho_o(q^2 / \varepsilon k_B T)\,\mathrm{e}^{-\Psi}. \tag{9.8}$$

The combination $q^2/\varepsilon k_B T$ in this equation has units of [*length*] and appears in the Bjerrum length ℓ_B:

$$\ell_B \equiv q^2 / 4\pi\varepsilon k_B T. \tag{9.9}$$

Through q and ε, the quantity ℓ_B depends upon the properties of the counterions and the medium, not on the charge density of the plate. For instance, the Bjerrum length of a single electron charge in air at room temperature is 58 nm, dropping to 0.7 nm in water where the permittivity is eighty times that of air. At this point, we introduce yet another parameter K, which has the units of [*length*]$^{-1}$, by

$$K^2 \equiv 2\pi\ell_B\rho_o, \tag{9.10}$$

to further simplify the appearance of the Poisson–Boltzmann equation:

$$\mathrm{d}^2\Psi / \mathrm{d}z^2 = -2K^2\mathrm{e}^{-\Psi}. \tag{9.11}$$

To solve Eq. (9.11), we first multiply both sides by $\mathrm{d}\Psi/\mathrm{d}z$ and then use the following two relations:

$$(\mathrm{d}\Psi/\mathrm{d}z) \cdot (\mathrm{d}^2\Psi/\mathrm{d}z^2) = (1/2) \cdot \mathrm{d}[(\mathrm{d}\Psi/\mathrm{d}z)^2]/\mathrm{d}z \tag{9.12}$$

$$\mathrm{e}^{-\Psi}\,(\mathrm{d}\Psi/\mathrm{d}z) = -\,\mathrm{d}(\mathrm{e}^{-\Psi})/\mathrm{d}z, \tag{9.13}$$

to rewrite the equation as

$$\mathrm{d}[(\mathrm{d}\Psi/\mathrm{d}z)^2]/\mathrm{d}z = 4K^2\,\mathrm{d}(\mathrm{e}^{-\Psi})/\mathrm{d}z. \tag{9.14}$$

Integrating this expression over z and taking the square root of the result gives

$$\mathrm{d}\Psi/\mathrm{d}z = 2K\,\mathrm{e}^{-\Psi/2}, \tag{9.15}$$

where the positive root is chosen to correspond to Fig. 9.4(c). The integration constant that should appear in Eq. (9.15) has been set equal to zero permitting the electric field, which is proportional to $\mathrm{d}\Psi/\mathrm{d}z$, to vanish at the large values of Ψ expected as $z \to \infty$. Equation (9.15) can be integrated easily by rewriting it as

$$\int \mathrm{e}^{\Psi/2}\,\mathrm{d}\Psi = 2K \int \mathrm{d}z, \tag{9.16}$$

which gives

$$\mathrm{e}^{\Psi/2} = K(z+\chi), \tag{9.17}$$

where the integration constants have been rolled into the factor χ.

We have now established that the functional form of $\mathrm{e}^{-\Psi}$ is $[K(z+\chi)]^{-2}$, allowing us to generate an expression for the charge density from $\rho = \rho_o\,\mathrm{e}^{-\Psi}$:

$$\rho(z) = \rho_o/[K(z+\chi)]^2 = 1\,/\,[2\pi\ell_B\,(z+\chi)^2]. \tag{9.18}$$

The integration constant is fixed by the value of the surface charge density σ_s. For the system to be electrically neutral, the integral over the positive charge density $\int q\rho\,\mathrm{d}z$ must equal σ_s in magnitude (recall ρ is a number per unit volume); in symbols

$$-\sigma_s = (q\,/\,2\pi\ell_B) \int \mathrm{d}z\,/\,(z+\chi)^2, \tag{9.19}$$

where the integral covers $0 \leq z \leq \infty$. The solution to this equation requires the distance χ to be

$$\chi = q\,/\,(-2\pi\ell_B\sigma_s) = 2\varepsilon k_B T\,/\,(-q\sigma_s), \tag{9.20}$$

which is positive because q and σ_s have opposite signs here. There are several features to note about the form of the number density of counterions. First, the density falls like the square of the distance from the plate, with half of the counterions residing within a distance χ of the plate. Second, from Eq. (9.20), the width of the distribution grows linearly with temperature, as the counterions venture further into the surrounding medium. Other properties of the solution are explored in Problem 9.16, confirming the expected behavior of the potential compared to that of a charged plate in the absence of counterions.

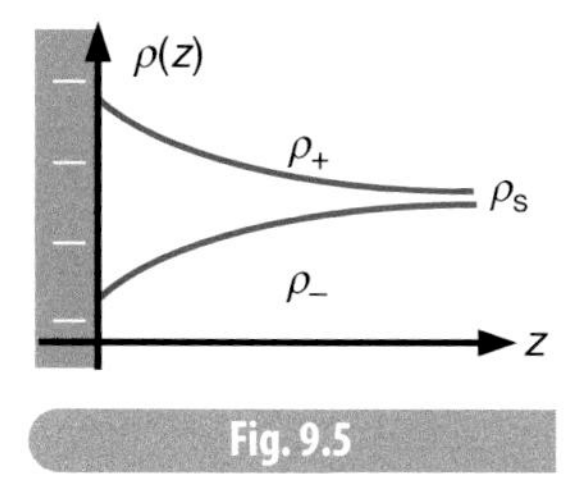

Fig. 9.5

Schematic representation of densities ρ_+ and ρ_- of positive and negative ions, respectively, as a function of distance z from a negatively charged plate (shaded region). The densities approach a common value ρ_s at large distance.

9.2.3 Charged plate in a salt bath

Both the interior of a cell and the environment surrounding it contain various organic compounds as well as ions released from salts such as NaCl, which we will refer to as "bulk" salts or ions. The effect that a negatively charged bilayer has on these bulk ions depends upon their charge: the density of positive ions is enhanced near the plate while negative mobile ions are depleted. The distributions are shown schematically in Fig. 9.5, where $\rho_+(z)$ and $\rho_-(z)$ are the number densities of positive and negative ions carrying charges $+q$ and $-q$, respectively. Only the case where the ions are monovalent ($q = e$) will be considered here. This situation is somewhat more complicated than the previous problem of a plate with counterions, but it can be approached with a suitably generalized version of the Poisson–Boltzmann equation.

Averaged over distances of tens of nanometers or more, the overall concentration of bulk salts generally exceeds that of the counterions. As a numerical example (see Problem 9.1), consider a negatively charged plate with the substantial surface charge density $\sigma_s = -0.3$ C/m^2, surrounded by water with permittivity 80 ε_o. If each counterion carries just one electron charge, their density may be in the molar range right at the interface with the plate, although they are largely confined to a region (χ) just 0.1–0.2 nm deep. Because the counterion number density ρ drops like the distance squared according to Eq. (9.18), then even at a distance of 10 nm, ρ has fallen by a factor of 10^4. For comparison, the concentration of ions in the cell's environment is about 0.2 M, and the salt concentration in the oceans themselves is 0.6 M. Thus, as far as the cell is concerned, the overall abundance of positive or negative ions in solution is dominated by that of the bulk salts. Now, in order to express ρ_+ and ρ_- according to the Boltzmann factor, a choice must be made for the position where the potential ψ vanishes. Given that $\rho_\pm$ approach their bulk value ρ_s as $z \to \infty$, a natural choice is $\psi \to 0$ as $z \to \infty$, and the Boltzmann factors are

$$\rho_+(z) = \rho_s \exp(-q\psi(z)/k_B T) \quad \text{and} \quad \rho_-(z) = \rho_s \exp(+q\psi(z)/k_B T), \tag{9.21}$$

leading to a net charge density of

$$\rho_{ch}(z) = q\rho_s [\exp(-q\psi(z) / k_B T) - \exp(+q\psi(z) / k_B T)]. \tag{9.22}$$

This expression may be substituted into Eq. (9.3) to yield

$$\nabla^2\psi = -(q\rho_s/\varepsilon) [\exp(-q\psi(z) / k_B T) - \exp(+q\psi(z) / k_B T)], \tag{9.23}$$

which can be generalized still further to include multiple ionic species.

The first steps in solving Eq. (9.23) are parallel to Eq. (9.6). To make the expression less visually cumbersome, the potential $\psi(z)$ is replaced by the dimensionless function $\Psi(z) \equiv q\psi(z)/k_B T$ defined in Eq. (9.7). The Bjerrum length appears unchanged at $\ell_B \equiv q^2/4\pi\varepsilon k_B T$, but we introduce a second length scale by

$$\ell_D^{-2} \equiv 8\pi\ell_B\rho_s. \tag{9.24}$$

Defined as the Debye screening length, ℓ_D depends only on the properties of the bulk medium, including its ionic content. Thus, the Poisson–Boltzmann equation for this situation becomes

$$d^2\Psi/dz^2 = -(1/2\ell_D^2)(e^{-\Psi} - e^{+\Psi}), \tag{9.25}$$

where we again assume that Ψ depends only on the distance from the plate, z. Now, the exponentials can be replaced by the hyperbolic function sinhΨ, yielding

$$d^2\Psi/dz^2 = +\ell_D^{-2}\sinh\Psi. \tag{9.26}$$

As with Eq. (9.11), we multiply both sides of Eq. (9.26) by $d\Psi/dz$, and then use Eq. (9.12) and $(d\Psi/dz)\bullet\sinh\Psi = d(\cosh\Psi)/dz$ to leave us with

$$(d\Psi/dz)^2 = (2\cosh\Psi - 2)/\ell_D^2, \tag{9.27}$$

where the additive factor of 2 from the integration constant permits $d\Psi/dz$ to vanish with Ψ at large distances. The identity $2\cosh\Psi - 2 = 4\sinh^2(\Psi/2)$ allows this expression to be simplified to

$$d\Psi/dz = \pm 2\sinh(\Psi/2)/\ell_D. \tag{9.28}$$

The potential is obtained by integrating this expression, as demonstrated in the problem set (or see Section 12.15 of Israelachvili, 1991).

Knowing the potential, the ion densities can be calculated from Eq. (9.21), although the resulting expressions may appear somewhat ungainly, given the exact solution for $\Psi(z)$ derived in Problem 9.17. However, if the potential is small, then we can replace sinhΨ by Ψ, leading to the Debye–Hückel approximation to Eq. (9.26):

$$d^2\Psi/dz^2 \cong \Psi/\ell_D^2. \tag{9.29}$$

This has the physical solution for the potential

$$\Psi(z) = \Psi_o \exp(-z/\ell_D), \tag{9.30}$$

which could also be obtained directly from the solution for $\Psi(z)$ in Problem 9.17 if Ψ_o is small. We see that ℓ_D characterizes the "screening" of the charged plate by the salt. Now, ℓ_D depends on the properties of the bulk medium, leaving us to obtain Ψ_o from the surface charge density σ_s, which is given by

$$\sigma_s = -q\int(\rho_+ - \rho_-)dz = +2q\rho_s\int\sinh\Psi\, dz, \tag{9.31}$$

if the system is electrically neutral. The second equality follows from Eq. (9.21); again, σ_s and q have opposite signs here. The integration variable can be changed from dz to dΨ using Eq. (9.28), leading to

$$\sigma_s = 2q\Psi_o \ell_D \rho_s = \varepsilon \psi_o / \ell_D, \tag{9.32}$$

after invoking $\sinh\Psi \cong \Psi$ for small Ψ and recalling its definition from Eq. (9.7).

What is the magnitude of the Debye length in a typical cellular environment? As stated above, the Bjerrum length of monovalent ions ($q = e$) in water is just 0.7 nm. Taking this value along with 0.2 M salt (for a number density $\rho_s = 1.2 \times 10^{26}$ m^{-3}), Eq. (9.24) yields $\ell_D = 0.7$ nm as well. In other words, the electric potential associated with a charged bilayer dies off rapidly over a distance of a few nanometers in a typical cellular electrolyte. Our results for the Debye length can be generalized to include several types of ions with charges q_i through the definition

$$\ell_D^{-2} = \Sigma_i \, \rho_{\infty,i} q_i^2 / \varepsilon k_B T, \tag{9.33}$$

where $\rho_{\infty,i}$ is the number density of ion type i in bulk. This relationship and other aspects of multicomponent systems are treated in Chapter 12 of Israelachvili (1991). Further studies of bilayer electrostatics have included the properties of asymmetrically charged bilayers (Ha, 2003); an introduction to the electrostatics of polymers can be found in Grønbech–Jensen *et al.* (1997) and Ha and Liu (1997) (see also Bloomfield, 1991).

In summary, we see that the presence of a negative charge on the lipid bilayer influences the distribution of mobile ions in its vicinity. Negative mobile ions are depleted near the bilayer while positive ions, both those that neutralize the bilayer and others from dissociated salts, are enhanced. For the typical cellular environment, the effects of a charged bilayer on mobile ions extend for several nanometers, as characterized by the Bjerrum and Debye lengths. In the following section, we bring two charged plates together and investigate the van der Waals force between them, as well as their interaction in an electrolyte.

9.3 van der Waals and electrostatic interactions

The components of the cell boundary – including the plasma membrane and extracellular structures – form a complex material of inhomogeneous elasticity and charge density. One aspect of this boundary region was treated in Section 9.2 by examining the distribution of mobile ions near a single charged plate. We now extend our analysis to consider two rigid plates interacting by:

- the van der Waals force between electrically neutral materials,
- the electrostatic force between charged objects in an electrolyte.

Our approach is to determine mathematically the strength of these interactions as a function of separation between the plates, and then uncover their domains of importance in the cell.

9.3.1 van der Waals forces

What is often labeled the van der Waals force between neutral atoms or molecules arises from a number of effects, including:

- the attractive interaction between electric dipole moments (Keesom),
- the attraction between permanent electric dipoles and induced dipoles in a neighboring molecule (Debye),
- the attraction between fluctuating dipole moments created by instantaneous movement of the electrons in an atom (London),

where the name in brackets is the person generally credited with calculating the effect (for a review, see Mohanty and Ninham, 1976). The potential energy associated with each of these angle-averaged contributions is proportional to r^{-6}, so they are often collectively written as

$$V_{\text{mol}}(r) = -C_{\text{vdw}} r^{-6}, \tag{9.34}$$

where r is the separation between molecules and C_{vdw} is a constant. The subscript mol serves as a reminder that Eq. (9.34) applies to molecules. The complete van der Waals potential between molecules incorporates a short-range repulsive term (proportional to $+r^{-12}$) complementary to the long-range attractive term ($-C_{\text{vdw}} r^{-6}$).

The interaction energy between aggregates such as sheets and spheres can be obtained by integrating V_{mol}, as performed for a variety of surface geometries in Chapter 10 of Israelachvili (1991). The calculations are easy enough to perform under the assumption that the force between extended objects is equal to a pairwise sum over individual molecules, neglecting correlations. The approach gives simple functions which capture a good fraction of the applicable physics, although the interpretation of macroscopic observables in terms of C_{vdw} is not always direct. Of interest for the cell boundary is the interaction energy per unit area between two rigid slabs separated by a distance D_s, as shown in Fig. 9.6. This situation is treated in Problem 9.14 for a molecular potential energy of the form $V_{\text{mol}}(r) = -C_{\text{vdw}}/r^n$ under the assumption of pairwise addition of forces. We show that the energy per unit area on one slab (B) due to its interaction with the other (A) is

$$V_{\text{slab}}(D_s)/A = -2\pi\rho^2 C_{\text{vdw}} \,/\, (n-2)(n-3)(n-4)D_s^{\,n-4}, \tag{9.35}$$

where ρ is the molecular density of the medium (i.e., the number of molecules per unit volume). The slabs are much thicker than the gap between them,

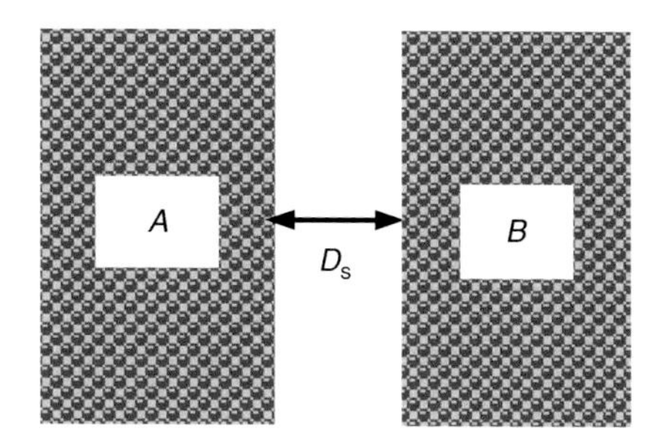

Fig. 9.6 Geometry of two rigid molecular slabs separated across a vacuum by a gap of width D_s. The slabs are semi-infinite in that they extend to infinity in directions parallel to the gap and their thickness is much greater than their separation.

and they extend to infinity in directions parallel to the gap. Substituting $n = 6$ of the van der Waals interaction into Eq. (9.35) gives

$$V_{\text{slab}}(D_s)/A = -\pi\rho^2 C_{\text{vdw}}/12D_s^2, \tag{9.36}$$

demonstrating that the energy density decays like the square of the separation. If D_s is not small compared to the thickness d_{sh} of the slab, Eq. (9.36) must be modified. As shown in Problem 9.15, the energy density of a thin sheet interacting with a semi-infinite slab is $-\pi C_{\text{vdw}}\rho^2 d_{\text{sh}}/6D_s^3$, while that of two thin, rigid sheets is $-\pi C_{\text{vdw}}\rho^2 d_{\text{sh}}^2/2D_s^4$. Comparing Eq. (9.36) with these two results shows that the energy density decreases faster with distance as the sheets become thinner: D_s^{-2}, D_s^{-3}, and D_s^{-4} for the situations considered (see also Section 5.3 of Safran, 1994).

Equation (9.36) includes interactions between molecules on separate slabs, but not between molecules within the same slab, nor does it allow for the presence of a medium in the gap between A and B. If the materials are dissimilar, then ρ^2 in Eq. (9.35) is replaced by the product of densities $\rho_A\rho_B$ from each medium. The combination $\pi^2 C_{\text{vdw}}\rho_A\rho_B$ is collectively referred to as the Hamaker constant (often denoted by the symbol A, the same as our notation for area, unfortunately), having a value of about 10^{-19} J or $25k_BT$ for many condensed phases interacting across a vacuum. The presence of an electrolyte solution between charged plates screens their interaction and reduces the energy density, as described further in Section 9.2. For the same reasons, the van der Waals force is also screened, although the reduction in the energy is not completely exponential in the distance, and the screening length is half the Debye length ℓ_D (see Section 11.8 of Israelachvili, 1991); for instance, the effective Hamaker constant in an electrolyte may be reduced by a factor of ten for distances of a few nanometers. Finite temperature and other aspects of the van der Waals interaction in continuous media can be found in Lifshitz (1956) and Ninham *et al.* (1970), as explained in Chapter 11 of Israelachvili (1991) and Section 5.4 of Safran (1994) (see also Mohanty and Ninham, 1976).

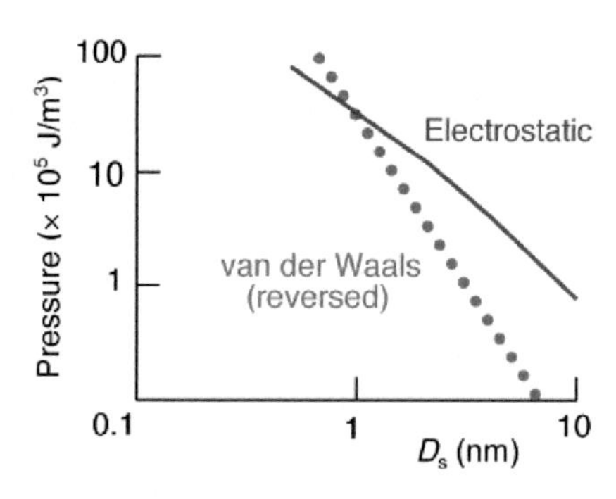

Fig. 9.7 Estimated pressure between two rigid plates as a function of their separation D_s, ignoring screening effects from the intervening medium. The van der Waals pressure (reversed in sign) is from Eq. (9.38) with parameters $C_{vdw} = 0.5 \times 10^{-77}$ J • m^6 and $\rho = 3.3 \times 10^{28}$ m^{-3}. The electrostatic pressure is a numerical solution of Eq. (9.56) with $\sigma_s = -0.1$ C/m^2, $q = e$ and $\varepsilon = 80\varepsilon_o$; screening reduces the calculated pressure.

The van der Waals interaction between plates results in a pressure P, which can be obtained from the change in the energy density $V(D_s)/A$ by

$$P = -\mathrm{d}(V(D_s)/A) \; / \; \mathrm{d}D_s, \qquad (9.37)$$

where $P < 0$ corresponds to attraction. As applied to Eq. (9.36), this yields

$$P = -\pi\rho^2 C_{vdw} \; / \; 6D_s^{\,3}, \qquad (9.38)$$

for two slabs; the attractive pressure decreases more rapidly than this for sheets whose thickness is much less than their separation (see Problem 9.15). To estimate the importance of the van der Waals interaction between bilayers, we consider two blocks of hydrocarbons, which we represent as aggregates of methyl groups with $C_{vdw} \sim 0.5 \times 10^{-77}$ J • m^6 and $\rho = 3.3 \times 10^{28}$ m^{-3} (from Table 11.1 of Israelachvili, 1991). With these values, the pressure from Eq. (9.38) is displayed in Fig. 9.7; for example, at $D_s = 5$ nm the pressure is 0.23×10^5 J/m^3 = 0.23 atm. We now contrast this interaction with the electrostatic force between charged plates.

9.3.2 Charged plates with counterions

Described in Section 9.2, phospholipids with serine head-groups are present in significant concentrations in some bilayers, awarding them a negative surface charge density that may range up to 0.3 C/m^2 in magnitude. As a first step towards understanding the forces between charged membranes, we calculate the electric field **E** between two rigid charged plates, as in Fig. 9.8(a) but in the absence of positive counterions. The plates, which extend to infinity, are separated by a distance D_s; they carry a charge density of σ_s and are immersed in a medium with permittivity ε, which equals the

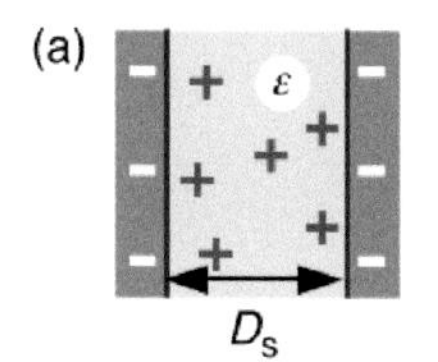

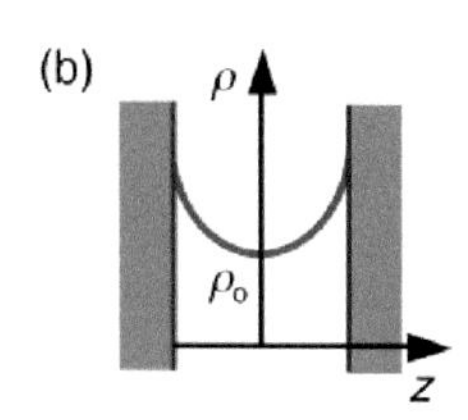

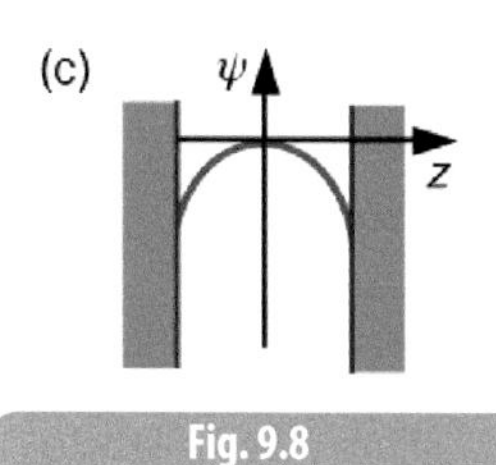

Fig. 9.8

(a) Mobile (positive) counterions spread through the region between two negatively charge plates. The counterion density ρ has a minimum (b) and the potential ψ has a maximum (c) at the midplane ($z = 0$), where their derivatives vanish. The vertical bars in (b) and (c) indicate the positions of the plate.

product of the dielectric constant and ε_o, the permittivity of free space. As established in Section 9.2, and most introductory physics textbooks, the magnitude of the electric field E from *one* plate is $E = \sigma_s/2\varepsilon$. For two parallel plates of the same charge density, *including the sign*, E vanishes between the plates and is equal to σ_s/ε outside the plates. That is, the fields from each plate cancel because they point in opposite directions between the plates. This is the reverse of a parallel plate capacitor, where the plates have opposite charges; there, the fields cancel outside the plates, but add between them, yielding $E = \sigma_s/\varepsilon$.

Now, let us introduce some (positive) counterions into the space between the plates, as illustrated in Fig. 9.8(a). If a single positive ion were placed between the plates, it would experience no net force because the electric field vanishes. However, a group of positive ions, initially spread throughout the space between the plates, would be driven towards the plates by their mutual repulsion. At finite temperature, the ions have thermal energy, allowing them to wander away from the plates and into the medium. The form of their density distribution $\rho(z)$ reflects a competition between energy and entropy: entropy favors a uniform distribution with the ions exploring all available configuration space, while energy favors the ions piled up against the plates to minimize their mutual repulsion. This cloud of counterions at the plate is referred to as the electric double layer.

Our next step is to obtain the potential $\psi(z)$ in Eq. (9.5), subject to constraints imposed by symmetry and the value of the surface charge density. The mirror symmetry of the system in Fig. 9.8 means that the potential and charge distribution must be symmetric about the midplane ($z = 0$), where we choose $\psi(0) = 0$. As seen in Figs. 9.8(b) and (c), the counterion density has a minimum and the potential has an extremum (depending on its sign) at the midplane: $\mathrm{d}\rho/\mathrm{d}z = 0$ and $\mathrm{d}\psi/\mathrm{d}z = 0$. Imposing the condition of overall electrical neutrality means that the integral of the counterion charge density $\rho_{\mathrm{ch}} = q\rho$ from $z = 0$ to $z = D_s/2$ must equal σ_s in magnitude:

$$\sigma_s = -q\int_0^{D_s/2} \rho \, \mathrm{d}z, \tag{9.39}$$

where the minus sign arises because σ_s and $q\rho$ have opposite signs. Poisson's equation, Eq. (9.3), can be used to replace $q\rho$ by $-\varepsilon\nabla^2\psi$, which is equal to $-\varepsilon\mathrm{d}^2\psi/\mathrm{d}z^2$ because the function depends only on coordinate z. The integral in Eq. (9.39) is thus

$$(-1)^2 \varepsilon \int_0^{D_s/2} (\mathrm{d}^2\psi/\mathrm{d}z^2)\, \mathrm{d}z = +\varepsilon(\mathrm{d}\psi/\mathrm{d}z)_{D_s/2}, \tag{9.40}$$

where we have used the condition $(\mathrm{d}\psi/\mathrm{d}z)_o = 0$ imposed by symmetry. Combined, Eqs. (9.39) and (9.40) yield

$$(\mathrm{d}\psi / \mathrm{d}z)_{D_s/2} = \sigma_s / \varepsilon, \tag{9.41}$$

which includes the correct signs. Being equal to $|(\mathrm{d}\psi/\mathrm{d}x)_{Ds/2}|$ from $\mathbf{E} = -\nabla\psi$ the magnitude of the electric field at the plate is $|\sigma_s|/\varepsilon$.

Armed with an expression for $(\mathrm{d}\psi/\mathrm{d}x)_{Ds/2}$, we can determine $\psi(z)$ from the Poisson–Boltzmann equation, Eq. (9.5), which, in its one-dimensional form, reads

$$\mathrm{d}^2\psi/\mathrm{d}z^2 = -(q\rho_o/\varepsilon)\exp(-q\psi/k_B T). \tag{9.42}$$

As in Eq. (9.7) of Section 9.2, we replace $\psi(z)$ by the dimensionless function

$$\Psi(z) = q\psi(z)/k_B T, \tag{9.43}$$

so that Eq. (9.42) becomes

$$\mathrm{d}^2\Psi/\mathrm{d}z^2 = -(q^2\rho_o/\varepsilon k_B T)\exp(-\Psi) = -2K^2\exp(-\Psi), \tag{9.44}$$

where

$$K^2 = q^2\rho_o/(2\varepsilon k_B T) = 2\pi\ell_B\rho_o, \tag{9.45}$$

and where ℓ_B is the Bjerrum length $q^2/4\pi\varepsilon k_B T$. Although symbolically identical, ρ_o is evaluated at different locations in Eqs. (9.10) and (9.45). Equation (9.44) has the solution

$$\Psi(z) = \ln(\cos^2[Kz]), \tag{9.46}$$

as can be verified by first demonstrating

$$\mathrm{d}\Psi/\mathrm{d}z = -2K\tan(Kz), \tag{9.47}$$

from which it follows that

$$\mathrm{d}^2\Psi/\mathrm{d}z^2 = -2K^2/\cos^2(Kz). \tag{9.48}$$

The boundary condition $(\mathrm{d}\psi/\mathrm{d}x)_{D_s/2} = q\sigma_s/\varepsilon k_B T$ from Eq. (9.41) fixes the value for K from Eq. (9.47):

$$-2K\tan(KD_s/2) = q\sigma_s/\varepsilon k_B T. \tag{9.49}$$

Note that $1/K$ has the units of [*length*]. The solution for $\Psi(z)$ in Eq. (9.46) vanishes at $z = 0$ and becomes negative for $|z| > 0$, as expected for negatively charged plates in Fig. 9.8(c).

Expressing Eq. (9.46) as $\exp[\Psi(z)] = \cos^2(Kz)$ permits the counterion density profile to be extracted easily from Eq. (9.4), namely

$$\rho(z) = \rho_o / \cos^2(Kz) = \rho_o + \rho_o\tan^2(Kz), \tag{9.50}$$

where the second equality follows from the trigonometric identity $1 + \tan^2\theta = 1/\cos^2\theta$. From Eq. (9.49), the value of $\tan^2(KD_s/2)$ at the plate is $[q\sigma_s/(2\varepsilon k_B T)]^2/K^2$, which simplifies to just $\sigma_s^2/(2\rho_o\varepsilon k_B T)$ when Eq. (9.45) is used for the definition of K^2. Thus, the counterion density at the plate obeys the particularly simple expression

$$\rho(D_s/2) = \rho_o + \sigma_s^2/(2\varepsilon k_B T). \qquad (9.51)$$

In other words, the number density of counterions is lowest at $z = 0$, from which it rises to $\rho(D_s/2)$ at the plates. From Eq. (9.51), the smallest value of $\rho(D_s/2)$ is $\sigma_s^2/(2\varepsilon k_B T)$ for a given charge density σ_s. For instance, the magnitude of σ_s could be as large as 0.3 C/m^2 for one charge per lipid in a bilayer, yielding $\rho(D_s/2) = 1.6 \times 10^{28}$ m^{-3} = 26 M for water with $\varepsilon = 80\varepsilon_o$. Equation (9.51) demonstrates that the counterion density at the plates declines with increasing temperature, as entropy encourages the counterions to explore new territory away from the plate boundaries.

The pressure P between charged plates in the absence of salts has the appealing form (see Section 12.7 of Israelachvili, 1991)

$$P = \rho_o k_B T = 2\varepsilon K^2 (k_B T/q)^2, \qquad (9.52)$$

where the second equality follows from Eq. (9.45). Except that ρ_o is the density of ions at the midplane, the first equality in this expression looks like the ideal gas law. Under what conditions do the counterions fill the gap between the plates and physically behave like a gas? The ions should spread away from the plates when the surface charge density is small, or the temperature is large. In this limit, K is small according to the right-hand side of Eq. (9.49), which then can be solved to yield

$$K^2 = -q\sigma_s / \varepsilon k_B T D_s. \qquad (9.53)$$

Using $\rho_o = 2\varepsilon k_B T K^2/q^2$ from Eq. (9.45), this region of K corresponds to

$$\rho_o = -2\sigma_s / qD_s, \qquad (9.54)$$

which is just the density expected if the counterions are spread evenly across the gap. The same expression for ρ_o applies at small gap width, where the product KD_s in Eq. (9.49) is proportional to $\sqrt{D_s}$. Under these conditions (large T; small σ_s or D_s), the system has the same pressure as an ideal gas of counterions

$$P = -2\sigma_s k_B T / qD_s, \qquad \text{(electrostatics, ideal gas limit)} \qquad (9.55)$$

which is inversely proportional to D_s. As usual, σ_s and q have opposite signs, making $P > 0$ and repulsive.

If the charges are more concentrated at the plates, Eq. (9.49) is easy enough to solve numerically by writing it as

$$y \tan y = -\zeta, \qquad (9.56)$$

where $y = KD_s/2$ and $\zeta = qD_s\sigma_s/ 4\varepsilon k_B T$, a constant of the configuration. At large values of ζ, the electrostatic interaction is approximately

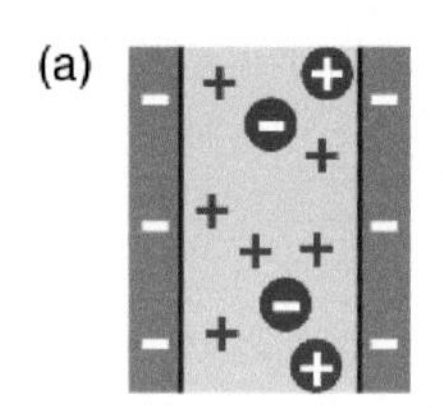

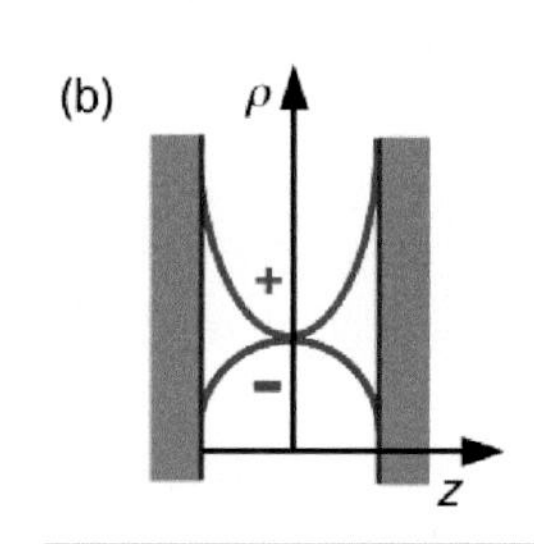

Fig. 9.9

(a) Ions from dissolved salts (black disks) may augment counterions between charged plates. (b) The number density of positive ions tends to be enhanced near a negatively charged plate, while negative ions are reduced. Vertical bars indicate the positions of the plates.

$$P = \pi k_B T / 2\ell_B D_s^2, \qquad \text{(electrostatics, large } D_s\text{, no screening)} \qquad (9.57)$$

an expression which depends only on the bulk properties of the medium (see Problem 9.21). Displayed in Fig. 9.7 is a sample calculation for $\sigma_s = -0.1$ C/m^2 in water; here, $\zeta = 1.4$ at $D_s = 1$ nm, so the solution to Eq. (9.56) approaches the ideal gas result and the large-D_s expression at its respective limits. Interested readers may try a similar calculation themselves in Problem 9.4. The figure demonstrates that the repulsive electrostatic pressure *without screening* dominates the attractive van der Waals pressure for $D_s \geq 1$ nm as it must: the electrostatic pressure decays like $1/D_s^2$ while the van der Waals declines more rapidly as $1/D_s^3$.

9.3.3 Charged plates in an electrolye

The electrostatic pressure between two plates separated by an electrolyte solution may be considerably less than that predicted by Eq. (9.57). The ion content of the medium for this case is illustrated in Fig. 9.9(a), and represented as the density function $\rho(z)$ in Fig. 9.9(b); for negatively charged plates, the density of positively charged ions is elevated near the plates while the negatively charged ions are suppressed. The presence of the bulk ions requires a modification of Eq. (9.52) for the electrostatic pressure between plates. The general case is treated in Section 12.17 of Israelachvili (1991); here, we treat only a monovalent electrolyte with charges $\pm e$. For this situation, the counterion density at the midplane ρ_o in Eq. (9.52) must be replaced with the difference in the total ion density at the midplane when the plates are separated by a distance D_s compared to when they are infinitely separated:

$$\rho_o \rightarrow [\textit{ionic density at separation } D_s]_m - [\textit{ionic density at infinite separation}]_m,$$

where the subscript indicates that the density is evaluated at the midplane. For each species of ion, the difference in densities at the midplane is $\rho_s\,[\exp(-q\psi_m/k_B T) - 1]$, where ψ_m is the potential evaluated at the midplane. Invoking the dimensionless potential $\Psi \equiv q\psi/k_B T$ from Eq. (9.43), the replacement for ρ_o becomes

$$\rho_o \rightarrow \rho_s\,\{[\exp(-\Psi_m) - 1] + [\exp(+\Psi_m) - 1]\},$$

for a monovalent electrolyte of positive and negative ions, permitting the pressure to be written as

$$P = 4\rho_s k_B T \sinh^2(\Psi_m/2). \qquad (9.58)$$

As expected, the pressure goes to zero as the potential vanishes.

The potential Ψ_m receives contributions from both plates, each a distance $D_s/2$ from the midplane. We now make the assumption that Ψ_m is small, and can be regarded as a sum of independent contributions from each plate, which we show in Problem 9.17 to be

$$\Psi(z) = 2 \ln[(1+\alpha)/(1-\alpha)], \tag{9.59}$$

where $\alpha = \tanh(\Psi_o/4) \exp(-z/\ell_D)$ and ℓ_D is the Debye screening length $\ell_D^{-2} = 8\pi\ell_B\rho_s$. For small values of α, this individual plate potential becomes $4 \bullet \tanh(\Psi_o/4) \bullet \exp(-z/\ell_D)$ leading to a pressure of

$$P \cong 64\rho_s k_B T \tanh^2(\Psi_o/4) \exp(-D_s/\ell_D), \qquad \text{(screened electrostatics)} \tag{9.60}$$

when the individual potentials are evaluated at $z = D_s/2$. The potential Ψ_o applies at the plates, not the midplane, and is given by $\Psi_o = q\sigma_s\ell_D / \varepsilon k_B T$ at low values of ψ_o according to Eq. (9.32). Thus, we are left with

$$P = (2\sigma_s^2/\varepsilon) \bullet \exp(-D_s/\ell_D), \qquad \text{(screened electrostatics, small } \psi_o) \tag{9.61}$$

after eliminating ρ_s with Eq. (9.24). We see from Eqs. (9.60) and (9.61) that the presence of the electrolyte screens the interaction and suppresses the repulsive pressure: the power-law decay in Eqs. (9.55) and (9.57) becomes an exponential decay in an electrolyte.

9.3.4 Combined interactions

Given the negative charge on phophatidylserine, the force between lipid bilayers may include both electrostatic and van der Waals components. At short distances, the van der Waals contribution dominates because of its stronger power-law dependence on $1/D_s$. At longer distances, the electrostatic interaction dominates if the electrolyte is dilute, resulting in a large Debye length. This is shown as the upper curve in Fig. 9.10. However, in a concentrated electrolyte with a small Debye length, the electrostatic interaction is rapidly extinguished with increasing distance, opening up the possibility that the electrostatic interaction is important only at intermediate separation, as illustrated by the lower curve in Fig. 9.10. In this situation, there are two minima in the potential energy density: a global (primary) minimum at short distance and a local (secondary) minimum at

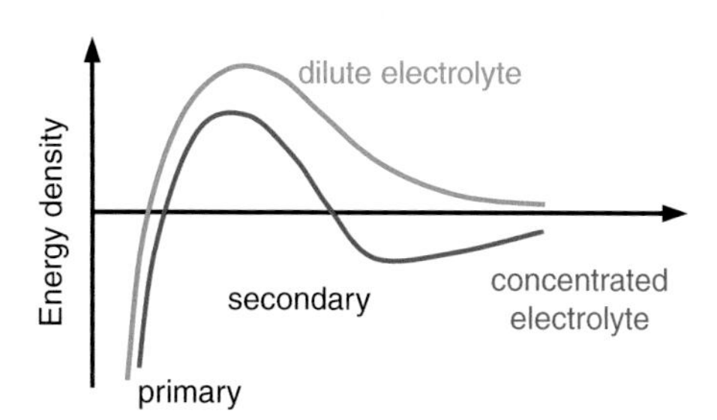

Fig. 9.10 van der Waals attraction dominates the interaction between charged plates at short distances; the behavior of the energy density at longer distances depends on the screening of the charges by the electrolyte solution.

intermediate distance. The combined electrostatic and van der Waals interaction, and its dependence on electrolyte concentration, forms the basis for the DLVO theory of colloids (Derjaguin and Landau, 1941; Verway and Overbeek, 1948; for reviews, see Derjaguin, 1949, or Section 12.18 of Israelachvili, 1991). If the temperature is low enough, a system may become trapped in the secondary minimum, the thermal fluctuations in its energy being insufficient to carry it into the global minimum on a reasonable time frame. Thus, individual colloidal particles may be stabilized in an electrolyte, rather than adhere to each other at close contact. Further studies of membrane interactions have included the effects of non-uniform charge distributions on the membranes (Russ *et al.*, 2003).

9.4 Entropic repulsion of sheets and polymers

Treating membranes as flat rigid sheets, as we have done in Sections 9.2 and 9.3, facilitates comparison of various membrane forces in a standard geometrical context. Yet we stress in Chapters 7 and 8 that biomembranes are flexible, such that thermal fluctuations in their energy result in visible undulations at room temperature (see Brochard and Lennon, 1975; Harbich *et al.*, 1976). These undulations generate a repulsive pressure from their entropic resistance to compression. Further, the extracellular matrix provides the cell with a fuzzy coat that also may inhibit the close approach of two cell envelopes. The transverse length scale of the undulations is not trivial: the root mean squared (rms) displacement of a membrane from a planar configuration is given by $(k_BTA / 4\pi^3\kappa_b)^{1/2}$ for a single gently undulating membrane, according to Eq. (8.82). For instance, taking a patch of membrane area A to be 1 μm^2 and the bending modulus κ_b to be 20 k_BT, we expect the rms displacement to be 20 nm, even though the surface orientation may change only slowly with distance along the membrane. In this section, we analyze the free energy of an undulating membrane stack and then briefly introduce the reader to the properties of polymer brushes.

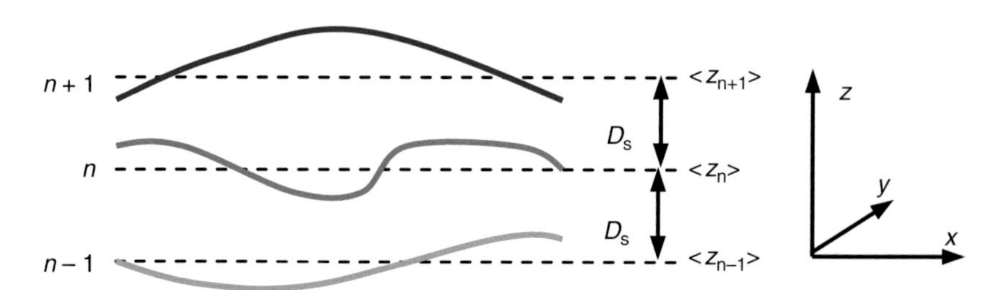

Fig. 9.11 An undulating membrane has a mean displacement $\langle z_n \rangle$ from the xy-plane and a mean separation D_s from a neighbor, where n denotes a specific membrane.

As a model system, consider a membrane stack lying parallel to the xy-plane, as displayed in Fig. 9.11. The location of a given membrane with label n is described by the function $z_n(x,y)$ with respect to the xy-plane. The mean separation between successive membranes is defined as D_s, or

$$\langle z_{n+1} \rangle - \langle z_n \rangle = D_s, \tag{9.62}$$

where the average is taken over the xy-coordinates of an individual surface. Following Chapter 8, we introduce a height function $h_n(x,y)$ which is the displacement in the z-direction of a location on the membrane from its mean position

$$h_n(x,y) = z_n(x,y) - nD_s. \tag{9.63}$$

The mean value of h_n vanishes

$$\langle h_n \rangle = 0, \tag{9.64}$$

for equally spaced membranes.

The free energy of our model system must capture the membrane bending resistance and steric repulsion between neighboring sheets. Omitting the Gaussian rigidity as in Section 7.4, the simplest expression for the bending energy density of a single membrane is given by Eq. (7.17) as $(\kappa_b/2) \bullet (C_1 + C_2)^2$ in the absence of spontaneous curvature. For gentle undulations, the sum of the principal curvatures C_1 and C_2 can be replaced by $h_{xx} + h_{yy}$ according to Eq. (8.19), where the subscripts on the function $h(x,y)$ refer to derivatives (i.e., $h_{xx} = \partial^2 h \,/\, \partial x^2$, etc.). The entropic contribution to the free energy density depends upon the local value of $h_{n+1} - h_n$, which is the difference between the membrane separation $(h_{n+1} - h_n + D_s)$ and its mean value D_s at a given (x,y) location. In its simplest form, this repulsive energy is quadratic in $h_{n+1} - h_n$, leading to a model for the free energy per unit area in the xy-plane of

$$\mathcal{F} = (\kappa_b/2)\; \Sigma_n \,(h_{n,xx} + h_{n,yy})^2 + (B/2)\; \Sigma_n \,(h_{n+1} - h_n)^2, \tag{9.65}$$

where the sum is over all membranes in the stack. The term B represents the stack's (entropic) compression resistance but, unlike the volume compression modulus, it has units of [*energy*]•[*length*]^4 and depends upon the separation D_s. This model energy was proposed by Helfrich (1978).

As in the single membrane situation studied in Chapter 8, Eq. (9.65) must be integrated over the membrane surfaces, and averaged over thermal fluctuations in their configurations, although the problem here is made complex by the presence of the stack. Using two slightly different approaches, Helfrich (1978) performed the thermal average by converting $h_n(x,y)$ to Fourier components in the z-direction as well as the xy-directions of Section 8.3. The derivation is somewhat lengthy for our context; instead, we quote and then analyze the result of the transformation, following closely the notation and approximations in Section 6.6 of Safran (1994). The Fourier components along the xy-axes introduce the two-dimensional wavevector **q** familiar from Section 8.3, and a second vector Q along the z-axis. The resulting free energy per unit volume $\Delta\mathcal{F}_v$ has the form

$$\Delta\mathcal{F}_v = (k_B T / 16\pi^3) \int_{-\pi/D_s}^{+\pi/D_s} dQ \int d\mathbf{q} \ln[1 + B(Q)/\kappa_b q^4], \tag{9.66}$$

where the function $B(Q) \equiv 2B(1 - \Delta\mathcal{F}_v \cos QD_s)$ depends only on Q and where the subscript reminds us that $\Delta\mathcal{F}_v$ is a three-dimensional density. This expression is the difference between the free energy of the stack and the membranes in isolation, vanishing with B, as appropriate.

It is not difficult to perform the integrals in Eq. (9.66). The **q** angle can be removed immediately such that $\int d\mathbf{q} = \pi \int dq^2$. By changing variables to $t^2 = \kappa_b q^4 / B(Q)$ and using $\int_0^\infty dt \, \ln(1 + 1/t^2) = \pi$, the integral over **q** in Eq. (9.66) becomes

$$\pi (B(Q)/\kappa_b)^{1/2} \int_0^\infty dt \ln(1 + 1/t^2) = \pi^2 (B(Q)/\kappa_b)^{1/2}, \tag{9.67}$$

yielding a free energy density of

$$\Delta\mathcal{F}_v = (k_B T / 16\pi) \int_{-\pi/D_s}^{+\pi/D_s} dQ [2B(1 - \cos QD_s)/\kappa_b]^{1/2}. \tag{9.68}$$

Collecting numerical constants and changing variables once again, this integral has the form

$$\Delta\mathcal{F}_v = (k_B T / 16\pi D_s) \bullet (2B/\kappa_b)^{1/2} \int_{-\pi}^{\pi} (1 - \cos\theta)^{1/2} \, d\theta, \tag{9.69}$$

which can be solved by the trigonometric substitution $(1 - \cos\theta)^{1/2} = \sqrt{2} \sin(\theta/2)$. Thus,

$$\Delta\mathcal{F}_v = (k_B T / 2\pi D_s) \bullet (B/\kappa_b)^{1/2}. \tag{9.70}$$

Helfrich (1978) uses a self-consistency argument to determine the functional dependence of B on D_s. Recall that for Hookean springs obeying $V(x) = k_{sp} x^2/2$, the spring constant is equal to the second derivative of the potential $V(x)$ with respect to the displacement from equilibrium x. In our case, the displacement variable is D_s and the relationship between B and the derivative of the energy is slightly modified to read

$$B = D_s \, \partial^2 \Delta\mathcal{F}_v / \partial D_s^2. \tag{9.71}$$

Helfrich proposed that B is a polynomial function of D_s, namely

$$B = \alpha D_s^n, \tag{9.72}$$

where α and n are constants to be determined. After some algebra, the right-hand side of Eq. (9.71) has the form

$$D_s \, \partial^2 \Delta \mathcal{F}_v \, / \, \partial D_s^2 = (k_B T \, / \, 2\pi) \bullet (\alpha / \kappa_b)^{1/2} (n/2 - 1)(n/2 - 2) D_s^{n/2-2}, \tag{9.73}$$

which can be equated with Eq. (9.72) to yield $n = -4$ and $\alpha = (6k_B T/\pi)^2/\kappa_b$, or

$$B(D_s) = (6k_B T/\pi)^2 \, / \, \kappa_b D_s^4. \tag{9.74}$$

Replacing B in Eq. (9.70) with this solution gives a free energy per unit area $\Delta\mathcal{F}$ of

$$\Delta\mathcal{F} = D_s \Delta\mathcal{F}_v = (3/\pi^2) \bullet (k_B T)^2 \, / \, \kappa_b D_s^2. \tag{9.75}$$

This expression, with a slightly different numerical prefactor, was obtained by Helfrich (1978); the argument has been extended by Evans and Parsegian (1986) to small separations. Multimembrane systems also have been investigated by computer simulation, which confirm the functional form of Eq. (9.75), but not the numerical prefactor. In light of this, it is more appropriate to write the free energy density as

$$\Delta\mathcal{F} = c_{fl} \, (k_B T)^2 \, / \, \kappa_b D_s^2, \tag{9.76}$$

where the constant c_{fl} is closer to 0.1 than $3/\pi^2 \cong 0.3$ of Eq. (9.75) (Gompper and Kroll, 1989; Janke and Kleinert, 1987; Netz and Lipowsky, 1995). The dynamics of fluctuating membranes are not covered in this book; pioneering work on this problem can be found in Milner and Safran (1987). Lastly, we mention that steric interactions between *polymerized* membranes (as opposed to *fluid* bilayers) have been studied via computer simulation by Leibler and Maggs (1989), who find that the energy density decreases like D_s^{-3}.

The behavior of the free energy per unit area according to Eq. (9.75) is much as we might expect. Being entropic, the free energy increases with temperature from a vanishing contribution at $T = 0$ where undulations are absent if $\kappa_b \neq 0$. In addition, the free energy falls as the mean spacing D_s rises, because the large excursions needed for adjacent membranes to collide at large D_s are uncommon. This dependence on D_s is the same as that displayed by the van der Waals expression for the attraction between two rigid slabs. Choosing $\pi^2 C_{vdw} \rho^2 = 10 \, k_B T$, the attractive van der Waals energy density is $-0.3 \, k_B T \, / D_s^2$ according to Eq. (9.36). In contrast, even if κ_b is as low as $10 \, k_B T$, the entropic contribution to the energy per unit area is $+0.03 \, k_B T \, / D_s^2$ according to Eq. (9.75), which is certainly lower than the van der Waals term, but not by many orders of magnitude. Now, the interaction between rigid slabs overestimates the interaction between thin sheets at large D_s: the energy density for sheets falls much faster like D_s^{-4}. Thus, there may be domains of D_s where the repulsive force from undulations dominates over other contributors.

Experimental evidence for the importance of membrane undulations comes from at least two sources: (i) measurements of the pressure between membranes and (ii) secondary effects of undulations on membrane adhesion. An example of item (i) is the observation by Evans and Parsegian

(1986) of an enhanced repulsive stress between electrically neutral egg lecithin membranes at $D_s \sim 2$–3 nm, which they attribute to undulations (data from Parsegian *et al.*, 1979). The physics behind item (ii) is described in Section 7.4: lateral tension suppresses undulations, thus increasing the possibility that two membranes may mutually adhere at close contact. Helfrich and coworkers proposed that this phenomenon underlies the onset of adhesion they observed for egg lecithin membranes under tension (Servuss and Helfrich, 1989; for earlier measurements, see Helfrich and Servuss, 1984).

Over time, our stack of model membranes separates because of entropic pressure, which can be obtained from Eq. (9.76) by the usual relation $P = -\partial \Delta \mathcal{F} / \partial D_s$. This exfoliation can be prevented by applying an external pressure P_{ext}, which generates an equilibrium spacing D_{eq} of

$$D_{eq}^{\ 3} = 2c_{fl}\,(k_B T)^2 / \kappa_b P_{ext}, \tag{9.77}$$

found by balancing the internal and applied pressures. The presence of a strong attractive force also may offset the entropic pressure between membranes. At low temperature, the attraction may be sufficiently strong to bind the membranes, while at high temperature, the entropic pressure leads to unbinding (Lipowsky and Leibler, 1986). To estimate the magnitude of this effect, we consider just two membranes subject to an attractive potential energy density of the step-function form

$$\begin{aligned} \mathcal{V}(z) &= \infty \qquad & z < 0 \\ \mathcal{V}(z) &= -\mathcal{V}_o \qquad & 0 \leq z \leq w \\ \mathcal{V}(z) &= 0 \qquad & z \geq w, \end{aligned} \tag{9.78}$$

where $\mathcal{V}_o$ and w are constants. Lipowsky (1994) provides a guide to membrane potentials capturing more physics than the simple form of Eq. (9.78). Mathematically, the behavior of two membranes of bending rigidity κ_1 and κ_2 interacting by $\mathcal{V}(z)$ is the same as that of a single membrane of rigidity $\kappa = \kappa_1\kappa_2 \,/\, (\kappa_1 + \kappa_2)$ interacting with a rigid wall via the same potential. Although the system is trapped by the potential to lie within the distance w at low temperature, the binding becomes ever less effective as the temperature rises. The approximate temperature scale T^* at which the system escapes from the potential well can be obtained by equating $\mathcal{V}_o$ with $\Delta\mathcal{F}$ in Eq. (9.76)

$$k_B T^* = (\kappa_b w^2 \mathcal{V}_o \,/\, 2c_{fl})^{1/2}, \tag{9.79}$$

for two identical membranes with bending modulus κ_b (such that the relevant bending rigidity is $\kappa_b/2$).

A variety of theoretical approaches and simulation studies confirm the existence of a distinct unbinding transition, although the transition is continuous, in the language of phase transitions, and not discontinuous as one might expect from a comparison of energies as in Eq. (9.79) (Lipowsky and Leibler, 1986). As determined in simulation studies (Lipowsky and

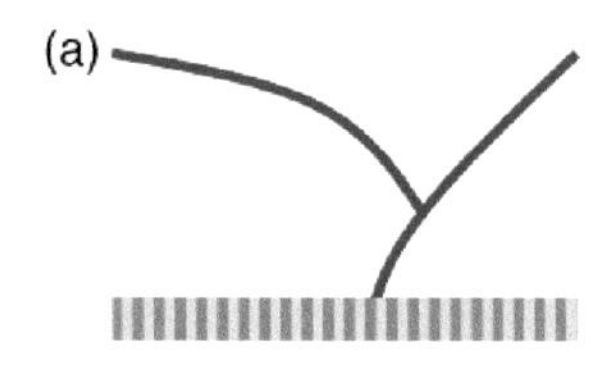

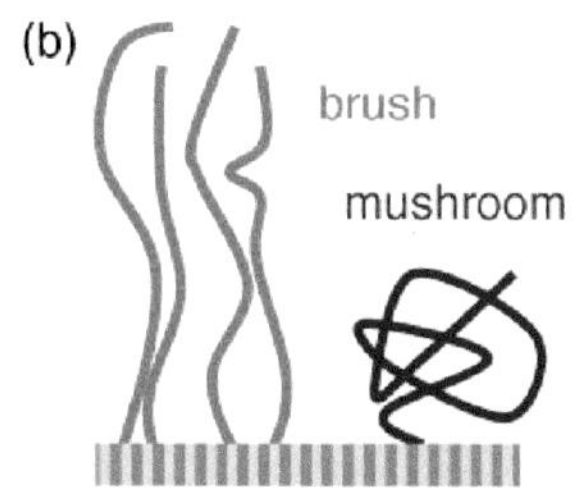

Fig. 9.12

Stiff polysaccharides appear like twigs on the bilayer length scale (a), while more flexible polymers resemble mushrooms or brushes, depending on their mean spacing (b). The entropic pressure of an attached polymer may cause the neighboring section of a flexible membrane to bend (c).

Zielinska, 1989), Eq. (9.79) sets the order of magnitude for the transition temperature, but the exact value depends on the short length scale of the simulation (i.e., the graininess of the discretized membrane). Further simulations have examined the unbinding transition of three or more membranes (Cook-Röder and Lipowsky, 1992; Netz and Lipowsky, 1993). The first experimental observation of a temperature-dependent unbinding transition was made by Mutz and Helfrich (1989) for digalactosyl-diacylglycerol (DGDG) in 0.1 M NaCl aqueous solution. We return to the general topic of membrane adhesion in Section 9.5.

The furry coat of polymers attached to the exterior of the plasma membrane also impedes cell adhesion. As displayed in Figs. 9.1 and 9.12(a), some attached polymers are relatively stiff polysaccharides (typical persistence length of 10 nm) and appear twig-like on the bilayer length scale of 4 nm. Other polymers are much more flexible, with persistence lengths of a few ångstroms, forming shapes imaginatively described as mushrooms or brushes in Fig. 9.12(b), at low or high surface density, respectively. The size of an unhindered polymer configuration like a mushroom, as measured by its end-to-end distance r_{ee} or radius of gyration R_g (see Problem 3.22), increases slowly with its contour length L_c, where L_c is the total length of the polymer chain, including branches. From Table 3.1, the radius of gyration of a branched polymer or an ideal chain (mathematically permitted to intersect itself) is proportional to $L_c^{1/2}$ in three dimensions, although the proportionality constant is inequivalent for these two polymer types. For instance, the end-to-end displacement of an ideal chain obeys

$$\langle \mathbf{r}_{ee}^{\,2} \rangle = 2\xi_p L_c - 2\xi_p^{\,2}\,[1 - \exp(-L_c/\xi_p)], \tag{9.80}$$

from Section 3.3, where ξ_p is the polymer persistence length; as an example, $\langle \mathbf{r}_{ee}^{\,2} \rangle^{1/2}$ is just 10 nm for a polymer with $L_c = 50$ nm and $\xi_p = 1$ nm. However, if the polymer has strong internal attraction, it may adopt condensed configurations whose size is proportional to $L_c^{1/3}$, more dense than ideal chains.

The effective volume of a polymer flopping about in solution, based upon Eq. (9.80), is much larger than the true physical volume of the molecule, such that there is entropic repulsion when the fuzzy coats of polymers on two cells are brought into contact. Such a repulsive interaction is well known in colloids, where polymer coatings may keep particles from adhering under some conditions (see Napper, 1983, or Van de Ven, 1989, for a guide to the extensive literature on this subject). Depending as it does on polymer density and solvent conditions, the functional form of the repulsive pressure is rather complex and is not presented here. The reader is directed to Chapter 14 of Israelachvili (1991) for an introduction to this field; entry points to more recent literature can be found in Milner (1991), Ross and Pincus (1992), Lai and Binder (1992), Grest and Murat (1993), Yeung *et al.* (1993) and Soga *et al.* (1995). Experimental measurements of the effects of grafted polymers on membrane adhesion can be found in Evans *et al.* (1996).

Lastly, we mention the effect that a polymeric mushroom has on the membrane to which it is attached. Just as a random chain has a compression resistance arising from its entropy, an attached chain may exert a steric pressure on its membrane base, as illustrated in Fig. 9.12(c). This pressure forces the membrane away from the attachment point, giving it an induced curvature. The magnitude of the curvature is reviewed in Lipowsky *et al.* (1998); being entropic, the induced curvature increases with temperature as $k_B T / \kappa_b$, where κ_b is the bending modulus. Of course, the presence of attractive interactions between polymer and membrane could modify the curvature, or even reverse its sign.

9.5 Adhesion

As determined in Sections 9.2–9.4, the pressure arising from several types of interactions between membranes decreases like an inverse power of D_s, the intermembrane separation. Crudely speaking, the pressure is 10^{-1} to 1 atm at $D_s \sim 2$ nm, built up from van der Waals (attractive), electrostatic (repulsive) and undulation (repulsive) terms. These contributions rise to 10^1 to 10^3 atm as D_s drops to 0.5 nm, where they are augmented by the steric repulsion between the extracellular matrix and other surface features. Finally, when the membranes are in very close contact, non-covalent binding is possible between specific molecules. We now describe these short-range couplings for biomembranes, before summarizing measurements of membrane adhesion energies and determining their effect on the shapes of adhering cells.

9.5.1 Site-specific interactions

Protein binding sites in immature cells, such as are present in an embryo, tend to be diffuse and may not display a locally organized structure in the plasma membrane. Organized adhesive structures appear as the cell matures, in some cases acting as junctions or switchyards for components of the cytoskeleton. In the epithelial cells shown schematically in Figs. 1.9 and A.2, for instance, desmosomes link bundles of intermediate filaments on adjacent cells, while hemidesmosomes join such filaments to the basal lamina. These structures provide local reinforcement to a bilayer-based membrane such that adhesion proteins are not simply ripped from the bilayer when the cell is subjected to stress. The largest families of proteins involved in cell adhesion are cadherins, integrins and selectins, with

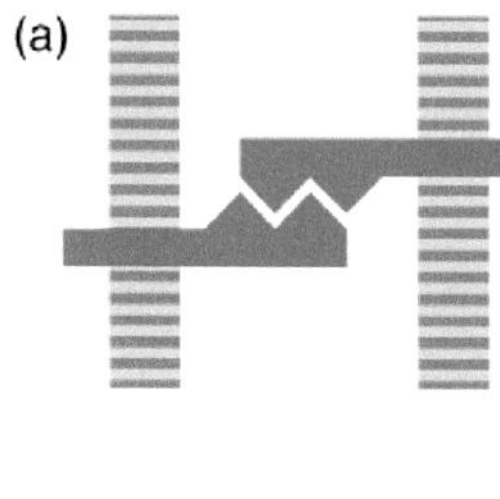

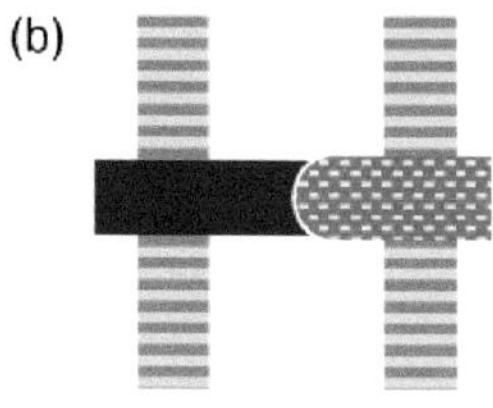

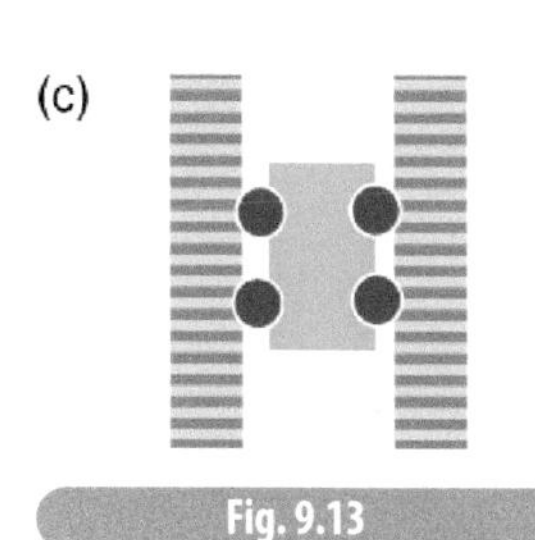

Fig. 9.13

Sample configurations of bound membranes. (a), (b) Direct binding between identical (homophilic) and non-identical (heterophilic) membrane proteins. (c) Extracellular protein linked to four small ligands (disks) attached to neighboring membranes. Bilayers are indicated by striped bars.

molecular masses around 80, 250 and 80 kDa, respectively, with some variation. The configurations adopted by proteins and their bound ligands (a generic term referring to any molecule capable of binding to a specified receiving molecule) fall into several categories. Simple binding geometries for pairs of identical or non-identical proteins are displayed in Figs. 9.13(a) and (b), corresponding to homophilic or heterophilic binding, respectively. For example, two desmosomes are bonded through their cadherin proteins as in (a), while the integrin proteins in a hemidesmosome are bonded to ligands on the extracellular matrix as in (b) (reviewed in Hynes, 1992). Another binding arrangement involves an extracellular protein attached to ligands resident on different cells, as in Fig. 9.13(c). Although less common than the other mechanisms, this is the structure of the biotin–avidin complex extensively used for cellular studies (see Meier and Fahrenholz, 1996). Avidin is a tetrameric glycoprotein of molecular mass 68 kDa; as expected, it has four sites for receiving the vitamin biotin, a small molecule of molecular mass 0.244 kDa.

The free energies of association have been measured for a number of protein–ligand pairs, starting as low as 5 k_BT per bond (for a review, see Weber, 1975). Binding is particularly strong for biotin–streptavidin at 32 k_BT, although other ligands bind to streptavidin with as little as 8 k_BT (Weber *et al.*, 1992). The free energy for binding biotin to avidin is even higher at 35 k_BT (see Green, 1975; the force–distance relationship of this binding is investigated in Wong *et al.*, 1997). However, even these strong associations are still weaker than the typical energy of a covalent bond at 180 k_BT for a C–H bond, or 150 k_BT for C–C.

Measurements also have been made of the force required to break an adhesive bond, using an atomic force microscope or a synthetic vesicle as a force transducer. As emphasized by Bell (1978), any constant force ultimately causes a bond to break at non-zero temperature; it's only a question of the time scale (see also Evans and Ritchie, 1997). This phenomenon is introduced in the context of membrane rupture under a lateral tension in Section 7.5; according to Fig. 7.21, the membrane configuration with the lowest free energy for any tension contains a hole, although the time required to cross an energy barrier to reach this ruptured state may be exceedingly long for low tensions. Two studies have confirmed the relation between applied force and bond lifetime for adhesion molecules. Merkel *et al.* (1999) have shown that the force required to break a biotin–streptavidin bond increases with the loading rate of the force: the faster the force is applied, the stronger the bond appears to be. The effect is dramatic, with the apparent rupture force growing from 5 to 170 pN when the loading rate increases by six orders of magnitude. Shao and Hochmuth (1999) have measured the related problem of removing an integrin or selectin from a neutrophil surface, observing that the time taken to extract the protein decays exponentially with the applied force; i.e., the protein could

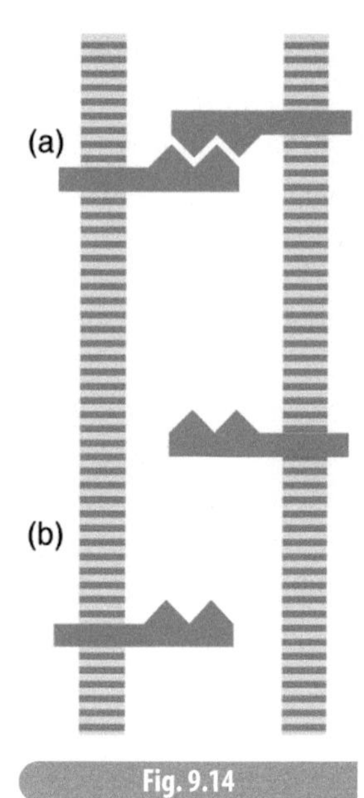

Fig. 9.14

Although some pairs of proteins on neighboring membranes may form an adhesive bond easily (a), others may have to diffuse through their membranes or reorient themselves to do so (b). Bilayers are indicated by the two striped bars.

be extracted by forces in the 20–100 pN range, but took the least time with the strongest force. Bearing this interpretational question in mind, the typical force required to break an adhesive bond on laboratory time scales is quoted at 50–150 pN (for example, Florin *et al.*, 1994, Dammer *et al.*, 1995); similar values are observed for the removal of proteins from a bilayer (Waugh and Bauserman, 1995; Shao and Hochmuth, 1999).

The formation of specific ligand–receptor bonds introduces additional time scales to the adhesion process beyond the dynamics of two membranes riding down a potential gradient that depends only on their separation D_s. As illustrated in Fig. 9.14, some interacting molecules may be close enough to easily form a bond (configuration (a)), while others may have to laterally diffuse along the bilayer to reach a suitable mate (b) or may have to change their conformational state to permit adhesion (Różycki *et al.*, 2006). Further, the participants may have to rotate in order to present the correct orientation for binding. Fundamental aspects of diffusion and bond formation have been incorporated into a set of rate equations developed by Bell (1978). This formalism involves several of the same diffusion concepts discussed in Section 11.2 on filament growth rates; namely, there are capture rate constants for the encounter of receptor–ligand pairs and dissociation constants for the breakup of bound complexes. The resulting rate equations interpret the binding process as a series of steps (e.g., initial association followed by local association to form a bound pair) and permit estimates of the rate constants on the basis of lateral diffusion constants and interaction distances (see Problem 9.5). The reaction rates have been measured for a number of bound systems; for recent examples, see Swift *et al.* (1998) or McKiernan *et al.* (1997).

9.5.2 Ligand–receptor binding

The weakest bonds that we have described have energies of less than 10 k_BT, meaning that some fraction of the bonds in a system will be disrupted by thermal fluctuations. To see the effect within a simple analysis, we consider two molecular species labeled L for ligand and R for receptor, that form a bond of energy ε_{bond} with respect to the molecules in isolation. In a fluid environment, even if the L and R molecules are not bound in an LR pair, they will nevertheless interact with their environment; the interaction energy of a single L molecule in solution is denoted ε_L while that of a single R molecule is ε_R.

Let's concentrate on the binding of a single R molecule surrounded by a solution of N_L ligand molecules that are free to diffuse within their environment. The probability of an LR pair being bound represents a competition between energy, which favors the bound state, and entropy, which favors the largest number of free molecules exploring configuration space. The entropic term is most important at high temperatures. Following the same pedagogical strategy as Phillips *et al.* (2009), the contribution of the entropic term will be calculated within a lattice representation of configuration space rather than the continuous representation of configuration space that we used for determining the critical aggregation threshold in Section 7.2. The lattice approach is somewhat friendlier mathematically.

The overall volume of the container in which the ligands reside, defined as V, is divided into Ω small boxes of equal volume V/Ω. According to statistical mechanics, the probability that an LR pair is bound involves a comparison of the number of states (times the appropriate Boltzmann weight) of the bound configuration against the total number of states available, bound and unbound (times the appropriate Boltzmann weight). Our LR system is chosen to be particularly simple: in the counting of configurations, either the receptor is bound or not.

If the receptor is unbound The N_L indistinguishable ligands can be distributed with equal probability throughout Ω locations in configuration space, and the total number of ways of doing this is

$$[\textit{number of free configurations}] = \Omega! \,/\, N_L!\,(\Omega - N_L)! \qquad (9.81)$$

If $\Omega >> N_L$, then

$$\Omega! \,/\, (\Omega - N_L)! = \Omega \bullet (\Omega - 1) \cdots (\Omega - N_L + 1) \cong \Omega^{N_L}, \qquad (9.82)$$

and Eq. (9.81) becomes

$$[\textit{number of free configurations}] = \Omega^{N_L} \,/\, N_L! \qquad (9.83)$$

Including the Boltzmann weight, which is the same for every configuration, yields

$$[\textit{weighted number of free configurations}] = (\Omega^{N_L}/N_L!)\exp(-\beta N_L \varepsilon_L - \beta \varepsilon_R), \qquad (9.84)$$

where $\beta = 1/k_B T$.

If the receptor is bound Here, $N_L - 1$ indistinguishable ligands can be distributed with equal probability throughout Ω locations in configuration space, and the number of ways of doing this is

$$[\textit{number of bound configurations}] = \Omega! \,/\, (N_L-1)!\,(\Omega - N_L+1)! \qquad (9.85)$$

Following the same reasoning leading to Eq. (9.82) for free configurations, Eq. (9.85) becomes

$$[\textit{number of bound configurations}] = \Omega^{N_L-1}/(N_L - 1)! \qquad (9.86)$$

Including the Boltzmann weight yields

$$[\textit{weighted number of bound configurations}] = [\Omega^{N_L-1}/(N_L - 1)!] \exp(-\beta[N_L - 1]\varepsilon_L - \beta\varepsilon_{bond}). \quad (9.87)$$

To find the probability p_{bound} that the receptor is in a bound state, we divide Eq. (9.87) by the sum of Eqs. (9.87) plus (9.84). Removing a factor of $\exp(-\beta N_L \varepsilon_L)\Omega^{N_L-1}/(N_L-1)!$, the expression for p_{bound} is

$$p_{bound} = \exp(\beta\varepsilon_L - \beta\varepsilon_{bond}) \,/\, \{(\Omega/N_L)\exp(-\beta\varepsilon_R) + \exp(\beta\varepsilon_L - \beta\varepsilon_{bond})\}. \quad (9.88)$$

With a few manipulations, this can be rewritten as

$$p_{bound} = \exp(\beta[\varepsilon_R + \varepsilon_L - \varepsilon_{bond}]) \,/\, \{(\Omega/N_L) + \exp(\beta[\varepsilon_R + \varepsilon_L - \varepsilon_{bond}])\}, \quad (9.89)$$

or

$$p_{bound} = (N_L/\Omega)\exp(-\beta\Delta\varepsilon) \,/\, \{1 + (N_L/\Omega)\exp(-\beta\Delta\varepsilon)\}, \quad (9.90)$$

where

$$\Delta\varepsilon \equiv \varepsilon_{bond} - (\varepsilon_R + \varepsilon_L). \quad (9.91)$$

That is, $\Delta\varepsilon$ represents the difference in energy of an LR pair when the molecules are bound compared to when they are unbound in solution.

As presented, both N_L and Ω are pure numbers, although they have very different origins. Dividing them both by the volume of the system yields the concentration of ligands $c_L = N_L/V$, and a standardization or reference concentration $c_o = \Omega/V$. The resulting equation for the probability goes under several names, but we will refer to it as the Hill function with Hill coefficient $n = 1$:

$$p_{bound} = (c_L/c_o)\exp(-\beta\Delta\varepsilon) \,/\, \{1 + (c_L/c_o)\exp(-\beta\Delta\varepsilon)\}, \quad \text{(Hill function, } n = 1) \quad (9.92)$$

Other variants of this equation are derived in the end-of-chapter problems. The general form of the Hill function is

$$p_{bound} = (c_L/c_o)^n\exp(-\beta\Delta\varepsilon) \,/\, \{1 + (c_L/c_o)^n\exp(-\beta\Delta\varepsilon)\} \quad \text{(Hill function)} \quad (9.93)$$

where n is the Hill coefficient. Receptors with more than one binding site have $n > 1$.

According to Eq. (9.91), if the LR pair can form a bound state, then $\Delta\varepsilon < 0$. The behavior predicted by Eqs. (9.92) and (9.93) is that at low temperature ($\beta\Delta\varepsilon << 0$) the probability of the receptor being bound approaches 100% ($p_{bound} \rightarrow 1$) for a fixed ligand concentration. In contrast, at high temperature ($\beta\Delta\varepsilon \sim 0$), $p_{bound} = [1 + (c_o/c_L)^n]^{-1}$ which vanishes if $c_L << c_o$. The behavior of p_{bound} as a function of concentration is analyzed in the end-of-chapter problems. A more thorough treatment of ligand–receptor binding can be found in Dill and Bromberg (2003). The formation of the adhesive state depends on the number density of receptors, as well as the ligand concentration just described; for an introduction to the experimental literature on this topic, see Sarda *et al.* (2004) and Selhuber-Unkel *et al.* (2010). The spatial pattern of the adhering region has been studied by Hategan *et al.* (2004).

9.5.3 Adhesion energies

A number of techniques have been employed to measure the adhesion energy density $\mathcal{W}_{ad}$ of pure bilayers, synthetic vesicles and simple cells. For instance, if two vesicles adhere as in Fig. 9.15(a), $\mathcal{W}_{ad}$ may be determined from the angle ϕ between the flat adhesion disk and the adjacent curved region (Bailey *et al.*, 1990). In the micromanipulation approach of Fig. 9.15(b), one vesicle is kept taut (B) while its contact area with a second flaccid vesicle (A) is controlled by the application of suction pressure, permitting $\mathcal{W}_{ad}$ to be extracted from the resulting dependence of pressure on contact area (Evans and Needham, 1988). Yet another approach uses interference patterns to obtain the height profile of a cell or vesicle near its contact region with a substrate as in Fig. 9.16(b), yielding a contact angle as in Fig. 9.15(a) (Rädler and Sackmann, 1993; Rädler *et al.*, 1995). Contact angles have also been measured for partially bound stacks of membranes (Servuss and Helfrich, 1989).

The adhesion energies of a number of pure bilayer systems are known experimentally, starting from the measurements by LeNeveu *et al.* (1975) on dimyristoyl phosphatidyl choline (DMPC); under many experimental conditions (see, however, Marra and Israelachvili, 1985), $\mathcal{W}_{ad}$ for DMPC lies close to 10^{-5} J/m^2. In 0.1 M NaCl buffer, SOPC vesicles (1-stearoyl-2-oleoyl phosphatidyl choline) display $\mathcal{W}_{ad} = 1.3 \times 10^{-5}$ J/m^2, while DGDG bilayers (digalactosyl diacyl glycerol) are ten times stickier at 2.2×10^{-4} J/m^2

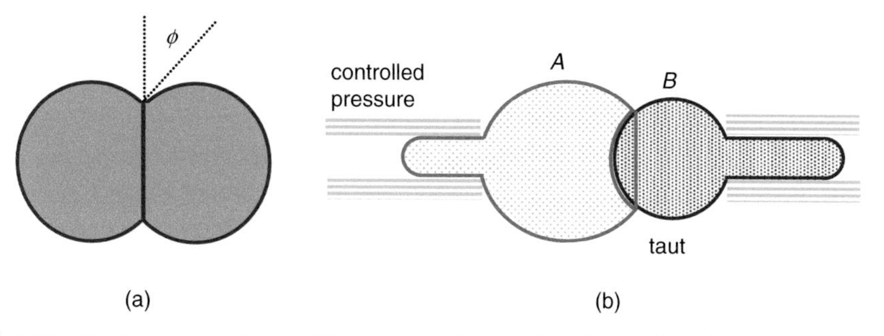

Fig. 9.15 (a) Two identical vesicles adhere, making an angle ϕ between the adhesion disk and the curved membrane. (b) Held in place by micropipettes, a flaccid vesicle (A) spreads onto a pressurized vesicle (B) in the shape of a sphere.

(Evans and Metcalfe, 1984; Evans and Needham, 1988). Introducing a polymer into the medium surrounding the SOPC vesicles raises W_{ad} linearly with concentration of the polymer; for instance, W_{ad} increases by a factor of ten in the presence of dextran at 10% by volume, in agreement with a mean field calculation of adhesive forces in polymeric solutions. In addition to pure bilayer vesicles, the adhesion of complete cells may be examined by interference microscopy; for example, W_{ad} for mutants of *Dictyostelium* has been found to lie in the range 0.6×10^{-5} to 2.2×10^{-5} J/m^2 (Simson *et al.*, 1998), and the adhesion energies of red blood cells are higher yet (Hochmuth and Marcus, 2002; Pierrat *et al.* 2004). Applied to synthetic model cells with surface receptors and a polymer glycocalyx, interference microscopy has established that the adhering region may phase separate, with local domains of strong contact separated by weakly associated membrane patches undergoing undulations (Albersdörfer *et al.*, 1997). The relationship between W_{ad} and other interfacial energies is established in Chapter 15 of Israelachvili (1991); note that the measured values of W_{ad} are several orders of magnitude lower than the surface tension of water with common alkanes, for example.

Let's compare the adhesion energy scales for a typical cell with a 10 μm^2 contact area. Consider two contributions:

- non-specific adhesion of a pure bilayer with $W_{ad} = 10^{-5}$ J/m^2, resulting in an adhesion energy of 10^{-16} J,
- specific protein binding at 15 k_BT per bond for 100 bonds / μm^2 (i.e., 10^4 copies over a surface area of 10^2 μm^2), yielding an energy of 0.6×10^{-16} J.

These two contributions are comparable in magnitude, and larger than the total membrane bending energy of 10^{-18} J for a spherical shell ($8\pi\kappa_b$) at $\kappa_b = 10\ k_BT$. Thus, a cell's shape may be strongly deformed upon adhering to a substrate.

9.5.4 Cell shape with adhesion

The shape of an adhering cell in Fig. 9.16(b) bears some resemblance to the liquid drop on a solid substrate of Fig. 9.16(a), although we caution that the bilayer has more complex elastic properties than the interface between structureless fluids. The adhering droplet serves as a good introduction to adhering cells, so we first derive the classic equation for its boundary, taking the medium above the substrate to be a vapor with the same composition as the droplet. We define the surface tension between the vapor and the solid substrate or liquid drop to be γ_S and γ_L, respectively, while the energy density for the substrate and droplet interface is γ_{SL}. Assuming that the shape of the droplet is a truncated sphere of radius R (proven in

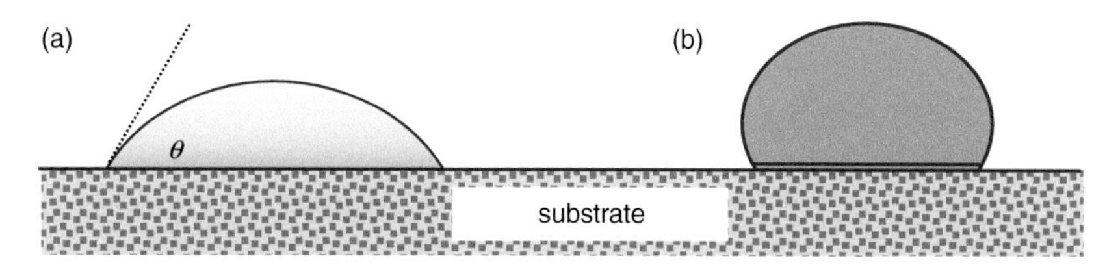

Fig. 9.16 (a) Droplet of liquid spreading on a rigid substrate with a contact angle θ at its boundary. (b) Flexible shell adhering to a rigid substrate.

Section 4.2 of Safran, 1994), then the area of the curved region is $A_{\text{curve}} = 2\pi R^2(1 - \cos\theta)$ while that of the flat adhesion region is $A_{\text{flat}} = \pi R^2 \sin^2\theta$, where θ is the contact angle shown in Fig. 9.16(a). Hence, the total interfacial energy from the adhering droplet is

$$E = (\gamma_{\text{SL}} - \gamma_{\text{S}})A_{\text{flat}} + \gamma_{\text{L}} A_{\text{curve}}, \tag{9.94}$$

where the γ_{S} term arises from the removal of substrate area in contact with the vapor. The droplet shape that minimizes this energy is characterized by a contact angle θ_{o} satisfying $\partial E / \partial A_{\text{flat}} = 0$. One only needs $\partial A_{\text{curve}} / \partial A_{\text{flat}} = \cos\theta$ from Problem 9.17 to prove

$$\cos\theta_{\text{o}} = (\gamma_{\text{S}} - \gamma_{\text{SL}}) / \gamma_{\text{L}}, \tag{9.95}$$

a result known as the Young equation (Young, 1805); the Young–Dupré equation, which involves a different parametrization for the energies, is discussed in Chapter 15 of Israelachvili (1991).

The value of the contact angle clearly depends on the relative magnitudes of the interfacial energies. For instance, taking water on wax as a typical situation, the relevant energy densities are $\gamma_{\text{L}} = 0.073$, $\gamma_{\text{S}} = 0.025$ and $\gamma_{\text{SL}} = 0.051$ J/m^2, yielding $\theta_{\text{o}} = 111°$. This confirms our experience that a water droplet forms a "bead" on a wax surface, rather than coating it ($\theta_{\text{o}} << \pi/2$) as in Fig. 9.16(a). The quantitative reason for this behavior is that $\gamma_{\text{SL}} > \gamma_{\text{S}}$: more energy is needed to create new area at a wax–water interface than at a wax–vapor interface. For less polar liquids than water, γ_{SL} may be much lower and consequently $\theta_{\text{o}} < \pi/2$.

The presence of a lipid bilayer surrounding a fluid interior may strongly influence the appearance of a cell adhering to a substrate. A bilayer resists area expansion through its area compression modulus K_{A} (although the deformation energy has a different dependence on area change than does a liquid with surface tension γ), and also resists bending through the rigidities κ_{b} and κ_{G}. The sharp change in orientation of the droplet interface at the boundary of Fig. 9.16(a) would involve severe bending of a membrane at a large energy cost; consequently, an ideal bilayer must meet the adhesion zone smoothly with a contact angle of $\theta_{\text{o}} = \pi$ when represented as a mathematically continuous surface. In such a representation, the principal curvatures at the contact point are $(2W_{ad}/\kappa_b)^{1/2}$ along a meridian of the vesicle and zero along the line of contact (Seifert and Lipowsky, 1990). Compared with an unconstrained cell or vesicle, the shape equations of an adhering cell have only a somewhat more complex mathematical form; however, the cell shape depends on more variables than κ_b and the reduced volume of the spontaneous curvature model of Sections 10.2 and 10.3, and models must recognize

- the work of adhesion W_{ad},
- K_A of the membrane,
- conservation of enclosed volume, if applicable.

Further, cell shape may also be influenced by the presence of chemically distinct domains in the substrate (Lipowsky *et al.*, 2005).

With such a large parameter space, it is not surprising that the phase diagram of adhering cell shapes is nowhere near complete. We mention two special situations of several investigated thus far.

- Bailey *et al.* (1990) have analyzed two adhering cells to determine the contact angle including only W_{ad} and K_A, but ignoring κ_b; expressions were obtained for θ_o under conditions of either fixed or variable volume.
- Seifert and Lipowsky (1990) have determined cell shape for $\kappa_b \neq 0$, assuming that the bilayer area is fixed but the enclosed volume is not. They established that, even at zero temperature, a minimum adhesion energy W_{ad} is required to bind a simple vesicle.

To gauge the importance of various contributions to the cell's energy, we estimate the minimum adhesion energy needed to bind a cell with a fixed surface area but unconstrained volume. Performed in Problem 9.8, the calculation yields a value for W_{ad} that is easily an order of magnitude less than W_{ad} found for some pure bilayer systems, indicating that the observed adhesion energy may be considerably larger than the change in membrane bending energy for some cells (supporting the approach of Bailey *et al.*, 1990). Nevertheless, the phenomenon of unbinding at weak adhesion is expected to be valid at zero temperature, and also at finite temperature where the net adhesion energy may be reduced by membrane undulations (Lipowsky and Seifert, 1991).

The weakness of the adhesion energy W_{ad} at zero temperature is not the only reason for membranes to unbind. At $T > 0$, steric repulsion between undulating membranes is present (Helfrich, 1978), and is predicted to cause unbinding at a well-defined transition (Lipowsky and Leibler, 1986; Lipowsky and Zielinska, 1989). In studies of egg lecithin membranes under mild tension, Servuss and Helfrich (1989) observe the apparent adhesion energy to increase linearly with lateral tension, reaching 0.6×10^{-6} J/m^2 at the highest tensions in the experiment, a magnitude not inconsistent with

$W_{ad} > 10^{-5}$ J/m^2 extracted from bilayers stiffened by strong lateral tension (e.g. Evans and Needham, 1988, Bailey *et al.*, 1990). More recent studies with reflection interference microscopy, while confirming the presence of undulations, yield a quadratic dependence on lateral tension for the related situation of a membrane adhering to a solid substrate (Rädler *et al.*, 1995). Further theoretical studies have shown that adhesion arising from the suppression of undulations may occur only under restricted conditions (Seifert, 1995). We also mention that the effect of membrane undulations on the formation of local molecular bonds has been examined by a number of researchers (Bruinsma *et al.*, 1994; Zuckermann and Bruinsma, 1995; Lipowsky, 1996; Weikl *et al.*, 2000). As with many topics in this text, we explore mainly equilibrium aspects of cell adhesion; the reader is directed to Dembo *et al.* (1988) for an introduction to the dynamics of adhesion, to Foty *et al.* (1994) for its effect on cell movement and to Brunk and Hammer (1997) and references therein for rolling adhesion. The adhesive interaction between the plasma membrane and its associated cytoskeleton is introduced in Sheetz (2001).

Summary

With their complex molecular structures and charge distributions, biological membranes experience forces from a variety of physical sources. In this chapter, we determine the pressure between membranes for several idealized models in order to identify their effects and assess their importance. Of the various contributors to intermembrane forces, the bulk of our attention is devoted to van der Waals, electrostatic and steric interactions from undulations. We first discuss the properties of charged membranes. Now, electrostatics in a cellular environment is more complex than that encountered in freshman physics: charged membranes and filaments may be surrounded by mobile counterions, as well as other ionic species. Further, the zero-temperature behavior of charges must be modified to accommodate the ambient temperature of cells, where entropy encourages ions to explore their surroundings in spite of the energy cost in doing so.

A good starting point for understanding such phenomena is the Poisson–Boltzmann equation, which reads $\nabla^2\psi = -(q\rho_o/\varepsilon)\exp(-q\psi(\mathbf{r})/k_BT)$, where $\psi(\mathbf{r})$ is the electrostatic potential experienced by a charge q at position $\mathbf{r}$. The number density of charged objects is ρ_o at the location where the potential is made to vanish; in general, the number density is given by $\rho(\mathbf{r}) = \rho_o \exp(-q\psi(\mathbf{r})/k_BT)$. As determined by the Poisson–Boltzmann equation, the spatial distribution of counterions lying a distance z to one side (only) of a large charged plate is $\rho(z) = [2\pi\ell_B (z + \chi)^2]^{-1}$, where the

Bjerrum length ℓ_B is a property of the medium through $\ell_B \equiv q^2/4\pi\varepsilon k_B T$. The length parameter χ is fixed by charge neutrality to be $\chi = 2\varepsilon k_B T \,/\, (-q\sigma_s)$, where σ_s is the surface charge density of the plate. In an electrolyte solution, the potential from a charged plate is reduced or *screened* by a factor $\exp(-z/\ell_D)$; for monovalent ions in solution at a number density ρ_s, the Debye screening length is $\ell_D^{-2} \equiv 8\pi\ell_B\rho_s$.

Both electrostatic and van der Waals forces may contribute to the interaction between membranes at separations D_s of a few nanometers or more. Unfortunately, the Poisson–Boltzmann equation must be solved numerically to determine the electrostatic pressure between charged plates, even in the absence of screening. However, there are two useful approximations to the exact result: when counterions are spread uniformly throughout the gap between the plates, the pressure is given by $P = -2\sigma_s k_B T \,/\, qD_s$, while at large separations, it becomes $P = \pi k_B T/2\ell_B D_s^2$. Noting that q and σ_s have opposite signs here, the pressure is always repulsive ($P > 0$). In the presence of an electrolyte, the electrostatic pressure is sharply reduced by Debye screening, as described above for a single charged plate. When the electrostatic potential is weak, the pressure decays as $P = (2\sigma_s^2/\varepsilon) \bullet \exp(-D_s/\ell_D)$; for instance, the Debye length of a 0.2 M solution of monovalent salt (e.g. NaCl) in water is 0.7 nm, showing that the pressure declines substantially over 5–10 nm.

The attractive part of the van der Waals potential for a pair of molecules falls rapidly with distance as $-C_{vdw}/r^6$, where C_{vdw} is commonly in the neighborhood of 5×10^{-78} J$\bullet$m^6 for hydrocarbons. This potential can be integrated easily for a variety of geometries to yield the energy density of bulk materials under the assumption that the pairwise molecular potentials add independently. For instance, semi-infinite slabs experience an attractive pressure of $P = -(\pi^2\rho^2 C_{vdw})/6\pi D_s^3$; the pressure between a slab and a sheet falls more rapidly at D_s^{-4}, changing to D_s^{-5} for two sheets. The combination $\pi^2\rho^2 C_{vdw}$ is called the Hamaker constant, having a value in the region of 10^{-20} J $\cong$ 2–3 $k_B T$ for common organics separated by water.

In the absence of an electrolyte, the electrostatic pressure falls more slowly with D_s than does the van der Waals pressure. Depending on numerical prefactors, there is a value of D_s below which the van der Waals dominates and above which electrostatics is more important, considering only these two contributions. This behavior may change in the presence of electrolytes because of screening, and it is possible that the electrostatic term decays so rapidly as to be relevant only at intermediate distances. With van der Waals attraction at both short and long separations, the electrostatic component may then just serve as an energy barrier between the global energy minimum at small separations, and a local minimum somewhat further away. This effect, which underlies the DLVO theory of colloid stability, allows systems at low temperatures to be trapped in the local minimum, rather than condense into the global one.

Other interactions between membranes include solvation forces at short distances, and steric interactions at intermediate and long distances. The fuzzy glycocalyx coating the cell often provides a repulsive force as the entropy of the polymeric fur resists compression when cells approach one another. Thermal undulations of the membrane boundary at long wavelengths generate an entropic pressure of $2c_{\text{fl}}(k_BT)^2/\kappa_bD_s^3$, where c_{fl} is a constant equal to about 0.1. For large separations, this pressure may be competitive with the attractive van der Waals pressure between membranes and consequently reduce cell adhesion; in some situations, placing the membranes under lateral tension to reduce their undulations is observed to encourage adhesion.

In close proximity, the association of protein receptors and ligands on neighboring membranes can also contribute to their adhesion. Although the binding energy of individual molecular pairs may range up to 35 k_BT, the bonds have less than one-quarter the strength of conventional covalent bonds. The probability p_{bound} of an individual receptor to be bound to a ligand at concentration c_L is described by the Hill function $p_{\text{bound}} = (c_L/c_o)^n\exp(-\beta\Delta\varepsilon)/\{1 + (c_L/c_o)^n\exp(-\beta\Delta\varepsilon)\}$, where n is the Hill coefficient and c_o is a reference concentration with $c_L << c_o$; Here, $\Delta\varepsilon$ is negative for bound systems. At typical receptor densities, these site-specific interactions may contribute up to about 10^{-5} J/m^2 to the short-distance attraction of membranes, which is competitive with an adhesion energy density W_{ad} of (1–2) $\times 10^{-5}$ J/m^2 observed for pure bilayers. Adhesion energies in this range are sufficient to deform the boundary of a cell, permitting it to bind to other cells or a rigid substrate at low temperature. However, thermal undulations of membranes at higher temperature reduce their binding strength, perhaps forcing them to become unbound if the adhesion energy W_{ad} is otherwise weak.

Problems

Applications

9.1. Consider a negatively charged plate with surface charge density σ_s = 0.3 C/m^2, immersed in water ($\varepsilon = 80\varepsilon_o$) which contains counterions on one side of the plate only. The counterions have charge $q = +e$ and the electrolyte is at $T = 300$ K.

(a) Find numerical values for ℓ_B, χ, and $\rho(z = 0)$ (quote ρ in M).
(b) Make a semi-logarithmic plot of ρ in the domain $0 \leq z \leq 5$ nm.
(c) At what value of z does ρ fall below 0.2 M?

(d) What is the magnitude of the electric field at $z = 1$ nm?

9.2. To one side of a single plate with surface charge density σ_s lies a monovalent electrolyte solution with $\rho_s = 0.2$ M. The potential at the plate is $\psi_o = -0.0125$ V.

(a) Find ℓ_B and ℓ_D for this system at $T = 300$ K.
(b) Calculate σ_s in the Debye approximation.
(c) Make a semilogarithmic plot of $\psi(z)$ for $0 \leq z \leq 5$ nm using both the exact expression from Problem 9.12 and the Debye approximation.

9.3. Using the potential energy density from Problem 9.20(b), derive the van der Waals pressure for two parallel sheets, each of thickness $d_{sh} = 4$ nm, separated by a distance D_s. Numerically compare your expression for the pressure to Eq. (8.38) for two slabs over the range $10 \leq D_s \leq 50$ nm. Assume $C_{vdw} = 0.5 \times 10^{-77}$ J•m^6 and $\rho = 3.3 \times 10^{28}$ m^{-3}.

9.4. Two parallel plates, separated by a distance D_s, form the boundaries of an aqueous medium containing only counterions, not dissolved salt. Their surface charge density is -0.2 C/m^2, and the medium is at 300 K.

(a) Calculate the electrostatic pressure between the plates for $0.2 \leq D_s \leq 5$ nm.
(b) Compare your result at $D_s = 0.2$ nm with the ideal gas expression for P.
(c) Compare your result at $D_s = 5$ nm with the large-D_s expression for P.

9.5. Plot (logarithmically) the magnitudes of the following contributions to the pressure between membranes over the range $2 \leq D_s \leq 10$ nm:

- van der Waals attraction between two semi-infinite slabs (small D_s approximation),
- van der Waals attraction between two sheets of thickness 4 nm, using the potential from Problem 9.15 to derive the corresponding pressure (large D_s approximation),
- repulsion from undulations for $\kappa_b = 25\, k_BT$.

At what value of D_s does the repulsion from undulations match the van der Waals attraction? Use $c_{fl} = 0.1$, $\pi^2 C_{vdw}\rho^2 = 2\, k_BT$ at $k_BT = 4 \times 10^{-21}$ J.

9.6. Proteins A and B move in parallel planes, producing a bound state C when they approach each other within a distance R_{AB} according to rate constants k_+ and k_-:

$$A + B \rightarrow C\ (k_+)$$

$$A + B \leftarrow C\,(k_-)$$

The applicable rate equations can be found in most introductory chemistry texts. If governed by diffusion, the rate constants are given by (see Bell, 1978)

$$k_+ = 2\pi(D_A + D_B) \qquad k_- = 2(D_A + D_B)/R_{AB}^2,$$

where D_A and D_B are the appropriate diffusion constants. Assume for proteins in a bilayer that $D_A = D_B = 10^{-14}$ m^2/s and $R_{AB} = 0.75$ nm.

(a) Find an expression for the equilibrium constant $K_{eq} = [C] / [A] \bullet [B]$ in terms of R_{AB} from the condition d$[C]$/dt = 0, where [...] represents concentration (here, in two dimensions). Evaluate your expression for our specific example.
(b) If the initial concentrations are $[A] = [B] = 5 \times 10^{14}$ m^{-2} and $[C]$ = 0, what is the initial value of d$[C]$/dt? What is this production rate for a 1 μm^2 membrane patch?
(c) If the total number of copies of each of A and B in a 1 μm^2 patch is 500, which of the following is the most appropriate for the membrane adhesion time by the formation of the C bound state: 10^{-2} s, 1 s, 10^2 s?

9.7. In bulk, cells may adhere to many of their neighbors, not just the pair-wise associations described in this chapter. Consider a cell with the unlikely shape of a rounded cube, each flat face of dimension $4R \times 4R$, for an overall width of $6R$. The principal curvatures are 0 and R^{-1} along the edges and R^{-1} at the corners.

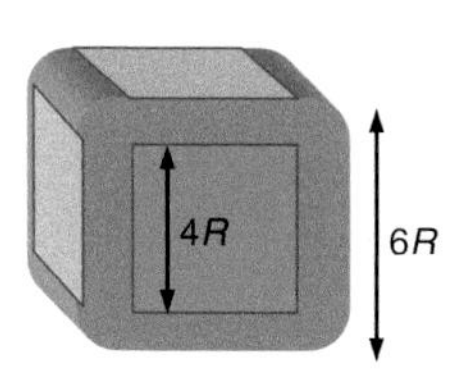

Evaluate the deformation energy ΔE of this cell compared to a sphere, and compare it with the adhesion energy per cell for $\kappa_b = 10^{-19}$ J, $W_{ad} = 10^{-5}$ J/m^2 and R = 3 μm. You may use the result from Problem 9.18 without proof.

9.8. A vesicle with the shape of an axially symmetric pancake adheres to a flat substrate. The vesicle has a fixed area A = 200 μm^2 and radius parameter R = 4 μm, but its volume is unconstrained.

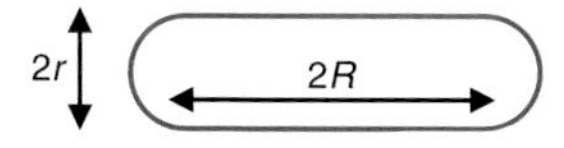

(a) What is the value of r?

(b) Find the energy $\Delta E/\kappa_b$ required to deform a spherical vesicle into this shape, using results from Table 7.1.

(c) What adhesion energy density W_{ad} is required to achieve this deformation if $\kappa_b = 20\ k_BT$? Quote your answer in J/m^2.

9.9. In deriving the Hill function for ligand–receptor systems, we stated that for most situations of interest, the ligand concentration c_L is much less than the reference concentration c_o. Taking $c_L = 1$ M and $\Delta\varepsilon/k_BT = -10$ for the binding energy, plot p_{bound} against c_L (in physical units) for $0 \leq p_{bound} \leq 1/2$ if $n = 1$ and if $n = 4$.

9.10. As a ligand–receptor system, consider two identical cells which adhere to each other along a flat interface; each cell has both ligands and receptors on its surface.

(a) If $\Delta\varepsilon/k_BT = -15$, what is the two-dimensional concentration c_L (in ligands/m^2) on one cell for $p_{bound} = 1/2$ on the other? Assume the binding probability is described by a Hill function with $n = 1$; take the reference concentration to be 2×10^{18} /m^2. You may use results from Problem 9.19 without proof.

(b) If the cell has a surface area of 100 μm^2, how many copies of the ligand are there on each cell?

Formal development and extensions

9.11. The electrostatic potential $\psi(z)$ and counterion distribution $\rho(z)$ are determined in Section 9.2 for a large plate with surface charge density σ_s adjacent to a bath of counterions, each with charge q, in a medium of permittivity ε.

(a) Find $\rho(z = 0)$ in terms of σ and ε.

(b) What fraction of the counterions are present in the region $0 \leq z \leq \chi$?

(c) Compare the behavior of the potential with that of a plate in the absence of counterions, particularly near $z = 0$ and $z \to \infty$.

9.12. Show that the dimensionless potential $\Psi(z)$ for a single charged plate in a salt medium is

$$\Psi(z)/2 = \ln([1+\alpha] / [1-\alpha]),$$

where $\alpha = \tanh(\Psi_o/4) \bullet \exp(-z/\ell_D)$ and $\Psi_o = \Psi(z = 0)$. Start from $d\Psi/dz = \pm 2\sinh(\Psi/2)/\ell_D$, as established in Section 9.2, and be cautious with the choice of sign. You may need to use $\int dx \,/\, \sinh x = \ln \tanh(x/2)$ from Section 2.423 of Gradshteyn and Ryzhik (1980).

9.13. Suppose that the potential energy between two molecules separated by a distance r has the power-law form $V(r) = -C_{vdw}/r^n$, where C_{vdw} and n are constants.

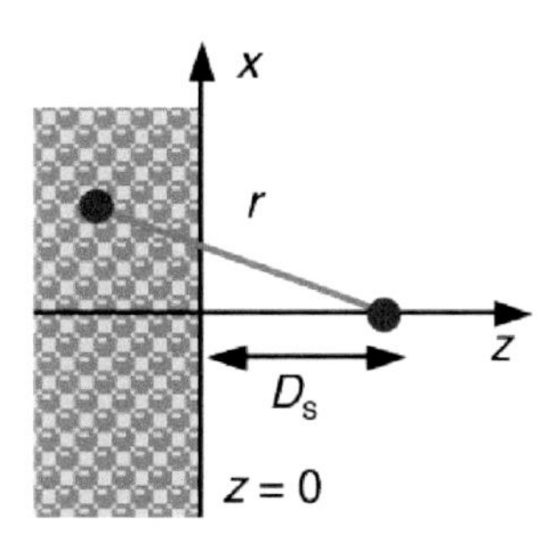

Show that a point molecule a distance D_s from a semi-infinite flat slab of material (infinite in the x- and y-directions, extending to $-\infty$ along the z-axis) experiences a potential energy given by

$$V_{point}(D_s) = -2\pi\rho C_{vdw}/(n-2)(n-3)D_s^{n-3}, \qquad \text{(for } n > 3)$$

where ρ is the number density of molecules (units of $[length]^{-3}$). Assume that the potential is additive. (*Hint: perform the integration using cylindrical coordinates.*)

9.14. Consider two molecules interacting with a potential energy $V(r) = -C_{vdw}/r^n$, as in Problem 9.13, where r is the separation between molecules. Show that the energy of

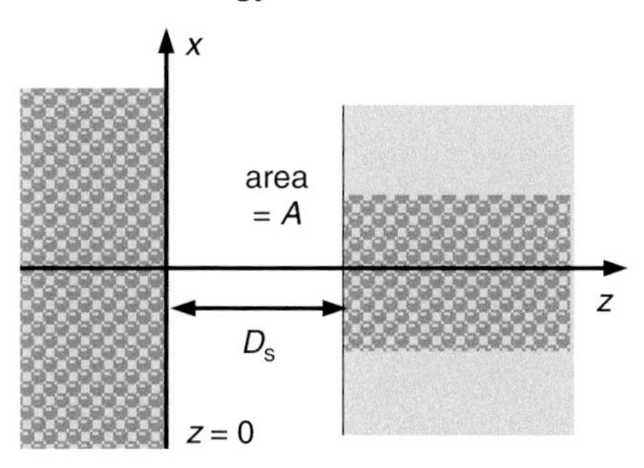

the molecules in the slab to the right (of the diagram, of area A) arising from their interaction with the molecules in the semi-infinite slab on the left is

$$V_{slab}(D_s)/A = -2\pi\rho^2 C_{vdw} / (n-2)(n-3)(n-4)D_s^{n-4}.$$

Both materials have the same molecular density ρ and are semi-infinite in extent. Assume that the potential energy of the molecules is additive and use the result from Problem 9.13.

9.15. Molecules interacting according to the van der Waals potential $-C_{vdw}/r^6$ are present in the sheets and slabs shown in the diagram below. In (a), a sheet of thickness d_{sh} interacts with a semi-infinite slab separated by a distance D_s, while in (b),

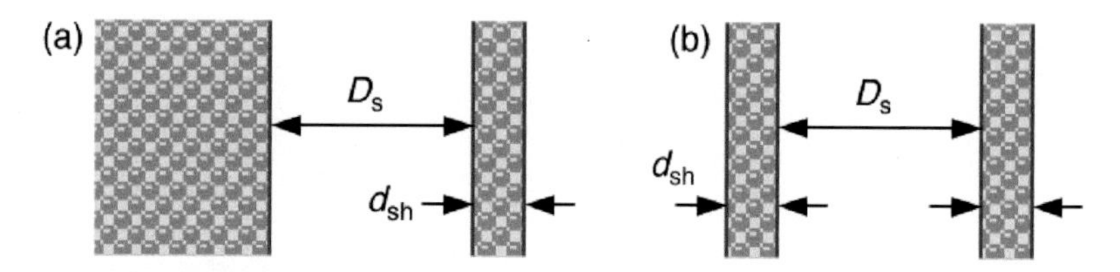

two identical sheets are a distance D_s apart. Starting with the result from Problem 9.13 for the point + slab combination, show that the energy per unit area of the sheet in configuration (a) is $V(D_s)/A = -\pi C_{vdw}\rho^2 d_{sh} / 6D_s^3$. Repeating Problem 9.13 for the point + sheet geometry, establish that the energy per unit area of a sheet in configuration (b) is $V(D_s)/A = -\pi C_{vdw}\rho^2 d_{sh}^2/2D_s^4$. In both cases, assume $D_s >> d_{sh}$.

9.16. Show that, in the limit of large separation D_s, the electrostatic pressure between two charged plates bounding a medium containing counterions (but no salt) is given by

$$P = \pi k_B T / 2\ell_B D_s^2.$$

You may start your proof from $-2K \tan(KD_s/2) = q\sigma_s / \varepsilon k_B T$.

9.17. Assume that the shape of a liquid droplet adhering to a planar substrate is a section of a sphere with radius R. The areas of the curved and flat regions of the droplet are defined as A_{curve} and A_{flat}, respectively, and the boundary has a contact angle θ.

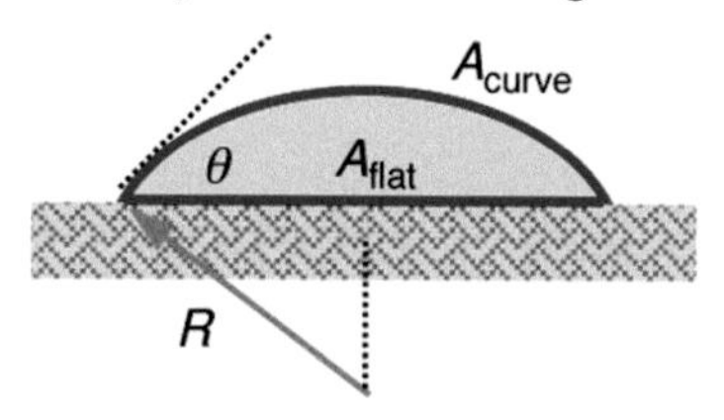

(a) Show that the areas and volume of the droplet are:

$$A_{flat} = \pi R^2 \sin^2\theta \qquad A_{curve} = 2\pi R^2(1 - \cos\theta)$$
$$V = (\pi R^3/3)\bullet(1 - \cos\theta)^2\bullet(2 + \cos\theta).$$

(b) Establish that $\partial A_{curve}/\partial A_{flat} = \cos\theta$ at fixed V by evaluating $\partial A_{curve}/\partial\theta$, etc. (where R is a function of θ, according to your results from part (a)).

9.18. As in Problem 9.7, consider a cell with the shape of a rounded cube, each flat face of dimension $4R \times 4R$, for an overall width of $6R$. The principal curvatures are 0 and R^{-1} along the edges and R^{-1} at the corners.

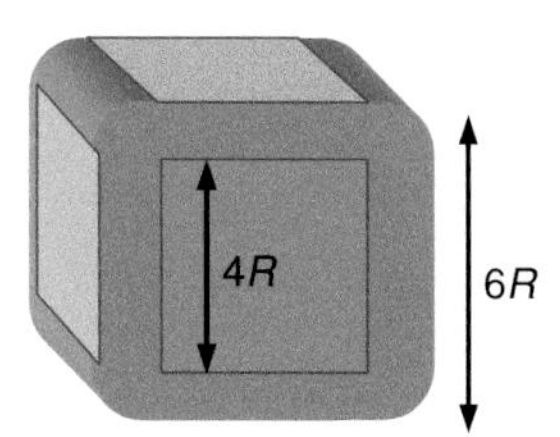

Taking into account only its bending resistance κ_b, show that the deformation energy $\Delta E/\kappa_b$ of this shape is 5/2 times that of a sphere.

9.19. For a ligand–receptor system described by the Hill function with coefficient n:

(a) Show that the relationship between concentration and temperature is given by

$$T = \Delta\varepsilon \,/\, k_B \ln\chi^n \qquad \chi = c_L/c_o$$

when the probability for a receptor to be bound is 1/2.

(b) If the ligand concentration were changed from c_L to c_L', at what new temperature T' would $p_{bound} = 1.2$ (calculate $1/T' - 1/T$)?

(c) Does this temperature difference increase or decrease with n?

9.20. In a ligand–receptor system described by the Hill function with arbitrary n, what is the rate of change of p_{bound} with T (that is, dp_{bound}/dT) when $p_{bound} = 1/2$? By examining the signs of the physical parameters in the system, is dp_{bound}/dT positive or negative? Assume the ligands are dilute.

PART III

THE WHOLE CELL

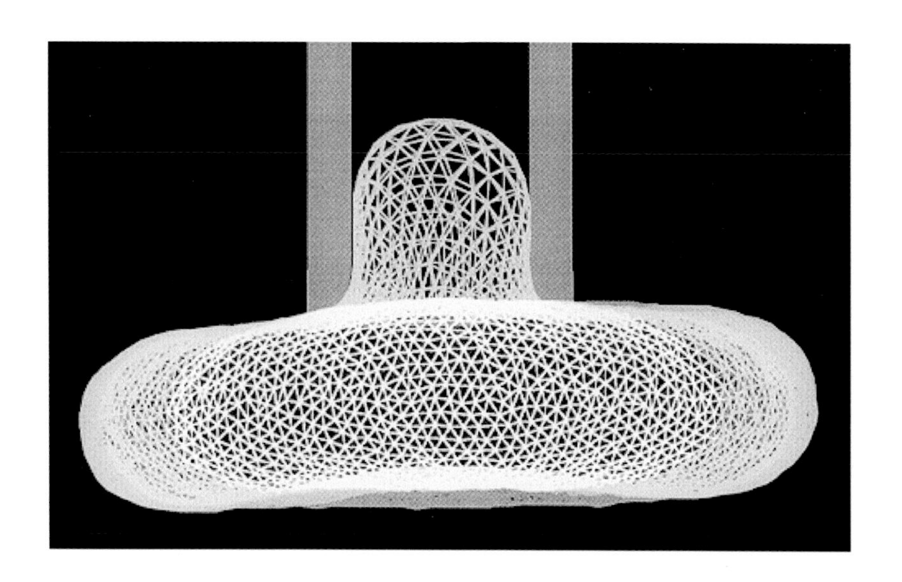

10 The simplest cells

Most cells have a complex internal structure of biological rods, ropes and sheets. In terms of the metaphors for the cell in Chapter 1 – a hot air balloon, a sailing ship – our task in Parts I and II of this text was to analyse the individual structural elements of the cell – the masts, rigging and hull – without regard to their interconnection. In Part III, we begin to assemble the units together to form composite systems. As the introduction to Part III, Chapter 10 treats the equilibrium shapes of the simplest systems, namely cells or vesicles without a nucleus or space-filling cytoskeleton. A thorough treatment of complex cells is beyond the reach of our analytical tools; rather, we mention a selection of their properties in Chapters 11 and 12. In Chapter 11, we describe mechanisms that cells have developed for changing their shape, as part of cell division or locomotion, for example. Cell division proper is the subject of Chapter 12, including changes to the division cycle over the history of the Earth. Lastly, in Chapter 13, we study methods that have evolved for control and organization in the cell, using specific situations as examples.

10.1 Cell shapes

The first figure in Chapter 1 of this book displays examples of the variety of shapes that cells can adopt, ranging from smooth cylindrical bacteria to branched and highly elongated neurons. Yet the primary mechanical elements of all cells are the same: one or more fluid sheets surround the cell and its internal compartments while filaments form a flexible scaffolding primarily, but not exclusively, confined within the cell. What design principles give rise to this wonderful collection of shapes? Let's begin with the lipid bilayer. If the two leaflets of a bilayer have the same homogeneous composition and are adjacent to equivalent environments, the equilibrium shape of the bilayer is flat at zero temperature. However, compositional differences between leaflets, such as the size or shape of their molecules, may cause a membrane to spontaneously bend; for example, the differing lipid head-groups in the schematic bilayer of Fig. 10.1(a) cause its equilibrium shape to have non-vanishing curvature. The rate of migration of lipids between leaflets is sufficiently slow (10^{-7} s^{-1} or less for some phospholipids;

(a)

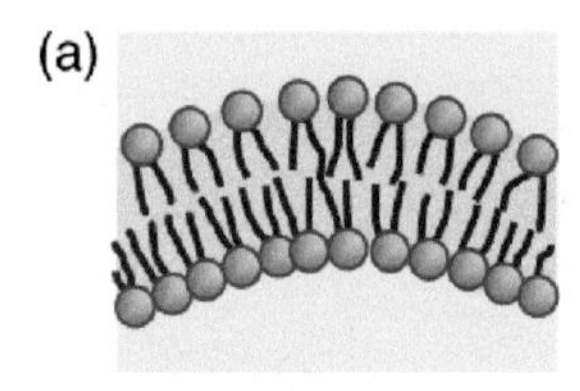

(b)

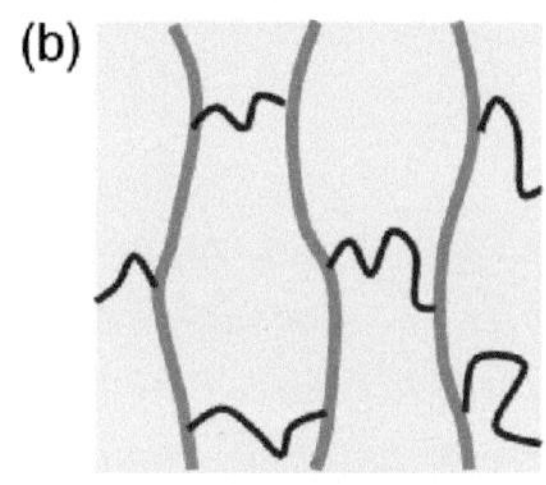

(c)

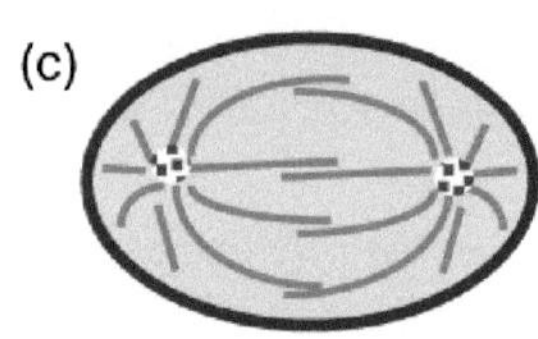

Fig. 10.1

Deformations of the cell boundary can arise from a variety of causes, including: (a) compositional inhomogeneity of the bilayer leaflets, (b) anisotropic network in a cell wall under lateral stress, and (c) pressure from structural elements, such as interacting microtubules during cell division. Note that the length scale of the drawings increases from (a) to (c).

see Vance and Vance, 1996) that it may not take much effort for this compositional imbalance to be maintained. Lateral phase separation may occur for leaflets with several molecular components, again opening the door to locally curved regions. Even if the bilayer is symmetric, spontaneous curvature may arise if the external environments of the leaflets are inequivalent.

Anisotropic structural elements, such as the network shown in Fig. 10.1(b), also can influence cell shape. For instance, networks may stretch more easily in one direction than another if the network components have different stiffness. Even when the network in the figure is placed under isotropic tension, it will exhibit a larger strain horizontally than vertically. Further, the magnitude of the lateral stress experienced by a membrane may be direction-dependent according to cell shape. For example, we show in Section 10.2 that the stress around the girth of a cylinder is twice the stress along its symmetry axis.

A third example of shape-changing forces is provided by interacting microtubules during cell division. As described in Chapter 11, microtubules radiate from a center near the cell's nucleus and are dynamic, in the sense that they grow and shrink continuously. During cell division, microtubules grow from two distinct centers, as shown in Fig. 10.1(c); filaments originating from opposite poles of the cell may interact, providing a force that pushes the centers apart and elongates the cell.

In this chapter, we describe several mechanical features of a cell that influence its shape, concentrating on those cells whose elastic elements are assembled at its boundary to form a composite membrane enclosing a structureless interior. In Section 10.2, a number of results from the continuum mechanics of thin shells are derived, and a methodology is presented for determining the shape of a system with axial symmetry. Only a limited number of structures can be explored analytically with this technique; in general, the equations governing cell shape must be solved numerically.

Biological ropes and sheets can be assembled into cells or cell-like objects with a hierarchy of complexity. Excluding viruses, the simplest structures are cargo-carrying vesicles with just a plasma membrane and diameters as little as 50 nm. Without a genetic blueprint, vesicles cannot reproduce; however, model vesicles can be manufactured in the laboratory from lipids and other molecules, allowing their shape to be determined systematically as a function of size and composition. These structures, along with red blood cells, are the subject of Sections 10.3 and 10.4. With dimensions of a few hundred nanometers, mycoplasmas have few mechanical elements and are also amenable to theoretical analysis; unfortunately, systematic quantification of their erratic shapes has not yet been performed.

Next to vesicles and erythrocytes, the cells whose mechanics have been most thoroughly examined are bacteria, the subject of Section 10.5. The elasticity of their cell boundaries is essentially two-dimensional: their construction involves one or two fluid membranes to which is attached a network of flexible links in the form of an extended macromolecule of peptidoglycan. Aspects of this network can be captured, in part, by the harmonic networks of Chapters 5 and 6, as demonstrated in the problem sets of those chapters. Here, we relate the properties of membranes and their associated networks to the shape of cells with axial symmetry; however, we caution that our understanding of these cells is not at all complete and new aspects of the bacterial cytoskeleton continue to be revealed.

10.2 Energetics of thin shells

The envelope of the cells of interest in this chapter ranges from about 5 to 50 nm in depth. The thinnest shells are found in small vesicles bounded by a lipid bilayer just 4 to 5 nm wide, which is about 10% of the vesicle radius. In contrast, the walls of bacteria are much thicker and the cells may be further wrapped in a regularly structured S-layer; the two principal designs for the bacterial envelope are as follows (see Fig. 1.7).

- Gram-negative bacteria have two fluid membranes mechanically connected by filaments spanning a periplasmic space often 20 nm wide (with considerable variation among cell types), a region which also contains a thin sheet of peptidoglycan.
- Gram-positive bacteria have a single membrane enclosed by multiple layers of peptidoglycan 20–80 nm across.

The boundary of eukaryotic cells such as the red blood cell or auditory outer hair cell have a protein-laden bilayer along with a membrane-associated cytoskeleton perhaps 30–40 nm across. For cells with complex boundaries, then, the thickness of the outer shell (often 40–60 nm) is about 5%–10% of the cell radius (~1/2 micron for common bacteria), permitting us to view the cell boundary as essentially two-dimensional for mechanical purposes. In developing a set of analytical tools to describe cell shape, we will first include the thickness of the shell as an explicit parameter before folding it into the stress tensor to obtain formal results for two-dimensional sheets.

Unlike animal cells, bacteria (and plant cells) may operate at an elevated internal pressure that is sustained by their cell wall. The tension created within the wall is proportional to its curvature and the pressure difference it experiences. To see how this arises, we first analyze the forces in a spherical shell defined by inner and outer radii r and $r + d_{sh}$, respectively, where d_{sh} is the shell thickness. Let's take an imaginary slice through the center of the sphere, dividing the shell in two and exposing the ring shown in the

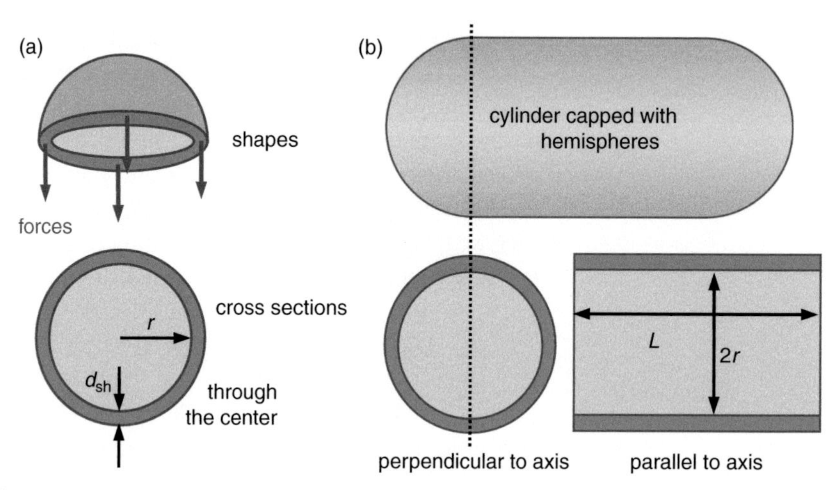

Fig. 10.2 Two shells of finite thickness d_{sh} and inner radius r; the appearance of the shell is shown on the top row, their cross sections are on the bottom. Pressurized shells bear forces perpendicular to the cross sections. (a) Half of a spherical shell, whose cross section is a thick ring. (b) Cylindrical shell, capped by hemispheres; the two cross sections are perpendicular and parallel to the symmetry axis.

bottom row of Fig. 10.2(a). An interior pressure P results in a force of $\pi r^2 P$ across the plane of the ring. Regarding the hemisphere as a free body, this "upward" force is balanced by the "downward" force across the ring itself, as indicated by the arrows in Fig. 10.2(a). Because the force is spread over the cross-sectional area of the ring, which is $\pi[(r + d_{sh})^2 - r^2] \cong 2\pi r d_{sh}$ at small d_{sh}/r, the mean stress $\langle\sigma\rangle$ in the shell is

$$\langle\sigma\rangle = [force] \,/\, [area]$$

$$\cong rP \,/\, 2d_{sh}. \qquad \text{(spherical shell).} \qquad (10.1)$$

An exact result can be obtained by using the full expression for the area of the ring. As a technical note, we could have evaluated the force by summing $P\,\mathrm{d}A$ over all area elements $\mathrm{d}A$ of the hemisphere, rather than considering the force on the midplane. However, we would then have to determine the component of the force in the direction normal to the midplane ($P\,\mathrm{d}A\cos\theta$) in calculating the overall force, leaving us to evaluate the integral $2\pi r^2 P \int \sin\theta \cos\theta \,\mathrm{d}\theta$, where the polar angle θ runs from 0 to $\pi/2$. Explicit evaluation gives the previous result: the force is $\pi r^2 P$.

Stresses in the cylindrical shells of many bacteria are only slightly more complex than those in spheres. The model bacterium in Fig. 10.2(b) consists of a cylindrical tube capped at each end by hemispherical shells to form what is sometimes referred to as a spherocylinder. The mean stress experienced by the endcaps is the same as that of the sphere in Fig. 10.2(a), namely $rP/2d_{sh}$. However, the stresses in the cylindrical section are not isotropic, but are different for area elements in directions perpendicular and parallel to the symmetry axis, which we will define as σ_z and σ_θ, respectively. A cross section through the cylinder perpendicular to the axis, has the shape of a ring, as in the bottom left corner of Fig. 10.2(b). Determination of the stress across an area element of the ring follows the same steps as the sphere and leads to the same conclusion, $\langle\sigma_z\rangle = rP / 2d_{sh}$. However, a slice through the tube containing the axis of cylindrical symmetry intersects the shell as shown in the lower right corner of Fig. 10.2(b): there are two parallel strips of length L and width d_{sh}.

To obtain the "hoop" stress $\sigma\theta$, we follow the same logic as for the sphere. Viewing one half of the cylinder as a free body, the pressure P exerts a force $2rLP$, over the area $2rL$ of the cross section. This force is balanced by the force borne by the shell, which is the product of the mean stress $\langle\sigma_\theta\rangle$ and the area of the two surfaces cut from the shell, $2Ld_{sh}$. Equating these expressions for the forces, the length of the cylinder cancels, leaving

$$\langle\sigma_\theta\rangle = rP/d_{sh} \quad \text{(cylindrical shell, hoop direction)} \quad (10.2a)$$

$$\langle\sigma_z\rangle = rP/2d_{sh}, \quad \text{(cylindrical shell, axial direction)} \quad (10.2b)$$

where Eq. (10.2b) is included for completeness. In other words, the stress is twice as great around the hoop direction than along the symmetry axis. That boiled hot dogs and sausages split more readily along their length, than around their girth, may reflect this difference axis (for further reading on stresses in shells, see Flügge, 1973; Fung, 1994; and Landau and Lifshitz, 1986).

The stress σ in Eqs. (10.1) and (10.2) is a three-dimensional quantity with dimensions of energy per unit volume. In Chapter 5, we introduced a two-dimensional tension τ, called a stress resultant, which carries dimensions of energy per unit area and is equal to the product of the mean stress and the shell thickness. Thus, Eq. (10.2) corresponds to the tension

$$\tau_\theta = rP \quad \text{(cylindrical shell, hoop direction)} \quad (10.3a)$$

$$\tau_z = rP / 2. \quad \text{(cylindrical shell, axial direction)} \quad (10.3b)$$

A relationship between the components of the tension is available for arbitrary radii of curvature, as we now obtain by analyzing the small section of the curved surface in Fig. 10.3(a). The rectangular shape of the section is taken to coincide with the principal directions 1 and 2 (of the curvature), along which lie arcs of length ds_1 and ds_2, resulting in an area element $dA = ds_1 \bullet ds_2$. The tension components τ_1 and τ_2 may be unequal, as may the radii of curvature R_1 and R_2.

Now consider a section through the surface along the 1-direction, as shown in Fig. 10.3(b), which is a somewhat exaggerated view illustrating R_1 and the angle $2\theta_1$ subtended by arc length ds_1. At each boundary, perpendicular to the plane of the diagram, the 1-component of the tension exerts

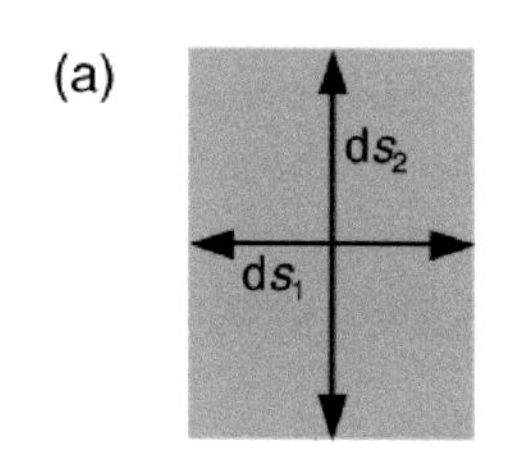

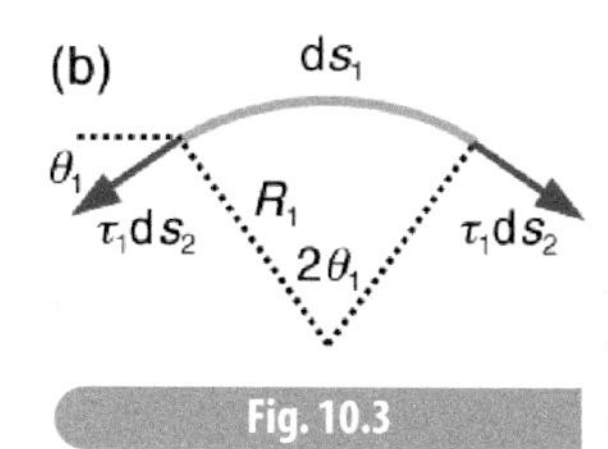

Fig. 10.3

(a) Face view of a small element of a curved shell with area $dA = ds_1 \bullet ds_2$ and principal curvatures $1/R_1$ and $1/R_2$. (b) Cross section through the shell, showing the tension at two edges directed into the plane.

a force $\tau_1 \bullet ds_2$ in a direction along the tangent to the surface at the edge. The vertical component of the force at each edge is $\tau_1 \bullet ds_2 \sin\theta_1$, which we may approximate by $\tau_1 \bullet ds_2 \bullet \theta_1$ at small angles. Thus, the total vertical force from τ_1 is $2\tau_1 \bullet ds_2 \bullet \theta_1$. The angle θ_1 is related to the arc length ds_1 by the radius of curvature, $2\theta_1 = ds_1/R_1$, permitting the vertical component of the force to be written as $\tau_1 \bullet ds_1 \bullet ds_2/R_1$. This argument can be repeated for a section in the 2-direction, yielding a second contribution to the vertical force of $\tau_2 \bullet ds_1 \bullet ds_2/R_2$. The sum of these forces is balanced by the force from the pressure difference across the surface, $P \bullet ds_1 \bullet ds_2$. Equating the forces and removing the common factor $ds_1 \bullet ds_2$ yields

$$\tau_1/R_1 + \tau_2/R_2 = P, \qquad (10.4)$$

one of several results obtained for thin shells in Chapter 2 of Flügge (1973). Equation (10.4) reduces to Eq. (10.3) upon substituting $R_1 = R\theta = r$ and $R_2 = R_z = \infty$. For fluid sheets, the tension is isotropic, and Eq. (10.4) becomes the Young–Laplace equation

$$\tau\,(1/R_1 + 1/R_2) = P, \qquad \text{(isotropic tension)} \qquad (10.5)$$

a result obtained by Laplace in his study of soap films. Note that bending energy considerations, which we have completely omitted, changes the form of these results (see Flügge, 1973).

Equations (10.3)–(10.5) relate the in-plane stretching forces on a thin sheet to a three-dimensional stress like the pressure. These relations augment the elastic properties of networks and membranes obtained in Chapters 5 and 7, where we introduced the two-dimensional shear modulus μ, the area compression modulus K_A, and the edge tension λ for membranes with free edges. There, we also expressed the bending energy of a membrane in terms of its local curvature and elastic parameters κ_b and κ_G, the bending rigidity and Gaussian bending rigidity, respectively. In other words, we can now write the energy of a thin shell using a two-dimensional description, without reference to its thickness. By minimizing this energy, the zero-temperature shape of the shell can be found, in principle, as a function of its elastic parameters and other physical characteristics such as its surface to volume ratio. In practice, however, finding energy minima is not always an easy numerical task, and the biological systems treated thus far usually have axial symmetry and shells without shear resistance (i.e. fluid membranes), as demonstrated in Section 10.3. In some applications, it is not overly important to work with the true ground state, and these situations can be treated by numerical simulation; in this vein, Section 10.4 contains examples of simple cells with surface shear resistance and arbitrary shape.

Before making recourse to computer simulations, let us push our analytical formalism a little harder and explore systems restricted to axially symmetric shapes, where the boundary is subject only to bending resistance.

Physical examples of such systems include pure lipid bilayer vesicles, which can be manufactured in the laboratory. The simplest model for the energy density $\mathcal{F}$ of a membrane depends quadratically on its local principal curvatures C_1 and C_2 (see Section 8.2) as

$$\mathcal{F} = (\kappa_b/2) \cdot (C_1 + C_2 - C_o)^2 + \kappa_G C_1 C_2, \tag{10.6}$$

where C_o is a parameter representing the spontaneous curvature that a bilayer may possess arising from compositional inhomogeneities in its two leaflets (a sphere with $C_1 + C_2 - C_o = 0$ has a radius of $2/C_o$). For a spherical shell, the enclosed volume V is related to the surface area A by $V = A^{3/2}/(6\sqrt{\pi})$, and the energy of the shell is uniquely

$$E = 4\pi(2\kappa_b + \kappa_G) \qquad \text{(spheres)}, \tag{10.7}$$

as found by integrating Eq. (10.6) over the surface. However, suppose now that the volume is less than $A^{3/2}/(6\sqrt{\pi})$, regarding A as fixed, such that the shell looks like a pancake, a cigar, or something else. How do we determine which shape has the minimum energy?

The energy of a closed surface involves an integral of the energy density $\mathcal{F}$, say the model energy in Eq. (10.6), over the area of the shell. Let's consider an axially symmetric shell, such as that displayed in Fig. 10.4. The shape of the shell can be described by a function $x = f(\ldots)$, where x is the distance from the symmetry axis to the surface, and the argument of f may be just the z-coordinate. In this shell, the area element dA of a ring of radius x and width ds can be easily written as

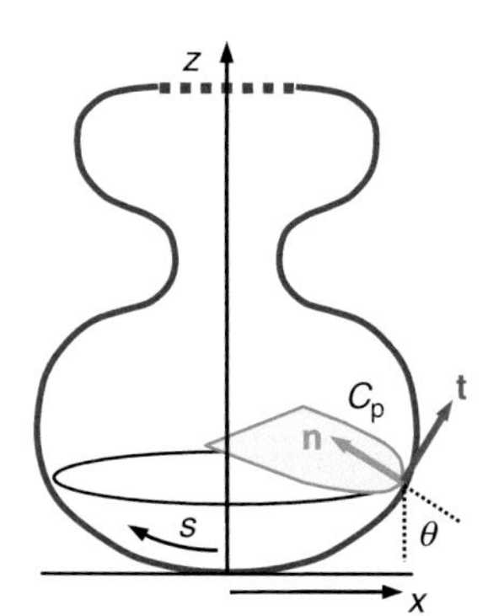

Fig. 10.4

Shape of a shell with axial or cylindrical symmetry. A position on the surface has normal and tangent vectors **n** and **t**, respectively, and principal curvatures C_m and C_p; C_m is defined by a plane containing the symmetry axis (such as the plane of the diagram) and C_p is defined by the shaded plane perpendicular to the plane of the diagram. The normal vector **n** makes an angle θ with respect to the z-axis; s denotes arc length.

$$dA = 2\pi x \, ds, \tag{10.8}$$

where s is the arc length. The tangent $\mathbf{t}$ and normal $\mathbf{n}$ to the surface can be obtained from derivatives of the position vector $\mathbf{r}(x, z)$ as described in Chapter 3 for polymers; we define θ as the angle between the interior normal and the z-axis. The curvature C_m defined by a plane containing the symmetry axis in Fig. 10.4 can be found from $\partial\mathbf{t}/\partial s = C_m\mathbf{n}$ along the boundary, where the subscript "m" indicates a meridian. In the diagram, $\mathbf{n} = (-\sin\theta, \cos\theta)$ and $\mathbf{t} = (\cos\theta, \sin\theta)$, so that $\partial\mathbf{t}/\partial s = (-\sin\theta, \cos\theta)\, d\theta/ds$. Taking the scalar product $\mathbf{n} \cdot (\partial\mathbf{t}/\partial s)$ yields the curvature

$$C_m = d\theta/ds. \tag{10.9}$$

The second principal curvature C_p is defined by a plane perpendicular to the plane of the diagram and contains the surface normal $\mathbf{n}$. Only if $\mathbf{n}$ were perpendicular to the z-axis would C_p be equal to $1/x$; for a general orientation of $\mathbf{n}$,

$$C_p = \sin\theta/x, \tag{10.10}$$

which is obtained by noting that surface normals on a parallel of latitude intersect at the symmetry axis a distance $x/\sin\theta$ from the surface. Parallels

Table 10.1 Bending energy, area and volume of five axisymmetric shells. The surface area of the arc and volume of the shaded region are integrated around the (vertical) symmetry axis. The ratio R /r is defined as ρ.

	Shell	Energy	Area	Volume
I	2r 2R	$\pi\kappa_b(8 + \pi\rho) + 4\pi\kappa_G$ $\rho >> 1$	$2\pi r^2(\pi\rho + 2)$	$2\pi r^3(\pi\rho/2 + 2/3)$
II	2r	$4\pi\kappa_b + 2\pi\kappa_G$	$2\pi r^2$	$2\pi r^3/3$
III	2r 2R	$\pi\kappa_b(-8 + \pi\rho) - 4\pi\kappa_G$ $\rho >> 1$	$2\pi r^2(\pi\rho - 2)$	$2\pi r^3(\pi\rho/2 - 2/3)$
IV	2r 2R	$\pi^2\kappa_b\rho^2 / (\rho^2 - 1)^{1/2}$	$2\pi^2 rR$	$\pi^2 r^2 R$
V	2r	$\pi\kappa_b L/r$	$2\pi rL$	$\pi r^2 L$

of latitude meet meridians at right angles, and hence the subscript "p" in C_p. Curvatures are positive when the center of curvature is in the same direction as the interior normal.

Combining Eqs. (10.8)–(10.10), the energy $E[S]$ of a particular surface S can be written as

$$E[S] = \pi\kappa_b \int x\,(\sin\theta/x + d\theta/ds - C_o)^2\, ds \;+\; 2\pi\kappa_G \int (d\theta/ds)\sin\theta\, ds, \tag{10.11}$$

and can be generalized to include shapes with holes (Boal and Rao, 1992b). Equation (10.11) is a model-dependent expression for the energy of a closed membrane without constraints on the shape other than axial symmetry. As applied to a biological cell, however, constraints such as fixed surface area and volume of the cell must be imposed on the search for energy-minimizing shapes. The search might include conditions such as the smoothness of the shape at the poles, which forces the surface to meet the symmetry axis at right angles (i.e., $\theta = 0$ at the poles). Applications of this approach to cell shape will be reported in Sections 10.3 and 10.4; however, because the solutions are ultimately numerical, we will not derive the intermediate algebraic equations here.

It is not difficult to evaluate Eq. (10.11) for the simplest shapes; as examples, several shells treated in the problem set are displayed in

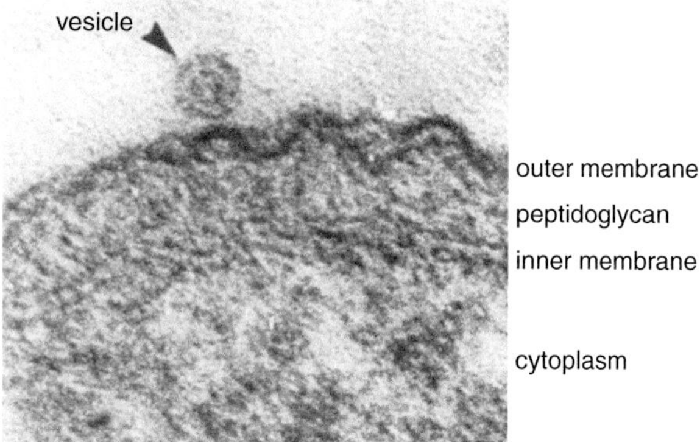

Fig. 10.5 Thin section showing a 50-nm diameter vesicle which has come off the outer membrane of the Gram-positive bacterium *Pseudomonas aeruginosa*. The dark areas in the cytoplasm are ribosomes (courtesy of Dr. Terry Beveridge, University of Guelph).

Table 10.1. The table includes the bending energy at $C_o = 0$, as well as the surface area and volume of the cross-hatched regions, both quantities integrated around the rotational axis. These results have familiar limiting situations: for example, the $\rho \to 0$ limit of shape I is a sphere, whose energy, volume and area are twice the hemisphere of shape II (the unitless parameter ρ characterizes the shape and is defined in the table). Similarly, shapes I and III can be joined to give a torus, a section of which is shape IV. We use these results in Section 10.3 to find the bending energies of a variety of simple cell shapes.

10.3 Pure bilayer systems

Mechanically simple cells, which may have a two-dimensional scaffolding attached to their plasma membrane, often adopt smooth symmetrical shapes such as the familiar biconcave disc of the red blood cell, or the spherocylinder of some bacteria. What features of cell shape arise from the properties of the membrane itself, without the presence of a cytoskeleton or cell wall? Controlled experimental studies on cell shape can be performed in the laboratory using bilayer systems called liposomes or artificial vesicles, which are structurally similar to the biological vesicles shown in Fig. 10.5, although liposomes may have diameters two orders of magnitude larger than a vesicle. In cells, vesicles are small cargo-carrying structures that may be absorbed at a membrane or pinch off from it, as described later in this section.

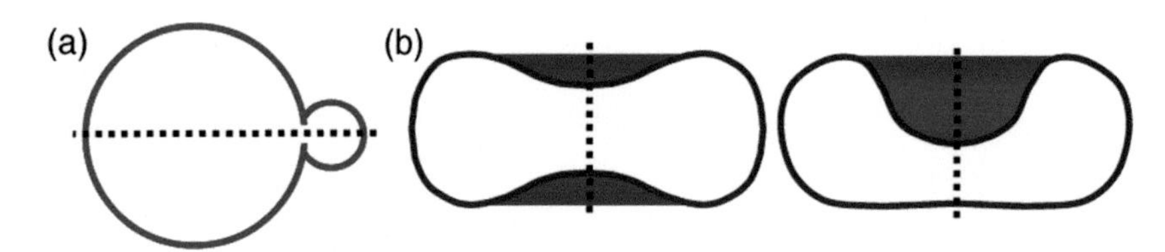

Fig. 10.6 Schematic cross sections of several observed vesicle shapes. For each shape, the axis of rotational symmetry is indicated by the dashed line.

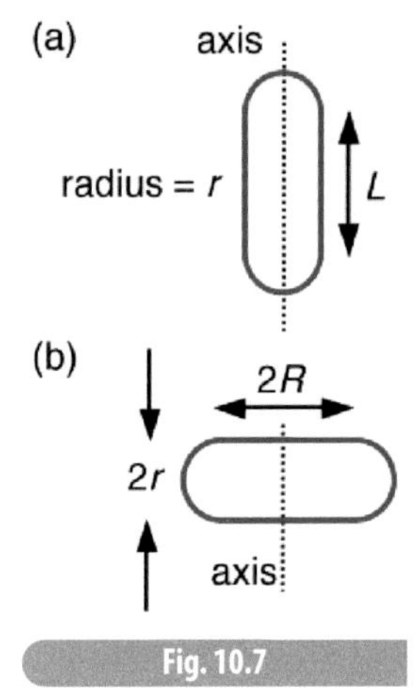

Fig. 10.7 The spherocylinder (a) and pancake (b) are axially symmetric shapes whose bending energy is easy to evaluate in the spontaneous curvature model.

A typical laboratory study of bilayer vesicles records their shape as the surface area is changed while holding the volume roughly constant (by adjusting the ambient temperature, for instance). Examples of axially symmetric vesicle shapes that may be observed experimentally are displayed in Fig. 10.6, where the rotational axis is indicated by the dashed vertical line. On the left, a bud is appended to a nearly spherical shell whose enclosed volume is not far below the maximum allowed by the vesicle surface area. The other two sample configurations have volumes well below the maximum value; these shapes are not ellipsoidal, where the curvature would have the same sign everywhere on the surface, but rather display both positive and negative curvature along meridians running from pole to pole. Note that not all of the configurations are "up–down" symmetric. In this section, we examine the origin of these phenomena in terms of the bending properties of fluid membranes.

We work within the spontaneous curvature model for bending, as represented by Eq. (10.11). The model has as its parameters the rigidities κ_b and κ_G, as well as the spontaneous curvature parameter C_o. We return to the physical origin of C_o later; for now, we treat it as a parameter and explore its effect on cell shape. With vanishing C_o, the bending energy of a spherical shell is $E_{sphere} = 4\pi(2\kappa_b + \kappa_G)$. How does this deformation energy change as the spherical shell is distorted, with a resulting loss of volume at fixed surface area? Which generic shape – prolate (like a rod) or oblate (like a disk) – has the lower energy for a given volume and surface area? To address this question, we use two configurations, the spherocylinder and the pancake, to represent prolate and oblate shapes. Referring to Fig. 10.7, these shapes are described by length parameters r, R, and L, and have energies

$$E_{spherocylinder} = 8\pi\kappa_b + \pi\kappa_b(L/r) + 4\pi\kappa_G \qquad (10.12a)$$

$$E_{pancake} = \pi\kappa_b(8 + \pi R/r) + 4\pi\kappa_G \qquad (R/r >> 1), \qquad (10.12b)$$

when $C_o = 0$, as can be constructed from Table 10.1. Equation (10.12b) is an approximation valid for $R/r >> 1$; the exact result is more cumbersome.

Displayed in Fig. 10.8 are the energies of the two representative shapes compared with E_{sphere}, as in $\Delta E_{pancake} = E_{pancake} - E_{sphere}$, where the exact result

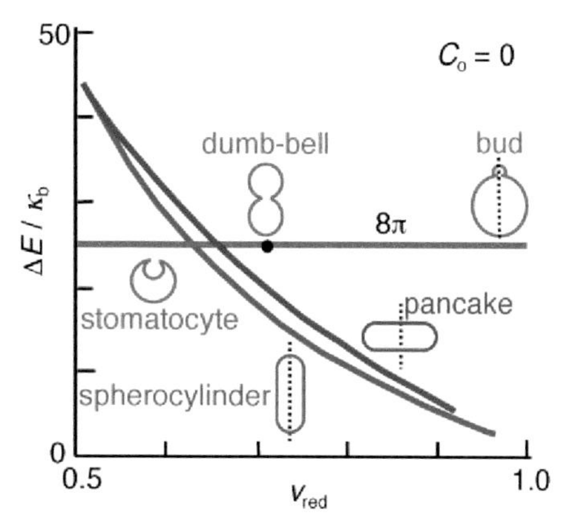

Fig. 10.8 Difference in bending energy ΔE relative to a sphere for three axisymmetric shapes – two touching spheres, a spherocylinder and a pancake, all with $C_o = 0$. The energies are displayed as a function of reduced volume v_{red} and normalized by κ_b. Dashed lines indicate axes of rotational symmetry.

is used for the pancake energy, not Eq. (10.12b). The energy difference has been divided by κ_b to produce a unitless observable. To quantify the volume, we introduce the unitless combination v_{red} given by

$$v_{red} = 6\sqrt{\pi}\, V / A^{3/2}, \tag{10.13}$$

which equals unity for a sphere. When $C_o = 0$, Eqs. (10.12) do not depend on the absolute size of the shell, only on the ratios L/r and R/r; hence, ΔE is unique for each v_{red} when $C_o = 0$. The figure shows that the spherocylinder is energetically favored over the pancake in the volume range $0.6 < v_{red} < 1$, with an energy difference as high as $3\kappa_b$ at $v_{red} \sim 0.7$. However, the energies converge and then reverse as v_{red} passes below 0.5, with the pancake becoming the favored shape for $v_{red} < 0.5$.

Also displayed on the figure is the energy of two touching spheres joined by a narrow neck with $E_{doublet} \sim 16\pi\kappa_b + 4\pi\kappa_G$, an approximation valid at $C_o = 0$ for two shells joined externally as a bud or internally as a pocket, illustrated by the two shapes in Fig. 10.9(a). Fourcade *et al.* (1994) have shown that the bending energy of the neck region may not be important because the principal curvatures have opposite sign (resulting in a small mean curvature). In fact, shape III of Table 10.1 could be regarded as a section of a neck whose bending energy can be made small by a suitable choice of curvatures. With $\Delta E_{doublet} / \kappa_b \sim 8\pi$, the energy of these configurations is independent of v_{red}; the reduced volume of the stomatocyte shape can range from 0 to 1, but the doublets obey $1/\sqrt{2} \leq v_{red} < 1$. Higher multiplet states with n externally connected spheres can attain lower values of v_{red} (down to $1/\sqrt{n}$) at the cost of adding another $8\pi\kappa_b$ to the bending energy for each new sphere, at vanishing C_o. Shapes with small buds are clearly not favored in Fig. 10.8, and hence processes

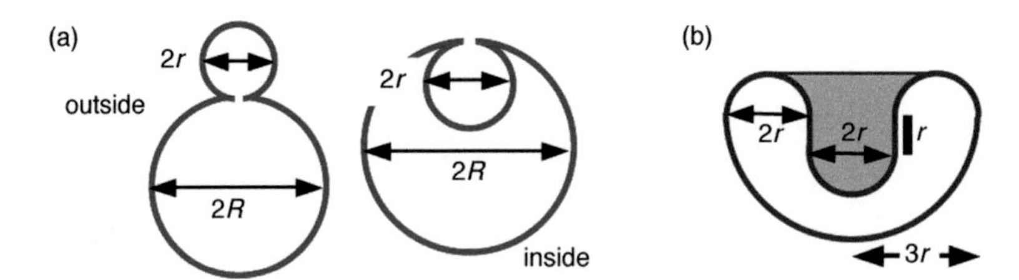

Fig. 10.9 Axially symmetric shapes whose bending energies are easy to calculate. (a) Doublet and extreme stomatocyte. (b) Stomatocyte with simple curvatures.

like the formation of a vesicle in Fig. 10.5 must involve more than just membranes with $C_o = 0$.

Figure 10.8 predicts that, of the shapes considered, the prolate spherocylinder is favored for $v_{red} > 0.63$ and the extreme stomatocyte (internal cavity) is favored for $v_{red} < 0.63$. However, these shapes may not be true energy minima, as they have been selected only for numerical convenience. Softening the curves may raise or lower the energy of the configurations, and affect which shape has the lowest energy. The gentle stomatocyte in Fig. 10.9(b), for instance, has a bending energy of $\pi\kappa_b(9 + 4\pi/\sqrt{3}) + 4\pi\kappa_G$ at $C_o = 0$, as established in the problem set. The energy difference between this shape and E_{sphere} is $\pi\kappa_b(1 + 4\pi/\sqrt{3}) = 8.255\ \pi\kappa_b$, barely more than the doublet energy, and has a reduced volume $v_{red} = 0.67$, close to the transition region in Fig. 10.8. Numerical searches have been performed within the spontaneous curvature model, and others, for the true minimal energy states, assuming axial symmetry and continuity of the curvature at the poles of the configuration. A selection of shapes from one such search at $C_o = 0$ are displayed in Fig. 10.10 as a function of v_{red} (Seifert *et al.*, 1991; specific configurations also have been considered by Canham, 1970, and Deuling and Helfrich, 1976). Although similar to our approximate boundary between stomatocytes and prolates at $v_{red} = 0.63$, the numerical results yield a narrow range of reduced volume around $0.59 \leq v_{red} \leq 0.65$ where the oblate shapes have the lowest energy. Further, these shapes are not ellipsoidal but rather biconcave, like the red blood cell.

What effect does a non-zero value for C_o have on the configuration energy? First, it introduces a length scale C_o^{-1}. For example, the bending energy of a spherical shell with radius r is $8\pi\kappa_b(1 - rC_o/2)^2 + 4\pi\kappa_G$ at $C_o \neq 0$. The first term vanishes at $r = 2/C_o$, with the result that the spherical shell with the lowest deformation energy has a particular size at $C_o \neq 0$. This conclusion is true for arbitrary shapes, meaning that the bending energy is a function of cell shape *and* size at $C_o \neq 0$. In addition, the sign of C_o influences the favored shape, as we can see by considering the two axisymmetric shells in Fig. 10.9(a). The doublet shape on the left has curvatures $1/r$ in the small bud and $1/R$ in the main body, whereas the extreme stomatocyte to

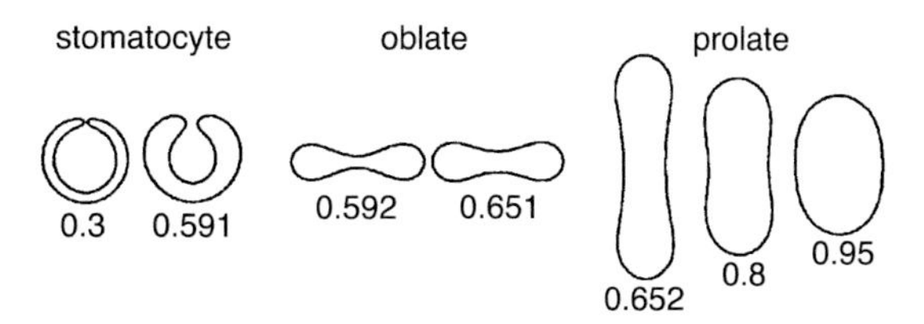

Fig. 10.10 Energy-minimizing shapes in the spontaneous curvature model with $C_o = 0$, shown for selected values of the reduced volume v_{red}. Note how the shape changes dramatically at $v_{red} = 0.59$ and 0.65. The axis of symmetry for all shapes is vertical (reprinted with permission from Seifert *et al.*, 1991; ©1991 by the American Physical Society).

its right has $-1/r$ in the cavity and $1/R$ on the exterior surface. As a result, the bending energies are simply

$$E_{\text{outside}} = 8\pi\kappa_b[(1 - rC_o/2)^2 + (1 - RC_o/2)^2] + 4\pi\kappa_G \tag{10.14a}$$

$$E_{\text{inside}} = 8\pi\kappa_b[(1 + rC_o/2)^2 + (1 - RC_o/2)^2] + 4\pi\kappa_G, \tag{10.14b}$$

where the labels refer to the position of the smaller shell and where the energy of the neck has been neglected. Let's take RC_o and rC_o to be the same for both configurations, meaning that the shells have the same areas but different enclosed volumes. The difference in their energies is then

$$E_{\text{outside}} - E_{\text{inside}} = -16\pi\kappa_b(r/R)RC_o, \tag{10.15}$$

indicating that the doublet configuration is favored ($E_{\text{outside}} < E_{\text{inside}}$) if $C_o > 0$ and the stomatocyte shape is favored if $C_o < 0$. In other words, negative values of C_o favor shapes with regions of negative curvature.

What determines the magnitude and sign of C_o? We recall from Section 7.2 how the molecular shape parameter was related to the stability of structures like micelles and bilayers: lipids with head-groups large in cross section compared to their hydrocarbon regions tend to form micelles which match the natural curvature of the lipids' molecular shape. A lipid monolayer, then, may have a spontaneous curvature arising from its molecular composition; measurements of this curvature can be found in Marsh (1996) or Chen and Rand (1997), for example. Similarly, spontaneous curvature can arise in lipid bilayers, either because the composition of each leaflet or its environment is inequivalent (see Döbereiner *et al.*, 1999, and references therein), or because of molecular packing arrangements within leaflets (Safran *et al.*, 1990). The "phase diagram" for vesicle shapes at non-zero C_o has been partially explored (Seifert *et al.*, 1991; Miao *et al.*, 1991) and is displayed in Fig. 10.11. The shapes along the $C_o = 0$ line correspond to Fig. 10.10, with the narrow domain of oblates visible at $v_{red} \sim 0.6$. Having regions of negative curvature, stomatocytes are more favored at negative C_o; in contrast, prolate ellipsoids and pears (doublets

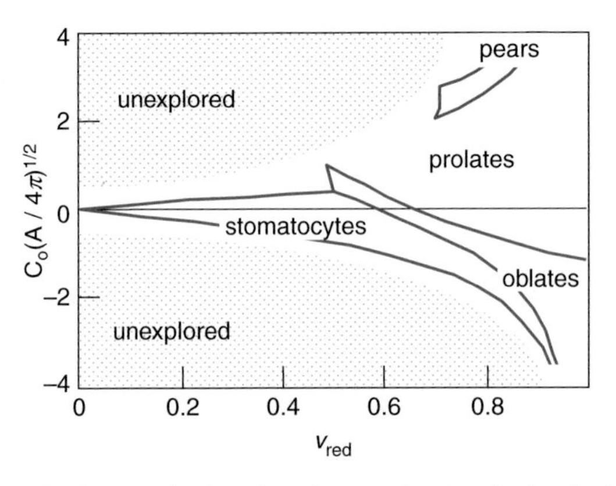

Fig. 10.11 Partial phase diagram for shapes with spherical topology as a function of reduced volume v_{red} and spontaneous curvature C_o, written as the unitless combination $C_o(A/4\pi)^{1/2}$, where A is the surface area (redrawn with permission from Seifert *et al.*, 1991; see original paper for exact position of phase boundaries; ©1991 by the American Physical Society).

with smooth necks) are favored at large positive C_o. If the composition of a leaflet is multicomponent, there is a possibility that C_o is not uniform over the shell; one can imagine that molecules with conical shapes might migrate to regions of high curvature, such as the folds in Fig. 1.11(b). With C_o being a function of position on the membrane, the energy minimization problem becomes more involved (Seifert, 1993; Kumar *et al.*, 1999).

The spontaneous curvature model recognizes that the inner and outer leaflets of a bilayer may be compositionally inequivalent, or may be embedded in differing environments. However, it does not recognize that the two leaflets may be mechanically decoupled, in the sense that they can slide past one another to relieve stress that would arise if the membrane were a single continuous sheet like a metal plate. Specifically, consider the inner and outer leaflets of the small vesicle in Fig. 10.5. The lipid head-groups in the outer leaflet form a spherical shell whose diameter is roughly double that of the inner shell of lipid head-groups. If the leaflets of this vesicle were mechanically coupled, such that they contained the same number of lipids, then the outer leaflet would have only a quarter the number of lipids per unit area as the inner leaflet, clearly a high-stress situation. The fact that the leaflets are not coupled reduces this stress, as extra material from the outer leaflet of the parent cell can be pulled onto the daughter vesicle largely independent of the make-up of the inner leaflet.

The difference in area of the leaflets ΔA can be written in a two-dimensional representation, so that it is easily incorporated into the spontaneous curvature model. Consider first the two arcs of a circle displayed in Fig. 10.12 (in bold) separated by a distance d_{bl} and having a common center of curvature; the curvature of the mid-line between the arcs is defined as

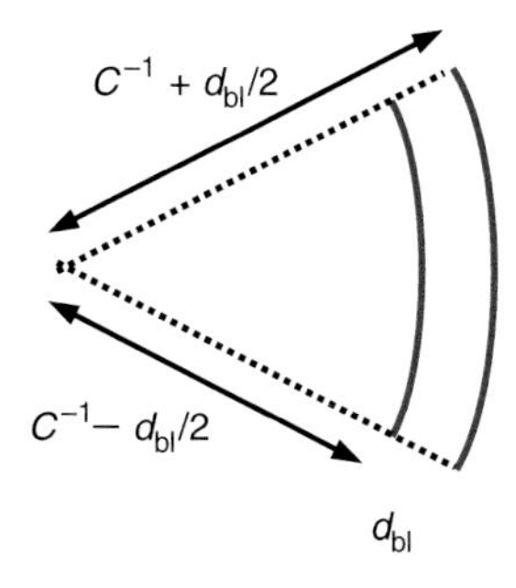

Fig. 10.12

Two arcs with a common center of curvature. The ratio of the outer to inner arc lengths, indicated by the bold lines, is equal to $(1 + Cd_{dl}/2)/(1 - Cd_{dl}/2)$, where C is the curvature of the mid-line between the arcs.

C. By geometry, the ratio of the outer arc length to the inner one is just $(1 + Cd_{bl}/2)/(1 - Cd_{bl}/2) \approx 1 + Cd_{bl}$. The same result holds in an orthogonal direction, so that the ratio of the corresponding outer and inner areas is just $A_{outer}/A_{inner} = (1 + C_1d_{bl})\bullet(1 + C_2d_{bl}) \approx 1 + C_1d_{bl} + C_2d_{bl}$, where C_1 and C_2 are the principal curvatures. Thus, the difference between the areas of the outer and inner boundaries of a membrane segment (of area dA) is approximately $(C_1d_{bl} + C_2d_{bl})dA$, and the total area difference, integrated over the surface, is

$$\Delta A \approx d_{bl} \int (C_1 + C_2)dA. \quad (10.16)$$

As expected, ΔA vanishes when $d_{bl} = 0$ or the integrated mean curvature vanishes. To be clear, even though it is obtained by summing over local differences in leaflet area, ΔA is a global quantity characteristic of the surface as a whole.

Now, ΔA for the vesicle shown in Fig. 10.5 is about half of the mean area of the inner and outer leaflets; in other words, ΔA need not vanish for an unstressed system. Thus, the energy associated with the leaflet area difference is proportional to the (square of the) *deviation* of ΔA from its unstressed value ΔA_o. The bending energy of the spontaneous curvature and area difference contributions can be parametrized as

$$E = (\kappa_b/2) \int (C_1 + C_2 - C_o)^2 \, dA + \kappa_G \int C_1C_2 \, dA$$
$$+ (\kappa_{nl}/2)\bullet(\pi / Ad_{dl}^2)\bullet(\Delta A - \Delta A_o)^2, \quad (10.17)$$

which is referred to as the ADE model (for *area difference elasticity*). The constant κ_{nl} is a non-local bending resistance, carrying units of energy, and can be related to the area compression modulus of the leaflets (Svetina *et al.*, 1985; Miao *et al.*, 1994). Two limiting cases of Eq. (10.17) include the spontaneous curvature model at $\kappa_{nl} = 0$ and the bilayer couple (or ΔA) model at $\kappa_{nl} / k_BT \rightarrow \infty$, where ΔA is driven to the fixed value ΔA_o. This latter approach was developed more than two decades ago and underlies the calculations appearing in Fig. 10.14 (Sheetz and Singer, 1974; Svetina and Zeks, 1985; see also Evans, 1974, and Helfrich, 1974a, b). One measurement of κ_{nl} (Waugh *et al.*, 1992) and estimates of κ_{nl} from the underlying bilayer deformation (Miao *et al.*, 1994) argue that κ_{nl}/κ_b is of order unity.

It can be shown that the set of stationary shapes in the ADE model (pears, ellipsoids, dumb-bells, etc.) is the same as the spontaneous curvature approach (see Miao *et al.*, 1994); however, the energy of a given configuration will vary according to the values of C_o, κ_{nl}/κ_b, etc. Thus, the phase diagram will be different from one model to the next. A section of the phase diagram for $\kappa_{nl}/\kappa_b = 4$ at $C_o = 0$ is displayed in Fig. 10.13, showing several shapes familiar from Fig. 10.10. The fixed area difference ΔA_o is plotted as the unitless quantity $\Delta A_o/[2d_{bl}(4\pi A)^{1/2}]$, which is equal to unity for a sphere. In bilayer language, adding excess lipid to the outer leaflet increases ΔA_o.

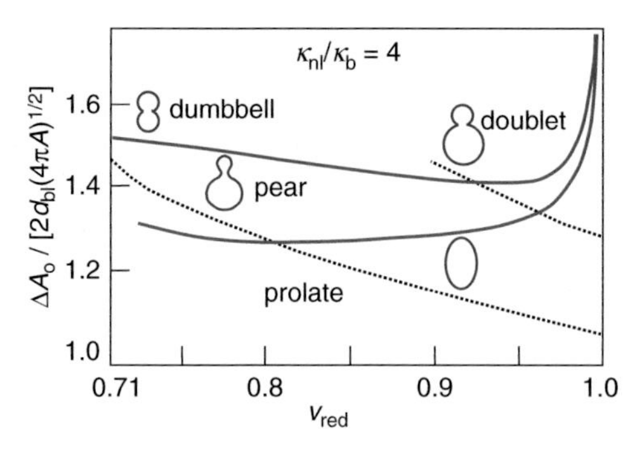

Fig. 10.13 Partial phase diagram in the ADE model for the ratio of bending parameters $\kappa_{nl}/\kappa_b = 4$, as a function of the unitless area difference and reduced volume v_{red}. At fixed v_{red}, the minimal energy shape goes from prolate to pear to multiplet as the area difference increases. The dashed lines are trajectories that a vesicle would follow as it is heated, causing v_{red} to rise (redrawn with permission from Miao *et al.*, 1994; ©1994 by the American Physical Society).

From the configurations in the diagram, one can see how the area of the outer leaflet increases with respect to the inner leaflet as the shape changes from prolate ellipsoids to pears to multiplets at fixed volume. The nature of the transition from one generic shape to another (e.g. ellipsoids to pears) also depends upon the model: the transition may be smooth (continuous) in some cases while abrupt (discontinuous) in others.

The spontaneous curvature model and its variants allow the prediction of energy-minimizing shapes for a given parameter set, as demonstrated in Figs. 10.8, 10.10, 10.11 and 10.13. In addition, one can analyze the shapes of specific cells, vesicles or liposomes and determine the relevant parameters describing the shape within a given model. Of course, not all cells of a given type have exactly the same shape: for instance, even the blood cells drawn from one person vary in size and shape, potentially requiring different values of C_o or ΔA for their description (see Bull, 1977, for examples). Does this mean that our models are nothing more than parametrizations without predictive power? Certainly not. Once the characteristics such as C_o and ΔA have been determined for a specific cell, other attributes such as the cell's reduced volume can be varied (by heating, for example) and the resulting change in shape can be both predicted and compared against experiment. As an illustration, Fig. 10.13 displays trajectories that specific cells should follow in configuration space under the ADE model (Miao *et al.*, 1994). Figure 10.14 contains a specific sequence from the ΔA model, plotted over the very narrow temperature range 43.8 °C to 44.1 °C; the predicted shapes reproduce the observed ones very well (Berndl *et al.*, 1990; see also Käs and Sackmann, 1991). The membrane area of a vesicle increases faster with temperature than does its enclosed volume, so that the reduced volume decreases with rising temperature. By following the

Fig. 10.14 Predicted shapes as a function of temperature within a curvature-based energy description of vesicle shape (reprinted with permission from Berndl *et al.*, 1990; ©1990 by EDP Sciences). The cross sections have axial symmetry and the reduced volume v_{red} increases from left to right.

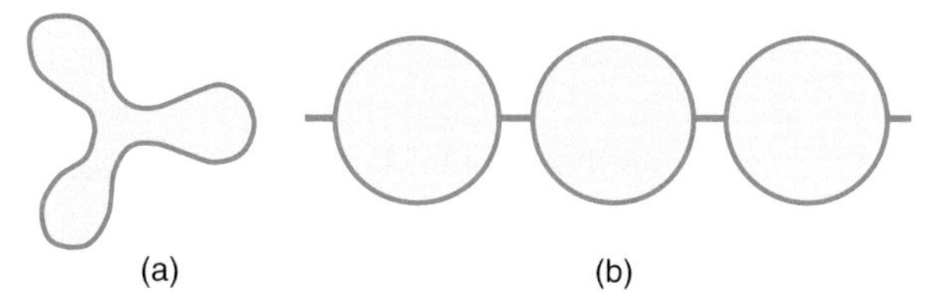

Fig. 10.15 (a) Stable starfish shape for $v_{red} = 0.5$ (redrawn from Wintz *et al.*, 1996). (b) Representation of pearls (spheres connected by narrow capillaries) appearing in tubular membranes under tension (after Bar-Ziv and Moses, 1994).

trajectories and determining the stability range of specific shapes and the nature of the changes between shapes, differences between models for the bending energy can be exposed; for instance, the spontaneous curvature model does not capture all of the physics of the shape transitions (for example, Döbereiner *et al.*, 1997, and Lim *et al.*, 2002).

At ever smaller values of reduced volume v_{red}, the menu of available shapes grows larger, and new phenomena appear. One of these is the loss of axial symmetry, as exemplified by the starfish shape in Fig. 10.15(a) (Wintz *et al.*, 1996; see also Bar-Ziv *et al.*, 1998). The configuration is a locally stable shape in a curvature-based model for the membrane energy, and it mirrors the observed shape very well. Within the spontaneous curvature model at vanishing C_o, this shape has an energy difference with respect to the sphere of $1.90 \bullet 8\pi\kappa_b$ and a reduced volume of $v_{red} = 0.5$, so that it lies close to the pancake and spherocylinder energies in Fig. 10.8 at the same v_{red}.

Vesicles with non-spherical topology, for instance doughnut shapes with one or more holes, also have been discovered (Mutz and Bensimon, 1991; Fourcade *et al.*, 1992; Michalet *et al.*, 1994). As determined in the problem set, an axially symmetric doughnut with a single hole (a torus) may have an energy as low as $4\pi^2\kappa_b$ in the spontaneous curvature model with $C_o = 0$, there being no Gaussian rigidity contribution to this topology. This is for a torus with a specific shape, called a Clifford torus, having a reduced volume $v_{red} = 0.71$; its bending energy is $(\pi/2)E_{sphere} = 1.57E_{sphere}$ when $\kappa_G = 0$. Referring to Fig. 10.8, the energy of the Clifford torus is comparable to, or lower than, the energies of the spherical topologies in the region $v_{red} \sim 0.7$, demonstrating that shapes with more exotic topologies than spheres are energetically accessible at lower values of v_{red}. Within the spontaneous curvature model, energy-minimizing shapes of toroidal vesicles have been investigated by Banavar *et al.* (1991) and Seifert (1991), including non-axisymmetric shapes; even more exotic topologies have been examined by Jülicher (1996).

Tensile stress also affects membrane behavior. Applied to tubular configurations like straws, where the membrane forms a fluid boundary, tension may induce "pearling", in which the fluid contents of a tube becomes segregated into regularly spaced spheres linked by narrow capillaries, as shown in Fig. 10.15(b) (Bar-Ziv and Moses, 1994). Although instabilities against undulations for tubes are expected from energy-minimizing shapes within the spontaneous curvature model (Bar-Ziv and Moses, 1994), the dynamic flow of material also contributes to the formation of pearls (Nelson *et al.*, 1995; Goldstein *et al.*, 1996).

To summarize, we have adopted the spontaneous curvature model as a simple representation for the bending energy of a fluid membrane and shown how energetics affect the equilibrium shape of vesicle-like systems. The model allows us to understand analytically many qualitative features of artificial vesicles manipulated under controlled conditions. However, the model does not capture all the features of shape changes observed experimentally, necessitating the inclusion of effects arising from, for example, the deviation from equilibrium of the mean area per lipid in each leaflet. In the next section, we investigate simple composite systems consisting of fluid sheets and shear-resistant networks such as the cytoskeleton or cell wall.

10.4 Vesicles and red blood cells

The membrane shapes introduced in Section 10.3 are among those adopted by cells and their compartments: red blood cell shapes include biconcave discs while some bacteria resemble spherocylinders. However, biological systems have more structural components than just a membrane, forcing us to ask which aspects of cell shape are determined by membrane bending energy alone, and which arise from other mechanical elements. In this section, we discuss the architecture of two simple systems – small structureless vesicles and mammalian red blood cells with a membrane-associated cytoskeleton. Only the shapes of cells in a quiescent fluid environment are considered here; for the effects of fluid flow on cell shape, see Kraus *et al.* (1996), Bruinsma (1996), Noguchi and Gompper (2005, 2007), Kantsler and Steinberg (2006), and Lebedev *et al.* (2007).

10.4.1 Vesicles

The biochemical factories of eukaryotic cells tend to be concentrated in localized industrial districts such as the extensive protein production facilities found on the endoplasmic reticulum. Transporting biochemical products from their point of manufacture to their usage sites involves several familiar steps:

(i) packaging the products to prevent loss during transport,
(ii) labelling the package so it can be recognized at its destination,
(iii) shipping the package along an efficient transportation route.

Here, we describe step (i), the packaging process itself. Item (ii) is beyond the scope of this book; however, we cover the third topic in Chapter 11 under molecular motors.

The transport containers of the cell are small membrane-bound vesicles, typically 100 nm in diameter. When produced, a vesicle pinches off from an existing membrane by the process of endocytosis, capturing material from the fluid environment on the opposite side of the parent membrane. Upon arrival at its destination, the vesicle may fuse with another membrane via exocytosis, releasing its contents to the medium on the opposite side; budding and fusion of vesicles are illustrated in Fig. 10.16. Both the vesicle's fluid interior and its membrane provide locations for carrying cargo to a new location. Looked upon solely as a bilayer, the bending energy of a spherical vesicle in the ADE model of Eq. (10.17) is

$$E = 8\pi\kappa_b + 4\pi\kappa_G + (\kappa_{nl}/2)\bullet(\pi / Ad_{bl}^2)\bullet(\Delta A - \Delta A_o)^2, \qquad (10.18)$$

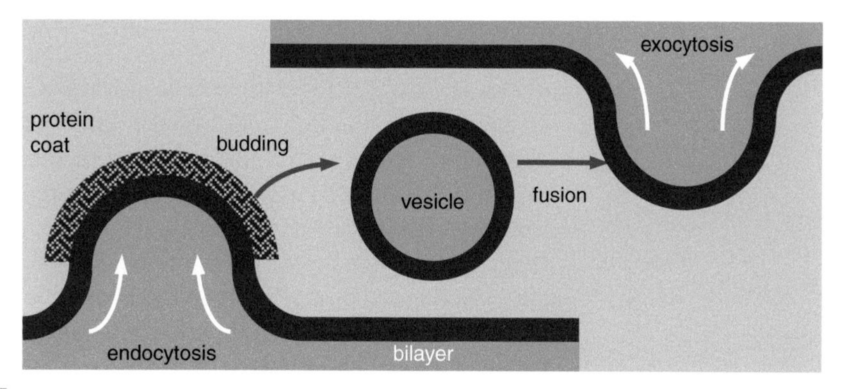

Fig. 10.16 Endocytosis and exocytosis involve the creation of a vesicle and its subsequent fusion after transport within a cell. Often, the budding process is aided by the attachment of coating proteins (e.g. clathrin or coatomer) to the parent membrane; clathrin, but not coatomer, is released once the vesicle pinches off.

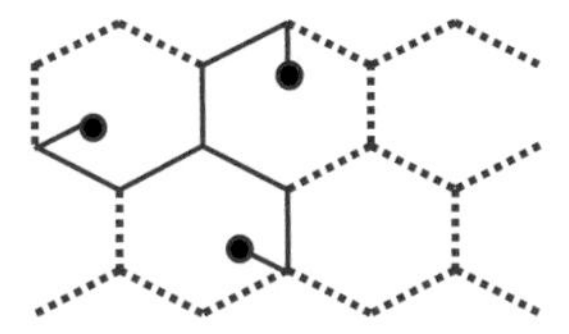

Fig. 10.17

The protein clathrin has a three-legged shape called a triskelion; six segments of these arms lie in a plane and are available to bond with neighboring molecules, while the three terminal segments lie below the plane of the drawing.

when the spontaneous curvature C_o vanishes. Depending on the flow of the inner and outer leaflets of the membrane during the budding process, the bending energy associated with individual leaflets may be augmented by the non-local contribution of the area difference $\Delta A - \Delta A_o$.

In the laboratory, the formation of buds has been studied for pure bilayer vesicles, where the bending energy is unquestionably important. When the reduced volume v_{red} is less than unity, budding appears in the phase diagram of Fig. 10.11 at sufficiently positive spontaneous curvature, and in Fig. 10.13 when the excess area in the outer leaflet is large enough (Miao *et al.*, 1991, 1994; Seifert *et al.*, 1991; Döbereiner *et al.*, 1997). In cells, the formation of vesicles is mediated by proteins such as clathrin, coatomer and others (see Chapter 13 of Alberts *et al.*, 2008). A three-armed protein with the appearance shown in Fig. 10.17 (Ungewickell and Branton, 1981), clathrin self-assembles to form regular two-dimensional arrays of hexagons about 25 nm across, with a small fraction of pentagons. There is one clathrin molecule per vertex in the hexagonal array (Crowther *et al.*, 1976); at six segments per molecule and 3/2 lattice "bonds" per lattice vertex, there must be four segments lying along each bond. Indirectly linked to the membrane by the protein adaptin, the curved clathrin coat deforms the parent membrane to produce buds of relatively uniform curvature and size (Heuser, 1980). The coat does more than induce curvature, however; adaptin is not attached to membrane lipids but rather to a transmembrane receptor protein capable of holding a cargo molecule on the opposite side of the bilayer. Thus, the coat both induces budding and also helps organize the contents of the vesicle. Once the vesicle is free of the parent membrane, the clathrin coat is released by hydrolysis of ATP. In contrast, the formation of coatomer coats is driven by ATP hydrolysis, and the coat is not shed until the vesicle arrives at its destination. The role of lattice defects in generating curvature of the clathrin meshwork is described in Mashi and Bruinsma (1998).

The curvature of the clathrin coat provides a natural length scale for the size of a vesicle budding from a planar parent membrane. Budding from a curved membrane, such as the perimeter of a pancake in the Golgi stack (see Figs. 1.8 and 1.9) may also involve a preferred vesicle size, even if C_o vanishes. Although the bending energy of a vesicle is independent of its radius in the spontaneous curvature model with $C_o = 0$, the energy of the path taken by the bud depends upon its curvature. In Fig. 10.18, we show a particular path for a bud at the edge of a pancake. The surface consists of two planar membrane regions (the top one is indicated by light shading in part (a)) connected by half a cylinder of radius r and length $4R$ (indicated by the darker shading); a cross-sectional view of this curved edge can be seen in the lower half of part (b). We let the bud push out in the same plane as the pancake, so its top and bottom sections remain flat, as seen by the top view in (c) and side view in (b); its edge is characterized by curvatures

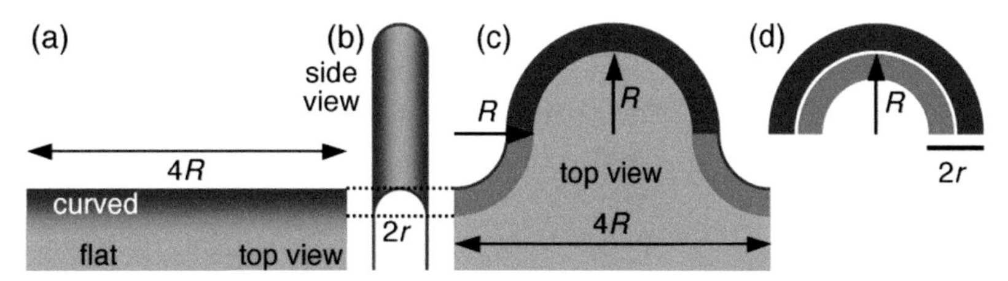

Fig. 10.18 (a) Top view of the edge of a pancake: the lightly shaded region is flat and the darker region is cylindrical with curvature r^{-1}, as illustrated by the lower part of the side view in (b). Both the flat (lightest) and curved regions (darker) are extended by the bud in (c), also shown in cross section (b). The light and dark curved regions in the bud can be assembled into half a torus with radii R and r as in (d).

$C_m = r^{-1}$ along a meridian and $C_p = R^{-1}$ in the plane, as indicated. The curved regions of the bud can be assembled into half a torus, with radii r and R, as seen in (d).

With their simple geometry, the bending energy of configurations (a) and (c) in Fig. 10.18 are easily calculated in the spontaneous curvature model with $C_o = 0$: as half a cylinder, initial configuration (a) has bending energy $2\pi\kappa_b\rho$, where $\rho = R/r$, while configuration (c) has half the energy of a torus, or $\pi^2\kappa_b\rho^2 / (\rho^2-1)^{1/2}$ (see problem set). The smallest difference in energy between these two configurations is about $10\kappa_b$ for bud radii near $R \sim 2r$. Although we have not followed the pathway to completion, nor have we considered arbitrary shapes, the bending energy suggests that buds and vesicles with radii on the order of the thickness of the pancake will be favored. In a cell, vesicles in this size range are observed to break off from edges of the Golgi stack; even here, however, coating proteins play a role in vesiculation, so bending energy may not be the primary factor in determining the size of these vesicles.

In the process of budding, a curved region forms on a membrane and eventually grows to encapsulate a small volume of the cytoplasm. A related process is vesicle fusion, where a small vesicle adheres to, and then fuses with, a large membrane surface, expelling its contents to the opposite side of the membrane. In cells, the processes are not strictly the inverse of each other, as the fusion process begins with the two membranes approaching and initially binding through ligand–receptor association. High-speed image capture (for example, Lei and MacDonald, 2003; Haluska *et al.*, 2006) reveals that the two bilayers of the membrane pair initially adhere, after which the two "outside" leaflets may join to form a transient hemifusion state. The central section of this hemifusion conformation has the thickness of two leaflets (namely the "inside" leaflets from each membrane) whilst outside of this small core domain, two complete bilayers adhere to one another. In short order, the central pair of leaflets also fuse and withdraw toward the perimeter of the fusion region, removing the barrier between the fluid contents of the cell/vesicle pair. This leads to completion of the fusion event. Membrane fusion has been studied by video microscopy for years and has established that the fusion time scale is very short, in the millisecond range. There have been many efforts to model the fusion sequence analytically or numerically; regretfully, these studies are too numerous to list here without doing injustice through omission to many articles in the field.

10.4.2 Red cell shapes

Their biological importance, their easy availability and their mechanical simplicity has made the mature human erythrocyte an attractive cell to study both experimentally and theoretically. In our circulatory system, a typical red cell has a surface area of ~135 μm^2 and a volume of ~95 μm^3 with 10%–20% variation (see Steck, 1989), corresponding to a reduced volume $v_{red} = 6\sqrt{\pi}\, V / A^{3/2} = 0.64$; note that red cells in other animals may be up to five times smaller or ten times larger in linear dimension than in humans. Referring to Figs. 10.8 and 10.10, this region of v_{red} is something of a cross roads in the phase diagram, with a variety of shapes having similar bilayer bending energy. It should be no surprise, then, that altering the environment of a blood cell can drive the cell away from its customary discocytic or biconcave shape. Catalogs of red cell shape exist (Bessis, 1973; see also Steck, 1989) and include the stomatocytes of Fig. 10.6(a) in addition to the more exotic echinocytes of Fig. 10.19(a), whose bumps may be gentle or rough, depending on conditions.

What factors contribute to this assembly of diverse shapes? First, the bending energy of the plasma membrane must be taken into account, recognizing that the molecular composition of its two leaflets is not symmetric – the inner leaflet tends to be richer in anionic phospholipids such as phosphatidylethanolamine (PE) and phosphatidylserine (PS), while the outer leaflet contains principally uncharged lipids such

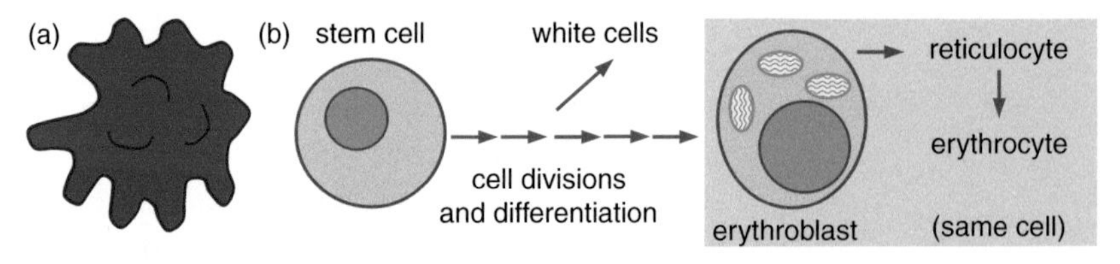

Fig. 10.19 (a) Among the shapes adopted by human erythrocytes at lower reduced volumes are lumpy echinocytes. (b) From common stem cells, blood cells differentiate by successive division; in the last step, an erythroblast loses its nucleus to become a reticulocyte, changing to an erythrocyte with the loss of other organelles.

as phosphatidylcholine (PC), among many other molecular components. Thus, it would not be surprising if the red cell membrane had a non-zero spontaneous curvature C_o. Canham (1970) analysed a collection of more than 30 human red cells (flaccid, swollen and stomatocytes) within the spontaneous curvature model with $C_o = 0$, confirming that these shapes were close to those which minimized the model bending energy at fixed volume. Of course, the mean size of a red cell is not predicted by the bending energy, but presumably reflects the ease of transporting the cell through the circulatory system, and the efficiency of its operation in a capillary. The cell volume is maintained osmotically through the active regulation of sodium and potassium ion concentrations across the plasma membrane (see Beck, 1991).

The membrane-associated cytoskeleton also provides deformation resistance in a red cell, although the structure of the cytoskeleton may evolve during the maturation of the cell. Blood cells have a sufficiently short lifetime that they must be continously and copiously produced in the body. Descendants of self-renewing stem cells, blood cells differentiate during successive cell divisions into families of white and red cells, as depicted in Fig. 10.19(b). Even the last cell in the sequence that yields an erythrocyte undergoes dramatic changes in its lifetime: born as a nucleated erythroblast, the cell expels its nucleus to produce an irregularly shaped reticulocyte. Shortly after abandoning the bone marrow for the circulatory system, the reticulocyte loses its ribosomes and remaining organelles to become a mature erythrocyte with its familiar discocytic shape (see Chapter 23 of Alberts *et al.*, 2008). The shape of the cytoskeleton is similar to the discocyte, although independent studies in which the plasma membrane is washed away by a detergent have found that the cytoskeleton may be somewhat smaller (Lange *et al.*, 1982; for a review, see Steck, 1989) or larger and more spherical, than the parent cell (Svoboda *et al.*, 1992). An extensive computational study of the large suite of red cell shapes has established the importance of cytoskeleton elasticity in determining the energy-minimizing surfaces of erythrocytes (Lim *et al.*, 2008); the behavior of a cytoskeleton at large deformation is treated formally by Dao *et al.* (2006).

10.4.3 Aspirated red cells

Irrespective of its contribution to the rest shape of the red cell or to elasticity under weak deformations, the cytoskeleton dominates the cell's resistance to strong deformation, as we now demonstrate by considering a specific situation commonly used in mechanical studies of the cell. Micropipette aspiration involves the application of a suction pressure to a cell via a micropipette 1 μm or so in diameter; experimental images of aspirated red cells can be seen in Figs. 1.15 and 5.24(a), and of an auditory outer hair cell in Fig. 5.25. The deformation in the figures principally involves bending the membrane, and stretching and shearing the network, while simultaneously preserving the overall areas of the cytoskeleton and membrane (the latter is relatively incompressible). To estimate the relative energies of bending and stretching, let's examine the tip of the aspirated segment shown in Fig. 10.20(a), where the network deformation is a dilation. In experiments, the stretched area at the tip may be twice as large as the unstressed area (Discher *et al.*, 1994). From Eq. (5.18), the energy density of a two-dimensional isotropic material under dilation is $(K_A/2)\bullet(u_{xx} + u_{yy})^2$, where K_A is the area compression modulus and u_{ij} is the strain tensor; the sum $u_{xx}+u_{yy}$ equals the relative change in area $\Delta A\,/A$. Choosing $u_{xx} + u_{yy} = 1/2$ and $K_A = 10^{-5}$ J/m^2 (see Section 5.5) as representative of moderate deformations, we expect the energy density of the stretched network to be ~ 10^{-6} J/m^2. Now, the energy density of the bending mode of the spherical cap in Fig. 10.20(a) is $(\kappa_b/2)(2/R_p)^2$, where R_p is the pipette radius (note that we have ignored the Gaussian and area-difference contributions). With $R_p = 1$ µm and $\kappa_b = 20k_BT$ as a typical experimental situation, the bending energy density is 1.6×10^{-7} J/m^2. The order of magnitude difference between stretching and bending depends upon our choice of geometry, etc., but is not atypical of the configuration as a whole; in general, we find that the bending energy of the membrane is not as important as the stretch and shear of the cytoskeleton at large deformations.

The pure dilation of the network at the tip of the aspirated segment is a special situation; in general, the network is subject to shear whose magnitude can be estimated from Fig. 10.20(b). When material is drawn up into the pipette, the remaining network in the plane outside is strained as it moves closer to the pipette entrance. In Fig. 10.20(b), for example, the

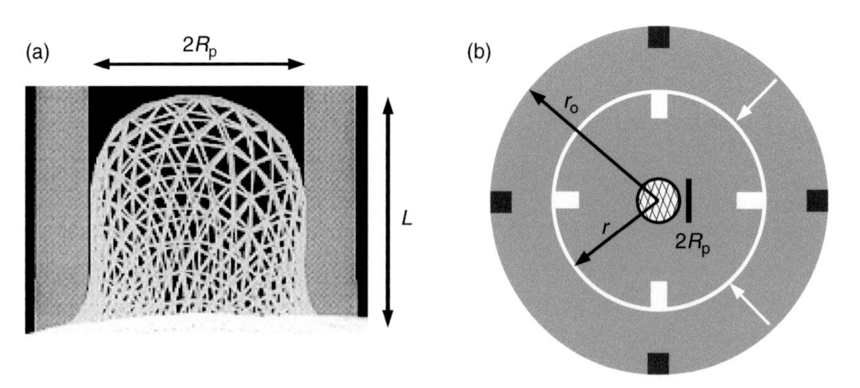

Fig. 10.20 (a) Simulation of the aspiration of a red cell, showing the dilution of the cytoskeleton at the tip of the aspirated segment. The micropipette is represented by the grey bars at the sides. (b) Face view of a schematic network being drawn up a pipette of radius R_p perpendicular to the plane. When aspirated, the shaded region of the radius r_o is reduced to an annulus of radius r, plus the cylinder and hemisphere inside the pipette.

material at the boundary of the shaded region is moved from a distance r_o to a distance r of the pipette center. This results in a strain because the distance between network vertices along the arc at r falls to r/r_o of its original, unstrained value. The strain along the arc is then $\Lambda - 1$, where $\Lambda = r_o/r$. The area of the network within r_o before aspiration is πr_o^2, but this section covers $\pi[r^2 + 2R_pL - R_p^2]$ afterwards (a sum of $\pi[r^2 - R_p^2]$ from the plane, $2\pi R_p[L - R_p]$ from the cylinder in the pipette and $2\pi R_p^2$ from the hemispherical cap). Taking the overall area to be unchanged by the deformation, then $\pi r_o^2 = \pi[r^2 + 2R_pL - R_p^2]$ and

$$\Lambda^2 = 1 + (R_p/r)^2(2L/R_p - 1). \tag{10.19}$$

As expected, this part of the strain vanishes (i.e., $\Lambda \rightarrow 1$) as $r \rightarrow \infty$, and has its maximum value $\Lambda^2 = 2L/R_p$ at the pipette entrance $r = R_p$. Demanding that area be conserved locally, the in-plane deformation is a shear, with network elements moving further apart in a radial direction, but closer together around a hoop.

By imposing a shear on the network, the aspiration deformation can be used to extract its shear modulus μ. From Laplace's law, Eq. (10.5), the force per unit length pulling the cylindrical section of the network up the pipette is $\Delta p R_p/2$, as visualized for a hemispherical cap in Fig. 10.2, where Δp is the pressure difference applied by the pipette. Balancing forces leads to the expression (Waugh and Evans, 1979; earlier work includes Evans, 1973a, b)

$$\Delta p = (2\mu/R_p)\int_{R_p}^{\infty}(\Lambda^2 - 1/\Lambda^2)r^{-1}\mathrm{d}r. \tag{10.20}$$

This integral can be evaluated exactly to yield

$$\Delta p = (\mu/R_p)[(2L/R_p - 1) + \ln(2L/R_p)]. \tag{10.21}$$

Several of the elastic moduli quoted in Chapter 5 were obtained by analyzing the aspiration length as a function of pressure using Eq. (10.21). Although this particular deformation mode is of more interest in the laboratory than in the circulatory system, it illustrates the importance of the cytoskeleton to the cell's deformation resistance.

10.5 Bacteria

Without an extensive cytoskeleton to distort their boundaries in peculiar ways as happens in eukaryotes, bacteria tend to choose from among a smaller set of basic cell shapes, although there are many exceptional cases. Common bacterial shapes are displayed in Fig. 10.21. Cells are

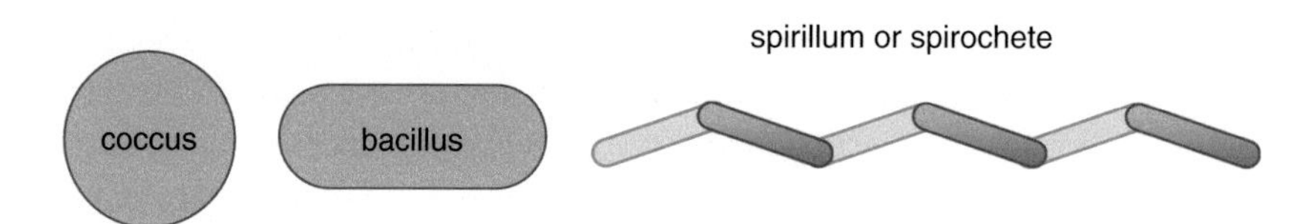

Fig. 10.21 Bacterial shapes include spheres (or cocci), rods (or bacilli) and spirals (spirilla and spirochaetes), among others. Typical dimensions for each shape are quoted in the text; the configurations are not drawn to the same scale.

approximately spherical in one shape category; referred to as *cocci* (plural of *coccus*) they typically have diameters of a few microns. Rods or *bacilli* (plural of *bacillus*) make up a second category; these are often 0.5–1 μm in diameter and several times that in length, although they may range up to 500 μm long for the cigar-shaped *Epulopiscium fishelsoni*. A third category includes corkscrew-shaped bacteria such as rigid *spirilla* and flexible *spirochaetes*, whose tubular cross section may have a diameter of just 0.1 μm and an overall length of a few microns (although, again, some may approach hundreds of microns). In addition to these main classifications, there are more exotic shapes such as flat plates or stalks with bulbous ends.

Bacterial shape is influenced by many factors (for a review, see Cabeen and Jacobs-Wagner, 2005). Most bacteria support an elevated osmotic pressure that may reach 10–20 atmospheres, easily sufficient to inflate the cell into a sphere if its envelope were isotropic. This pressure difference P across the envelope generates a two-dimensional tension within it which, from Eq. (10.3), is equal to the product RP, give or take a factor of two, where R is the radius of the sphere or cylindrical cross section. For instance, with $P = 10^6$ J/m^3 (10 atm) and $R = 10^{-6}$ m, the tension is 1 J/m^2, well above the bilayer rupture tension of ~10^{-2} J/m^2 on laboratory time scales (see Section 5.3). As a consequence, the bilayer of micron-sized bacteria under pressure must be mechanically reinforced by a cell wall. Notable exceptions are tiny mycoplasmas just 0.3 μm in diameter, which could operate at half an atmosphere without a cell wall by virtue of their small radius (and hence small value of RP). As can be seen from the image in Fig. 1.6, the shape of a mycoplasma is more irregular than the pressurized *E. coli* in Fig. 1.1(c).

The two-dimensional tension τ borne by the cell boundary is isotropic in spherical cocci, but depends on direction for rod-shaped bacilli: from Eq. (10.3), τ at the end caps and along the axial direction is $RP/2$, but τ_θ around the girth of the cell (hoop direction) is RP, twice as large. Not surprisingly, a cell can accommodate this anisotropic stress with an anisotropic cell wall: Yao *et al.* (1999) find the Young's modulus of the cell wall is twice as large in the hoop direction as it is in the longitudinal direction, as expected. The molecular composition of the bacterial cell wall (plant walls are different!) are chains of sugar rings which are thought to run in parallel, linked

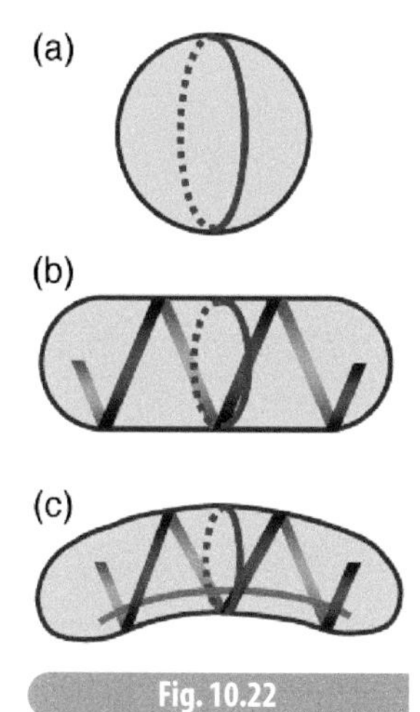

Fig. 10.22

Cytoskeletal proteins in bacteria. FtsZ (blue) creates a ring at the midplane of most bacteria and is the primary filamentous element known in cocci (a) at present. MreB and Mlb (magenta) form helical structures in both rod-shaped (b) and crescent-shaped (c) cells. Crescentin fibers (green) lie along one side of *Caulobacter crescentus* and give this bacterium its characteristic shape.

transversely by floppy peptide strings, to form peptidoglycan. Certain long tubular sheaths enclosing strings of cells are known to be constructed from a series of hoops linked laterally to each other (Xu *et al.*, 1996). A model for the linkage of peptidoglycan is shown in Figs. 7.24 and 7.25; note that the short peptide side-chains radiating from the glycan strands are chiral: in the diagrams, the strands are left-handed. Thus, only every fourth side-chain is oriented correctly to link to a given neighboring glycan, such that the network is more flexible than what might be naïvely expected.

Although the cell wall and plasma membrane have primary roles in the mechanics of the bacterial cell boundary and the flow of material in and out of the cell, other structural elements are present, similar to the membrane-associated cytoskeleton of eukaryotic cells. At least three filament-forming proteins are known to be important.

FtsZ

Having a molecular conformation very much like individual monomers of α- and β-tubulin, the protein FtsZ is the tubulin homolog in the bacterial cytoskeleton. It forms polymeric filaments and, again like tubulin, the GTP that it carries will hydrolyze once FtsZ is polymerized, resulting in a conformational change. As will be discussed momentarily, filaments of FtsZ feature dynamic growth and shrinkage as part of the so-called Z-ring at the bacterium's midplane, shown in Fig. 10.22(a); the half-life of the filaments is a brief 30 s (Stricker *et al.*, 2002).

MreB, Mbl

There are several bacterial homologs of actin, of which MreB and Mbl form filaments that run in cell-enclosing spirals near the plasma membrane as shown in Fig. 10.22(b). Like actin, individual monomers of MreB and Mbl carry ATP that hydrolyzes upon polymerization, resulting in a dynamic filament whose half-life is a few minutes. *B. subtilis* contains both MreB and Mbl (Jones *et al.*, 2001) while *E. coli* contains only MreB (Shih *et al.*, 2003).

Crescentin

This filamentous protein displays structural similarities to intermediate filaments in eukaryotes (Ausmees *et al.*, 2003). When organized into 1–2 μm long fibers, it causes *Caulobacter crescentus* to adopt a crescent shape whose radius of curvature is comparable to the length of the cell, as in Fig. 10.22(c).

Each of these types of proteins plays a different role in the shape of a bacterium and its growth. Individual filaments of FtsZ are arranged in parallel to form the Z-ring at the equator of a coccus or midcell of an elongated bacterium. While the Z-ring is dynamic, it is not like the contractile ring of a eukaryotic cell which is an actin–myosin composite generating a contraction force. Rather, the force originating from the Z-ring may have to do with conformational changes of FtsZ itself (Erickson *et al.*, 1996; Allard and Cytrynbaum, 2009). As the septum is formed in the bacterial cell cycle, the Z-ring shrinks to zero when the cell divides, but begins to assemble in each of the two daughter cells before division is complete (Sun and Margolin, 1998; Miyagishima *et al.*, 2001). The growth and shrinkage dynamics of FtsZ (Gonzáles *et al.*, 2003) and another actin analog ParM (Garner *et al.*, 2004) are not dissimilar to actin itself, as described further in Section 12.3.

Filaments of MreB and Mbl may be part of the bacterial cytoskeleton, but not universally so. Although these proteins have been observed in rod-shaped bacteria, neither of them is present in spherical cocci, suggesting that MreB and Mbl have a primary role in the elongation of cells, rather than the formation of the septum; this conclusion is supported through studies of cells in which these proteins were modified or deleted (Jones *et al.*, 2001). Further, MreB has not been found to play a role in chromosome segregation for at least one genus of cyanobacteria (Hu *et al.*, 2007). MreB and Mbl both trace out helical paths along the cell boundary. In Gram-positive *Bacillus subtilis*, MreB forms right-handed helices with a pitch of 0.73 ± 0.12 μm compared to a cell width of 0.8 μm and length of 1.5 to 3.5 μm; in the same cells, Mbl also forms right-handed helices with a much longer pitch of 1.7 ± 0.28 μm (Jones *et al.*, 2001). Thus, the helices complete one to two turns within the cylindrical section of the cell; in addition, visual evidence suggests that Mbl may form two intersecting helices. It is possible that the helices follow energy-minimizing paths in the confined environment of the rod, and this has been explored by Andrews and Arkin (2007) and by Allard and Rutenberg (2009).

Although filaments like MreB and Mbl are not as copious in bacteria as is peptidoglycan, they may still have an effect on the cells' elastic behavior for deformations which place them under tension. An analogous situation would be a suspension bridge where the tension-bearing cables may not be as massive as the compression-bearing towers. In a study of the MreB network in *E. coli*, Wang *et al.* (2010a) found that removing MreB lowered the flexural rigidity κ_f of the cell by about 30%, largely independent of its initial rigidity. In a population of *E. coli* cells whose individual κ_f values ranged from $1\text{–}5 \times 10^{-20}$ N•m², the mean κ_f decreased by a similar fraction from $(2.8 \pm 0.4) \times 10^{-20}$ N•m² to $(2.0 \pm 0.4) \times 10^{-20}$ N•m².

Returning to the cellular length scale of microns, examples of cells linked to form linear or branched filaments are known for both bacteria and eukaryotes. Two examples of this were shown in Fig. 2.1 from the world of cyanobacteria. The indentations between adjacent cells is barely

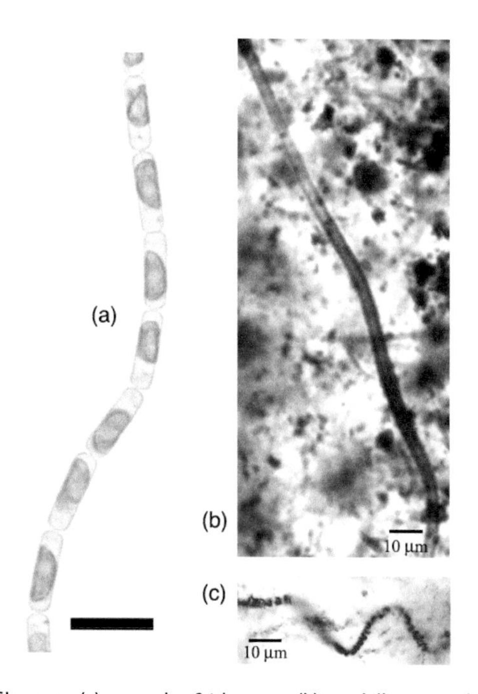

Fig. 10.23 Examples of cellular filaments: (a) green alga *Stichococcus*; (b) two-billion year-old microfossil *Gunflintia grandis* (Boal and Ng, 2010); (c) 2-billion-year-old microfossil *Halythrix* (Hofmann, 1976). Scale bar is 10 μm in all images.

visible in these examples, but such is not always the case. For both cyanobacteria (for example, *Pseudanabaena*) and algae, there are examples of filamentous morphologies that are more like a string of sausages, such as shown in Fig. 10.23(a), for the green alga *Stichococcus*. Without an encompassing sheath, such structures may be fragile and easily disrupted by mild agitation. As a cell design, filaments appeared early in the Earth's history (Cloud, 1965; Barghoorn and Schopf, 1966; Hofmann, 1976; Walsh and Lowe, 1985; Schopf and Packer, 1987; Schopf, 1993). Two examples of 2-billion-year-old filamentous structures are displayed in Fig. 10.23: panel (b) is *Gunflintia grandis* (Barghoorn and Tyler, 1965) and panel (c) is *Halythrix* (Schopf, 1968). Even older examples of filamentous structures include pyritic replacement filaments with ages exceeding three billion years (Rasmussen, 2000).

To date, the mechanical properties of filaments have not been as well characterized as those of simple polymers or biofilaments like microtubules or F-actin. Part of the reason for this is the inapplicability of techniques for obtaining the bending resistance κ_b or torsion resistance κ_{tor} that are based on thermal fluctuations (Sections 3.2 and 4.3, respectively): the fluctuations are too small to be measured. Another reason is interpretational: the filaments are structurally composite and therefore less easily analyzed. Although we do not have absolute measurements of κ_b and κ_{tor}, nevertheless their relative behavior can be examined semi-quantitatively by extracting their tangent correction length ξ_t.

The procedure is the same as that used for polymers in Section 3.2. The tangent correlation function $C_t(\Delta s)$,

$$C_t(\Delta s) \equiv \langle \mathbf{t}(s_1) \bullet \mathbf{t}(s_2) \rangle, \tag{10.22}$$

is constructed from the tangent vectors $\mathbf{t}(s_1)$ and $\mathbf{t}(s_2)$ at all pairs of points at arc lengths s_1 and s_2 subject to the constraint that $|s_2 - s_1|$ is equal to a particular Δs specified on the left-hand side of the equation. As with polymers, the behavior of $C_t(\Delta s)$ is described by exponential decay in Δs:

$$C_t(\Delta s) = \exp(-\Delta s / \xi_t), \tag{10.23}$$

where ξ_t is the tangent correlation length. For filaments with shapes governed by thermal fluctuations in their deformation energy, the correlation length ξ_t of Eq. (10.23) is the same as the persistence length of Eq. (3.21). The correlation function depends on the dimensionality of the system: the true correlation length ξ_3 of a filament in three-dimensional space is related to the correlation length ξ_{2p} of the same filament whose shape is projected into two dimensions via $\xi_3 = (3\pi/8)\ \xi_{2p}$.

Tangent correlation lengths are displayed in Fig. 10.24 for three genera of cyanobacterial filaments that represent three very different cell geometries (*Geitlerinema*, *Pseudanabaena*, and *Oscillatoria*, from the largest to the smallest cell length-to-width ratio). The correlation length is labeled ξ_2 because the filaments have been forced to lie in the two-dimensional plane of a microscope slide. Both *Geitlerinema* and *Oscillatoria* exhibit values of ξ_2 that rise with filament diameter D among species within the genus as $D^{5.1\pm1}$ and $D^{3.3\pm1}$, respectively. This power-law dependence is what one would expect if the tangent correlation length were proportional to the filament stiffness, assuming that all species of a given genus have similar construction. That is, Eqs. (3.12) and (3.13) show that the flexural rigidity κ_f scales as D^4 for solid cylinders and D^3 for hollow ones, for a fixed value of the Young's modulus of the material, and it would be reasonable to suppose that the correlation length is proportional to κ_f, just as the thermal persistence length is proportional to κ_f. Thus, Fig. 10.24 supports a mechanical interpretation of the tangent correlation length, although the conclusion would be more compelling if more species were measured.

The behavior of modern filamentous cyanobacteria can be compared with the measured correlation lengths of microfossils such as those displayed in Fig. 10.23. For a selection of 2-billion-year-old filamentous microfossils, ξ_{2p} lies in the 300–700 μm range for populations with apparent diameters of 1–4 μm, the same as the modern filamentous bacteria (Boal and Ng, 2010). However, this similarity by itself does not imply that the microfossil taxa must be cyanobacteria, as the correlation lengths of

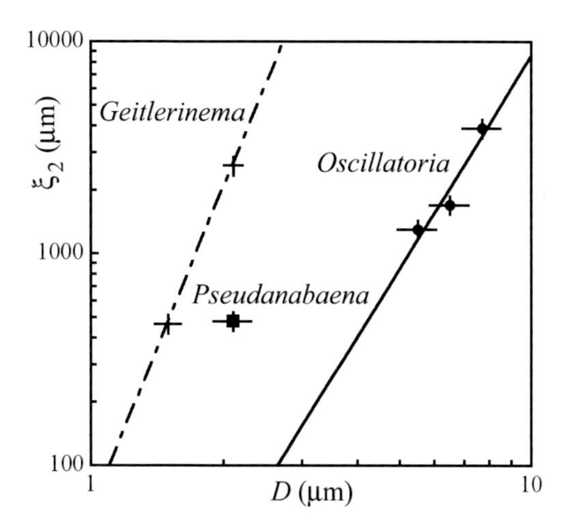

Fig. 10.24 Measured ξ_2 (in μm) for filamentous cyanobacteria as a function of their mean diameter D. The correlation functions are approximately described by $4.3 \cdot D^{3.3\pm1}$ for *Oscillatoria* and $62 \cdot D^{5.1\pm1}$ for *Geitlerinema* (solid and dot-dashed lines, respectively; both D and the result are in μm). The cyanobacteria are *Geitlerinema* (crosses), *Pseudanabaena* (square), and *Oscillatoria* (disks) (Boal and Ng, 2010).

eukaryotic green algae with similar diameters also lie in this range; for example, $\xi_2 = 900 \pm 100$ μm for the green alga *Stichococcus* with a mean diameter of 3.5 ± 0.2 μm (Forde and Boal, unpublished data).

Summary

The mechanical elements of a simple cell or vesicle are just the plasma membrane and perhaps a cell wall or cytoskeleton forming a compound envelope, whose elastic moduli (including bending, compression and shear resistance) strongly influence cell shape. A pressure difference P across the envelope gives rise to a lateral tension whose magnitude and direction-dependence is determined by the local curvature of the surface and which obeys the Young–Laplace equation $\tau_1/R_1 + \tau_2/R_2 = P$, where R_1 and R_2 are the local radii of curvature along directions 1 and 2. For example, the tension around the ring direction of a cylinder of radius R is RP, while the corresponding tension along the axis is $RP/2$. The bending energy of many shapes with constant, but not isotropic, local curvature can be evaluated analytically for simple models of the energy density. The most thoroughly investigated approach is the spontaneous curvature model introduced in Section 6.2, with a free energy density given by $\mathcal{F} = (\kappa_b/2) \cdot (C_1 + C_2 - C_o)^2 + \kappa_G C_1 C_2$, where κ_b and κ_G are the bending rigidity and Gaussian rigidity, respectively. Unrestricted in sign, the spontaneous curvature parameter C_o may arise from the chemical inequivalence of the bilayer leaflets or their environment. Integrated values of $\mathcal{F}$ for a selection of surfaces are summarized in Table 10.1; these expressions can be used to construct the bending energies of a variety of surfaces with axial symmetry.

Our understanding of membrane mechanics is applied to several situations: pure bilayer vesicles, the red blood cell, rod-shaped bacteria and cellular filaments. Synthetic vesicles of micron dimensions have been well-studied experimentally, although their membranes do not possess the protein components of 100-nm vesicles in cells. The phase diagrams of spherical or toroidal vesicles have been evaluated within both the spontaneous curvature and area-difference elasticity models. Although the true energy-minimizing shapes usually must be found numerically, analytical approximations are available which allow us to determine the dependence of vesicle shape on its reduced volume $v_{red} = 6\sqrt{\pi}\ V/A^{3/2}$, where A and V are the vesicle's surface area and volume, respectively. In the spontaneous curvature model with $C_o = 0$, the energetically favored state with axial symmetry is prolate (cigar-shaped) as v_{red} initially decreases from unity, changing to a biconcave disk near v_{red} = 0.65, then becoming a stomatocyte at v_{red} less than 0.59. Interestingly, shapes such as dumb-bells and doughnuts, which have been observed experimentally, also become energetically competitive around $v_{red} \sim 0.65$. We note that the red cell operates close to this volume, permitting shape change with little cost in energy. The phase diagram of vesicle shapes has been investigated experimentally by following trajectories in v_{red} induced by changing the temperature of synthetic vesicles. The shapes of vesicles as they pinch off from a membrane can be studied within the spontaneous curvature model, and we show how the preferred vesicle radius is of the order of the inverse curvature of the parent membrane, in the absence of other factors. Of course, other factors abound, including the presence of coating proteins such as clathrin (some of whose properties are examined in the problem set).

The deformation resistance of a red cell is strongly influenced by the presence of its cytoskeleton, many aspects of which are covered in Chapter 3. Here, we describe the deformation of a compound membrane induced by micropipette aspiration, which can be used to determine the cytoskeleton shear modulus. As a further example, we examine the shapes and mechanical properties of the bacterial cell wall. For micron-sized cells, the plasma membrane is able to sustain a pressure difference of just a tenth of an atmosphere, and must be strengthened by a cell wall to withstand internal pressures of 3–5 atmospheres in a Gram-negative bacterium (thin wall) or more than 20 atmospheres in a Gram-positive bacterium (thick wall). The wall material is peptidoglycan, an anisotropic fabric consisting of short glycan ropes (5–10 dissacharide units) laid out in parallel and cross-linked at 1 nm intervals by strings of amino acids. These cross-linking peptides have a contour length of 4.2 nm, but a much shorter end-to-end displacement of 1.3 nm.

This simple picture is not the whole story, in that the fabric in a rod-shaped bacterial cell wall must be at least partially aligned to accommodate its anisotropic stress, and to encourage cell growth along the cylindrical axis. Further, bacteria have a membrane-associated cytoskeleton which, although nowhere near as extensive as a eukaryotic cytoskeleton, is related to cell shape and division. FtsZ, MreB, Mbl and crescentin are filamentous proteins that assemble into fibers that are microns in length:

- the equatorial ring of FtsZ is associated with septum formation and division,
- MreB and Mbl form helices in rod-like cells and influence their shape,
- cresentin causes *Caulobacter crescentus* to bend into a crescent.

Many but not all linked chains of individual bacteria are enclosed and strengthened by a permanent or semi-permanent sheath; the mechanical properties of such composite systems are not well studied as yet.

Problems

Applications

10.1. Suppose that we have a synthetic solid material with volume compression modulus $K_V = 3 \times 10^9$ J/m^3, out of which we wish to make membranes to enclose

(i) a spherical cell of radius 10 μm,
(ii) a spherical weather balloon of radius 10 m.

Both membranes must support a pressure difference of one atmosphere (10^5 J/m^3). Assuming that the membrane fails if the lateral tension exceeds 5% of its area compression modulus K_A, what minimum membrane thicknesses are required for the cell and balloon? (*An approximate relationship between K_A and K_V for solids is given at the end of Section 7.3.*)

10.2. The inner and outer membranes of the nuclear envelope are joined by circular, protein-lined pores with the approximate dimensions indicated in the cross section in the figure.

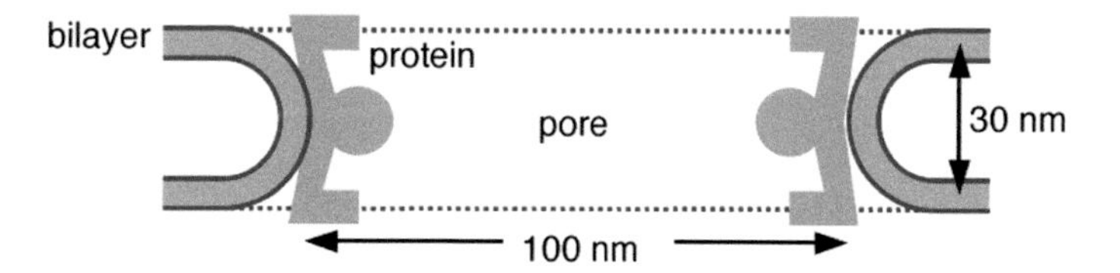

The pore has axial symmetry with the bilayer joined smoothly around the ring, as shown by the dashed lines (see also Fig. 5.2). For a nucleus with 3500 pores, typical of a mammalian cell, what is the total bending energy associated with the pores (in $k_B T$; use Table 10.1 and take $\kappa_b = 20 k_B T$, $\kappa_G = 0$). If this energy were provided by the hydrolysis of ATP to ADP, how many ATP molecules would be required?

10.3. Shown below is a cross section through a simplified model for the three-dimensional stacks in the Golgi apparatus (see also Fig. 1.8).

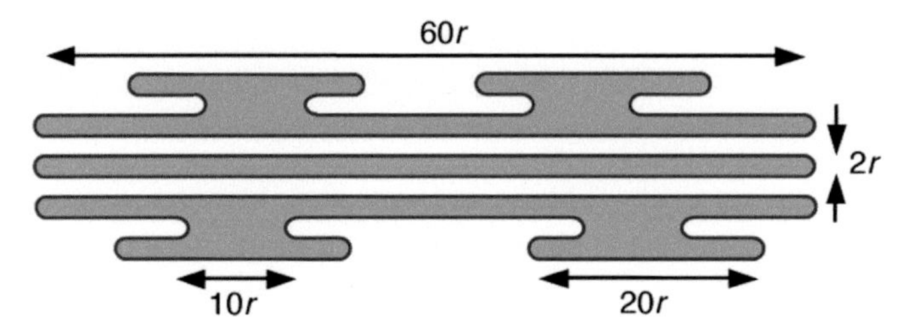

The length scale is set by the radius of curvature r in the plane of the drawing (i.e. along meridians). Each long horizontal region is a pancake perpendicular to the plane linked by axially symmetric necks. There are four pancakes at the top and bottom of diameter $20r$ and three pancakes in the middle with diameter $60r$. The four necks have diameter $10r$. Using results from Table 10.1, determine the bending energy of this structure in the spontaneous curvature model with $C_o = 0$. Assuming $\kappa_b = 20 k_B T$ and $\kappa_G = 0$, express this energy in $k_B T$ at $T = 300$ K.

10.4. Some cells can adhere to a substrate by spreading, as shown here in the idealized shape.

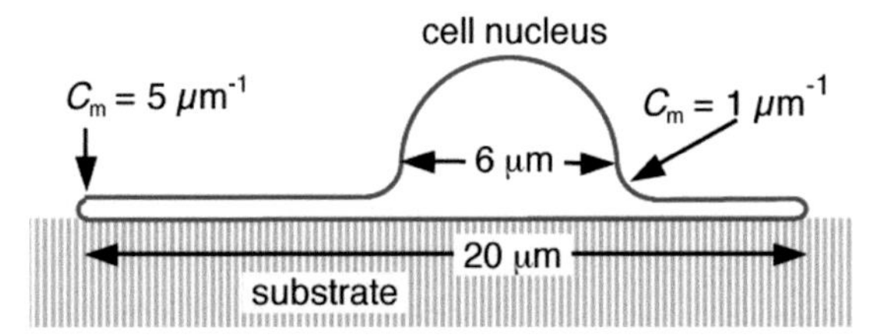

Assume that the thin region adhering to the substrate is a circular pancake of diameter 20 μm. Note that for many cells, the curvature at the edge may be a factor of two higher than indicated.

(i) Using the dimensions and curvatures along the meridians (C_m) in the diagram, what is the membrane bending energy in terms of κ_b and κ_G (use Table 10.1).

(ii) What is the energy difference $\Delta E / k_B T$ between this shape and a sphere, assuming $\kappa_b = 20 k_B T$?

10.5. The clathrin network in a cell has a honeycomb structure with dimensions as indicated in the figure.

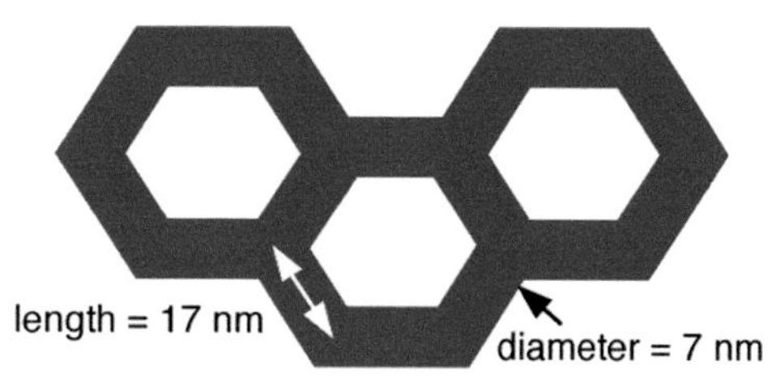

Take the network elements to be cylinders of diameter 7 nm with a Young's modulus of $Y = 10^9$ J/m^3.

(i) Find the effective spring constant of the network elements using the result from Problem 5.29.
(ii) What is the area compression modulus K_A of this network according to Problem 10.17? Compare your result with $K_A = 10^{-5}$ J/m^2 observed for the human red cell cytoskeleton.

10.6. Suppose that the area expansion of a thin layer of peptidoglycan is described by the honeycomb network in Problem 10.17, with $k_{sp} = 10^{-2}$ J/m^2 and $s_o = 1$ nm. At what tension τ_{exp} does the network expand without limit? If a spherical bacterium of radius 2 μm is enclosed by this network, what is its interior pressure relative to its environment at τ_{exp} (quote your answer in J/m^3 and in atmospheres)?

10.7. The area compression modulus K_A of a plate is related to its volume compression modulus K_V by the approximate relationship $K_A = dK_V$, where d is the plate thickness. Assuming that $K_V = 2 \times 10^7$ J/m^3 for peptidoglycan, find the change in relative area of a cell wall encasing a sphere of diameter 1 μm sustaining 5 atm internal pressure. Plot the area strain as a function of wall thickness for $5 \leq d \leq 25$ nm. Areal stress and strain are described in Section 7.3.

10.8. Plot the bending energy (in units of κ_b) of an axially symmetric torus in the spontaneous curvature model as a function of the reduced volume v_{red}. Use results from Problem 10.11 for the area and volume, and obtain the bending energy from Problem 10.14(i). Over what range of v_{red} are doublet shapes with $E = 16\pi\kappa_b$ favored over tori (see Fig. 10.8)?

10.9. A cellular filament with diameter 4 μm grows in a straight line, bounded at each end by immovable walls. Treating the filament as a uniform rod with a Young's modulus of 10^9 J/m^3, find the buckling force on the filament under two situations: the distance between the walls is (a) 20 μm and (b) 100 μm.

Formal development and extensions

10.10. The pancake shape shown in the cross section (where the dashed line is the axis of rotational symmetry)

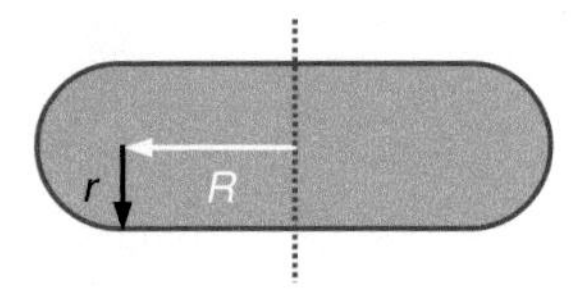

is described by two geometrical parameters r and R. Prove that the surface area A and enclosed volume V of this shape are given by

$$A = 2\pi r^2 \, [(R/r)^2 + \pi R/r + 2]$$

$$V = 2\pi r^3 \, [(R/r)^2 + (\pi/2)R/r + 2/3].$$

10.11. The torus displayed in the cross section (where the dashed line is the axis of rotational symmetry)

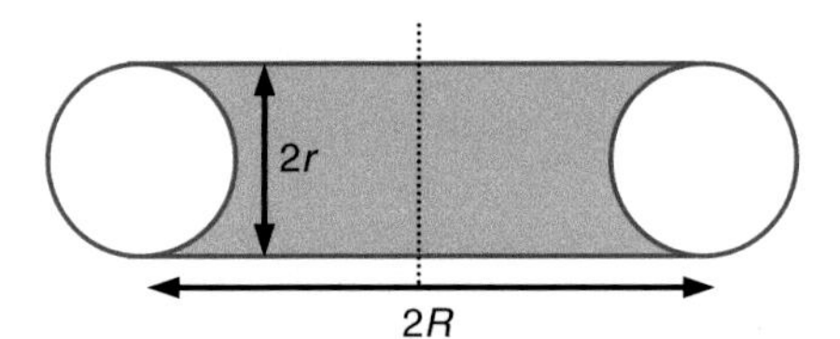

is described by two geometrical parameters r and R.

(i) Prove that the surface area A and enclosed volume V of this shape are given by

$$A = 4\pi^2 r R$$

$$V = 2\pi^2 r^2 R.$$

(ii) Express the reduced volume v_{red} in terms of r/R, and find its allowed range.

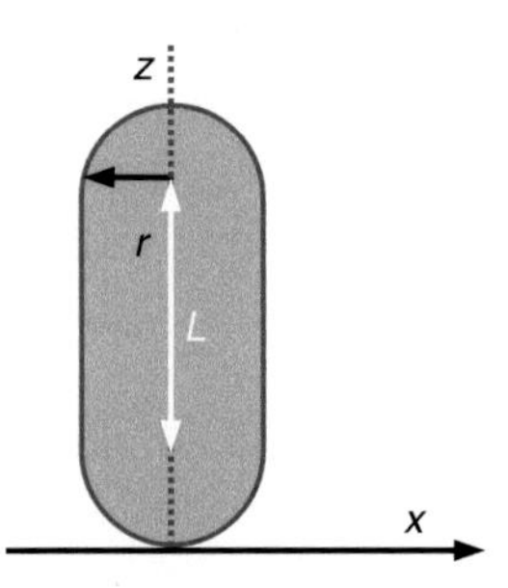

10.12. As a simple application of Eqs. (10.9) to (10.11), calculate the bending energy of a spherocylinder of radius r and total length $L + 2r$.

(i) For the lower hemisphere, obtain $\theta(s)$ and $x(s)$ with the coordinate origin at the south pole. Prove that the principal curvatures are $C_m = C_p = 1/r$ by evaluating Eqs. (10.9)–(10.10). Integrate Eq. (10.11) to obtain the energy of the south pole.

(ii) Repeat the steps in part (i) for the cylindrical section in the middle of the shell.

(iii) Add the sections together to obtain

$$E = 2\pi\kappa_b(2 - rC_o)^2 + \pi\kappa_b(L/r)\bullet(1 - rC_o)^2 + 4\pi\kappa_G.$$

Note that the Gaussian term is the same for a sphere or a pancake (Problem 10.13).

10.13. For the pancake shape shown in Problem 10.10, establish that the bending energy is given by

$$E_{\text{pancake}} = 8\pi\kappa_b + 4\pi\kappa_b\rho^2 (\rho^2\text{–}1)^{-1/2} \{\arctan[(\rho+1)(\rho^2\text{–}1)^{-1/2}] - \arctan[(\rho^2\text{–}1)^{-1/2}]\} + 4\pi\kappa_G \qquad (\rho > 1),$$

in the spontaneous curvature model when $C_o = 0$ (the relevant integral can be found in Section 2.552 of Gradshteyn and Ryzhik, 1980). Here, we have put $\rho = R/r$. Show that this energy becomes

$$E_{\text{pancake}} = \pi\kappa_b(8 + \pi\rho) + 4\pi\kappa_G$$

in the limit where ρ is large.

10.14. Apply the spontaneous curvature model to the torus in Problem 10.11 without resorting to Table 10.1.

(i) Prove that the bending energy is given by

$$E_{\text{torus}} = 2\pi^2\kappa_b\,(R/r)^2 \,/\, [(R/r)^2\text{–}1]^{1/2} \qquad (R > r)$$

when $C_o = 0$. Confirm that the Gaussian contribution to the bending energy vanishes.

(ii) Show that the minimum value of E_{torus} occurs at $R/r = \sqrt{2}$, a special geometry called the Clifford torus. Calculate $E_{\text{torus}}/E_{\text{sphere}}$ (at $\kappa_G = 0$) for this shape and determine its reduced volume v_{red} using results from Problem 10.11.

10.15. Consider the specific stomatocyte shape which is assembled from sections of spheres, cylinders and tori.

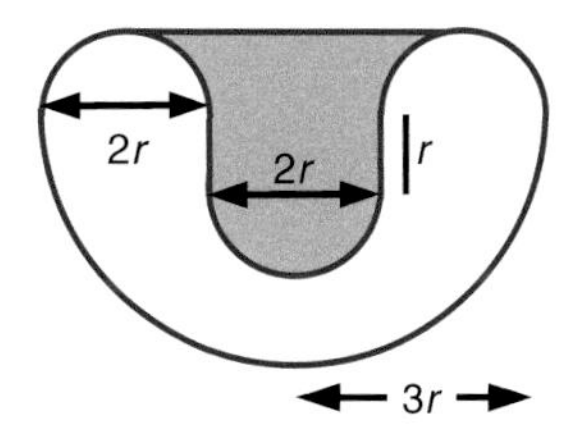

To simplify the algebra, the curvatures along the meridians are chosen to be 0, r^{-1}, or $(3r)^{-1}$.

(i) Prove that the area A and volume V are given by

$$A = 2\pi(11 + 2\pi)r^2$$

$$V = (49\pi/3 + 2\pi^2)r^3.$$

(ii) What is the reduced volume v_{red} of this configuration?

(iii) Using results from Table 10.1, show that the bending energy of this shape in the spontaneous curvature model is

$$E_{special} = \pi\kappa_b(9 + 4\pi/\sqrt{3}) + 4\pi\kappa_G$$

when $C_o = 0$. What is the ratio of this energy to that of a sphere if $\kappa_G = 0$?

10.16. Consider the uniformly curved surface joining two flat regions, as might appear in the neck of the doublet of spheres in Fig. 10.8.

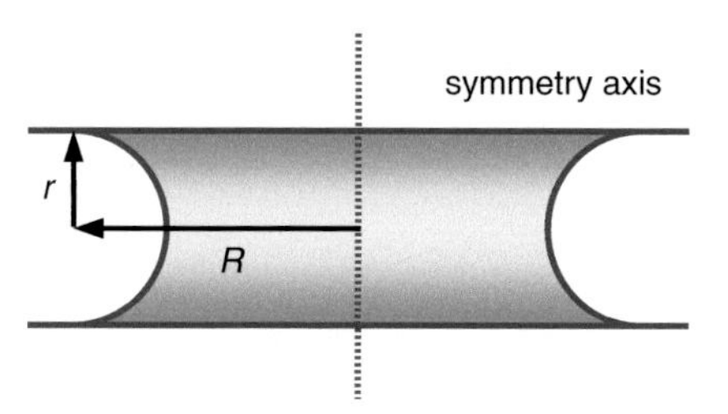

(i) Prove that the area of the curved region is $2\pi r^2\,[(\pi R/r) - 2]$.

(ii) Defining $\rho = R/r$, establish that the bending energy of the curved region is

$$E_{pore} = -8\pi\kappa_b + 4\pi\kappa_b\rho^2\,(\rho^2-1)^{-1/2}\,\{\arctan[(\rho-1)\bullet(\rho^2-1)^{-1/2}] + \arctan[(\rho^2-1)^{-1/2}]\} - 4\pi\kappa_G$$

when $\rho > 1$. Show that this energy becomes

$$E_{pore} = \pi\kappa_b(-8 + \pi\rho) - 4\pi\kappa_G$$

in the limit where ρ is large.

10.17. The clathrin network of a cell has the appearance of a two-dimensional honeycomb with three-fold connectivity, as in the diagram here.

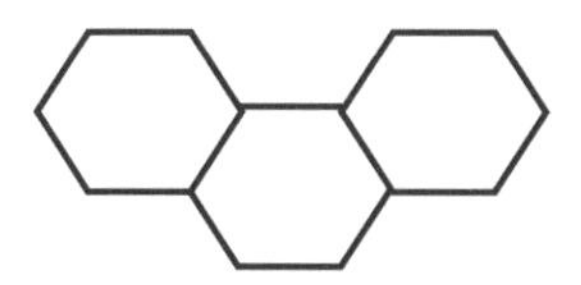

Suppose that each network element is a spring with force constant k_{sp} and rest length s_o. If the network is placed under a tension τ, show that

(i) the enthalpy per network vertex is

$$H_v = (3/4)k_{sp}(s - s_o)^2 - (3\sqrt{3}\ /\ 4)\tau s^2,$$

(ii) the spring length under tension is $s_\tau = s_o\ /\ (1 - \sqrt{3}\ \tau\ /k_{sp})$,
(iii) the area compression modulus K_A is given by

$$K_A = (k_{sp}\ /2\sqrt{3}) \bullet (1 - \sqrt{3}\tau/k_{sp}).$$

10.18. A single strand of a cytoskeletal filament, connected end-to-end to form a continuous loop, resides within a bacterium in the shape of a spherocylinder of radius R and overall length $L + 2R$.

(a) Draw the path taken by the loop for the following situations in all of which the filament passes through the each pole of the spherocylinder. In your diagrams, "unroll" the cylindrical section of the spherocylinder and start the path at the lower left-hand corner, taking the symmetry axis of the cell to be vertical.

(i) The filament runs straight up one side of the cylinder, across the hemisphere at the top, straight down the other side, and across the hemisphere at the bottom.

(ii) Similar to part (i), but twist the path across the top pole clockwise by half a turn (or π radians). What is the handedness of the helical path?

(iii) Similar to part (ii), but twist the path across the top pole counter-clockwise by a full turn (or π radians). What is the handedness of the helical path?

(b) In terms of its flexural rigidity κ_f, what is the bending energy of the filament along the cylindrical section of the cell for all of the configurations in part (a)?

11 Dynamic filaments

Cells are more than just passive objects responding to external stresses: they can actively change shape or move with respect to their environment. A very familiar example of cellular shape change is the contraction of our muscle cells. Less familiar, but very important to our health, is the locomotion of cells such as macrophages, which work their way through our tissues to capture and remove hostile cells and material. Another example of cell movement is propulsion in a fluid medium, which in some cells is provided by the rotation of flagella (Latin plural for the noun whip*) extending from the cell body. Flagella can be seen from several cells of the bacterium* Brevundimonas alba *in Fig. 11.1. Structurally related to flagella are cilia (Latin plural for* eyelash*), which occur on the surfaces of some cells and wave in synchrony like tall grass in the wind, creating currents in their fluid environment. What microscopic mechanisms underlie a cell's movement or* motility*?*

In Section 11.1, we review several aspects of cell movement and trace their origin to the molecular level. Dynamic biofilaments alone can be responsible for certain shape changes because they can generate a force by increasing their length, as described in Section 11.2. However, contractile forces in muscles, or transport along a microtubule, arise from specialized motor proteins capable of crawling along a biofilament. A number of models have been proposed for the operation of such molecular motors, and these are explained in Section 11.3. Measurements of motile forces, and the effect of dynamic structural elements on cell shape are presented in Section 11.4. Lastly, several mechanisms that can provide cell movement are analyzed in Section 11.5.

11.1 Movement in the cell

Actin and tubulin are dynamic polymers: their fundamental protein building blocks (G-actin or the tubulin heterodimer) can both polymerize and depolymerize, depending on the conditions, changing the length of the filament in the process. Further, each end of the filament grows and shrinks at a different rate: the rapidly growing end is called the plus end, while the slowly growing end is minus. This is illustrated in

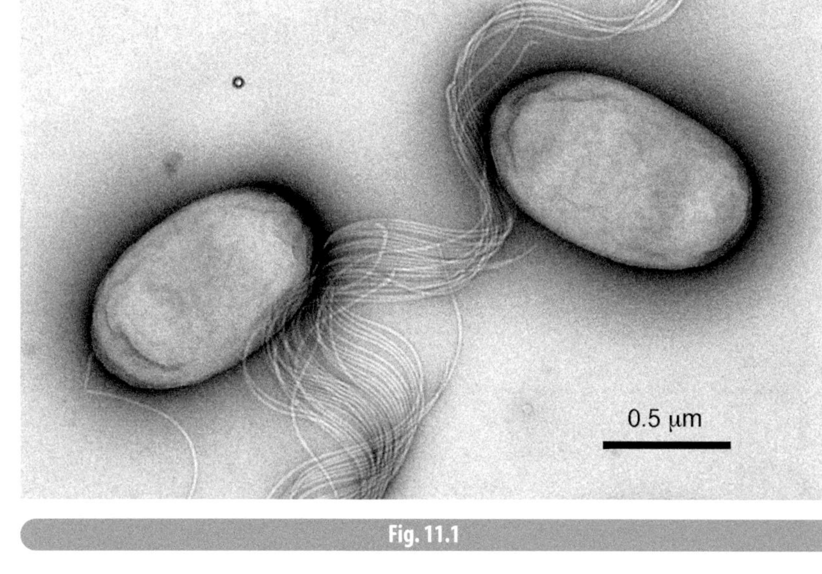

Fig. 11.1

Electron micrograph of *Brevundimonas alba*, showing the flagella from several bacteria (bar is 0.5 μm in length; courtesy of Dr. Cezar Khursigara and Robert Harris, University of Guelph).

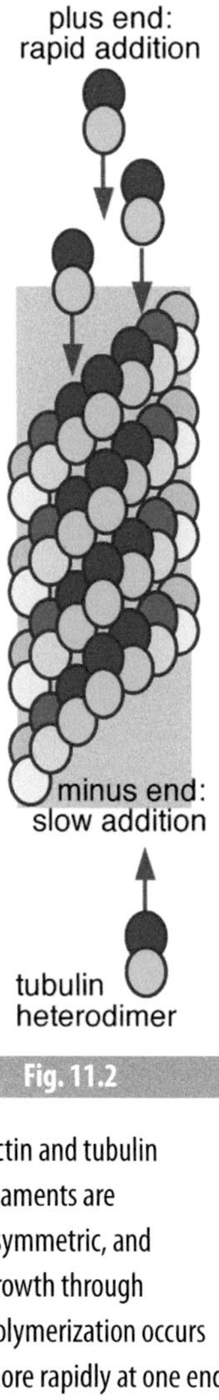

plus end:
rapid addition

minus end:
slow addition

tubulin
heterodimer

Fig. 11.2

Actin and tubulin filaments are asymmetric, and growth through polymerization occurs more rapidly at one end than the other (called the plus and minus ends, respectively). Depolymerization also can occur at either end.

Fig. 11.2 for tubulin; it has been argued that the α-tubulin units of the heterodimer face towards the plus end of the filament (Oakley, 1994). For actin, electron microscopy shows the physical appearance of each filament end to be different: the plus end resembles the feathered end of an arrow (hence the "barbed" end) while the minus end looks like the arrowhead (or "pointed" end). A representation of the three-dimensional structure of an individual actin monomer in its globular form is displayed in Fig. 11.3.

Once a unit has attached to the end of the filament, a further compositional change occurs by hydrolysis. In the case of tubulin, each α- and β-unit contains GTP (guanosine triphosphate) binding sites; GTP in the β-unit hydrolyzes to GDP (guanosine diphosphate) shortly after polymerization and the tubulin has been incorporated into the filament. Similarly, actin monomers contain ATP (adenosine triphosphate), which hydrolyzes to ADP (adenosine diphosphate) after polymerization. In both cases, the effect of the hydrolysis is to weaken the polymeric bonds, making depolymerization easier. Depending on the environment, growth of the filament may be so rapid that hydrolysis lags behind the addition of new units, resulting in a region near the growing end of the filament where hydrolysis has yet to occur. For tubulin, this region is called the GTP cap, as illustrated in Fig. 11.4, and is *more* stable against depolymerization than the bulk filament itself. The growth and shrinkage rates of these filaments are discussed at length in Section 11.2.

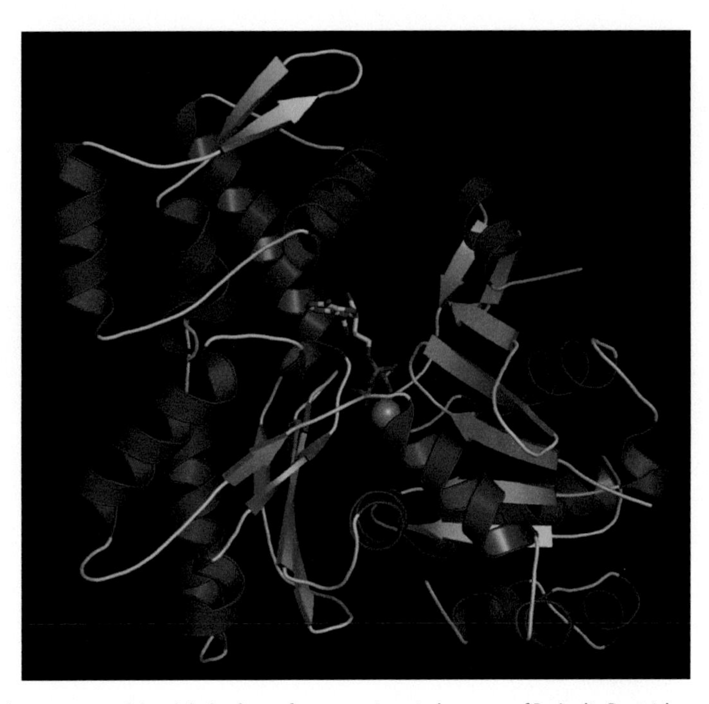

Fig. 11.3 Secondary structure of the globular form of monomeric actin (courtesy of Dr. Leslie Burtnick, University of British Columbia; data from Wang *et al.*, 2010b).

The growth of dynamic filaments such as microtubules affects cell shape and motility. As shown in Fig. 11.5(a), the centrosome of most animal cells provides a nucleation region (or microtubule organizing center, MTOC) from which literally hundreds of microtubules radiate at any given instant. The filaments grow and shrink continuously, with their rapidly growing plus ends extending out towards the cell periphery. Thus, the filaments can exert a collective pressure on the plasma membrane in some cases, and, as a result, can push the nucleation region towards the center of the cell. During cell division, the centrosome replicates, providing two centers for microtubule nucleation and growth. Imagine two copies of Fig. 11.5(a) placed beside each other. Some microtubules from each centrosome approach each other plus-to-plus, meeting near the midplane of the dividing cell; these can link to form long bridges between the centrosomes. Other filaments can attach to the cell's chromosomes lying between the centrosomes. In one mechanism for cell division, the bridging microtubules force the centrosomes apart, dragging sister chromatids along with them by means of the attached microtubules. Further aspects of microtubule organization are described by Dustin (1978) and Tuszynski (2008).

Filament growth also plays a role in cell locomotion. For example, fibroblasts and other cells can move along a substrate, adhering to it

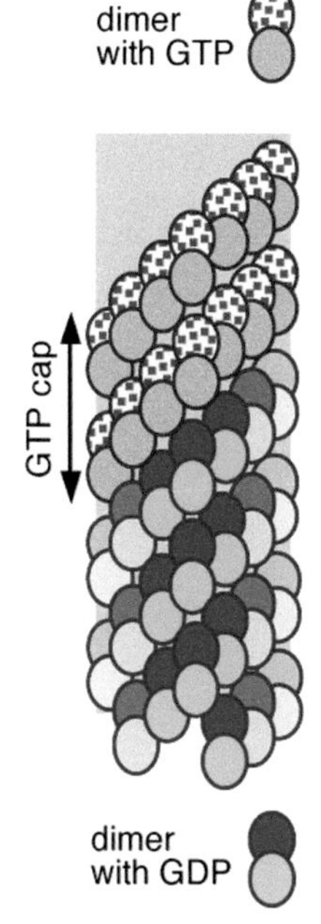

Fig. 11.4

When tubulin heterodimers are added to a microtubule faster than the rate of GTP hydrolysis, the filament acquires a GTP-rich cap. Here, growth is to the top.

by spreading a sheet-like structure (the lamellipodium) across the surface, as illustrated schematically in Fig. 11.5(b). Other types of cells create extensions called pseudopods (literally false feet) to enable them to crudely walk over a surface. Observations of fluorescently labeled filaments show that the leading edge of such cells is actin-rich, with filament growth occurring at the cell boundary. In a study of keratocytes, the actin filaments move back relative to the cell body at a speed approximately equal to the crawling speed of the cell with respect to the substrate (up to 0.1 μm/s), such that a given position on a filament remains roughly stationary with respect to the substrate (Theriot and Mitchison, 1991; see also Cao *et al.*, 1993, Lin and Forscher, 1993; actin growth rates of 0.05–0.1 μm/s are observed for neuronal growth cones by Forscher and Smith, 1988). At a speed of 0.1 μm/s, newly incorporated actin can be carried along a filament in the lamellipodia of these cells in about 20–30 s.

Actin and microtubules need not exist as unconnected entities in the cell. We have already described in Section 6.1 a number of actin-binding proteins (ABPs) capable of linking actin filaments into networks, and there also exist microtubule-associated proteins (MAPs) for fabricating bundles of microtubules. Further, specialized proteins can slide along actin and tubulin filaments, the energy for their motion being provided by ATP hydrolysis. Although their specific amino-acid sequence varies with the cell type, these so-called motor proteins can be grouped into three families: myosins associate with actin, while kinesins and dyneins associate

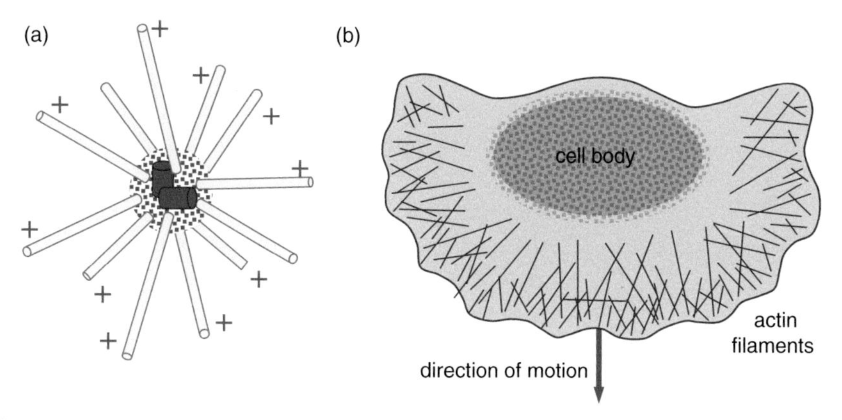

Fig. 11.5

(a) Long microtubules radiate from the cell's centrosome (fuzzy region around two cylindrical centrioles), their plus ends extending towards the cell boundary. (b) Actin filaments occur at high density within several microns of the leading edge of a moving cell on a substrate; in fish scale keratocytes, the filaments often meet near right angles (Small *et al.*, 1995). Seen in cross section, the thin lamellipodium hugs the substrate and only the cell body has an appreciable thickness.

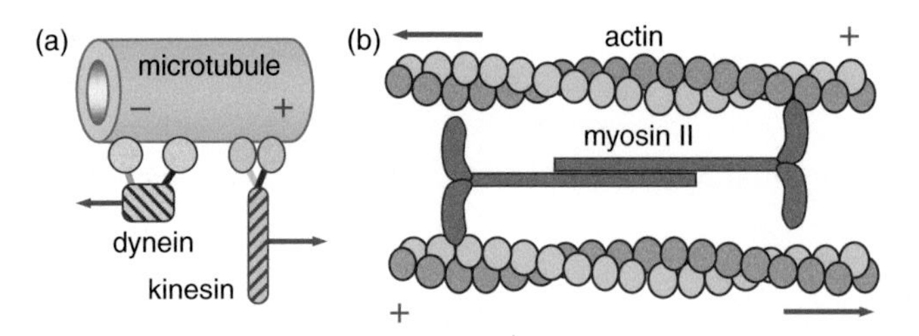

Fig. 11.6 (a) Kinesin and dynein walk towards the plus and minus ends of a microtubule, respectively. (b) Myosin walks towards the plus end of actin, so that the minus ends of the filaments are pulled toward each other.

with microtubules. Samples of movements possible with motor proteins are shown in Fig. 11.6.

Neither actin filaments nor microtubules are symmetric end-to-end, and the motor proteins have a preferred direction to their movement.

- Microtubules: kinesins generally walk towards the plus end, cytoplasmic dyneins walk towards the minus end.
- Actin: myosins walk towards the plus end.

Given that the plus ends of microtubules radiate from the centrosome, kinesin and dynein together provide a mechanism for transport in either direction from the nucleus with modest speeds at cellular length scales. Motors on axonal microtubules can walk at speeds up to 2–5 μm/s towards the end of the axon, a rate which permits a chemical cargo of neurotransmitter, for example, to be transported in 2–6 days from a production site in the brain to the end of a motor neuron a meter away.

Actin filaments and bundles of myosin may be organized into highly cooperative structures in our muscles. Figure 11.7 is a schematic representation of skeletal muscle, which is the common muscle of our biceps, etc. (the other types of muscle cells are described in Appendix A). The thick filaments are bundles of myosin with a mean diameter of about 30 nm, separated from adjacent myosin bundles by 40–50 nm. Walking towards the plus end of the actin, as shown in Fig. 11.6, mysosin pulls the minus ends of the filament towards one another, contracting the muscle along the horizontal direction in Fig. 11.7.

Molecular motors attached to parallel microtubules generate the whip-like motion of cilia and some flagella. The core of these thin structures is the axoneme, a bundle of microtubules about 225 nm in diameter, consisting of two microtubules in its center, surrounded by nine microtubule doublets. Each doublet has a single microtubule (13 protofilaments) to which is attached an incomplete second tubule (11 protofilaments),

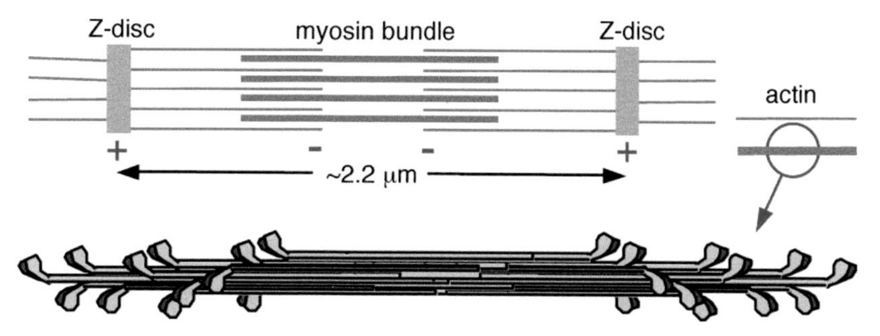

Fig. 11.7 Arrangement of actin filaments and myosin bundles in a skeletal muscle cell. The myosin bundles contain more than one hundred individual myosin-II filaments, as shown in the inset (the average lateral separation between myosin bundles is 40–50 nm). Myosin "walks" along actin towards its plus end, causing the muscle to contract. The Z-discs define the boundary of a single sarcomere about 2.2 μm long (after Alberts *et al.*, 2008; reproduced with permission; ©2008 by Garland Science/Taylor and Francis LLC).

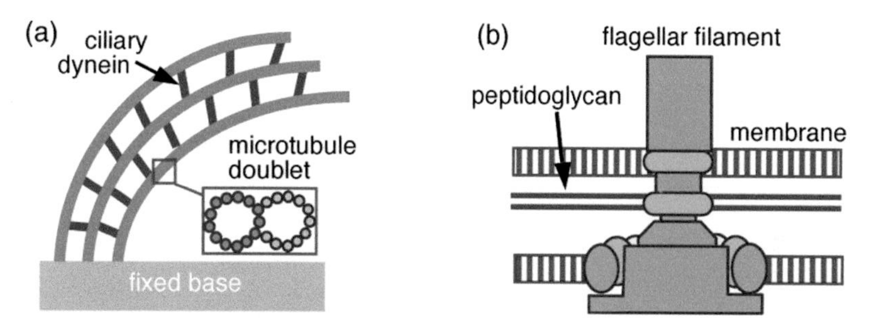

Fig. 11.8 (a) The bending of a cilium results from the motor protein ciliary dynein walking along a neighboring doublet of microtubules. If the cross-linking structure at the base of this diagram were removed, the filaments would slide past one another without bending. Insert shows a cross section of a microtubule doublet. (b) Highly schematic representation of the motor region of a (Gram-negative) bacterial flagellum (based upon material drawn from Francis *et al.*, 1992, and Kubori *et al.*, 1992; © 1992 by Academic Press).

forming a structure whose cross section looks like the letters CO; see inset in Fig. 11.8(a). The doublets are cross-linked at various points to form a bundle, and are also linked by a form of dynein called ciliary dynein (as opposed to cytoplasmic dynein described above) at 24 nm intervals. Because the microtubules are cross-linked, the effect of the dynein motors attempting to walk along the filaments causes them to bend, as demonstrated in Fig. 11.8(a). The period of the wave motion is 0.1–0.2 s.

Flagella attached to sperm cells undulate transversely, although their motion is not identical to that of cilia. In contrast, the flagella of the bacterium in Fig. 11.1 rotate about their axis, driven by a much more complicated mechanism than that of sliding filaments. A schematic drawing of the rotary mechanism for a Gram-negative bacterium is shown in Fig. 11.8(b), although the drawing does not do justice to its complexity. Just as with a conventional electric motor, there is a rotor surrounded by a stator, in this case consisting of proteins. Seals must be present as well, as the motor passes through two membranes of the bacterium. These flagella are capable of 150 rotations per second, and provide a means of propelling the bacterium (see Chapter 6 of Berg, 1983, and Berg, 2000). We now describe several molecular aspects of motility, before returning in Section 11.4 to the forces and torques generated within the cell.

11.2 Polymerization of actin and tubulin

Actin and tubulin filaments arise from the polymerization of subunits, which undergo a chemical change upon incorporation into a filament. Actin monomers contain an ATP molecule that hydrolyzes to ADP shortly after polymerization; similarly, the β-unit of the tubulin heterodimer contains a GTP molecule that hydrolyzes to GDP after polymerization. As the subunits in the filament are chemically inequivalent to free monomers, the filaments are capable of two phenomena – treadmilling and dynamic instability – not present in the simplest form of polymerization where free and bound subunits are equivalent. Our pedagogical approach to understanding this complex behavior is to present a simple model for polymerization first, followed by discussions of treadmilling and dynamic instability.

11.2.1 Simple polymerization

Consider the time rate of change, dn/dt, of the number of monomers n in a single filament. In principle, monomers can be both captured and released by a filament, as illustrated in Fig. 11.9, the two processes generally having different rates. The rate equation for the capture of monomers by a single filament has the form

$$dn/dt = +k_{on}\,[M], \qquad \text{(capture)} \qquad (11.1)$$

where $[M]$ is the concentration of free monomer in solution and k_{on} is the capture rate constant, with units of $[concentration \cdot time]^{-1}$. That is, the number of monomers captured per unit time is proportional to the number of monomers available for capture. In contrast, the release rate from

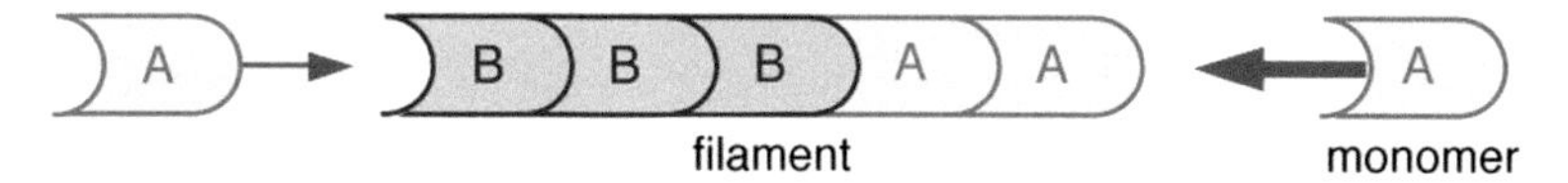

Fig. 11.9 The capture (or release) rates at each end of a filament need not be equal if there is a chemical change of free monomer *A* to monomer *B* upon polymerization, as occurs for actin and tubulin. In the simplest polymerization process, *A* and *B* are identical.

a single filament need not depend upon the free monomer concentration, and may be governed by the simple expression

$$\mathrm{d}n/\mathrm{d}t = -k_{\mathrm{off}}, \qquad \text{(release)} \qquad (11.2)$$

where k_{off} has units of $[time]^{-1}$. Taking both processes into account, we have

$$\mathrm{d}n/\mathrm{d}t = +k_{\mathrm{on}}\,[M] - k_{\mathrm{off}}, \qquad (11.3)$$

a form proposed by Oosawa and Asakura (1975). Note than Eq. (11.3) applies only once a nucleation site for filament growth is available.

The linear dependence of the elongation rate on monomer concentration is observed experimentally for both actin and tubulin filaments, although their behavior is slightly more subtle than Eq. (11.3) because of hydrolysis. The two rate constants k_{on} and k_{off} are easily extracted from a plot of $\mathrm{d}n/\mathrm{d}t$ against $[M]$, as shown in Fig. 11.10; k_{on} is the slope of the line and $-k_{\mathrm{off}}$ is the y-intercept. Negative values of $\mathrm{d}n/\mathrm{d}t$ mean that the filament is shrinking, with the result that the minimum concentration for filament growth $[M]_{\mathrm{c}}$ (often called the *critical concentration*) is

$$[M]_{\mathrm{c}} = k_{\mathrm{off}} / k_{\mathrm{on}}, \qquad (11.4)$$

arising when $\mathrm{d}n/\mathrm{d}t$ in Eq. (11.3) vanishes.

11.2.2 Effects of hydrolysis

The protein subunits of actin filaments and microtubules undergo more complex polymerization reactions than the simple situation summarized in Fig. 11.10. First, the subunits are not symmetric and add to each other with a preferred orientation, giving rise to an oriented filament as well. Second, the free subunits carry a triphosphate nucleotide (ATP for actin, GTP for tubulin) which is hydrolyzed to a diphosphate (ADP for actin, GDP for tubulin) after polymerization. Being chemically inequivalent, the ends of the filament polymerize at different rates, with the faster-growing (slower-growing) end refered to as the plus (minus) end. The various combinations of filament end (plus or minus), nucleotide state (di- or triphosphate) and reaction process (capture or release) mean that a minimum of $2^3 = 8$ rate constants are needed to describe the simplest generalization of Eq. (11.3),

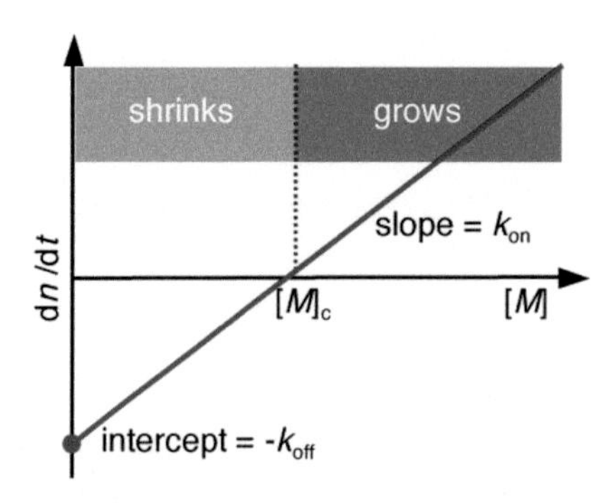

Fig. 11.10 Filament growth rate dn/dt as a function of free monomer concentration $[M]$ expected from Eq. (11.3). The slope of the line is k_{on} and the y-intercept is $-k_{off}$. Filaments grow only if $[M]$ exceeds the critical concentration $[M]_c$ defined by Eq. (11.4).

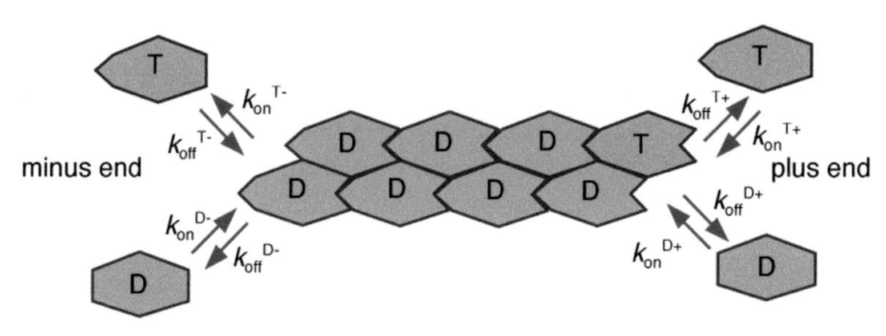

Fig. 11.11 The subunits of actin filaments and microtubules are asymmetric and contain a triphosphate nucleotide (T) that hydrolyzes after polymerization to a diphosphate (D). The capture and release of these subunits is described by a minimum of eight inequivalent rate constants. Note the stylized difference in conformations of the free and polymerized subunits.

as illustrated in Fig. 11.11. Accommodating the hydrolysis of ATP and factors such as local chemical composition of the filament requires even more terms in the rate equations (Pantaloni *et al.*, 1985).

The rate constants are obtained by observing the time-evolution of filaments by a variety of methods (for a review, see Chapters 3 and 7 of Amos and Amos, 1991). Usually, a mechanism must be devised for identifying the plus and minus ends of the filament in order to isolate the plus and minus rate constants. For example, growth from a particular nucleating center may allow the orientation of the filament to be distinguished (e.g. Bergen and Borisy, 1980; Bonder *et al.*, 1983; Pollard, 1986). Alternatively, a specific protein might be used to cap one end of the filament and restrict further changes to it (see Korn *et al.*, 1987); in actin filaments, for example, phalloidin inhibits disassembly, while cytochalasin inhibits assembly at the plus end. The lengths of the filaments themselves can be obtained directly through electron microscopy (Bergen

Table 11.1 Samples of measured values for rate constants of actin filaments (Pollard, 1986) and microtubules (Walker *et al.*, 1988); units of $(\mu M \cdot s)^{-1}$ for k_{on}, s^{-1} for k_{off}, and μM for $[M]_c$

Monomer in solution	$k_{on}+$ (plus end)	$k_{off}+$	k_{on}^- (minus end)	k_{off}^-	$[M]_c+$	$[M]_c^-$
Actin						
ATP-actin	11.6±1.2	1.4±0.8	1.3±0.2	0.8±0.3	0.12±0.07	0.6±0.17
ADP-actin	3.8	7.2	0.16	0.27	1.9	1.7
Microtubules						
growing (GTP)	8.9±0.3	44±14	4.3±0.3	23±9	4.9±1.6	5.3±2.1
rapid disassembly	0	733±23	0	915±72	not applicable	

Notes. The actin measurements vary by a factor of two or more with the solute concentration (for a more complete compilation, see Sheterline *et al.*, 1998). The growth mode of microtubules is measured with GTP-tubulin, while rapidly retreating microtubules are presumably GDP-tubulin.

and Borisy, 1980; Bonder *et al.*, 1983; Mitchison and Kirschner, 1984; Pollard, 1986) and optical microscopy (Horio and Hotani, 1986; Walker *et al.*, 1988; Verde *et al.*, 1992) or indirectly via light scattering or spectroscopically (Lal *et al.*, 1984).

The dependence of the rate constants on the nucleotide composition of the monomer is obtained by observing filaments in pure solutions of ATP- or ADP-monomers; the rates also may depend strongly on other solutes. Sample values for the rate constants obtained for both actin and tubulin are given in Table 11.1. The first column on the left specifies the monomer in solution, while the headings above the data indicate the filament end at which the reaction occurs. The actin results quoted here are similar to those found with electron microscopy by Bonder *et al.* (1983), but are larger than other measurements summarized in Korn *et al.* (1987). The stated microtubule rates are somewhat larger than Bergen and Borisy (1980). The data establish that:

- for both triphosphate and diphosphate monomer, the capture and release rates are almost always larger at the plus end than at the minus end
- the capture rates of the triphosphate are larger than the diphosphate at both ends.

In addition, the rate constants depend on the composition of the solvent; for example, altering the salt content of a solution at fixed monomer concentration may change some ks by more than a factor of two for actin (see collection of measurements in Lal *et al.*, 1984, and Pollard, 1986). The ratio k_{off}/k_{on} of the rate constants gives the critical concentrations shown in the two columns at the right, as expected from Eq. (11.4). Other determinations of the critical concentrations for ATP-actin (Lal *et al.*, 1984) are generally higher than the range given in the table.

Microtubules appear to grow only in the presence of GTP-tubulin (Walker *et al.*, 1988). Once incorporated into the microtubule and hydrolyzed, the resulting GDP-tubulin is capable of very rapid disassembly: k_{off} is one or two orders of magnitude larger for tubulin than it is for actin. Other studies of microtubules using dark field microscopy (Horio and Hotani, 1986) and electron microscopy (Mitchison and Kirschner, 1984) also yield values of hundreds of dimers per second for k_{off}. Note that the critical concentrations $[M]_c$ for GTP-tubulin are the same at each filament end to within experimental uncertainties. We return to the dynamics of microtubule growth after describing the phenomenon of treadmilling observed for actin.

11.2.3 Treadmilling

The fact that actin and tubulin subunits undergo hydrolysis leads to an interesting dynamical mode called treadmilling (Wegner, 1976; Bonder *et al.*, 1983). We can see how this arises by inspecting the rate equations for each end of the filament, having the approximate form

$$dn^+/dt = k_{on}^+ [M] - k_{off}^+ \quad (11.5a)$$

$$dn^-/dt = k_{on}^- [M] - k_{off}^-, \quad (11.5b)$$

where +/– refer to the filament end. The triphosphate version of free actin and tubulin is copiously present *in vivo*, and clearly dominates the rate constants in Table 11.1; hence we take $[M]$ to be the concentration of free triphosphate proteins. Viewed independently, each of these equations may lead to a different critical concentration

$$[M]_c^+ = k_{off}^+ / k_{on}^+ \qquad [M]_c^- = k_{off}^- / k_{on}^-. \quad (11.6)$$

Examples of the critical concentrations for solutions containing ATP-actin and GTP-tubulin are shown on the two right-hand columns of Table 11.1. In the special case $[M]_c^+ = [M]_c^-$, which is approximately obeyed by tubulin, the rate of growth or shrinkage has the schematic form in Fig. 11.12(a): at a given free monomer concentration, both ends grow or both ends shrink simultaneously, although the *rates* of growth or shrinkage may be different.

The dynamics are more intriguing for the general situation $[M]_c^+ \neq [M]_c^-$ displayed in Fig. 11.12(b). At high or low values of $[M]$, both ends grow or both ends shrink simultaneously, as in Fig. 11.12(a). However, in the intermediate concentration range $[M]_c^+ < [M] < [M]_c^-$, the plus end grows at

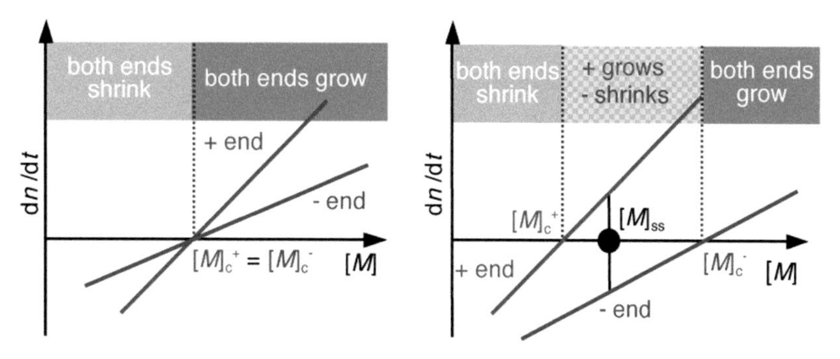

Fig. 11.12 (a) If $[M]_c^+ = [M]_c^-$, both filament ends grow or shrink simultaneously. (b) If $[M]_c^+ \neq [M]_c^-$, there is a region where one end grows while the other shrinks. The vertical line indicates the steady-state concentration $[M]_{ss}$ where the filament length is constant.

the same time as the minus end shrinks. A special case occurs when the two rates have the same magnitude (but opposite sign): the total filament length remains the same although monomers are constantly moving through it. Setting $dn^+/dt = -dn^-/dt$ in Eqs. (11.5), this steady-state dynamics occurs at a concentration $[M]_{ss}$ given by

$$[M]_{ss} = (k_{off}^+ + k_{off}^-) / (k_{on}^+ + k_{on}^-). \quad (11.7)$$

Here, we have assumed that there is a source of chemical energy to phosphorylate, as needed, the diphosphate nucleotide carried by the protein monomeric unit; this means that the system reaches a steady state, but not an equilibrium state.

The behavior of the filament in the steady-state condition is called treadmilling, as illustrated in Fig. 11.13. Inspection of Table 11.1 tells us that treadmilling should not be observed for microtubules since the critical concentrations at the plus and minus ends of the filament are the same; that is, $[M]_c^+ = [M]_c^-$ and the situation in Fig. 11.12(a) applies. However, $[M]_c^-$ is noticeably larger than $[M]_c^+$ for actin filaments, and treadmilling should occur. If we use the observed rate constants in Table 11.1 for ATP-actin solutions, Eq. (11.7) predicts treadmilling is present at a steady-state actin concentration of 0.17 μM, with considerable uncertainty. A direct measure of the steady-state actin concentration under not dissimilar solution conditions yields 0.16 μM (Wegner, 1982). At treadmilling, the growth rate from Eqs. (11.5) is

$$dn^+/dt = -dn^-/dt = (k_{on}^+ \bullet k_{off}^- - k_{on}^- \bullet k_{off}^+) / (k_{on}^+ + k_{on}^-), \quad (11.8)$$

corresponding to $dn^+/dt = 0.6$ monomers per second for $[M]_{ss} = 0.17$ μM.

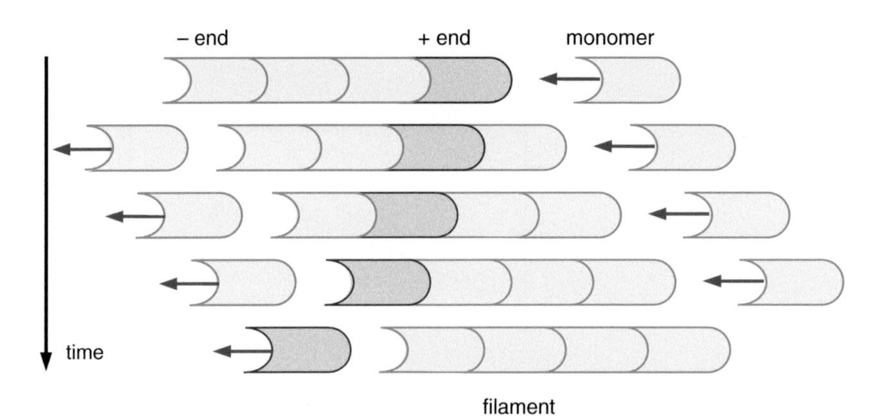

Fig. 11.13 Time evolution of a filament undergoing treadmilling. The magenta monomer is seen to move along the filament during succesive additions, while the length of the filament remains constant. Whether the filament remains stationary depends upon its mechanical environment.

11.2.4 Dynamic instability

Rate equations such as Eq. (11.3) generally represent average behavior, perhaps an average over a number of systems, or perhaps a time average over a single system. Clearly, in a molecular picture, the capture or release of a monomer at a filament end is stochastic, such that $\mathrm{d}n/\mathrm{d}t$ varies with time. Over how long a time interval must the rate be measured in order to appear relatively constant? Observations of a single filament using optical microscopy, such as the plot shown in Fig. 11.14, indicate that the growth rate is uniform when sampled at time intervals of fractions of a second. However, for microtubules, growth is interrupted at seemingly random locations where dramatic shortening of the filament occurs. That is, growth does not appear to continue forever (unbounded), in spite of the presence of tubulin monomer; rather, the filaments alternately lengthen and shrink. The figure shows that disassembly events occur at both the plus and minus ends of the filament. This intermittent disassembly of the microtubule, referred to as dynamic instability, can be suppressed by the presence of a number of microtubule associated proteins (or *MAPS*; see Horio and Hotani, 1986).

Time scales for the growth and retreat phases were first measured by Horio and Hotani (1986) and more systematically by Walker *et al.* (1988). The frequency of disassembly is found to decrease linearly with monomer concentration above $[M]_c$, becoming very small above $[M] \sim 15$ μM. Similarly, the frequency of recovery during a disassembly event rises linearly with monomer concentration for $[M] > [M]_c \sim 5$ μM. That is, disassembly events occur less frequently and are of shorter duration as the concentration of monomers rises.

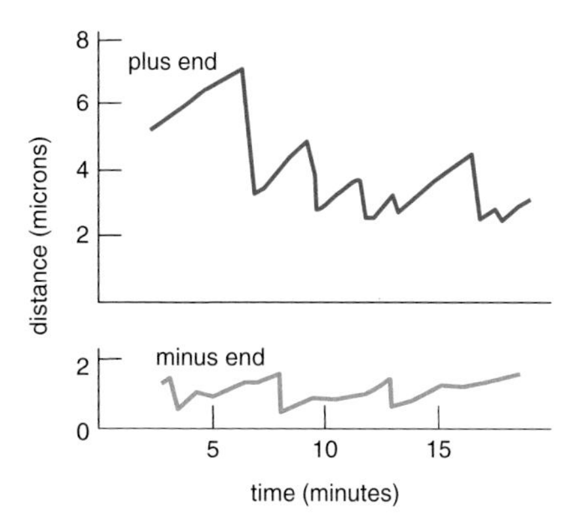

Fig. 11.14 Growth of the plus and minus ends of a single microtubule as seen with optical microscopy (redrawn with permission from Horio and Hotani, 1986; ©1986 by Macmillan Magazines Limited).

It is generally thought that the instability is associated with the hydrolysis of GTP at the microtubule tip (Mitchison and Kirschner, 1984; see also Carlier and Pantaloni, 1981). As illustrated in Fig. 11.4, the GTP-bearing tubulin hydrolyzes to GDP-tubulin once the molecule is polymerized. If the growth rate of the tip is fast enough, the hydrolysis reaction lags behind monomer capture and a GTP-rich cap builds up at the filament tip. From Table 11.1, GTP has a much lower off-rate k_{off} than GDP-tubulin, so the GTP cap is more stable than the main body of the filament. However, if growth slows for whatever reason (e.g. fluctuations in monomer concentration) the cap may be shortened or lost, permitting the filament to disassemble. Thus, this picture predicts that at high monomer concentration $[M]$:

- growth is rapid, the GTP cap is long, and disassembly is infrequent,
- disassembly is more likely to be reversed because $k_{\mathrm{on}}[M]$ is large.

These model expectations are in general agreement with measurements of growth and shrinkage reversal by Walker *et al.* (1988).

A simple set of rate equations that include a parametrization of the random switching between growth and shrinkage epochs has been solved numerically or analytically for some special cases (Verde *et al.*, 1992; Dogterom and Leibler, 1993). This model exhibits a phase transition between unbounded filament growth and the bounded behavior of Fig. 11.14 when

$$f_{-+}\,\mathrm{d}n^{+}/\mathrm{d}t = f_{+-}\,\mathrm{d}n^{-}/\mathrm{d}t, \tag{11.9}$$

where f_{+-} is the frequency of changing from growth to shrinkage, and f_{-+} is the inverse. Although the rate equations do not include a microscopic prediction for parameters such as f_{-+}, the model permits a data analysis whose internal consistency can be checked through observables such as the mean filament length.

Dynamic instability is an efficient growth strategy for microtubules. Among their other roles, microtubules function as probes, which may seek out the cell boundary or organize components of the cell. In its "seeking phase", a microtubule should grow only at a modest rate as it extends through the cell; but once its job is complete (even if unsuccessful), there is no need for the microtubule to retreat gracefully. Rapid disassembly releases the contents of the microtubule to the cytoplasm and permits it, or nearby microtubules, to start searching afresh in a different direction. In some sense, the process involves similar principles as fly fishing, where placing a fishing lure in the water is a fast process, but retrieving the lure is done slowly to increase its likelihood of attracting a fish.

11.2.5 Benchmarks for k_{on} and k_{off}

We have used the measured values of k_{on} and k_{off} to understand dynamic behavior such as treadmilling and catastrophic disassembly, without attempting to interpret their numerical values within a molecular context (for a review of microscopic approaches to filament growth, see Bayley *et al.*, 1994). Molecular diffusion is one component of the overall reaction mechanism and is easily estimated, as we now show. Diffusion-based benchmarks for k_{on} and k_{off} permit us better to visualize the process of monomer capture and release in filament formation.

The disassembly process could be viewed as sequential if the average monomer randomly diffuses away from the filament before the next monomer is released. If the mean time between release is much smaller than this diffusion time scale, the disassembly is, in a sense, more cooperative. Now, the mean square displacement $\langle r^2 \rangle$ of a randomly diffusing particle in three dimensions increases linearly with time t according to

$$\langle r^2 \rangle = 6Dt, \tag{11.10}$$

where D is the diffusion constant as presented in Section 2.4. The diffusion constant for a monomer depends on its size and the viscosity η of the medium; the Stokes–Einstein relation (Section 2.4; see Chapter VI of Landau and Lifshitz, 1987)

$$D = k_B T \,/\, 6\pi\eta R, \tag{11.11}$$

for spherical particles of radius R, provides a starting point for estimating the applicable D. The viscosity of the cytoplasm should be higher than water ($\eta = 10^{-3}$ kg/m•s) but less than glycerine ($\eta = 830 \times 10^{-3}$ kg/m•s), such that Eq. (11.11) predicts D in the range 6×10^{-14} to 5×10^{-11} m^2/s for proteins with a 4 nm radius. For an experimental comparison, lactate dehydrogenase (from dogfish) with $R \sim 4$ nm has $D = 5 \times 10^{-11}$ m^2/s in water (see Chapter 7 of Creighton, 1993). We choose $D \sim 10^{-12}$ m^2/s to be representative of cellular conditions, noting that D for actin monomers in endothelial cells is observed at $3\text{–}5 \times 10^{-12}$ m^2/s (McGrath *et al.*, 1998).

Let's use the filament diameter to set $\langle r^2 \rangle$ of the diffusion process, recognizing that Eq. (11.10) averages over a distribution of paths, including those where the particle has not moved very far. The corresponding time scale for a protein to diffuse over a distance $\langle r^2 \rangle^{1/2} = 25$ nm is 10^{-4} s with our choice of D in Eq. (11.10). Thus, if each successive monomer unbinds once its former neighbor has moved away by a filament diameter, the monomers can be released at a rate k_{off} of order 10^4 s^{-1} for microtubules. The off-rates of tubulin in Table 11.1 are an order of magnitude lower than this benchmark, indicating that disassembly is moderately fast on a molecular length scale but not so fast as to be cooperative. Actin monomers, in contrast, are released much more slowly, presumably reflecting their tighter binding to the filament.

Diffusion also provides us with an estimate for k_{on}. The growing tip of a filament depletes monomers from its neighborhood, such that the local concentration of free monomers is less than the bulk concentration $[M]_\infty$ some distance away. As a result, free monomers flow into the depleted region, driven by the concentration gradient $\text{d}[M]/\text{d}r$, where r is the distance between the monomer and the capture site. The flux of monomers (the number per unit area per unit time) is simply $-D\text{d}[M]/\text{d}r$ from Fick's first law of diffusion, where the minus sign indicates that the flow is in the opposite direction to the concentration gradient (see Section 12.5 of Reif, 1965). The gradient can be integrated to give a local concentration profile, rising smoothly to $[M]_\infty$ from zero at $r = R$, where all monomers are captured. From the number of monomers passing into the capture region, one obtains (see problem set)

$$k_{\text{on}} = 4\pi DR, \tag{11.12}$$

for a stationary reaction region (see Section 6.7 of Laidler, 1987, or Section 6.3 of Weston and Schwarz, 1972; for other biological applications, see Berg and von Hippel, 1985). The expression for k_{on} depends upon the trapping mechanism; e.g., for the bimolecular reaction $A + B \rightarrow AB$ of two mobile species, D is replaced by the sum of the individual diffusion constants $D_A + D_B$. Let us apply Eq. (11.12) to microtubules by taking $D = 10^{-12}$ (m$^2 \bullet$ s)$^{-1}$, as usual, and R to be the sum of the microtubule radius (12 nm) and tubulin size (about 6 nm). With these values, Eq. (11.12) gives $k_{\text{on}} = 2.3 \times 10^{-20}$ m^3/s = 138 (μM $\bullet$ s)$^{-1}$, about an order of magnitude larger than the results in Table 11.1. Given the uncertainty in D and the omission of rotational diffusion when the monomer docks at the filament, this estimate is appropriately larger than the observed values. In other words, the diffusion benchmarks for k_{on} and k_{off} provide a rough value for the capture rate and establish that disassembly is more random than cooperative along the length of the filament.

11.3 Molecular motors

The specialized proteins capable of walking or crawling along a cellular filament are referred to as motor proteins, of which there are separate families for actin and tubulin. The principal motor proteins associated with actin are myosin-I and myosin-II, having the distinct structures shown in Fig. 11.15 (the myosin family has more than two members, by the way). Myosin-I has a single globular "head" of mass ~80 kDa, containing the actin-binding site, and a tail of mass ~50 kDa, giving a typical overall mass in the 150 kDa range. The tail is attached to another structure such as a vesicle, which acquires movement as the myosin head crawls along an actin filament. Myosin-II is a dimer with two actin-binding heads attached to a flexible, but not floppy, tail (see problem set). The head region, at 95 kDa, is augmented by two light chains to bring the total mass to ~150 kDa, while each component of the tail has a mass of ~110 kDa. For both myosins, the linear dimension of the heads is about 15–20 nm. In skeletal muscle, more than one hundred myosin-II dimers can form a bipolar bundle with a length in the micron range, as illustrated in Fig. 11.7. These myosin motors pull themselves towards the plus end of actin filaments.

Also shown in Fig. 11.15 is myosin-V, which has a two-headed structure but with the axis of its "tail" perpendicular to the actin filament. The two "legs" of myosin-V move like feet in a normal walk, in the sense that one leg passes the other in taking a step. In the diagram, the left leg detaches from the actin filament and then reattaches itself a little more than 70 nm to the right, a distance which is double the 36 nm separation between legs. This motion is not like a sideways or crab walk, where the left leg would only move up to the current location of the right leg, after which the right leg would move forward a similar distance.

Two motor proteins associated with microtubules are kinesin and dynein. Structurally, the kinesin dimer has the same general appearance as myosin-II, except that the tail region is shorter (60–70 nm at ~50 kDa per strand) and the heads are smaller (10–15 nm at 45 kDa). Each strand of

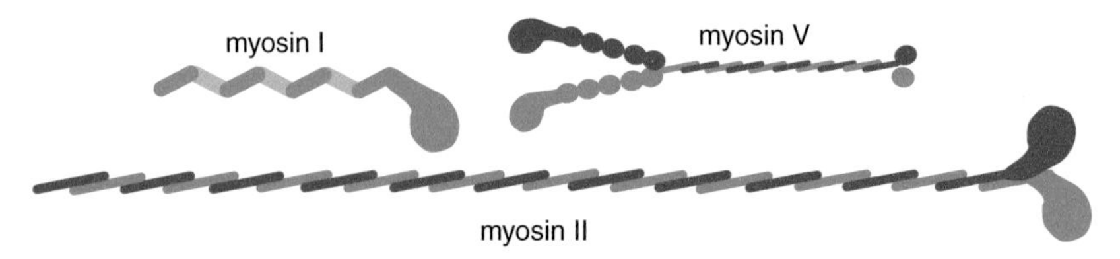

Fig. 11.15 Schematic representation of the motor proteins myosin-I, myosin-II, and myosin-V; the globular ends are their actin-binding sites. These proteins walk on actin filaments towards their plus end. The proteins are all drawn to different scales.

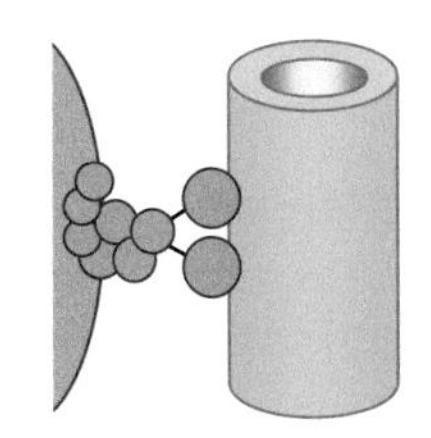

Fig. 11.16

Schematic representation of cytoplasmic dynein showing its attachment to a microtubule (right) and a section of a vesicle (left). It has been suggested that the heads may be linked to microtubules by thin stalks about 10 nm long (see Chapter 8 of Amos and Amos, 1991). Dynein usually walks towards the minus end of a microtubule and kinesin towards the plus end.

the tail terminates in several intermediate length chains, bringing the overall mass to about 180 kDa for each element of the dimer. Both heads of kinesin attach to a single microtubule and slide parallel to a protofilament, with the tail perpendicular to the axis of the microtubule, as shown in Fig. 11.6. Having a mass in the range 1200–1300 kDa, cytoplasmic dynein is much heavier than the other motor proteins and is structurally different, as seen in Fig. 11.16. The globular heads are spherical with diameters of 9–12 nm; cytoplasmic dynein is two-headed whereas ciliary dynein may have between one and three heads. The orientation of cytoplasmic dynein as it links a microtubule to a cellular component like a vesicle is displayed in Fig. 11.16 as well. A more detailed exposition of these proteins can be found in Amos and Amos (1991) and Howard (2001).

The energy source of the motors is ATP, which induces configurational changes during its capture, hydrolysis and release. A proposed mechanism for the structural changes of the myosin motor, based upon a determination of its head geometry from X-ray scattering, is shown in Fig. 11.17 (Rayment *et al.*, 1993a, b). Five configurations are shown in the figure:

(i) tight binding of the myosin head to the actin filament in the "rigor" position (so named from *rigor mortis*),
(ii) release of the filament upon capture of ATP; the head and filament remain in close proximity because of neighboring connection sites,
(iii) configurational change to the cocked position during hydrolysis of ATP; as a result, the head moves along the length of the filament,
(iv) weak binding of the head to the filament in a new position accompanying release of a phosphate group,

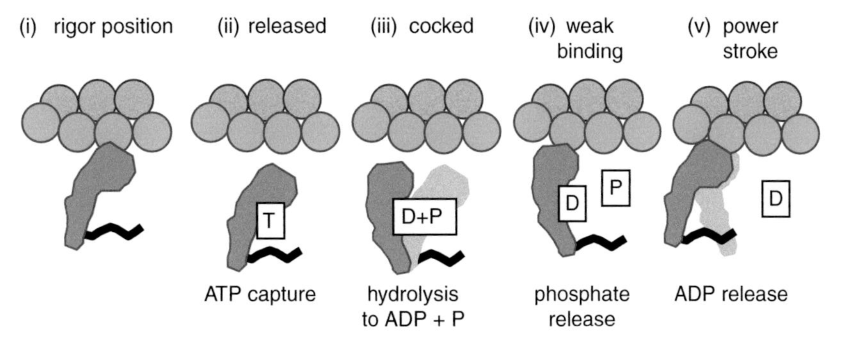

Fig. 11.17

Postulated mechanism for the movement of a myosin head (redrawn with permission after Rayment *et al.*, 1993b; ©1993 by the American Association for the Advancement of Science). Steps (i)–(v) are described in the text; in (iii) and (v) the initial configuration during the step is indicated by the light shaded region and the final configuration is dark. The tail of the myosin molecule, indicated by the wiggly line at the bottom of each frame, would be buried in a myosin bundle in Fig. 11.7. The symbols T, D and P represent ATP, ADP and phosphate, respectively.

(v) configurational change during the power stroke, initiated by the release of ADP.

In the configuration changes of steps (iii) and (v), the position of the head in the previous step is indicated by the light shaded region, while its position after the change is indicated by the shape with the dark border. Typical displacement of the head is 5 nm, occuring with a repetition rate which may be well above a cycle per second. The measurements in support of the directed motion model of Fig. 11.17 have been reviewed by Howard (1995).

Directed motion may be the most energy-efficient means of generating linear movement, and there is evidence in favor of directed motion for myosins-I, -II and -V, even though the motor may step in the reverse direction on occasion. Nevertheless, other mechanisms based on biased random motion may have been important in earlier stages of evolution, or for different types of cellular locomotion. To gain insight into their operation and efficiency, we now analyze a thermal ratchet model for the step-wise movement of motor proteins. Later, in Section 11.5, we examine a variant of this mechanism, in which the actively growing ends of an ensemble of polymerizing filaments exert a force on an object.

11.3.1 Thermal ratchets

A simple, but inefficient, means of generating motion is a thermal ratchet, which makes use of a spatially asymmetric potential that oscillates with time, as sketched out in Fig. 11.18 (for general properties of thermal ratchets, see Ajdari and Prost, 1992; Doering and Gadoua, 1992; Magnasco, 1993; Bier and Astumian, 1993; Prost *et al.*, 1994). The potential $V(x)$ is taken to be periodic in the x-direction, with a peak-to-trough difference in part (a) that is large compared with k_BT; the potential is also periodic in time, alternately turned on and off in the sequence (a)–(c). In part (a), a particle subject to this potential lies near the minimum of $V(x)$, where the probability distribution for its location is sharply peaked. However, when the potential is turned off (b), the particle undergoes a random walk, such that the probability of finding it in a given location is described by a Gaussian distribution centered at the original potential minimum with a dispersion that increases with time. In part (c), the potential is turned back on again, and the particle tumbles towards the nearest potential minimum, which may not be its original position. Because the potential is asymmetric, the probability of moving to the right may be different from that of moving to the left. Thus, application of an oscillating asymmetric potential can generate a net movement in one direction, although the movement is not steady and may reverse direction from one step to the next. Clearly, if the potential is completely symmetric, a particle may jump between local

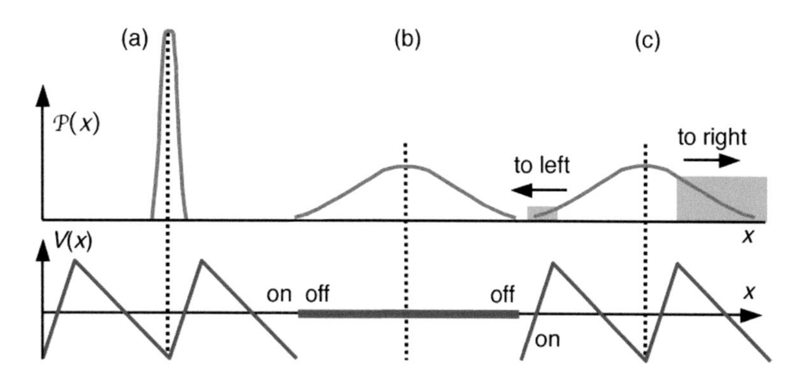

Fig. 11.18 Probability distribution (top row) of a particle subject to an asymmetric potential (bottom row) that is alternately switched on and off from (a) to (c). Initially very sharply peaked at the potential minimum, the probability distribution broadens when the potential is off (b). Restoration of the potential in (c) results in a greater probability that the particle will move to the right than to the left.

potential minima, but with equal probability in either direction, resulting in no preferred direction of travel over long times. The form of the potential energy function is reminiscent of a mechanical ratchet, hence the model is referred to as a thermal or Brownian ratchet, after random or Brownian motion.

Why might this mechanism be important for cells (Astumian and Bier, 1994; Prost *et al.*, 1994)? Suppose that there is a charge distribution along a line of monomers in a filament as in Fig. 11.19. If it were charged, the head of a motor molecule could bind to this filament electrostatically; in the figure, the head on the left side carries a positive charge, and experiences the potential energy shown at the bottom. Upon capture of a charged molecule like ATP, the net charge on the head is reduced or eliminated, rendering the potential energy featureless and releasing the head from the filament. Lastly, when ATP is hydrolyzed and ultimately released, the charge is restored and the head binds to the filament once again. As we now show, this mechanism is not 100% efficient for generating motion, and may not be the method of choice in today's cells. However, because it is operationally simple and may have been present earlier in evolution, we develop an approximate model for thermal ratchets that permits us to probe their efficiency and flux.

We invoke a number of approximations to get an analytically simple form for the thermal ratchet model. First, the rounded peaks and troughs of the potential in Fig. 11.19 are sharpened to give the sawtooth form in Fig. 11.20, where the repeat distance of $V(x)$ is defined as b and the shortest peak-to-trough distance along the x-axis is ab. Second, the one-dimensional Gaussian probability distribution $\mathcal{P}(x) = (2\pi\sigma^2)^{-1/2} \exp(-x^2/ 2\sigma^2)$ is replaced by a triangular form of the form $\mathcal{P}(x) = (1 - |x|/w) /w$, such that

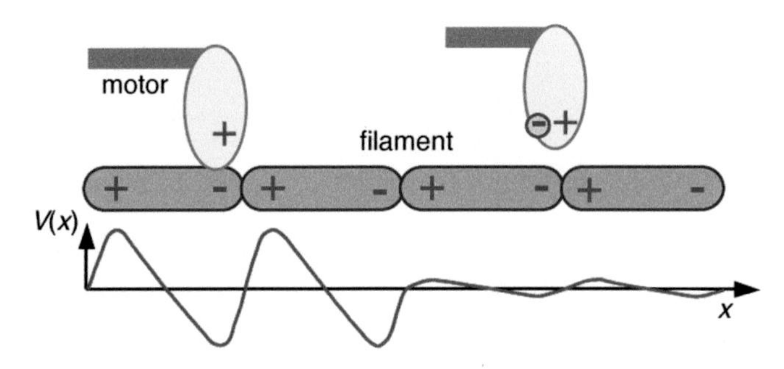

Fig. 11.19 If the monomers of a biofilament have a dipolar charge distribution, the head of a motor molecule carrying a local positive charge experiences a potential energy with a soft sawtooth form (left). This potential is largely eliminated if the head captures an oppositely charged molecule like ATP from solution (right).

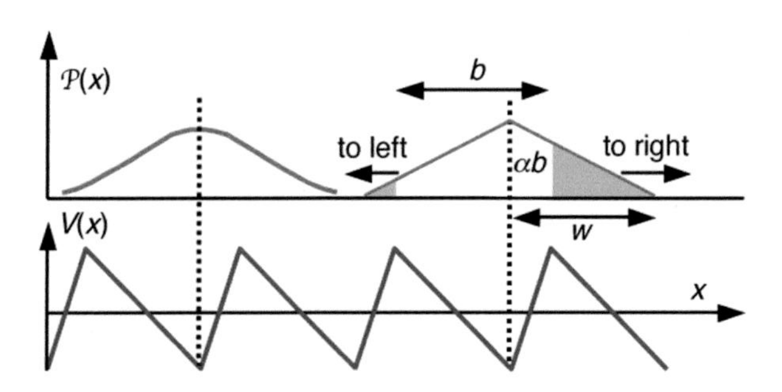

Fig. 11.20 In a simplified model for a thermal ratchet, the potential energy has a sharply peaked sawtooth form with period b and minimum peak-to-trough separation αb. We replace the Gaussian probability distribution by a triangular one with base $2w$ to calculate the probability of motion.

$\int \mathcal{P}(x)\, dx = 1$. Equating the dispersions of the two distributions, namely $\langle x^2 \rangle = \sigma^2$ and $\langle x^2 \rangle = w^2/6$, gives the relation

$$w = \sqrt{6}\, \sigma \cong 2.5\, \sigma, \tag{11.13}$$

as one might intuitively expect from the shapes of the distributions. The accuracy of the triangle approximation is tested further in the problem set.

To begin our calculation of motion in the ratchet model, we refer to Fig. 11.20 and determine the probability of a particle hopping to the right. If the potential is turned off for too short a time, a particle diffuses locally and the probability distribution is very narrow, such that w is less than αb. In such situations, a particle simply returns to its original position once the potential is turned on again. For somewhat longer diffusion times with $w > \alpha b$, the particle may have moved so far that it seeks out a neighboring minimum once the potential is restored. Now, we will assume that the potential is turned on so fast that any particle moves rapidly down its potential gradient to a local energy minimum and does not diffuse into other regions of the system. The probability of movement to the right is then equal to the area of the right-hand shaded triangle in Fig. 11.20, or $\int \mathcal{P}(x)\, dx$ integrated over $ab \le x \le w$. With the form $\mathcal{P}(x) = (1 - |x|/w)\,/w$, the probability of movement to the right is $P_R = (1 - ab/w)^2/2$, for $w > ab$. This same argument can be repeated for motion to the left. If $w < (1-a)b$, the distribution is sufficiently narrow that the probability of motion to the left is zero in the triangle approximation: $P_L = 0$. However, once the distribution is wide enough, the probability of a hop to the left is $P_L = [1 - (b - ab)/w]^2/2$. The net probability of motion, P_{net}, is the difference between P_R and P_L

$$P_{net} = P_R - P_L, \tag{11.14}$$

and can be summarized by:

$$P_{net} \begin{cases} = 0 & 0 < w < ab & (11.15a) \\ = (1 - ab/w)^2/2 & ab < w < (1-a)b & (11.15b) \\ = (b/w)\bullet[1 - (b\,/\,2w)]\bullet(1-2a) & (1-a)b < w. & (11.15c) \end{cases}$$

Let's examine the properties of P_{net} as a function of a. When $a = 1/2$, the potential is symmetric and there is equal probability of movement to the left or right; hence, $P_{net} = 0$ in Eq. (11.15) (note that Eq. (11.15b) has no domain at $a = 1/2$). At the other extreme, when $a = 0$, one wall of the potential is vertical and the probability distribution is split right through the center when the potential is turned on. This gives an upper bound on P_{net} of 1/2, again as can be verified in Eqs. (11.15). At intermediate values of a, the maximum value of P_{net} is less than 1/2, as shown for the case $a = 0.1$ in Fig. 11.21. The probability rises steadily from zero once $w > ab$, reaching a maximum value of 2/5 at $w/b = 1$ before slowly falling towards zero at large w/b. By solving the condition $dP_{net}/dw = 0$ at $w/b > a$, it is easy to show that the maximum probability for net motion P_{max} occurs at $w/b = 1$, and has the value

$$P_{max} = (1-2a)/2. \tag{11.16}$$

This expression yields $P_{max} = 0.4$ for the parameter choice in Fig. 11.21.

We were forced to make a number of strong approximations in deriving Eqs. (11.15), and each of these approximations leads to an overestimate of P_{net} for molecular motors. For example, the physical potential does not have a sharp peak and is likely to appear bumpy in all regions because of the charge distribution and other effects. In addition, the time for switching the potential on and off is not instantaneous, but is at least on the order of the time to capture ATP or release ADP. Both

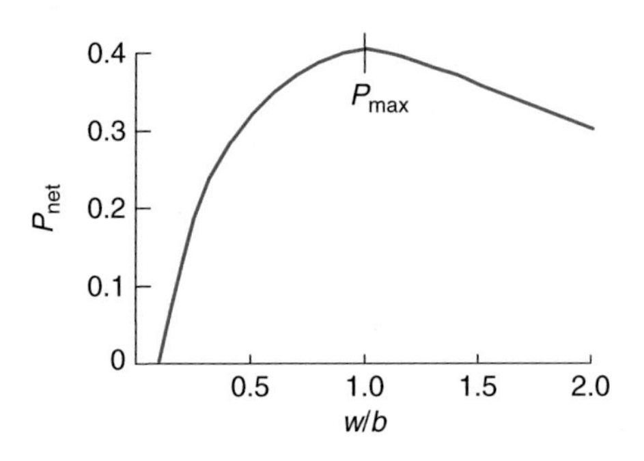

Fig. 11.21 The probability of net motion to the right P_{net} as a function of w/b for a sawtooth potential with α = 0.1, according to Eqs. (11.15); $P_{max} = 0.4$ at $w/b = 1$.

of these effects give the motor molecule more time to diffuse forward or backward along the filament, thus reducing the probability of motion in a specific direction. More accurate evaluation of the properties of simple potentials has been performed by Astumian and Bier (1994) and Prost *et al.* (1994), who find that P_{max} may be 0.25 or much less, depending on the potential.

Let us estimate the flux of thermal ratchets to determine if they are even relevant for molecular motors in the cell. First, we use the diffusion equation in one dimension to estimate the time scale for the step. The trajectory of a diffusing particle obeys the usual Gaussian distribution with $\sigma^2 = 2Dt$, where D is the diffusion constant and t is the time, such that $w = \sqrt{6}\,\sigma = (12Dt)^{1/2}$ according to Eq. (11.13). Now, in Section 11.2, our order of magnitude estimate for D was 10^{-12}–10^{-14} m^{2}s, indicating that t must be 5×10^{-6} to 5×10^{-4} s if we impose the condition for P_{max}: $w = b = 8$ nm. This time scale is rather small: if the period for one cycle of the motion is twice this time, the motor could step at a rate of 10^3 to 10^5 steps per second. Of course, the local chemical environment of the motor may not be immediately ready for a new cycle at the end of the power stroke, increasing the time between completed steps; for instance, myosin is observed to move at less than 200 steps per second (Finer *et al.*, 1994). At 10^3 attempts per second and 8 nm per step at a success rate of, say $P_{net} = 1/10$, the motor could advance at 800 nm/s. This is clearly in the experimental range ~0.5 μm/s observed for kinesin motors by Svoboda *et al.* (1993, 1994).

The primary drawback of the ratchet mechanism, for an advanced cell, is efficiency: it may take the hydrolysis of 10 ATP molecules to complete one step (if, for example, $P_{net} = 0.1$). Certainly, this efficiency is much lower than the directed motion of conformational changes such as Fig. 11.17, which could complete one step for every ATP hydrolysis. This question has been addressed experimentally by Svoboda *et al.* (1994) who measured the fluctuations, as a function of time, in the position of a kinesin motor sliding on a microtubule. Having established that the movement occurs in 8 nm steps (Svoboda *et al.*, 1993), they show that the probability of completing a step per ATP hydrolysis is 0.5, which lies at the upper limit of the ratchet model developed here. Realistic ratchet models have efficiencies that range from 0.02 to 0.25 as considered by Astumian and Bier (1994) or Prost *et al.* (1994). The implications of these observations for two-step models are described by Svoboda *et al.* (1994). What we have demonstrated, then, is that the speeds of simple thermal ratchets are not unreasonable, but their efficiency is low, which may restrict their importance to earlier stages in the evolutionary development of motor mechanisms.

11.4 Forces from filaments

The previous two subsections are devoted to quantifying filament dynamics at the microscopic level: Section 11.2 analyzes the rates of growth and shrinkage of filaments in solution, while Section 11.3 examines the motion of molecular motors as they slide, dance or walk along a filament. The speeds observed for several types of motion associated with filaments are summarized in Table 11.2. The important feature to note is that the benchmark speed is 1 μm/s for many processes, although cellular movements such as slow axonal transport admittedly are much less. We have already established that these speeds could be achieved by a variety of mechanisms, although we have not ranked the importance, or even the relevance, of such mechanisms in actual cells. In this section, we place dynamic filaments in a cellular environment: our discussion is organized by filament type (actin first, then microtubules) and complexity of system (isolated filaments first, then composite mechanical systems). The mechanisms of bacterial propulsion are the subject of Section 11.5, and the forces involved in DNA replication and mitosis are treated in Chapter 12.

11.4.1 Actin filaments

The growth rates of individual actin filaments depend on the concentration of actin monomer $[M]$, which varies from cell to cell, and even within regions of a given cell. The total actin concentration of many cells falls in the 1–5 mg/ml range, although much of this actin may be bound in filaments rather than free in solution. Given a molecular mass of 42 000 Da, a 1 mg/ml actin solution is equivalent to 24 μM, for example. At this concentration, the plus end of an actin filament would grow at a rate of 280 monomers per second, according to the rate constants of Table 11.1. Each

Table 11.2 Summary of speeds observed in various types of cellular movement. The monomer concentration in solution is denoted by $[M]$. Fast axonal transport involves kinesin or dynein moving along microtubules

Motion	Typical speed (μm/s)	Example
actin filament growth	10^{-2}–1	0.3 μm/s at $[M]$ = 10 μM (Table 11.1)
actin-based cell crawling	10^{-2}–1	fibroblasts move at ~10^{-2} μm/s
myosin on actin	10^{-2}–1	0.1–0.5 μm/s common in muscles
microtubule growth	up to 0.3	0.03 μm/s at $[M]$ = 10 μM (Table 11.1)
microtubule shrinkage	0.4–0.6	0.5 μm/s (Table 11.1)
fast axonal transport	1–4	
slow axonal transport	10^{-3}–10^{-1}	

monomer adds 2.73 nm to the filament length (as can be verified from a mass per unit filament length of 16 000 Da/nm from Table 3.2), such that a rate of 300 monomers per second at the plus end corresponds to 0.8 μm/s in length. However, free actin concentrations in endothelial cells, for example, are 1/2 to 2/3 of total actin concentrations (McGrath *et al.*, 1998), pushing the expected growth rate down to about 0.3 μm/s. Thus, filament growth rates approaching 1 μm/s are achievable for actin, but only if the monomer concentration is rather large (see also Abraham *et al.*, 1999).

One mechanism for cell movement, shown in Fig. 11.4(b), involves the growth and disassembly of actin filaments at the leading edge of a lamellipodium. Is the rate of filament growth consistent with the movement of the cell? Although there are cells capable of crawling at a micron per second, typical speeds of crawling fibroblasts are closer to 10^{-2} μm/s, easily within the range of actin filament growth. Of course, the motion of a lamellipodium involves more than filament elongation at the cell boundary: materials such as monomeric actin and other proteins must be transported to the leading edge (where the plus-end of the filament resides) and each of these transport processes has a time frame. Several models have been proposed for the cooperative machinery of a crawling cell, and the reader is directed to the literature for more details (a sampling would include Dembo and Harlow, 1986; Dembo, 1989; Lee *et al.*, 1993; Evans, 1993; Lauffenburger and Horwitz, 1996; Mitchison and Cramer, 1996).

Does the behavior of actin filaments in lamellipodia arise from unassisted treadmilling? In isolated actin filaments, treadmilling is a relatively slow process: we showed from Eq. (11.8) that dn^+/dt is just 0.6 s^{-1} at steady state, or about 0.002 μm/s in filament length for the particular values of the rate constants in Table 11.1 (see also Selve and Wegner, 1986, and Korn *et al.*, 1987). This speed is rather low compared to most processes in Table 11.2. In some locomoting cells, actin filaments do have a roughly constant filament length even though the cell boundary may be moving at 0.1 μm/s; actin filament formation in neuronal growth cones also proceeds at 0.05–0.1 μm/s (Forscher and Smith, 1988). In such situations, the apparent treadmilling may be caused by actin-related proteins that sever the filament away from the leading edge of the cell, exposing regions of ADP-type actin that are more prone to disassembly.

Listeria monocytogenes and *Shigella flexneri* (Gram positive and negative, respectively) are two bacteria whose locomotion makes use of actin filaments within a host cell but *outside* the bacterium itself (Dabiri *et al.*, 1990; Theriot *et al.*, 1992; Goldberg and Theriot, 1995; and references therein). The surface of the bacterium contains a protein that helps organize free actin into a comet-like tail of cross-linked filaments pointing away from the direction of motion, as in Fig. 11.22. The plus end of the filaments is at the bacterium (Sanger *et al.*, 1992) and grows at rates up to 0.2–0.4 μm/s. Although composed of shorter actin segments, the tails themselves range up to about 10 μm in length and are stationary with respect to the host cell (Dabiri *et al.*, 1990; Theriot *et al.*, 1992). Newly incorporated monomers reside at their local position in the tail for about 30–40 s on average, independent of the speed of the bacterium. Thus, the tail length increases with the speed of the bacterium: the separation between ends equals the product of the bacterial speed with the (constant) lifetime of the tail. Theoretical approaches to this motion have concentrated on mechanisms by which actin can insert at the plus end against the bacterial surface (Mogilner and Oster, 1996; for an earlier version, see Peskin *et al.*, 1993). Thermal fluctuations in the gap between the bacterium and filament end, arising in part from its bending motion (see Kroy and Frey, 1996), are sufficient to explain the observed filament growth rates.

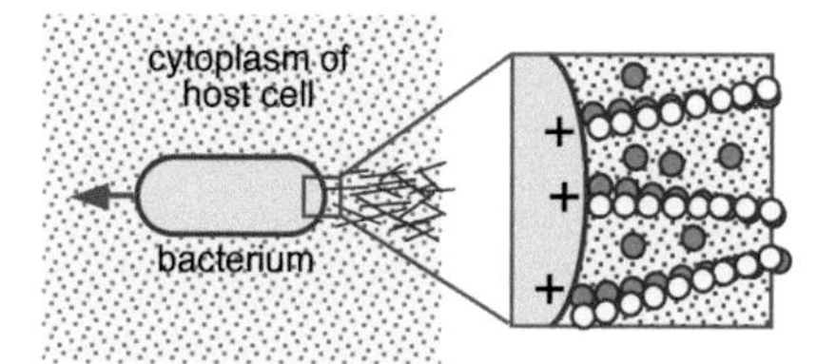

Fig. 11.22 Several bacteria are capable of sequestering actin monomers from a host cell into filaments with their plus ends at the bacterium. The filaments are stationary with respect to the host cell, and the capture of new monomers at the plus end generates a locomotive force on the bacterium.

11.4.2 Actin/myosin systems

The relative speed of actin and myosin falls in a range from 10^{-2} to 1 μm/s, depending on conditions in the cell. As discussed in Section 11.3, this range is anticipated by a variety of mechanical models, even the noisy motion of thermal ratchets. If the myosin head in Fig. 11.17 moves forward at about 5 nm per completed step (Veigel *et al.*, 1998), and if 10^1–10^2 steps can be completed per second (a range permitted by diffusion and also observed in optical trap experiments by Finer *et al.*, 1994) then the motor protein can advance at a rate of 0.05 to 0.5 μm/s. How is this speed related to the macroscopic speeds of our muscles? From Fig. 11.7, the relative motion of myosin on actin causes the two Z-discs (defining the boundary of the sarcomere) to advance towards each other at a speed twice that of the individual myosin head, i.e. 0.1–1 μm/s in this calculation. Given that the sarcomere in the figure has a length in its relaxed state of 2.2 μm, the muscle unit may contract by 5%–45% per second. This admittedly broad relative change is consistent with the motion of the long muscles in our arms or legs; taking 30 cm as a typical muscle length, a 40% change corresponds to 12 cm per second, or 1 cm in a tenth of a second. Recalling the lever-like mechanical structure of our arms (i.e., our hand swings through perhaps 10 cm when our biceps contracts by 1 cm), the rapid contraction of our muscles has an order of magnitude of 10 cm/s.

The force generated by a single myosin motor has been measured by several techniques. Early experiments by Hill (1981) found a force of 8 pN (pN = 10^{-12} N) and observations using optical traps by Finer *et al.* (1994) yields 3–4 pN. We can estimate the force generated in a single step knowing that the energy provided by the hydrolysis of an ATP molecule is about 20 k_BT. If 50% of the energy released results in a single 5 nm step of the motor, the resulting force must be $0.5 \cdot 20k_BT / 5$ nm = 8 pN from [*work*] = [*force*] • [*distance*]. This estimate is consistent with the measured values.

11.4.3 Microtubules

The concentration of tubulin dimer in the cell, expressed as a mass per unit volume, is in the same general range as G-actin, although the corresponding molarity is a factor of two lower because of the difference in molecular mass – 42 kDa for G-actin and 100 kDa for tubulin dimers. The total tubulin concentration in a common cell like a fibroblast is 2 mg/ml, or 20 μM, of which half is free and half is bound in filaments, a similar ratio of free- to-bound monomer as found for actin. Thus, the typical concentration of free tubulin dimer is 10 μM. From the rate constants in Table 11.1, we expect the plus end of a microtubule to grow at ~50 units per second at 10 μM and to catastrophically disassemble at 800 s^{-1}, this latter value independent of tubulin concentration. At 13 protofilaments per microtubule, a single 8 nm long tubulin unit adds just 8/13 = 0.62 nm to the overall length of a filament, so the growth and shrinkage speeds of a microtubule are 0.03 and 0.5 μm/s, respectively. Unlike actin, the crawling motion of cells may not require the presence of microtubules (Euteneuer and Schliwa, 1984).

11.4.4 Motor/microtubule systems

The motor proteins kinesin and dynein walk towards the plus and minus ends of microtubules, respectively, at a similar range of speeds, from 1 to 4 μm/s. Given that the polarity of microtubules in most cells has the plus end towards the cell boundary, the two molecular motors permit the transport of cargo-laden vesicles or organelles to and from the protein manufacturing sites near the cell's nucleus. Vesicle transport over long distances is particularly important for nerve cells, where chemical neurotransmitters produced in the cell body must be delivered to a synapse (anterograde motion) and waste products must be returned to the cell body for recycling (retrograde motion). Parenthetically, we note that the microtubules of the axonal highways are continuous only over distances of a few hundred microns on average, which is much less than the length of some axons. Both anterograde and retrograde motion have a fast component of a few microns per second; their cargo may range from small vesicles to mitochondria, which provide ATP to molecular motors as they work their way along axonal microtubules. In addition to this *fast axonal transport* in both directions, there is *slow axonal transport* in the anterograde direction, with much lower speeds of 10^{-3} to 10^{-1} μm/s (axonal transport is treated in more detail in Amos and Amos, 1991). The speed of the fast component *in vivo* is consistent with, and perhaps faster than, the motion of molecular motors as we described in Section 11.3.

The force generated by a motor on a single microtubule has been measured by several independent experiments to be ~2.6 pN (Ashkin *et al.*, 1990), 1.9±0.4 pN (Kuo and Sheetz, 1993), 4–5 pN (Hunt *et al.*, 1994) and 4–6 pN (Svoboda and Block, 1994). Our general expectation of the force, based on nucleotide hydrolysis as we discussed for myosin, is 5 pN for 20 k_BT released per hydrolysis resulting in a step of 8 nm at 50% efficiency. This force range is consistent with the drag force on a vesicle, as described in the problem set.

11.5 Cell propulsion

What we have described in Sections 11.2 to 11.4 is the movement of materials within cells or the change in cell shape resulting from molecular motors or the dynamic polymerization of filaments. Broadly speaking, the focus has been on changes within the cell, although we recognize that the polymerization of actin in the lamellipodium is a mechanism for the crawling motion of certain cells. In this section, we move outside the cell and address the propulsion of bacteria in a fluid environment. We examine three different aspects of this motion:

- the general mechanisms that lead to net movement in swimming,
- the rotary motion of flagella driven by an ionic turbine,
- the role of external thermal ratchets to create propulsion.

We begin this section by developing our intuition about movement in a viscous environment.

11.5.1 Swimming in molasses

Rowboats have appeared in idyllic paintings for centuries: two rigid oars move toward the stern of the boat, generating a reaction force that propels the boat forward. In a low viscosity fluid, the boat continues to glide forward while the oar blades are lifted from the water and moved back to their original positions pointing towards the bow of the boat. Some swimming strokes have the same conceptual basis: each arm provides a power stroke while it is underwater, and it is then lifted from the water as it returns to its original orientation. Let us imagine the motion of the rowboat if two aspects of its propulsion were changed:

(i) the water was replaced with molasses so viscous that the boat came to a complete stop as soon as the oars ceased moving,
(ii) the oars had to be kept in the water as they were moved forward, and the vertical orientation of their blades could not be changed.

We assume that the oars are moved in complete synchrony – both are moving backward or both are moving forward at the same time. Under these conditions, the boat might move forward and backward as the oars were moved (at considerable effort to the rower because the molasses is so viscous), but its net overall motion would be zero: at the end of each complete cycle of the oars, the boat would return to where it started.

Cells face an environment much like the rowboat in molasses. As discussed in Chapter 2, their movement through water is completely dominated by viscous damping: we showed in Example 2.1 that a bacterium will abruptly come to rest in less than an atomic diameter as soon as its propulsion units are stopped. In the language of fluid dynamics, cells move in the low Reynolds number regime: they experience essentially no inertial drift in the absence of a propulsive force. Cells stop dead unless there is a force to keep them moving.

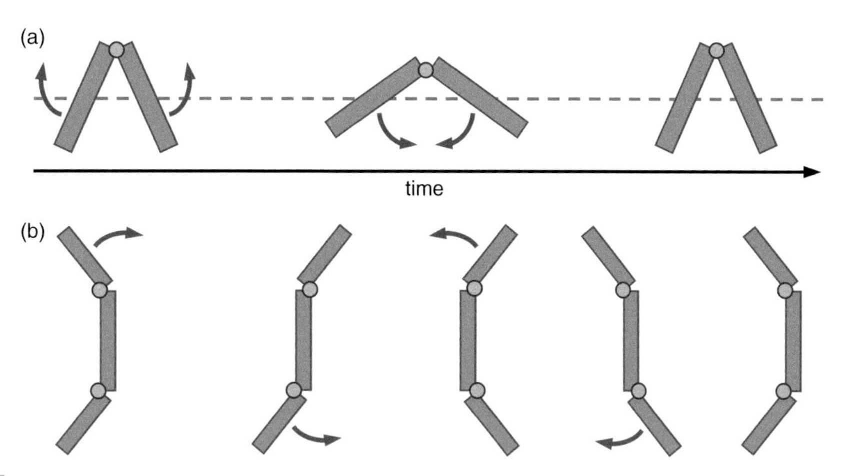

Fig. 11.23 (a) The motion of a simple object: two plates connected by a single hinge (shown in cross section). In a viscous environment, the object returns to its original position after one complete cycle of movement. (b) Configurations of a more complex object: three plates connected by two hinges. This object is capable of net motion in a viscous environment (after Purcell, 1977).

Purcell (1977) provides a classic semi-quantitative analysis of several important features of cell movement; here, we include the mathematical arguments presented in Phillips *et al.* (2009). To see why the rowboat is dysfunctional in molasses, let it be represented by two identical plates joined along one side by a single hinge as in Fig. 11.23(a) . The planes of the plates make an angle of 2θ between them. As described in Chapter 2, movement at low Reynolds number is smooth and non-turbulent, and the drag force is proportional to the velocity of the moving object. Thus, each plate in Fig. 11.23(a) experiences a torque due to drag that is proportional to its angular velocity, $\mathrm{d}\theta/\mathrm{d}t$. This pair of torques from the oars generates a force F to propel the boat; once again because of drag, the velocity $\mathrm{d}x/\mathrm{d}t$ that can be achieved by the boat is linearly proportional to F and does not grow with time as it would under drag-free conditions. In other words,

$$\mathrm{d}x/\mathrm{d}t = [constant] \bullet \mathrm{d}\theta/\mathrm{d}t, \tag{11.17}$$

where the magnitude of the proportionality constant need not concern us. The sign of the motion is also accounted for in Eq. (11.17): if $\mathrm{d}\theta/\mathrm{d}t > 0$, then $\mathrm{d}x/\mathrm{d}t$ and the boat moves forward, but if $\mathrm{d}\theta/\mathrm{d}t < 0$ the motion is reversed. Now, to find the overall movement between t_1 and t_2, all we need to do is integrate Eq. (11.17) to obtain

$$x(t_2) - x(t_1) = [constant] \bullet [\theta(t_2) - \theta(t_1)]. \tag{11.18}$$

Thus, for this object, over one complete cycle of the motion when θ returns to its original value, $\theta(t_2) = \theta(t_1)$ and consequently $x(t_2) = x(t_1)$: the object hasn't moved.

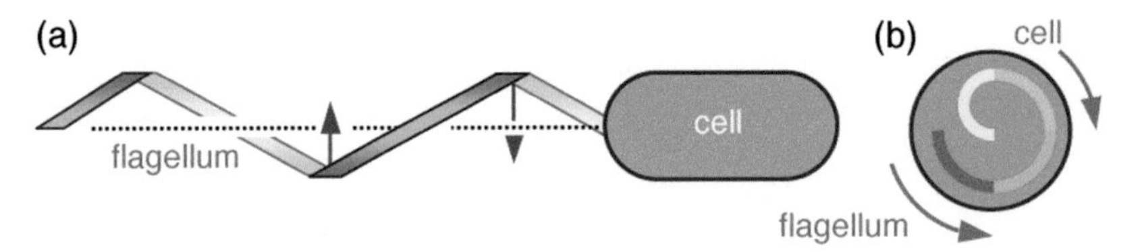

Fig. 11.24 (a) Rotating about the symmetry axis of a bacterium, a flagellum adopts a helical shape. (b) Looking along the axis towards the cell, the flagellum rotates counter-clockwise to provide thrust, and the cell slowly rotates clockwise in response. Darker regions are closer to the viewer.

It's clear from Eq. (11.18) that it doesn't matter how fast the plates swing – for example a slow opening speed and a rapid close of the plates – the motion over one complete cycle will be the same. Purcell refers to this as reciprocating motion: the configurational change of the object in the first half of the cycle is identically reversed in the second half. He argues that any system with a reciprocating cycle cannot generate propulsion. As a simple example of a non-reciprocating object, consider the three-plate system in Fig. 11.23(b). If the two end-plates moved up and down in synchrony, the cycle would be reciprocating. But if one plate moves down, followed by the other, and then the second plate moves up, followed by the other, the motion is non-reciprocating: when the right-hand plate moves down, the left-hand plate is in a different position than it is when the right-hand plate moves up.

In the cell, two non-reciprocating cycles involve flagella and cilia. Flagella adopt a helical shape as they rotate around an axis as shown in Fig. 11.24: there is no reversal of direction during one complete cycle and the motion is clearly non-reciprocating. Cilia, on the other hand, do not rotate but rather wave back and forth. What differentiates their movement from a reciprocating cycle is their flexible shape: they bend away from the direction of motion such that the shape on the initial stroke in one direction is reversed from the stroke in the opposite direction. We now examine the motion of flagella in more detail.

11.5.2 Flagella

Many bacteria and complex single-celled organisms that inhabit aqueous environments have developed ways of locomoting in fluids. We described in Section 11.4 the actin-based movement of *listeria*, which we will model as a thermal ratchet in a moment. However, a more common bacterial propulsion mechanism employs whiplike flagella, which can be seen extending from both ends of the cell displayed in Fig. 11.1. Flagella have a typical length of 10 μm, although examples ten times this length are known, such that their length is usually several times that of the main body of the bacterium. Both torque and thrust are generated by a flagellum as it rotates about its axis, driven by a rotary motor schematically illustrated in Fig. 11.8(b). Embedded in the bacterial membrane and cell well, the motor consists of several protein-based components, including:

- bushings to seal the cell membrane,
- a circular stator, attached to the cell,
- a rotor, with a typical radius of 15 nm, attached to the flagellum.

Early observations favored eight force-generation units in the stator (see references in Berg, 1995; but see also Berry and Berg, 1999). A curved segment or "hook" separates the motor from the main length of the filament, such that the filament is bent away from the normal to the membrane over a distance of tens of nanometers. Thus, the filament executes a helical motion as its rotates, and acts like a propeller.

As illustrated in Fig. 11.24(a), the flagellar helix moves through a fluid at an angle with respect to the symmetry axis of the bacterium. The resistive force from the fluid can be resolved into two components: one generates a thrust along the symmetry axis while the other results in a torque around the axis. As shown in Fig. 11.24(b), the rotation of the flagellum is balanced by the slower counter-rotation of the cell proper. The figure displays the rotation of *E. coli*: normally the filament rotates counter-clockwise (CCW) as viewed looking along the axis towards the cell. When the appropriate flagella rotate CCW cooperatively, the cell can move forward at speeds of 20 μm/s in a mode of motion called a "run". However, the motor has a switch permitting it to run in reverse, or clockwise (CW); flagella rotate CW independently, resulting in no net thrust on the cell such that it "tumbles", losing its orientation. For *E. coli* under common conditions, a typical run lasts 1 s, and a typical tumble lasts 0.1 s, although the duration of a given mode is exponentially distributed.

As a bacterium swims, its flagella may rotate at 100 revolutions per second (Hz) or more. Such rotational rates are comparable to an automobile engine, which runs comfortably at 30 Hz (or 2000 rpm) and reaches its operating limit at ~100 Hz; also like a car engine, flagellar motors fail catastrophically if driven too hard. A single flagellum attached to a fixed substrate can cause the cell body to rotate at 10 Hz, a lower rotation rate than the free flagellum because of viscous drag on the cell. Extensive measurements have been made of the torque generated by the *E. coli* flagellar motor (see Iwazawa *et al.*, 1993, Berg and Turner, 1993, and references therein; the behavior of motors forced to run in reverse has been revisited by Berry and Berg, 1999); the torque decreases monotonically to zero at several hundred revolutions per second, depending on conditions. As estimated in the problem set, the magnitude of the torque lies in the range 2–6×10^{-18} N•m in a number of cells investigated to date.

Flagella are not driven directly by ATP hydrolysis; rather, the motion is generated by ions running down a potential gradient across the membrane – electrical and pH gradients are both possible (see Khan and Macnab, 1980, Manson *et al.*, 1980, and references therein). Recent observations (Berry and Berg, 1999) favor viewing the motor as a proton turbine (Berry, 1993; Elston and Oster, 1997). The free energy release for each ion that moves across the membrane has two components.

(i) A contribution ΔG_{coul} from the change in electrostatic energy across the membrane. As discussed in Chapter 13, the cytoplasmic surface of the membrane often has an electrostatic potential energy ΔV that is 70–90 mV lower than the exterior surface, which results in an energy release ΔG_{coul} of

$$\Delta G_{coul} = e\,\Delta V \tag{11.19}$$

when a proton with elementary charge $e = 1.6 \times 10^{-19}$ C crosses into the cell.

(ii) A contribution ΔG_{conc} from the concentration gradient across the membrane. Also, as discussed in Chapter 13, the free energy released per particle for a given atomic or molecular species in moving from outside the cell at concentration c_{out} to concentration c_{in} inside is given by

$$\Delta G_{conc} = k_B T \ln(c_{out}/c_{in}). \tag{11.20}$$

The sum $\Delta G_{coul} + \Delta G_{conc}$ is the total energy released by an ion as it passes through a rotary motor. In turn, this free energy release is equal to the product of the torque $\mathcal{T}$ generated by the motor and the angular step size $\Delta\theta$ associated with the passage of the ion:

$$\mathcal{T}\,\Delta\theta = \Delta G_{coul} + \Delta G_{conc}. \tag{11.21}$$

Typical changes in free energy are in the several $k_B T$ range, and typical torques are many tens of pN•nm, or 10^{-19} N•m.

11.5.3 Thermal ratchets

As applied to a growing filament, the idea behind the thermal ratchet approach is that thermal fluctuations in the position of the cell boundary create locations where a monomer can be added to an existing filament. If the binding of the monomer is strong, then the process is like a ratchet, in that the filament lengthens irreversibly and exerts a force on the cell boundary as a consequence. Let's analyze the growth of the filament displayed in Fig. 11.25, where the left-hand end of the filament is fixed and the right-hand end is growing against a fluctuating barrier. There is an action–reaction couplet of forces F between the filament tip and the barrier; we assume that F is not so large as to cause the filament to buckle (see Section 3.5). The tip of the filament is not in constant contact with the

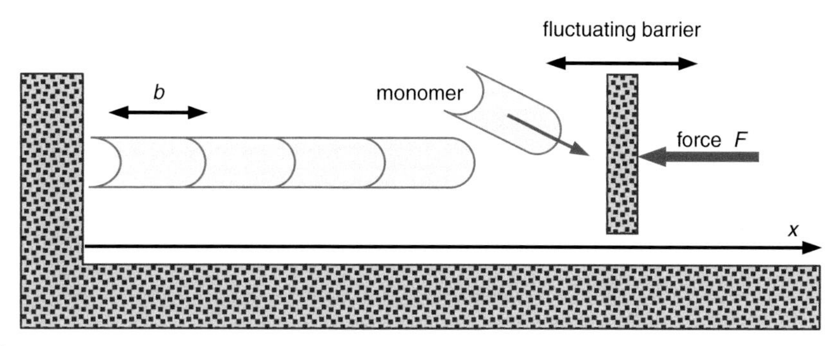

Fig. 11.25 Thermal ratchet model for a filament growing against a fluctuating barrier subject to a force F. When added to the filament, each monomer of length b does work Fb on the barrier.

barrier: as the boundary fluctuates to the left and right along the x-axis, a gap large enough to accommodate an additional monomer may open up from time to time.

The binding of a new monomer does work Fb on the barrier because it is displaced one monomer length b to the right. As a result of this work, the likelihood of the polymerization reaction taking place is reduced according to the Boltzmann factor $\exp(-Fb/k_BT)$ compared to the likelihood in the absence of the barrier, the reason being that the net free energy gain from the reaction has been reduced because of the work done in shifting the barrier. The Boltzmann factor shows up in the dissociation constant for the reaction as follows.

The addition of the monomer at the tip of the filament can be described in terms of the reaction

$$AB \leftrightarrow A + B, \tag{11.22}$$

that has the usual on and off rate constants k_{on} and k_{off} that we have already introduced. The dissociation constant K_D for the reaction can be written in terms of the equilbrium concentrations of each reactant as

$$K_D = [A]_{eq} \bullet [B]_{eq} / [AB]_{eq}, \tag{11.23}$$

where $[X]_{eq}$ is the equilibrium concentration of species X. Thus, if we denote the dissociation constant at zero force and at force F by $K_D(0)$ and $K_D(F)$, respectively, then the dissociation constant will be *increased* by the Boltzmann factor, which favors the *depolymerization* of the filament:

$$K_D(F) = K_D(0) \exp(Fb/k_BT). \tag{11.24}$$

Running through the same argument as in Section 11.2 (see end-of-chapter problems), it can be shown that

$$K_D = k_{off}/k_{on}, \tag{11.25}$$

and this equation can be combined with Eq. (11.6) to yield

$$K_D(0) = [M]_c, \tag{11.26}$$

where $[M]_c$ is the critical concentration of the polymerization reaction. Further, for a given monomer concentration $[M]$, there is a maximum force F_{max} beyond which the filament cannot lengthen without buckling. At F_{max}, the net filament growth is zero and once again

$$K_D(F_{max}) = [M]. \tag{11.27}$$

Combining both of these expressions for the dissociation constant gives

$$F_{max} = (k_B T/b) \ln([M] / [M]_c), \tag{11.28}$$

after a little rearrangement.

Our intuition about the behavior of F_{max} is borne out in Eq. (11.28). The force on the barrier is higher due to increased polymerization when: (i) more monomers are available for polymerization, and (ii) more gaps appear at the filament end because of fluctuations at higher temperatures. The effect of the step length b on F_{max} can be seen by comparing actin to tubulin polymerization. Both of these filaments can be viewed as composites of several protofilaments in parallel, such that the step length b is less than the monomer length: b is just 1/2 of the monomer length for actin with 2 protofilaments, or 1/13 the monomer length for a microtubule with 13 protofilaments. Thus, even if the actin and tubulin monomers had the same lengths (they are not hugely different) the force from a growing microtubule would be about six times that of an actin filament. Many other features of thermal ratchets in the context of translational motion have been examined and tested experimentally; for a more thorough treatment, including the effects of diffusive time scales on filament growth, see Chapter 16 of Phillips *et al.* (2009).

Summary

Cells use a variety of mechanisms to change their shape, move through their environment and internally transport chemical cargo and cellular subunits. The simplest dynamic mechanism involves growth and shrinkage of the cytoskeletal filaments actin and tubulin. The monomers in these filaments have asymmetric shapes and, once a monomer is captured, a triphosphate nucleotide it carries is hydrolyzed to a diphosphate. As a result, the ends of the filament are chemically and structurally inequivalent, with the faster (slower) growing end referred to as the plus (minus) end. In many situations, the plus ends of the filaments extend towards the cell boundary. The overall length of both types of filaments changes at a concentration-dependent rate: under common cellular conditions, actin filaments lengthen at 10^{-2} to 1 μm/s while microtubules grow up to 0.3 μm/s and shrink at 0.4–0.6 μm/s.

The simplest representation of the change in the number of monomers n in a filament is $dn/dt = k_{on} \cdot [M] - k_{off}$, where $[M]$ is the concentration of free monomers and k_{on}, k_{off} are rate constants. Because these filaments are in dynamic equilibrium, net loss of monomer occurs if $[M]$ is below the critical concentration $[M]_c = k_{off}/k_{on}$. For actin filaments and microtubules, k_{on} is often in the range 5–10 $(\mu M \cdot s)^{-1}$, depending on conditions; k_{off} is in the range 1–10 s^{-1} for actin but dramatically larger at ~800 s^{-1} for rapidly disintegrating microtubules. These rates can be interpreted in terms of monomer diffusion; for example, in a very simple reaction model where a monomer is captured if it strays within a distance R of a filament tip, we expect $k_{on} \sim 4\pi DR$, where D is the diffusion constant of the monomer in solution. Cytoskeletal filaments have two interesting dynamical characteristics. With inequivalent values of $[M]_c$ at each end, an actin filament may grow at one end while it simultaneously shrinks at the other; in fact, the two rates are equal at a unique monomer concentration, resulting in a filament of constant length, a phenomenon known as treadmilling. Microtubules, for which treadmilling is not an important process, are subject to dynamic instability where disassembly is so rapid that one tubulin dimer has barely diffused away from the filament before another is released.

Filament growth can exert internal pressure on a cell boundary causing it to move, as in the lamellipodium of migrating cells, or can be harnessed to exert an external pressure on the wall of a bacterial invader like *listeria*. The maximum force F_{max} exerted by a growing filament on a barrier whose position undergoes thermal fluctuations at a temperature T is given by $F_{max} = (k_B T/b) \ln([M] / [M]_c)$, where b is the change in filament length associated with monomer capture, and $[M]$ is the monomer concentration. If the filament is a composite structure with several protofilaments in parallel, b is less than the individual monomer length: b is just 1/2 of the monomer length for actin with 2 protofilaments, or 1/13 the monomer length for a microtubule with 13 protofilaments.

More complex machinery for cell movement and transport is provided by special motor proteins capable of walking along actin filaments or microtubules, summarized in Table 11.3. These motors have one or more globular-shaped "heads" about 5–10 nm in size, connected to a longer "tail" that can be attached to other motor proteins or to small vesicles or organelles. Present in both one- and two-headed varieties, the motor protein myosin forms a mechanical complex with actin, a widespread example being our muscle cells. Myosin commonly attains a speed of 10^{-2} to 1 μm/s as it works its way towards the plus end of an actin filament. Two distinct motor proteins travel along microtubules: kinesin generally walks towards the plus end while cytoplasmic dynein walks towards the minus end (there are a few examples of kinesin moving in the minus direction). Again, the

Table 11.3 General attributes of four motor proteins found in the cell. Omitted from the list is ciliary dynein, which displays a wider range of mass than the other proteins

Motor protein	Associated filament	Mass (kDa)	Usual direction of travel
myosin-I	actin	~150	plus
myosin-II	actin	~ 2 × 260	plus
kinesin	microtubules	~ 2 × 180	plus
cytoplasmic dynein	microtubules	1200–1300	minus

typical speeds of these motors may range up to 1–4 μm/s. The mechanism by which the motor proteins travel along their respective filaments involves conformational changes of the "head" region, during the capture, hydrolysis and release of ATP and its reaction product ADP. The most energy-efficient mechanism for movement of the heads involves deliberate steps like a walk; it is less efficient for a motor to use a mechanism called a thermal ratchet, in which trial steps are made subject to a potential bias, not every step being successful. Current experiment leans towards the direct motion picture, although thermal ratchets may have been important in the evolutionary development of today's molecular motors. In their movement, motor proteins generate a force of a few piconewtons, consistent with the chemical energy released during the hydrolysis of ATP as the motor makes steps of 5–8 nm, depending upon the system.

The rotary motors driving bacterial flagella are the most complex source of cellular motion: many different proteins are used to fabricate the stator, rotor and bushings of this cellular engine. The analogue with an everyday mechanical engine is not inappropriate, as both engines have operating ranges of 100 Hz or more, and both fail catastrophically if overdriven. However, the torque produced by a flagellar motor is in the range 10^{-18} N•m, rather less than an automobile engine (1–3 × 10^2 N•m), but sufficient to provide a few piconewtons of thrust and propel a bacterium at up to a few tens of microns per second. Most flagellar motors can be described as protein turbines, driven by the flow of protons down a potential gradient across the cell boundary.

Problems

Applications

11.1. Find the time taken for the plus end of a microtubule to grow 5 μm from the centrosome of a hypothetical cell to its boundary. How long does it take for the filament to shrink to zero length if it undergoes rapid depolymerization upon reaching the boundary? Assume $[M] = 10\ \mu\text{M}$ and take rate constants from Table 11.1 (a tubulin dimer is about 8 nm long).

11.2. Suppose that we have an ensemble of microtubules which are constrained to grow/shrink only at their plus ends. If their mean length is constant, what fraction of their time is spent growing and what fraction shrinking? Use rate constants from Table 11.1 and assume $[M] = 15\ \mu\text{M}$ (tubulin dimers are about 8 nm long).

11.3. Evidence suggests that the hydrolysis of the GTP cap of a microtubule occurs at a rate constant k_{hydro} that is independent of monomer concentration and has units of $[time]^{-1}$, just like k_{off}. Let's evaluate the growth of the GTP cap for a specific tubulin concentration.

(a) Show that the number of tubulin dimers in the cap grows like

$$\mathrm{d}n^{+}_{\text{cap}}/\mathrm{d}t = \mathrm{d}n^{+}/\mathrm{d}t - k_{\text{hydro}},$$

assuming that the monomer solution is pure GTP-tubulin.

(b) Calculate $\mathrm{d}n^{+}/\mathrm{d}t$ from Table 11.1 if $[M] = 20\ \mu\text{M}$.

(c) Assume, without experimental justification, that $k_{\text{hydro}} = 30\ \text{s}^{-1}$. Plot the filament length, and the cap length, as a function of time until the filament reaches 10 μm assuming that the negative end does not change with time (a tubulin dimer is about 8 nm long).

11.4. The "straight" tail of myosin-II consists of two parallel segments, each an α-helix of mass of 110 kDa, wound together into a coil of contour length 150 nm. Find the flexural rigidity κ_f and persistence length ξ_p of the tail using results from Fig. 3.16. Find $\langle r_{ee}^{2}\rangle^{1/2}$ for the tail using the continuum filament model of Section 3.3.

11.5. Using Stokes' relation for the diffusion constant, Eq. (11.11), determine the diffusion model prediction for k_{on} in the capture of a protein of radius R_p by a filament according to Eq. (11.12). Assume that the capture occurs when the centers of the protein and filament are separated by $3R_p$.

(a) What is the value for k_{on} if the viscosity of the medium is 10^{-1} kg/m•s; quote k_{on} in units of $(\mu\text{M}\cdot\text{s})^{-1}$? Does k_{on} depend on R_p?

(b) Compare your calculation for k_{on} with that of actin from Table 11.1. Comment on the difference between these values.

11.6. The viscous drag force exerted by a stationary fluid on a spherical object of radius R is, according to Stokes' law,

$$F = 6\pi\eta vR,$$

where η is the viscosity of the medium and v is the object's speed. Let's apply this to a spherical vesicle moving at 0.5 μm/s in a medium with $\eta = 10^{-2}$ kg/m•s, which is ten times the viscosity of pure water.

(a) Plot the drag force as a function of vesicle radius for the range $20 \leq R \leq 200$ nm.

(b) Find the radius at which the drag force equals 2 pN, the force typical of a molecular motor.

11.7. What force is needed to propel a spherical bacterium of radius $R = 1$ μm at a speed v of 20 μm/s in water. Take the drag force to be $6\pi\eta Rv$ with $\eta = 10^{-3}$ kg/ m•s (quote your answer in pN). What power must be supplied to keep the cell moving (quote your answer in watts)?

11.8. In a viscous medium, an object rotates at an angular frequency ω when subject to a torque $\mathcal{T}$ according to

$$\mathcal{T} = \omega f_{drag}$$

where f_{drag} is the drag coefficient. For a sphere of radius R, f_{drag} is given by

$$f_{drag} = 8\pi\eta R^3,$$

where η is the viscosity of the medium. Find $\mathcal{T}$ if the rotational speed is 10 revolutions per second, $R = 1$ μm, and $\eta = 10^{-3}$ kg/ m•s (water).

11.9. It is shown in Problem 11.18 that the free energy released for a proton traversing a membrane with a potential difference ΔV and pH difference ΔpH is

$$\Delta G = e\,\Delta V + 2.303\,k_B T\,\Delta \text{pH},$$

where Δa means $a_{inside} - a_{outside}$. Applying this free energy to drive an ionic turbine in the membrane, find the torque generated at room temperature if ΔpH = 0.75, $\Delta V = 90$ mV, and 10 protons move through the turbine in one cycle. Quote your answer in pN•nm.

11.10. The reversible dissociation reaction $AB \leftrightarrow A + B$ involves the same rate constants k_{off} and k_{on} as the polymerization reaction for a filament. The dissociation constant K_D is written in terms of the equilibrium concentrations of the reactants as $K_D = [A]_{eq}[B]_{eq}/[AB]_{eq}$.

(a) Write the equation for $d[AB]/dt$ and express K_D in terms of k_{off} and k_{on}.

(b) Calculate K_D for the plus and minus ends of ATP-actin. Is the binding of actin stronger or weaker than the biotin–avidin protein–ligand pair with $K_D \sim 10^{-15}$ M?

Formal development and extensions

11.11. The rate expressions of Fig. 11.11(b) are one example of the general case where $k_{on}^{+} \neq k_{on}^{-}$ and $k_{off}^{+} \neq k_{off}^{-}$.

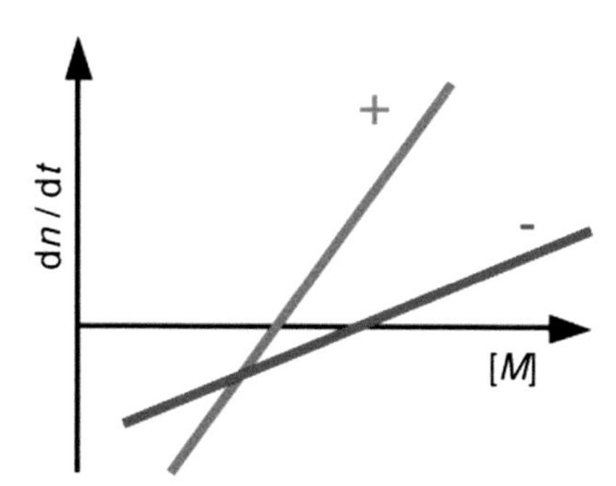

(a) What is the significance of the intersection point of the lines?
(b) What are the growth characteristics of a filament where the two lines are parallel?
(c) Suppose that the lines intersect when $dn/dt > 0$. What would be the properties of such a filament?

11.12. A filament, starting from a small seed, grows in a monomer solution of initial concentration M_o in volume V_o. Suppose that no new monomer is added to the solution as the filament grows, and that the rate equation for the number of monomers in the filament is governed by the usual $dn/dt = k_{on}[M] - k_{off}$.

(a) Sketch $[M]$ as a function of time.
(b) What is the equation of this plot at $t \sim 0$?
(c) What is the asymptotic value of $[M]$ as $t \to \infty$?

11.13. Perform the following calculation to confirm how well the triangle distribution in Fig. 11.19 approximates the Gaussian distribution $\mathcal{P}(x) = (2\pi\sigma^2)^{-1/2} \exp(-x^2/ 2\sigma^2)$.

(a) Determine the functional form of $\mathcal{P}(x)$ in the triangle approximation (as parametrized by w) and evaluate the dispersion $\langle x^2 \rangle$. Equate this to the dispersion of the Gaussian distribution to establish $w = \sqrt{6}\,\sigma$.
(b) Compare $\mathcal{P}(0)$ and the full width at half maximum for both distributions, with $w = \sqrt{6}\,\sigma$ from part (a).

11.14. From their functional form alone, find the value of w/b where the expressions for P_{net} in Eqs. (11.15b) and (11.15c) are identical (you may *not* assume $w/b = 1 - \alpha$). Are the derivatives dP_{net}/dw also the same at this location?

11.15. For the bimolecular reaction $A + B \to AB$, the diffusion constant in Eq. (11.12) is replaced by $D_A + D_B$ if both species are mobile (Chandrasekhar, 1943). If A and B are viewed as hard spheres of radius R_s, establish that

$$k_{on} = 8k_BT / 3\eta$$

in the diffusion model.

11.16. Derive Eq. (11.12) for the capture of monomers in a solution. From Fick's law, the current of monomers (J, a number per unit time) inbound through a spherical shell of area A is $J = DA\, dc(r)/dr$, where $c(r)$ is the concentration of monomers at radius r (note the lack of a minus sign with this definition).

(a) Assuming that J is independent of r (why?), show that $c(r)$ must satisfy

$$c(r) = [M] - J/4\pi Dr,$$

where $[M]$ is the monomer concentration at infinite separation.

(b) Sketch $c(r)$ as a function of r. What is the physical significance of R, the radius at which the concentration vanishes?

(c) Use R to obtain an r-independent expression for J, from which you can deduce

$$k_{on} = 4\pi DR.$$

11.17. A spherical object of radius R moving in a stationary fluid of viscosity η experiences a drag force given approximately by Stokes' law

$$\mathbf{F} = -6\pi\eta R\, \mathbf{v},$$

where $\mathbf{v}$ is the object's velocity and the minus sign signifies that $\mathbf{F}$ and $\mathbf{v}$ point in opposite directions. The velocity of an object subject only to drag steadily decreases from its initial value v_o.

(a) Starting from Newton's law, establish that the velocity decays with time t as

$$v(t) = v_o \exp(-t/t_{visc}),$$

where the characteristic time $t_{visc} = m/6\pi\eta R$, m being the object's mass.

(b) Find t_{visc} for a cell with the density of water (10^3 kg/m^3) and radius $R = 3$ μm moving in a fluid with $\eta = 10^{-3}$ kg/m•s (water).

(c) By integrating the velocity, show that the cell comes to rest in a distance equal to $v_o t_{visc}$. If the cell is initially moving at v_o = 10 μm/s, find the distance over which it comes to a stop (after Chapter 6 of Berg, 1983).

11.18. A proton is transported across a membrane from concentration c_{out} to c_{in}, simultaneously experiencing a potential difference ΔV.

(a) Show that the total free energy released per proton by this process is

$$\Delta G = e\,\Delta V + 2.303\, k_B T\, \Delta \mathrm{pH},$$

where $\Delta\mathrm{pH} = \mathrm{pH}_{inside} - \mathrm{pH}_{outside}$.

(b) Find an algebraic expression for ΔpH at which there is no free energy released during transport.
(c) What is ΔpH in part (b) if $\Delta V = -80$ mV?

11.19. Find the maximum pressure exerted on the boundary of a eukaryotic cell of radius 3 μm by a collection of 50 microtubules growing in random directions from the cell's center (quote your answer in atmospheres). Treat the growth mechanism to be a thermal ratchet and assume that the free tubulin dimer concentration in the cell is 100 μM. Is the maximum force on each microtubule large enough to cause it to buckle (see Section 3.5)?

11.20. A flagellum experiences a thrust axis because of the difference in drag forces parallel and perpendicular to its length. Consider an element of the flagellum of length ℓ rotating at a speed v about the flagellum's helical axis, where the flagellum proper makes an angle ϕ with respect to the helical axis.

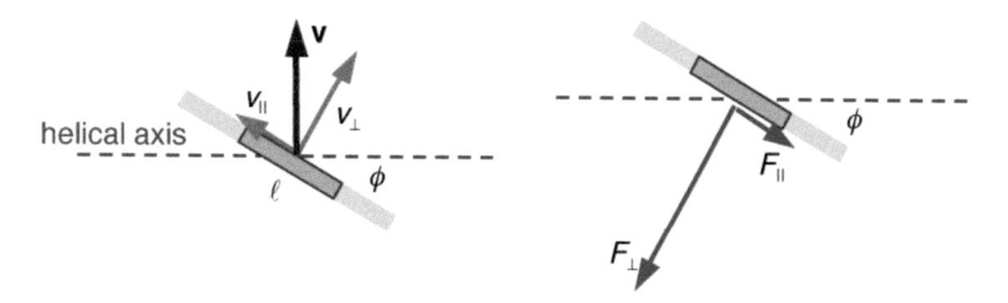

The perpendicular and parallel components of the drag force each have the form $F_k = C_k \eta \ell v_k$, where $C_{\perp} \approx 2C_{\parallel}$ owing to the different orientations of the filament ($C_{\perp} \approx 4\pi$, $C_{\parallel} \approx 2\pi$).

(a) Show that $F_{\parallel}/F_{\perp} = (C_{\parallel}/C_{\perp})\tan\phi$. What does this imply about the directions of **F** and **v**?
(b) Show that the propulsive force, which is the net drag force to the left in the diagram, is given by $F_{\text{net}} = (C_{\perp} - C_{\parallel})\eta \ell v \sin\phi \cos\phi$ for this segment. What is F_{net} for the whole flagellum of length L?
(c) If the drag force along the flagellum as a whole is $C_{\parallel}\eta L V$, where V is the speed of the bacterium, show that

$$V = (C_{\perp}/C_{\parallel} - 1)v \sin\phi \cos\phi.$$

11.21. Find the speed of a bacterium driven by a single flagellum of length 10 μm if the flagellum rotates at 50 Hz. The helical shape assumed by the flagellum has a diameter of 0.5 μm and a pitch of 2 μm (see Section 4.2 for definitions). Take $C_{\perp}/C_{\parallel} = 2$ and quote your answer in μm/s. You may use results from Problem 11.20 without proof.

12 Growth and division

More often than not, life on Earth exists in an environment that is both comfortable yet competitive. It is comfortable in that molecular construction materials are available in a fluid medium at a suitable temperature for self-assembly and appropriate functionality. It is competitive in that cell populations are unregulated and cell design is subject to random changes, permitting better-adapted designs to flourish at the expense of less-adapted designs. Cells have no choice but to go forth and multiply if their design is to survive: individual cells must duplicate their blueprint for use by successive generations and they must grow, although not without limit. The maximum size to which an individual cell may grow within a sustainable population may be influenced by many factors, including the time scale for the transport of nutrients from external sources and the surface stresses on the cell's membranes and walls, which generally scale with cell size. Further, as cell design becomes more complex for advanced or specialized cells, the functionality and perhaps structural arrangements of its components may be forced to change during its lifetime, which in turn requires the cell to develop means of regulating its functionality.

In this chapter, we describe the gross structural changes that a cell undergoes during the cell cycle, and the mechanism by which its blueprint is duplicated. Section 12.1 contains a survey of the main events of prokaryotic and eukaryotic cell cycles, which is then quantitatively examined for bacteria and simple eukaryotes (such as green algae) in Section 12.2. This section includes influences on the cell cycle that have a physical origin, such as mechanical stresses and diffusion. Although the simplest models portray the cell cycle as having a doubling time with division into identical daughter cells, in reality both doubling time and daughter size are distributed around a mean; asymmetry and asynchrony of the cycle are the subjects of Section 12.3. Two successive sections of this chapter (12.4 and 12.5) are devoted to microscopic issues in the cell cycle, namely the forces and mechanisms of DNA replication and segregation; additional biomolecular aspects of transcription are treated later in Section 13.4. For an introduction to the mechanical aspects of cell growth in the context of tissues, see Shraiman (2005).

12.1 Overview of the division cycle

We begin by reviewing the sizes and structures of the simplest cells in order to understand some of the structural aspects of cell division. As described in Chapter 10, common bacterial shapes include cocci (simple spheres), diplococci (fused spherical caps), bacilli (rod-like cells or spherocylinders) and spirals; these fundamental shapes have been present on Earth for billions of years. For example, Fig. 12.1 displays the fossilized remnants of a few taxa of 2-billion-year-old microfossils from the Belcher Island, Canada (Hofmann, 1976; Bennett *et al.*, 2007). In clockwise order from the upper left: bacillus-like *E. moorei* and diplococci *EB* (unclassified), *S. parvum*, and *E. belcherensis capsulata*; although displayed in isolation, these cells can be found in colonies, as will be described below. Among their other notable features, microfossils have linear dimensions that correspond to a variety of modern cells, namely diameters of a few microns. Given the

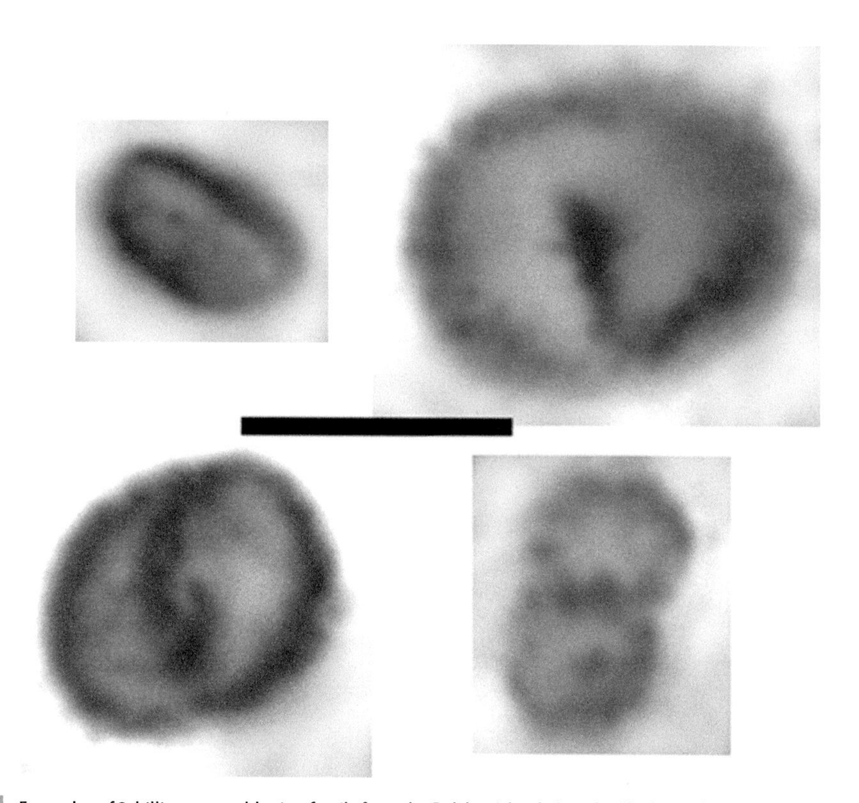

Fig. 12.1 Examples of 2-billion-year-old microfossils from the Belcher Island, Canada. Clockwise from the upper left: bacillus-like *Eosynechococcus moorei* and diplococcus-like *EB* (unclassified colony), *Sphaerophycus parvum*, and *Eoentophysalis belcherensis capsulata* (scale bar is 5 microns; Bennett *et al.*, 2007).

huge number cell divisions that have taken place in the last two billion years since these taxa were present, does the relative constancy of size and shape indicate that there are optimization or other principles at play in the selection of designs for viable cell types? Let's examine some generic aspects of cell size.

The energy of the exposed membrane boundary and the energy of membrane bending scale in different ways with the physical size of the membrane, with the result that a flat membrane must reach a minimum size before it becomes energetically favorable for the membrane to reduce the length of its exposed boundary by closing up into a sphere. According to Eq. (7.23), the minimum radius of resulting vesicle is

$$R_V^* = (2\kappa_b + \kappa_G) / \lambda, \tag{12.1}$$

if the energy is governed only by membrane bending elastic constants κ_b and κ_G and the edge tension of the boundary, λ. Neglecting κ_G and using typical values of $\kappa_b = 5 \times 10^{-20}$ J (from Table 7.3) and $\lambda = 10^{-11}$ J/m, Eq. (12.1) yields $R_V^* = 10$ nm. Although not a terribly useful lower bound on cell size, this result shows that from the mechanical point of view, cells or cell-like structures could be so small as to have radii comparable to the bilayer thickness.

In contrast to this lower bound, the sizes of cells with an elevated internal pressure are subject to an upper bound imposed by the tensile strength of the cell boundary. For a sphere of radius R, the (two-dimensional) surface stress Π resulting from a pressure difference P across the membrane is given by

$$\Pi = PR/2. \tag{12.2}$$

When subjected to a surface stress, a membrane first stretches and then ruptures: depending on their composition, lipid bilayers typically rupture at Π around 1×10^{-2} J/m^2 on laboratory time scales (Needham and Hochmuth, 1989). For a spherical cell of radius $R = 1$ μm and no cell wall, rupture occurs at a fairly low osmotic pressure: Eq. (12.2) predicts that the pressure at a failure stress of $\Pi = 10^{-2}$ J/m^2 would be 2×10^4 J/m^3 = 0.2 atm. Thus, a bacterium requires a cell wall to support an osmotic pressure of several atmospheres, which is more than the lipid bilayer of the plasma membrane can withstand. However, such is not the case for smaller cells: the same type of calculation shows that a pure bilayer vesicle just 100 nm in radius (or 0.2 μm in diameter) could operate at an osmotic pressure of 2 atm without needing a cell wall for additional strength. The general behavior of the maximal pressure sustainable by a bilayer is shown in Fig. 12.2, which is a plot of $P = \Pi / R$ with $\Pi = 10^{-2}$ J/m^2 at failure.

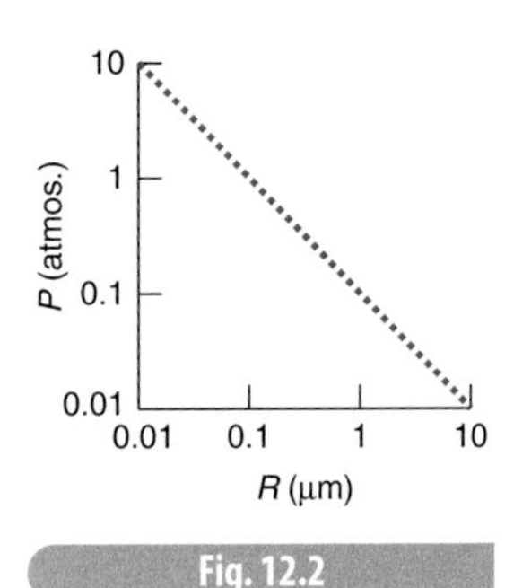

Fig. 12.2

Rough estimate of the maximal pressure sustainable for a bilayer-bounded cell as a function of radius R, assuming a (two-dimensional) rupture tension of 10^{-2} J/m^2.

Equations (12.1) and (12.2) provide crude boundaries beyond which the cell architecture needs to be modified to avoid mechanical failure. The specific strategy adopted by a cell to accommodate stress varies considerably.

As described in Section 10.5, most bacteria support an internal pressure through a network of peptidoglycan: the thin network in the Gram-negative cell boundary sustains a modest internal pressure, whereas the thick network of Gram-positive cells can support a substantial pressure. Similarly, layers of cellulose permit plant cells to operate at large internal pressures. In contrast, eukaryotic cells do not operate at elevated internal pressures, so their bilayer-based plasma membranes need little further support to retain their mechanical integrity.

What structural issues are faced by the cell boundary during growth and division? In all cells, bilayer-based membranes can grow through the addition of newly manufactured lipids to one leaflet of the bilayer; material is most commonly added to the inner leaflet. With a temperature-dependent rate constant of about 10^{-5} s^{-1} (Chapter 4 of Gennis, 1989; see for example Anlgin *et al.*, 2007), the flip-flop of lipids between leaflets is small but it can be speeded up through the presence of a class of enzymes called flippases. This permits the redistribution of lipids between leaflets, reducing mechanical stress as the bilayer grows. Other than the incorporation of new lipids, membrane-associated proteins and other molecular components of the plasma membrane, cell-boundary growth in eukaryotic cells is not ridden with mechanical problems.

Unfortunately, this is less true of bacteria. The issue with the growth of the bacterial cell wall is that the wall is often under lateral stress, meaning that new material must be incorporated into the wall without creating a void that would cause the existing network to fail. It is thought that the mixed peptide and glycan composition of individual strands of peptidoglycan permits an incoming strand to cross-link to an existing strand via floppy peptide chains before the existing peptide cross-links are broken: in essence, a "make before break" strategy (Höltje, 1993, 1998) illustrated in Fig. 12.3. In Gram-positive bacteria, new peptidoglycan is added to the inside surface of the cell wall, and it gradually makes its way outward as the wall material is replenished.

The growth and division mechanism of a cell recognizes the need for synchronization such that DNA replication is complete before it is sequestered and the cell cleaves into two daughters. We delay a detailed analysis

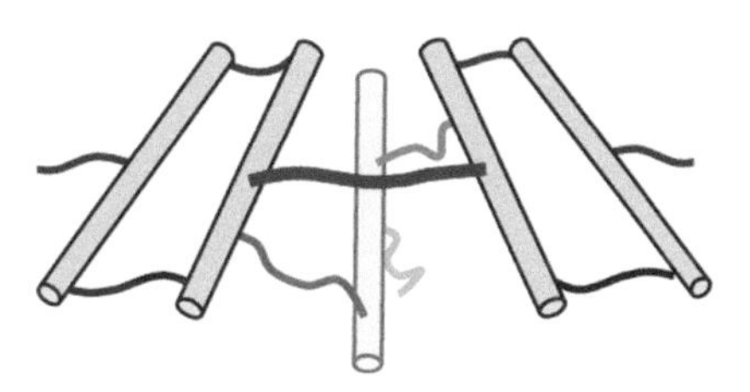

Fig. 12.3 "Make before break" strategy of bacterial wall growth. A new strand of peptidoglycan approaches an existing network from below, creating new links with it (green) before an existing link (red) is severed.

of the process and regulation of DNA replication and transcription until Section 13.4; here we examine the packaging of DNA in the cell, and the steps that must be taken during division to separate the macromolecules once they have been duplicated. As a double-stranded helix, DNA is considerably stiffer than a simple flexible polymer like a saturated alkane, such that for eukaryotic cells, the packaging of DNA inside the cell is a challenge. From Section 3.3, a filament with persistence length ξ_p undergoes random changes in direction all along its contour length L_c, such that the end-to-end displacement $\mathbf{r}_{ee}$ of the filament obeys Eq. (3.33):

$$\langle r_{ee}{}^2 \rangle = 2\xi_p L_c. \tag{12.3}$$

A related measure of the size of a flexible filament is its radius of gyration, R_g, which is defined by $\langle R_g{}^2 \rangle = N^{-1} \Sigma_{i=1,N} \mathbf{r}_i{}^2$, where the filament has been appropriately sampled at N points with displacements $\mathbf{r}_i$ from the center-of-mass position of the filament. If the physical overlap of remote sections of the filament is permitted, then randomly oriented filaments are governed by

$$\langle R_g{}^2 \rangle = \langle r_{ee}{}^2 \rangle / 6 = \xi_p L_c / 3. \tag{12.4}$$

Figure 12.4 shows a log–log plot of Eq. (12.4) for DNA with $\xi_p \sim 53$ nm (Table 3.2), and a contour length of $L_c = 0.34$ nm • [*no. of base pairs*]. For example, $\langle R_g{}^2 \rangle^{1/2} = 5.3$ mm for the DNA of *E. coli* (4.7 million base pairs) and $\langle R_g{}^2 \rangle^{1/2} = 2.2$ mm for mycoplasma (800 000 base pairs) if they were open strings. In both cases, the effective size of a ball of their DNA has roughly the same linear dimension as the cell itself. However, this is not true for eukaryotic cells: human DNA is very long, such that it would occupy far more volume as a random coil than bacterial DNA. Consequently, advanced cells have developed a packaging technique in which their DNA is wrapped around barrel-shaped proteins called histones, with a diameter of 11 nm, in order to organize and sequester their long genetic blueprints.

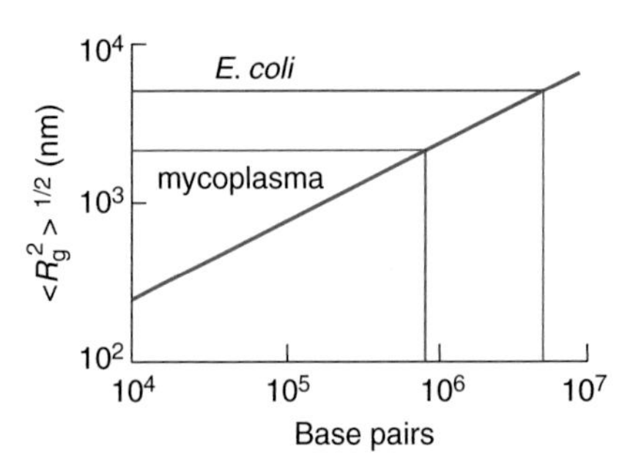

Fig. 12.4 Root-mean-square radius of gyration $\langle R_g{}^2 \rangle^{1/2}$ for a random polymer having the persistence length of DNA. The number of base-pairs in mycoplasma and *E. coli* are indicated for comparison.

In the discussion above, we have identified some broad aspects of cell size that are relevant to cell growth and division. The minimal cell size permitted by mechanical considerations is well below what is required for genome storage: the dimensions of the smallest cells are appropriate to the radius of gyration of their DNA configured simply as a random chain. Complex cells with longer DNA (including non-coding regions), must organize their genetic blueprints into compact packages to keep the cell to a reasonable size; while this strategy conserves materials so that cell size remains in the many micron range, it adds considerable complexity to the segregation of the two copies of DNA once the cell is ready to divide. Another problem for large cells is the surface stress arising when the interior of the cell is pressurized, which can be accommodated by the addition of a stress-bearing cell wall. Once again, this modification of the simplest cell architecture introduces the twin challenges of building a wall with specific curvature, and of adding material to the wall without causing it to fail.

Let's now examine the segregation of genetic material during division. At its simplest level, the division cycle only requires that the mean volume V and surface area A of the cell double independently during the cycle once a steady-state growth pattern has been achieved. These changes in the area and volume of the cell can be accommodated through the use of a (two-dimensional) fluid boundary such as a lipid monolayer or bilayer, which has the flexibility to change shape as needed with a minimal cost in deformation energy. An illustration of a division cycle that any fluid membrane can support is displayed in Fig. 12.5 (Boal and Jun, unpublished): the membrane grows by the random addition of new molecules and ultimately forms wavy arms that can pinch off to form new cells, assuming that the cell's genetic blueprint has been replicated on the same time scale as the doubling of cell volume. The main physical principle that drives this model cycle is one that nature adores: the maximization of entropy (Jun

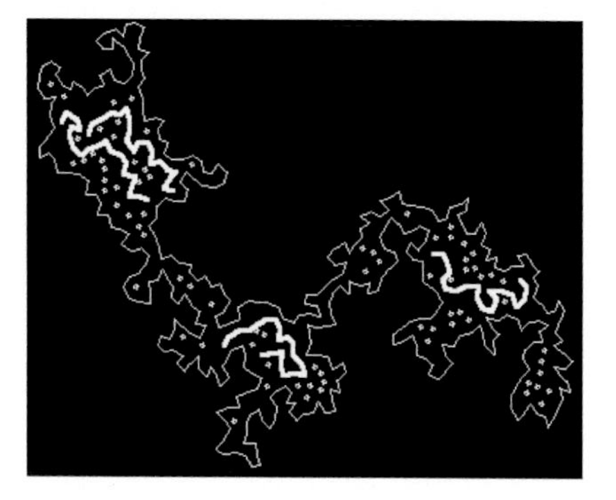

Fig. 12.5 Computer simulation of entropy-driven cell division in two dimensions. Enclosed within this cell are four genetic polymers (linked spheres) as well as numerous solvent spheres. Entropy-laden arms appear only when the perimeter of the cell becomes large (Boal and Jun, unpublished).

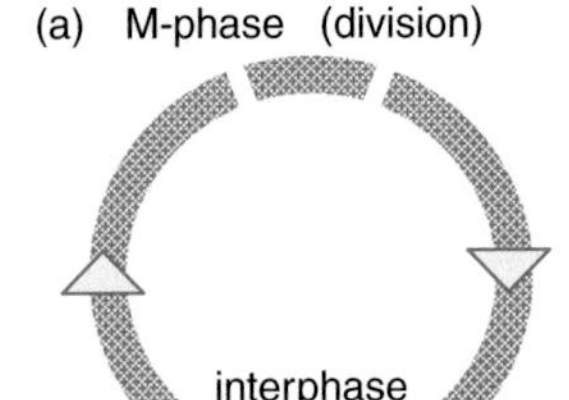

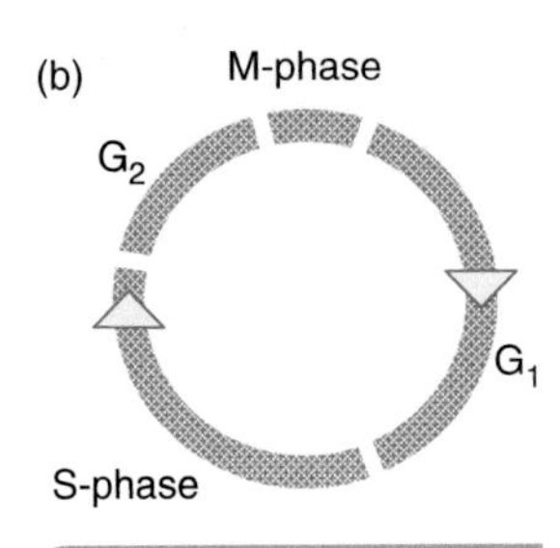

Fig. 12.6

Schematic representation of the division cycle for eukaryotic cells. The interphase period of panel (a) can be subdivided into S-phase (DNA synthesis) and the two gap phases G_1 and G_2 in panel (b). Arrows indicate the direction of time.

and Mulder, 2006; see Luisi, 2006, and Zhu and Szostak, 2009, for overviews of related experimental work).

The cell shape displayed in Fig. 12.5 belongs to a family of random shapes that obey branched polymer scaling, where the surface area is proportional to the enclosed volume; this is unlike a spherical balloon where $A \sim V^{2/3}$ (see Section 8.4). However, these shapes are not efficient in the usage of materials: it isn't so much that the surface area is forced to grow so fast, but rather that the cell must be sufficiently large before (i) the branched polymer shapes emerge, and (ii) its genetic blueprints have replicated and separated. Given that cells must expend metabolic energy to produce the molecules composing the cell boundary, other routes to cell division may be more appropriate. One possible design is based on the use of molecules that, perhaps because of their spatial conformation, generate a membrane that is naturally deformed. For instance, suppose a membrane is made from molecules that favor surfaces that spontaneously deform to some particular curvature C_o, thus imparting a preferred radius of $1/C_o$ to the cell. As more molecules are added to this membrane, it grows at constant curvature in the form of two overlapping spherical caps linked together at a ring with radius less than $1/C_o$, until the ring closes, leaving just two touching spheres. Now, this design is not flawless, in that the membrane curvature in the intersection region of the ring has the wrong sign – if the surface is concave (inwardly curved) over most of the linked spheres, it is convex (outwards) along the intersection ring itself, like the shape of an old-fashioned hour-glass.

Although it has been shown that entropy may play a large role in the division of bacteria, other factors must be at work in more complex cells, as we now describe. The events of the eukaryotic cell cycle are, at their simplest, grouped into interphase, during which the cell grows and its DNA is replicated, and M-phase (M for mitosis) where the DNA separates and the cell divides. The trajectory of the cycle is often represented by panel (a) of Fig. 12.6, where the arrows indicate the direction in time followed by the trajectory. Interphase itself can be further partitioned into three subsections as the S-phase (S for synthesis, where DNA is replicated), and the so-called gap phases G_1 and G_2 that define the beginning and end of interphase; this is illustrated by panel (b) of Fig. 12.6. Although the cell grows and performs various functions during G_1 and G_2, DNA replication is absent. Depending on environmental and other issues, cells may also suspend their growth or other activities during G_1 if necessary. In a typical cell with a one-day doubling period, half the time is spent in G_1, and a quarter in each of S and G_2 phases; mitosis accounts for a mere 30 minutes.

The cell's most visually spectacular activity is displayed in M-phase, the mechanical division of the cell. Based on the appearance of the cell's chromosomes, the events of M-phase are grouped into mitosis (five stages) and cytokinesis (one stage), as displayed in Fig. 12.7. Most of the features of

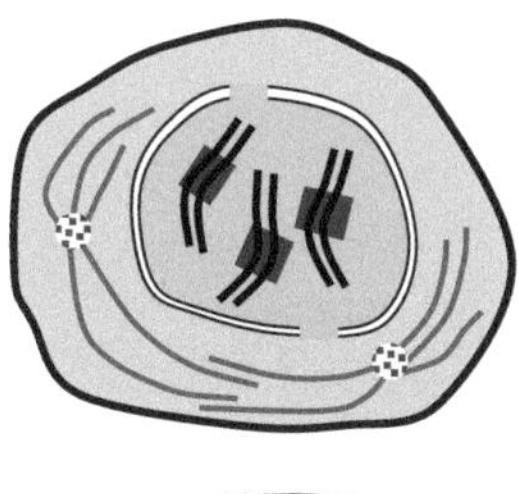

prophase chromosomes slowly condense (two sister chromatids each); centrosomes have duplicated and begun to form spindle poles; the nuclear envelope is still intact (kinetochores are depicted in black on the chromosomes)

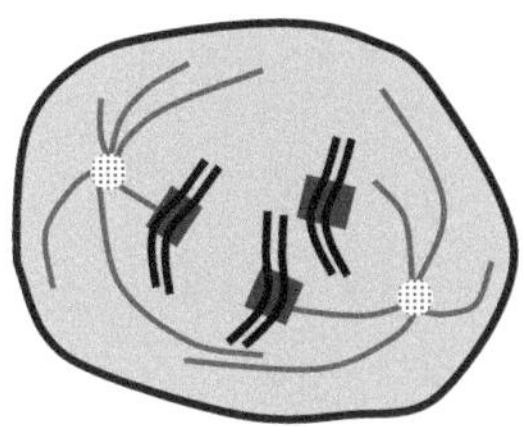

prometaphase the nuclear envelope breaks up, allowing microtubules from the separated centrosomes to seek out the kinetochore on each sister chromatid; *polar* microtubules extend towards midplane

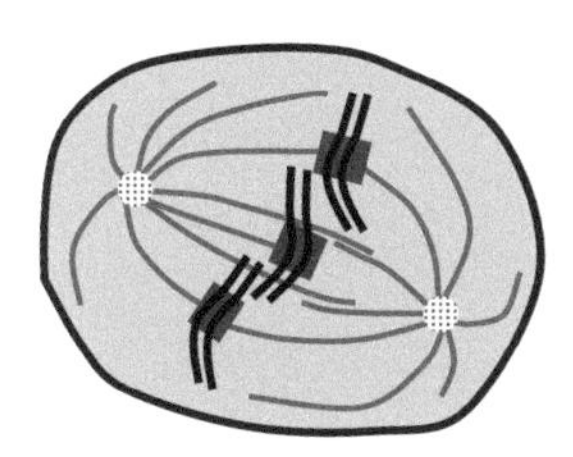

metaphase once connected to each other, polar microtubules provide a force opposing that from the kinetochore microtubules; this creates a tension on the chromosomes, pulling them towards the midplane

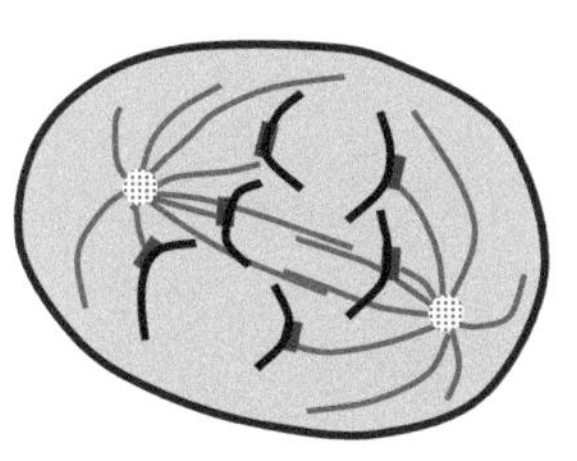

anaphase chromosomes break into pairs of chromatids, which are dragged by shortening kinetochore microtubules to the spindle poles; pressure from polar microtubules begins to push spindle poles apart

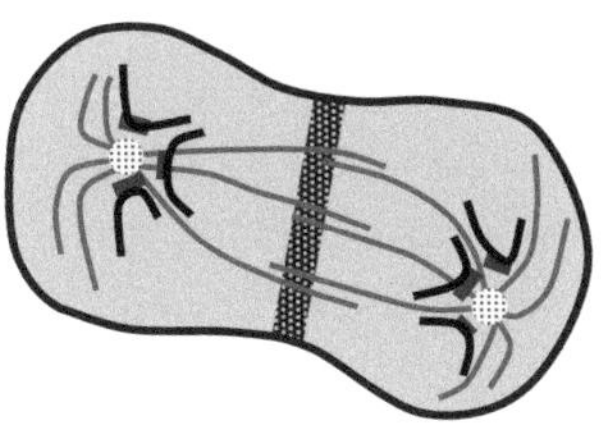

telophase the kinetochore microtubules disappear; chromatids decondense; a contractile ring begins to constrict around the equator; the nuclear envelope begins to reform as mitosis ends

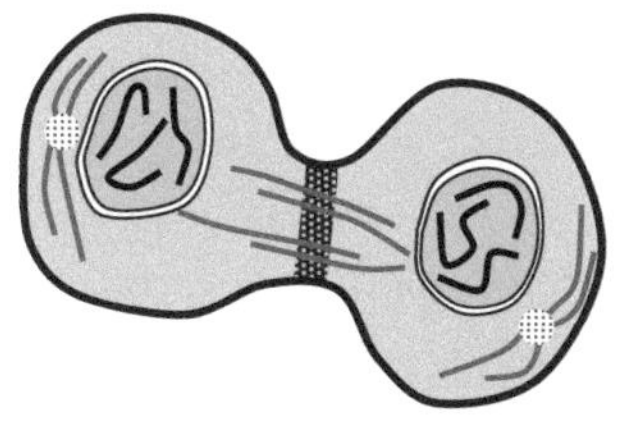

cytokinesis nuclear envelopes complete; contractile ring causes cleavage furrow and leads to cell division

Fig. 12.7 Representational features of cell division for eukaryotic animal cells. Plant cells may face the additional challenge of extending the cell wall during division.

each phase are introduced in the figure and their introductory description need not be repeated here. There are a variety of intriguing biochemical and biophysical aspects to the division process: of greatest interest in this book are the forces from microtubules that organize and segregate chromosomes and also the force generated in the actin/myosin contractile ring as it drives cytokinesis. These will be treated in greater detail in Sections 12.4 and 12.5.

Although these drawings represent a typical sequence for mitosis in eukaryotic cells, it is by no means the only sequence, nor the most studied one. For example, the doubling time of yeasts is competitive with bacteria and, given their low cost as lab specimens and their compact DNA, their cell cycle has been well characterized. Two different sequences are observed in yeasts: fission yeasts divide symmetrically, while budding yeasts divide into unequal daughter cells. In the next section of this chapter, we examine generic features of cell growth in bacteria, following which we explore asymmetric division, and variations in doubling time, in Section 12.3.

12.2 Growth of rods and cocci

The first aspect of the cell cycle that we investigate in detail is the growth rate of structurally simple cells such as bacteria. Our aim here is to examine whether the growth is uniform in time or whether the overall growth rate changes even as the cell's molecular contents continue to evolve. A rather primitive mechanism for cell division has been presented in Section 12.1, namely the creation in large cells of irregularly shaped entropy-rich arms that can pinch off and create new daughter cells. Owing to its large surface-to-volume ratio, this mechanism is not an efficient usage of materials; hence, it will not be analyzed further here. Instead, several models based on growth at constant surface curvature are introduced and their applicability to simple cells is evaluated.

As described in Section 10.5, most genera of bacteria have multi-component boundaries: a cell wall, one or two bilayer-based membranes plus a limited network of membrane-associated filaments. For the moment, let's view the boundary as effectively a single-component system. In Fig. 12.8(a), the surface of a spherical coccus with radius R and curvature $C = 1/R$ grows with unchanging curvature into two intersecting spherical caps until its surface area and volume have both doubled to produce two identical daughter cells. The growth of rod-shaped cells in Fig. 12.8(b) is only slightly more complex, and can be viewed as two distinct stages where growth is at constant curvature throughout, but with different values of C_1 and C_2 after the cylindrical section of the bacterium has doubled: $C_1 = 1/R$

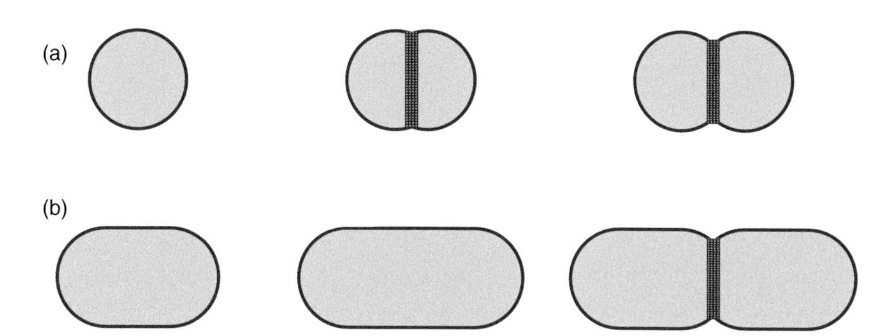

Fig. 12.8 Growth or elongation at constant curvature for cocci (a) or rod-like cells (b). The dark band at the intersection region represents a contractile belt.

and $C_2 = 0$ during extension of the cylinder and $C_1 = C_2 = 1/R$ as the mid-plane shrinks to the division point. Thus, the rod-like bacterium in Fig. 12.8(b) needs a sensor to indicate when cylinder elongation is complete, necessitating a change in C_2 of the newly added material.

Before describing this one-component system mathematically, we pause to consider the growth of cells whose boundaries have two mechanical components: these components need not possess distinct molecular composition, but rather need to move independently of one another over some range of shapes. Even a bilayer can be considered a two-component system if the lipid flip-flop rate between leaflets is low. Consider the series of shapes shown in Fig. 12.9: we assume that the inner component is a fluid membrane while the outer one is a network like a cell wall. If the area of the network is relatively constant, then material initially added to the membrane does not increase its area, resulting in an increase in its molecular density and, correspondingly, its state of strain. This strain can be relieved through buckling, which adds a ring of material to the membrane, as shown in cross section in Fig. 12.9(b). Although there is deformation energy associated with the region where the ring joins the spherical shape of the membrane, for the most part the membrane neck formed by the ring is flat and can be extended at no cost in deformation energy; in fact, the "hole" in the flat part of the membrane created by the ring possesses an edge tension that favors the contraction of the ring to form two separate chambers as in Fig. 12.9(c). Of course, this is not a complete cell cycle, as the configuration in Fig. 12.9(c) still has the original volume and network area as Fig. 12.9(a); in fact, even the area of the membrane has not doubled yet. However, if the transfer rate of material to the network is not too slow, the outer layer will grow, allowing the enclosed volume to do likewise as in Fig. 12.9(d).

There is a particular time dependence for the length, area and volume [$L(t)$, $A(t)$, and $V(t)$] of the cell shape within each model for the division cycle described above. For living cells, this time dependence can be measured

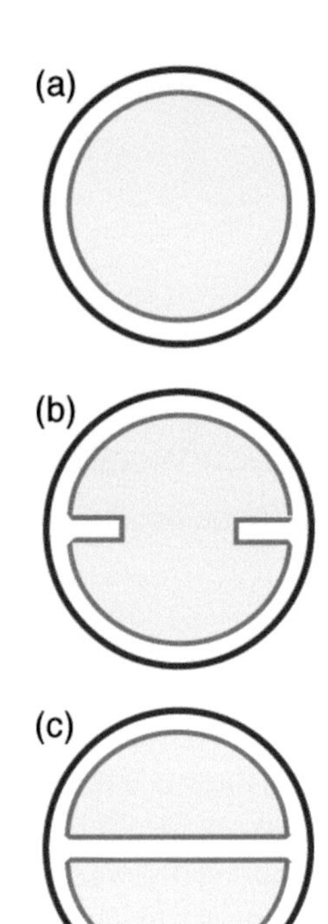

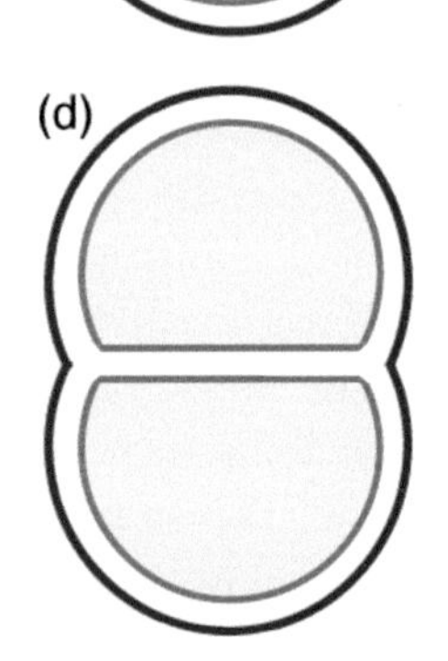

Figure 12.9

One possible model for the division cycle of a cell with a boundary having two mechanical components. In panels (a) to (c), the inner layer grows and buckles to form two separate chambers, then both layers grow at constant curvature as in panel (d) until the original area and volume have doubled.

in the lab by imaging the growth of single cells or it can be obtained from the measurement of an ensemble of cells undergoing steady-state growth. Suppose that we have an ensemble of n_{tot} cells whose shape we measure one by one; examples of such colonies are shown in Fig. 12.10 for the modern green alga *Stichococcus* in panel (a) and the microfossil taxon *Eoentophysalis belcherensis capsulata* in panel (b). We assume that each cell started to grow at a random initial time, so the shapes of the cells in the sample are uncorrelated. Choosing a particular variable φ (for example, L, A, V...) we count dn_φ cells having a value of φ between φ and $\varphi + d\varphi$. Now, dn_φ is a number, which necessarily depends on the total size of the sample n_{tot}. This dependence on the sample size can be removed by constructing the probability density $\mathcal{P}_\varphi$ (the probability per unit φ) from the definition

$$dn_\varphi = n_{\text{tot}}\, \mathcal{P}_\varphi\, d\varphi. \tag{12.5}$$

By integrating Eq. (12.5) over φ, one finds that $\mathcal{P}_\varphi$ is normalized to unity: $\int \mathcal{P}_\varphi\, d\varphi = 1$. Note that $\mathcal{P}_\varphi$ has units of φ^{-1}, whereas dn_φ is simply a number.

The link between $\mathcal{P}_\varphi$ and the time-dependence $\varphi(t)$ is that under steady-state conditions, $\mathcal{P}_\varphi$ is given by

$$\mathcal{P}_\varphi = (\partial\varphi \,/\, \partial t)^{-1} \,/\, T_2, \tag{12.6}$$

where T_2 is the doubling time of the cell cycle. For illustration, suppose the cell has the shape of a uniform cylinder that increases in length L from ℓ to 2ℓ (at fixed radius) whereupon it divides symmetrically at time T_2. If $L(t)$ grows linearly with time as $L(t) = (1 + t/T_2)\ell$, then $\mathcal{P}_\ell = 1/\ell$. In words, the physical meaning of Eq. (12.6) is that the more rapidly the quantity φ changes (i.e. larger $\partial\varphi/\partial t$), the less time the cell spends in that range of φ because $(\partial\varphi/\partial t)^{-1}$ is small. This is familiar in the simple pendulum, which moves the fastest through its vertical position and slowest through its turning points, such that it is least likely to be found in the vertical position and most likely to be found at the turning points.

Let's turn now to measurements of bacteria shape. In Fig. 12.11 is a scatter plot of the observed cell length $L_{\parallel}$ and width $L_{\perp}$ for colonies of the diplococcus *Synechocystis* PCC 6308 and rod-like *Synechococcus* PCC 6312, both of which are cyanobacteria. The projected length $L_{\parallel}$ is less than or equal to the true length, while the projected width $L_{\perp}$ is the true width and consequently equal to twice the cell radius. The cells in these colonies (as well as the 2-billion year-old microfossils of Fig. 12.1) possess widths that are relatively constant compared to their lengths: the standard deviation in cell width is less than 10% of the width itself. Similar results are obtained for other bacterial genera by Sharpe *et al.* (1998) and by Trueba and Woldringh (1980), among many others. If the width increased as the cell grew, the scatter plot would drift to the right with increasing $L_{\perp}$. To a good approximation then, the diplococcus and bacillus shapes grow at constant width $2R$, and this is true for many other genera that have been

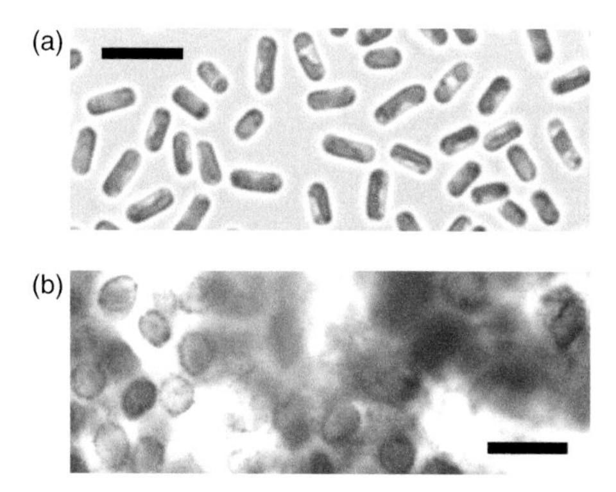

Figure 12.10 Samples from colonies of (a) rod-like green alga *Stichococcus* and (b) *Eoentophysalis belcherensis capsulata*. The cells in panel (a) lie mostly parallel to the focal plane, while in panel (b) they have random orientations. Scale bars are 10 μm.

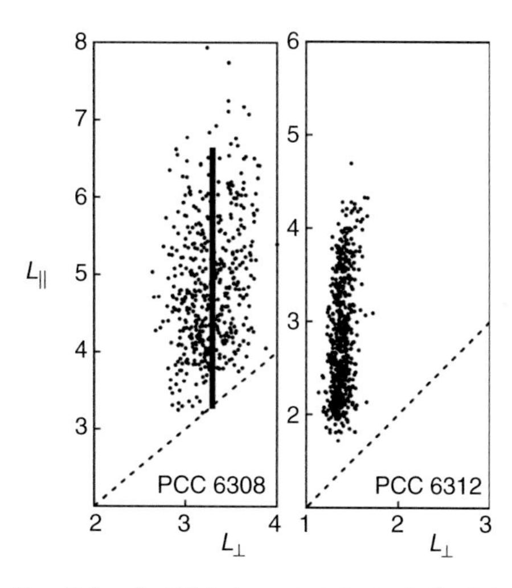

Fig. 12.11 Scatter plot of measured length $L_{||}$ and width $L_{\perp}$ (as projected onto the focal plane of the microscope objective) for diplococcus *Synechocystis* PCC 6308 and rod-like *Synechococcus* PCC 6312, both of which are cyanobacteria. The vertical line is an idealized trajectory assuming growth at constant curvature. Units are μm. (Forde and Boal, unpublished data.)

studied; measurements of their surface curvatures also appear to be close to $1/R$ (see Itan *et al.*, 2008, for a detailed analysis of *E. coli* shapes). With the cell radius fixed, the length, area and volume of this family of shapes depend on only one geometrical quantity, which we can choose to be the separation s between the centers of the intersecting spheres. Through the division cycle, the diplococcus grows from $s = 0$ (a single spherical cell) to $s = 2R$ (two spheres in contact), with the length L, area A, and volume V of the cell depending on s as

$$L = 2R(1+\beta) \tag{12.7a}$$

$$A = 4\pi R^2(1+\beta) \tag{12.7b}$$

$$V = (4\pi R^3/3)\cdot[1 + \beta(3 - \beta^2)/2)], \tag{12.7c}$$

where $\beta \equiv s / 2R$. Similar expressions can be obtained for rod-like cells, as performed in the end-of-chapter problems. Once the time dependence of just one of L, A, or V is known, the time dependence of the remainder is determined and the probability densities $\mathcal{P}$ can be calculated from Eq. (12.6) as follows (Bennett *et al.*, 2007).

Linear volume increase A linear increase with time implies $\mathrm{d}V/\mathrm{d}t$ is constant; here $\mathrm{d}V/\mathrm{d}t = (4\pi R^3/3) / T_2$. Thus:

$$\mathcal{P}_\beta = 3(1 - \beta^2) / 2. \qquad \text{(linear in } t\text{)} \tag{12.8}$$

Exponential volume increase Here, the rate at which the volume increases is proportional to the instantaneous value of the volume, or $dV/dt = V \ln 2/T_2$. From this,

$$\mathcal{P}_\beta = [3(1 - \beta^2)/\ln 2] / [2 + \beta(3 - \beta^2)] \qquad \text{(exponential in } t\text{)} \tag{12.9}$$

In both of these expressions, T_2 from Eq. (12.6) has cancelled out. The conversion from $\mathcal{P}_V$ to $\mathcal{P}_\beta$ is done via the chain rule of calculus.

The probability density $\mathcal{P}_\beta$ is shown in Fig. 12.12 for two species of the diplococcus *Synechocystis* from the Pasteur Culture Collection, PCC 6804 and PCC 6714 (Boal and Forde, 2011). As a result, $\mathcal{P}_\beta$ is non-vanishing in the smallest measurable range of β. In fact, $\mathcal{P}_\beta$ is peaked around $\beta = 0$, from which it declines and eventually vanishes as $\beta \to 1$. This behavior of $\mathcal{P}_\beta$ has the physical interpretation that the cell grows most slowly at the start of its division cycle (large $\mathcal{P}_\beta$) and most rapidly at its end (small $\mathcal{P}_\beta$).

How do the measurements of $\mathcal{P}_\beta$ in Fig. 12.12 compare with models of cell growth? Exponential growth of cell length or area should obey $\mathcal{P}_\beta = [\ln 2\ (1 + \beta)]^{-1}$, a function that decreases from 1.44 at $\beta = 0$ to 0.72 at $\beta = 1$ (dotted line); these values do not agree with the data in the figure. The predicted $\mathcal{P}_\beta$ based upon volume growth with linear (dashed line) or exponential (solid line) time dependence is also plotted in the figure. The differences between the theoretical curves are obviously not large, which is expected because the exponential function e^x is approximately linear in x at small x. The intercept of $\mathcal{P}_\beta$ at $\beta = 0$ is predicted to be $3/(2 \ln 2) = 2.16$ for exponential volume growth and 3/2 for linear volume growth. The higher (exponential) value of $\mathcal{P}_\beta$ ($\beta = 0$) is mildly preferred by experiment in both panels of the figure, but linear volume growth is not ruled out. This preference for exponential growth in volume is seen for many other cyanobacteria of differing topologies as well as *Bacillus cereus* (Collins and Richmond, 1962), *Bacillus subtilis* (Sharpe *et al.*, 1998) and *Escherichia coli* (Koppes

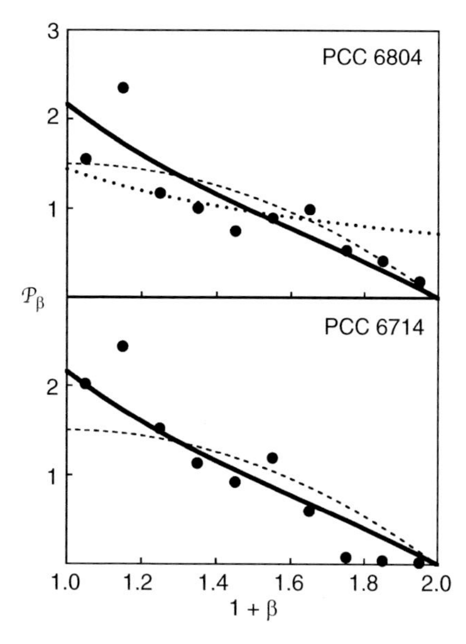

Fig. 12.12 Probability density $\mathcal{P}_\beta$ for diplococci *Synechocystis* PCC 6804 and *Synechocystis* PCC 6714, where $\beta \equiv s/2R$. Shown for comparison are predictions from models based on exponential (solid curve) or linear (dashed curve) volume growth, as well as exponential growth of cell length or area (dotted curve, top panel only). The statistical uncertainty is about 9% per individual datum. (Boal and Forde, 2011.)

and Nanninga, 1980; Grover *et al.*, 1987); see Cooper (1991) for a review. More precise observations on *E. coli* that follow the growth of single cells find that three linear growth regimes with different rates provide a somewhat better description than exponential (Reshes *et al.*, 2008).

Let's now return to the principles behind models for cell growth. Models where the growth of a cell is linear in time assume that change occurs at the same rate throughout the division cycle no matter what the contents of the cell. Examples of linear models can be found in some eukaryotic cells, where cell mass grows linearly with time (Killander and Zetterberg, 1965). In contrast, a study of mouse lymphoblasts found that the rate of cell volume increase was not constant, but proportional to cell volume itself (Tzur *et al.*, 2009). For bacteria, the only linear model not immediately ruled out by data is the linear rise in volume, for which agreement with data is marginal. Exponential growth may arise from several different mechanistic origins. Exponential growth in area corresponds to new surface being created at a rate proportional to the area available to absorb new material – a logical possibility but not supported by Fig. 12.12. Lastly, exponential growth in volume arises if new volume is created at a rate proportional to the cell's contents, which is the only scenario to comfortably describe the data. Not only do modern rod-like bacteria and diplococci show exponential increase in volume with time, the fossilized bacteria of Fig. 12.1 also display this behavior, implying that exponential volume growth has very likely been a feature of bacterial growth for at least two billion years.

For most bacteria to accommodate cell growth, their cell walls must be able to continuously incorporate new strands of peptidoglycan (see Goodell, 1985). This is a rather delicate operation for Gram-negative cell walls, where the lateral stress may be borne by just a single layer of peptidoglycan. A number of possible mechanisms for wall growth have been advanced (see reviews by Höltje, 1993; Koch, 1993; and Thwaites, 1993), which generally recognize the importance of forming new network bonds before breaking old ones (see Koch, 1990, for an overview of wall growth issues). New strands may be added with random orientation to spherical cocci, as the network is presumably isotropic. However, strands must be added with a preferred direction to rod-shaped bacteria, both to accommodate the anisotropic tension and to direct growth along the cylindrical direction of the cell. That is, rod-shaped bacteria do not expand like a balloon, but rather elongate and divide like a string of sausages.

For cocci, the addition of new wall material occurs at the furrow (septum) that will ultimately become the division plane. That is, a daughter cell can be thought of as two hemispheres, one of which is its original peptidoglycan wall and the other is new material added through growth, as shown in Fig. 12.13(a). Rod-like cells display two different growth strategies. For cells like *E. coli* (Gram-negative) or *B. subtilis* (Gram-positive), insertion of new peptidoglycan occurs first in the cylindrical sidewalls of the cell but not at its end-caps (Schlaeppi *et al.*, 1982; Mobley *et al*, 1984; De Pedro *et al.*, 2003; Scheffers *et al.*, 2004). Once the cylindrical section has doubled in length, formation of the septum proceeds through the building of new wall, as illustrated in Fig. 12.13(b). However, cells such as *Corynebacterium diptheriae* elongate through their end-caps, following which the growth site switches to the septum, as in Fig. 12.13(c) (Umeda and Amako, 1983; Daniel and Errington, 2003).

In virtually all eubacteria (wall-bearing bacteria), formation of the septum is associated with the presence of the so-called Z-ring containing the tubulin-like protein FtsZ. As the name implies, the ring is an equatorial hoop circling the midplane of the cell. With time, the ring shortens from an initial circumference of $2\pi R$ to zero, where $2R$ is the width of the cell (for cylinders or spheres). The Z-ring is on the inside of the cell boundary and is not a belt around the outside of the cell wall. We can test the model time dependence of the shortening of the Z-ring by repeating the same line of reasoning that leads to Eqs. (12.8) and (12.9). For diplococci, the radius r of the ring is related to the separation s between the curvature centers of the end-caps simply by Pythagoras' Theorem,

$$r^2 = R^2 - (s/2R)^2. \tag{12.10}$$

Thus, if the radius decreases linearly with time according to

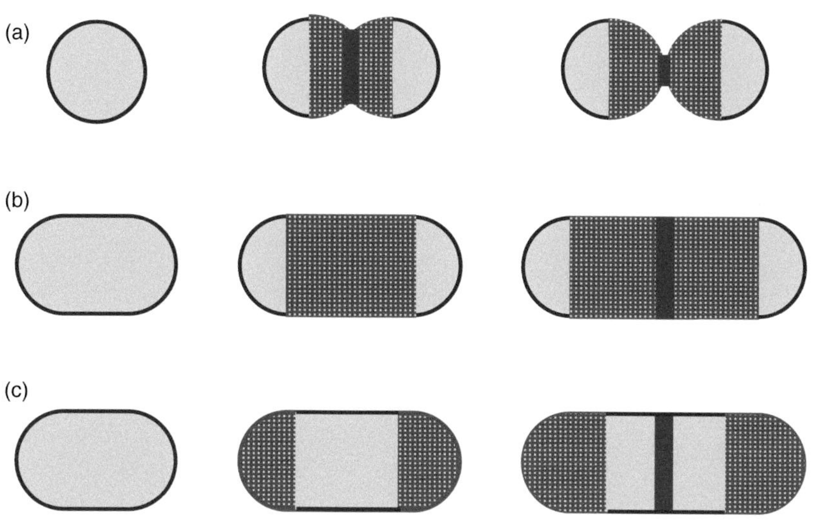

Fig. 12.13 New cell wall (in red) is usually added to the septum of spherical cells (a) or to the sidewalls and then the septum of rod-like bacteria (b). In some rod-like bacteria, it is added to the end-caps before the septum (c). The contractile ring is shown in blue.

$$r(t) = R\,(1 - t/T_2), \qquad (12.11)$$

then the probability density $\mathcal{P}_\beta$ is found to be

$$\mathcal{P}_\beta = \beta\,/\,(1 - \beta^2)^{1/2}, \qquad (12.12)$$

as derived in the end-of-chapter problems. This behavior of $\mathcal{P}_\beta$ is rather different than Eqs. (12.8) and (12.9), in that Eq. (12.12) increases with β, rather than decreases, and does not properly describe the data in Fig. 12.12. Thus, the time-dependence of the Z-ring in these cells is more complex than simply linear shrinkage. See Biron *et al.* (2005) for an assessment of the dynamics of the contractile ring in animal cells.

12.3 Asymmetry and asynchrony in cell division

In an idealized division cycle for growth at constant cell width $L_\perp$, the length $L_\parallel$ of a coccus would increase to two cell diameters at the division point, and the length of a rod-like cell would double from its initial value. If all cells of a given species had exactly the same width, then a scatter plot of $L_\parallel$ against $L_\perp$ in a colony of cells would simply show a vertical line: e.g., the shapes of a diplococcus would run from $(L_\perp, L_\parallel) = (2R, 2R)$ to $(2R, 4R)$, where R is the radius of the coccus when the cell cycle begins. This ideal trajectory is indicated by the vertical bar in Fig. 12.11 for the diplococcus-shaped cyanobacterium *Synechocystis* PCC 6308. Similar behavior would be observed in a scatter plot of rod-like cells, except that the smallest cell would not lie on the $L_\perp = L_\parallel$ locus (the dashed line in Fig. 12.11) but would be displaced along the y-axis to the initial length of the cell.

Taken from colonies of cells growing under steady-state conditions, the ($L_\perp$, $L_\parallel$) scatter plots of *Synechocystis* PCC 6308 and bacillus-like *Synechococcus* PCC 6312 shown in Fig. 12.11 are not far from idealized growth trajectories, but beyond what would arise solely from experimental uncertainties in analyzing the shapes of the cells. What is the source of the deviation from the ideal trajectory? Perhaps each cell produces identical twin daughters upon division, but the division occurred before or after the daughters were identical to their parent in size. Alternatively, it's possible that the doubling time is fixed, but the division is asymmetric and produces non-identical daughters. Both of these explanations have been investigated experimentally; here, we will discuss asymmetric division at length before presenting asynchronous division.

An example of explicit asymmetric division can be found in yeasts, which display two different cell cycles. Although they are eukaryotes, fission yeasts grow like the rod-like cells in Fig.12.13(c) – they elongate until they divide into identical daughters. In contrast, budding yeasts grow asymmetrically like the series of dumbbell shapes in Fig. 10.8 except that budding yeast cells are slightly elongated to begin with: the shapes are not intersecting spheres and when division occurs, the daughter cells are not identical. However, the distribution of cell shapes in Fig. 12.11 is narrower and has a more subtle origin than gross asymmetry in the cell cycle; instead, the figure represents the effects of the slightly off-center location of the division plane.

The origin of the asymmetry we're interested in is not the deliberate division strategy of budding yeast, but is rather the statistical uncertainty of locating the midplane of the cell, or in the determination of the division time in the case of asynchronous division. To quantify the asymmetry, more information is needed than just the distribution of cell shapes as in Fig. 12.11; rather, the correlation in cell sizes for daughter pairs must be measured. For *E. coli*, the location of the division plane is at 0.5 ± 0.013 of the cell length, which implies an uncertainty of 2%–3% in the location of the septum, and a larger range for the distribution of the length of daughter cells.

The variable location of the midplane in an otherwise symmetric system may have several causes (statistical limits on the accuracy of positional information is discussed in Gregor *et al.*, 2007). In eukaryotic cells, microtubules provide a centering mechanism by pushing on the plasma membrane, and on each other when linked together. However, there is an inherent randomness to the orientation of the microtubules, so one does not expect this centering mechanism to be too exact. Possessing only a modest membrane-associated cytoskeleton, bacteria must have an alternate centering mechanism than what is available to eukaryotic cells. Two such mechanisms can be imagined using only repulsion effects.

Buckling Figure 12.9 shows how the differential growth of a boundary with two mechanical components can lead to the buckling of the inner component, whose lowest stress state will be an equatorial ring.

Elongation of the Z-ring If the Z-ring elongates as it forms, then it will maximize its length before it succumbs to buckling, producing an equatorial ring in a coccus.

These mechanisms, if implemented, would only apply to cocci: the first mechanism is probably ineffective for rod-shaped cells, and the second mechanism would produce a ring that ran longitudinally from pole to pole, not around the equator. Although a pole-to-pole loop could generate a helical path if it were deformed, it would not look like the Z-ring, and its specific handedness would need an alternate explanation.

The challenge, then, is to find a mechanism that works for rod-like cells. First, it is known that the Z-ring avoids formation in the region of the bacterial nucleoid, which is not unexpected. Next, the recruitment of the Z-ring at midcell involves several proteins in Gram-negative *E. coli*. Collectively called the Min proteins, their role in broad terms is as follows.

MinC This protein has a negative function in that it inhibits the condensation of FtsZ into filaments.

MinD Loosely associated with the bacterial plasma membrane only when in its ATP-bound form, MinD recruits MinC to its location and increases the latter's concentration. The combined MinC/MinD complex suppresses the formation of the Z-ring near the plasma membrane.

MinE Also playing a negative role, MinE blocks the action of MinC/MinD and permits the Z-ring to form where MinE is locally present.

Both MinC and MinD are necessary for proper cell division: if either protein is absent, then cell division occurs near a pole of the cell, resulting in highly asymmetric division having one small cell without a chromosome. If MinE is absent, then MinC/MinD are present over the entire membrane surface and prohibit the Z-ring from forming on it. The necessity of the "double negative" regulation mechanism is now clear: one global mechanism (MinC/MinD) prevents the Z-ring from assembling where it is not wanted and another mechanism (MinE) places the ring in the correct site. But what specifies the location of MinE?

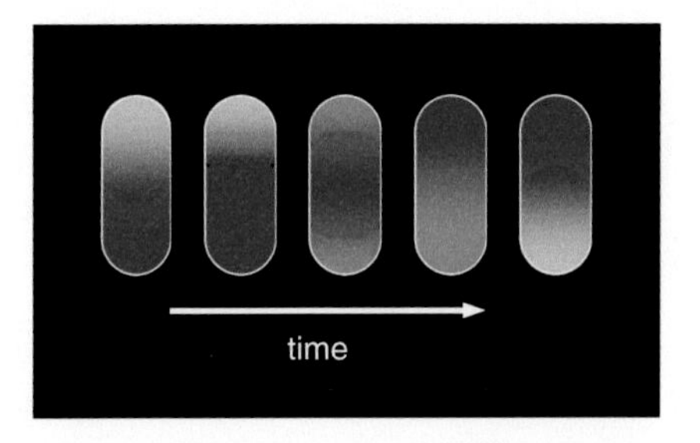

Fig. 12.14 Time evolution at equal intervals of membrane-associated MinD in *E. coli* as observed at low resolution, shown as a concentration profile for half the period of the oscillation (period is about one minute). Projected images of the cell will show higher concentrations toward the cell boundaries because of the surface curvature (interpretation of Hale *et al.*, 2001).

Owing to the small cell size of *E. coli*, the earliest imaging of fluorescently labeled Min proteins *in vivo* was understandably relatively low resolution. Time-sequence images of a given cell showed an oscillatory behavior in which MinD initially occupies a region near only one pole of the cell. The region then retreats toward the pole, as a new region of Min D begins to form at the opposite pole, as shown in Fig. 12.14. After all of the MinD at the first pole has disappeared, the process starts all over again at the second pole (Raskin and de Boer, 1999). A complete cycle is relatively fast with a minimum period of about 30 s, depending on both the total amount of MinD and MinE in the cell: the period is directly proportional to the amount of MinD, and inversely proportional to MinE. Although it may be found throughout the cytoplasm, the density of MinC at the cell membrane follows the behavior of MinD associated with the cell boundary (Hu and Lutkenhaus, 1999). At the plasma membrane, MinE is present at the edge of the MinD polar region and propagates along with it as the MinD region shrinks away from the midcell plane (Hale *et al.*, 2001).

More detailed imaging of the Min system has revealed that a MinD region is not uniformly covered but has the same type of membrane-associated helical filaments as observed for MreB and Mbl (Shih *et al.*, 2003). Thus, the fuzzy areas of MinD near the poles in the images of Fig. 12.14 are actually helices when seen at higher resolution, and the ring of MinE that encircles the MinD region is also canted like a helix with respect to the cell's long axis. The picture that emerges of the Min oscillations is then one where MinE travels up the tail of the MinD helix, starting at midcell, releasing MinD into the cytoplasm as it goes. Cytoplasmic MinD then regroups at the opposite pole, from which it forms a new helix growing towards midcell. In a given cell, there are far fewer copies of MinE than MinD, so the MinE ring is relatively short. The behavior of the MinD filament and MinE ring is shown schematically in Fig. 12.15.

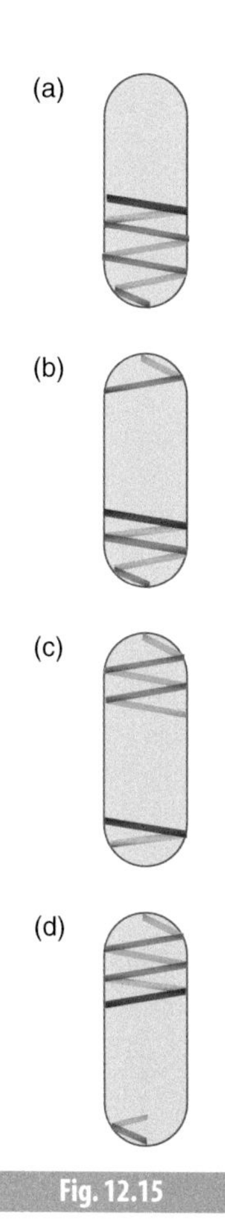

Fig. 12.15

Schematic representation of the growth of MinD filaments (teal) and MinE ring (red) in a rod-shaped cell. In panels (a)–(c), MinE works along an existing MinD helix, releasing its individual molecules into the cytoplasm. In panels (b) and (c), a new MinD helix has begun to grow from the opposite pole. Once MinE has finished its work in panel (c), it moves to midcell and begins to release MinD from the upper filament.

Thus, there are two helical components to the cytoskeleton of rod-like bacteria: MreB (and Mbl where present) influence the shape of the rod and MinCDE help define the cell's midplane for septum formation. An obvious question is whether the two groups of helices are structurally coincident. The answer appears to be negative, as the the pitch of MreB helices is about twice that of MinD; in addition, MinE may not entirely dismantle the MinD helix, but leave a residual track at each pole for the next construction phase in the Min cycle (Shih *et al.*, 2003).

The low-resolution data were initially interpreted in terms of reaction–diffusion models in which the Min proteins are described only in terms of their concentration as a function of their location along the symmetry axis of the bacterium (Howard *et al.*, 2001; Meinhardt and de Boer, 2001). That is, the Min proteins were not treated as dynamic filaments. To their credit, for some regions of their parameter space, the models show oscillatory behavior for MinD and MinE, and indicate a region near the bacterial midcell where MinD is depleted and MinE enhanced on average, thus specifying the location of the Z-ring. Although many features of the time-dependence of the Min concentration profiles are correctly captured, the models are incomplete (Huang *et al.*, 2003) and may be regarded as effective representations of a more complete theory.

To give a flavor of the mathematics involved in these calculations, we reproduce the set of equations from the model by Howard *et al.* (2001):

$$\partial\rho_D/\partial t = D_D(\partial^2\rho_D/\partial x^2) - \sigma_1\rho_D/(1 + \sigma'_1\rho_e) + \sigma_2\rho_e\rho_d, \tag{12.13a}$$

$$\partial\rho_d/\partial t = \sigma_1\rho_D/(1 + \sigma'_1\rho_e) - \sigma_2\rho_e\rho_d, \tag{12.13b}$$

$$\partial\rho_E/\partial t = D_E(\partial^2\rho_E/\partial x^2) - \sigma_3\rho_D\rho_E + \sigma_4\rho_e/(1 + \sigma'_4\rho_D), \tag{12.13c}$$

$$\partial\rho_e/\partial t = \sigma_3\rho_D\rho_E - \sigma_4\rho_e/(1 + \sigma'_4\rho_D), \tag{12.13d}$$

where

- ρ_D and ρ_E are the number densities of MinD and MinE in the cytoplasm,
- ρ_d and ρ_e are the number densities of MinD and MinE at the cytoplasmic membrane.

The diffusion of MinD and MinE in the cytoskeleton is the basis for the terms in Eqs. (12.13a) and (12.13c) that have the form of Fick's second law of diffusion, $\partial\rho/\partial t = D(\partial^2\rho/\partial x^2)$ expressed in Eq. (2.53). The other terms are reactions among MinD and MinE in the cytoplasm and on the membrane. Looking at Eq. (12.13a) for example, the term $-\sigma_1\rho_D$ represents the removal of MinD from the cytoplasm, and the term $+\sigma_2\rho_e\rho_d$ represents the release of MinD into the cytoplasm caused by the interaction of MinD and MinE in the membrane. The other terms have similar origins. In addition to the diffusion constants D_D and D_E, this model contains six parameters (σ) that must be estimated from data to test its true validity.

Table 12.1 Samples of measured values for rate constants for FtsZ (Gonzáles *et al.*, 2003) and ParM (Garner et al., 2004); units of (μm • s)−1 for k_{on}, s−1 for k_{off}, and μm for $[M]_c$

Monomer	k_{on}	k_{off}	$[M]_c$
FtsZ			1.25
ATP-ParM	5.3 ± 1.3	—	0.55–0.68
ADP-ParM	—	64 ± 20	100

Reaction–diffusion models are capable of capturing many of the observations of Min transport in *E. coli* seen at low resolution, although their detailed success varies according to the content of the model (Huang *et al.*, 2003). To probe the dynamics more deeply, however, a model must incorporate the localization effects of the MinD molecules when in their filamentous form and this has been the focus of recent work (Drew *et al.*, 2005; Cytrynbaum and Marshall, 2007). In addition, better knowledge of the rate constants σ is required; a few examples of the data are given in Table 12.1.

The other aspect of the division cycle of interest in this section is asynchrony – the idea that a group of cells which start their cycle at the same time may complete the cycle at slightly different times. The two approaches that we introduce here for measuring the distribution and correlations in division times are based on different ensembles. In one case, doubling times are repeatedly measured for a single cell through successive generations, while in the other case, the division state of many progeny of a given cell are measured simultaneously, but just once. An example of the single-cell approach is a study of the doubling time of *Caulobacter crescentus* by Siegel-Gaskins and Crosson (2008) in an experimental setup illustrated in Fig. 12.16. A mother cell attaches itself to a substrate by means of a stalk, and grows in a moving fluid. Upon each division, the daughter cell is removed by the fluid and the cycle begins again. The doubling times for the mother cell with a stalk already in place, averaged over many division cycles and over several individual cells, show a distribution with a mean value of 58.3 min and a standard deviation of 9.5 min. However, the complete cell cycle for a daughter cell includes the growth of the stalk, which is not present at birth; thus, the whole cycle has a doubling time of 68.7 ± 8.6 min. The long doubling time means that the separation point can be determined accurately, with an error less than the standard deviation of the division time; as usual, the division time will depend upon the growth environment.

Does the variation in division time mean that a given cell has a fixed division time, but there is a distribution of this time among the mother cells in the sample? This possibility can tested by measuring the correlation

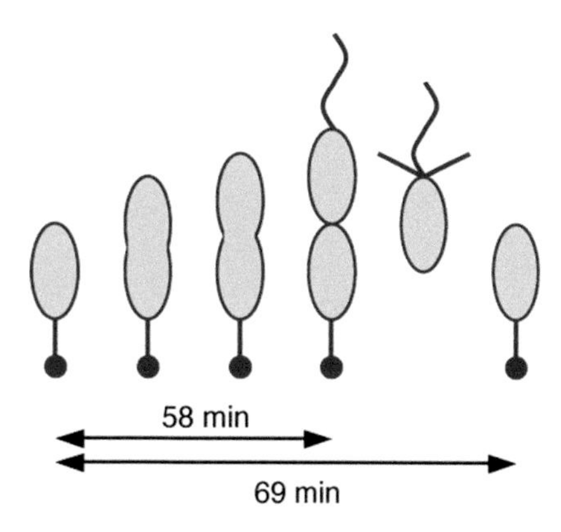

Fig. 12.16 In one study of the division time of *Caulobacter crescentus*, the mother cell attaches itself to a substrate and grows in a moving fluid; upon division, the daughter cell is removed by the fluid.

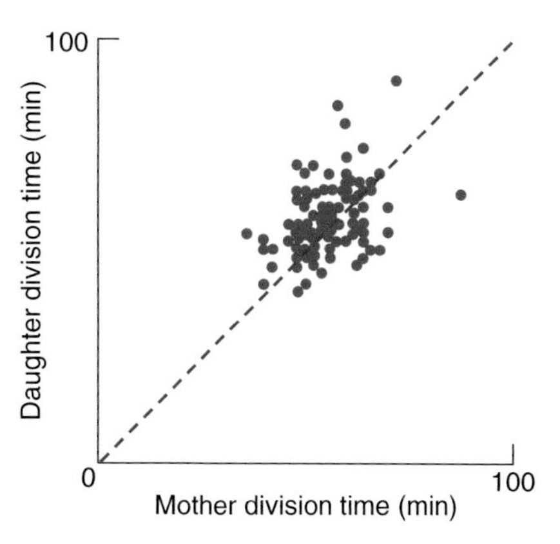

Fig. 12.17 Scatter plot showing the correlation between the division time of a daughter cell with that of its mother for *Caulobacter crescentus* (data from Siegel-Gaskins and Crosson, 2008).

between the division time of a daughter cell with that of its mother, as in Fig. 12.17. If the division time of the parent were exactly hereditary by the daughter, the scatter plot would be a straight line of unit slope. Clearly, the data are not so tightly correlated, but a trend along the diagonal line is apparent and statistically supported. Thus, the lifetime of a daughter cell is correlated with its parent, but there will be variation of the division time from one generation to the next.

A different approach to asynchronous division applicable to filamentous cells is based on simultaneous measurement of a large sample. Starting from a single cell, if all successive daughters have exactly the same division time, then all cells in a colony grown from a single parent will be observed at the same point in their cell cycle whenever a measurement is made of the colony. In a filament, this means that all cells will have the same size. However, if there is variation in the division time, then there will be variation in the cell size along the filament because the cells in each generation will gradually lose their synchrony: after just a few generations, neighboring cells are likely to have the same size, but after many generations, each cell will have one neighbor (its sister at birth) of approximately the same size, and another neighbor chosen randomly from within the general distribution of sizes. Simply taking a snapshot of a filament will not disclose which neighbor is the sister of a given cell. The effect is visible by inspecting even a moderately short strand of some genera like the alga *Stichococcus* in Fig. 10.23(a), where pairs of small cells are noticeable. Quantitative measurements of one hundred *Stichococcus* cells in sequence are shown in Fig. 2.16, and again, groups of cells of approximately the same size can be seen extending over half-a-dozen to a dozen cells.

12.4 Transcription and replication

Our approach to cell growth and division has been to examine micron-scale issues first, such as the time evolution of cell shape, and then probe cellular processes at ever smaller length scales. In this section, we complete this task by describing the operation of the cell's genetic blueprint on the molecular level – the structure of the DNA blueprint and its related macromolecules is presented in Appendix B. Only a limited sample of topics from the vast field of genomics and proteomics is visited here, where we provide an overview of DNA transcription and replication. Chromosome segregation in cell division is treated in Section 12.5, and the theoretical framework of genetic control is constructed in Sections 13.4 and 13.5. This section concludes with a summary of the molecular-level forces at work during DNA transcription and replication.

As carried by DNA, the genetic blueprint of the smallest modern cells contains approximately 800 000 base pairs, compared to human DNA with 3.2 billion base pairs (the sum of all chromosomes) and the amoeba *Polychaos dubium* with an astounding 670 billion base pairs. Each protein is encoded in roughly 1000–1200 base pairs on either string of the DNA double helix. The smallest cells are very efficient and almost all their DNA is *coding* in the sense that it contains the genetic code for a particular protein. However, the DNA in organisms such as ourselves is notoriously inefficient: the 20 000–25 000 human protein-coding genes can be coded in a few tens of millions of base pairs, not 3 billion. Worse yet, the thousand-base-pair sequence for a given protein may be broken up by the presence of non-coding *introns* (or intervening sequence) which effectively lengthens the stretch of DNA occupied by a gene. We return to the question of gene structure in Section 13.4; here we are interested in how a particular region of DNA is *transcribed* for use in protein production, and how the whole blueprint is *replicated* in preparation for cell division.

12.4.1 DNA transcription

An early step in the process of transcribing a DNA gene is the creation of a strand of messenger RNA, or *mRNA*. For this to occur, the DNA double helix must be unwound so that its base pair sequence can be read. Although a strand of mRNA may be a few thousand base pairs long (it must carry more information than just the thousand bases of the gene) only a small portion of the DNA helix is unwound at any given time. RNA polymerase (mass ~ 500 kDa) is the enzyme that directs production of RNA. In bacteria, following a collision at a random point on the DNA macromolecule, RNA polymerase sticks lightly to the DNA helix and slides along it. When it encounters a *promoter* sequence on the DNA, RNA polymerase then binds tightly and begins to unwind the helix as it moves along, as illustrated in Fig. 12.18. It reads the DNA sequence and assembles a strand of RNA from the appropriate chemical building blocks available in its environment. Finally, it rewinds the helix continuously as it leaves the parts it has transcribed.

The ends of each strand of DNA are inequivalent, and are labeled 3′ or 5′ according to the unbonded link on the sugar ring at the end of the strand; this is described further in Appendix B. Because of their handedness, the two strands in the helix run in opposite directions, so to speak, such that a truncated end of a double helix is the 3′ end of one strand and the 5′ end of the other, as illustrated in Fig. 12.19(a). Each strand of DNA

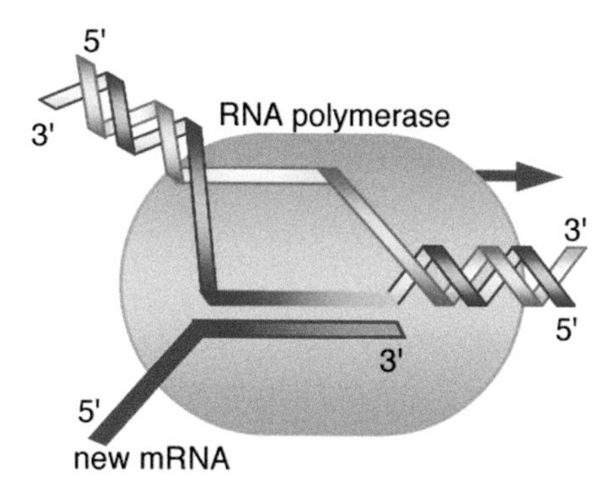

Fig. 12.18 Schematic representation of DNA transcription. RNA polymerase reads a single strand of DNA from its 3′ to 5′ ends, and produces mRNA from its 5′ to 3′ ends. The polymerase moves to the right.

(a) DNA strand 1 5' AATTTGCGCGTTAGAGACCTG 3'
DNA strand 2 3' TTAAACGCGCAATCTCTGGAC 5'

(b) mRNA copy 5' UUAAACGCGCAAUCU
DNA template 3' AATTTGCGCGTTAGA

(c) coding strand 5' NNNNNNGTTGACANNNNNNNNNNNNNNNNNTATAATNNNNNNA
template strand 3' NNNNNNCAACTGTNNNNNNNNNNNNNNNNNATATTANNNNNNT
-35 region -10 region +1

(d) mRNA copy 5' CCCACAGCCGCCAGUUCCGCUGGCGGCAUUUUAACUUUC
DNA template 3' GGGTGTCGGCGGTCAAGGCGACCGCCGTAAAATTGAAAG

(e) 5' CCCACAGCCGCCAGUU
||||||||||||||||||| C
3' AUUUUACGGCGGUCGC

Fig. 12.19 Processes involved in the transcription of DNA to produce messenger RNA (mRNA). (a) Each strand of DNA is paired with the other according to A ↔ T and G ↔ C binding. The ends of each strand are inequivalent, with a 3′ sugar on one strand and a 5′ sugar on the other (see Appendix B for details). (b) The mRNA made from a DNA template is paired with it according to A ↔ U and G ↔ C binding. The 3′ end of the DNA template corresponds to the 5′ end of the mRNA copy. (c) Two strands of bacterial DNA for which the attachment sequence for RNA polymerase on the template strand is indicated in green. The first base of the gene is the last character (T) at the right-hand end of the strand. Even though it is not read, the upper strand is called the coding strand because it corresponds to the mRNA sequence. The symbol N represents any nucleotide that is properly paired. (d) Sequence in bacterial DNA that marks the end of transcription; lower strand is the DNA template and the upper strand is the mRNA copy. (e) The mRNA sequence in panel (d) can double-back on itself via internal base-pair bonds to form a loop that terminates the mRNA strand.

is paired with the other according to A ↔ T and G ↔ C complementarity. DNA and RNA have slightly different quartets of bases (CGAT for DNA and CGAU for RNA) so the base pairing between DNA and RNA, namely A ↔ U and G ↔ C, differs from that between the complementary strands of a DNA helix. Further, just as the unbound links of the sugar rings at each end of a DNA strand are inequivalent, the same is true of RNA. The enzyme RNA polymerase reads a single strand of DNA from its 3′ to 5′ ends, and therefore produces mRNA from its 5′ to 3′ ends, as demonstrated in Fig. 12.19(b). Note: although RNA polymerase reads the template strand, the quoted gene sequence is that of the complementary strand (called the coding strand) because it is the same as the RNA sequence with the usual U/T replacement.

Superficially, the arrangement of the base-pair letters in a strand of DNA has been likened to a book written with no spaces and no punctuation: how does one know where each sentence begins and how to parse the letters into words? We'll address the punctuation question first, and leave the parsing of words until later. Imagining a gene on a strand of

DNA to be a sentence with more than a thousand letters, the start point of the sentence is several small groups of base-pair letters that tell RNA polymerase where to bind and start reading the DNA template. The short sequences are shown in Fig. 12.19(c) for both the DNA template and its complementary sequence on the other strand of the helix; the figure displays the consensus sequence in the promoter region for bacterial DNA, where the consensus represents an average over many genes. The lower strand is read from left to right (3′ to 5′) with the AACTGT sequence around 35 base pairs before the actual gene (called the −35 region); the first character of the gene is indicated by the last base on the right (T). At 0.34 nm per base pair, the −35 region is 10 nm from the +1 site, which is about the length of DNA polymerase.

The consensus sequence in Fig. 12.19(c) indicates where RNA polymerase should start transcription, but what signal tells it where to stop? For bacteria, this sequence is displayed in Fig. 12.19(d): the lower strand is the DNA template and the upper strand is the mRNA copy. Close inspection of the two longer RNA sequences in red (that is, not the UUUU quartet) reveals that they are complementary to each other but in reverse order. Consequently, these two sequences can bind as RNA base pairs to form a hairpin loop as shown in Fig. 12.19(e), thus tidying up the dangling end of the mRNA very nicely. Transcription can be stopped by the application of a force to RNA polymerase in a direction opposed to its movement; in *E. coli*, this arrest force is in the range 8–14 pN depending on temperature, with a maximum value of 14 ± 1 pN at 21 °C (Mejia *et al.*, 2008).

Genes may be encoded on either strand of the DNA double helix – there is no preferred strand – but the template regions do not overlap. Because of their differing 3′ and 5′ orientations, RNA polymerase reads the strands in opposite directions, transcribing about 20 nucleotides per second in eukaryotes, but higher at 50 s^{-1} in *E. coli*. Generating a typical strand of primary RNA in a eukaryotic cell involves the transcription of 3000–4000 nucleotides including introns, and takes about 3 minutes. This primary string contains far more than the 1200± nucleotides needed to make a protein, so the unwanted surplus nucleotides are removed in the nucleus before release of the mRNA. In eukaryotes, the "stop work" order to RNA polymerase may not take effect immediately even once the the primary RNA has been released, such that the RNA polymerase may continue to generate useless RNA from non-coding DNA for hundreds of nucleotides. In the process of trimming down the primary RNA strand in the nucleus, a cap is added to its 5′ end and about 200 adenine residues are added to the 3′ end. Many molecules of RNA polymerase can work on the same gene simultaneously, such that upward of one thousand transcripts of a given gene can be produced per hour.

12.4.2 Protein production

Messenger RNA provides a compact genetic blueprint for a specific protein, one of several steps in the process of assembling the protein from amino acids. Let's describe protein manufacturing in general and then return to specific ways mRNA must be modified for its role in this process. The challenge is to link up the appropriate amino acids in the order specified by mRNA. There are four letters in the RNA alphabet, but 20 types of amino acids commonly found in a protein, so the translation from RNA to protein requires more than a one-to-one association of nucleotide and amino acid. Even a code based on two nucleotides won't work, as there are only $4^2 = 16$ nucleotide pairs, which is not quite enough to do the translation and may also be more error prone. The translation code is therefore based on combinations of three nucleotides, which has the benefit of increasing the accuracy of the construction process, even though it has more information capacity than is needed ($4^3 = 64$ triplets).

The amino acid itself does not recognize its three-letter sequence on an mRNA string and bind directly to it. Rather, an amino acid binds to an intermediate piece of RNA called tRNA (for transfer RNA) and it is tRNA that reads mRNA. As can be seen from Table 12.2, there are usually two or more three-letter RNA sequences that code for the same amino acid; often, the first two letters in this group of sequences are the same. Often having only 80 nucleotides, tRNA is very short compared to mRNA, and its structure has been compared to a three-leaf clover, as shown in Fig. 4.17. The central leaf contains a nucleotide triplet that is the anticodon of the code given in Table 12.2; that is, it contains a triplet of bases that are complementary to the three-nucleotide codon on mRNA. Being complementary bases, the two triplets on tRNA and mRNA easily form hydrogen bonds with each other. The stem of the tRNA clover is a pair of nucleotide strings, one of which (the 3′ end) is longer than the other. It is the long string that binds the amino acid. Thus, tRNA molecules read mRNA and line up their amino acid cargo in the correct order for the protein. As expected, this process doesn't happen all by itself, and a molecule called a ribosome shepherds the whole process, as illustrated in Fig. 12.20(a). Having a molecular mass of 2–4 $\times 10^6$ Da, a ribosome is a large complex of rRNA (r for ribosomal) and protein; its two main structural parts move with respect to one another during protein synthesis.

The decoding process does not start right at the end of an mRNA molecule; rather, the end orients the ribosome as it attempts to find the start point, which is defined by an AUG nucleotide triplet (which is the codon for methionine). However, there may there be many "valid" AUG triplets along the mRNA strand in bacteria, and there may be a huge number of "accidental" AUG

Table 12.2 RNA coding sequences for amino acids and start/stop commands.

Amino acid or instruction	RNA sequence					
Ala	GCA	GCC	GCG	GCU		
Arg	AGA	AGG	CGA	CGC	CGG	CGU
Asp	GAC	GAU				
Asn	AAC	AAU				
Cys	UGC	UGU				
Glu	GAA	GAG				
Gln	CAA	CAG				
Gly	GGA	GGC	GGG	GGU		
His	CAC	CAU				
Ile	AUA	AUC	AUU			
Leu	UUA	UUG	CUA	CUC	CUG	CUU
Lys	AAA	AAG				
Met	AUG					
Phe	UUC	UUU				
Pro	CCA	CCC	CCG	CCU		
Ser	AGC	AGU	UCA	UCC	UCG	UCU
Thr	ACA	ACC	ACG	ACU		
Trp	UGG					
Tyr	UAC	UAU				
Val	GUA	GUC	GUG	GUU		
start	AUG (met)					
stop	UAA	UAG	UGA			

triplets arising from juxtaposed codons for other amino acids. How does the ribosome decide which AUG is the first one in the coding sequence?

Eukaryotes While in the nucleus, a cap is added to the 5′ end of mRNA and about 200 adenine residues are added to the 3′ end. A ribosome recognizes the 5′ cap, then walks along mRNA until it finds an AUG. Only one protein is coded per mRNA in eukaryotes.

Prokaryotes There is no 5′ cap. Further, several proteins are coded per mRNA, so that there are several appropriate AUG starting points.

The situation is illustrated in Fig. 12.20(b), which shows the organization of mRNA sequences at the "start" end of the template. In both prokaryotes and eukaryotes, a ribosome reads mRNA from its 5′ to 3′ ends, generating a protein from its N-terminus to its C-terminus.

Formation of a peptide bond between the amino acids carried by adjacent tRNAs is a complex process. Among many roles, a ribosome can accept up

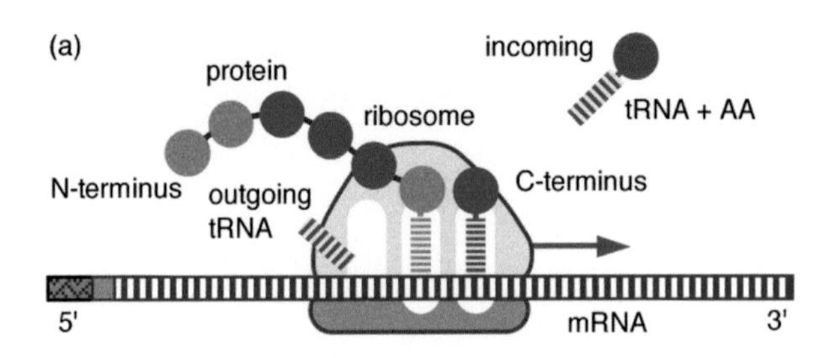

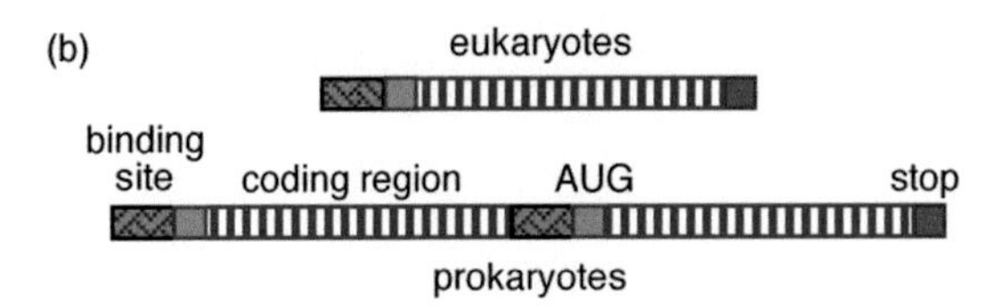

Fig. 12.20 (a) Protein synthesis as a ribosome reads mRNA from left (5′) to right (3′). Incoming amino acids carried by tRNA arrive at the right-hand pocket of the ribosome, while outgoing tRNA leaves from the left pocket. Once a new polypeptide bond is formed that lengthens the protein at its C-terminus, the "empty" tRNA is released and the top unit of the ribosome shifts to the right, corresponding to the two captured tRNA units shifting internally to the left. Subsequently, the bottom ribosome unit shifts to the right as well, matching the shift of the top unit. (b) Eukaryotic mRNA consists of a binding site, a start codon (AUG), a long coding region of about a thousand nucleotides, and a stop codon. Prokaryotic mRNA has similar construction, but may code for more than one protein.

to three tRNAs at a time as illustrated schematically in Fig. 12.20(a), where an incoming tRNA is accepted at the right-hand side of the ribosome, and released at the left after its amino acid has been added to the protein chain. Once the left-most tRNA is released, the top unit of the ribosome shifts to the right (so the two tRNAs already in the ribosome shift relatively leftward), opening up a slot for a new tRNA to join the queue. Subsequently, the bottom unit of the ribosome shifts to the right by the same distance as the top unit. There is more chemistry involved with the initiation and polymerization steps than is shown; for further details, the interested reader is directed to standard sources such as Chapter 6 of Alberts *et al.* (2008) or Chapter 2 of Snyder and Champness (2007)

12.4.3 DNA replication

In general, replication rates are higher than transcription rates. In bacteria, the replication rate can be 500–1000 nucleotides per second (vs. 50 s^{-1} for transcription) and in mammals it is a factor of ten lower at 50 nucleotides per second (vs. ~20 s^{-1} for transcription). Thus, the circular DNA of *E. coli* at 4.6×10^6 base pairs could be replicated in 80 minutes at 1000 nucleotides per second, but human DNA at 3×10^9 base pairs would require two years. These times are greater than the known replication times of the cells (e.g. 40 min for *E. coli*), which indicates that DNA replication is not done linearly from one end of the strand to the other, but may involve simultaneous replication at several if not many locations.

The replication strategy for both prokaryotic and eukaryotic cells is that special initiator proteins open up a section of the double helix at a particular sequence called a *replication origin*. Subsequently, a ring-shaped *helicase* enzyme wraps around one of the exposed strands of DNA, creating a *replication fork* as illustrated in Fig. 12.21(a). The exposed region in what is otherwise a DNA zipper is referred to as a bubble. On the circular DNA of bacteria, there is just one replication origin, but in eukaryotic cells there may be many on each chromosome. Both replication forks in each bubble are active, and many bubbles may be active simultaneously. The helicase travels along the strand, opening up the helix at rates of up to 1000 nucleotide pairs per second. About 20 different helicases are known in *E. coli*, and each type moves along the strand in one direction only; in Fig. 12.21(b), the helicase moves toward the 3′ end.

As with transcription, further enzymes (DNA polymerase) then replicate the exposed section of the helix. Just as with RNA polymerase, DNA polymerase reads in one direction only, from the 3′ end to the 5′ end, and this creates a challenge for replication. It's easy for DNA polymerase to attach itself to the 3′ end of a strand and follow along as the bubble opens up: the polymerase simply works toward the 5′ end. However, at the replication fork, the complementary strand is exposed at its 3′ end, and DNA polymerase does not read toward the 3′ end. As a result, what happens on the second strand is that the polymerase must attach itself near the helicase at the replication fork and then work away from it, producing only a short segment of a single strand. Then it must shift to the replication fork again and work backwards along a short stretch of newly exposed DNA to produce another segment, stopping once it reaches the DNA it synthesized in the previous step.

Replication along the 3′ to 5′ strand is fast and continuous, so that the 3′ strand is referred to as the *leading* strand. In contrast, replication on the 5′ to 3′ strand is slower and discontinuous, and giving it the moniker of the *lagging* strand. The spatial arrangement of the helicase and polymerase on the lagging strand is complex, as described further in Alberts *et al.* (2008). The DNA fragments produced on the lagging strand are referred to as *Okazaki fragments*, and the small gaps between them must be filled in after the fact. In eukaryotes, Okazaki fragments are 100–200 nucleotides in length, while in bacteria, they are ten times as long, at 1000–2000 nucleotides. Each exposed strand in the bubble becomes the template for the addition of a new complementary strand, such that each strand in the "parent" DNA helix ends up in a different "daughter" helix. Unfortunately, creating

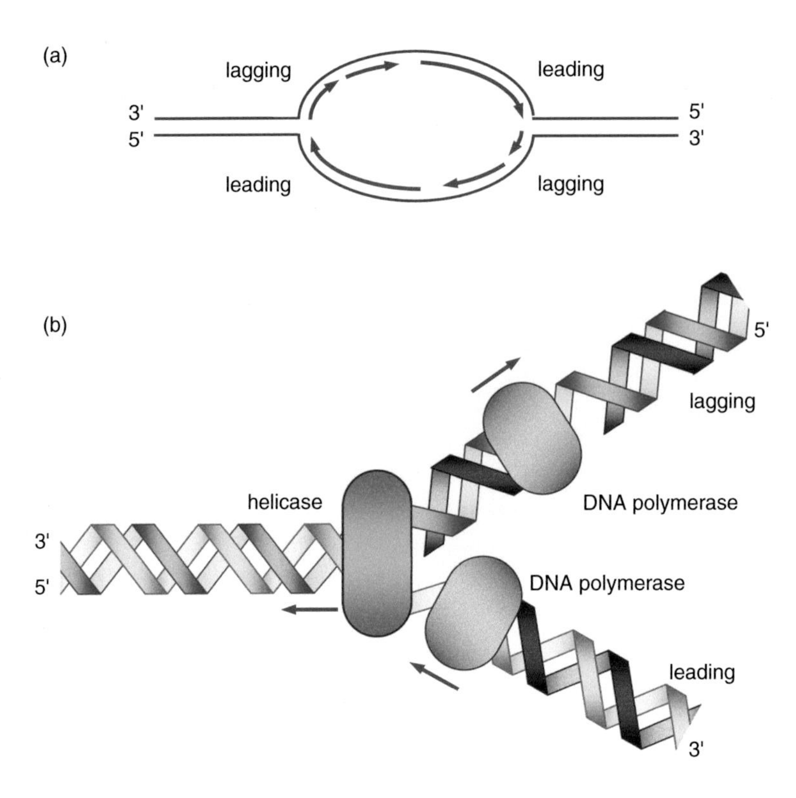

Fig. 12.21 Schematic overview of DNA replication. (a) Helicase enzymes pry apart the strands of a DNA helix at active replication forks that define a bubble. Newly synthesized DNA is shown in red; the arrow indicates the direction of synthesis. Each strand is a leading strand at one fork, and a lagging strand at the other (for simplicity, the double helix is represented by two parallel lines). (b) A replication fork in more detail. Replication of DNA is continuous on the 3′ to 5′ strand (leading strand), but not on the 5′ to 3′ strand (lagging strand), where the 3′ to 5′ reading of DNA polymerase produces short sections of new DNA called Okazaki fragments.

and replicating the bubbles can exert a torque on the DNA helix and cause supercoiling, which is discussed at length in Section 4.2. The enzyme DNA topoisomerase I performs the delicate task of snipping a strand of DNA, passing it across its complementary strand to relieve the torsional stress, and then finally reconnecting it and restoring the helix to its unstressed state.

12.4.4 Quality control

Mistakes happen. And in multi-step processes like transcription and replication, errors can arise from the mis-orientation of a molecule owing to thermal and other noise, from the need to read a potentially fragmented gene, from the breakage of a macromolecule under stress, or from a variety of other sources. Not all mistakes are harmful, of course; in DNA, some mutations are beneficial and may lead to more efficient operation of the cell. Describing the mechanisms by which the cell identifies and repairs mistakes that involve the genetic blueprint is beyond the scope of this text, but a few issues are worth mentioning.

In spite of all the cell's efforts at quality control, mutation rates are not trivial; unsuccessful mutations lead to death of the cell or organism, or its removal from a competitive population by natural selection. The mutation rate is thought to limit an organism to 50 000 genes, which is a factor of two higher than the number of genes in human DNA. Successful mutation rates are about 1 mutation per gene per 200 000 years, as discovered by comparing DNA or protein sequences between existing species. By using the fossil record to identify when the species diverged in evolution, the rate at which they successfully mutated can be determined; note that evolutionary time is twice the elapsed time, since both species evolve. Such studies reveal that some proteins "evolve" faster than others, in the sense that they can tolerate a higher variation without the host dying. For example, the tail end of fibrinogen is functionally discarded, and therefore it can tolerate greater variation without compromising its operation. The successful mutation rate extracted by this technique is consistent with laboratory studies that conclude there is 1 replication error per 10^9 base pairs per cell generation for bacteria. The error rate for RNA synthesis and for translating mRNA sequences to protein sequences is much higher (1 error per 10^5 nucleotides) because these processes lack the error correction mechanisms of DNA replication.

12.5 Forces arising in cell division

The process by which a eukaryotic cell physically divides involves several steps, as outlined in Fig. 12.7: the sequestering of chromosomes into two separate nuclei during mitosis is subdivided into five stages, following which the cytoplasm is partitioned into distinct cells during cytokinesis. One of these steps, anaphase, is recalled in Fig. 12.22(a), which illustrates the organization of the cytoskeleton in separating sister chromatids marshaled at the cell's midplane, after the nuclear membrane has been disassembled and two copies of the centrosome have moved to opposite ends of the cell, defining two spindle poles. During cytokinesis, shown in Fig. 12.22(b), an actin–myosin contractile ring assembles just inside the plasma membrane, and cinches the cell into a dumb-bell shape by shortening. In this section, we describe the forces carried by microtubules and the contractile ring during these two stages; the forces arising in DNA transcription are reviewed in Section 12.4.

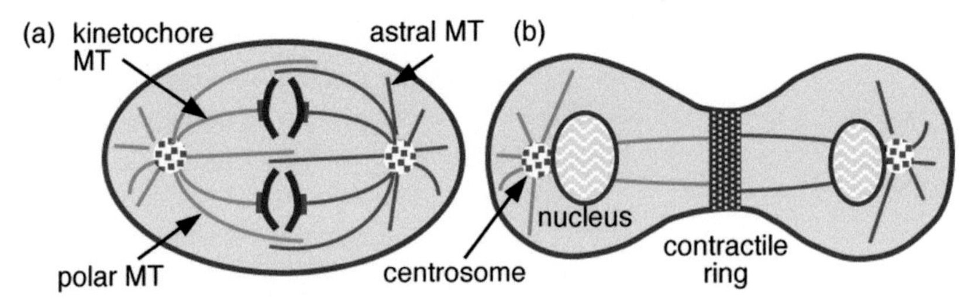

Fig. 12.22 (a) Having replicated, centrosomes (containing two centrioles in most animal cells) provide two nucleating sites for microtubules (MT), called astral, polar and kinetochore according to their function in separating sister chromatids, seen at the center. (b) Later in the cycle, an actin–myosin contractile ring assembles, ultimately forming a cleavage furrow as it shortens.

Microtubules are labeled according to their function or appearance during mitosis. Without considering the long preparation time involved in DNA replication, the onset of the visible part of mitosis is accompanied by a dramatic rise in the rate of growth and shrinkage of microtubules, as they seek out the attachment point (kinetochore) on a chromosome; indicated by the two small disks on the chromosome in Fig. 12.23(b), one kinetochore is present on each of the two sister chromatids and may be capable of attaching more than one microtubule, depending on cell type. *Kinetochore* microtubule is the label given to a filament that has successfully hit its target; the attachment process involves more steps than we have indicated here (see Chapter 17 of Alberts *et al.*, 2008). If they are growing in the correct direction, some microtubules will snag, not a chromosome, but another microtubule radiating from the opposite centriole. Meeting plus end to plus, these microtubules may be linked by other proteins to form a dynamic bond, and are referred to as *polar* microtubules, running between the spindle poles defined by the centrosomes. Growing away from the cell center, a third set of filaments has no luck at all in fishing for chromosomes, and are termed *astral* microtubules. The labels are displayed on Fig. 12.22.

Several aspects of filament growth and molecular motors introduced in Chapter 11 are present during mitosis. The plus end of a microtubule (the rapidly growing end) is connected to the kinetochore of a chromatid in an arrangement that has been likened to an open collar, rather than a cap, as illustrated in Fig. 12.23(a); this structure permits new tubulin dimers to be added to, or lost from, the plus end according to dynamics established in Section 11.2. The minus end of the filament is buried in the centrosome, and loses tubulin with time, such that a given tubulin dimer moves along a microtubule from plus to minus, as displayed in Fig. 12.23(b). This behavior is reminiscent of the phenomenon of treadmilling introduced in Section 11.2, where one end of a filament shrinks as the other grows (see Fig. 11.13). Although there is a reasonable range of

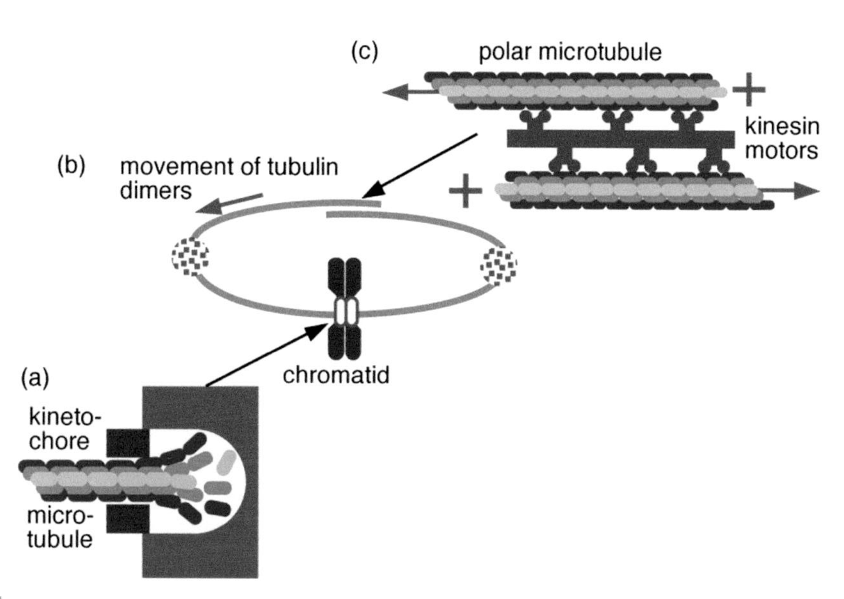

Fig. 12.23 (a) Cross section of a possible structure of a kinetochore collar surrounding the plus end of a microtubule as it depolymerizes. (b) Movement of newly incorporated tubulin dimers along a polar microtubule towards the centrosome. (c) Possible arrangement of kinesin motors linking polar microtubules with opposing plus ends; the movements of the microtubules in reaction to the kinesin forces are indicated by the arrows. Each of (a)–(c) is drawn to a different scale.

monomer concentrations under which actin undergoes treadmilling, this is not true for tubulin. The steady-state treadmilling condition where the growth rate at the plus end of the filament is exactly equal to the loss rate at the minus end occurs at a monomer concentration $[M]_{ss}$ that lies between the zero growth concentrations $[M]_c^+$ and $[M]_c^-$ at the plus and minus ends, respectively (see Fig. 11.12). According to Table 11.1, $[M]_c^+$ and $[M]_c^-$ for tubulin have the same value within statistical errors, arguing that treadmilling occurs in a rather restrictive concentration regime, if it is present at all. Further, the treadmilling rates for microtubules extracted from experiment (Bergen and Borisey, 1980; Hotani and Horio, 1988) are of the order microns per hour, far slower than the migration rate of tubulin during mitosis. In other words, even though there is tubulin loss at the centrosome, the concentration of free tubulin monomer is sufficiently high that the plus end of a microtubule grows well above the rate for the steady-state maintenance of filament length.

Let's now quantify some aspects of chromosome segregation. From visual recordings of mitosis, it is observed that kinetochore microtubules can drag a chromatid towards a centrosome at a rate of about 2 μm per minute. Taking this movement to arise from the shortening of kinetochore microtubules, the reduction in filament length corresponds to a release of about 50 tubulin heterodimers per second per microtubule, given that there are 13 dimers for every 8 nm of filament length (Sawin and Mitchison, 1991; Mitchison and Salmon, 1992). About 60%–80% of the tubulin loss occurs at the plus end and the remainder at the centrosome (Mitchison and Salmon, 1992). However, the loss rate at the plus end is not as catastrophic as permitted by the values quoted in Table 11.1, where the loss rate for rapid disassembly (k_{off}) is ~700 dimers per second, not 50.

A variety of forces are at play during chromatid separation: drag experienced by chromatids, tension within kinetorchore microtubules, and compression sustained by pairs of polar microtubules. The drag force experienced by an individual chromatid arising from its motion through the cytoplasm is relatively small. In the problem set, we find a theoretical benchmark of 10^{-2} pN for a hypothetical chromatid in a dilute environment, but more realistic estimates are an order of magnitude higher at 10^{-1} pN (for a review, see Nicklas, 1988). Of course, a chromatid in a crowded environment is subject to more forces than just the viscous drag of the cytoplasm, and the sum of these forces must be carried by the kinetochore microtubules.

Lest it lose mechanical contact, the collar around the microtubule at the kinetochore must move along the filament towards the centrosome in advance of the plus end of the microtubule as it depolymerizes. The combination of collar movement and tubulin loss at the centrosome results in a tensile force of several hundred piconewtons applied to the attached chromatid. However, given that tens of microtubules are attached to each kinetochore, the force per filament is correspondingly less, closer to 50 pN (Nicklas, 1983). What mechanism is responsible for this force?

One approach is to assume that the force is generated by molecular motors attached to the kinetochore collar, walking along the microtubule toward its minus end. At about 5 pN per motor (see Section 11.5), a small collection of molecular motors would be required to produce a force of 50 pN, although this is far more than what is needed to overcome just the viscous drag on a chromatid. Presumably, each motor in the collar would not run at full speed, given that the speed of a typical motor protein is microns per second, not the microns per minute at which a chromatid is pulled.

A second approach is based on a simple repulsive interaction between the collar and the depolymerizing end of the microtubule. Electromicrographs have provided images of the plus ends of microtubules during the depolymerization process, revealing that the tubulin protofilaments splay into short strands like the frayed end of a rope. As sketched in Fig. 12.23(a), the splayed end is restrained from simply flopping out of the kinetochrore collar, but at the same time, the loss of tubulin forces the filament to continually shorten. If it followed the depolymerization front of the protofilaments, the speed of the collar would be determined by the off-rate k_{off} of the plus end, which is 40–50 dimers per second according to Table 11.1. This is close to the observed tubulin loss as calculated above for a filament retreating at 2 μm per minute.

As expected from Newton's Third Law, a reaction force must act in opposition to the force pulling on the chromosomes, or else the two centrosomes will be pulled towards each other by the ensemble of kinetochore microtubules, thus defeating the operation of cell division. The reaction force is provided by pairs of opposing polar microtubules that are thought to overlap at a junction complex containing the molecular motor kinesin. A candidate structure is displayed in Fig. 12.22(c), reminiscent of the actin–myosin complex in muscles of Fig. 11.6 and 11.7, where a motor pulls opposing filaments towards one another. In mitosis, the force would be generated by coupled kinesin motors walking towards the plus ends of the filaments, thus generating compression on the microtubule and driving the spindle poles to move apart; in addition, polar microtubules lengthen during the later part of anaphase. If the junction contains several kinesin motors, then the compressive force on the pair of polar microtubules will lie in the 10–20 pN range needed to balance the force from kinetochore microtubules. This force range is around the threshold force that will cause buckling in a microtubule. From Eq. (3.69), the buckling force F_{buckle} for a filament of length L_c is equal to $\pi^2\kappa_f / L_c^2$, where the flexural rigidity of a microtubule is in the range 2×10^{-23} J•m according to Table 3.2 (we have combined the relation $\kappa_f = k_B T \xi_p$ with $\xi_p = 5 \times 10^{-3}$ m for microtubules). If $L_c = 5$ μm, F_{buckle} is then estimated to be 8 pN.

The mechanism for drawing the membrane together at the midplane is provided by the contractile ring, which in animal cells is comprised of some 20 actin filaments. The molecular motor myosin-II is thought to provide the driving force to shorten the ring, just as it draws actin filaments together in a muscle cell. The operation of the ring is somewhat like a belt, in that it cinches down the middle region of the dividing cell, ultimately forming a cleavage furrow where the cell splits. Unlike a belt, however, surplus actin is released from the ring as the neck narrows, and the ring disappears before the daughter cells separate, each carrying off their half of the cytoplasm and its organelles.

In contrast to Chapter 10, which focused on the bending of membranes and cell walls as factors in determining cell shape, the emphasis in Chapter 12 has been placed on the cytoskeleton both as a means of locating the cell's center and of segregating copies of its genetic blueprint. Experimental study of the structure and mechanical capacity of the cytoskeleton has been aided by the development of synthetic model systems – a bilayer encapsulating a medium containing actin filaments or microtubules (Cortese *et al.*, 1989; Miyata and Hotani, 1992; Bärman *et al.*, 1992; Elbaum *et al.*, 1996). Some aspects of the microtubule-bearing vesicles were described in Section 3.5, so we briefly review just the actin systems. The strategy is to prepare a liposome (a bilayer vesicle) in a medium containing unpolymerized G-actin and perhaps other proteins. Once the liposome has closed into a seemless shell, polymerization of monomers can be induced by one of several means, such as increasing the temperature or altering the ion concentration. Shape changes of the liposome resulting from the synthetic cytoskeleton can be recorded microscopically. Cross-linking can be induced by means of the actin cross-linker filamin; the mean filament length can be controlled by introducing gelsolin as a nucleation site, giving mean filament lengths of 0.07–5.4 μm in one study (Cortese *et al.*, 1989). In some cases, F-actin is found to assemble into stiff bundles that force the vesicle into a highly prolate shape with the filaments running parallel to the axis of symmetry (Miyata and Hotani, 1992).

Summary

Although certainly varied, the sizes of cells such as bacteria have remained remarkably uniform for literally billions of years. Part of the reason for this must reflect an optimal design for collecting resources and reproducing in a competitive environment. Part must also be structural: what cell designs are robust and efficient given the biochemical building blocks at hand? A mechanical analysis of some of the material properties of bacteria reveals a few constraints on cell size. For example, if a cell is too small, it is more energetically favorable for a closed surface with the topology of a sphere to open up into a flat sheet. For spherical membranes, this will occur when the radius is less than $R^* = (2\kappa_b + \kappa_G) / \lambda$, where κ_b and κ_G are the bending and Gaussian rigidities, respectively, and λ is the edge tension of the membrane. At larger length scales, if the cell supports an internal pressure P, it will rupture if the (two-dimensional) tension Π within its membrane exceeds a material-specific threshold Π^*. For bilayer-based membrane, Π^* is about 10^{-2} J/m^2, and the pressure-dependent radius at which a spherical cell fails is $R = 2\Pi^*/P$. Another challenge to cell size is the packaging of its genetic blueprint. Treating a strand of DNA with length L_c as a random coil, its mean squared radius of gyration is given by $\langle R_g^2 \rangle = \xi_p L_c/3$, where ξ_p is the DNA persistence length of about 50 nm. For a bacterium, the radius of gyration of its DNA is comparable to the cell size; for eukaryotic DNA, $\langle R_g^2 \rangle$ is much larger than the cell dimension, which forces eukaryotic cells to find strategies for storing their DNA.

For a mean cell shape to be unchanging from one generation to the next, the cell's surface area A and volume V must double on average during its growth and division cycle. Branched polymer shapes, with their irregular entropy-rich arms, satisfy this criterion in that their surface area is proportional to their volume, averaged over an ensemble of shapes. Division cycles based on branched polymers have been produced in the lab and may have been a cell design early in evolution, although the design is not particularly efficient in the use of materials. Although A and V must double, there is no *a priori* condition on their time evolution to do so. The trajectory of shapes taken by a cell during its lifetime can be investigated by monitoring the growth of individual cells, or by taking a snapshot of an ensemble of cells under steady-state conditions. In the latter approach, the probability density $\mathcal{P}_\varphi$ for observing a geometrical characteristic φ (such as A or V) is given by $\mathcal{P}_\varphi = (\partial\varphi / \partial t)^{-1} / T_2$, where T_2 is the doubling time. Thus, the distribution $\mathcal{P}_\varphi$ can be obtained from the time dependence $\varphi(t)$ through its derivative. For example, exponential volume growth of a coccus leads to $\mathcal{P}_\beta = [3(1 - \beta^2)/\ln 2] / [2 + \beta(3 - \beta^2)]$, where the dimensionless constant β is equal to $s/2R$, with R the radius of the coccus and s the distance between the centers of curvature in the diplococcus shape. Experimentally, the bacterial division cycle can be described by exponential growth, or three regimes of linear growth that approximate an exponential function. This growth pattern was established early on in evolution, and has been present among cyanobacteria for at least two billion years.

The models for cell growth and division that are described above focus on the behavior of the mean cell shape in the absence of fluctuations. Yet we know experimentally that the division cycle is asymmetric and asynchronous: not all cells divide into identical daughters nor do all cells have exactly the same doubling time. In some cases, asymmetric division is an intrinsic part of the cell cycle, while in others, the mechanism for finding the cell's midplane, while robust, is not entirely accurate. An example of the former is budding yeast; an example of the latter is *E. coli*, whose division plane is at 0.5 ± 0.013 of the cell length, a variation of a few percent. Similarly, the standard deviation of the division time in an ensemble of *Caulobacter crescentus* is around 10%, a variation well beyond experimental uncertainty. In bacteria, such fluctuations can be interpreted within detailed models for the division process that recognize the role of the helical cytoskeleton based on the Min, MreB and Mbl proteins, and the role of the Z-ring constructed from FtsZ.

In almost all cell types, the forces at play in the division cycle include those from the membrane-associated cytoskeleton and mid-cell ring; in eukaryotic cells, forces from microtubules contribute to the movement of chromosomes. The biochemical issues involved with DNA replication and the production of proteins are explained in Section 12.4, but will not be repeated here. The behavior of dynamic filaments (Chapter 11) is well suited to the capture of chromosomes by the ensemble of microtubules radiating from opposing centrosomes during mitosis. At the tubulin dimer concentrations present in the cytoplasm, microtubules are able to grow at a suitable rate as they search for kinetochores on pairs of sister chromatids, and then disassemble catastrophically when a search fails in a particular direction. Once captured by the collar of a kinetochore, a microtubule shortens through depolymerization, and its splayed end may transmit a tension to the collar as it is pulled against it. In addition, force at a kinetochore may be generated directly through the presence of molecular motors. Were it not for the compression borne by linked pairs of polar microtubules, the shortening of kinetochore microtubules would just result in centrosomes being pulled toward the chromosomes assembled at the division plane. However, myosin motor proteins form a structural link between polar microtubules, and these motors walk in opposite directions toward the plus (radiating) end of the microtubules and generate forces of several pN per motor. At ten or more pN per filament, this compressive force is in the range of the filament's buckling force, so it is not surprising that pairs of polar microtubules adopt the shape of a bow. Molecular motors are also active in the actin–myosin contractile ring of eukaryotic cells. Composed of roughly 20 actin filaments in animal cells, the contractile ring provides the force that creates the cleavage furrow where these cells divide.

Problems

Applications

12.1. Suppose that the tubulin (heterodimer) concentration $[M]$ in a hypothetical cell of radius 10 μm is 10 μM.

(i) What total length of microtubule could be made from this amount of protein if each heterodimer is 8 nm long?
(ii) If all filaments radiate from the center of the cell to its boundary, what is the average membrane area per microtubule?
(iii) At a force of 5 pN per filament, what would be the total pressure exerted on the cell membrane (quote your answer in J/m^3 and atmospheres)?

12.2. Plot the minimal cell wall thickness d for a spherical cell to sustain a pressure difference of one atmosphere according to the bound in Section 12.1. Take the cell radius R to lie between 0.05 and 1.0 μm and assume the wall material to be peptidoglycan with $K_V = 3 \times 10^7$ J/m^3. Does $d < 1$ nm have a physical meaning? Given a rupture tension of 10^{-2} J/m^2, show the range in R for which a lipid bilayer alone can support this pressure.

12.3. Approximate the shape of the *E. coli* bacterium in Fig. 1.1(c) to be a cylinder 2.0 μm long and 1.0 μm in diameter. Equate the volume of this cylinder to that of a sphere to determine an equivalent radius of gyration R_g for the sphere. If the DNA of this bacterium obeyed ideal scaling, what must be its persistence length in order to have the same R_g? How does this compare with the measured persistence length of DNA? (*Note: the contour length of E. coli DNA is given in the text.*)

12.4. For the hypothetical division cycle shown in Fig. 12.9, suppose that the area of the inner layer grows linearly with time t over the entire doubling time T_2. Plot the areas of the inner and outer boundaries of this cycle over $0 \leq t \leq T_2$. On the same graph, plot the area of the outer layer if it increases exponentially over $0 \leq t \leq T_2$. Define the initial area as A_o.

12.5. Suppose that a spherical cell were so small that the separation of its chromatids during cell division could be driven by actin filaments, rather than microtubules. What would be the radius of the cell such that an actin filament spanning the cell would not buckle when subject to a force of 5 pN? (*Note: choose a value for the actin persistence length in the middle of the range quoted in Table 3.2.*)

12.6. Analogous to the elastic constant for a spring, the bending stiffness for a uniform rod of length L is given by $k = 3\kappa_f/L^3$, where κ_f is the flexural rigidity defined in Eq. (3.10). For *E. coli*, the bending stiffness is measured to be about 0.22 pN/nm (Wang *et al.*, 2010a).

(a) Determine κ_f if the bendable length of the bacteria in the experiment is 3 μm (quote your answer in J • m).
(b) If the elasticity arises from the boundary alone (the cell wall plus the MreB cytoskeleton), what is its thickness if its Young's modulus is 10^7 J/m^3? Take the diameter of the cell to be 0.5 μm, and assume that the thickness t is much less than the radius R.

12.7. Estimate the viscous drag force experienced by a chromatid of length L = 5 μm as it is dragged through a fluid of viscosity η = 0.002 kg/m • s at a speed of v = 2 μm per minute. Take the drag force to be equal to $4\pi\eta Lv$. If the chromatid were dragged parallel to its cylindrical axis, what would the force be? (See Problem 11.20.)

Formal development and extensions

12.8. A longitudinal compressive force F is applied uniformly across the ends of a hollow cylinder of length L and radius R, as in the following diagram.

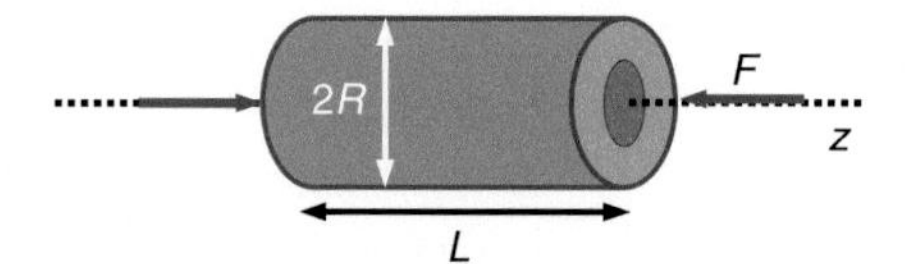

The hollow core has a radius of $R/2$. If F is less than the buckling force $F_{\text{buckle}} = \pi^2 Y\mathcal{I}/L^2$ (see Section 3.5), the rod has a non-zero strain in the z-direction, where the axis of symmetry lies along the z-axis. (a) Find u_{zz} in terms of R and L just at the buckling point F_{buckle} (calculate the stress σ_{zz} first). (b) What is u_{zz} at the buckling point for such a filament of radius 12 nm and length 4 μm?

12.9. Suppose that a set of filaments are lined up in a spherical cell of radius R as shown: all filaments are uniformly distributed and point along the z-axis.

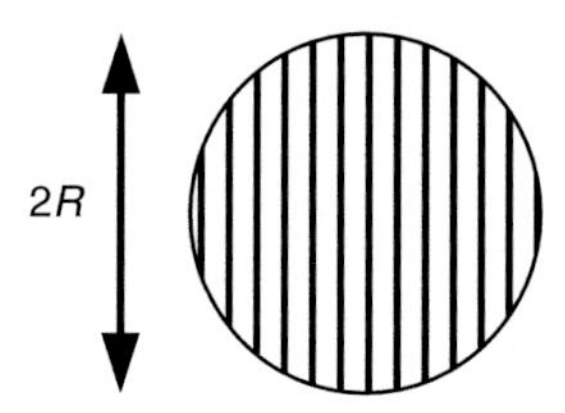

Show that the average filament length is $4R/3$.

12.10. Establish that the length L, surface area A and enclosed volume V of the diplococcus shape of width $2R$ is given by Eq. (12.7):

$$L = 2R(1+\beta)$$

$$A = 4\pi\, R^2(1+\beta)$$

$$V = (4\pi\, R^3/3) \bullet [1 + \beta\,(3 - \beta^2)/2)],$$

where $\beta \equiv s\,/\,2R$ and s is the separation between centers of the intersecting spheres. Verify the limits of these expressions when $\beta = 0$ and 1.

12.11. For the diplococcus shape, show that $\mathcal{P}_\beta$ has the same functional form for a given growth function $F(t)$, whether it is the length or the area that obeys $F(t)$.

12.12. Generalize the analysis of the diplococcus shape in Eqs. (12.7) and (12.8) to a spherocylinder whose cylindrical core section initially has a length s_o.

(a) Find the analogous expressions to Eq. (12.7) when $s_o \neq 0$.

(b) If the length increases linearly with time, show

$$\mathcal{P}_\beta = 1\,/\,(1 + \beta_o).$$

(c) If the length increases exponentially with time, show

$$\mathcal{P}_\beta = 1\,/\,[\ln 2 \bullet (1 + \beta)].$$

In your expressions, retain s as the distance between the centers of the spherical end-caps, but now s runs from s_o initially to $2s_o + 2R$ at the division point. As in Eq. (12.7), $\beta = s/2R$ but in addition $\beta_o = s_o/2R$.

12.13. Generalize the analysis of the diplococcus shape in Eqs. (12.7) and (12.8) to a spherocylinder whose cylindrical core section initially has a length s_o.

(a) Find the analogous expressions to Eq. (12.7) when $s_o \neq 0$.

(b) If the volume increases linearly with time, show

$$\mathcal{P}_\beta = 1 / (2/3 + \beta_o) \qquad \text{for } \beta_o \leq \beta \leq 2\beta_o,$$

$$\mathcal{P}_\beta = (1 - \Delta\beta^2) / (2/3 + \beta_o) \qquad \text{for } 2\beta_o \leq \beta \leq 2\beta_o + 1.$$

(c) If the length increases exponentially with time, show

$$\mathcal{P}_\beta = 1 /[(\ln 2) \bullet (2/3 + \beta)] \qquad \text{for } \beta_o \leq \beta \leq 2\beta_o,$$

$$\mathcal{P}_\beta = (1 - \Delta\beta^2) / 2\ln 2 \bullet \{[1 + \Delta\beta(3 - \Delta\beta^2)/2]/3 + \beta_o\} \qquad \text{for } 2\beta_o \leq \beta \leq 2\beta_o + 1.$$

In your expressions, retain s as the distance between the centers of the spherical end-caps, but now s runs from s_o initially to $2s_o$ + $2R$ at the division point. As in Eq. (12.7), $\beta = s/2R$ but in addition $\beta_o = s_o/2R$.

12.14. The radius r of the intersection ring between the two spherical caps of a diplococcus decreases with time from $r = R$ to $r = 0$ during the division cycle, where R is the radius of the initial coccus. Suppose that $r(t)$ decreases linearly with time according to $r(t) = R \bullet (1 - t/T_2)$, where T_2 is the doubling time. Show that the expression for the probability density P_β is given by $P_\beta = \beta / (1 - \beta^2)^{1/2}$, where $\beta = s/2R$, and s is the distance between the centers of curvature of the two elements of the diplococcus. You may find the algebra easier using the dimensionless time $\tau = t/T_2$.

13 Signals and switches

The cell is a dynamic object that grows, divides and, in some cases, locomotes. Even if its growth is the simple exponential volume increase observed for some genera in Chapter 12, various biochemical pathways continually open and close as required during the division cycle. Further, multicellular organisms as a whole display cell differentiation and cooperative behavior that occur on macroscopic time scales; an example of the latter might be the response of a muscle to the receipt of sensory input. These types of processes imply the transmission of information over a variety of length scales, and the existence of switches that can activate or deactivate processes within cells.

In this chapter, we illustrate some of the mechanisms that underlie switching and signal transmission. The classic example is signal transmission along the axon of a nerve: here, we describe the behavior and physical basis of the axon's action potential in Section 13.1 and then establish a theoretical framework for its analysis in Section 13.2. The framework was developed by Hodgkin and Huxley (1952) based on their own experimental findings, which were obtained in part with B. Katz. Following this, Section 13.3 takes a mathematical look at the structure of equations that possess switch-like features. There are many examples of switches in both the genetic and the mechanical functioning of the cell, and two of these are treated in Section 13.4.

13.1 Axons and the action potential

Compared to the rotund cell shapes of many bacteria, the shapes of nerve cells are exotic. From the main cell body extend branched dendrites that gather sensory input and an elongated axon that delivers a signal dispatched from the cell body (introduced schematically in Fig. 1.1). An example of a neuron is shown in the two panels of Fig. 13.1, where the staining in the right-hand panel corresponds to nuclear DNA (blue), the Golgi apparatus (yellow), dendrites (red), and secretory vesicles (green). Axons may range up to a few meters in length in larger animals, and have diameters that are as high as a millimeter in squid, although diameters of a micron

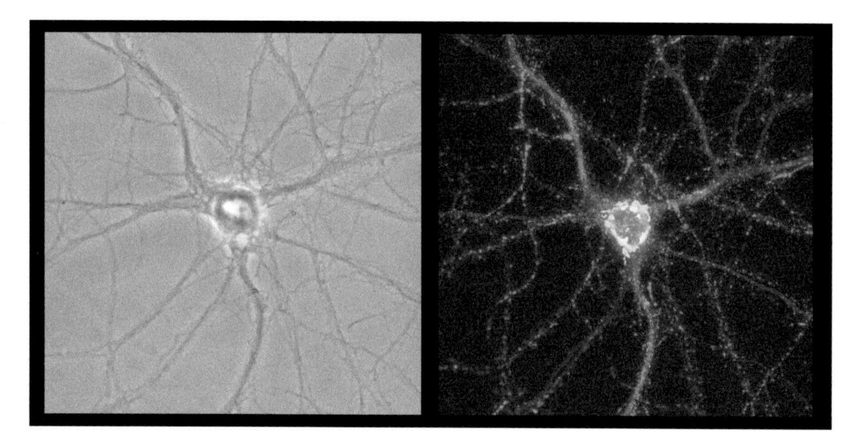

Fig. 13.1 Images of a primary cultured hippocampal neuron. On the right-hand panel, the stains correspond to nuclear DNA (blue), the Golgi apparatus (yellow), dendrites (red), and secretory vesicles (green) (courtesy of Dr. Pouya Mafi and Dr. Michael Silverman, Simon Fraser University).

or so (in the absence of a myelin sheath) to 10–20 microns for myelinated axons are not uncommon. The role of the axon is both to propagate a signal as well as to deliver chemicals to its terminus at a synapse. The transport of chemicals is accomplished through secretory vesicles in the axon's interior, imaged in green in Fig. 13.1; the transport of vesicles is treated in Chapter 11. In contrast, the signal is propagated along the axon's surface, as we will now describe in detail.

The transmission of a signal along the axon involves time scales of fractions of a second, and macroscopic distances of up to a few hundred centimeters. To understand the physics at work in nerve signals, let's compare a benchmark speed for the signal of 30 m/s (that is, 30 cm in 10 ms) against what we might expect from diffusion or from transport by a motor protein. As shown in Chapter 2, the mean square end-to-end displacement $\langle r_{ee}^2 \rangle$ of a random trajectory in one dimension is $\langle r_{ee}^2 \rangle = 2Dt$, where D is the diffusion constant and t is the elapsed time. Taking $D \sim 10^{-12}$ m^2/s of lipids in a bilayer, we find $\langle r_{ee}^2 \rangle^{1/2} = 0.14$ μm for $t = 10$ ms, which is about a million times smaller than the benchmark. Similarly tiny distances are found for transport by molecular motors in a cell: at a speed of 2 μm/s, a motor could travel a distance of 0.02 μm in 10 ms. These examples show that neither diffusion nor molecular transport is fast enough to account for the speed of signals in nerves.

In this and the following sections, we discuss signal transmission as a phenomenon involving the propagation of an ionic concentration gradient along an axon. Along with the environment surrounding the cell, the fluid medium within the cell (the cytosol) contains a variety of ions of physiological importance. Because of their hydrophobic interiors, pure lipid bilayers are relatively impermeable to charged molecules, even if they are physically small. Consequently, most ionic species will diffuse in and out of the cell only slowly unless a strategy is adopted by the cell to hasten this process. Specific transport proteins embedded within the plasma membrane are capable of pumping ions across it, which permits the ion concentration in the interior of the cell to be different from that of the surrounding medium. Thus, a uniform concentration gradient can be maintained over the cell's surface – the cell's rest state – even though it is not an equilibrium state. However, if a local inhomogeneity in concentration develops at the plasma membrane for some reason, then the resulting concentration gradient may not be static but rather evolve with time; the mathematical formalism for this is described in Chapter 2.

A variety of transmembrane proteins are associated with the passage of molecules and ions across a membrane; the two classes of interest here are *channel* proteins and *transporters*. When it is open, a channel protein creates a narrow aqueous link across the bilayer as in Fig. 13.2(a). The movement of a molecular or ionic species through the channel is driven solely by an electrochemical gradient across the membrane; in that sense, the movement is *passive*. Gradients may arise from electric fields or from differences in concentration; for example, the cytosol-facing side of most plasma membranes is negatively charged with respect to the extracellular side, and the resulting electric field favors the migration of positively charged ions into the cell. The migrating species is only loosely associated with the interior core of a channel protein and the movement is like that of a fish in a stream.

In contrast to the structure of channels, transporter proteins carefully shepherd their cargo through the hydrophobic interior of the bilayer by undergoing a conformational change. The change of shape takes time, so that the rate at which a molecule is carried along by a transporter is slower than that in a channel protein. Transporters employ several mechanisms to get molecules to their destination; in each case, the migrating species

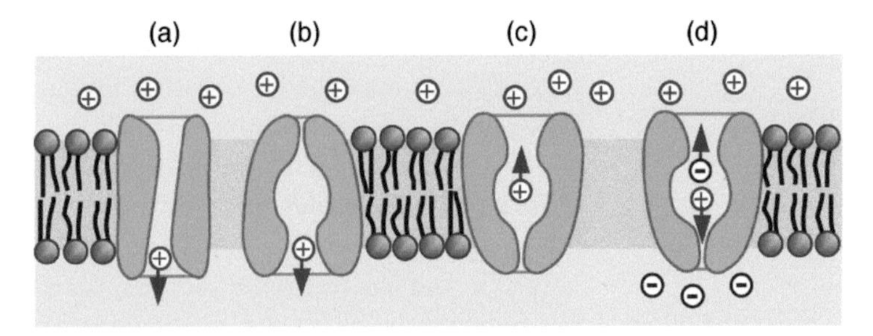

Fig. 13.2 A cross section through a bilayer-based membrane illustrating the mechanisms for transporting ionic species; the concentration of positive ions is higher above the membrane. Panels (a) and (b) show passive movement along a concentration gradient for a channel protein (a) and a transporter (b). Panels (c) and (d) show active transport against the gradient (c) and in two opposing directions (d).

is closely associated with the protein through attachment to one of what might be several binding sites in its interior. One transport mechanism is passive, where the transporter functions as a gate that opens and closes randomly, as in Fig. 13.2(b). Although the binding site may select which molecules are to be carried, the net flow of the molecules follows the electrochemical gradient.

In contrast, *active* transport involves the movement of a species against its electrochemical gradient. Such a process requires an energy input, either directly to the transporter or to an associated transporter protein that is part of a multi-step process. Transporters that pump their cargo uphill against an electrochemical gradient may use light as an energy source (mainly found in bacteria and archaea) or use ATP hydrolysis, as in Fig. 13.2(c). Although some transporters move a single species in a single direction (called *uniporters*), this need not be true in general. *Coupled* transporters move two species at once and come in two variants: symporters move two species in the same direction (in or out) while antiporters are bidirectional in that they carry one species into the cell and a different species in the opposite direction, as in Fig. 13.2(d). For example, the sodium–potassium pump removes 3 Na^+ ions from the cell and brings 2 K^+ ions into the cell per ATP hydrolysis; both ionic species are moved against their electrochemical gradient. Although the relevant electrochemical gradients may favor the movement of both species, it may also be the case that the change in the concentration of one of the species is unfavorable to the cell; in this case a second transporter may be required to replenish or remove the unfavored species in a direction opposed by its electrochemical gradient. Thus, even if the bidirectional transporter can operate without an energy source, the second transporter in this example must be an energy-driven pump. In animal cells, Na^+ is often the co-transported ion. Although many types of pumps require only one ATP hydrolysis per step, some require two. More detailed descriptions of these transporters can be found in sources such as Alberts *et al.* (2008) or Hille (2001).

Typical concentrations of the primary ionic species in cells are summarized in Table 13.1. Of course, life forms have adapted to an extraordinary range of conditions, so the concentrations quoted in the table are only chosen for illustration. For example, although many strains of bacteria and eukaryotic algae thrive in ordinary freshwater and marine environments, others are found in more extreme conditions of both temperature and salt concentration. As a benchmark for assessing the ion concentrations commonly found in the cytosol, Table 13.1 also displays the corresponding concentrations in sea water, from which it can be seen that the salinity of cells, while not miniscule, is noticeably lower than sea water (which is about 3.5%, with variation). Note that there are more charged objects in the cell than just solvated ions – DNA and a number of phospholipids are also charged. As a consequence, the concentration of ions listed in Table

Table 13.1 Concentrations of ions commonly found in the cytosol and the surrounding environment for mammalian cells (Alberts *et al.*, 2008; Hille, 2001). Approximate concentrations of these ions found in sea water are shown for comparison (Turekian, 1976); SO_4^{2-} is present in sea water at higher concentrations than Ca^{2+}. Units are milliMolar

	Intracellular concentration	Extracellular concentration	Sea water
Na^+	5–15	145	470
K^+	140–155	4–5	10
Mg^{2+}	0.5	1–2	50
Ca^{2+}	10^{-4}	1–2	10
Cl^-	5–15	110–120	550

13.1 need not sum to zero algebraically for the cell to be overall electrically neutral.

The local charge density of each species can be calculated using the mathematical framework outlined in Sections 9.2 and 9.3. From the definition of electrostatic potentials, the energy of an ion of charge ze when placed in an electric potential V is zeV, where e is the elementary unit of charge (1.6×10^{-19} C) and z is the number of charges on the ion, including its sign. Suppose now that the potential varies from one place to another as $V(x)$. Then the probability of finding the ion at location x is equal to the usual Boltzmann factor $\exp(-zeV / k_BT)$ multiplied by a normalization constant. Comparing the probabilities P_1 and P_2 at two locations 1 and 2, the ratio of the probabilities must be

$$P_1/P_2 = \exp(ze[V_2 - V_1] / k_BT), \tag{13.1}$$

where the proportionality constant has been eliminated. Now, the concentration of an ionic species is proportional to the probability P, so Eq. (13.1) can be rewritten as

$$V_2 - V_1 = (k_BT/ze) \ln(c_1/c_2). \quad \text{Nernst equation} \tag{13.2}$$

This expression, called the Nernst equation, relates concentration ratios at two locations to their potential differences. The prefactor k_BT/ze sets the physical scale for the potential difference, and is equal to 25 mV for $z = 1$ and $k_BT = 4 \times 10^{-21}$ J at room temperature. Examination of the ionic concentrations of Na^+, K^+ and Cl^- in Table 13.1 shows that potential differences V are commonly in the −60 to −80 mV range according to Eq. (13.2) where $V = V_{inside} - V_{outside}$. A minus sign for V means that the inside location is more negative than the outside; because of the complexity of the equations later in this section, we use V for the potential difference across the membrane, not ΔV.

Under steady-state conditions, ions may move through the membrane, being driven by their concentration gradients (in part established by ion pumps) and the potential difference across the membrane arising from all of the ionic species. If the pumps are turned off, the immediate effect on the potential difference is small, and it is straightforward to calculate the quasi-steady-state potential V_{qss} applicable in this circumstance. At V_{qss}, the net current transiting the membrane for all species must be zero, even if the current for an individual species is non-zero. For an ionic species with the label α, its current density (current per unit area of the membrane) $\mathcal{I}_\alpha$ is given by $\mathcal{I}_\alpha = \gamma_\alpha (V_\alpha - V_{qss})$, where γ_α is its conductivity through the membrane and V_α is the Nernst potential for species α according to Eq. (13.2). The conductivity γ_α is the (two-dimensional) conductance per unit area with units of $\Omega^{-1}m^{-2}$, such that the conductance G of a membrane patch of area A is

$$G = \gamma_\alpha A, \tag{13.3}$$

for a single ionic species. The total conductivity of a number of species transiting a membrane is just the sum of the individual contributions

$$\gamma_{tot} = \Sigma_\alpha \gamma_\alpha. \tag{13.4}$$

Ion channels can be added to this expression as needed. Recall that the conductance is the reciprocal of the resistance R

$$G = 1/R. \tag{13.5}$$

The condition that there is no net current density then reads

$$\Sigma_\alpha \gamma_\alpha (V_\alpha - V_{qss}) = 0. \tag{13.6}$$

Restricting our attention to the movement of only three ionic species, Na^+, K^+ and Cl^-, Eq. (13.6) can be rearranged to yield

$$V_{qss} = (\gamma_{Na} V_{Na} + \gamma_K V_K + \gamma_{Cl} V_{Cl}) \,/\, (\gamma_{Na} + \gamma_K + \gamma_{Cl}). \tag{13.7}$$

Note that V_{qss} is the mean value of the individual potential differences only when all the conductivities are equal; otherwise, the sum is weighted by each γ_α. Knowing the ionic concentrations maintained by the ion pumps, Eq. (13.7) can be solved for V_{qss}. This is performed in the end-of-chapter problems, and yields $V_{qss} = -74$ mV for the mid-range of the concentrations given in Table 13.1; this estimate is very close to the measured potential of the membrane.

Let's now briefly describe what happens during the propagation of a signal along an axon; we describe the situation in which the axon is not wrapped in a myelin sheath. Under resting conditions, concentration gradients across the plasma membrane are maintained by transporters such that the concentration of Na ions is much lower inside the cell than its exterior (for many cell types), while that of K ions is much higher inside

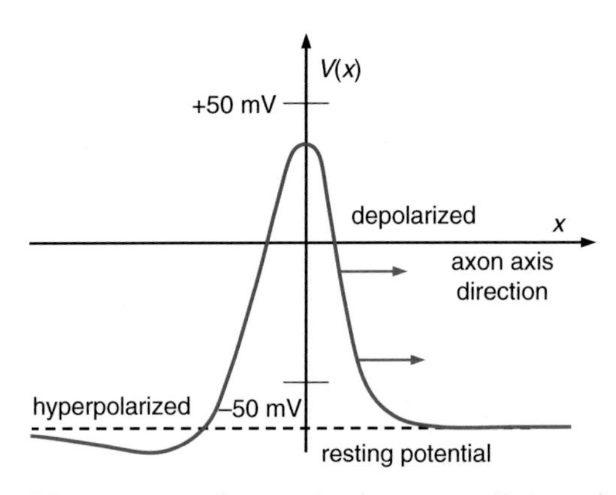

Fig. 13.3 Schematic features of the action potential propagating along an axon. Moving to the right in the diagram, the potential first swings from its rest value to a peak as Na^+ ions flood into the axon (depolarized); following this, the potential then drops as K^+ flows out, resulting in the potential becoming slightly more negative than its rest value (hyperpolarized), before ultimately recovering.

the cell (see Table 13.1). When a signal passes along the axon, Na and K ions are (unequally) exchanged across the plasma membrane in a local region and the membrane potential swings from its rest value by roughly 100 mV, depending on the cell, to +30 mV or more, before falling back toward the rest value as the axon recovers. The swing in the membrane potential is referred to as the action potential, and it is shown schematically in Fig. 13.3. The time scale for the change in the potential is of the order a millisecond or less for sodium-based potentials, but much longer for calcium-based ones (which may be 100 ms or longer).

Such fast movement of ions across the membrane does involve transmembrane proteins, but not transporter proteins which are relatively slow to operate because they must undergo a conformational change to capture and then release their cargo. This time scale is low compared to the flow rate of ions during the transit of an action potential, so the mechanism for depolarization of the membrane must come from channel proteins. Now, there are many different types of channel proteins, and they are sensitive to a variety of triggers. Among the most common are mechanically-gated channels, ligand-gated channels and, of importance here, voltage-gated channels. This last-mentioned channel is one that opens when the voltage across it lies in a specific operating range; the effect of a voltage-dependent conductance within the theoretical description of signal propagation is presented in Section 13.2.

In order to be useful in a thermal environment, the channel must require a minimum voltage in order to open; otherwise thermal fluctuations in the local ion concentration would cause the channel to open at random, not a satisfactory situation. The typical threshold for the change in potential is about 15 mV; below this threshold, a stimulus does not propagate along the axon but simply decays away, also on a time scale of milliseconds. Initially, the channels that open are those that strongly favor the movement of sodium ions from outside the cell to inside the cell (see Table 13.1). The membrane potential becomes less negative because of the flow of positive ions, which in turn causes additional (voltage-sensitive) sodium ion channels to open, further increasing the flow of Na^+. As a result of this positive feedback, the change in the sodium concentration gradient is very rapid.

Examination of Table 13.1 shows that potassium ions also possess a strong concentration gradient across the plasma membrane, but in the opposite sense to the sodium gradient. Not surprisingly, K^+ ions may move through voltage-sensitive ion channels as well, but from inside the cell to outside, opposing the flow of Na^+ and eventually driving the membrane potential negative again. However, the passage of K^+ ions occurs slightly later than Na^+, with the overall result that the potential rapidly swings positive from Na flow, and then returns to its resting value by means of K flow. Once the membrane potential has swung positive, sodium channels close, helping prevent the back-propagation of the signal.

13.2 Signal propagation in nerves

The propagation speed of a signal along an axon depends on both a time scale and a length scale. We examine each of these quantities within the framework of electrical circuit theory from introductory physics courses. Our starting point is the observation is that the plasma membrane may be regarded as a parallel plate capacitor with the charges Q on each side of the membrane supporting a potential difference V via

$$Q = CV, \tag{13.8}$$

where C is the capacitance of the membrane. Potentially, there are several contributions to the membrane capacitance; in the problem set, we examine an idealized membrane to show that the capacitance per unit area $\mathcal{C}$ is expected to be around 1×10^{-2} F/m^2, which is in the range measured experimentally. For a membrane potential of -70 mV, a capacitance per unit area of $\mathcal{C} = 1 \times 10^{-2}$ F/m^2 corresponds to a charge density Q/A of 0.7×10^{-3} C/m^2.

A single loop circuit with just a resistor and capacitor in series will discharge exponentially with time. This can be seen by substituting Ohm's Law $V = IR$ into Eq. (13.8) to obtain a differential equation for $Q(t)$

$$\mathrm{d}Q/\mathrm{d}t = -Q \,/\, RC, \tag{13.9}$$

where the current I is dQ/dt; the minus sign on the right-hand side reflects the signs of I and V. The solution to Eq. (13.9) is $Q(t) = Q_o \exp(-t/\tau)$, where the decay time τ is

$$\tau = RC. \tag{13.10}$$

Expressed in terms of conductivity and capacitance per unit area, Eq. (13.10) can be rewritten as

$$\tau = \mathcal{C} / \gamma. \tag{13.11}$$

For example, the conductivity is about 5 $\Omega^{-1}\text{m}^{-2}$ (or less) for the passage of individual ionic species across a membrane; see end-of-chapter problems. Taking this value and $\mathcal{C} = 1 \times 10^{-2}$ F/m^2 leads to a decay time τ of 2 ms according to Eq. (13.11). For future reference, we rearrange Eq. (13.9) in terms of the change in the membrane potential as

$$dV / dt = -V/RC, \tag{13.12}$$

where the decay time for $V(t)$ from this expression is the same as that in Eq. (13.9).

The analysis leading up to Eq. (13.12) considers only the time variation of V but ignores its spatial variation along the axis of the neuron. We now make V a function of position as well as time, and consider the behavior of a small ring of membrane surface with area $A = 2\pi r\ dx$, where r is the radius of the axon and dx is a distance along its length, assumed to lie parallel to the x-axis. The flow of ions through this membrane patch can be viewed using the circuit diagram of Fig. 13.4, where the movement of three principal ions (Na^+, K^+ and Cl^-) is considered. Each ionic species is shown as having its own path within the circuit, as there is a separate Nernst potential (shown by the battery symbol) and conductivity (shown by the resistor symbol) for each ionic species. The capacitance doesn't depend on the individual species, so only one capacitive element is required for this section of the axon. The horizontal wire at the top of the circuit represents the current outside of the axon while the wire at the bottom represents the current in the interior of the axon.

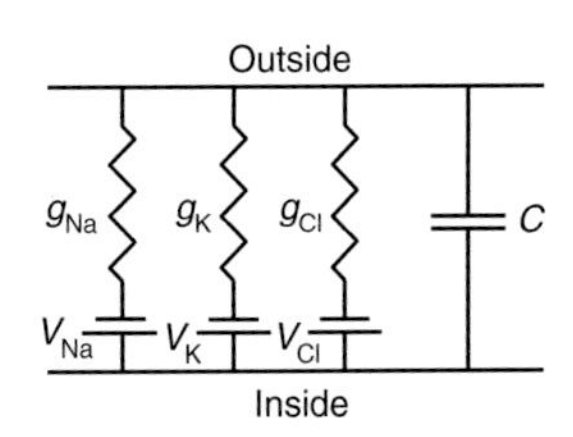

Fig. 13.4

Circuit diagram representing several aspects of ion flow through a membrane patch of area A. Three ionic species are considered (Na^+, K^+ and Cl^-), each having a unique Nernst potential V_α and conductivity γ_α. The capacitance of the patch as a whole is equal to C. The length L of the axon segment is assumed to be short enough that the membrane potential does not vary strongly in the x-direction.

The circuit in Fig. 13.4 describes the flow of several ionic species in one section of an axon. However, the effective circuit element can be simplified if we are primarily interested in the potential difference $V(x,t)$, rather than the flow of individual species. The individual Nernst/resistive elements can be combined into a single element for the overall current:

- the conductance for the overall current is just the sum of the individual conductances, so only one resistive element is required, with conductivity $\gamma = \gamma_{Na} + \gamma_K + \gamma_{Cl}$,
- the Nernst potentials for each ion can be replaced with the quasi-steady-state potential V_{qss} defined in Eq. (13.7).

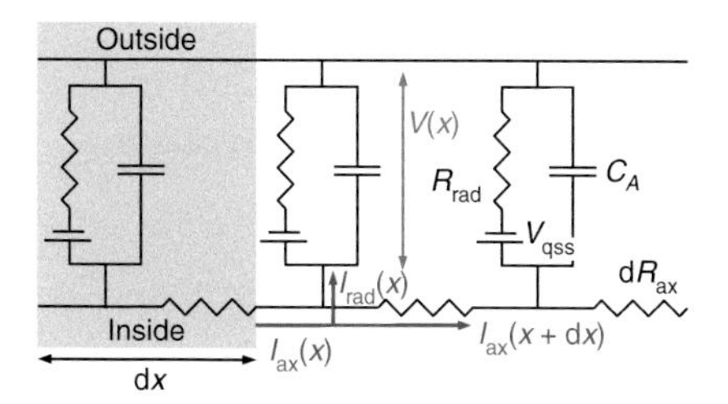

Fig. 13.5 Effective circuit diagram for the flow of current along an axon (bottom wire), through the plasma membrane (vertical wires) and along the axon's exterior (top wire). The resistance of the exterior fluid is assumed to be relatively small and has been omitted. Each circuit loop represents a distance Δx along the axon; between loops, the resistance of the axon's interior along its axis is ΔR_{ax}. Encountering a new loop, a current $I_{ax}(x)$ loses an amount $I_{rad}(x)$ in the radial direction, part of which is collected by the capacitor C and part of which leaves the axon after passing through the effective Nernst potential V_{qss} and membrane resistance R_{rad} in series. The quantities V_{qss} and R_{rad} represent the aggregate behavior of the three ionic species (Na^+, K^+ and Cl^-) in Fig. 13.4.

Thus, the circuit for a section of the axon with length dx along its axis (in the x-direction) can be simplified to the loop within the box on the left-hand side of Fig. 13.5. The new element added to this diagram is the resistance of the axonal fluid dR_{ax} which opposes the flow of current in the longitudinal direction of the axon; a similar resistance should appear in the top wire of the diagram representing current flow outside of the axon, but because the exterior fluid region is so large, this resistance is small compared to the interior value dR_{ax}.

Let's examine what happens as the current I_{ax} flows to the right along the bottom wire and encounters the loop circuit at location x linking the interior of the axon to the exterior. The current will divide at this point, such that its value after the loop $I_{ax}(x + \mathrm{d}x)$ is less than its value before, $I_{ax}(x)$. The difference in these values,

$$I_{rad} = I_{ax}(x) - I_{ax}(x + \mathrm{d}x), \tag{13.13}$$

is the amount of current lost to the loop. Now, the radial current I_{rad} can take two routes through the loop as it moves from the axonal fluid to the plasma membrane: it can either leak through the membrane via V_{qss} (subject to the resistance in the radial direction R_{rad}) or it can add to the charge on the membrane at the capacitive element C_A. The current that leaks out is equal to the radial current density $\mathcal{I}_{rad}(x)$ multiplied by the area of the membrane at location x, namely $2\pi r$ dx. The charge dQ that accumulates at the capacitor is just d$Q = C_A$dV, which arises from the current dQ / dt = d/dt ($C_A V$) in time interval dt. Thus, the radial current is

$$I_{rad}(x) = \mathcal{I}_{rad}(x) \bullet 2\pi r\, \mathrm{d}x + \mathrm{d}/\mathrm{d}t\, (C_A V). \tag{13.14}$$

In this expression, the capacitance of the membrane element C_A is equal to the product of the capacitance per unit area $\mathcal{C}$ multiplied by the area element $2\pi r\,\mathrm{d}x$; $\mathcal{C}$ is assumed to be constant here. Thus, Eq. (13.14) can be rewritten as

$$I_{rad}(x) = 2\pi r\,\{\mathcal{I}_{rad}(x) + \mathcal{C}\,(\mathrm{d}V/\mathrm{d}t)\}\,\mathrm{d}x. \tag{13.15}$$

Substituting Eq. (13.15) for $I_{rad}(x)$ into Eq. (13.13) yields

$$I_{ax}(x) - I_{ax}(x + \mathrm{d}x) = 2\pi r\,\{\mathcal{I}_{rad}(x) + \mathcal{C}\,(\mathrm{d}V/\mathrm{d}t)\}\,\mathrm{d}x, \tag{13.16}$$

or

$$\mathrm{d}I_{ax}\,/\,\mathrm{d}x = -2\pi r\,\{\mathcal{I}_{rad}(x) + \mathcal{C}\,(\mathrm{d}V/\mathrm{d}t)\}. \tag{13.17}$$

The minus sign comes from the reversed order of the currents on the left-hand side of Eq. (13.16) compared to the conventional definition of a derivative $\mathrm{d}f/\mathrm{d}x$. The axial current can be related to the decrease in potential from x to $x + \mathrm{d}x$ through Ohm's Law:

$$V(x + \mathrm{d}x) - V(x) = -I_{ax}(x)\,\mathrm{d}R_{ax}(x), \tag{13.18}$$

where $\mathrm{d}R_{ax}(x)$ is the resistance of the interior of the axon over a distance $\mathrm{d}x$. The minus sign arises because current flows in the direction of decreasing potential; here, a positive value for $\mathrm{d}V/\mathrm{d}x$ means that the potential increases with x, which drives the current toward negative x, the opposite direction than what is assumed in Fig. 13.5. The resistance $\mathrm{d}R_{ax}$ of the axon segment is equal to the length of the segment $\mathrm{d}x$ divided by the product of its cross-sectional area πr^2 and the axial conductivity κ_{ax} (in three dimensions; units of $\Omega^{-1}\mathrm{m}^{-1}$)

$$\mathrm{d}R_{ax} = \mathrm{d}x\,/\,\pi r^2 \kappa_{ax}. \tag{13.19}$$

Thus, Eq. (13.18) can be rewritten as

$$I_{ax}(x) = -(\pi r^2 \kappa_{ax}) \bullet (\mathrm{d}V(x)\,/\,\mathrm{d}x). \tag{13.20}$$

Equations (13.17) and (13.20) are a couplet that can be combined to yield a differential equation for the voltage $V(x,t)$.

The first step in deriving such an expression is to take the spatial derivative of Eq. (13.20) to obtain $\mathrm{d}I_{ax}/\mathrm{d}x$, and then substitute the result into Eq. (13.17). After some algebraic manipulations,

$$(r\kappa_{ax}/2) \bullet (\mathrm{d}^2 V(x)\,/\,\mathrm{d}x^2) = \mathcal{I}_{rad}(x) + \mathcal{C}\,(\mathrm{d}V/\mathrm{d}t). \tag{13.21}$$

This is one version of the cable equation, so named because it was derived by Lord Kelvin in the nineteenth century to understand the properties of signals in undersea electrical cables. The next step is to relate the radial current density $\mathcal{J}_{rad}(x)$ to the potential. To obtain Eq. (13.7) for the quasi-steady-state potential V_{qss}, we imposed the condition that there was no net current density out of the parallel circuits in Fig. 13.4. However, if the applied potential V deviates from V_{qss}, then there will be a current density $\mathcal{J}$ governed by Ohm's Law $\mathcal{J} = \gamma \Delta V$, where the relevant ΔV is $V(x,t) - V_{qss}$ and the relevant (two-dimensional conductivity) γ is γ_{tot} because the current is the sum over all ionic contributions:

$$\gamma_{tot} = \Sigma_\alpha \, \gamma_\alpha. \tag{13.22}$$

That is, the radial current per unit area is

$$\mathcal{J}_{rad}(x) = \gamma_{tot} \, (V - V_{qss}). \tag{13.23}$$

Placing this into Eq. (13.21) yields

$$(r\kappa_{ax}/2) \bullet (\mathrm{d}^2 V / \mathrm{d}x^2) = \gamma_{tot} \, (V - V_{qss}) + \mathcal{C} \, (\mathrm{d}V / \mathrm{d}t). \tag{13.24}$$

As a last step, we shift the voltages by defining the local voltage as $v(x,t)$ as

$$v(x,t) = V(x,t) - V_{qss}, \tag{13.25}$$

to extract the linear cable equation:

$$\lambda^2 \, (\mathrm{d}^2 v / \mathrm{d}x^2) - \tau \, (\mathrm{d}v / \mathrm{d}t) = v. \qquad \text{linear cable equation} \tag{13.26}$$

By comparison with Eq. (13.24) the length and time scales are

$$\lambda = (r\kappa_{ax} / 2\gamma_{tot})^{1/2} \qquad \tau = \mathcal{C} / \gamma_{tot}. \tag{13.27}$$

The time scale τ has appeared before in Eq. (13.11) for *RC* circuits.

How do we interpret the cable equation and what are its properties? First, we note that if the potential v does not depend on x, Eq. (13.26) reduces to Eq. (13.12) for an *RC* circuit, with τ still providing the characteristic time scale for the exponential decay of $v(t)$. Second, the left-hand side of the cable equation has the same form as Fick's second law of diffusion introduced in Chapter 2 via Eq. (2.53); the difference between Fick's diffusion equation and the linear cable equation is the presence of the linear term on the right-hand side of the latter. If the initial disturbance applied to the axon is small, meaning that $V - V_{qss}$ is small, then the linear cable equation is not far from Fick's Law and predicts that the disturbance will evolve diffusively. That is, if the disturbance is sub-threshold for the action potential, then it will spread out with time but will not propagate intact. The predicted time scale for the disturbance to dissipate is around a millisecond, as discussed in the numerical example following Eq. (13.11). This feature is explored in the end-of-chapter problems.

To make the cable equation a proper description of signal propagation, it must incorporate the effects of voltage-gated ion channels. To accomplish this, the two-dimensional conductivities γ_α in Eq. (13.22) must be modified to reflect the experimental observation that sodium channels open as the membrane potential rises at the leading edge of the action potential. When channels open in the membrane, the conductivity increases in proportion to the number of open channels per unit area, n_{open}, as in

$$\gamma_\alpha(V) = n_{\text{open}} G_{\text{channel}} + \gamma_\alpha(0), \qquad (13.28)$$

where G_{channel} is the conductance of a single channel and where $\gamma_\alpha(V)$ is the conductivity of species α at voltage V. Given that more channels open as the voltage increases, then γ_{tot} is voltage-dependent. Considering only the sodium and potassium conductivities, Eq. (13.24) can be rearranged to read

$$(r\kappa_{\text{ax}}/2\gamma_{\text{K}}) \cdot (\text{d}^2V / \text{d}x^2) - (\mathcal{C}/\gamma_{\text{K}}) \cdot (\text{d}V/\text{d}t) = (\gamma_{\text{Na}}/\gamma_{\text{K}})\,(V - V_{\text{Na}}) + (V - V_{\text{K}}). \qquad (13.29)$$

Now, the conductivities of the primary ionic species in a squid axon under resting conditions are $\gamma_{\text{Na}} = 0.11\ \Omega^{-1}\text{m}^{-2}$, $\gamma_{\text{K}} = 3.7\ \Omega^{-1}\text{m}^{-2}$, and $\gamma_{\text{Cl}} = 3.0\ \Omega^{-1}\text{m}^{-2}$. For the two-species (Na and K) system here, $\gamma_{\text{tot}} \cong \gamma_{\text{K}}$ and the two prefactors in Eq. (13.29) are approximately equal to λ^2 and τ in Eq. (13.26)

$$\lambda \cong (r\kappa_{\text{ax}} / 2\gamma_{\text{K}})^{1/2} \qquad \tau \cong \mathcal{C}/\gamma_{\text{K}}, \qquad (13.30)$$

so long as the number of open sodium channels is small. Hence, the left-hand sides of Eqs. (13.26) and (13.29) are very similar, while the right-hand side of Eq. (13.29) reveals the presence of the voltage-dependent sodium conductivity $\gamma_{\text{Na}}(V)$.

Taking the potassium conductivity to be constant at the leading edge of the action potential, the $\gamma_{\text{Na}}{:}\gamma_{\text{K}}$ ratio changes from 1:25 in the resting state to 20:1 at the peak of the potential. This dramatic variation is sufficient to completely change the behavior of Eq. (13.29) from signal dissipation below the threshold voltage to signal propagation above it. The speed of the propagating signal is proportional to λ/τ, which are the length and time scales of the equation. In the discussion following Eq. (13.11), we have already estimated τ to be about 1–2 ms, depending on the values used for the conductivities. To estimate the length scale, we take the (three-dimensional) conductivity of the axonal fluid to be $\kappa_{\text{ax}} = 1\ \Omega^{-1}\text{m}^{-1}$; this choice is motivated by the observed range of conductivities from less than $10^{-10}\ \Omega^{-1}\text{m}^{-1}$ for insulators to 10^7 or $10^8\ \Omega^{-1}\text{m}^{-1}$ for metals, with sea water having $5\ \Omega^{-1}\text{m}^{-1}$. Taking $r = 10\ \mu\text{m}$ as a midrange of axon radii and $\gamma = 5\ \Omega^{-1}\text{m}^{-2}$ yields $\lambda = 10^{-3}$ m. Thus, the propagation speed of the action potential with these parameter values should be about 0.5–1 m/s within the cable equation approach. Although measured values of large axons tend to be higher than this, our estimate is in the correct range. Note that λ is proportional to the square root of the axon radius, so the estimated propagation speed for the giant squid axon will be much higher, given its radius of $> 200\ \mu\text{m}$.

13.3 Rate equations: switches and stability

As described in the previous two sections, signal propagation along an axon displays switch-like behavior in that there are two stable states: the membrane potential is close to either the potassium Nernst potential (the resting potential) or the sodium Nernst potential (the depolarized state near the peak of the action potential). As the action potential arrives at a particular location, the membrane potential changes from one state to another, including a change in its sign. This behavior arises because the overall membrane conductivity depends on voltage, so the cable equation, Eq. (13.29), is dominated by one or other of the Nernst potentials. In Section 13.4, we discuss other manifestations of switches in the context of gene regulation. However, we now pause for a mathematical interlude on the properties of differential equations that arise in the analysis of genetic networks.

One of the simplest control circuits is a single negative-feedback loop, where the presence of a particular protein (which we give the label R) inhibits its own production. An example of how this might work in the cell is the situation where a protein can bind to its own gene. When the concentration of R is low, then its small number of copies in the cell does little to inhibit the transcription of its mRNA. In contrast, when R is abundant, transcription is blocked by the binding of R to its gene. Thus, R builds to a certain critical concentration and then is held fixed at that level through negative feedback. If for some reason the concentration of R falls below this critical value, the production of R will resume until the critical concentration is reached once again. In this situation, the production rate of R might be described by a single differential equation with a form like $\mathrm{d}u/\mathrm{d}t = -u + u_{ss}$, where u_{ss} is the steady-state value of u.

A more general form of a regulatory system involves the variation of two quantities, which we denote by u and v, with a time dependence governed by the coupled equations

$$\mathrm{d}u/\mathrm{d}t = -u + \alpha\,/\,(1 + v^n) \tag{13.31a}$$

$$\mathrm{d}v/\mathrm{d}t = -v + \alpha\,/\,(1 + u^n). \tag{13.31b}$$

Here, u and v might be concentrations of proteins, scaled by appropriate rate constants to make them dimensionless. For our applications in the cell, the exponent n obeys $n > 1$. If the parameter $\alpha = 0$, then the equations decouple and their solution is just exponential decay of u and v with time. Within the cell, this might be the case if a protein starts at a fixed concentration and decreases with time. More interesting behavior arises when $\alpha \neq 0$.

To understand the generic properties of Eq. (13.31), we start with the solution under steady-state conditions where the time derivatives on the left-hand side of the equations vanish. Unlike the $\alpha = 0$ case, now the equations remain coupled in u and v:

$$u_{ss} = \alpha / (1 + v_{ss}^{\ n}) \tag{13.32a}$$

$$v_{ss} = \alpha / (1 + u_{ss}^{\ n}). \tag{13.32b}$$

These equations resemble the Hill functions of Eq. (9.92) and (9.93) of Section 9.5, where they were introduced in the context of receptor–ligand binding; there, n is the Hill coefficient. The ss subscripts have been introduced to identify u_{ss} and v_{ss} as steady-state solutions, whose properties we now analyze following the same approach as Phillips *et al.* (2009).

First, we consider the situation when $\alpha >> 1$. There are three distinct functional regimes present:

Case 1 Assuming u_{ss} is small, then Eq. (13.32b) gives the form $v_{ss} = \alpha$, which can be substituted into Eq. (13.32a) to yield a consistent solution for u_{ss}. Thus,

$$u_{ss} = \alpha^{1-n} \qquad v_{ss} = \alpha. \tag{13.33}$$

Case 2 Next, assume v_{ss} is small and follow the same steps as Case 1 to obtain

$$u_{ss} = \alpha \qquad v_{ss} = \alpha^{1-n}. \tag{13.34}$$

Case 3 Now, u_{ss} and v_{ss} cannot be small simultaneously if $\alpha >> 1$ as inspection of Eq. (13.32) confirms. Thus, the only other possibility left is that they are both large simultaneously; solving Eq. (13.32) for this situation yields

$$u_{ss} = v_{ss} = \alpha^{1/(1+n)}. \tag{13.35}$$

Next, consider the opposite range of α, where $\alpha << 1$. Again, we start by assuming u_{ss} is small, so that Eq. (13.32b) yields $v_{ss} = \alpha$, from which $u_{ss} = \alpha$ according to Eq. (13.32a). Thus, one possible solution is

$$u_{ss} = v_{ss} = \alpha. \tag{13.36}$$

However, proposing that one of u_{ss} or v_{ss} is large does not yield a consistent solution upon substitution into Eq. (13.32). So, the regime with $\alpha << 1$ has only one solution, Eq. (13.36), not the three solutions present in Eqs. (13.33)–(13.35) when $\alpha >> 1$.

Equations (13.33)–(13.36) are the asymptotic steady-state solutions to Eq. (13.31) in two limits of the parameter α: there is one solution at small α and three solutions at large α. The next step is to find which solutions

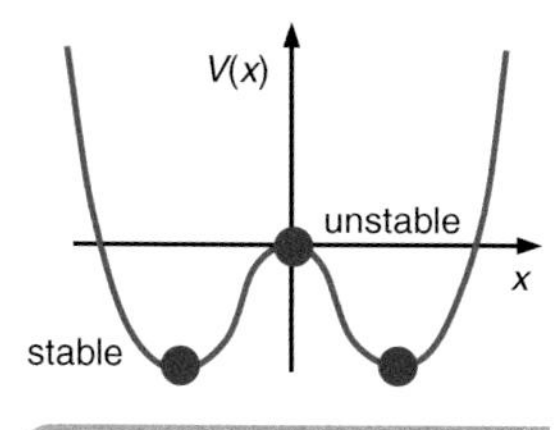

Fig. 13.6

Potential energy function $V(x) = -k_2x^2/2 + k_4x^4/4$ showing the three solutions to the condition $dV/dx = 0$, namely $x = 0$, $\pm(k_2/k_4)^{1/2}$ (blue dots). The $x = 0$ solution, with $d^2V/dx^2 < 0$, is unstable, while the other two solutions with $d^2V/dx^2 > 0$ are stable.

are stable. The problem to be resolved can be illustrated by the potential energy function $V(x) = -k_2x^2/2 + k_4x^4/4$, where k_2 and k_4 are positive constants, as shown in Fig. 13.6. At large values of x, the quartic term dominates and V becomes very large and positive, while at $x = 0$, V vanishes. However, at small x, the function is negative. If one uses the condition $\mathrm{d}V/\mathrm{d}x = 0$ to determine the extrema of V, three solutions are obtained, namely the derivative vanishes at $x = 0$ as well as $x = \pm(k_2/k_4)^{1/2}$. The solution at $x = 0$ is therefore unstable, in the sense that there are two solutions at $\pm(k_2/k_4)^{1/2}$ with lower energy than $V(0)$ and it is these latter two solutions that are the lowest energy states.

A mathematical test of the stability of solutions to a potential energy function $V(x)$ is that its second derivative must be positive: $\mathrm{d}^2V/\mathrm{d}x^2 > 0$ for stability. Graphically, a positive second derivative means that the shape of the potential energy curve around the solution is concave up and therefore stable. In Eq. (13.31), we are interested in the time-dependence of u and v around the values of u_{ss} and v_{ss}: if u is displaced slightly from the steady-state value, does it oscillate around u_{ss} as happens with $V(x)$ for $x = \pm(k_2/k_4)^{1/2}$, indicating stability? Or does u move away from u_{ss} like $V(x)$ does at $x = 0$, indicating instability? By introducing the small quantities $\delta_u(t)$ and $\delta_v(t)$ via the equations

$$u(t) = u_{ss} + \delta_u(t) \tag{13.37a}$$

$$v(t) = v_{ss} + \delta_v(t), \tag{13.37b}$$

the time dependence of the perturbations can be corralled into a set of equations for $\delta_u(t)$ and $\delta_v(t)$. To simplify the notation, Eq. (13.31) is rewritten as

$$\mathrm{d}u/\mathrm{d}t = -u + g(v) \tag{13.38a}$$

$$\mathrm{d}v/\mathrm{d}t = -v + g(u), \tag{13.38b}$$

with

$$g(x) = \alpha / (1 + x^n), \tag{13.39}$$

so that the steady-state solutions obey $u_{ss} = g(v_{ss})$ and $v_{ss} = g(u_{ss})$. Combining Eqs. (13.38) with the series expansion $g(u) = g(u_{ss}) + g'(u_{ss})\delta_u$ [with a similar form for $g(v)$], yields

$$\mathrm{d}\delta_u /\mathrm{d}t = -\delta_u + g'(v_{ss})\delta_v \tag{13.40a}$$

$$\mathrm{d}\delta_v /\mathrm{d}t = -\delta_v + g'(u_{ss})\delta_u, \tag{13.40b}$$

where $g'(x)$ is the derivative of $g(x)$ with respect to x.

Assuming that unstable states diverge from their steady-state solutions exponentially with time, we assign $\delta_u(t)$ and $\delta_v(t)$ the functional forms

$$\delta_u(t) = \delta_{uo} \exp(\lambda t) \qquad (13.41a)$$

$$\delta_v(t) = \delta_{vo} \exp(\lambda t), \qquad (13.41b)$$

where δ_{uo} and δ_{vo} are constants and λ is a rate constant. If $\lambda > 0$, the perturbation grows with time (unstable) whereas if $\lambda < 0$, it decays with time (stable); we have assumed that the same rate constant applies to u and v. Substituting Eq. (13.41) into Eq. (13.40) gives the set of coupled equations

$$(1 + \lambda)\, \delta_{uo} = g'(v_{ss})\, \delta_{vo} \qquad (13.42a)$$

$$(1 + \lambda)\, \delta_{vo} = g'(u_{ss})\, \delta_{uo}, \qquad (13.42b)$$

which can be combined to yield

$$1 + \lambda = \pm\, [g'(u_{ss})g'(v_{ss})]^{1/2}. \qquad (13.43)$$

The stability condition $\lambda < 0$ then imposes the requirement

$$g'(u_{ss})g'(v_{ss}) < 1, \qquad \text{stable solutions} \qquad (13.44)$$

and that $g'(u_{ss})g'(v_{ss})$ be positive.

We now apply this stability analysis to the steady-state solutions in Eqs. (13.33)–(13.36). Considering first the regime $\alpha << 1$, the solutions in Eq. (13.36) give $g'(u_{ss})g'(v_{ss}) = n^2\alpha^{2n}$. For small α and n greater than unity, α^{2n} must be much less than 1, so the single symmetric solution ($u_{ss} = v_{ss}$) is stable. However, in the regime $\alpha >> 1$, the solution in Case 3 leads to $g'(u_{ss})g'(v_{ss}) = n^2$, which must be larger than unity because $n > 1$. Thus, the symmetric solution $u_{ss} = v_{ss}$ is *unstable* at $\alpha >> 1$ even though the symmetric solution is stable at $\alpha << 1$. However, the remaining two solutions at $\alpha >> 1$ are both stable, as shown in the end-of-chapter problems. The overall behavior of the stable solutions to Eqs. (13.32) is that there is a single, symmetric solution at small values of the parameter α, and two asymmetric solutions at large values of α. The large-α solutions have the properties of a switch, and the transition from a single-solution regime to the switch regime occurs at a value of α that depends on n.

In this section, we have searched for switch-like behavior in the properties of coupled differential equations, without explaining why these equations are relevant in the cell. After introducing the molecular basis for genetic regulation in Section 13.4, we will interpret Eq. (13.31) in terms of chemical reactions and demonstrate their applicability in describing switch behavior in the cell cycle. More complex genetic pathways are possible than Eq. (13.31), and one of these, the repressilator circuit, is introduced in Section 13.4 as well.

13.4 Molecular basis of regulation

As described in Chapter 12, the cell cycle involves changes in cell shape and activity, which means that the concentrations of various proteins or other molecular constituents of the cell may change with time, particularly for eukaryotes. As a result, mechanisms must be available to control the production rate for many of the cell's molecular products, both individually and as members of a group involved in a collective activity. Further, there are times at which events must occur with some degree of synchronization. In yeasts, an example is the Start event near the end of G_1 phase: once the conditions are right, DNA replication begins in S-phase and the task cannot be left incomplete without serious consequences for the cell, such as the incorrect proportion of genes on partly duplicated DNA. The cell employs various techniques to regulate the production, distribution and functioning of proteins, although some of these methods are exclusive to eukaryotes. In this section, we examine the molecular basis of regulation, focusing on the transcription of DNA and the role of the *cyclin* family of proteins in eukaryotic cells.

13.4.1 Transcription control in prokaryotes

At the molecular level, the first step in the transcription process is the attachment of RNA polymerase to double-stranded DNA at a *promoter* site, whose location is shown diagrammatically in Fig. 13.7(a). For transcription to occur, RNA polymerase first encompasses the double helix at the promoter, then opens the helix and begins to read along the gene (moving to the right in the diagram; see Section 12.4). The promoter is much shorter than the gene itself; a consensus sequence for the promoter region in bacteria is displayed in Fig. 12.19(c). In some situations in bacteria, a single promoter is the start point for a series of related genes, all transcribed sequentially onto the same mRNA; in such instances, the promoter and associated genes are collectively called an *operon*. Transcription can be influenced in both a negative and positive fashion by the attachment of *regulatory proteins* to the DNA helix. The regulatory proteins serve as sensors to their environments: for example, they may be able to bind a particular chemical compound or *ligand*, enabling them to sense the abundance of that compound and, as a result, influence how many copies of mRNA the cell needs to make in order to maintain the desired concentration of the ligand at that point in the cell cycle (assuming the ligand is produced by the cell's proteins).

Under *negative* control, a *gene repressor protein* (also called a *transcriptional repressor*) attaches to a site called the *operator* (which is within or

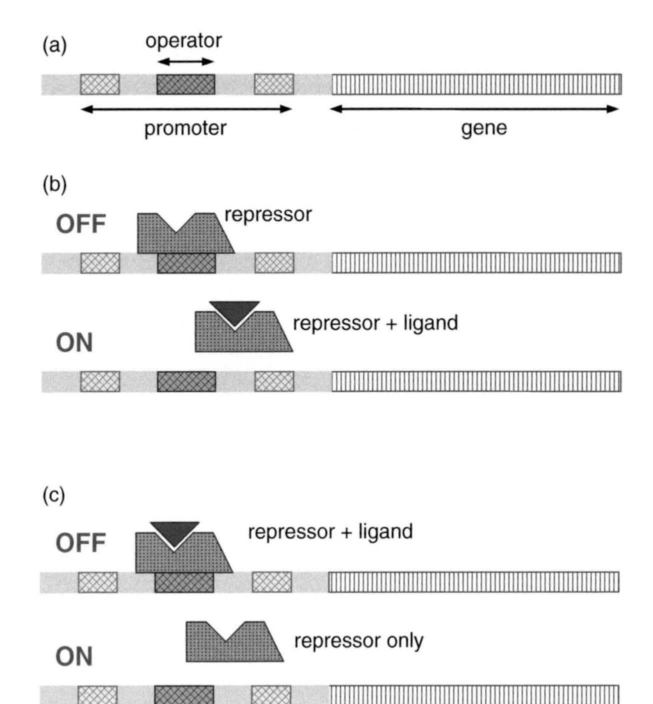

Fig. 13.7 (a) Arrangement of the operator region with respect to the promoter for negative control, where the gene is turned off when a protein is bound to the DNA helix and blocks the attachment of RNA polymerase. In panel (b), the gene is turned off when the repressor alone is bound, whereas in panel (c), it is turned off when a repressor–ligand complex is bound.

spatially near the promoter) and is able to block the attachment of RNA polymerase. In bacteria, the operator is usually smaller than the promoter. Once again, the operator provides a ligand-sensitive switch that can function in one of two ways.

Example 1 – negative The repressor binds in the absence of the ligand, turning the gene off. Then, when the ligand binds to the repressor, the pair is released from the DNA helix and the gene turns on, as shown diagrammatically in Fig. 13.7(b).

Example 2 – negative The repressor binds in the presence of the ligand, turning the gene off. Then, when the ligand is absent, the repressor is released from the helix and the gene turns on. This is illustrated in Fig. 13.7(c).

In the discussing negative control, the binding of the repressor protein to DNA is portrayed as undesirable, in that the repressor blocks the binding of RNA polymerase. However, there are situations in which a bound

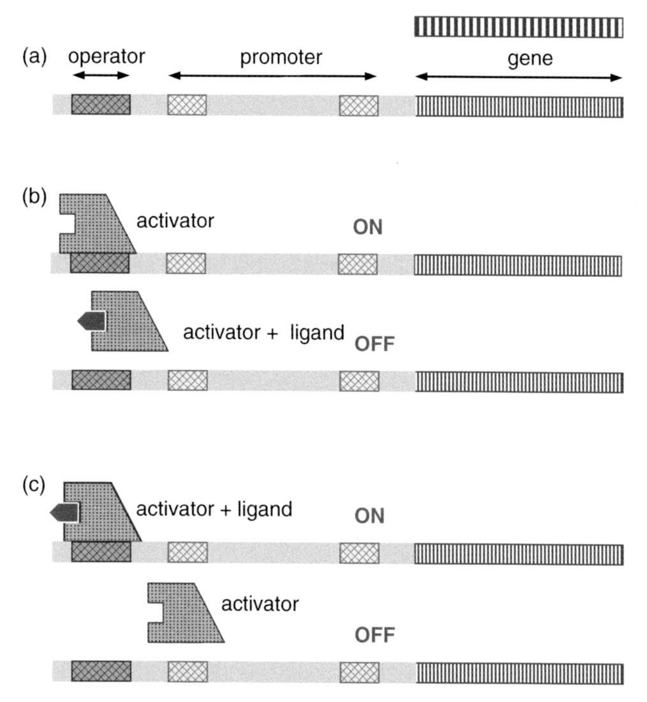

Fig. 13.8 (a) Arrangement of the operator region with respect to the promoter for positive control, where the gene is turned on when an activator protein is bound to the DNA helix and assists the attachment of RNA polymerase. In panel (b), the gene is turned on when the activator alone is bound, whereas in panel (c), it is turned on when a repressor–ligand complex is bound.

protein can have a positive effect by enhancing the function of RNA polymerase, for example by helping it open the DNA helix. Such proteins are called *gene activator proteins* (or *transcriptional activators*), and they too can function as switches if their ability to bind to DNA is affected by the presence/absence of a ligand. Several possibilities arise when the activator binds outside of the promoter region, as shown in Fig. 13.8(a), including the following:

Example 3 – positive The activator binds to DNA in the absence of the ligand, turning the gene on. Then, when the ligand binds to the activator, the pair is released from the helix and the gene turns off or the frequency of its transcription is dramatically curtailed. This is displayed in Fig. 13.8(b).

Example 4 – positive The activator binds to DNA in the presence of the ligand, turning the gene on. When the ligand is not present, the activator is released from the helix and the gene turns off, as in Fig. 13.8(c).

Note the reversal of the on–off states in Fig. 13.8 compared to the negative control situations in Fig. 13.7.

Armed with these basic switch mechanisms, one can imagine ever more complex situations in which activator or repressor proteins are sensitive to the presence or absence of more than one ligand: for example, a switch turning the gene on when ligand A is present and ligand B is absent concurrently. Indeed, eukaryotic switches may be much more complex than the examples above, and may involve several binding sites spread along the DNA helix some distance away from the gene. Even though binding sites may be well separated along the helical contour, they may be spatially close if the helix forms a loop; see Müller *et al.* (1996) and references therein.

We now begin to describe the regulation mechanisms in Examples 1–4 in terms of the mathematical formalism developed in Section 13.3 for coupled equations. Suppose that we can isolate the production and degradation of two particular proteins A and B from the remaining biochemical pathways in the cell. If these proteins are present at some initial concentration in the cell, but are not produced or destroyed during the cell cycle, then their concentration will decline as the cell volume expands. The rate at which the concentration falls is proportional to the concentration itself at any given time, so that $dc_A/dt = -\lambda c_A$. This equation will also apply if the protein is degraded by additional means, but the rate constant λ will be different and may depend on the concentration of the species, protein or otherwise, that removes protein A from the system. Next, let protein A also be produced in the cell at a rate γ, independent of the cell volume. As a result, the rate equation will be modified to read $dc_A/dt = -\lambda c_A + \gamma$, where the plus sign indicates A is being produced, not destroyed.

As a further extension, let another protein B act as a repressor to A, so that the rate of production γ will be reduced according to the probability p_B that B can bind to the operator site of the gene that codes for protein A. To accommodate the repressor, the equation governing dc_A/dt becomes

$$dc_A/dt = -\lambda c_A + \gamma(1 - p_B), \tag{13.45a}$$

where the factor $(1 - p_B)$ means that the more likely the binding of repressor B, the smaller the production rate $\gamma(1 - p_B)$. The possibility of switch-like behavior arises if there is another pathway in which A acts as a repressor to the production of B, so that the time evolution of species B obeys

$$dc_B/dt = -\lambda c_B + \gamma(1 - p_A), \tag{13.45b}$$

where the same degradation rate λ and production rate γ have been used in both equations for simplicity.

The last step in the puzzle is to determine the functional form of the binding probabilities p_A and p_B. For this, we return to Section 9.5, where we showed that the probability of a ligand at concentration c_L to bind to a receptor was given by the Hill function of Eq. (9.93)

$$p_{\text{bound}} = (c_L/c_o)^n \exp(-\beta\Delta\varepsilon) \,/\, \{1 + (c_L/c_o)^n \exp(-\beta\Delta\varepsilon)\}, \tag{13.46}$$

where n is the Hill coefficient. As applied to the two-species transcription problem we have formulated, the reference concentration c_o, receptor–ligand binding energy $\Delta\varepsilon$, and inverse temperature β are all fixed, so they can be wrapped together into a single constant K as

$$K = \exp(-\beta\Delta\varepsilon) \,/\, c_o^{\,n}, \tag{13.47}$$

permitting us to rewrite p_{bound} as

$$p_{\text{bound}} = Kc_L^{\,n} \,/\, (1 + Kc_L^{\,n}). \tag{13.48}$$

Thus,

$$1 - p_{\text{bound}} = 1 \,/\, (1 + Kc_L^{\,n}), \tag{13.49}$$

and Eq. (13.45) can be expressed as

$$\mathrm{d}c_A/\mathrm{d}t = -\lambda c_A + \gamma/(1 + Kc_B^{\,n}) \tag{13.50a}$$

$$\mathrm{d}c_B/\mathrm{d}t = -\lambda c_B + \gamma/(1 + Kc_A^{\,n}). \tag{13.50b}$$

Once again, the combination of terms in K is taken to be the same for species A and B to simplify the mathematics.

Leaving aside the Hill coefficient, there are three parameters in Eq. (13.50), two of which can be absorbed into the definition of the concentration c and time t, leaving only a single combination $\gamma K^{1/n}/\lambda$ which takes the place of α in Eq. (13.31). Thus, Eq. (13.50) has the same coupled structure as Eq. (13.31) of Section 13.3. There, we showed that under weak coupling (small values of $\gamma K^{1/n}/\lambda$ here), the equations permitted only a single stable solution, whereas at large coupling, there were two stable solutions and the system possessed switch-like behavior. Translating the u_{ss} and v_{ss} language of Section 13.3, the stable solutions in the switch regime are asymmetric, in which one of the protein concentrations is large and the other is small, depending on the initial conditions of the system. An experimental realization of this switch can be found in Gardiner *et al.* (2000).

The genetic switch that we have described above is not the same as an oscillator; rather, it is a system that can be driven between two different states. However, a simple extension of the two-component switch model to include a third repressor leads to oscillatory behavior in some ranges of its parameter space; further, this network has been realized experimentally in *E. coli* (Elowitz and Leibler, 2000), just like the toggle-switch network of Eq. (13.50) has. The network, dubbed a repressilator, involves three repressor proteins interacting in a loop:

- protein A represses the expression of protein B,
- protein B represses the expression of protein C,
- protein C represses the expression of protein A.

To describe the concentrations of each protein and its corresponding mRNA requires six coupled rate equations of the generic form:

$$\mathrm{d}m_j/\mathrm{d}t = -\lambda_\mathrm{m} m_j + \gamma/(1 + Kp_{j-1}{}^n) + \gamma_\mathrm{o} \tag{13.51a}$$

$$\mathrm{d}p_j/\mathrm{d}t = -\lambda_\mathrm{p} p_j + \zeta m_j. \tag{13.51b}$$

The index $j = 1$–3 refers to species A, B and C periodically (that is, $j = 0$ is species C). Here, all proteins have the same rate parameters, as do all types of mRNA: the rates do not depend on the protein species A, B or C. However, the degradation rates (λ_m and λ_p) and production rates (γ and ζ) are different for mRNA and proteins. Other than the fact that twice as many equations are required for determining the concentrations of both protein and mRNA, the only new term to appear here is γ_o, which allows for the production of each protein even when its associated repressor concentration is at saturation.

The appearance of these equations can be improved with the usual conversion to dimensionless variables, as done in reducing Eq. (13.50) to Eq. (13.31). The six equations have a larger parameter space than Eq. (13.31), but the starting point for their solution is the same as that outlined in Section 13.3: find the steady-state solutions and then evaluate their stability and other properties. Some of the steps in the process are easy, others require numerical evaluation. For example, imposing $\mathrm{d}p_j/\mathrm{d}t = 0$ on Eq. (13.51b) immediately yields

$$\lambda_\mathrm{p} p_{j,\mathrm{ss}} = \zeta m_{j,\mathrm{ss}}, \tag{13.52}$$

for the steady-state concentrations of each of the three protein–mRNA pairs. To find the individual concentrations, we impose $\mathrm{d}m/\mathrm{d}t = 0$ on Eq. (13.51a) to obtain

$$\lambda_\mathrm{m} m_j = \gamma/(1 + Kp_{j-1}{}^n) + \gamma_\mathrm{o}. \tag{13.53}$$

Substituting Eq. (13.52) gives

$$m_j = (\gamma/\lambda_\mathrm{m})/[1 + K(\zeta m_{j-1}/\lambda_\mathrm{p})^n] + (\gamma_\mathrm{o}/\lambda_\mathrm{m}). \tag{13.54}$$

Each m_j is related to its neighbor through an equation like this, so that m satisfies the iterative equation $m_j = H\{H[H(m_j)]\}$ for a function $H(m)$. Hence, $m_j = H(m_j)$ is an allowed solution for monotonically decreasing $H(m)$, and the steady-state solution satisfies

$$m = (\gamma/\lambda_\mathrm{m})/[1 + K(\zeta m/\lambda_\mathrm{p})^n] + (\gamma_\mathrm{o}/\lambda_\mathrm{m}). \tag{13.55}$$

The stability analysis of the solution can be performed as in Section 13.3, although there are now six functions whose time dependence must be evaluated, not just u and v of Eq. (13.38).

The stability analysis reveals that there are large regions of parameter space where the steady-state solutions become unstable and, as a result, oscillatory (for example, see Chapter 19 of Phillips *et al.* (2009). This is in contrast to the two-protein switch model where at least one stable steady state was present for any value of the parameter α. Thus, the addition of the extra repressor changes the nature of the solutions and provides a means for cells to express proteins in an oscillatory manner when needed.

Other important examples of control in eukaryotic cells are the checkpoints at which an incorrect decision to proceed further in the division cycle could have disastrous consequences. These are points where a number of conditions must be simultaneously satisfied and/or where several processes must proceed in concert.

Start checkpoint. Cell size, operation, composition and environment must be appropriate for the commencement of DNA replication in S phase.

G_2/M checkpoint. DNA replication must be complete and all other conditions appropriate for the organization of chromosomes in the early components of M phase must be met.

Metaphase-to-anaphase transition. Sister chromatids must be ready for concurrent separation followed by segregation during cytokinesis.

Many issues are involved with the control mechanisms at these checkpoints and the genetic circuits are potentially complex because of the conditions that need to be simultaneously satisfied.

One group of proteins that oversee the cell cycle are the *cyclin-dependent kinases* or *Cdk*s, which bind with another group of protein regulators called *cyclins* to form *cyclin–Cdk* complexes. Although present throughout the cell cycle, often in relatively constant concentrations, the Cdks themselves are not active without cyclin, meaning that activity in the cycle is governed more by changes in cyclin concentration than by Cdk abundance. Not expectedly, there may be more than one kind of Cdk in the cell, particularly within complex organisms such as mammals. A cell whose cycle has been studied extensively, budding yeast (*Saccharomyces cerevisiae*) is a pleasantly simple system with only one Cdk, namely Cdk1. However, it possesses several types of cyclin.

G_1–Cdk complex:	Cln3
G_1/S–Cdk complex:	Cln1, Cln2
S–Cdk complex:	Clb5, Clb6
M–Cdk complex:	Clb1, Clb3, Clb3, Clb4.

This means, for example, that near the G_1/S boundary, two cyclin–Cdk complexes may be present, namely Cln1–Cdk1 and Cln2–Cdk1; during M-phase, four such complexes may be prominent.

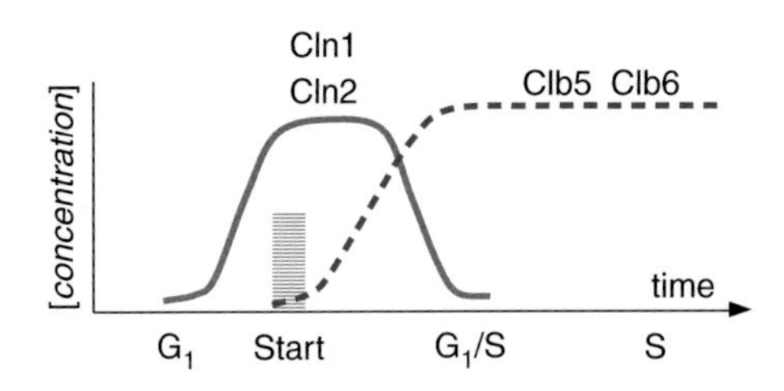

Fig. 13.9 Qualitative representation of representative cyclin concentrations during G_1 and S phases of the cell cycle.

The relative concentrations of several cyclin proteins in budding yeast are displayed in Fig. 13.9 for the G_1 and S phase transition region. Noteworthy is the fact that the changes in concentration are switch-like: Cln1 and Cln2 appear around the Start checkpoint, but then retreat as S phase is approached, to be replaced by Clb5 and Clb6. Other cyclins may be present at reduced concentrations throughout the division cycle; for example, Cln3 is also important in Start. The regulation of cyclin production has been extensively studied in the G_1/S region, and also at the transition to anaphase. Although the biochemical pathways involved are not trivial to map, there is good evidence for the presence of feedback loops in regulatory networks. Positive feedback has been observed in single-cell measurement of Start in budding yeast (Skotheim *et al.*, 2008, and references therein); negative feedback is seen in the initiation of anaphase of *Xenopus laevis* eggs (Pomerening *et al.*, 2005). Switch-like behavior arises in many other circumstances in the cell, such as the change in the rotational direction of the bacterial flagellum (Bai *et al.*, 2010).

Summary

With billions of years of evolution embedded in their design, modern cells employ a variety of techniques to obtain sensory input and analyze that input through chemical circuitry to affect the cell's functioning. The production rates of some proteins can be controlled by a circuit with only a few elements, but the commitment to, and orchestration of, the cell division process may involve highly complex, multicomponent circuits. In this chapter, we have provided only a very modest sampling of signalling and control: signal propagation along an axon in Sections 13.1 and 13.2, the switch-like properties of certain coupled rate equations in Section 13.3, and their application to DNA transcription in Section 13.4. An introduction to the mathematics of receptor–ligand binding, which is also part of the signaling process, is introduced earlier in Section 9.5.

The subject as a whole is much broader than the limited region we have explored here.

Signal propagation along an axon involves speeds that range within a factor of two about 30 m/s, which is orders of magnitude faster than what can be achieved by diffusion or transport via molecular motors. Instead, the signal reflects a change in the transmembrane potential: in many cells the cytosol-facing side of the plasma membrane is held negative by transporter proteins that function as ion pumps. The dramatic change in the membrane potential $V = V_{\text{inside}} - V_{\text{outside}}$ arises by the flow of ions through voltage-gated ion channels. For a single ionic species, the concentration gradient so created can be described by the Nernst potential in Eq. (13.2)

$$V_2 - V_1 = (k_B T / ze) \ln(c_1 / c_2),$$

where c_1 and c_2 are concentrations at different locations and the prefactor $k_B T/ze$ is equal to 25 mV for $z = 1$. When more than one species is present, the overall potential under quasi-steady-state conditions V_{qss} is given by Eq. (13.6)

$$\Sigma_\alpha \gamma_\alpha (V_\alpha - V_{\text{qss}}) = 0,$$

where γ_α is the conductivity (in two dimensions) of species α and V_α is its Nernst potential from Eq. (13.2). This expression applies to homogeneous membranes and may be modified in the presence of ion channels. Typical values for V_{qss} are more negative than −50 mV, often hovering around −70 to −80 mV.

The Hodgkin–Huxley description of axonal signal propagation uses circuit theory to follow ionic current as it travels longitudinally along the axon and radially through the plasma membrane (which also acts as a parallel plate capacitor). Omitting the voltage dependence of the membrane conductivity, the membrane potential $v(x,t)$ obeys the linear cable equation (13.26), which has the form $\lambda^2 (d^2v / dx^2) - \tau (dv / dt) = v$, where the x-axis lies along the axon and v is the difference between the local membrane potential and V_{qss} from Eq. (13.6). The two constants in this equation are the length and time scales of a disturbance to the potential according to Eq. (13.27): $\lambda = (r\kappa_{\text{ax}} / 2\gamma_{\text{tot}})^{1/2}$ and $\tau = \mathcal{C} / \gamma_{\text{tot}}$, where r is the axon radius, κ_{ax} is its conductivity (in 3D), γ_{tot} is the total conductivity (in 2D) of the membrane, and $\mathcal{C}$ is the membrane capacitance per unit area. Unfortunately, this equation, which bears a resemblance to Fick's law for diffusion, does *not* describe signal propagation; rather it describes the dissipation of a low-level signal that is not strong enough to propagate. By permitting the ionic conductivities γ_α (at the membrane) to depend on the membrane voltage v, the cable equation admits a richer suite of solutions, including the propagation of a strong voltage disturbance. The speed at which the disturbance travels is roughly λ/τ, which has values in the meter/second range for common choices of r, κ_{ax} γ_{tot} and $\mathcal{C}$. Note the $\sqrt{r}$ dependence of the speed on the axon radius: signals travel faster along larger axons, all other things being equal.

The presence of voltage-gated ion channels causes the membrane potential to be switch-like, in that it swings from its rest value of about −75 mV through an impressive 100 mV or more to the peak of the action potential as sodium ions flood into the cell once their associated ion-gated channels open. That is, the potential jumps rapidly from one state to another, before retreating to the first state again as potassium ions flow out of the cell (ultimately, ion pumps restore the ionic concentrations to their values before the action potential arrived). Switch-like behavior is present in many other situations in the cell, an example being protein synthesis, where the transcription of a particular gene is turned on or off according to conditions in the cell. Thus, some molecular processes in the cell may obey classes of equations that possess switch-like behavior. An example of such equations are the coupled differential equations in Eq. (13.31) for the time dependence of two quantities u and v: $du/dt = -u + \alpha / (1 + v^n)$ and $dv/dt = -v + \alpha / (1 + u^n)$, where $n > 1$ for many situations. When the parameter α is small, these equations possess only the steady state solution $u_{ss} = v_{ss} = \alpha$, and this solution is stable in the sense that small perturbations away from it dissipate with time. On the other hand, at large values of α, the symmetric solution under steady-state conditions becomes $u_{ss} = v_{ss} = \alpha^{1/(1+n)}$, and this solution is unstable. When perturbed, the unstable solution decays into one of two asymmetric states ($u_{ss} = \alpha^{1-n}$; $v_{ss} = \alpha$) and ($u_{ss} = \alpha$; $v_{ss} = \alpha^{1-n}$), both of which are stable. Thus, the equations at large α describe a system with switch-like behavior: the system can be in either of two inequivalent stable solutions.

At the molecular level, the cell employs a variety of strategies to regulate transcription and, consequently, the abundance of each type of protein. The strategies may be negative or positive: a repressor protein may bind to a gene and block the gene's transcription (negative) or an activator protein may bind near a gene and help initiate transcription (positive). The rate equations that describe the concentrations of molecules participating in a genetic network can be very complex; however, there are several naturally occurring and synthetic networks that are amenable to the analysis developed in this chapter. One situation involves two proteins (say A and B) that repress each other's expression. The rate equations for their concentration have the form of Eq. (13.31), demonstrating that their genetic circuit has switch-like properties. Adding a third repressor C to the system (A represses B; B represses C; C represses A) results in a network known as a repressilator, a circuit that may display oscillatory behavior. Circuits with even more elements and pathways may be part of the control mechanism for coordinating major events during the division cycle.

Problems

Applications

13.1. (a) Using concentration values from the mid-range of those quoted in Table 13.1, find the transmembrane potential V for each of Na^+, K^+ and Cl^- assuming the extracellular concentrations given. (b) Now replace the extracellular concentrations with those of sea water and repeat the calculation in part (a). Make sure to include the sign of V. (c) For parts (a) and (b), solve for the quasi-steady-state potential V_{qss} using for the conductivities $\gamma_{Na} = 0.11\ \Omega^{-1}m^{-2}$, $\gamma_K = 3.7\ \Omega^{-1}m^{-2}$, and $\gamma_{Cl} = 3.0\ \Omega^{-1}m^{-2}$ for a squid axon under resting conditions.

13.2. (a) Find the range of capacitance per unit area $\mathcal{C}$ for membrane systems of thickness $d = 4$–5 nm and dielectric constant $D = 3$–6 (experimentally, D is around 2.5 for some pure bilayers, 4 for oil, and 6 for physical membranes). (b) For a membrane potential of –70 mV, what is the charge density σ on opposite sides of the membrane expected for your results from part (a)? Quote your answer in C/m^2 and also in elementary charges ($e = 1.6 \times 10^{-19}$ C) per lipid head-group, assuming a membrane area per lipid of 0.5 nm^2.

13.3. Estimate the signal propagation speed along an axon as a function of its radius r. Use λ and τ from Eq. (13.27) for the characteristic length and times scales; for parameters, use $\mathcal{C} = 10^{-2}$ F/m^2, $\gamma_{tot} = 5.0\ \Omega^{-1}m^{-2}$ and $\kappa_{ax} = 2\ \Omega^{-1}m^{-1}$. Plot your results for the range 2 μm $\leq r \leq$ 1 mm.

13.4. Determine the conversion factors between the physical concentration c and time t of Eq. (13.50) and the dimensionless u, v and τ of $du/d\tau = -u + \alpha/(1 + v^n)$ in Eq. (13.31).

13.5. An expression is obtained in Problem 13.7 for the dissipation of a sub-threshold signal $w(x,t)$ according to the linear cable equation. The spatial part of the disturbance is Gaussian in form, centered on $x = 0$; at any given time, define x_e to be the value of x where $w(x_e,t) = (1/e)\, w(0,t)$. To gauge how the signal dissipates, plot x_e against time t over the range 10^{-3} s $\leq t \leq$ 1 s for the parameter choice $\lambda = 1$ mm and $\tau = 2$ ms. Quote your answers for x_e in mm.

Formal development and extensions

13.6. Estimate the amount of charge that passes through an axon during the passage of an action potential and compare it with the charge present within the axon's cytoplasm. We break the problem up into the following steps by examining a section of an axon 1 μm long and 5 μm in radius. (a) When the membrane potential V_{mem} changes from −70 mV to +30 mV, what current must flow across the membrane? Use $\mathcal{C} = 10^{-2}$ F/m^2 for the capacitance per unit area, and quote your answer in elementary charges e. (b) In which direction does the current flow? (c) Find the ratio of the ion flow from part (a) to the number of Na^+ ions in the axon cytoplasm, taking the sodium concentration to be 10^{-2} molar.

13.7. Determine the functional form of the potential $v(x,t)$ in the linear cable equation through the following steps.

(a) Show that Eq. (13.26) can be recast as

$$\lambda^2 (\mathrm{d}^2 w / \mathrm{d}x^2) - \tau (\mathrm{d}w / \mathrm{d}t) = 0$$

under the substitution

$$w(x,t) = v(x,t) \exp(t/\tau).$$

(b) Verify that the solution to this equation is

$$w(x,t) = (1 / \sqrt{t}) \exp(-x^2\tau / 4\lambda^2 t).$$

13.8. Using Eq. (13.44), determine the stability of all four steady-state solutions of Eq. (13.31); obtain expressions for $g'(u_{ss})g'(v_{ss})$ as a function of α and n in the appropriate ranges of these parameters.

13.9. Find the steady-state solutions to Eq. (13.31) for $\alpha = 2$ when the Hill coefficient $n = 2$ using the following steps.

(a) Show that the equation governing u_{ss} has the form

$$(u_{ss}^2 - \alpha u_{ss} + 1) \bullet (u_{ss}^3 + u_{ss} - \alpha) = 0.$$

(b) Find the zeroes of the first factor. What is the corresponding v_{ss}?

(c) Find the zeroes of the second factor. What is the corresponding v_{ss}?

Appendix A Animal cells and tissues

Focusing on the mechanical operation of a cell, this text tends to emphasize aspects of structure and organization common to all cells. However, cells differ in many details according to their function, evolution and environment, such that, within a multicellular organism such as ourselves, there may be 100 or 200 different cell types. Because frequent reference is made to these cell types in the text, we provide in this appendix a brief survey of animal cells and their organization in tissues. Readers seeking more than this cursory introduction to cell biology should consult one of many excellent textbooks, such as Alberts *et al.* (2008), Goodsell (1993), or Prescott *et al.* (2004).

A1.1 Tissues

Cells act cooperatively in a multicellular system, and are hierarchically organized into tissues, organs and organ systems. Tissues contain the cells themselves, plus other material such as the extracellular matrix secreted by the cells. Several different tissues may function together as an organ, usually with one tissue acting as the "skin" of the organ and providing containment for other tissues in the interior. Lastly, organs themselves may form part of an organ system such as the respiratory system. Animal tissues are categorized as epithelial, connective, muscle or nervous tissue, whose relationship within an organ (in this case the intestinal wall) is illustrated in Fig. A1.1. The intestine is bounded on the inside and outside by epithelial cell sheets (or epithelia), wherein epithelial cells are joined tightly to each other in order to restrict the passage of material out of the intestinal cavity. Connective tissue, which lies inside the boundaries provided by the epithelia, consists of fibroblastic cells that secrete an extracellular matrix which bears most of the stress in connective tissues. Lastly, several layers of smooth muscle tissue lie on the exterior of the connective tissue, in which they circle around the intestine (circular fibers) or run along its length (longitudinal fibers). Even a given cell type may exhibit several different forms, depending on the specific role that it plays. We now describe in more detail the four principal animal tissues.

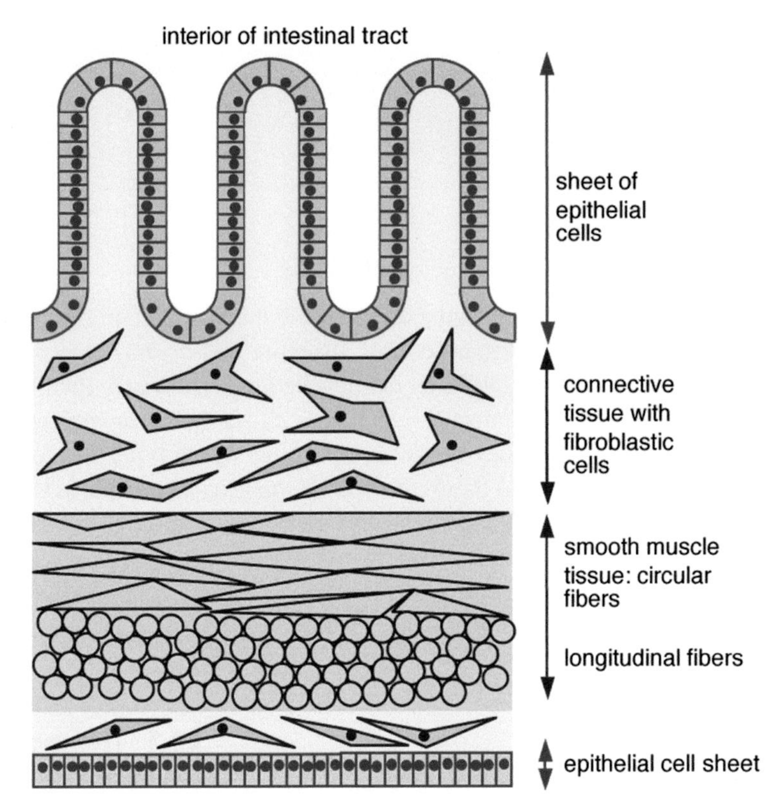

Fig. A1.1 Schematic cross section of the intestine, seen transverse to its cylindrical axis, showing three types of tissue: epithelial, connective and smooth muscle tissue, including both circular and longitudinal fibers (after Alberts *et al.*, 2008; reproduced with permission; ©2008 by Garland Science/Taylor and Francis LLC).

A1.2 Epithelial cells

The passage of material across the boundary of an organ must be selectively controlled, a task performed by epithelial cells. Figure A1.2 displays one subtype of epithelial cell, namely an absorptive cell which possesses microvili to increase the cell's effective area. The surfaces of the cell, denoted according to their location as apical, lateral or basal in Fig. A1.2, are attached by a variety of junctions both to other cells and to the basal lamina; the latter underlies the epithelial cell sheet and is composed of the protein collagen. Linking the cells near their apical surface are tight junctions, which are not so much globs of glue as they are many strands of junction proteins which encircle the cell, such that a barrier of several strands faces any molecule attempting to slip by the cell along its lateral surface.

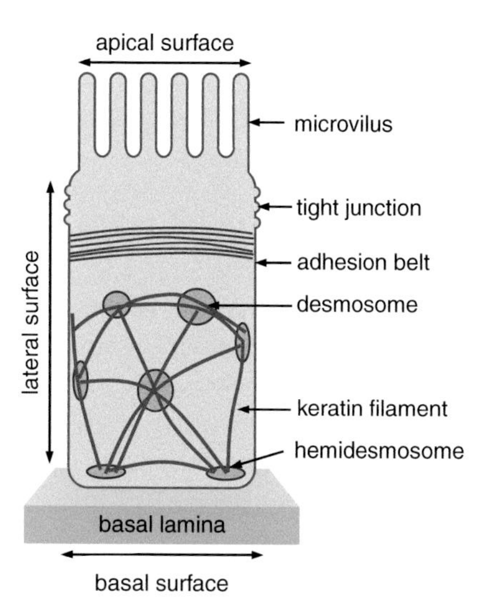

Fig. A1.2 Principal mechanical components of an absorptive epithelial cell. The apical surface faces into the intestine, while the basal surface is attached to the basal lamina. The view is quasi-perspective and the cell's organelles have been omitted for clarity. Typical cell width is about 25 μm.

The mechanical stress on epithelial cells is borne largely by the cytoskeleton and anchoring junctions, of which there are several different types. Filaments of the protein actin are attached at adherens junctions, while thicker intermediate filaments are attached at desmosomes and hemidesmosomes. Bundles of actin filaments form an adhesion belt that encircles the cell; in Fig. A1.2, the belt lies just below the ring of tight junctions. Adhesion belts on neighboring cells are aligned with each other, and are thought to be linked across the lateral surfaces by cell–cell adhesion molecules called cadherins. Epithelial cells also are linked to the basal lamina through cell–matrix adherens junctions. Not shown in the figure are gap junctions, which are regions in which the plasma membranes of adjacent cells lie close to each other, separated by a distance of about 3 nm. Gap junctions act as pores linking the cells and allowing small molecules to pass between them.

A1.3 Fibroblasts and connective tissue

There may be several different cell types present in connective tissue, including nerve cells, capillaries lined with endothelial cells, and also macrophages. Fibroblastic cells, which are largely responsible for secreting the extracellular matrix, are also present. Lastly, connective tissue is usually criss-crossed by fibers such as collagen, secreted by the fibroblasts and forming a network whose density is reflected in the strength of the tissue. Aside from the usual organelles, fibroblasts contain stress fibers, which are bundles of the proteins actin and myosin, as illustrated in Fig. A1.3. One end of a stress fiber terminates in a focal contact (or adhesion plaque) in the plasma membrane, by means of which a fibroblast can attach itself to a substrate. The other end of the fiber may be attached to a network of intermediate filaments surrounding the nucleus, or to a second focal contact. As their name implies, stress fibers can generate and bear a tensile stress, permitting the fibroblast to pull on the extracellular matrix of the connective tissue. Individually, stress fibers are not permanent but form when the cell is subject to tension. As in most cells, the cytoskeleton of fibroblasts also contains a network of microtubules.

A1.4 Muscle cells

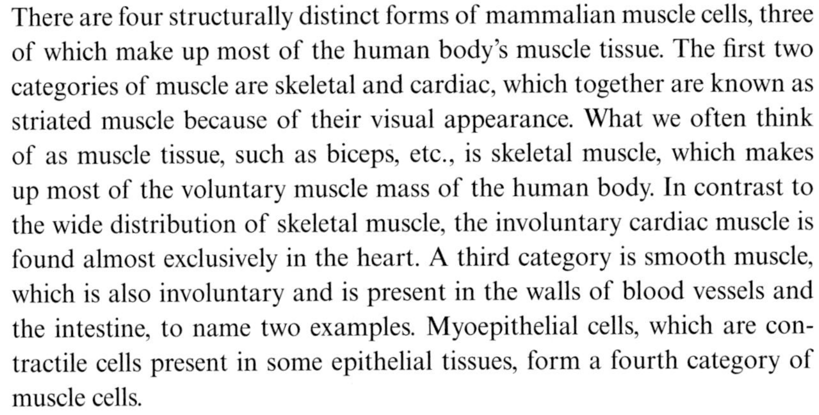

There are four structurally distinct forms of mammalian muscle cells, three of which make up most of the human body's muscle tissue. The first two categories of muscle are skeletal and cardiac, which together are known as striated muscle because of their visual appearance. What we often think of as muscle tissue, such as biceps, etc., is skeletal muscle, which makes up most of the voluntary muscle mass of the human body. In contrast to the wide distribution of skeletal muscle, the involuntary cardiac muscle is found almost exclusively in the heart. A third category is smooth muscle, which is also involuntary and is present in the walls of blood vessels and the intestine, to name two examples. Myoepithelial cells, which are contractile cells present in some epithelial tissues, form a fourth category of muscle cells.

Fig. A1.3

Schematic representation of a fibroblast attached to a substrate in culture, illustrating the network of actin stress fibers and microtubules; the focal contacts in this cell would be near the edges where the stress fibers terminate. Organelles are omitted for clarity. Grown in culture, a fibroblast may have a length in the 50–100 μm range when attached to a substrate.

Skeletal muscle fibers are highly elongated, multinuclear cells: in humans, a skeletal muscle fiber may range up to 100 μm in diameter (although 50 μm is more typical) and up to half a meter long, a length-to-width ratio of 10^4! As displayed in Fig. A1.4, many cylindrical myofibrils about 1–2 μm in diameter are packed together in a single skeletal muscle fiber. The multinuclear nature of the cells arises when mononuclear myoblasts fuse together early in the development of a vertebrate; their nuclei lie near the plasma membrane and do not reproduce once the skeletal muscle cell is formed. Individual multinuclear cells are organized into bundles and surrounded by connective tissue. Having a less cylindrical structure than skeletal muscle cells, the mononuclear smooth muscle cells are nevertheless highly elongated: the typical diameter of a smooth muscle cell in its relaxed state is 5–6 μm, while its length is 0.2 mm.

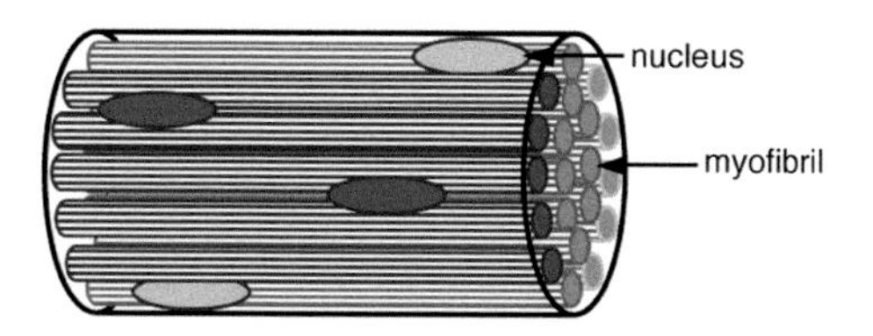

Fig. A1.4 Schematic drawing of a multinuclear skeletal muscle cell, or muscle fiber, the interior of which is packed with myofibrils often running the length of the cell. Only a very short section of the cell is displayed. The envelope encasing the myofibrils is made to appear transparent.

A1.5 Nerve cells

The human nervous system is composed of perhaps 10^{11} nerve cells or neurons, and although their visual appearance covers a range of shapes, neurons share common structural features, as illustrated in Fig. A1.5. The central part of the cell is called the soma, a region with a diameter of 10–80 μm which contains the nucleus and is the main site of protein synthesis. Radiating from the soma are fine dendrites that collect information from sensory cells and other neurons, before passing it on to the soma in the form of an electrical impulse travelling along the plasma membrane. The highly branched structure of dendrites allows them to receive up to 10^5 inputs from other cells. The collective information-processing activities of the dendrites and soma may result in an impulse being sent along an extension of the cell body called the axon. Typically a few microns in diameter, the length of an axon can range from a few millimeters in the brain, up to a meter for large motor neurons; further, there are examples of neurons with more than one axon. At its remote end, an axon divides into branches, each with an axon terminal from which chemical neurotransmitters are released into the synaptic cleft between the axon and an adjacent cell. Because the neurotransmitters and other cell components must be delivered to the axon terminals from their distant production site in the soma, the axon contains microtubules to provide highways for the transport of chemically laden vesicles.

A1.6 Blood cells

The final group of cells that we mention are the red and white blood cells of the human circulatory system. By far the most numerous cells in human blood, red blood cells (or erythrocytes) have the task of carrying oxygen and carbon dioxide from and to the lungs. A variety of white blood cells (or leukocytes) play a prime role in fighting infections, although they are

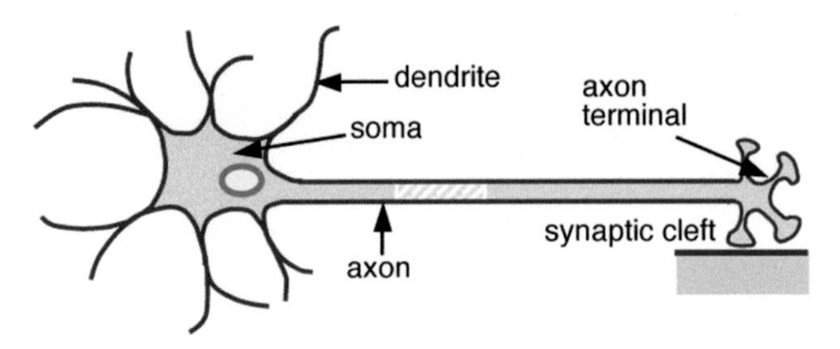

Fig. A1.5 Schematic drawing of a neuron. The soma is typically 10–80 μm in diameter, while the axon and dendrites have diameters of a few microns. The axon is not drawn to scale, as it may be more than 10 000 times as long as it is wide.

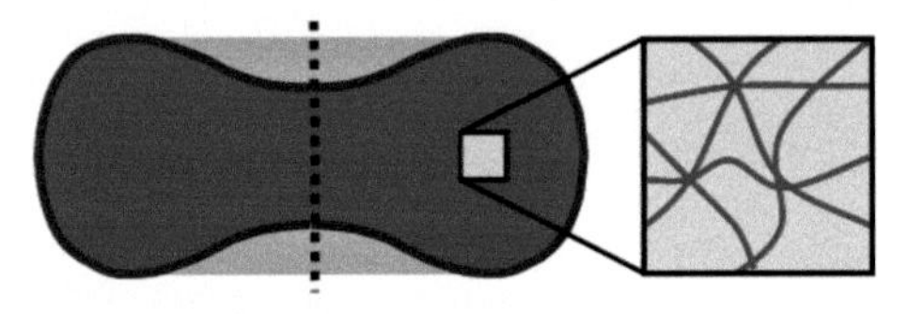

Fig. A1.6 Biconcave shape of the flaccid red cell, with a mean diameter of about 8 μm. The enlargement illustrates the membrane-associated cytoskeleton.

outnumbered by red cells by about 500 to 1. Small incomplete cell fragments called platelets help repair damaged blood vessels, and are present in the blood at about one platelet for every 12–14 erythrocytes. Leukocytes can slip by the endothelial cells lining a capillary and, after passing through the basal lamina, can move into connective tissue. Within a day or two of entering such tissue, monocytes differentiate into macrophages, which can attack and remove foreign cells and debris.

With a biconcave shape approximately 8 μm in mean diameter, erythrocytes have a remarkably simple internal structure. The nucleus and other organelles that are present in red cells during their development are expelled before and shortly after the cells are released into the circulatory system, leaving the mature cells with no internal structural components other than the membrane-associated cytoskeleton. As represented in Fig. A1.6, this cytoskeleton is a loose, two-dimensional network composed principally of tetrameric chains of the protein spectrin, with each tetramer having one or perhaps more attachment sites to the plasma membrane.

B

Appendix B The cell's molecular building blocks

The purpose of this book is to explore the mesoscale mechanics of the cell: its structure and behavior on length scales of tens of nanometers to microns. This expedition cannot be undertaken without a biochemical lexicon, in part because the mechanical properties of some structures are a direct reflection of their molecular composition, and in part because the text draws constantly from the language of biochemistry. In this appendix, we describe in more detail the nomenclature and composition of several classes of compounds referred to in the main text, including:

membrane components: fatty acids and phospholipids,
biopolymers: sugars, amino acids and proteins,
the genetic blueprint: DNA and RNA.

For a more thorough treatment of biomolecular building blocks, the reader is directed to books such as Alberts *et al.* (2008), Gennis (1989), and Lehninger *et al.* (1993).

B1.1 Fatty acids and phospholipids

One of the principal constituents of cellular membranes is the family of dual-chain lipids formed from *fatty acids*, which are carboxylic acids of the form RCOOH, where R represents a hydrocarbon chain. In cells, fatty acids usually are not found in their free state, but rather are components of covalently bonded molecules such as phospholipids or triglycerides. A major component of cell membranes, phospholipids contain two fatty acids linked to a glycerol backbone. The remaining OH group in the glycerol is replaced by a phosphate group PO_4 that, in turn, is linked to yet another group, usually referred to as the polar head group; the spatial arrangement is shown in Fig. B1.1(a). Samples of the fatty acids found in the cell's membranes are given in Table B1.1, including the position of any C=C double bonds. Hydrocarbon chains containing double bonds adopt a *cis* or *trans* form at the double bond: the upper chain in Fig. B1.1(a) is *cis* and the lower chain would be all *trans* if double bonds were present.

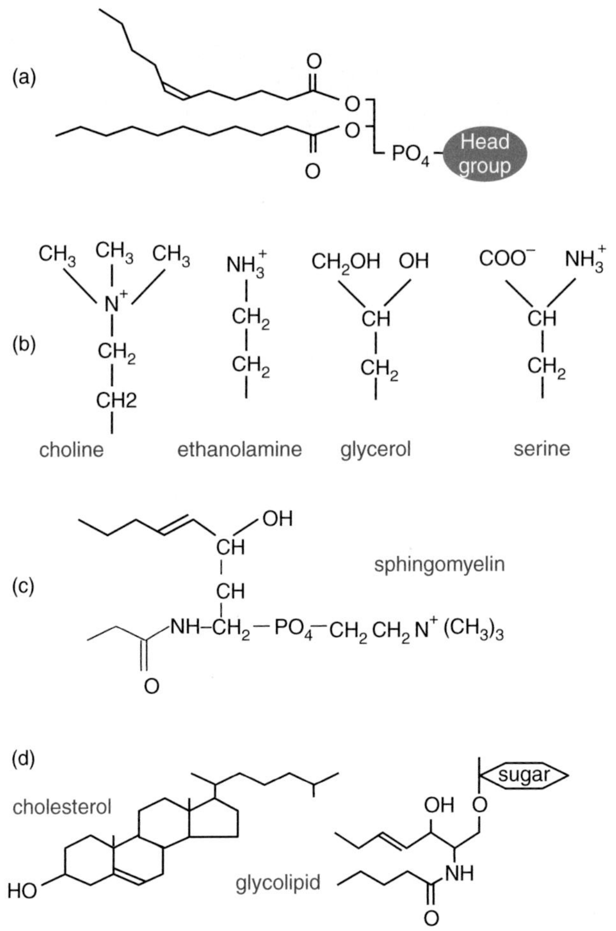

Fig. B1.1 (a) Example of a dual-chain phospholipid with inequivalent hydrocarbon chains. (b) Polar head groups commonly found in phospholipids in the plasma membrane. (c) Sphingomyelin is a phospholipid not based on acyl chains. (d) Examples of membrane lipids that are not phospholipids.

The polar head groups of phospholipids may be chosen from a variety of organic compounds, such as shown in Fig. B1.1(b). Because the phosphate group has a negative charge, serine and glycerol phospholipids are negative while choline and ethanolamine phospholipids are neutral. The naming convention for a phospholipid mirrors, in part, its fatty acid composition. For example, a phospholipid containing two myristic acids (14 carbons each) and an ethanolamine group is called dimyristoyl phosphatidylethanolamine. Common abbreviations (italics provided for ease of pronounciation) include:

phosphatidyl*choline*	PC	phosphatidyl*ethanolamine*	PE
phosphatidyl*glycerol*	PG	phosphatidyl*serine*	PS.

Table B1.1 Some fatty acids commonly found in membrane lipids

Acid name	Total number of carbon atoms	Number of double bonds	Position of double bond
lauric	12	0	
myristic	14	0	
palmitic	16	0	
palmitoleic	16	1	9-*cis*
stearic	18	0	
oleic	18	1	9-*cis*
linoleic	18	2	9-*cis*, 12-*cis*
arachidonic	20	4	9-*cis*, 8-*cis*, 11-*cis*, 14-*cis*

Note. Column 4 displays the C–C bond numbers on which any double bonds occur (see Fig. B1.1).

Not all phospholipids are based solely on fatty acids. For example, sphingomyelin has the structure shown in Fig. B1.1(c). Further, there are other lipids present in cellular membranes which are not phospholipids at all, such as cholesterol and glycolipids, displayed in Fig. B1.1(d). Cholesterol is a member of the steroid family and is frequently found in the membranes of eukaryotic cells. Glycolipids are similar to phospholipids in that they have a pair of hydrocarbon chains, but the phosphate group is replaced by a sugar residue (based on a ring of five carbons and one oxygen).

B1.2 Sugars

Sugar molecules play an obviously important role in metabolism, and also are one component of DNA and of the peptidoglycan network of the bacterial cell wall. A single sugar molecule has the chemical formula $(CH_2O)_n$, the most biologically important sugars having $n = 5$ or 6. Examples of the conformations available for the glucose molecule ($n = 6$) are shown in Fig. B1.2. Part (a) shows the molecule as a linear chain, illustrating that there are several chemically inequivalent functional groups. Five of the oxygens are part of –OH groups while the sixth is double-bonded as an aldehyde. The double-bonded oxygen can be placed at one of several different positions on the chain, each corresponding to an inequivalent, yet related, molecule. The chain can be closed into a ring using one of the oxygens in a hydroxyl group [not the oxygen in the aldehyde group (RCOH)] as illustrated in Fig. B1.2(b). As drawn, the ring in part (b) appears planar, in spite of the

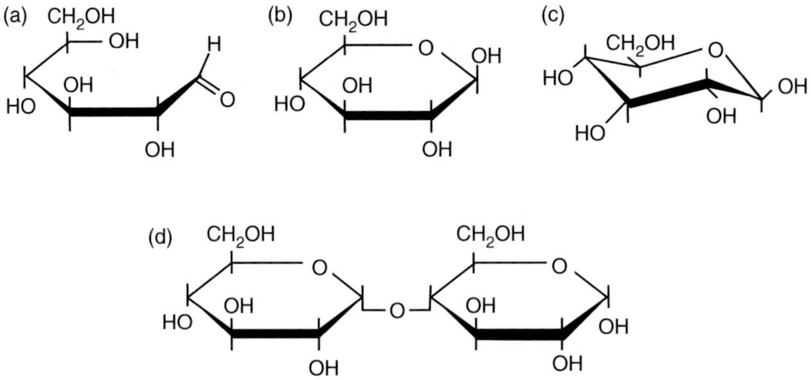

Fig. B1.2 The sugar glucose can be a linear chain (a) or a six-member ring (b). Although often drawn as a planar ring, in fact glucose adopts configuration (c). Two glucose molecules may combine to form the disaccharide maltose, liberating H_2O as a product (d).

lack of in-plane double bonds such as are present in planar compounds like benzene. In fact, the actual configuration of a glucose ring is the bent form in part (c), just as it is in the single-bonded ring of cyclohexane. A single sugar molecule in isolation is referred to as a *monosaccharide*. But sugar molecules can polymerize through reactions in which two alcohol groups (one on each ring) combine to give a single bond between rings, liberating H_2O as a product. For example, glucose molecules may combine to form the disaccharide maltose, as shown in Fig. B1.2(d), or longer polysaccharides.

B1.3 Amino acids and proteins

A principal component of the cytoskeleton and present in most membranes, proteins are linear chains of amino acids, which are a family of organic compounds containing an amino group ($-NH_3^+$) and a carboxyl group ($-COO^-$). Only 20 specific amino acids appear in protein construction, none of which has a high molecular mass, as can be seen from Fig. B1.3. Amino acids can join together to form chains through an amide linkage (–OC–N–) referred to as a peptide bond. The reaction liberates H_2O, and has the general form displayed in Fig. B1.4. A protein chain of high molecular mass will be built up when this reaction occurs repeatedly; for example, the two inequivalent spectrin proteins in the human erythrocyte have molecular masses of ~220 000 and ~230 000 Da. Amino acids appear in a protein with varying relative abundance, and some, such as tryptophan, are uncommon. In a large protein, the average molecular mass of an amino acid residue is 115 Da. This means that one spectrin monomer in the human red blood cell having a molecular mass of 230 000 Da would be composed of approximately 2000 amino acids. Each residue contributes 0.38 nm to the contour length of the polypeptide.

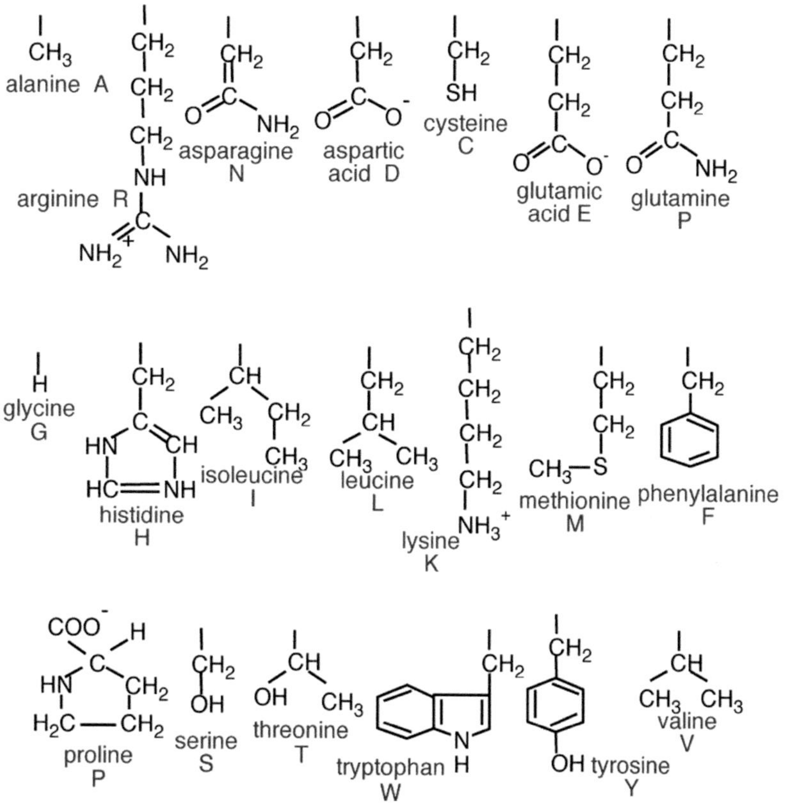

Fig. B1.3 Summary of the 20 amino acids that commonly appear in proteins. All but one of the acids (proline) have the form NH_3^+–RCH–COO^- at pH 7. The name is given first, followed by the abbreviation.

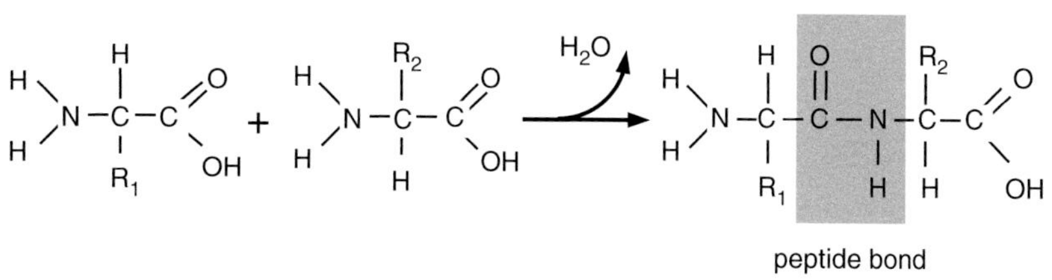

Fig. B1.4 Formation of a peptide bond through the reaction of two amino acids.

B1.4 Nucleotides and DNA

Ribonucleic acid, or RNA, directs protein synthesis in the cell, whereas **deoxy**ribonuclei acid, or DNA, is the carrier of genetic information. The elementary chemical units of DNA and RNA are nucleotides, which have a more complex structure than the amino acid building blocks of proteins. Firstly, nucleotides are themselves composed of subunits, namely a sugar, an organic base and a phosphate group. The two different sugars found in nucleotides, ribose in RNA and deoxyribose in DNA, are five-membered rings differing from each other by only one oxygen atom as in Fig. B1.5(a). In either molecule, the OH groups are potential reaction sites for addition of a base. Five organic bases are found in nucleotides, and one can see from Fig. B1.5(b) that they fall into two chemically similar groups – purines and pyrimidines. Only four of the five bases are present in a given DNA or RNA molecule, and the one "missing" base is different for each:

RNA: adenine, guanine, cytosine, uracil,
DNA: adenine, guanine, cytosine, thymine.

The reaction of a sugar with a base releases water (an –OH from the sugar plus an H from the base) and produces a sugar–base combination called a *nucleoside*. Addition of a phosphate to a nucleoside releases water and produces a *nucleotide*, two of which are shown in the chain of Fig. B1.6(a). The nucleotides themselves can polymerize to form DNA and RNA, through a linkage between a sugar from one nucleotide and a phosphate from another, as schematically illustrated in Fig. B1.6(a). In the double-stranded helix of DNA, the bases lie in the interior of the helix, and hold the helix together through hydrogen bonding between base-pairs. As illustrated in Fig. B1.6(b), each matching base pair on the opposing strands consists of one purine and one pyrimidine: adanine/thymine and guanine/cytosine. The two ends of a single RNA or DNA strand are inequivalent and are labelled 3′ or 5′ according to the last covalent bond on the sugar ring; it is somewhat easier to remember that the 3′ end terminates with a sugar and the 5′ end terminates with a phosphate.

B1.5 ADP and ATP

Chemical processes in cells involve changes in energy. In almost all examples, the cell captures or delivers energy by chemical means, minimizing the energy loss through heat or light. The energy economy of the cell has two components: a currency for local energy exchange, and a

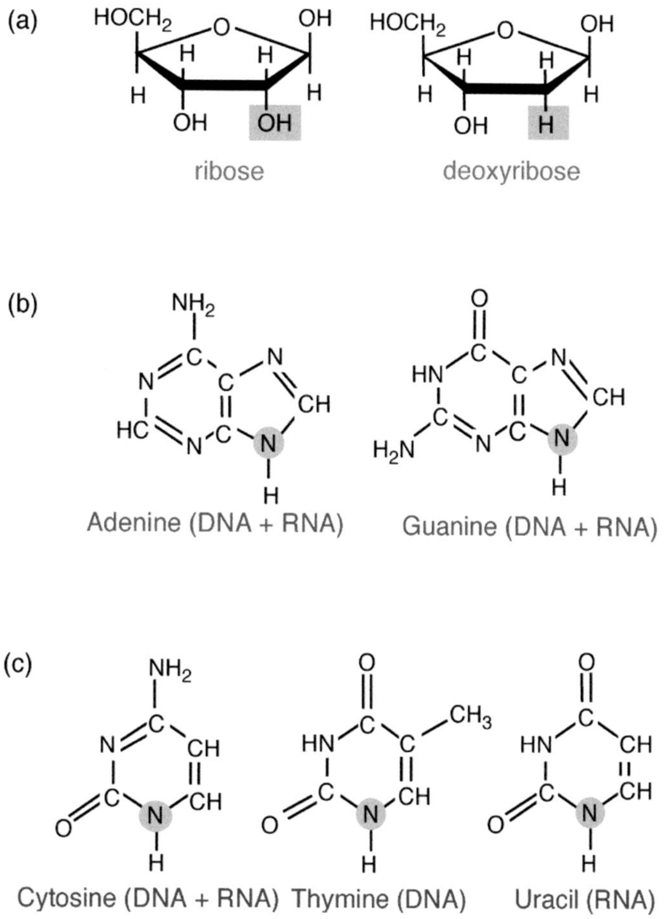

Fig. B1.5 (a) Ribose and deoxyribose are the sugars present in RNA and DNA. (b) The five organic bases found in DNA and RNA fall into two structural families. The shaded nitrogen is the reactive site of the base that joins onto the sugar of the nucleic acid.

mechanism for long-term storage. The concept of an energy currency has parallels with a monetary currency: the amount of currency in circulation need not be nearly as large as the total economy, and long-term storage of economic output does not involve hoarding the currency itself. Long-term energy storage can be accomplished with triglycerides, whose structure is mentioned at the beginning of this appendix. However, the currency itself is based upon a sugar–base–phosphate nucleotide whose elementary components are the same as RNA, a very efficient use of materials. The sugar of the currency is ribose of RNA, and the most common base is adenine (or, much less frequently, guanine). The currency differs from RNA by the presence of more than one phosphate group, as illustrated in Fig. B1.7. The lower energy state of the pair has two phosphate groups (adenosine

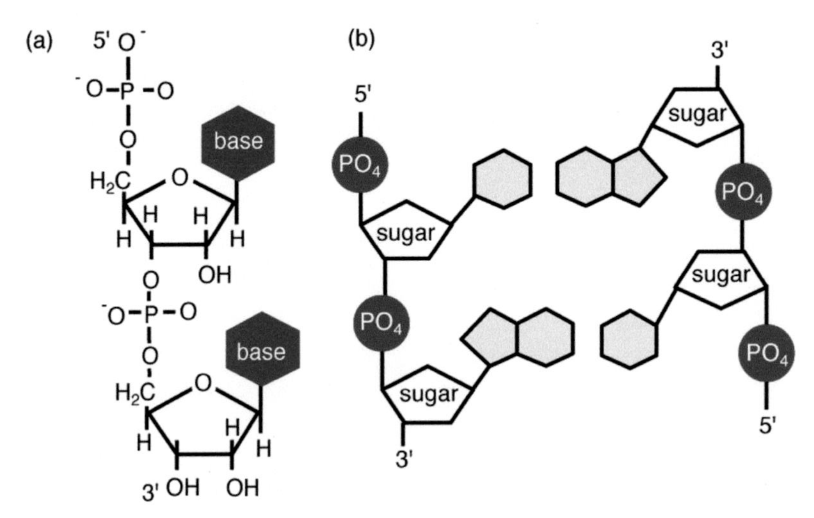

Fig. B1.6 (a) Schematic representation of the sugar/phosphate linkages in a single strand of DNA. (b) Hydrogen bonding between bases on different strands contributes to the interactions that stabilize the double helix structure of DNA.

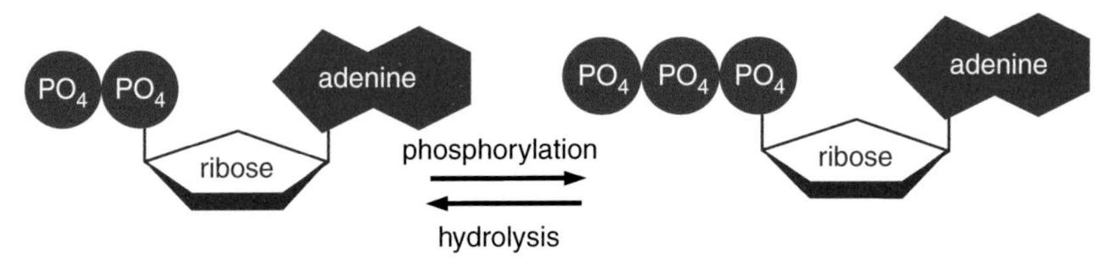

Fig. B1.7 The most common denominations of the cell's energy currency are ADP (adenosine diphosphate) and ATP (adenosine triphosphate).

diphosphate, or ADP), while the higher energy state has three (adenosine triphosphate, or ATP). The energy-consuming process of adding a phosphate is called phosphorylation, while the release of a phosphate is hydrolysis. Depending on conditions, the energy change for a single phosphate group is 11–13 kcal/mol, corresponding to 8×10^{-20} J = $20k_BT$ per reaction.

C Appendix C Elementary statistical mechanics

Soft materials, such as the filaments and membranes of the cell, may be subject to strong thermal fluctuations and sample a diverse configuration space. Each of these accessible configurations, of varying shape and energy, contribute to the ensemble average of an observable such as the mean length of a polymer chain. Statistical mechanics provides a formalism for describing the thermal fluctuations of a system; in this appendix, we review several of its cornerstones – entropy, the Boltzmann factor and the partition function – following a compact approach laid out in Reif (1965). As applications, the thermal fluctuations of a Hooke's law spring and the entropy of an ideal gas are calculated.

C1.1 Temperature and entropy

Any configuration available to a system, be it a polymer or a gas in a container, can be characterized by its energy, although more than one configuration may have the same energy. At zero temperature, the system seeks out the state of lowest energy, but at non-zero temperature the system samples many configurations as it exchanges energy with its environment. To determine the magnitude of these energy fluctuations, let's examine two systems, L and S, in thermal contact only with each other, as in Fig. C1.1. System S is the subject of our observations, while L is the reservoir; L has far more energy than S. Because L and S exchange energy only with each other, their total energy, E_{TOT}, is a constant:

$$E_{TOT} = E_L + E_S. \tag{C1.1}$$

We now specify E_{TOT} to be a particular value E_o, say $E_{TOT} = E_o = 10^6$ Joules. We are interested in a particular state r of the small system with energy E_r, such that

$$E_S = E_r \quad \text{and} \quad E_L = E_o - E_r. \tag{C1.2}$$

Just to make sure that the notation is clear: E_S, E_L and E_{TOT} are general parameters, while E_r and E_o represent a specific choice of energies.

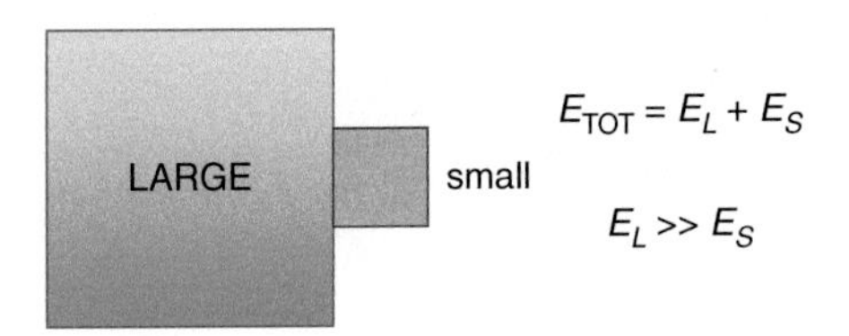

Fig. C1.1 Two subsystems (small and LARGE) in thermal contact can exchange energy; their energy is dominated by the large system.

We define the number of states of the large system having energy E_L to be

$$\Omega_L(E_L) = [\textit{number of states of large system with } E_L]. \tag{C1.3}$$

The number of states of the large system $\Omega_L(E_L)$ varies, perhaps even rapidly, with E_L, but its logarithm varies more slowly, and can be expanded in a series around the specific value $E_L = E_o$:

$$\ln \Omega_L(E_o - E_r) = \ln \Omega_L(E_o) - [\partial \ln \Omega_L / \partial E_L]_o\, E_r + \ldots, \tag{C1.4}$$

where the derivative of $\ln \Omega_L$ with respect to E_L is evaluated at $E_L = E_o$. The minus sign in front of the derivative arises because the energy E_L of the large system decreases by E_r when the energy of the small system increases by E_r. Higher order terms can be neglected since $E_r << E_o$. The derivative $[\partial \ln\Omega_L / \partial E_L]_o$ characterizes the large system around $E_L = E_o$, and does not depend on E_r; hence, it is notationally convenient to replace this derivative with a single symbol

$$[\partial \ln\Omega_L / \partial E_L]_o \equiv \beta. \tag{C1.5}$$

It can be shown that β^{-1}, bearing units of energy, has the properties of a temperature; the physical temperature scale is set through $T = 1 / k_B\beta$, where k_B is Boltzmann's constant ($k_B = 1.38 \times 10^{-23}$ J/K).

Returning now to Eq. (C1.4), the derivative on the right-hand side can be replaced by β

$$\ln \Omega_L(E_o - E_r) = \ln \Omega_L(E_o) - \beta E_r + \ldots,$$

$$\Rightarrow \Omega_L(E_o - E_r) = \Omega_L(E_o)\exp(-\beta E_r). \tag{C1.6}$$

The number of configurations available to a system affects its behavior at finite temperature. For instance, the molecules of a gas are more likely to be found scattered throughout the volume of a container than collected together in one of its corners, all other things being equal. Mathematically, we say that a system at fixed volume minimizes its free energy $E - TS$, rather than just minimizes its energy E, where S denotes entropy; an even more general expression for the free energy is needed if the volume or the number of particles is not fixed. Entropy increases logarithmically with the number of states Ω accessible to the system,

$$S = k_B \ln \Omega(\langle E \rangle), \tag{C1.7}$$

where Ω is a function of the mean energy $\langle E \rangle$ of the system (see Section 6.6 of Reif, 1965). Because TS enters the free energy with a minus sign, the free energy of a system falls as its entropy rises, a process that can occur spontaneously.

C1.2 Boltzmann factor

We have selected a particular state r of the small system with energy E_r. The probability $\mathcal{P}_r$ of the small system being in this state is proportional to the number of states of the *large* system having the appropriate value of E_L; that is, $\mathcal{P}_r \propto \Omega_L(E_o - E_r)$ or

$$\mathcal{P}_r \equiv A\Omega_L(E_0 - E_r), \tag{C1.8}$$

where the proportionality constant A is a characteristic of the *large* system. The value of A can be determined through the condition that the small system must always occupy an available state, although it may occupy different states as time passes:

$$\Sigma_r \mathcal{P}_r = 1, \tag{C1.9}$$

where the sum is over all of the states r available to the small system. We replace $\Omega_L(E_o - E_r)$ in Eq. (C1.8) by its functional dependence on E_r in Eq. (C1.6) to obtain

$$\mathcal{P}_r = A\Omega_L(E_0 - E_r) = [A\Omega_L(E_0)] \exp(-\beta E_r). \tag{C1.10}$$

This is the Boltzmann factor, and it shows that the probability of the small system being in a specific state r with energy E_r is a function of the energy of the state r and of the temperature of the large system with which it is in thermal contact. The two factors in the square braces of Eq. (C1.10) are both constants, and can be rolled into one as

$$\mathcal{P}_r = Z^{-1} \exp(-\beta E_r) \tag{C1.11}$$

with Z determined from

$$Z = \Sigma_r \exp(-\beta E_r), \tag{C1.12}$$

so that $\mathcal{P}_r$ is properly normalized to unity. The quantity Z involves a sum over the states available to the small system, and is defined as the partition function. Note that if several configurations have the same E_r, then *each configuration* has the same probability of occurence $\mathcal{P}(E_r)$. Equation (C1.11) tells us that if T is small, or β is correspondingly large, then the system is much more likely to be found with E_r near zero than with E_r

much greater than zero [for which $\exp(-\beta E_r)$ is tiny], all other things being equal.

The probability also can be related to the entropy by manipulation of Eq. (C1.11). Multiplying the logarithm of Eq. (C1.11) by $\mathcal{P}_r$ yields $-\mathcal{P}_r \ln \mathcal{P}_r = \beta \mathcal{P}_r E_r + \mathcal{P}_r \ln Z$, or

$$- \Sigma_r \mathcal{P}_r \ln \mathcal{P} = \beta \langle E \rangle + \ln Z, \tag{C1.13}$$

after summing over all states r and defining $\langle E \rangle \equiv \Sigma_r \mathcal{P}_r E_r$. Now, it can be established from thermodynamics that

$$S = k_B(\beta \langle E \rangle + \ln Z), \tag{C1.14}$$

(see Section 6.6 of Reif, 1965, for example). Thus, we expect

$$S = -k_B \, \Sigma_r \, \mathcal{P}_r \ln \mathcal{P}_r, \tag{C1.15}$$

by comparing Eqs. (C1.13) and (C1.14).

C1.3 Example: harmonic oscillator

As an application of the Boltzmann factor, we consider the one-dimensional motion of a particle in the quadratic potential $V(x)$ characteristic of Hooke's law for springs,

$$V(x) = k_{sp} x^2 / 2, \tag{C1.16}$$

where k_{sp} is the spring constant and x is the displacement from equilibrium. For a given value of x, Eq. (C1.16) is the corresponding potential energy, and the Boltzmann factor $\exp(-\beta k_{sp} x^2/2)$ provides the likelihood that the particle can be found with that energy.

Now, x is a continuous variable, so one talks of the probability $\mathcal{P}(x)\mathrm{d}x$ of the particle having a displacement between x and $x + \mathrm{d}x$. The continuous versions of Eqs. (C1.11) and (C1.12) become

$$\mathcal{P}(x)\mathrm{d}x = \mathrm{d}x \bullet \exp(-\alpha x^2) / \int \exp(-\alpha x^2)\, \mathrm{d}x, \tag{C1.17}$$

where the combination $\beta k_{sp}/2$ has been replaced with a single constant α:

$$\alpha \equiv \beta k_{sp}/2. \tag{C1.18}$$

Equation (C1.17) can be used to evaluate the fluctuations in x about $x = 0$:

$$\begin{aligned} \langle x^2 \rangle &= \int x^2 \, \mathcal{P}(x)\, \mathrm{d}x \\ &= \alpha^{-1} \int z^2 \exp(-z^2)\, \mathrm{d}z / \int \exp(-z^2)\, \mathrm{d}z \\ &= \alpha^{-1} (\sqrt{\pi} / 2) / \sqrt{\pi} = 1 / 2\alpha, \end{aligned} \tag{C1.19}$$

whence

$$\langle x^2 \rangle = 1 / \beta k_{sp} = k_B T / k_{sp}. \qquad (C1.20)$$

Equation (C1.20) shows that $\langle x^2 \rangle$ increases with temperature T, as expected

- high temperature $\Rightarrow$ large fluctuations in x
- low temperature $\Rightarrow$ small fluctuations in x
- zero temperature $\Rightarrow$ no fluctuations in x (ground state at $x = 0$).

C1.4 Partition function

The partition function introduced in Eq. (C1.12) allows many thermodynamic quantities of interest to be written in a compact form. As an example, we determine the ensemble average of the system's energy, $\langle E \rangle$, which is the sum of the energy of each state according to the probability of its occurence: $\langle E \rangle = \Sigma_r E_r \mathcal{P}_r$, or

$$\langle E \rangle = Z^{-1} \Sigma_r E_r \exp(-\beta E_r). \qquad (C1.21)$$

Each term in the sum, $E_r \exp(-\beta E_r)$, can be written as the (negative of the) derivative of $\exp(-\beta E_r)$ with respect to β, so that

$$\begin{aligned} \langle E \rangle &= -Z^{-1} \Sigma_r (\partial / \partial \beta) \exp(-\beta E_r) \\ &= -Z^{-1} (\partial / \partial \beta) \Sigma_r \exp(-\beta E_r) = -Z^{-1} (\partial Z / \partial \beta), \end{aligned} \qquad (C1.22)$$

or, equivalently,

$$\langle E \rangle = -\partial \ln Z / \partial \beta. \qquad (C1.23)$$

Similarly compact forms occur for other observables.

The sum-over-states in Eq. (C1.12) involves a discrete set of states r. What happens in a classical system where there are many particles, each described by a position vector $\mathbf{r}$ and a momentum vector $\mathbf{p}$? In most systems, the potential energy of a particle depends upon its position, and the kinetic energy depends upon its momentum. If these variables are continuous, the sum over r must be replaced by a continuous integral over $\mathbf{r}_i$ and $\mathbf{p}_i$ for each of the N particles in the system, as in

$$\exp[-\beta E(\mathbf{r}_1, \mathbf{r}_2, \ldots \mathbf{r}_N, \mathbf{p}_1, \mathbf{p}_2, \ldots \mathbf{p}_N)] \, d\mathbf{r}_1 \, d\mathbf{r}_2 \ldots d\mathbf{r}_N \, d\mathbf{p}_1 \, d\mathbf{p}_2 \ldots d\mathbf{p}_N.$$

As it stands, this expression has dimensions, whereas the partition function should be dimensionless. The complete expression for the partition function must therefore include a normalization factor reflecting the density of states: the number of states per unit volume of phase space (each point in *phase space* represents both a position and a momentum of a particle). We parametrize the phase space volume occupied by one state as h^3, without proving that the parameter h is actually Planck's constant, so that the normalized partition function becomes

$$Z = (N!)^{-1} h^{-3N} \int\int \cdots \int \exp[-\beta E(\mathbf{r}_1, \mathbf{r}_2, \ldots \mathbf{r}_N, \mathbf{p}_1, \mathbf{p}_2, \ldots \mathbf{p}_N)] \, \mathrm{d}\mathbf{r}_1 \, \mathrm{d}\mathbf{r}_2 \cdots \mathrm{d}\mathbf{r}_N \, \mathrm{d}\mathbf{p}_1 \, \mathrm{d}\mathbf{p}_2 \ldots \mathrm{d}\mathbf{p}_N. \quad \text{(C1.24)}$$

The factor of $N!$ in Eq. (C1.24) arises from the fact that, at the molecular level, the particles are indistinguishable, and the $N!$ possible permutations implied by the integral do not correspond to physically distinguishable states (see Section 7.3 of Reif, 1965, for further details).

C1.5 Example: entropy of an ideal gas

For most situations of interest in this text, the potential energy depends only on the positions of the particles, not their momenta. In other words, the conventional problem that one has to solve is described by a total energy, E, with

$$E = (p_1^2/2m + p_2^2/2m + \cdots + p_N^2/2m) + V(\mathbf{r}_1, \mathbf{r}_2, \ldots \mathbf{r}_N), \quad \text{(C1.25)}$$

where m is the mass of an individual particle (taken to be identical) and $V(\mathbf{r}_1, \mathbf{r}_2, \ldots \mathbf{r}_N)$ is the total potential energy. Since the exponential of a sum is equal to the product of individual exponentials, the partition function then can be written as a product of $3N$ identical momentum integrals Z_p along with a $3N$-dimensional spatial integral Z_c:

$$Z = (N!)^{-1} Z_p^{3N} Z_c, \quad \text{(C1.26)}$$

where

$$Z_p = h^{-1} \int \exp(-\beta p^2/2m) \, \mathrm{d}p \quad \text{(C1.27)}$$

$$Z_c = \int\int \cdots \int \exp[-\beta V(\mathbf{r}_1, \mathbf{r}_2, \ldots \mathbf{r}_N)] \, \mathrm{d}\mathbf{r}_1 \, \mathrm{d}\mathbf{r}_2 \ldots \mathrm{d}\mathbf{r}_N. \quad \text{(C1.28)}$$

The $3N$ identical momentum integrals arise from the fact that the momentum of each particle is independent and has three Cartesian components ($p^2 = p_x^2 + p_y^2 + p_z^2$).

One can obtain Z_p using the expression $\int \exp(-x^2) dx = \sqrt{\pi}$, where the integration extends from $-\infty$ to $+\infty$. Changing variables in Eq. (C1.27) to make the integral dimensionless, we find

$$Z_p = h^{-1} (2\pi m/\beta)^{1/2}. \quad \text{(C1.29)}$$

For an ideal gas, all interactions between particles vanish, $V(\mathbf{r}_1, \mathbf{r}_2, \ldots \mathbf{r}_N) = 0$, so that the coordinate-space partition function Z_c just involves N integrals over the volume available to each particle:

$$Z_c = \int_V \mathrm{d}\mathbf{r}_1 \int_V \mathrm{d}\mathbf{r}_2 \ldots \int_V \mathrm{d}\mathbf{r}_N = V^N. \quad \text{(C1.30)}$$

Thus, the complete partition function is

$$Z = (N!)^{-1} V^N h^{-3N} (2\pi m/\beta)^{3N/2}. \quad (C1.31)$$

The quantities that we wish to calculate involve the logarithm of Z,

$$\ln Z = N [\ln V - 3\ln h - 3/2 \ln\beta + 3/2 \ln(2\pi m)] - \ln N! \quad (C1.32)$$

This expression can be simplified by using Sterling's approximation for the logarithm of the factorial at large N,

$$\ln N! = N \ln N - N, \quad (C1.33)$$

which is a familiar tool in statistical mechanics, and leads to

$$\ln Z = N [\ln(V/N) - 3/2 \ln\beta + 3/2 \ln(2\pi m/h^2) + 1]. \quad (C1.34)$$

Equation (C1.34) can be substituted into Eq. (C1.33) to obtain the average energy

$$\langle E \rangle = -\partial \ln Z / \partial\beta = 3/2\ N/\beta = 3/2\ Nk_BT, \quad (C1.35)$$

which tells us that the average kinetic energy per particle is $3/2\ k_BT$. The entropy of the gas can be found by returning to Eq. (C1.14), $S = k_B(\ln Z + \beta\langle E \rangle)$. Substituting Eq. (C1.34) and $\beta\langle E \rangle = 3/2\ N$ from Eq. (C1.35), we obtain

$$S = k_BN[\ln(V/N) + 3/2 \ln T + 3/2 \ln(2\pi mk_B/h^2) + 5/2], \quad (C1.36)$$

for the entropy of an ideal gas.

D Appendix D Elasticity

Biological materials are generally soft, meaning that the shapes of objects made from them, such as strings and sheets, are moderately deformable under conditions commonly arising in the cell. Various aspects of the mathematical representation of deformations have been introduced in several chapters of this text (notably Sections 3.2, 4.2, 5.2, 6.2 and 8.2), although not in as much detail as would be found in a standard text on elasticity. Consequently, several calculations and fundamental results from continuum mechanics were quoted in this book without proof. To rectify this somewhat unsatisfactory situation, we develop in Appendix D:

- the symmetry relations among elastic constants for materials with four-fold and six-fold symmetry in two dimensions,
- the symmetry relations among elastic constants for materials with four-fold symmetry in three dimensions,
- the relation between shape fluctuations and elastic constants.

There is considerable variation in the notation and analytic approaches adopted by the spectrum of disciplines that make use of continuum mechanics; the notational convention adopted here is close to that of Landau and Lifshitz (1986).

D1.1 Deformations and the strain tensor

As introduced in Sections 5.2 and 6.2, the deformation of an object can be described by means of a displacement vector **u**, which is the difference in position of an element of the object arising from the deformation. The displacement is not universal and **u** is a function of position **x** on the object. A non-vanishing value of **u** does not immediately imply that the object is deformed; for example, constant **u** is just a translation of the object, not a deformation. Thus, it is the rate of change of **u** with **x** that is a measure of the deformation, not the magnitude of **u** itself.

In Cartesian coordinates, **u** has as many components u_x, u_y... as there are dimensions of the system, and each of these components can vary with each of x, y... independently. For example, a simple shear in two dimensions involves the variation of u_y with x or u_x with y. Thus, one way to represent a general deformation is through a tensor quantity u_{ij} (the strain tensor) that involves all first-order derivatives of **u** with respect to **x**. The general definition of u_{ij} is given in Chapter 5, but for many situations of interest, the deformations are small and the strain tensor can be approximated by

$$u_{ij} = 1/2\ (\partial u_i/\partial x_j + \partial u_j/\partial x_i). \tag{D1.1}$$

Just as the displacement **u** varies with position, so does the strain tensor u_{ij}.

The magnitude of the deformation, and hence the magnitude of the strain tensor, depends on both the forces experienced by the object as well as its material characteristics. Further, account must taken of the direction of an external force with respect to the orientation of the surface to which it is applied, as well as the surface area over which it is spread. In Chapter 6, the stress tensor σ_{ij} was introduced as a means of accounting for these two aspects of the force: σ_{ij} is the applied force per unit area, taking into account the direction of the force with respect to the surface. The component of the force in the i-direction, F_i, is given in terms of σ_{ij} by

$$F_i = \Sigma_j\ \sigma_{ij}\ a_j. \tag{D1.2}$$

The sum is over the components of the surface area vector **a**, which has a direction perpendicular to the surface and a magnitude equal to the surface area. The stress tensor has units of energy density and is symmetric in indices i and j.

At small deformations, the stress is linearly proportional to the strain, which is the same form as Hooke's Law for ideal springs: there, the restoring force F (the stress) is proportional to the displacement from equilibrium x (the strain). However, the general form of Hooke's law involves tensors, so that the simple spring constant k_{sp} present in $F = k_{sp}x$ must be replaced by more complex proportionality constants. There is a choice of conventions for defining the generalized spring constants: one can write

$$u_{ij} = \Sigma_{k,l}\ S_{ijkl}\sigma_{kl}, \tag{D1.3}$$

where the constants S_{ijkl} are called the elastic compliance constants (or elastic constants), or one can choose

$$\sigma_{ij} = \Sigma_{k,l}\ C_{ijkl}u_{kl}, \tag{D1.4}$$

where the constants C_{ijkl} are called the elastic stiffness constants (or elastic moduli). In this text, we use the elastic moduli C_{ijkl} which have units of [*energy*]/[*area*] in two dimensions or [*energy*]/[*volume*] in three dimensions. The elastic moduli determine the energy associated with a deformation: the change in the free energy density $\Delta\mathcal{F}$ of a continuous object under deformation is quadratic in the strain tensor u_{ij}:

$$\Delta\mathcal{F} = 1/2\ \Sigma_{i,j,k,l}\ C_{ijkl}\ u_{ij}\ u_{kl}. \tag{D1.5}$$

This is similar to the potential energy of a Hooke's Law spring, which is quadratic in the square of the displacement from equilibrium.

The elastic moduli are necessarily more complex than the single k_{sp} of an isolated spring: collectively, C_{ijkl} of a deformable object must be able to accommodate situations where the applied stress in one direction generates a strain in another, like the squeezing of jelly in one's hand. However, the number of indices on C_{ijkl} should not intimidate the reader: symmetry considerations greatly reduce the number of independent components from the $3^4 = 81$ terms naively expected in three dimensions, or $2^4 = 16$ expected in two dimensions. To show this, we first consider symmetry relations arising from Eq. (D1.5) that can be applied to all materials. Because u_{ij} is symmetric under exchange of i and j, then according to Eq. (D1.5), C_{ijkl} can be defined such that it is pairwise symmetric under exchange of i and j, or k and l; that is,

$$C_{ijkl} = C_{jikl} = C_{ijlk}. \tag{D1.6}$$

Further, because the product $u_{ij}u_{kl}$ in Eq. (D1.5) is symmetric under exchange of the pairs of indices ij and kl, the moduli must obey

$$C_{ijkl} = C_{klij}. \tag{D1.7}$$

The set of conditions in Eqs. (D1.6) and (D1.7) decreases the number of independent values of C_{ijkl} to 21 in three dimensions, and 6 in two dimensions. Specifically, in two dimensions, the six independent moduli are

$$\begin{aligned}
&C_{xxxx} \quad C_{yyyy} \quad C_{xxyy} = C_{yyxx} \\
&C_{xyxy} = C_{xyyx} = C_{yxyx} = C_{yxxy} \\
&C_{xxxy} = C_{xxyx} = C_{xyxx} = C_{yxxx} \\
&C_{yyyx} = C_{yyxy} = C_{yxyy} = C_{xyyy}.
\end{aligned} \tag{D1.8}$$

Symmetries of the material itself further reduce the number of elastic moduli, as we now demonstrate for two-dimensional networks with four-fold and six-fold connectivity. Although the word "network" is used to describe the materials in the following examples, the results apply to any system with the symmetries of the examples.

D1.2 Six-fold networks in 2D

The two-dimensional network illustrated in Fig. D1.1(a) possesses six-fold axes of rotational symmetry through the vertices. To determine the effects of this symmetry, we follow Landau and Lifshitz (1986) and change from Cartesian coordinates x and y to complex coordinates ξ and η, where

(a)
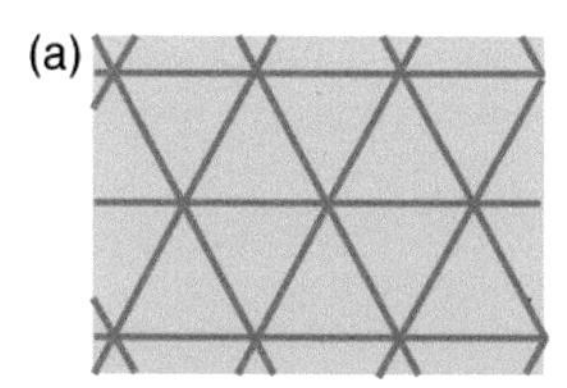

(b)
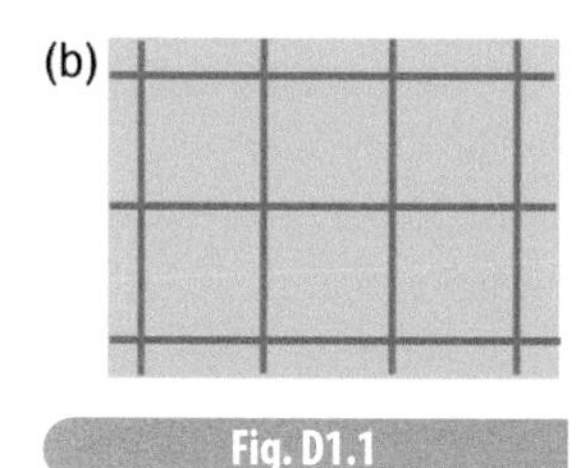

Fig. D1.1

Two-dimensional networks with six-fold (a) and four-fold symmetry (b).

$$\xi \equiv x + \mathrm{i}y \qquad \eta \equiv x - \mathrm{i}y, \tag{D1.9}$$

such that the free energy involves products of the form $C_{\xi\xi\eta\eta}u_{\xi\xi}u_{\eta\eta}$ and others. Now, a rotation about the xy-coordinate origin by an angle θ changes the coordinates (x, y) of a given position to $(x\cos\theta - y\sin\theta, x\sin\theta + y\cos\theta)$ or, equivalently, changes the coordinates (ξ, η) to

$$\xi \rightarrow \xi\exp(\mathrm{i}\theta) \qquad \eta \rightarrow \eta\exp(-\mathrm{i}\theta). \tag{D1.10}$$

Specifically, six-fold symmetry demands that the moduli be invariant under rotations through $\theta = \pi/3$, or $\xi \rightarrow \xi\exp(\mathrm{i}\pi/3)$ and $\eta \rightarrow \eta\exp(-\mathrm{i}\pi/3)$. The only non-zero components of C_{ijkl} that remain unchanged by this transformation must contain ξ and η the same number of times, since $\exp(\mathrm{i}\pi/3)\bullet\exp(-\mathrm{i}\pi/3) = 1$. Hence only two components of C_{ijkl} are invariant under the symmetry of the network, namely the pair $C_{\xi\xi\eta\eta}$ and $C_{\xi\eta\xi\eta}$ and terms related by symmetry of the indices.

Expressed in the $\xi\eta$ representation, the change in the free energy density $\Delta\mathcal{F}$ from Eq. (D1.5) becomes

$$\Delta\mathcal{F} = 2C_{\xi\eta\xi\eta}u_{\xi\eta}u_{\xi\eta} + C_{\xi\xi\eta\eta}u_{\xi\xi}u_{\eta\eta}, \tag{D1.11}$$

where the first term results from four combinations of $C_{\xi\eta\xi\eta}$ and the second term is from two combinations of $C_{\xi\xi\eta\eta}$; the expression includes the normalization factor of 1/2 from Eq. (D1.5). To replace $u_{\xi\xi}$, etc., in Eq. (D1.11) by the Cartesian components of the strain tensor, we use the fact that the components of a tensor transform as the products of the corresponding coordinates. That is, because $\xi^2 = (x + \mathrm{i}y)^2 = x^2 - y^2 + 2\mathrm{i}xy$, then

$$u_{\xi\xi} = u_{xx} - u_{yy} + 2\mathrm{i}u_{xy}, \tag{D1.12}$$

and similarly with $u_{\xi\eta}$ and $u_{\eta\eta}$, permitting Eq. (D1.11) to be written as

$$\Delta\mathcal{F} = 2C_{\xi\eta\xi\eta}\,(u_{xx} + u_{yy})^2 + C_{\xi\xi\eta\eta}\,\{(u_{xx} - u_{yy})^2 + 4u_{xy}^{\ 2}\}. \tag{D1.13}$$

As introduced in Chapter 5, the trace of the strain tensor is equal to the relative change in area associated with the deformation (in two dimensions) or the relative change in volume (in three dimensions). Thus, $u_{xx} + u_{yy}$ in Eq. (D1.13) represents an expansion or compression and $C_{\xi\eta\xi\eta}$ must be proportional to the area compression modulus K_A. Similarly, the combinations of the strain tensor in the second term of Eq. (D1.13) represent shear deformations, so $C_{\xi\xi\eta\eta}$ must be proportional to the two-dimensional shear modulus μ. The exact relations between C_{ijkl} and the compression and shear moduli are

$$K_A = 4C_{\xi\eta\xi\eta} \qquad \mu = 2C_{\xi\xi\eta\eta}, \tag{D1.14}$$

so that Eq. (D1.13) becomes

$$\Delta\mathcal{F} = (K_A/2)(u_{xx} + u_{yy})^2 + \mu\,\{(u_{xx} - u_{yy})^2/2 + 2u_{xy}^{\ 2}\} \qquad \text{(six-fold symmetry).} \tag{D1.15}$$

This is the same expression as that for isotropic materials; that is, in two dimensions, isotropic materials and those with six-fold symmetry are both described by two elastic moduli at small deformations.

D1.3 Four-fold networks in 2D

As we established above from general symmetry considerations, two-dimensional materials with four-fold symmetry, as displayed in Fig. D1.1(b), have, at most, the six independent moduli given by Eq. (D1.8). This number of independent moduli is reduced by additional symmetries, including invariance under each of the coordinate inversions $x \rightarrow -x$ and $y \rightarrow -y$ independently. Hence, any component of C_{ijkl} with an odd number of x or y indices changes sign under a single application of the inversions, because the components of a tensor transform as products of the corresponding coordinates. Now, the mechanical properties of a material should not change sign upon reflection, so that those components with an odd number of x or y indices must vanish, reducing the number of independent components to four. Lastly, clockwise rotations by an angle of $\pi/2$ about the four-fold symmetry axis results in $x \rightarrow -y$ and $y \rightarrow x$, which implies that $C_{xxxx} = C_{yyyy}$. The moduli for pure deformations can be written in terms of these three C_{ijkl} by

$$K_A = (C_{xxxx} + C_{xxyy}) / 2$$

$$\mu_p = (C_{xxxx} - C_{xxyy}) / 2 \qquad \text{(pure shear)} \qquad \text{(D1.16)}$$

$$\mu_s = C_{xyxy} \qquad \text{(simple shear)},$$

so that the free energy density $\Delta\mathcal{F}$ in Eq. (D1.5) becomes

$$\Delta\mathcal{F} = (K_A/2)(u_{xx} + u_{yy})^2 + (\mu_p/2)(u_{xx} - u_{yy})^2 + 2\mu_s u_{xy}^2 \qquad \text{(four-fold symmetry).} \qquad \text{(D1.17)}$$

There are two shear moduli present in Eq. (D1.17), μ_p and μ_s, reflecting the resistance against pure shear and simple shear, respectively, introduced in Chapter 5.

D1.4 Four-fold networks in 3D

In three dimensions, four-fold (or cubic) systems require only three moduli, just as four-fold systems do in two dimensions. In fact, the expression for the free-energy density is a straightforward generalization of the two-dimensional result, becoming (Landau and Lifshitz, 1986, p. 35)

$$\Delta\mathcal{F} = (C_{xxxx}/2) \bullet (u_{xx}^2 + u_{yy}^2 + u_{zz}^2) + C_{xxyy}(u_{xx}u_{yy} + u_{xx}u_{zz} + u_{yy}u_{zz})$$
$$+ 2C_{xyxy}(u_{xy}^2 + u_{xz}^2 + u_{yz}^2). \quad \text{(D1.18)}$$

Although this equation clearly expresses its historical roots in Eq. (D1.5), it does not directly reflect simple deformations. By taking combinations of elastic moduli and strain tensor components, Eq. (D1.18) can be rewritten

$$\Delta\mathcal{F} = (K_V/2) \bullet (u_{xx} + u_{yy} + u_{zz})^2 + (\mu'/3)[(u_{xx} - u_{yy})^2 + (u_{xx} - u_{zz})^2 + (u_{yy} - u_{zz})^2]$$
$$+ 2\mu_s(u_{xy}^2 + u_{xz}^2 + u_{yz}^2), \quad \text{(D1.19)}$$

where

$$K_V = (C_{xxxx} + 2C_{xxyy})/3 \quad \mu' = (C_{xxxx} - C_{xxyy})/2 \qquad \mu_s = C_{xyxy}. \quad \text{(D1.20)}$$

With this definition of the elastic moduli, Eq. (D1.19) reduces to the isotropic situation in Section 6.2 when $\mu' = \mu_s = \mu$.

D1.5 Fluctuations

The strain tensor describes a deformation from one specific configuration to another. However, systems at finite temperature exchange energy with their surroundings, allowing their shapes and volumes to fluctuate. Thus, the deformation of an object under stress at non-zero temperature reflects changes in the average positions of its elements, owing to thermal fluctuations. The magnitude of the shape fluctuations is inversely proportional to the corresponding elastic modulus of the object, as can be seen by considering fluctuations in volume. The volume compression modulus (at constant temperature) K_V is related to the change in volume V with respect to pressure P via

$$K_V^{-1} = -V^{-1} \, (\partial V / \partial P)_T. \quad \text{(D1.21)}$$

That is, the smaller the change in volume for a given change in pressure (i.e., $\partial V / \partial P$ small), the larger the compression modulus. In terms of fluctuations, this means that a system with a large K_V undergoes only small fluctuations in volume if held at a fixed pressure as illustrated in Fig. D1.2.

We obtain a relation between the compression modulus and volume fluctuations following Reif (1965). Consider a small system at pressure P in thermal and mechanical contact with a large reservoir at pressure P_R, as displayed in Fig. D1.3. Being in thermal contact, both systems have the same temperature T, a result which we do not bother to prove. The small system has a volume V, which it is free to exchange with the large system by moving a piston connecting the systems. As the piston moves, the pressure of the small system may change, but the pressure of the reservoir does

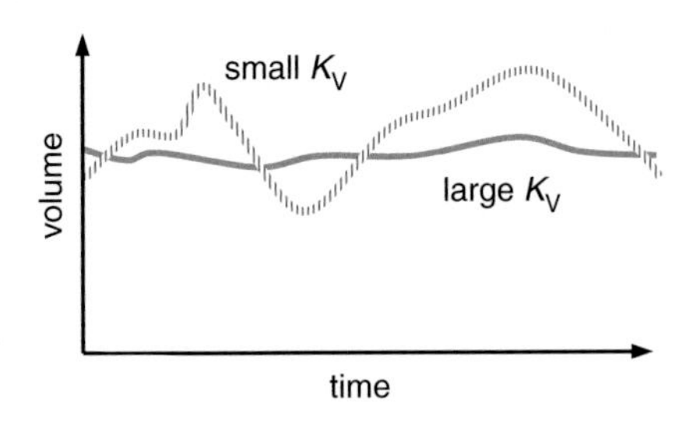

Fig. D1.2 Fluctuations in volume as a function of time for different compression moduli.

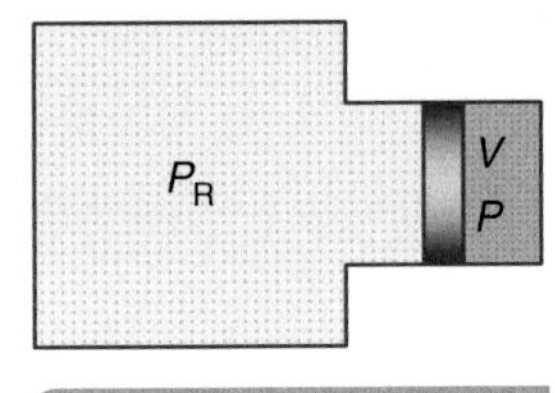

Fig. D1.3

A small system with volume V at pressure P in contact with a large reservoir at fixed pressure P_R. The piston is able to slide and conduct heat, allowing the small system to exchange energy and volume with the reservoir. Both systems have the same temperature.

not. We define a Gibbs free energy for the small system, G_S, in terms of the temperature and pressure of the reservoir,

$$G_S = E + P_R V - TS, \tag{D1.22}$$

where G_S is a function of V at fixed T.

For a given pressure and temperature of the reservoir, the volume of the small system fluctuates about a value V_o that minimizes G_S at a value G_{min}. That is, writing the free energy as a function of the volume, $G_S(V)$, then

$$G_{min} = G_S(V_o). \tag{D1.23}$$

For small fluctuations around V_o, the free energy can be expanded as a power series in the change in volume ΔV,

$$G_S(V) = G_{min} + (\partial G_S / \partial V)_T \, \Delta V + 1/2 \, (\partial^2 G_S / \partial V^2)_T \, (\Delta V)^2, \tag{D1.24}$$

where

$$\Delta V = V - V_o, \tag{D1.25}$$

and the notation $(\ldots)_T$ indicates that the derivative is taken at constant temperature. To find G_{min}, we must evaluate derivatives of G_S with respect to V, obtaining

$$(\partial G_S / \partial V)_T = (\partial E / \partial V)_T + P_R - T(\partial S / \partial V)_T, \tag{D1.26}$$

from Eq. (D1.22). However, the fundamental relation of thermodynamics

$$T \, dS = dE + P \, dV, \tag{D1.27}$$

applied to the small system gives

$$T \, (\partial S / \partial V)_T = (\partial E / \partial V)_T + P, \tag{D1.28}$$

so that the first derivative of G_S becomes

$$(\partial G_S / \partial V)_T = P_R - P, \tag{D1.29}$$

when Eq. (D1.28) is substituted into Eq. (D1.26).

For G_S to be a minimum at V_o, the first derivative of G_S must vanish, and hence $P_R = P$, according to Eq. (D1.29). Just as importantly, Eq. (D1.29) can be differentiated to yield an expression for the second derivative of G_S,

$$(\partial^2 G_S / \partial V^2)_T = -(\partial P / \partial V)_T. \tag{D1.30}$$

The first and second derivatives of G_S now can be substituted into Eq. (D1.24) to give

$$\Delta G_S = G_S - G_{min} = -1/2\, (\partial P / \partial V)_T\, (\Delta V)^2, \tag{D1.31}$$

or, in terms of the volume compression modulus from Eq. (D1.21).

$$\Delta G_S = +K_V\, (\Delta V)^2 / 2V_o. \tag{D1.32}$$

Knowing the change in free energy with volume allows us to determine the relative probability for the system to have a particular volume; as discussed in Appendix C on statistical mechanics, the relative probability $\mathcal{P}(V)$ for the system to have a volume V is

$$\mathcal{P}(V) \sim \exp(-\beta \Delta G_S) \sim \exp\{-\beta K_V\, (\Delta V)^2 / 2V_o\}. \tag{D1.33}$$

Having a Gaussian form, the probability distribution can be integrated easily to give the dispersion in the volume. The steps are the same as those in evaluating the mean square displacement of the harmonic oscillator in Appendix C, and yield

$$\langle (\Delta V)^2 \rangle = V_o / \beta K_V, \tag{D1.34}$$

or in the more familiar form

$$1/\beta K_V = \langle (\Delta V)^2 \rangle / V_o. \tag{D1.35}$$

As expected, this expression shows that the fluctuations in V rise as the compression modulus falls.

Glossary

α-actinin rod-shaped actin-binding protein which links parallel actin filaments into bundles.

actin protein of mass 42 kDa; forms a polymeric filament about 8 nm in diameter; mechanical element of the cytoskeleton, ubiquitous from yeasts to humans; major component of muscle tissue.

actin-binding protein type of protein capable of binding to actin, often to form a network or bundle.

ADP (adenosine diphosphate; see ATP).

amino acid organic molecule containing both carboxyl and amino groups attached to the same carbon atom (see Appendix B for a summary of the 20 amino acids commonly found in proteins).

amphiphile molecule with both polar and non-polar regions; simultaneously attractive to immiscible fluids such as oil and water.

anisotropic a material or medium whose properties are direction-dependent.

ankyrin protein that links β-spectrin filaments of the red cell cytoskeleton to the protein band 3 embedded in the plasma membrane.

archaebacteria a subkingdom of bacteria capable of thriving in environments of extreme acidity or temperature.

arc length (s) distance along a curve following its contour.

ATP (adenosine triphosphate) along with GTP, a fundamental unit of energy currency of the cell, releasing energy upon hydrolysis to ADP.

avidin tetrameric protein of mass 68 kDa capable of forming adhesive bonds with ligands on adjacent membranes (similar to **streptavidin**).

Avogadro's number (N_A) number of particles in a mole, 6.022×10^{23} mol^{-1}.

axisymmetric unchanged by rotation about an axis (e.g. a cylinder).

axon long, string-like section of a nerve cell extending from the soma (which contains the cell's nucleus).

bacteria single-celled micro-organisms without a nucleus, classified into the subkingdoms of archaebacteria and eubacteria.

basal lamina thin mat of interconnected proteins adjacent to an epithelial sheet, separating it from connective tissue (see Appendix A).

base pair complementary organic bases linking two strands of DNA by means of hydrogen bonds.

bending rigidity (κ_b) elastic constant characterizing a membrane's resistance to a change in its mean curvature; commonly has a value of 10^{-19} J for biomembranes.

bilayer quasi-two-dimensional structure composed of two monolayers, each an extended array just a single molecule in thickness.

biotin a vitamin of molecular mass 0.244 kDa; binds tightly (but not covalently) to the protein avidin.

Bjerrum length (ℓ_B) temperature-dependent length scale for a medium of permittivity ε containing ions of charge q; $\ell_B \equiv q^2 / 4\pi\varepsilon k_B T$.

Boltzmann constant (k_B) fundamental physical constant equal to 1.38×10^{-23} J/K; classically, $k_B T/2$ is the mean thermal energy of each independent mode of motion.

Boltzmann factor [$\exp(-\Delta E/k_B T)$] relative probability for a system at temperature T to occupy a specific state having energy ΔE with respect to a reference state (derived in Appendix C).

branched polymer polymeric structure whose chains have multiple branch points.

cadherin adhesion protein present in desmosomes of animal cells.

cell wall stress-bearing fabric enclosing the plasma membranes of most plant cells and bacteria, but not animal cells; composition varies according to cell type.

cellulose principal component of the plant cell wall, consisting of bundles of linear polysaccharides.

centriole cylindrical structure present (usually in pairs) in the centrosome of a cell.

centrosome diffuse region of the cell near the nucleus, from which radiate microtubule components of the cell's cytoskeleton.

chloroplast organelle found in plant and algae cells; site of photosynthesis.

cholesterol a type of sterol; often present in cell membranes.

clathrin a three-armed protein that assembles into a two-dimensional coat on a local region of a membrane during the formation of a vesicle in endocytosis.

collagen fibrous protein found in connective tissue; the most abundant protein in animals.

compression deformation mode characterized by a change in area or volume, depending on the dimensionality of the system; magnitude of deformation is determined by a compression modulus K_A or K_V in two or three dimensions, respectively.

connective tissue heterogeneous tissue in animal cells providing mechanical support, among its other roles.

contour length (L_c) total length of a filament as measured along its contour.

critical micelle concentration (CMC) in aqueous solutions of amphiphiles, the minimum concentration for the formation of condensed phases such as micelles.

curvature (C) measure of the rate of change with arc length of a unit tangent vector to a line or normal vector to a surface; at any location on a surface, there are two **principal curvatures**, C_1 and C_2, which may be combined as the **mean curvature** ($C_1 + C_2$) or **Gaussian curvature** (C_1C_2).

cytoplasm interior contents of the cell, including the cytosol and most organelles but excluding the nucleus.

cytoskeleton filamentous network permeating the cell's interior; composed of microtubules, intermediate filaments, actin and sometimes spectrin, depending on the cell.

cytosol heterogeneous fluid component of the cell's cytoplasm, excluding organelles.

dalton (Da) elementary unit of atomic mass, defined as 1/12 of the mass of the ^{12}C-atom; equal to 1.66054×10^{-27} kg.

Debye length (ℓ_D) length scale characterizing the exponential decay of the electrostatic potential in an electrolyte; for a neutral electrolyte of monovalent ions, $\ell_D^{-2} \equiv 8\pi\ell_B\rho_s$, where ℓ_B is the Bjerrum length and ρ_s is the number density of the ions.

dendrites branched structures emanating from the main cell body of a neuron; collect signals from adjacent cells and propagate them to the cell body.

desmosome local junction between adjacent cells (as in epithelia), mechanically linking their cytoskeletons.

diagonal matrix a matrix m_{ij} with vanishing off-diagonal elements ($i \neq j$).

diffusion constant (D) proportionality constant relating the mean square displacement $\langle \mathbf{r}^2 \rangle$ of a randomly moving object to the elapsed time t of its movement; $\langle \mathbf{r}^2 \rangle = 2dDt$, where d is the embedding dimension.

dimer a composite of two molecular subunits, which, in the case of **heterodimers**, need not be identical.

DLVO theory theory of colloids based on electrostatic and van der Waals interactions; formulated by Derjaguin, Landau, Verway, and Overbeek.

DNA (deoxyribonucleic acid) genetic material of the cell consisting of a linear chain whose elemental unit is a troika of a sugar derivative (deoxyribose), an organic base, and a phosphate group; pairs of chains intertwine as a double helix.

echinocyte a conformation of the red blood cell adopted under unusual conditions and characterized by the appearance of bumps on its surface.

edge energy energy associated with the free boundary of a surface; in its simplest form, equal to $\lambda\Gamma$, where Γ is the length of the boundary and λ is the material-specific edge tension (energy per unit length).

elastic moduli (C_{ijkl}) general set of elastic constants relating stress to strain; for Hooke's law materials, $\sigma_{ij} = \Sigma_{k,l} C_{ijkl}\, u_{kl}$, where σ_{ij} and u_{kl} are the stress and strain tensors, respectively.

electrolyte a solution or molten state that conducts electricity owing to the presence of mobile ions.

electrostatic potential (ψ) for a charged object, $\psi = V/q$, where V is the potential energy of the object and q is its charge.

endocytosis process in which material enters a cell via the formation of vesicles by invagination of the plasma membrane.

endoplasmic reticulum an organelle with an extensive membrane area, attached to the nuclear envelope; location of protein synthesis.

end-to-end displacement ($\mathbf{r}_{ee}$) displacement vector between the ends of a linear polymer.

enthalpy (H) in thermodynamics, the sum of the internal energy E of a system and the work of deformation under pressure P ($H = E + PA$ or $H = E + PV$ in two or three dimensions, respectively).

entropy (S) in statistical mechanics, a measure of the number of configurations Ω available to a system; $S = k_B \ln\Omega$, where k_B is Boltzmann's constant (see Gibbs free energy).

epithelium a sheet-like tissue, such as the skin or intestinal wall, isolating the body from its environment.

equilateral having sides of equal length, as in equilateral triangle.

erythrocyte blood cell containing hemoglobin, responsible for oxygen transport in the circulatory system.

eukaryote cell with a distinct nucleus containing most, but not all, of its DNA; all cells are eukaryotic except bacteria.

exocytosis process by which material is expelled from the cell by the fusion of vesicles with its plasma membrane.

extracellular matrix stress-bearing fabric secreted by cells, composed of proteins and polysaccharides.

fatty acid organic molecule consisting of a single hydrocarbon chain terminating at a carboxyl group (−COOH).

fibroblast cell commonly found in connective tissue; function includes secretion of extracellular matrix.

filamin as a V-shaped dimer, a protein capable of cross-linking actin filaments at wide angles.

fimbrin short protein that links parallel actin filaments into dense bundles.

flagellum long, whip-like structure extending from a cell; undulations of the flagellum may generate cell locomotion.

flexural rigidity (κ_f) material parameter characterizing the bending resistance of a filament; $\kappa_f = \pi Y R^4/4$ for a solid cylindrical rod of radius R and Young's modulus Y.

flux (j) the number of particles crossing a plane per unit area per unit time.

freely jointed chain model for a polymer in which linked segments may pivot about their junctions without restriction.

freely rotating chain model for a polymer in which the polar angle between neighboring straight segments is fixed, although the segments are free to rotate azimuthally (like an alkane chain).

FtsZ protein that forms a ring at the bacterial midplane in preparation for cell division

Gaussian distribution in statistics, the normalized probability distribution for an observable x given by $(2\pi\sigma^2)^{-1/2} \exp(-[x - \mu]^2/2\sigma^2)$, where σ^2 is the variance and μ is the mean.

Gaussian rigidity (κ_G) elastic constant characterizing a membrane's resistance to a change in its Gaussian curvature.

GDP (guanosine diphosphate; see GTP).

gene a region of DNA generally more than a thousand base pairs long that carries the genetic code for a protein.

Gibbs free energy (G) in thermodynamics, the free energy of a system at fixed temperature T and pressure P; in three dimensions, $G = E + PV - TS$, where E, V, and S are the internal energy, volume and entropy of the system, respectively.

glucose a specific sugar molecule (one of many with the composition $C_6H_{12}O_6$) used in cell metabolism.

glycocalyx sugar-based coat surrounding most eukaryotic cells.

Golgi apparatus organelle with extensive folded membranes, whose function includes protein sorting.

Gram-negative bacterium bacterium whose thin cell wall is sandwiched between the inner (plasma) membrane and an outer membrane.

Gram-positive bacterium bacterium with a single membrane encapsulated by a thick cell wall.

GTP (guanosine triphosphate) one element of the cell's energy currency, releasing energy during hydrolysis to GDP.

Hooke's Law behavior of materials for which stress is linearly proportional to strain; the force **F** experienced by a Hookean spring obeys $\mathbf{F} = -k_{sp}\mathbf{x}$, where k_{sp} is an elastic parameter and $\mathbf{x}$ is the displacement from equilibrium.

hydrocarbon molecule, or section thereof, containing exclusively hydrogen and carbon atoms.

hydrolysis chemical reaction involving the breakage of covalent bonds by the addition of water.

hydrophilic literally, water-loving; in a molecule, usually an electrically polar region with an attraction to water.

hydrophobic literally, water-avoiding; in a molecule, generally a non-polar region.

ideal gas gas of N point-like constituents obeying the ideal gas law $PV = Nk_BT$, where P, V, and T are pressure, volume, and temperature, respectively.

ideal scaling particular relationship between the effective size of a polymer and its contour length L_c; the end-to-end displacement $\mathbf{r}_{ee}$ of ideal polymers obeys $\langle \mathbf{r}_{ee}^2 \rangle \propto L_c$.

intermediate filament one of the common filaments of the cytoskeleton; a bundle of intertwined and cross-linked proteins about 10 nm in diameter.

isosceles a triangle with two sides of equal length.

isotropic describing a medium or material whose physical properties are independent of direction.

keratin one of the fibrous proteins capable of forming intermediate filaments.

kinase an enzyme that transfers a phosphate group from a molecule like ATP to another protein.

kinesin a type of molecular motor that (generally) moves towards the plus end of a microtubule.

lamellipodium thin, sheet-like extension of a cell spread on a substrate; generally rich in actin filaments and part of the cell locomotion apparatus.

lamin one of the proteins capable of forming intermediate filaments; present in the nuclear lamina.

linking number (Lk) in topology the algebraic sum of the twist and writhe that characterize a helical shape that may itself be wound into larger loops or helices.

lipids broad class of organic molecules that are insoluble in water; phospholipids and cholesterol are the principal lipid constituents of cellular membranes.

ligand a molecule capable of binding to a specific receptor molecule.

lysis mechanical failure of the plasma membrane under tension.

macrophage specialized white blood cell responsible for removing bacteria and cell debris from the body.

macroscopic associated with large length scales, such as a millimeter or more, visible to the unaided eye.

matrix mathematically, a two-dimensional array of elements m_{ij}.

mean value mathematically, for a set of N elements x_i, the mean value is $(1/N)\ \Sigma_i x_i$.

membrane in the cell, a sheet-like structure that forms the boundaries of a cell and its organelles; derives its two-dimensional structure from a lipid bilayer, which may also contain cholesterol and embedded proteins.

mesophase phase of matter with properties intermediate between solids and isotropic liquids; examples include some phases of liquid crystals.

mesoscopic having a length scale larger than molecular sizes yet smaller than macroscopic dimensions.

micelle an aggregate of amphiphilic molecules in a fluid, having roughly spherical or cylindrical shapes.

microfossil in paleontology, fossilized remains of lifeforms less than or equal to a few mm in size.

microtubule polymeric protein filament composed of tubulin monomers, present in the cytoskeleton, as well as flagella and cilia, with a hollow structure 25 nm in diameter.

Min a family of proteins that form a membrane-associated cytoskeleton in a bacterium, and help determine the location of the cell's midplane in preparation for cell division.

minus end slow-growing or unstable end of a polymeric protein filament such as actin or a microtubule.

mitochondrion one of two organelles responsible for producing ATP; a few nanometers in length and bounded by a double membrane.

moment of inertia of cross section ($\mathcal{I}$) geometrical quantity influencing the energy needed to bend a filament; equal to $\pi R^4/4$ for a cylinder of radius R.

monolayer a quasi-two-dimensional array just one atom or molecule thick.

monomer chemical building block capable of associating into polymeric structures such as filaments.

motor protein a class of proteins (dynein, kinesin, and myosin) capable of movement along a cytoskeletal filament, often for the purpose of vesicle transport or cell contraction.

MreB, Mbl protein components of dynamical filaments forming the helical cytoskeletons in bacteria; Mbl is not present in all bacteria.

muscle tissue individual or composite cells capable of producing movement through contraction; muscle cells are classified as skeletal, smooth, cardiac or myoepithelial (see Appendix A).

mycoplasma class of small prokaryotic cells lacking a cell wall and having diameters of 0.8 μm or less.

myosin a family of motor proteins that move along actin filaments.

nematic a mesophase characterized by long-range orientational order without positional order.

neuron a cell of the nervous system capable of receiving and transmitting electrical signals; principal components of a neuron are dendrites, soma and axon.

normal vector (**n**) unit vector locally perpendicular to a surface or line.

nuclear lamina connected mat of intermediate filaments attached to the inner surface of the inner nuclear membrane.

nucleotide a subunit of DNA and RNA consisting of a sugar ring to which are separately attached an organic base and phosphate group;

polymerization occurs between the sugar and phosphate of successive nucleotides.

nucleus unique to eukaryotic cells, an organelle containing most, but not all, of a cell's DNA.

oblate axially symmetric shape like a pancake (short along the axis of rotation).

operator in genetics, a site on DNA to which a repressor may bind and stop or reduce the transcription of the associated gene by RNA polymerase.

operon a functional region of DNA that contains a gene as well as the promoter and operator sites that are associated with its expression and regulation.

organelle membrane-bound compartment of a eukaryotic cell; examples include the nucleus, mitochondria, etc.

organic acid an organic compound containing the carboxyl group (−COOH).

organic base in chemistry, an organic compound which is often basic by virtue of accepting a proton from aqueous solution, leaving OH^-.

peptide bond a covalent bond formed by the linkage of carboxyl (−COOH) and amino ($-NH_2$) groups, eliminating H_2O.

peptidoglycan material of the bacterial cell wall, composed of linear polysaccharides cross-linked by polypeptide chains.

percolation in physics, a concentration threshold above which a continuous path traverses a network or lattice.

persistence length (ξ_p) a measure of the distance scale over which tangent vectors to a curve, or normal vectors to a surface, become uncorrelated.

phantom membrane a mathematical surface which is permitted to intersect itself.

phosphate the chemical group PO_4.

phosphorylation the covalent addition of a phosphate group to a molecule.

pipette in chemistry, a glass tube used for transferring liquids; laboratory study of cells may involve micropipettes with a tapered end just a micron in diameter.

pitch the distance along the symmetry axis corresponding to one complete turn of a helix.

Planck's constant (h) fundamental unit in quantum mechanics equal to 6.626×10^{-34} J•s; the energy of a photon is the product of its frequency and Planck's constant.

plaquette elementary geometrical unit of a surface.

plasma membrane semi-permeable membrane surrounding a cell and controlling the passage of molecular material to and from its interior.

plectoneme a looped structure in the shape of a double helix extending laterally from a flexible rod that that has been overwound.

plus end rapidly growing end of a polymeric filament such as actin or a microtubule.

Poisson–Boltzmann equation an equation describing the electrostatic potential experienced by a charge at non-zero temperature (see Section 9.2).

Poisson ratio for an object under uniaxial stress, the ratio of the fractional contraction in the transverse direction compared to the fractional extension in the direction of the applied stress.

polar head-group polarized terminal group attached to the phosphate group of a phospholipid.

polymer a structure formed through the association of monomeric subunits; bonding need not be covalent, and the structure may be dynamic, continually adding and releasing monomers.

polysaccharide linear or branched polymer of linked sugar molecules.

promoter a region of DNA to which RNA polymerase attaches in preparation for opening the DNA helix and transcribing a gene.

prokaryote a mechanically simple cell or aggregate without a nucleus; bacteria are prokaryotic cells.

prolate axially symmetric shape like a cigar (long along the axis of rotation).

protein linear polymer of amino acids, covalently linked by peptide bonds.

purine a specific family of organic bases with a molecular structure of two rings (five and six members each) sharing a common bond; present in DNA and RNA.

pyrimidine a specific family of organic bases in DNA and RNA comprising a single six-member ring.

radius of curvature reciprocal of the curvature of a line or surface.

radius of gyration (R_g) for a collection of N objects with positions $\mathbf{r}_i$ with respect to the center-of-mass, $R_g^2 = \Sigma_i r_i^2/N$.

receptor protein a protein that binds to specific ligands, often as part of a signaling mechanism.

replication the process in which a cell's DNA is duplicated by DNA polymerase during the cell cycle.

repressor a protein that binds to the operator region of DNA and blocks the attachment of RNA polymerase, thus regulating the production of mRNA in a nearby gene.

Reynolds number (R) a dimensionless number proportional to the ratio of the inertial forces to the viscous forces experienced by an object while moving in a fluid.

ribose specific sugar with a five-member ring, present in RNA; a related compound, deoxyribose, is present in DNA.

ribosome complex cellular body constructed from protein and ribosomal RNA; site of protein synthesis.

RNA (ribonucleic acid) molecular chain of linked nucleotides where ribose is the sugar unit; plays a variety of roles in protein synthesis.

root mean square for a set of N elements x_i, the root mean square is $(\Sigma_i x_i^2/N)^{1/2}$.

self-avoiding walk an otherwise random trajectory which does not cross itself.

self-avoiding membrane a mathematical surface which is not permitted to intersect itself.

shear a force applied tangentially to a surface; the **shear modulus** (μ) of isotropic materials is the shear stress divided by the angle of deformation (in radians).

shear storage and loss moduli (G' and G'') frequency-dependent mechanical parameters of viscoelastic materials; in the limit where the deformation frequency ω vanishes, $G' \to \mu$, and $G'' \to \eta/\omega$, where η is the viscosity.

spectrin two closely related proteins (α- and β-spectrin) of mass about 220 kDa; as αβ-heterodimers, form a filament with a contour length of 200 nm; principal component of the human erythrocyte cytoskeleton.

strain deformation of an object in response to a stress; a dimensionless quantity, the **strain tensor** (u_{ij}) is given by $(\partial u_i/\partial x_j + \partial u_j/\partial x_i)/2$ for small displacements **u**, where **x** is the coordinate system.

stomatocyte cup-shaped conformation of a red blood cell.

stress in three dimensions, a force per unit area; the **stress tensor** (σ_{ij}) is defined in Appendix D.

sugar a family of compounds with the chemical formula $(CH_2O)_n$; for some values of n, the molecule may adopt either a linear chain or ring configuration.

supercoil in topology, structures such as helices and loops that form when a filament is overwound or underwound as a result of an applied torque.

surface tension (γ) energy required to increase a surface by a unit area.

surfactant a chemical compound that reduces the surface tension at an interface of a liquid (from surface active agent).

synapse junction between a cell and the axon of a nerve cell, across which a signal may by carried chemically or electrically.

tangent vector (t) unit vector locally tangent to a line or surface.

tensile strength stress required to cause mechanical failure of a material.

thermal ratchet model for molecular motion in which thermally activated movement is biased in one direction.

tissue an extended collection of cells and associated materials acting cooperatively with other tissues as part of an organ (see Appendix A for classification of tissues).

torsional rigidity (κ_{tor}) the resistance of a filament to deformation by a torque applied around its axis.

trace (tr) mathematically, the sum of the diagonal elements of a matrix; $\mathrm{tr}m = \Sigma_i m_{ii}$.

transcription the process by which the code carried on a single strand of DNA is read by RNA polymerase to produce a string of mRNA.

translation the process by which the code carried by mRNA is read by a ribosome to create a linear polypeptide, or protein.

treadmilling dynamic state of a polymer filament occurring when the addition rate of monomers at one end equals the release rate at the other.

tubulin monomeric component of microtubules; the proteins α- and β-tubulin form a heterodimer 8 nm in length.

twist (Tw) in topology, the number of turns made by a filament wound into the shape of a helix, including the winding of the filament about its own axis.

vacuole enclosed space within plant or animal cells often used for storage; may occupy a sizeable fraction of the volume of a plant cell.

van der Waals forces a class of attractive forces arising from the interaction of permanent and instantaneous electric dipole moments; corresponding potential energy falls like $1/r^6$ for large separations r.

variance in statistics, for a set of N measured values x_i, $\mathrm{Var}(x_1 \ldots x_N) = (N-1)^{-1}\,\Sigma_i(x_i - X)^2$, where the mean value is $X = \Sigma_i x_i\,/N$.

vesicle small membrane-bounded sphere used for transport in the cytoplasm.

vimentin a protein of mass 54 kDa capable of forming intermediate filaments.

virus parasitic life-form consisting of a core of tightly packed DNA or RNA protected by a proteinaceous capsid; can replicate inside a host cell but is otherwise inert.

viscosity (η) a measure of the resistance of a fluid to deformation under stress.

writhe (Wr) in topology, the number of times a filament crosses itself when deformed into one or more loops

Young equation relationship between the contact angle of a droplet on a substrate and the interfacial energies of the materials.

Young's modulus (Y) an elastic property of materials; ratio of stress to strain for a uniaxial deformation.

References

Abraham, F. F. and Kardar, M. (1991). Folding and unbinding transitions in tethered membranes. *Science*, 252, 419–422.

Abraham, F. F. and Nelson, D. R. (1990). Diffraction from polymerized membranes. *Science*, 249, 393–397.

Abraham, V. C., Krishnamurthi, V., Taylor, D. L. and Lanni, F. (1999). The actin-based nanomachine at the leading edge of migrating cells. *Biophys. J.*, 77, 1721–1732.

Abramowitz, M. and Stegun, I. A. (1970). *Handbook of Mathematical Functions*. New York, NY: Dover.

Aebi, U., Cohn, J., Buhle, L. and Gerace, L. (1986). The nuclear lamina is a meshwork of intermediate-type filaments. *Nature*, 323, 560–564.

Ajdari, A. and Prost, J. (1992). Mouvement induit par un potential périodique de basse symétrie: diélectrophorèse pulsée. *C. R. Acad. Sci. Paris Ser. II.*, 315, 1635–1639.

Albersdörfer, A., Feder, T. and Sackmann, E. (1997). Adhesion-induced domain formation by interplay of long-range repulsion and short-range attraction force: a model membrane study. *Biophys. J.*, 73, 245–257.

Alberts, B., Johnson, A., Lewis, J. *et al.* (2008). *Molecular Biology of the Cell*, 5th edn. New York, NY: Garland.

Allard, J. and Cytrynbaum, E. N. (2009). Force generation by a dynamic Z-ring in *Escherichia coli* cell division. *Proc. Nat. Acad. Sci. USA*, 106, 145–150.

Allard, J. and Rutenberg, A. D. (2009). Pulling helices inside bacteria: imperfect helices and rings. *Phys. Rev. Lett.*, 102, 158105.

Allende, D., Simon, S. A. and McIntosh, T. J. (2005). Melittin-induced bilayer leakage depends on lipid material properties: evidence for toroidal pores. *Biophys. J.*, 88, 1828–1837.

Amit, R., Oppenheim, A. B. and Stavans, J. (2003). Increased bending rigidity of single DNA molecules by H-NS, a temperature and osmolarity sensor. *Biophys. J.*, 84, 2467–2473.

Amos, L. A. and Amos, W. B. (1991). *Molecules of the Cytoskeleton*. London: MacMillan.

Andrews, S. S. and Arkin, A. P. (2007). A mechanical explanation for cytoskeletal rings and helices in bacteria. *Biophys. J.*, 93, 1872–1884.

Anglin, T. C., Liu, J. and Conboy, J. C. (2007). Facile lipid flip-flop in a phospholipid bilayer induced by Gramcidin A measured by sum-frequency vibrational spectroscopy. *Biophys. J.*, 92, L01–L03.

Arnoldi, M., Fritz, M., Bäuerlein, E. *et al.* (2000). Bacterial turgor pressure can be measured by atomic force microscopy. *Phys. Rev. E*, 62, 1034–1044.

Aronovitz, J. A. and Lubensky, T. C. (1988). Fluctuations of solid membranes. *Phys. Rev. Lett.*, 60, 2634–2637.

Ashkin, A., Schütze, K., Dziedzic, J. M., Euterneuer, U. and Schliwa, M. (1990). Force generation of organelle transport measured *in vivo* by an infrared laser trap. *Nature*, 348, 346–348.

Aström, J. A., Kumar, P. B. S., Vattulainen, I. and Karttunen, M. (2008). Strain hardening, avalanches, and strain softening in dense cross-linked actin networks. *Phys. Rev. E*, 77, 051913.

Astumian, R. D. and Bier, M. (1994). Fluctuation driven ratchets: molecular motors. *Phys. Rev. Lett.*, 72, 1766–1769.

Ausmees, N., Kuhn, J. R. and Jacobs-Wagner, C. (2003). The bacterial cytoskeleton: an intermediate filament-like function in cell shape. *Cell*, 115, 705–713.

Bai, F., Branch, R. W., Nicolau, D. V. Jr. *et al.* (2010). Conformal spread as a mechanism for cooperatively in the bacterial flagellar switch. *Science*, 327, 685–689.

Bailey, S. M., Chiruvolu, S., Israelachvili, J. N. and Zasadzinski, J. A. N. (1990). Measurements of forces involved in vesicle adhesion using freeze-fracture electron microscopy. *Langmuir*, 6, 1326–1329.

Baillie, C. F. and Johnston, D. A. (1993). Freezing in a fluid random surface. *Phys. Rev. D*, 48, 5025–5028.

Banavar, J. A., Maritan, A. and Stella, A. (1991). Geometry, topology and universality of random surfaces. *Science*, 252, 825–827.

Barghoorn, E. S. and Schopf, J. W. (1966). Microorganisms three billion years old from the Precambrian of South Africa. *Science,* 152, 758–763.

Barghoorn, E. S. and Tyler, S. A. (1965). Microorganisms from the Gunflint chert. *Science*, 147, 563–577.

Bärmann, M., Käs, J., Kurzmeier, H. and Sackmann, E. (1992). A new cell model: actin networks encaged by giant vesicles. In *The Structure and Conformation of Amphiphilic Membranes*, eds. R. Lipowsky, D. Richter and K. Kremer, pp. 137–143. Springer-Verlag, Berlin.

Bar-Ziv, R. and Moses, E. (1994). Instability and "pearling" states produced in tubular membranes by competition of curvature and tension. *Phys. Rev. Lett.*, 73, 1392–1395.

Bar-Ziv, R. Moses, E. and Nelson, P. (1998). Dynamic excitations in membranes induced by optical tweezers. *Biophys. J.*, 75, 294–320.

Batey, S., Scott, K. A. and Clarke, J. (2006). Complex folding kinetics of a multidomain protein. *Biophys. J.*, 90, 2120–2130.

Baumgärtner, A. (1982). Statistics of self-avoiding ring polymers. *J. Chem. Phys.*, 76, 4275–4282.

Baumgärtner, A. (1991). Does a polymerized membrane crumple? *J. Phys. I France*, 1, 1549–1556.

Baumgärtner, A. and Ho, J.-S. (1990). Crumpling of fluid vesicles. *Phys. Rev. A*, 41, 5747–5750.

Bausch, A. R., Ziemann, F., Boulbitch A. A., Jacobson, K. and Sackmann, E. (1998). Local measurements of viscoelastic parameters of adherent cell surfaces by magnetic bead microrheology. *Biophys. J.*, 75, 2038–2049.

Bausch, A. R., Möller, W. and Sackmann, E. (1999). Measurement of local viscoelasticity and forces in living cells by magnetic tweezers. *Biophys. J.*, 76, 573–579.

Bayley, P. M., Sharma, K. K. and Martin, S. R. (1994). Microtubule dynamics *in vitro*. In *Microtubules*, eds. J. S. Hyams and C. W. Lloyd, pp. 111–137. New York, NY: Wiley-Liss.

Beck, W. S., ed. (1991). *Hematology*, 5th edn. Boston, MA: MIT Press.

Bell, G. I. (1978). Models for the specific adhesion of cells to cells. *Science*, 200, 618–627.

Bennett, S., Boal, D. H. and Ruotsalainen, H. (2007). Growth modes of 2-Ga microfossils. *Paleobiology*, 33, 382–396.

Berg, H. C. (1983). *Random Walks in Biology*. Princeton, NJ: Princeton University Press.

Berg, H. C. (1995). Torque generation by the flagellar rotary motor. *Biophys. J.*, 68, 163s–167s.

Berg, H. C. (2000). Motile behavior of bacteria. *Physics Today*, 53, 24–29.

Berg, H. C. and Turner, L. (1993). Torque generated by the flagellar motor of *Escherichia coli*. *Biophys. J.*, 65, 2201–2216.

Berg, O. G. and von Hippel, P. H. (1985). Diffusion-controlled macromolecular interactions. *Ann. Rev. Biophys. Biophys. Chem.*, 14, 131–160.

Bergen, L. G. and Borisy, G. G. (1980). Head-to-tail polymerization of microtubules *in vitro*. *J. Cell Biol.*, 84, 141–150.

Berndl, K., Käs, J., Lipowsky, R., Sackmann, E. and Seifert, U. (1990). Shape transformations of giant vesicles: extreme sensitivity to bilayer asymmetry. *Europhys. Lett.*, 13, 659–664.

Berry, R. M. (1993). Torque and switching in the bacterial flagellar motor: an electrostatic model. *Biophys. J.*, 64, 961–973.

Berry, R. M. and Berg, H. C. (1999). Torque generated by the flagellar motor of *Escherichia coli* while driven backward. *Biophys. J.*, 76, 580–587.

Bessis, M. (1973). *Living Blood Cells and Their Ultrastructure*. New York, NY: Springer-Verlag.

Bhasin, N., Law, R., Liao, G. *et al.* (2005). Molecular extensibility of mini-dystrophins and a dystrophin rod construct. *J. Mol. Biol.*, 352, 795–806.

Bier, M. and Astumian, R. D. (1993). Matching a diffusive and a kinetic approach for escape over a fluctuating barrier. *Phys. Rev. Lett.*, 71, 1649–1652.

Biron, D., Alvarez-Lacalle, E., Tlusty, T. and Moses, E. (2005). Molecular model of the contractile ring. *Phys. Rev. Lett.*, 95, 098102.

Bloom, M., Evans, E. and Mouritsen, O. G. (1991). Physical properties of the fluid lipid-bilayer component of cell membranes: a perspective. *Quart. Rev. Biophys.*, 24, 293–397.

Bloomfield, V. A. (1991). Condensation of DNA by multivalent ions: considerations on mechanism. *Biophys. J.*, 31, 1471–1481.

Boal, D. H. (1991). Phases of two-dimensional vesicles under pressure. *Phys. Rev.*, A43, 6771–6777.

Boal, D. H. (1994). Computer simulation of a model network for the erythrocyte cytoskeleton. *Biophys. J.*, 67, 521–529.

Boal, D. H. and Boey, S. K. (1995). Barrier-free paths of directed protein motion in the erythrocyte plasma membrane. *Biophys. J.*, 69, 372–379.

Boal, D. H. and Forde, C. (2011). Evolution of the cell's mechanical design. In *The Minimal Cell: The Biophysics of Cell Compartment and the Origin of Cell Functionality*, eds. P. L. Luisi and P. Stano., Chap. 2. New York: Springer.

Boal, D. H. and Ng, R. (2010). Shape analysis of filamentous Precambrian microfossils and modern cyanobacteria. *Paleobiology*, 36, 555–572.

Boal, D. H. and Rao, M. (1992a). Scaling behavior of fluid membranes in three dimensions. *Phys. Rev. A*, 45, R6947–R6950.

Boal, D. H. and Rao, M. (1992b). Topology changes in fluid membranes. *Phys. Rev. A*, 46, 3037–3045.

Boal, D. H., Seifert, U. and Shillcock, J. C. (1993). Negative Poisson ratio in two-dimensional networks under tension. *Phys. Rev. E*, 48, 4274–4283.

Bockelmann, U., Thomen, Ph., Essevaz-Roulet, B., Viasnoff, V. and Helsot, F. (2002). Unzipping DNA with optical tweezers: high sequence sensitivity and force flip. *Biophys. J.*, 82, 1537–1553.

Bonder, E. M., Fishkind, D. J. and Mooseker, M. S. (1983). Direct measurement of critical concentrations and assembly rate constants at the two ends of an actin filament. *Cell*, 34, 491–501.

Bouchiat, C. and Mézard, M. (1998). Elasticity model of a supercoiled DNA molecule. *Phys. Rev. Lett.*, 80, 1556–1559.

Boulbitch, A., Quinn, B. and Pink, D. (2000). Elasticity of the rod-shaped Gram-negative eubacteria. *Phys. Rev. Lett.*, 85, 5246–5249.

Bowick, M. J. (2004). Fixed connectivity membranes. In *Statistical Mechanics of Membranes and Surfaces*, eds. D. Nelson, T. Piran and S. Weinberg, pp. 323–357. Singapore: World Scientific.

Bowick, M., Falcioni, M. and Thorleifsson, G. (1997). Numerical observation of a tubular phase in anisotropic membranes. *Phys. Rev. Lett.*, 79, 885–888.

Brangwynne, C. P., Koenderink, G. H., Barry, E. *et al.* (2007). Bending dynamics of fluctuating biopolymers probed by automated high-resolution filament tracking. *Biophys. J.*, 93, 346–359.

Braun, V., Gnirke, U., Henning, U. and Rehn, K. (1973). Model for the structure of the shape-maintaining layer of *Escherichia coli* cell envelope. *J. Bacteriol.*, 114, 1264–1270.

Brett, C. T. (2000). Cellulose microfibrils in plants: biosynthesis, deposition, and integration into the cell wall. *Int. Rev. Cyto.*, 199, 161–199.

Brewster, R. and Safran, S. A. (2010). Line active hybrid lipids determine domain size in phase separation of saturated and unsaturated lipids. *Biophys. J.*, 98, L21–L23.

Brochard, F. and Lennon, J. F. (1975). Frequency spectrum of the flicker phenomenon in erythrocytes. *J. Phys. Paris*, 36, 1035–1047.

Brotschi, E. A., Hartwig, J. H. and Stossel, T. P. (1978). The gelation of actin by actin-binding protein. *J. Biol. Chem.*, 253, 8988–8993.

Brown, F. L. H. (2003). Regulation of protein and mobility via thermal membrane undulations. *Biophys. J.*, 84, 842–853.

Brownell, W. E., Spector, A. A., Raphael, R. M. and Popel, A. S. (2001). Micro- and nanomechanics of the cochlear outer hair cell. *Ann. Rev. Biomed. Eng.*, 3, 169–194.

Bruinsma, R. (1996). Rheology and shape transitions of vesicles under capillary flow. *Physica A*, 234, 249–270.

Bruinsma, R., Goulian, M. and Pincus, P. (1994). Self-assembly of membrane junctions. *Biophys. J.*, 67, 746–750.

Brunk, D. K. and Hammer, D. A. (1997). Quantifying rolling adhesion with a cell-free assay: E-selectin and its carbohydrate ligands. *Biophys. J.*, 72, 2820–2833.

Brutzer, H., Luzzietti, N., Klaue, D. and Seidel, R. (2010). Energetics at the DNA supercoiling transition. *Biophys. J.*, 98, 1267–1276.

Bryant, Z., Stone, M. D., Gore, J. *et al.* (2003). Structural transitions and elasticity from torque measurements on DNA. *Nature*, 474, 338–341.

Bull, B. S. (1977). The resting shape of normal red blood cells is determined by bending and ___? *Blood Cells*, 3, 321–323.

Burge, R. E., Fowler, A. G. and Reaveley, D. A. (1977). Structure of the peptidoglycan of bacterial cell walls. I. *J. Mol. Biol.*, 117, 927–953.

Bustamante, C., Marko, J. F., Siggia, E. D. and Smith, S. (1994). Entropic elasticity of λ-phage DNA. *Science*, 265, 1599–1600.

Byers, T. and Branton, D. (1985). Visualizations of the protein associations in the erythrocyte membrane skeleton. *Proc. Natl. Acad. Sci. USA*, 82, 6153–6157.

Cabeen, M. T. and Jacobs-Wagner, C. (2005). Bacterial cell shape. *Nature Rev. Microbiol.*, 3, 601–610.

Canham, P. B. (1970). The minimum energy of bending as a possible explanation of the biconcave shape of the human red blood cell. *J. Theor. Biol.*, 26, 61–81.

Cao, L., Fishkind, D. J. and Wang, Y.-L. (1993). Localization and dynamics of nonfilamentous actin in cultured cells. *J. Cell Biol.*, 123, 173–181.

Carlier, M.-F. and Pantaloni, D. (1981). Kinetic analysis of guanosine 5′-triphosphate hydrolysis associated with tubulin polymerization. *Biochemistry*, 20, 1918–1924.

Cardini, G., Bareman, J. P. and Klein, M. L. (1988). Characterization of a Langmuir–Blodgett monolayer using molecular dynamics calculations. *Chem. Phys. Lett.*, 145, 493–498.

Cave, I. D. (1968). The anisotropic elasticity of the plant cell wall. *Wood Sci. and Tech.*, 2, 268–278.

Cevc, G. and Marsh, D. (1987). *Phospholipid Bilayers. Physical Principles and Models*. New York, NY: Wiley-Interscience.

Chaikin, P. M. and Lubensky, T. C. (1995). *Principles of Condensed Matter Physics*. Cambridge: Cambridge University Press.

Chan, H. S. and Dill, K. A. (1989). Compact polymers. *Macromolecules*, 22, 4559–4573.

Chan, H. S. and Dill, K. A. (1991). "Sequence space soup" of proteins and polymers. *J. Chem. Phys.*, 95, 3775–3787.

Chan, H. S. and Dill, K. A. (1994). Transition state and folding dynamics of proteins and heteropolymers. *J. Chem. Phys.*, 100, 9238–9257.

Chandrasekhar, S. (1943). *Rev. Mod. Phys.*, 15, 1.

Chen, Z. and Rand, R. P. (1997). The influence of cholesterol on phospholipid membrane curvature and bending elasticity. *Biophys. J.*, 73, 267–276.

Chernomordik, L. V. and Chizmadzhev, Y. A. (1989). Electrical breakdown of lipid bilayer membranes: phenomenology and mechanism. In *Electroporation and Electrofusion in Cell Biology*, eds. E. Neumann, A. E. Sowers and C. A. Jordan, pp. 83–95. New York, NY: Plenum.

Chyan, C.-L., Lin, F.-C., Peng, H. *et al.* (2004). Reversible mechanical unfolding of single ubiquitin molecules. *Biophys. J.*, 87, 3995–4006.

Cirulis, J. T., Keeley, F. W. and James, D. F. (2009). Viscoelastic properties and gelation of an elastin-like polypeptide. *J. Rheol.*, 53, 1215–1228.

Claessens, M. M. A. E., van Oort, B. F., Leemakers, F. A. M., Hoekstra, F. A. and Cohen Stuart, M. A. (2004). Charged lipid vesicles: effects of salts on bending rigidity, stability and size. *Biophys. J.*, 87, 3882–3893.

Cloud, P. E., Jr. (1965). Significance of the Gunflint (Precambrian) microflora. *Science*, 148, 27–35.

Coleman, T. R., Fishkind, D. J., Mooseker, M. S. and Morrow, J. S. (1989). Functional diversity among spectrin isoforms. *Cell Motil. Cytoskeleton*, 12, 225–247.

Collins, J. F. and Richmond, M. H. (1962). Rate of growth of *Bacillus cereus* between divisions. *J. Gen. Microbiol.*, 28, 15–33.

Cook-Röder, J. and Lipowsky, R. (1992). Adhesion and unbinding for bunches of fluid membranes. *Europhys. Lett.*, 18, 433–438.

Cooper, S. (1991). *Bacterial Growth and Division: Biochemistry and Regulation of Prokaryotic and Eukaryotic Division Cycles*. San Diego: Academic Press.

Coppin, C. M. and Leavis, P. C. (1992). Quantitation of liquid-crystalline ordering in F-actin solutions. *Biophys. J.*, 63, 794–807.

Cortese, J. D., Schwab, B. III, Freiden, C. and Elson, E. L. (1989). Actin polymerization induces a shape change in actin-containing vesicles. *Proc. Natl. Acad. Sci. USA.*, 86, 5773–5777.

Courty, S., Gornall, J. L. and Terentjev, E. M. (2006). Mechanically induced helix-coil transition in biopolymer networks. *Biophys. J.*, 90, 1019–1027.

Creighton, T. E. (1993). *Proteins: Structures and Molecular Properties*, 2nd edn. New York, NY: Freeman.

Crick, F. H. C. and Hughes, A. F. W. (1950). The physical properties of cytoplasm. A study by means of magnetic particle method part I. Experimental. *Exp. Cell Res.*, 1, 37–80.

Cros, S., Garnier, C., Axelos, M. A. V., Imberty, A. and Pérez, S. (1996). Solution conformations of pectin polysaccharides: determination of chain characteristics by small angle neutron scattering, viscometry, and molecular modeling. *Biopolymers*, 39, 339–352.

Crowther, R. A., Finch, J. T. and Pearse, B. M. F. (1976). On the structure of coated vesicles. *J. Mol. Biol.*, 103, 785–798.

Culver, H. B. (1992). *The Book of Old Ships*. New York, NY: Dover.

Cuvelier, D., Derényi, I., Bassereau, P. and Nassoy, P. (2005). Coalescence of membrane tethers: experiments, theory and applications. *Biophys. J.*, 88, 2714–2726.

Cytrynbaum, E. N. and Marshall, B. D. L. (2007). A multistranded polymer model explains MinDE dynamics in *E. coli* cell division. *Biophys. J.*, 93, 1134–1150.

Dabiri, G. A., Sanger, J. M., Portnoy, D. A. and Southwick, F. S. (1990). *Listeria monocytogenes* moves rapidly through the host-cell cytoplasm by inducing directional actin assembly. *Proc. Natl. Acad. Sci. USA*, 87, 6068–6072.

Dahl, K. N., Kahn, S. M., Wilson, K. L. and Discher, D. E. (2004). The nuclear envelope lamina network has elasticity and a compressibility limit suggestive of a molecular shock absorber. *J. Cell Sci.*, 117, 4779–4786.

Dahl, K. N., Engler, A. J., Pajerowski, J. D. and Discher, D. E. (2005). Power-law rheology of isolated nuclei with deformation mapping of nuclear substructures. *Biophys. J.*, 89, 2855–2864.

Dammer, U., Popescu, O., Wagner, P. *et al.* (1995). Binding strength between cell adhesion proteoglycans measured by atomic force microscopy. *Science*, 267, 1173–1175.

Daniel, R. A. and Errington, J. (2003). Control of cell morphogenesis in bacteria: two distinct ways to make a rod-shaped cell. *Cell*, 113, 767–776.

Dao, M., Li, J. and Suresh, S. (2006). Molecularly based analysis of deformation of spectrin network and human erythrocyte. *Mat. Sci. Eng. C*, 26, 1232–1244.

David, F. (1989). Geometry and field theory of random surfaces and membranes. In *Statistical Mechanics of Membranes and Surfaces*, eds. D. Nelson, T. Piran and S. Weinberg, pp. 157–223. Singapore: World Scientific.

Deam, R. T. and Edwards, S. F. (1976). The theory of rubber elasticity. *Phil. Trans. R. Soc. Lond.*, A280, 317–353.

de Duve, C.R. (1995). *Vital Dust*. New York, NY: Basic Books.

de Gennes, P.-G. (1979). *Scaling Concepts in Polymer Physics*. Ithaca, NY: Cornell University Press.

de Gennes, P. G. and Prost, J. (1993). *The Physics of Liquid Crystals*, 2nd edn. Oxford: Oxford University Press.

de Gennes, P. G. and Taupin, C. (1982). Microemulsions and the flexibility of oil/water interfaces. *J. Phys. Chem.*, 86, 2294–2304.

Dembo, M. (1989). Mechanics and control of the cytoskeleton in *Amoeba proteus*. *Biophys. J.*, 55, 1053–1080.

Dembo, M. and Harlow, F. (1986). Cell motion, contractile networks, and the physics of interpenetrating reactive flow. *Biophys. J.*, 50, 109–121.

Dembo, M., Torrey, D. C., Saxman, K. and Hammer, D. (1988). The reaction-limited kinetics of membrane-to-surface adhesion and detachment. *Proc. R. Soc. Lond.*, B234, 55–83.

Deniz, A. A., Mukhopadhyay, S. and Lemke, E. A. (2007). Single-molecule biophysics: at the interface of biology, physics and chemistry. *J. R. Soc. Interface*, 5, 15–45.

De Pedro, M. A., Schwarz, H. and Koch, A. L. (2003). Patchiness of murein insertion in the sidewall of *Escherichia coli*. *Microbiol.*, 149, 1753–1761.

Derjaguin, B. V. (1949). *Theory of Stability of Colloids and Thin Films*, trans. R.K. Johnston. New York, NY: Consultants Bureau.

Derjaguin, B. V. and Landau, L. (1941). *Acta Physicochim. URSS*, 14, 633–662.

Derrida, B. and Stauffer, D. (1985). Corrections to scaling and phenomenological renormalization for 2-dimensional percolation and lattice animal problems. *J. Phys. (Paris)*, 46, 1623–1630.

Deryagin, B. V. and Gutop, Yu. V. (1962). Theory of the breakdown (rupture) of free films. *Kolloidnyi Zh.*, 24, 431–437.

Deuling, H. J. and Helfrich, W. (1976). The curvature elasticity of fluid membranes: a catalogue of vesicle shapes. *J. Physique*, 37, 1335–1345.

Dewey, T. G. (1993). Protein structure and polymer collapse. *J. Chem. Phys.*, 98, 2250–2257.

Di Cola, E., Waigh, T. A., Trinick, J. *et al.* (2005). Persistence length of titin from rabbit skeletal muscles measured with scattering and microrheology techniques. *Biophys. J.*, 88, 4095–4106.

Dietrich, C., Bagatolli, L. A., Volovyk, Z. N. *et al.* (2001). Lipid rafts reconstituted in model membranes. *Biophys. J.*, 80, 1417–1428.

Dill, K. A. and Bromberg, S. (2003). *Molecular Driving Forces: Statistical Mechanics in Chemistry and Biology*. New York, NY: Garland.

Dill, K. A., Bromberg, S., Yue, K. *et al.* (1995). Principles of protein folding – a perspective from simple exact models. *Protein Sci.*, 4, 561–602.

Discher, D. E., Mohandas, N. and Evans, E. A. (1994). Molecular maps of red cell deformation: hidden elasticity and *in situ* connectivity. *Science*, 266, 1032–1035.

Discher, D. E., Boal, D. H. and Boey, S. K. (1997). Phase transitions and anisotropic responses of planar triangular nets under large deformation. *Phys. Rev. E*, 55, 4762–4772.

Discher, D. E., Boal, D. H. and Boey, S. K. (1998). Simulations of the erythrocyte cytoskeleton at large deformation. II. Micropipette aspiration. *Biophys. J.*, 75, 1584–1597.

Döbereiner, H.-G., Evans, E., Kraus, M., Seifert, U. and Wortis, M. (1997). Mapping vesicle shapes into the phase diagram: a comparison of experiment and theory. *Phys. Rev. E*, 55, 4458–4474.

Döbereiner, H.-G., Selchow, O. and Lipowsky, R. (1999). Spontaneous curvature of fluid vesicles induced by trans-bilayer sugar asymmetry. *Eur. Biophys. J.*, 28, 174–178.

Doering, C. R. and Gadoua, J. C. (1992). Resonant activation over a fluctuating barrier. *Phys. Rev. Lett.*, 69, 2318–2321.

Dogterom, M. and Leibler, S. (1993). Physical aspects of the growth and regulation of microtubule structures. *Phys. Rev. Lett.*, 70, 1347–1350.

Doi, M. and Edwards, S. F. (1986). *The Theory of Polymer Dynamics*. Oxford: Oxford University Press.

Drew, D., Osborn, M. and Rothfield, L. (2005). A polymerization-depolymerization model that accurately generates the self-sustained oscillatory system involved in bacterial division site placement. *Proc. Nat. Acad. Sci. USA*, 102, 6114–6118.

Dustin, P. (1978). *Microtubules*. Berlin: Springer.

Duplantier, B. (1990). Exact fractal area of two-dimensional vesicles. *Phys. Rev. Lett.*, 64, 493.

Duwe, H. P., Kaes, J. and Sackmann, E. (1990). Bending elastic moduli of lipid bilayers: modulation by solutes. *J. Phys. France*, 51, 945–962.

Egberts, E. and Berendsen, H. J. C. (1988). Molecular dynamics simulation of a smectic liquid crystal with atomic detail. *J. Chem. Phys.*, 89, 3718–3732.

Egelhaaf, S. U. and Schurtenberger, P. (1999). Micelle-to-vesicle transition: a time-resolved structural study. *Phys. Rev. Lett.*, 82, 2804–2807.

Elbaum, M., Fygenson, D. K. and Libchaber, A. (1996). Buckling microtubules in vesicles. *Phys. Rev. Lett.*, 76, 4078–4081.

Elowitz, M. B. and Leibler, S. (2000). A synthetic oscillatory network of transcriptional regulators. *Nature*, 403, 335–338.

Elson, E. L. (1988). Cellular mechanics as an indicator of cytoskeletal structure and function. *Ann. Rev. Biophys. Chem.*, 17, 397–430.

Elston, T. C. and Oster, G. (1997). Protein turbines I: the bacterial flagellar motor. *Biophys. J.*, 73, 703–721.

Eppenga, R. and Frenkel, D. (1984). Monte Carlo study of the isotropic and nematic phases of infinitely thin hard platelets. *Molec. Phys.*, 52, 1303–1334.

Erickson, H., Taylor, D., Taylor, K. and Bramhill, D. (1996). Bacterial cell division protein FtsZ assembles into protofilament sheets and minirings, structural homologs of tubulin polymers. *Proc. Nat. Acad. Sci. USA*, 93, 519–523.

Euteneuer, U. and Schliwa, M. (1984). Persistent, directional motility of cells and cytoplasmic fragments in the absence of microtubules. *Nature*, 310, 58–61.

Evans, E. A. (1973a). A new material concept for the red cell membrane. *Biophys. J.*, 13, 926–940.

Evans, E. A. (1973b). New membrane concept applied to the analysis of fluid shear- and micropipette-deformed red blood cells. *Biophys. J.*, 13, 941–954.

Evans, E. A. (1974). Bending resistance and chemically induced moments in membrane bilayers. *Biophys. J.*, 14, 923–931.

Evans, E. A. (1983). Bending elastic modulus of red blood cell membrane derived from buckling instability in micropipet aspiration tests. *Biophys. J.*, 43, 27–30.

Evans, E. (1993). New physical concepts for cell amoeboid motion. *Biophys. J.*, 64, 1306–1322.

Evans, E. and Metcalfe, M. (1984). Free energy potential for aggregation of giant, neutral lipid bilayer vesicles by van der Waals attraction. *Biophys. J.*, 46, 423–426.

Evans, E. and Needham, D. (1987). Physical properties of surfactant bilayer membranes: thermal transitions, elasticity, rigidity, cohesion and colloidal interactions. *J. Phys. Chem.*, 91, 4219–4228.

Evans, E. and Needham, D. (1988). Attraction between lipid bilayer membranes in concentrated solutions of nonadsorbing polymers: comparison of mean-field theory with measurements of adhesion energy. *Macromolecules*, 21, 1822–1831.

Evans, E. A. and Parsegian, V. A. (1986). Thermal-mechanical fluctuations enhance repulsion between bimolecular layers. *Proc. Natl. Acad. Sci. USA*, 83, 7132–7136.

Evans, E. and Rawicz, W. (1990). Entropy-driven tension and bending elasticity in condensed-fluid membranes. *Phys. Rev. Lett.*, 64, 2094–2097.

Evans, E. and Ritchie, K. (1997). Dynamic strength of molecular adhesion bonds. *Biophys. J.*, 72, 1541–1555.

Evans, E. and Ritchie, K. (1999). Strength of a weak bond connecting flexible polymer chains. *Biophys. J.*, 76, 2439–2447.

Evans, E. and Skalak, R. (1980). *Mechanisms and Thermodynamics of Biomembranes*. Boca Raton, FL: CRC Press.

Evans, E. A. and Waugh, R. (1977a). Osmotic correction to elastic area compressibility measurements on red cell membrane. *Biophys. J.*, 20, 307–313.

Evans, E. A. and Waugh, R. (1977b). Mechano-chemistry of closed vesicular membranes. *J. Coll. Interfac. Sci.*, 60, 286–298.

Evans, E. A. and Waugh, R. E. (1980). Mechano-chemical study of red cell membrane structure *in situ*. In *Erythrocyte Mechanics and Blood Flow*. New York, NY: Liss.

Evans, E., Klingenberg, D. J., Rawicz, W. and Szoka, F. (1996). Interactions between polymer-grafted membranes in concentrated solutions of free polymer. *Langmuir*, 12, 3031–3037.

Evans, E., Heinrich, V., Ludwig, F. and Rawicz, W. (2003). Dynamic tension spectroscopy and strength of biomembranes. *Biophys. J.*, 85, 2342–2350.

Evans, J., Gratzer, W., Mohandas, N., Parker, K. and Sleep, J. (2008). Fluctuations of the red blood cell membrane: relation to mechanical properties and lack of ATP dependence. *Biophys. J.*, 94, 4134–4144.

Everaers, R. and Kremer, K. (1996). Topological interactions in model polymer networks. *Phys. Rev. E*, 53, R37–R40.

Falcioni, M., Bowick, M. J., Guitter, E. and Thorleifsson, G. (1997). The Poisson ratio of crystalline surfaces. *Europhys. Lett.*, 38, 67–72.

Falvo, M. R., Washburn, S., Superfine, R. *et al.* (1997). Manipulation of individual viruses: friction and mechanical properties. *Biophys. J.*, 72, 1396–1403.

Farago, O. and Kantor, Y. (2002). Entropic elasticity at the sol-gel transition. *Europhys. Lett.*, 57, 458–463.

Farago, O. and Pincus, P. (2003). The effect of thermal fluctuations on Schulman area elasticity. *Eur. Phys. J. E*, 11, 399–408.

Faucon, J. F., Mitov, M. D., Méléard, P., Bivas, I. and Bothorel, P. (1989). Bending elasticity and thermal fluctuations of lipid membranes. Theoretical analysis and experimental requirements. *J. Phys. France*, 50, 2389–2414.

Felgner, H., Frank, R. and Schliwa, M. (1996). Flexural rigidity of microtubules measured with the use of optical tweezers. *J. Cell Sci.*, 109, 509–516.

Feneberg, W., Aepfelbacher, M. and Sackmann, E. (2004). Microviscoelasticity of the apical cell surface of human umbilical vein endothelial cells (HUVEC) within confluent monolayers. *Biophys. J.*, 87, 1338–1350.

Feng, S. and Sen, P. N. (1984). Percolation on elastic networks: new exponent and thresholds. *Phys. Rev. Lett.*, 52, 216–219.

Feng, S., Thorpe, M. F. and Garboczi, E. (1985). Effective-medium theory of percolation on central-force elastic networks. *Phys. Rev. B*, 31, 276–280.

Ferry, J. D. (1980). *Viscoelastic Properties of Polymers*, 3rd edn. New York, NY: Wiley.

Feynmann, R. P., Leighton, R. B. and Sands, M. (1964). *The Feynmann Lectures on Physics*, vol. II. Reading, MA: Addison-Wesley.

Finer, J. T., Simmons, R. M. and Spudich, J. A. (1994). Single myosin molecule mechanics: piconewton forces and nanometre steps. *Nature*, 368, 113–119.

Florin, E.-L., Moy, V. T. and Gaub, H. E. (1994). Adhesion forces between individual ligand-receptor pairs. *Science*, 264, 415–417.

Flory, P. J. (1953). *Principles of Polymer Chemistry*. Ithaca, NY: Cornell University Press.

Flory, P. J. (1969). *Statistical Mechanics of Chain Molecules*. New York, NY: Wiley.

Flory, P. J. (1976). Statistical thermodynamics of random networks. *Proc. R. Soc. Lond.*, A351, 351–380.

Flügge, W. (1973). *Stresses in Shells*, 2nd edn. New York, NY: Springer-Verlag.

Forscher, P. and Smith, S. J. (1988). Actions of cytochalasins on the organization of actin filaments and microtubules in a neuronal growth cone. *J. Cell Biol.*, 107, 1505–1516.

Foty, R. A., Forgacs, G., Pfleger, C. M. and Steinberg, M. S. (1994). Liquid properties of embryonic tissues: measurement of interfacial tensions. *Phys. Rev. Lett.*, 72, 2298–2301.

Fourcade, B., Mutz, M. and Bensimon, D. (1992). Experimental and theoretical study of toroidal vesicles. *Phys. Rev. Lett.*, 68, 2551–2554.

Fourcade, B., Miao, L., Rao, M., Wortis, M. and Zia, R. K. P. (1994). Scaling analysis of narrow necks in curvature models of fluid lipid-bilayer vesicles. *Phys. Rev. E*, 49, 5276–5286.

Fournier, L. and Joos, B. (2003). Lattice model for the kinetics of rupture of fluid bilayer membranes. *Phys. Rev. E*, 67, 051908.

Francis, N. R., Irikura, V. M., Yamaguchi, S., DeRosier, D. J. and Macnab, R. M. (1992). Localization of the *Salmonella typhimurium* flagella switch protein FliG to the cytoplasmic M-ring face of the basal body. *Proc. Natl. Acad. Sci. USA*, 89, 6304–6308.

Fromhertz, P. (1983). Lipid-vesicle structure: size control by edge-active agents. *Chem. Phys. Lett.*, 94, 259–266.

Fromhertz, P., Röcker, C. and Rüppel, D. (1986). From discoid micelles to spherical vesicles: the concept of edge activity. *Faraday Discuss. Chem. Soc.*, 81, 39–48.

Fudge, D. S. and Gosline, J. M. (2004). Molecular design of the α-keratin composite: insights from a matrix-free model, hagfish slime threads. *Proc. Roy. Soc. Lond. B*, 271, 291–299.

Fujimoto, B. S. and Schurr, J. M. (1990). Dependence of the torsional rigidity of DNA on base composition. *Nature*, 344, 175–178.

Fujimoto, B. S., Brewood, G. P. and Schurr, J. M. (2006). Torsional rigidities of weakly strained DNAs. *Biophys. J.*, 91, 4166–4179.

Fuller, N., Benatti, C. R. and Rand, R. P. (2003). Curvature and bending constants for phosphatidylserine-containing membranes. *Biophys. J.*, 85, 1667–1674.

Fung, Y. C. (1994). *A First Course in Continuum Mechanics*. Englewood Cliffs, NJ: Prentice-Hall.

Furukawa, R., Kundra, R. and Fechheimer, M. (1993). Formation of liquid crystals from actin filaments. *Biochem.*, 32, 12 346–12 352.

Gardel, M. L., Valentine, M. T., Crocker, J. C., Bausch, A. R. and Weitz, D. A. (2003). Microrheology of entangled F-actin solutions. *Phys. Rev. Lett.*, 91, 158302.

Gardner, T. S., Cantor, C. R. and Collin, J. J. (2000). Construction of a genetic toggle switch in *Escherichia coli*. *Nature*, 403, 339–342.

Garner, E. C., Campbell, C. S. and Mullins, R. D. (2004). Dynamic instability in a DNA-segregating prokaryotic actin homolog. *Science*, 306, 1021–1025.

Gennis, R. B. (1989). *Biomembranes: Molecular Structure and Function*. New York, NY: Springer-Verlag.

Gevorkian, S. G., Allahverdyan, A. E., Gevorgyan, D. S. and Simonian, A. L. (2009). Thermal (in)stability of type I collagen fibrils. *Phys. Rev. Lett.*, 102, 048101.

Gittes, F. and MacKintosh, F. C. (1998). Dynamic shear modulus of a semiflexible polymer network. *Phys. Rev. E*, 58, R1241–R1244.

Gittes, F., Mickey, B., Nettleton, J. and Howard, J. (1993). Flexural rigidity of microtubules and actin filaments measured from thermal fluctuations in shape. *J. Cell Biol.*, 120, 923–934.

Glaus, U. (1988). Monte Carlo study of self-avoiding surfaces. *J. Stat. Phys.*, 50, 1141–1166.

Gō, N. and Taketomi, H. (1978). Respective roles of short- and long-range interactions in protein folding. *Proc. Nat. Acad. Sci.*, 75, 559–563.

Goetz, A. (1970). *Introduction to Differential Geometry*. Reading, MA: Addison-Wesley.

Goetz, R. and Lipowsky, R. (1998). Computer simulations of bilayer membranes: self-assembly and interfacial tension. *J. Chem. Phys.*, 108, 7397–7407.

Goetz, R., Gompper, G. and Lipowsky, R. (1999). Mobility and elasticity of self-assembled membranes. *Phys. Rev. Lett.*, 82, 221–224.

Goldberg, M. B. and Theriot, J. A. (1995). *Shigella flexneri* surface protein IcsA is sufficient to direct actin-based motility. *Proc. Natl. Acad. Sci. USA.*, 92, 6572–6576.

Goldstein, R. E., Nelson, P., Powers, T. and Seifert, U. (1996). Front propagation in the pearling instability of tubular vesicles. *J. Phys. II*, 6, 767–796.

Gompper, G. and Kroll, D. M. (1989). Steric interactions in multimembrane systems: a Monte Carlo study. *Europhys. Lett.*, 9, 59–64.

Gompper, G. and Kroll, D. M. (1991). Fluctuations of a polymerized membrane between walls. *J. Phys. France*, 1, 1411–1432.

Gompper, G. and Kroll, D. M. (1995). Phase diagram and scaling behavior of fluid vesicles. *Phys. Rev. E*, 51, 514–525.

Gompper, G. and Kroll, D. M. (1997). Freezing flexible vesicles. *Phys. Rev. Lett.*, 78, 2859–2862.

Gompper, G. and Kroll, D. M. (2004). Triangulated-surface models of fluctuating membranes. In *Statistical Mechanics of Membranes and Surfaces*, eds. D. Nelson, T. Piran and S. Weinberg, pp. 359–426. Singapore: World Scientific.

Gompper, G. and Schick, M. (1994). Self-assembling amphiphilic systems. In *Phase Transitions and Critical Phenomena*, Vol. 16, eds. C. Domb and J. L. Lebowitz. London: Academic Press.

González, J. M., Jiménez, M., Vélez, M. *et al.* (2003). Essential cell division protein FtsZ assembles into one monomer-thick ribbons under conditions resembling the crowded intracellular environment. *J. Biol. Chem.*, 278, 37 664–37 671.

Goodell, E. W. (1985). Recycling of murein by *Escherichia coli*. *J. Bacteriol.*, 163, 305–310.

Goodsell, D. S. (1993). *The Machinery of Life*. New York, NY: Springer.

Goulian, M., Lei, N., Miller, J. and Sinha, S. K. (1992). Structure factor for randomly oriented self-affine membranes. *Phys. Rev. A*, 46, R6170–R6173.

Gov, N., Zilman, A. G. and Safran, S. (2003). Cytoskeleton confinement and tension of red blood cell membranes. *Phys. Rev. Lett.*, 90, 228101.

Gov, N., Zilman, A. G. and Safran, S. (2005). Red blood cell membrane fluctuations and shape controlled by ATP-induced cytoskeletal defects. *Biophys. J.*, 88, 1859–1874.

Gradshteyn, I. S. and Ryzhik, I. M. (1980). *Table of Integrals, Series and Products*, 4th edn. New York, NY: Academic Press.

Green, N. M. (1975). Avidin. *Adv. Prot. Chem.*, 29, 85–133.

Gregor, T., Tank, D. W., Wieschaus, E. F. and Bialek, W. (2007). Probing the limits to positional information. *Cell*, 130, 153–164.

Grest, G. S. and Murat, M. (1993). Structure of grafted polymeric brushes in solvents of varying quality: a molecular dynamics study. *Macromolecules*, 26, 3108–3117.

Griffith, A. A. (1921). The phenomena of rupture and flow in solids. *Phil. Trans. R. Soc. Lond.*, A221, 163–198.

Griffith, L. M. and Pollard, T. D. (1982). The interaction of actin filaments with microtubules and microtubule-associated proteins. *J. Biol. Chem.*, 257, 9143–9151.

Grønbech-Jensen, N., Mashl, R. J., Bruinsma, R. F. and Gelbart, W. M. (1997). Counterion-induced attraction between rigid polyelectrolytes. *Phys. Rev. Lett.*, 78, 2477–2480.

Gross, D. J. (1984). The size of random surfaces. *Phys. Lett.*, 138B, 185–190.

Grover, N. B., Woldringh, C. L. and Koppes, L. J. (1987). Elongation and surface extension of individual cells of *Escherichia coli* B/r: comparison of theoretical and experimental size distributions. *J. Theor. Biol.*, 129, 337–348.

Guitter, E. and Palmeri, J. (1992). Tethered membranes with long-range interactions. *Phys. Rev. A*, 45, 734–744.

Guitter, E., David, F., Leibler, S. and Peliti, L. (1989). Thermodynamical behavior of polymerized membranes. *J. Phys. France*, 50, 1787–1819.

Gutsmann, T., Fantner, G. E., Kindt, J. H., Venturoni, M. and Danielsen, S. (2004). Force spectroscopy of collagen fibers to investigate their mechanical properties and structural organization. *Biophys. J.*, 86, 3186–3193.

Ha, B.-Y. (2003). Effects of divalent counterions on asymmetrically charged lipid bilayers. *Phys. Rev. E*, 67, 030901(R).

Ha, B.-Y. and Liu, A. J. (1997). Counterion-mediated attraction between two like-charged rods. *Phys. Rev. Lett.*, 79, 1289–1292.

Hale, C. A., Meinhardt, H. and de Boer, P. A. J. (2001). Dynamic localization cycle of the cell division regulator MinE in *Escherichia coli*. *EMBO*, 20, 1563–1572.

Haluska, C. K., Riske, K. A., Marchi-Artzner, V. *et al.* (2006). Time scales of membrane fusion revealed by direct imaging of vesicle fusion with high temporal resolution. *Proc. Nat. Acad. Sci. USA*, 103, 15 841–15 846.

Harbich, W., Servuss, R. M. and Helfrich, W. (1976). Optical studies of lecithin-membrane melting. *Phys. Lett.*, 57A, 294–296.

Harley, R., James, D., Miller, A. and White, J. W. (1977). Phonons and the elastic moduli of collagen and muscle. *Nature*, 267, 285–287.

Hategan, A., Law, R., Kahn, S. and Discher, D. E. (2003). Adhesively tensed cell membranes: lysis kinetics and atomic force microscopy probing. *Biophys. J.*, 85, 2746–2759.

Hategan, A., Sengupta, K., Kahn, S., Sackmann, E. and Discher, D. E. (2004). Topographical pattern dynamics in passive adhesion of cell membranes. *Biophys. J.*, 87, 3547–3560.

Heinrich, V., Ritchie, K., Mohandas, N. and Evans, E. (2001). Elastic thickness compressibility of the red cell membrane. *Biophys. J.*, 81, 1452–1463.

Heins, S., Wong, P. C., Müller, S. *et al.* (1993). The rod domain of NF-L detemines neurofilament architecture, whereas the end domains specify filament assembly and network formation. *J. Cell Biol.*, 123, 1517–1533.

Helfrich, W. (1973). Elastic properties of lipid bilayers: theory and possible experiments. *Z. Naturforsch.*, 28c, 693–703.

Helfrich, W. (1974a). The size of bilayer vesicles generated by sonication. *Phys. Lett.*, 50A, 115–116.

Helfrich, W. (1974b). Blocked lipid exchanges in bilayers and its possible influence on the shape of vesicles. *Z. Naturforsch.*, 29c, 510–515.

Helfrich, W. (1975). Out-of-plane fluctuations of lipid bilayers. *Z. Naturforsch.*, 30c, 841–842.

Helfrich, W. (1978). Steric interaction of fluid membranes in multilayer systems. *Z. Naturforsch.*, 33a, 305–315.

Helfrich, W. (1986). Size distributions of vesicles: the role of the effective rigidity of membranes. *J. Phys. France*, 47, 321–329.

Helfrich, W. (1998). Stiffening of fluid membranes and entropy loss of membrane closure: two effects of thermal undulations. *Eur. Phys. J. B*, 1, 481–489.

Helfrich, W. and Servuss, R.-M. (1984). Undulations, steric interaction and cohesion of fluid membranes. *Nuovo Cimento*, 3D, 137–151.

Hellan, K. (1984). *Introduction to Fracture Mechanics*. New York: McGraw-Hill.

Heller, H., Schaefer, M. and Schulten, K. (1993). Molecular dynamics simulation of a bilayer of 200 lipids in the gel and in the liquid crystal phase. *J. Phys. Chem.*, 97, 8343–8360.

Hénon, S., Lenormand, G., Richert, A. and Gallet, F. (1999). A new determination of the shear modulus of the human erythrocyte membrane using optical tweezers. *Biophys. J.*, 76, 1145–1151.

Henriksen, J., Rowat, A. C., Brief, E. *et al.* (2006). Universal behavior of membranes with sterols. *Biophys. J.*, 90, 1639–1649.

Herrmann, H., Häuer, M., Brettel, M., Ku, N.-O, and Aebi, U. (1999). Characterization of distinct early assembly units of different intermediate filament proteins. *J. Mol. Biol.*, 286, 1403–1420.

Heuser, J. (1980). Three-dimensional visualization of coated vesicle formation in fibroblasts. *J. Cell Biol.*, 84, 560–583.

Heuser, J. E. (1983). Procedure for freeze-drying molecules adsorbed to mica flakes. *J. Cell Biol.*, 169, 155–195.

Heussinger, C. and Frey, E. (2006). Floppy modes and nonaffine deformations in random fiber networks. *Phys. Rev. Lett.*, 97, 105501.

Higgins, M. J., Sader, J. E. and Jarvis, S. P. (2006). Frequency modulation atomic force microscopy reveals individual intermediates associated with each unfolded I27 titin domain. *Biophys. J.*, 90, 640–647.

Hill, T. L. (1981). Microfilament or microtubule assembly or disassembly against a force. *Proc. Natl. Acad. Sci. USA.*, 78, 5613–5617.

Hille, B. (2001). *Ion Channels of Excitable Membranes,* 3rd edn. Sunderland, Mass.: Sinauer.

Hinner, B., Tempel, M., Sackmann, E., Kroy, K. and Frey, E. (1998). Entanglement, elasticity, and viscous relaxation of actin solutions. *Phys. Rev. Lett.*, 81, 2614–2617.

Hirokawa, N. (1982). Cross-linker system between neurofilaments, microtubules and membranous organelles in frog axons revealed by the quick-freeze, deep etching method. *J. Cell Biol.*, 94, 129–142.

Hochmuth, R. M. and Marcus, W. D. (2002). Membrane tethers formed from blood cells with available area and determination of their adhesion energy. *Biophys. J.*, 82, 2964–2969.

Hochmuth, R. M. and Needham, D. (1990). The viscoelasticity of neutrophils and their transit times through small pores. *Biorheol.*, 27, 817–828.

Hodgkin, A. L. and Huxley, A. F. (1952). A quantitative description of membrane current and its application to conduction and excitation in nerve. *J. Physiol.*, 117, 500–544.

Hofmann, H. J. (1976). Precambrian microflora, Belcher Islands, Canada: significance and systematics. *J. Paleontology*, 50, 1040–1073.

Hohenadl, M., Storz, T., Kirpal, H., Kroy, K. and Merkel, R. (1999). Desmin filaments studied by quasi-elastic light scattering. *Biophys. J.*, 77, 2199–2209.

Holley, M. C. and Ashmore, J. F. (1990). Spectrin, actin and the structure of the cortical lattice in mammalian cochlear outer hair cells. *J. Cell Sci.*, 96, 283–291.

Höltje, J.-V. (1993). "Three-for-one" – a simple growth mechanism that guarantees a precise copy of the thin rod-shaped murein sacculus of *Escherichia coli*. In *Bacterial Growth and Lysis*, eds. M. A. de Pedro, V.-J. Höltje and W. Löffelhardt. New York, NY: Plenum.

Höltje, J.-V. (1998). Growth of stress-bearing and shape-maintaining murein sacculus of *Escherichia coli*. *Microbiol. Mol. Biol. Rev.*, 62, 181–203.

Horio, T. and Hotani, H. (1986). Visualization of the dynamic instability of individual microtubules by dark-field microscopy. *Nature*, 321, 605–607.

Horowitz, D. S. and Wang, J. C. (1984). Torsional rigidity of DNA and length dependence of the free energy of DNA supercoiling. *J. Mol. Biol.*, 173, 75–91.

Hotani, H. and Horio, T. (1988). Dynamics of microtubules visualized by darkfield microscopy: treadmilling and dynamic instability. *Cell Motil. Cytoskeleton*, 10, 229–236.

Howard, J. (1995). The mechanics of force generation by kinesin. *Biophys. J.*, 68, 245s–255s.

Howard, J. (2001). *Mechanics of Motor Proteins and the Cytoskeleton*. Sunderland, Mass.: Sinauer.

Howard, M., Rutenberg, A. D. and de Vet, S. (2001). Dynamic compartmentalization of bacteria: accurate division in *E. coli*. *Phys. Rev. Lett.*, 87, 278102.

Hu, B., Yang, G., Zhao, W., Zhang, Y. and Zhao, J. (2007). MreB is important for cell shape but not for chromosomal segregation of the filamentous cyanobacterium *Anabaena* sp. PCC7120. *Mol. Microbiol.*, 63, 1640–1652.

Hu, Z. and Lutkenhaus, J. (1999). Topological regulation of cell division in *Escherichia coli* involves rapid pole to pole oscillation of the division inhibitor MinC under the control of MinD and MinE. *Mol. Microbiol.*, 34, 82–90.

Huang, C. and Mason, J. T. (1978). Geometrical packing constraints in egg phosphatidylcholine vesicles. *Proc. Natl. Acad. Sci. USA*, 75, 308–310.

Huang, K. C., Meir, Y. and Wingreen, N. S. (2003). Dynamic structures in *Escherichia coli*: spontaneous formation of MinE rings and MinD polar zones. *Proc. Nat. Acad. Sci. USA*, 100, 12 724–12 728.

Hudson, N. E., Houser, J. R., O'Brien III, E. T., Taylor II, R. M. and Superfine, R. (2010). Stiffening of individual fibrin fibers equitably distributes strain and strengthens networks. *Biophys. J.*, 98, 1632–1640.

Huisman, E. M., Storm, C. and Barkema, G. T. (2008). Monte Carlo study of multiply crosslinked semiflexible polymer networks. *Phys. Rev. E*, 78, 051801.

Hunt, A. J. and Howard, J. (1993). Kinesin swivels to permit microtubule movement in any direction. *Proc. Nat. Acad. Sci. USA*. 90, 11 653–11 657.

Hunt, A. J., Gittes, F. and Howard, J. (1994). The force extended by a single kinesin molecule against a viscous load. *Biophys. J.*, 67, 766–781.

Hwang, W. C. and Waugh, R. E. (1997). Energy of dissociation of lipid bilayer from the membrane skeleton of red blood cells. *Biophys. J.*, 72, 2669–2678.

Hynes, R. O. (1992). Integrins: versatility, modulation and signalling in cell adhesion. *Cell*, 69, 11–25.

Ingber, D. E. (1997). Tensegrity: the architectural basis of cellular mechanotransduction. *Ann. Rev. Physiol.*, 59, 575–599.

Isambert, H. and Maggs, A. C. (1996). Dynamics and rheology of actin solutions. *Macromolecules*, 29, 1036–1040.

Isambert, H., Venier, P., Maggs, A. C. *et al.* (1995). Flexibility of actin filaments derived from thermal fluctuations: effect of bound nucleotide, phalloidin, and muscle regulatory proteins. *J. Biol. Chem.*, 270, 11 437–11 444.

Israelachvili, J. N. (1991). *Intermolecular and Surface Forces*, 2nd edn. London: Academic Press.

Israelachvili, J. N., Mitchell, D. J. and Ninham, B. W. (1976). Theory of self-assembly of hydrocarbon amphiphiles into micelles and bilayers. *J. Chem. Soc. Faraday Trans.*, II 72, 1525–1568.

Itan, E., Carmon, G., Rabinovitch, A., Fishov, I. and Feingold, M. (2008). Shape of nonseptated *Escherichia coli* is asymmetric. *Phys. Rev. E*, 77, 061902.

Iwazawa, J., Imae, Y. and Kobayasi, S. (1993). Study of the torque of the bacterial flagellar motor using a rotating electric field. *Biophys. J.*, 64, 925–933.

James, H. M. and Guth, E. (1943). *J. Chem. Phys.*, 11, 470.

Janke, W. and Kleinert, H. (1987). Fluctuation pressure of a stack of membranes. *Phys. Rev. Lett.*, 58, 144–147.

Janmey, P. A., Hvidt, S., Lamb, J. and Stossel, T. P. (1990). Resemblence of actin-binding protein/actin gels to covalently crosslinked networks. *Nature*, 345, 89–92.

Janmey, P. A., Euteneuer, U., Traub, P. and Schliwa, M. (1991). Viscoelastic properties of vimentin compared with other filamentous biopolymer networks. *J. Cell Biol.*, 113, 155–159.

Janmey, P. A., Hvidt, S., Käs, J. *et al.* (1994). The mechanical properties of actin gels. Elastic modulus and filament motions. *J. Biol. Chem.*, 269, 32 503–32 513.

Janson, M. E. and Dogterom, M. (2004). A bending mode analysis for growing microtubules: evidence for a velocity-dependent rigidity. *Biophys. J.*, 87, 2723–2736.

Jarzynski, C. (1977). Nonequilibrium equality for free energy differences. *Phys. Rev. Lett.*, 78, 2690–2693.

Jeppesen, C. and Ipsen, J. H. (1993). Scaling properties of self-avoiding surfaces with free topology. *Europhys. Lett.*, 22, 713–716.

Jiang, F.-Y., Bouret, Y. and Kindt, J. T. (2004). Molecular dynamics simulations of the lipid bilayer edge. *Biophys. J.*, 87, 182–192.

Jones, L. J. F., Carballido-López, R. and Errington, J. (2001). Control of cell shape in bacteria: helical, actin-like filaments in *Bacillus subtilis*. *Cell*, 104, 913–922.

Jülicher, F. (1996). The morphology of vesicles of higher topological genus: conformal degeneracy and conformal modes. *J. Phys. II France*, 6, 1797–1824.

Jun, S. and Mulder, B. (2006). Entropy-driven spatial organization of highly confined polymers: lessons for the bacterial chromosome. *Proc. Nat. Acad. Sci. USA*, 103, 12388–12393.

Kahn, J. D., Yun, E. and Crothers, D. M. (1994). Detection of localized DNA flexibility. *Nature,* 368, 163–166.

Kang, H., Wen, Q., Janmey, P. A. *et al.* (2009). Nonlinear elasticity of stiff filament networks: strain stiffening, negative normal stress and filament alignment in fibrin gels. *J. Phys. Chem. B*, 113, 3799–3805.

Kantor, Y. (2004). Properties of tethered surfaces. In *Statistical Mechanics of Membranes and Surfaces*, eds. D. Nelson, T. Piran and S. Weinberg, pp. 111–130. Singapore: World Scientific.

Kantor, Y. and Nelson, D.R. (1987). Crumpling transition in polymerized membranes. *Phys. Rev. Lett.*, 58, 2774–2777.

Kantor, Y., Kardar, M. and Nelson, D. R. (1986). Statistical mechanics of tethered surfaces. *Phys. Rev. Lett.*, 57, 791–794.

Kantsler, V. and Steinberg, V. (2006). Transition to tumbling and two regimes of tumbling motion of a vesicle in shear flow. *Phys. Rev. Lett.*, 96, 036001.

Karatekin, E., Sandre, O., Guitouni, H. *et al.* (2003). Cascades of transient pores in giant vesicles: line tension and transport. *Biophys. J.*, 84, 1734–1749.

Käs, J. and Sackmann, E. (1991). Shape transitions and shape stability of giant phospholipid vesicles in pure water induced by area-to-volume changes. *Biophys. J.*, 60, 825–844.

Käs, J., Strey, H., Tang, J. X., *et al.* (1996). F-actin, a model polymer for semiflexible chains in dilute, semidilute and liquid crystalline solutions. *Biophys. J.*, 70, 609–625.

Kauzmann, W. (1966). *Kinetic Theory of Gases*. New York, NY: Benjamin.

Kasza, K. E., Broedersz, C. P., Koenderink, G. H. *et al.* (2010). Actin filament length tunes elasticity of flexibly cross-linked actin networks. *Biophys. J.*, 99, 1091–1100.

Keller, D., Swigon, D. and Bustamante, C. (2003). Relating single-molecule measurements to thermodynamics. *Biophys. J.*, 84, 733–738.

Kessler, D. A. and Rabin, Y. (2003). Effect of curvature and twist on the conformations of a fluctuating ribbon. *J. Chem. Phys.*, 118, 897–904.

Khan, S. and Macnab, R. M. (1980). The steady-state counterclockwise / clockwise ratio of bacterial flagellar motors is regulated by protonmotive force. *J. Mol. Biol.*, 138, 563–597.

Kikumoto, M., Kurachi, M., Tosa, V. and Tashiro, H. (2006). Flexural rigidity of individual microtubules measured by a buckling force with optical traps. *Biophys. J.*, 90, 1687–1696.

Killander, D. and Zetterberg, A. (1965). A quantitative cytochemical investigation of the relationship between cell mass and initiation of DNA synthesis in mouse fibroblasts *in vitro*. *Exp. Cell Res.*, 40, 12–20.

Kirkwood, J. G. and Auer, P. L. (1951). The visco-elastic properties of solutions of rod-like macromolecules. *J. Chem. Phys.*, 19, 281–283.

Kis, A., Kasas, S., Babíc, B. *et al.* (2002). Nanomechanics of microtubules. *Phys. Rev. Lett.*, 89, 248101.

Kittel, C. (1966). *Introduction to Solid State Physics,* 3rd edn. New York, NY: Wiley.

Koch, A. L. (1990). Growth and form of the bacterial cell wall. *Amer. Scientist*, 78, 327–341.

Koch, A. L. (1993). Stresses on the surface stress theory. In *Bacterial Growth and Lysis*, eds. M. A. de Pedro, V.-J. Höltje and W. Löffelhardt. New York, NY: Plenum.

Koch, A. L. (2006). The exocytoskeleton. *J. Mol. Microbiol. Biotechnol.*, 11, 115–125.

Koch, A. L. and Woeste, S. (1992). Elasticity of the sacculus of *Escherichia coli*. *J. Bacteriol.*, 174, 4811–4819.

Koenig, B. W., Strey, H. H. and Gawirsch, K. (1997). Membrane lateral compressibility determined by NMR and X-ray diffraction: effect of acyl chain unsaturation. *Biophys. J.*, 73, 1954–1966.

Komura, S. and Baumgärtner, A. (1991). Tethered vesicles at constant pressure: Monte Carlo study and scaling analysis. *Phys. Rev. A*, 44, 3511–3518.

Koppes, L. J. H. and Nanninga, N. (1980). Positive correlation between size at initiation of chromosome replication in *Escherichia coli* and size at initiation of cell constriction. *J. Bacteriol.*, 143, 89–99.

Korn, E. D., Carlier, M.-F. and Pantaloni, D. (1987). Actin polymerization. *Science*, 238, 638–644.

Kraus, M., Wintz, W., Seifert, U. and Lipowsky, R. (1996). Fluid vesicles in shear flow. *Phys. Rev. Lett.*, 77, 3685–3688.

Kreyszig, E. (1959). *Differential Geometry*. Toronto, ON: University of Toronto Press.

Kroll, D. M. and Gompper, G. (1992). The conformation of fluid membranes: Monte Carlo simulations. *Science*, 255, 968–971.

Kroy, K. and Frey, E. (1996). Force–extension relation and plateau modulus for worm-like chains. *Phys. Rev. Lett.*, 77, 306–309.

Kubori, T., Shimamoto, N., Yamaguchi, S., Namba, K. and Aizawa, S.-I. (1992). Morphological pathway of flagellar assembly in *Salmonella typhimurium*. *J. Mol. Biol.*, 266, 433–446.

Kuhn, W. and Grün, F. (1942). *Kolloid Z.*, 101, 248.

Kumar, P. B. S., Gompper, G. and Lipowsky, R. (1999). Modulated phases in multicomponent fluid membranes. *Phys. Rev. E*, 60, 4610–4618.

Künneke, S., Krüger, D. and Janshoff, A. (2004). Scrutiny of the failure of lipid membranes as a function of headgroups, chain length, and lamellarity measured by scanning force microscopy. *Biophys. J.*, 86, 1545–1553.

Kuo, S. and Sheetz, M. P. (1993). Force of single kinesin molecules measured by optical tweezers. *Science*, 260, 232–234.

Kurz, J. C. and Williams, R. C. Jr. (1995). Microtubule-associated proteins and the flexibility of microtubules. *Biochemistry*, 34, 13 374–13 380.

Kusumi, A. and Sako, Y. (1996). Cell surface organization by the membrane skeleton. *Curr. Opin. Cell Biol.*, 8, 566–574.

Kusumi, A., Shirai, Y. M., Koyama-Honda, I., Suzuki, K. G. N. and Fujiwara, T. K. (2010). Hierarchical organization of the plasma membrane: investigations by single-molecule tracking *vs.* fluorescence correlation spectroscopy. *FEBS Letter*, 584, 1814–1823.

Kwok, R. and Evans, E. (1981). Thermoelasticity of large lecithin bilayer vesicles. *Biophys. J.*, 35, 637–652.

Lai, P.-Y. and Binder, K. (1992). Structure and dynamics of polymer brushes near the Θ-point: a Monte Carlo simulation. *J. Chem. Phys.*, 97, 586–595.

Laidler, K. J. (1987). *Chemical Kinetics*, 3rd edn. New York, NY: Harper and Row.

Lal, A. A., Korn, E. D. and Brenner, S.L. (1984). Rate constants for actin polymerization in ATP determined using cross-linked actin trimers as nuclei. *J. Biol. Chem.*, 259, 8794–8800.

Lammert, P. and Discher, D. E. (1998). Tethered networks in two dimensions: a low temperature view. *Phys. Rev. E*, 57, 4386–4374.

Landau, L. D. and Lifshitz, E. M. (1986). *Theory of Elasticity*. Oxford: Pergamon Press.

Landau, L. D. and Lifshitz, E. M. (1987). *Fluid Mechanics*, 2nd edn. Oxford: Pergamon Press.

Lange, Y., Hadesman, R. A. and Steck, T. L. (1982). Role of the reticulum in the stability and shape of the isolated human erythrocyte membrane. *J. Cell Biol.*, 92, 714–721.

Lau, K. F. and Dill., K. A. (1989). A lattice statistical mechanics model of the conformational and sequence space of proteins. *Macromolecules*, 22, 3986–3997.

Lauffenburger, D. A. and Horwitz, A. F. (1996). Cell migration: a physically integrated molecular process. *Cell*, 84, 359–369.

Law, R., Carl, P., Harper, S. *et al.* (2003a). Cooperativity in forced unfolding of tandem spectrin repeats. *Biophys. J.*, 84, 533–544.

Law, R., Liao, G., Harper, S. *et al.* (2003b). Pathway shifts and thermal softening in temperature-coupled forced unfolding of spectrin domains. *Biophys. J.*, 85, 3286–3293.

Leake, M. C., Wilson, D., Gautel, M. and Simmons, R. M. (2004). The elasticity of single titin molecules using a two-bead optical tweezers assay. *Biophys. J.*, 87, 1112–1135.

Lebedev, V. V., Turitsyn, K. S. and Vergeles, S. S. (2007). Dynamics of nearly spherical vesicles in an external flow. *Phys. Rev. Lett.*, 99, 218101.

Le Doussal, P. and Radzihovsky, L. (1992). Self-consistent theory of polymerized membranes. *Phys. Rev. Lett.*, 69, 1209–1212.

Lee, C.-H., Lin, W.-C. and Wang, J. (2001). All-optical measurements of the bending rigidity of lipid-vesicle membranes across structural phase transitions. *Phys. Rev. E*, 64, 020901(R).

Lee, J., Ishihara, A., Theriot, J. A. and Jacobson, K. (1993). Principles of locomotion for simple-shaped cells. *Nature*, 362, 167–171.

Lehninger, A. L., Nelson, D. L. and Cox, M. M. (1993). *Principles of Biochemistry*. New York, NY: Worth.

Lei, G. and MacDonald, R. C. (2003). Lipid bilayer vesicle fusion: intermediates captured by high-speed microfluorescence spectroscopy. *Biophys. J.*, 85, 1585–1599.

Leibler, S. (1989). Equilibrium statistical mechanics of fluctuating films and membranes. In *Statistical Mechanics of Membranes and Surfaces*, eds. D. Nelson, T. Piran and S. Weinberg, pp. 45–103. Singapore: World Scientific.

Leibler, S. and Maggs, A. C. (1989). Entropic interactions between polymerized membranes. *Phys. Rev. Lett.*, 63, 406–409.

Leibler, S., Singh, R. R. P. and Fisher, M. E. (1987). Thermodynamic behavior of two-dimensional vesicles. *Phys. Rev. Lett.*, 59, 1989–1992.

Leidy, C., Kaasgaard, T., Crowe, J. H., Mouritsen, O. G. and Jorgensen, K. (2002). Ripples and the formation of anisotropic lipid domains: imaging two-component supported double bilayers by atomic force microscopy. *Biophys. J.*, 83, 2625–2633.

LeNeveu, D. M., Rand, R. P. and Parsegian, V. A. (1975). Measurement of forces between lecithin bilayers. *Nature*, 259, 601–603.

Lenormand, G., Hénon, S., Richert, A., Siméon, J. and Gallet, F. (2001). Direct measurement of the areaa expansion and shear moduli of the human red blood cell membrane skeleton. *Biophys. J.*, 81, 43–56.

Lenormand, G., Hénon, S., Richert, A., Siméon, J. and Gallet, F. (2003). Elasticity of the human red cell cytoskeleton. *Biorheology*, 40, 247–251.

Leontiadou, H., Mark, A. E. and Marrink, S. J. (2004). Molecular dynamics simulations of hydrophilic pores in lipid bilayers. *Biophys. J.*, 86, 2156–2164.

Levinson, E. A. (1991). Monte Carlo studies of crumpling for Sierpinski gaskets. *Phys. Rev. A*, 43, 5233–5239.

Lewis, B. A. and Engelman, D. A. (1983). Lipid bilayer thickness varies linearly with acyl chain length in fluid phosphatidylcholine vesicles. *J. Mol. Biol.*, 166, 211–217.

Li, B., Madras, N. and Sokal, A. D. (1995). Critical exponents, hyperscaling, and universal amplitude ratios for two- and three-dimensional self-avoiding walks. *J. Stat. Phys.*, 80, 661–754.

Li, C., Ru, C. Q. and Mioduchowski, A. (2006a). Length-dependence of flexural rigidity as a result of anisotropic elastic properties of microtubules. *Biochem. Biophys. Res. Comm.*, 349, 1145–1150.

Li, H., Helling, R., Tang, C. and Wingreen, N. (1996). Emergence of preferred structures in a simple model of protein folding. *Science*, 273, 666–669.

Li, L., Wetzel, S., Plückthun, A. and Fernandez, J. M. (2006b). Stepwise unfolding of ankyrin repeats in a single protein revealed by atomic force microscopy. *Biophys. J.*, 90, L30–L32.

Lifshitz, E. M. (1956). Theory of molecular attractive forces between solids. *Soviet Physics – JETP*, 2, 73–83.

Lim, G. H. W., Wortis, M. and Mukhopadhyay, R. (2002). Stromatocyte–discocyte–echinocyte sequence of the human red blood cell: evidence for the bilayer-couple hypothesis from membrane mechanics. *Proc. Nat. Acad. Sci. (USA)*, 99, 16 766–16 769.

Lim, G. H. W., Wortis, M. and Mukhopadhyay, R. (2008). Red blood cell shapes and shape transformations: Newtonian mechanics of a composite membrane. In Gompper, G. and Schick, M., eds., *Soft Matter, Vol. 4: Lipid Bilayers and Red Blood Cells*. Weinheim, Germany: Wiley-VCH Verlag.

Lin, C.-H. and Forscher, P. (1993). Cytoskeleton remodeling during growth cone-target interactions. *J. Cell Biol.*, 121, 1369–1383.

Liphardt, J., Onos, B., Smith, S. B., Tinoco, I. and Bustamante, C. (2001). Reversible unfolding of single RNA molecules by mechanical force. *Science*, 292, 733–737.

Lipowsky, R. (1994). Generic interactions of flexible membranes. In *Structure and Dynamics of Membranes*, eds. R. Lipowsky and E. Sackmann. Amsterdam: Elsevier.

Lipowsky, R. (1996). Adhesion of membranes *via* anchored stickers. *Phys. Rev. Lett.*, 77, 1652–1655.

Lipowsky, R. and Girardet, M. (1990). Shape fluctuations of polymerized or solid-like membranes. *Phys. Rev. Lett.*, 65, 2893–2896.

Lipowsky, R. and Leibler, S. (1986). Unbinding transitions of interacting membranes. *Phys. Rev. Lett.*, 56, 2541–2544; 59, 1983(E).

Lipowsky, R. and Seifert, U. (1991). Adhesion of membranes: a theoretical perspective. *Langmuir*, 7, 1867–1873.

Lipowsky, R. and Zielinska, B. (1989). Binding and unbinding of lipid membranes: a Monte Carlo study. *Phys. Rev. Lett.*, 62, 1572–1575.

Lipowsky, R., Brinkmann, M., Dimova, R. *et al.* (2005). Droplets, bubbles, and vesicles at chemically structured surfaces. *J. Phys.: Condens. Matter*, 17, S537–S558.

Lipowsky, R., Döbereiner, H.-G., Hiergeist, C. and Indrani, V. (1998). Membrane curvature induced by polymers and colloids. *Physica*, A249, 536–543.

Litster, J. D. (1975). Stability of lipid bilayers and red blood cell membranes. *Phys. Lett.*, 53A, 193–194.

Liu, D. and Plischke, M. (1992). Monte Carlo studies of tethered membranes with attractive interactions. *Phys. Rev. A*, 45, 7139–7144.

Liu, S.-C., Derick, L. H. and Palek, J. (1987). Visualization of the hexagonal lattice in the erythrocyte membrane skeleton. *J. Cell Biol.*, 104, 527–536.

López-García, P. (1998). DNA topoisomerases, temperature, adaptation and early diversification of life. In *Thermophiles: The Keys to Molecular Evolution and the Origin of Life?* J. Wiegel and M. W. W. Adams. London: Taylor and Francis.

Lorrain, P., Corson, D. R. and Lorrain, F. (1988). *Electromagnetic Fields and Waves*, 3rd edn. New York, NY: Freeman.

Love, A. E. H. (1944). *A Treatise on the Mathematical Theory of Elasticity*. New York: Dover.

Luisi, P. L. (2006). *The Emergence of Life: From Chemical Origins to Synthetic Biology*. Cambridge: Cambridge University Press.

Ly, H. V. and Longo, M. L. (2004). The influence of short-chain alcohols on interfacial tension, mechanical properties, area/molecule, and permeability of fluid lipid bilayers. *Biophys. J.*, 87, 1013–1033.

MacKintosh, F. C., Käs, J. and Janmey, P. A. (1995). Elasticity of semiflexible biopolymer networks. *Phys. Rev. Lett.*, 75, 4425–4428.

Maggs, A. C., Leibler, S., Fisher, M. E. and Camacho, C. J. (1990). The size of an inflated vesicle in two dimensions. *Phys. Rev. A*, 42, 691–695.

Magnasco, M. O. (1993). Forced thermal ratchets. *Phys. Rev. Lett.*, 71, 1477–1481.

Maniotis, A. J., Chen, C. S. and Ingber, D. E. (1997). Demonstration of mechanical connections between integrins, cytoskeletal filaments, and nucleoplasm that stabilize nuclear structure. *Proc. Natl. Acad. Sci.*, 94, 849–854.

Manson, M. D., Tedesco, P. M. and Berg, H. C. (1980). Energetics of flagellar rotation in bacteria. *J. Mol. Biol.*, 138, 541–561.

Marcelja, S. (1974). Chain ordering in liquid crystals II. Structure of bilayer membranes. *Biochim. Biophys. Acta*, 367, 165–176.

Marcelli, G., Parker, K. H. and Winlove, C. P. (2005). Thermal fluctuations of red blood cell membrane via constant-area particle dynamics model. *Biophys. J.*, 89, 2473–2480.

Marko, J. F. and Siggia, E. D. (1994). Fluctuations and supercoiling of DNA. *Science*, 265, 506–508.

Marko, J. F. and Siggia, E. D. (1995). Stretching DNA. *Macromolecules*, 28, 8759–8770.

Marra, J. and Israelachvili, J. N. (1985). Direct measurement of forces between phosphatidylcholine and phosphatidylethanolamine bilayers in aqueous electrolyte solutions. *Biochemistry*, 24, 4608–4618.

Marsh, D. (1990). *CRC Handbook of Lipid Bilayers*. Boca Raton, FL: CRC Press.

Marsh, D. (1996). Intrinsic curvature in normal and inverted lipid structures and in membranes. *Biophys. J.*, 70, 2248–2255.

Marsh, D. (1997). Renormalization of the tension and area expansion modulus in fluid membranes. *Biophys. J.*, 73, 865–869.

Marszalek, P. E., Oberhauser, A. F., Li, H. and Fernandez, J. M. (2003). The force-driven conformations of heparin studied with single molecule force microscopy. *Biophys. J.*, 85, 2696–2704.

Maruyama, K., Kaibara, M. and Fukuda, E. (1974). Rheology of F-actin II: effect of tropomyosin and troponin. *Biochim. Biophys. Acta*, 371, 30–38.

Mashi, R. J. and Bruinsma, R. (1998). Spontaneous curvature theory of clathrin coated membranes. *Biophys. J.*, 74, 2862–2875.

Maxwell, J. C. (1864). *Philos. Mag.*, 27, 294.

McGough, A. M. and Josephs, R. (1990). On the structure of erythrocyte spectrin in partially expanded membrane skeletons. *Proc. Natl. Acad. Sci. USA*, 87, 5208–5212.

McGrath, J. L., Tardy, Y., Dewey, C. F. Jr., Meister, J. J. and Hartwig, J. H. (1998). Simultaneous measurements of actin filament turnover, filament friction, and monomer diffusion in endothelial cells. *Biophys. J.*, 75, 2070–2078.

McIntosh, T. J. and Simon, S. A. (2007). Bilayers as protein solvents: role of bilayer structure and elastic properties. *J. Gen. Physiol.*, 130, 225–227.

McIntosh, T. J., Vidal, A. and Simon S. A. (2003). Sorting of lipids and transmembrane peptides between detergent-soluble bilayers and detergent-resistant rafts. *Biophys. J.*, 85, 1656–1666.

McKenzie, D. S. (1976). Polymers and scaling. *Phys. Rep.*, 27, 35–88.

McKeon, F. D., Kirschner, M. W. and Caput, D. (1986). Homologies in both primary and secondary structure between nuclear envelope and intermediate filament proteins. *Nature*, 319, 463–468.

McKiernan, A. E., MacDonald, R. I., MacDonald, R. C. and Axelrod, D. (1997). Cytoskeleton protein binding kinetics at planar phospholipid membranes. *Biophys. J.*, 73, 1987–1998.

Meier, T. and Fahrenholz, F. (1996). *A Laboratory Guide to Biotin-labelling in Biomolecule Analysis*. Basel: Birkhäuser Verlag.

Meinhardt, H. and de Boer, P. A. J. (2001). Pattern formation in *Escherichia coli*: a model for the pole-to-pole oscillations of Min proteins and the localization of the division site. *Proc. Nat. Acad. Sci. USA*, 98, 14202–14207.

Mejia, Y. X., Mao, H., Forde, N. R. and Bustamante, C. (2008). Thermal probing of *E. coli* RNA polymerase off-pathway mechanisms. *J. Mol. Biol.*, 382, 628–637.

Méléard, P., Gerbeaud, C., Pott, T. *et al.* (1997). Bending elasticities of model membranes: influences of temperature and sterol content. *Biophys. J.*, 72, 2616–2629.

Merkel, R., Nassoy, P., Leung, A., Ritchie, K. and Evans, E. (1999). Energy landscapes of receptor-ligand bonds explored with dynamic force spectroscopy. *Nature*, 397, 50–53.

Miao, L., Fourcade, B., Rao, M., Wortis, M. and Zia, R. K. P. (1991). Equilibrium budding and vesiculation in the curvature model of fluid lipid vesicles. *Phys. Rev. A*, 43, 6843–6856.

Miao, L., Seifert, U., Wortis, M. and Döbereiner, H.-G. (1994). Budding transitions of fluid-bilayer vesicles: the effect of area-difference elasticity. *Phys. Rev. E*, 49, 5389–5407.

Michalet, X., Bensimon, D. and Fourcade, B. (1994). Fluctuating vesicles of nonspherical topology. *Phys. Rev. Lett.*, 72, 168–171.

Mickey, B., and Howard, J. (1995). Rigidity of microtubules is increased by stabilizing agents. *J. Cell Biol.*, 130, 909–917.

Milner, S. T. (1991). Polymer brushes. *Science*, 251, 905–914.

Milner, S. T. and Safran, S. A. (1987). Dynamical fluctuations of droplet microemulsions and vesicles. *Phys. Rev. A*, 36, 4371–4379.

Mitaku, S., Ikegami, A. and Sakanishi, A. (1978). Ultrasonic studies of lipid bilayer. Phase transition in synthetic phosphatidylcholine liposomes. *Biophys. Chem.*, 8, 295–304.

Mitchison, T. J. and Cramer, L. P. (1996). Actin-based cell motility and cell locomotion. *Cell.*, 84, 371–379.

Mitchison, T. and Kirschner, M. (1984). Dynamic instability of microtubule growth. *Nature*, 312, 237–242.

Mitchison, T. and Salmon, E. D. (1992). Poleward kinetochore fiber movement occurs during both metaphase and anaphase-A in newt lung cell mitosis. *J. Cell Biol.*, 119, 569–582.

Miyagishima, S., Takahara, M., Mori, T. *et al.* (2001). Plastid division is driven by a complex mechanism that involves differential transition of the bacerial and eukaryotic division rings. *Plant Cell*, 13, 2257–2268.

Miyata, H. and Hotani, H. (1992). Morphological changes in liposomes caused by polymerization of encapsulated actin and spontaneous formation of actin bundles. *Proc. Natl. Acad. Sci. USA*, 89, 11 547–11 551.

Mobley, H. L. T., Koch, A. L., Doyle, R. J. and Streips, U. N. (1984). Insertion and the fate of the cell wall in *Bacillus subtilis*. *J. Bacteriol.*, 158, 169–179.

Mohanty, J. and Ninham, B. W. (1976). *Dispersion Forces.* London: Academic Press.

Moligner, A. and Oster, G. (1996). Cell motility driven by actin polymerization. *Biophys. J.*, 71, 3030–3045.

Moroz, D. and Nelson, P. (1997a). Dynamically stabilized pores in bilayer membranes. *Biophys. J.*, 72, 2211–2216.

Moroz, D. and Nelson, P. (1997b). Torsional directed walks, entropic elasticity, and DNA twist stiffness. *Proc. Nat. Acad. Sci. USA*, 94, 14 418–14 422.

Morse, D. C. (1998). Viscoelasticity of tightly entangled solutions of semiflexible polymers. *Phys. Rev. E*, 58, R1237–R1240.

Morse, D. C. and Milner, S. T. (1995). Statistical mechanics of closed fluid membranes. *Phys. Rev. E*, 52, 5918–5945.

Mouritsen, O. G. (1984). *Computer Simulations of Phase Transitions and Critical Phenomena.* Berlin: Springer-Verlag.

Mukhopadhyay, R., Emberly, E., Tang, T. and Wingreen, N. S. (2003). Statistical mechanics of RNA folding: importance of alphabet size. *Phys. Rev. E*, 68, 041904.

Müller, J., Oehler, S. and Müller-Hill, B. (1996). Repression of *lac* promoter as a function of distance, phase and quality of an auxiliary *lac* operator. *J. Mol. Biol.*, 257, 21–29.

Muroga, Y., Yamada, Y., Noda, I. and Nagasawa, M. (1987). Local conformation of polysaccharides in solution investigated by small-angle X-ray scattering. *Macromolecules*, 20, 3003–3006.

Mutz, M. and Bensimon, D. (1991). Observation of toroidal vesicles. *Phys. Rev. A*, 43, 4525–4527.

Mutz, M. and Helfrich, W. (1989). Unbinding transition of a biological model membrane. *Phys. Rev. Lett.*, 62, 2881–2884.

Mutz, M. and Helfrich, W. (1990). Bending rigidities of some biological model membranes as obtained from the Fourier analysis of contour sections. *J. Phys. France*, 51, 991–1002.

Nagle, J. and Tristram-Nagle, S. (2000). Structure of lipid bilayers. *Biochim. Biophys. Acta*, 1469, 159–195.

Nagy, A., Grama, L., Huber, T. *et al.* (2005). Hierarchical extensibility in the PEVK domain of skeletal-muscle titin. *Biophys. J.*, 89, 329–336.

Napper, D. H. (1983). *Polymeric Stabilization of Colloidal Dispersions*. London: Academic Press.

Needhan, D. and Hochmuth, R. M. (1989). Electromechanical permeabilization of lipid vesicles. *Biophys. J.*, 55, 1001–1009.

Needham, D., and Nunn, R. S. (1990). Elastic deformation and failure of lipid bilayer membranes containing cholesterol. *Biophys. J.*, 58, 997–1009.

Nelson, D. (2004). Theory of the crumpling transition. In *Statistical Mechanics of Membranes and Surfaces*, eds. D. Nelson, T. Piran and S. Weinberg, pp. 131–148. Singapore: World Scientific.

Nelson, D. R. and Halperin, B. I. (1979). Dislocation-mediated melting in two dimensions. *Phys. Rev. B*, 19, 2457–2484.

Nelson, D. R. and Peliti, L. (1987). Fluctuations in membranes with crystalline and hexatic order. *J. Physique*, 48, 1085–1092.

Nelson, P. C. (2003). *Biological Physics*. New York: Freeman.

Nelson, P., Powers, T. and Seifert, U. (1995). Dynamical theory of the pearling instability in cylindrical vesicles. *Phys. Rev. Lett.*, 74, 3384–3387.

Netz, R. R. and Lipowsky, R. (1993). Unbinding of symmetric and asymmetric stacks of membranes. *Phys. Rev. Lett.*, 71, 3596–3599.

Netz, R. R. and Lipowsky, R. (1995). Stacks of fluid membranes under pressure and tension. *Europhys. Lett.*, 29, 345–350.

Netz, R. R. and Schick, M. (1996). Pore formation and rupture in fluid bilayers. *Phys. Rev. E*, 53, 3875–3885.

Neuman, K. C., Lionnet, T. and Allemand, J.-F. (2007). Single molecule micromanipulation techniques. *Ann. Rev. Mat. Res.*, 37, 33–67.

Neumann, D. (2003). On the precise meaning of extension in the interpretation of polymer-chain stretching experiments. *Biophys. J.*, 85, 3418–3420.

Nicklas, R. B. (1983). Measurements of the force produced by the mitotic spindle in anaphase. *J. Cell Biol.*, 97, 542–548.

Nicklas, R. B. (1988). The forces that move chromosomes in mitosis. *Ann. Rev. Biophys. Biophys. Chem.*, 17, 431–449.

Nielsen, M. (1999). Ph. D. thesis (McGill).

Nienhuis, B. (1982). Exact critical point and critical exponents of $O(n)$ models in two dimensions. *Phys. Rev. Lett.*, 49, 1062–1065.

Ninham, B. W., Parsegian, V. A. and Weiss, G. H. (1970). On the macroscopic theory of temperature-dependent van der Waals forces. *J. Stat. Phys.*, 2, 323–328.

Noguchi, H. and Gompper, G. (2005). Shape transitions of fluid vesicles in capillary flows. *Proc. Nat. Acad. Sci. USA*, 102, 14 159–14 164.

Noguchi, H. and Gompper, G. (2007). Swinging and tumbling of fluid vesicles in shear flow. *Phys. Rev. Lett.*, 98, 128103.

Oakley, B. R. (1994). γ-Tubulin. In *Microtubules*, eds. J. S. Hyams and C. W. Lloyd, pp. 33–45. New York, NY: Wiley-Liss.

Oberhauser, A. F., Marszalek, P. E., Erickson, H. P. and Fernandez, J. M. (1998). The molecular elasticity of the extracellular matrix protein tenascin. *Nature*, 393, 181–185.

Oghalai, J. S., Patel, A. A., Nakagawa, T. and Brownell, W. F. (1998). Fluorescence-imaged microdeformation of the outer hair cell lateral wall. *J. Neuroscience*, 18, 48–58.

Olbrich, K., Rawicz, W., Needham, D. and Evans, E. (2000). Water permeability and mechanical strength of polyunsaturated lipid bilayers. *Biophys. J.*, 79, 321–327.

Onsager, L. (1949). The effects of shape on the interaction of colloidal particles. *Ann. N.Y. Acad. Sci. USA*, 51, 627–659.

Oosawa, F. and Asakura, S. (1975). *Thermodynamics of the Polymerization of Proteins*. New York, NY: Academic Press.

Osborn, M., Webster, R. and Weber, K. (1978). Individual microtubules viewed by immunofluorescence and electron microscopy in the same PtK2 cell. *J. Cell Biol.*, 77, R27–R34.

Palmer, A., Mason, T. G., Xu, J., Kuo, S. C. and Wirtz, D. (1999). Diffusing wave spectroscopy microrheology of actin filament networks. *Biophys. J.*, 76, 1063–1071.

Pampaloni, F., Lattanzi, G., Jonás, A. *et al.* (2006). Thermal fluctuations of grafted microtubules provide evidence of a length-dependent persistence length. *Proc. Nat. Acad. Sci. (USA)*, 103, 10 248–10 253.

Panorchan, P., Lee, J. S. H., Kole, T. P., Tseng, Y. and Wirtz, D. (2006). Microrheology and ROCK signalling of human endothelial cells embedded in a 3D matrix. *Biophys. J.*, 91, 3499–3507.

Pantaloni, D., Hill, T. L., Carlier, M. F. and Korn, E. D. (1985). A model for actin polymerization and the kinetic effects of ATP hydrolysis. *Proc. Natl. Acad. Sci. USA*, 82, 7202–7211.

Parsegian, V. A. (1966). Theory of liquid-crystalline phase transitions in lipid + water systems. *Trans. Faraday Soc.*, 62, 848–860.

Parsegian, V. A., Fuller, N. and Rand, R. P. (1979). Measured work of deformation and repulsion of lecithin bilayers. *Proc. Natl. Acad. Sci. USA*, 76, 2750–2754.

Pastor, R. W. (1994). Molecular dynamics and Monte Carlo simulations of lipid bilayers. *Curr. Opin. Struct. Biol.*, 4, 486–492.

Peliti, L. and Leibler, S. (1985). Effects of thermal fluctuations on systems with small surface tension. *Phys. Rev. Lett.*, 54, 1690–1693.

Perkins, T. T., Smith, D. E., Larson, R. G. and Chu, S. (1995). Stretching of a single tethered polymer in a uniform flow field. *Science*, 268, 83–87.

Pesen, D. and Hoh, J. H. (2005). Micromechanical architecture of the endothelial cell cortex. *Biophys. J.*, 88, 670–679.

Peskin, C. S., Odell, G. M. and Oster, G. (1993). Cellular motion and thermal fluctuations: the Brownian ratchet. *Biophys. J.*, 65, 316–324.

Petsch, I. B. and Grest, G. S. (1993). Molecular dynamic simulations of the structure of closed tethered membranes. *J. Phys. I France*, 3, 1741–1754.

Peterson, M. A., Strey, H. and Sackmann, E. (1992). Theoretical and phase contrast microscope eigenmode analysis of erythrocyte flicker: amplitudes. *J. Phys. II France*, 2, 1273–1285.

Petrov, A. G. (1999). *The Lyotropic State of Matter: Molecular Physics and Living Matter Physics*. Amsterdam: Gordon and Breach.

Phillips, R., Kondev, J. and Theriot, J. (2009). *Physical Biology of the Cell*. New York: Garland.

Pierrat, S., Brochart-Wyart, F. and Nassoy, P. (2004). Enforced detachment of red blood cells adhering to surfaces: statics and dynamics. *Biophys. J.*, 87, 2855–2869.

Pink, D. A., Green, T. J. and Chapman, D. (1980). Raman scattering in bilayers of unsaturated phophatidylcholines. Experiment and theory. *Biochemistry*, 19, 349–356.

Plischke, M. (2006). Critical behavior of entropic shear rigidity. *Phys. Rev. E*, 73, 061406.

Plishcke, M. (2007). Rigidity of disordered networks with bond-bending forces. *Phys. Rev. E*, 76, 021401.

Plischke, M. and Bergersen, B. (1994). *Equilibrium Statistical Physics*, 2nd edn. Singapore, World Scientific.

Plischke, M. and Boal, D. H. (1988). Absence of a crumpling transition in strongly self-avoiding tethered membranes. *Phys. Rev. A*, 38, 4943–4945.

Plischke, M. and Fourcade, B. (1991). Monte Carlo simulation of bond-diluted tethered membranes, *Phys. Rev. A*, 43, 2056–2058.

Plischke, M. and Joos, B. (1998). Entropic elasticity of diluted central force networks. *Phys. Rev. Lett.*, 80, 4907–4910.

Podolski, J. L. and Steck, T. L. (1990). Length distribution of F-actin in *Dictyostelium discoideum*. *J. Biol. Chem.*, 265, 1312–1318.

Pollard, T. D. (1986). Rate constants for the reactions of ATP- and ADP-actin with the ends of actin filaments. *J. Cell Biol.*, 103, 2747–2754.

Pomerening, J. R., Kim, S. Y. and Ferrell, J. E. Jr. (2005). Systems-level dissection of the cell-cycle oscillator: bypassing positive feedback produces damped oscillations. *Cell*, 122, 565–578.

Pope, L. H., Bennink, M. L., van Leijenhorst-Groener, K. A. *et al.* (2005). Single chromatin fiber stretching reveals physically distinct populations of disassembly events. *Biophys. J.*, 88, 3572–3583.

Powers, D. L. (1999). *Boundary Value Problems*, 4th edn. London: Academic Press.

Prescott, L. M., Harley, J. P. and Klein, D. A. (2004). *Microbiology*, 6th edn. Dubuque, IA: Brown.

Prochniewicz, E., Zhang, Q., Howard, E. C. and Thomas, D. D. (1996). Microsecond rotational dynamics of actin: spectroscopic detection and theoretical simulation. *J. Mol. Biol.*, 255, 446–457.

Prost, J., Chauwin, J.-F., Peliti, L. and Ajdari, A. (1994). Asymmetric pumping of particles. *Phys. Rev. Lett.*, 72, 2652–2655.

Puech, P.-H., Borghi, N., Karatekin, E. and Brochard-Wyatt, F. (2003). Line thermodynamics: adsorption at a membrane edge. *Phys. Rev. Lett.*, 90, 128304.

Purcell, E. M. (1977). Life at low Reynolds number. *Am. J. Phys.*, 45, 3–11.

Putzel, G. G. and Schick, M. (2009). Theory of raft formation by the cross-linking of saturated or unsaturated lipids in model lipid bilayers. *Biophys. J.*, 96, 4935–4940.

Rädler, J. and Sackmann, E. (1993). Imaging optical thickness and separation distances of phospholipid vesicles at solid surfaces. *J. Phys. II France*, 3, 727–747.

Rädler, J., Feder, T. J., Strey, H. H. and Sackmann, E. (1995). Fluctuation analysis of tension-controlled undulation forces between giant vesicles and solid substrates. *Phys. Rev. E*, 51, 4526–4536.

Radmacher, M., Fritz, M., Kacher, C. M., Cleveland, J. P. and Hansma, P. K. (1996). Measuring the viscoelastic properties of human platelets with the atomic force microscope. *Biophys. J.*, 70, 556–567.

Radzihovsky, L. (2004). Anisotropic and heterogeneous polymerized membranes. In *Statistical Mechanics of Membranes and Surfaces*, eds. D. Nelson, T. Piran and S. Weinberg, pp. 275–321. Singapore: World Scientific.

Radzihovsky, L. and Toner, J. (1995). A new phase of tethered membranes: tubules. *Phys. Rev. Lett.*, 75, 4752–4755.

Ragsdale, G. K., Phelps, J. and Luby-Phelps, K. (1997). Viscoelastic response of fibroblasts to tension transmitted through adherens junctions. *Biophys. J.*, 73, 2798–2808.

Rand, R. P. and Parsegian, V. A. (1989). Hydration forces between phospholipid bilayers. *Biochim. Biophys. Acta*, 988, 351–376.

Raskin, D. M. and de Boer, P. A. J. (1999). Rapid pole-to-pole oscillation of a protein required for directing division to the middle of *Escherichia coli*. *Proc. Nat. Acad. Sci. USA*, 96, 4971–4976.

Rasmussen, B. (2000). Filamentous microfossils in a 3.235-million-year-old volcanogenic massive sulphide deposit. *Nature*, 405, 676–679.

Rawicz, W., Olbrich, K. C., McIntosh, T., Needham, D. and Evans, E. (2000). Effects of chain length and unsaturation on elasticity of lipid bilayers. *Biophys. J.*, 79, 328–339.

Rayment, I., Rypniewski, W. R., Schmidt-Bäse, K. *et al.* (1993a). Three-dimensional structure of myosin subfragment-1: a molecular motor. *Science*, 261, 50–58.

Rayment, I., Holden, H. M., Whittaker, M. *et al.* (1993b). Structure of the actin–myosin complex and its implications for muscle contraction. *Science*, 261, 58–65.

Reif, F. (1965). *Fundamentals of Statistical and Thermal Physics*. New York, NY: McGraw-Hill.

Reshes, G., Vanounou, S., Fishov, I. and Feingold, M. (2008). Cell shape dynamics in *Escherichia coli*. *Biophys. J.*, 94, 251–264.

Rief, M., Gautel, M., Oesterhelt, F., Fernandez, J. M. and Gaub, H. E. (1997). Reversible unfolding of individual titin immunoglobulin domains by AFM. *Science*, 276, 1109–1112.

Rief, M., Gautel, M., Schemmel, A. and Gaub, H. E. (1998). The mechanical stability of immunoglobulin and fibronectin III domains in the muscle protein titin measured by atomic force microscopy. *Biophys. J.*, 75, 3008–3014.

Rief, M., Pascual, J., Saraste, M. and Gaub, H. E. (1999). Single molecule force spectroscopy of spectrin repeats: low unfolding forces in helix bundles. *J. Mol. Biol.*, 286, 553–561.

Riveline, D., Wiggins, C. H., Goldstein, R. E. and Ott, A. (1997). Elastohydrodynamic study of actin filaments using fluorescence microscopy. *Phys. Rev. E*, 56, R1330–R1333.

Rodionov, V. I., Nadezhdina, E. and Borisy, G. G. (1999). Centrosomal control of microtubule dynamics. *Proc. Natl. Acad. Sci. USA*, 96, 115–120.

Root, D. D., Yadavalli, V. K., Forbes, J. G. and Wang, K. (2006). Coiled-coil nanomechanics and coiling and unfolding of the superhelix and α-helices of myosin. *Biophys. J.*, 90, 2852–2866.

Ross, R. S. and Pincus, P. (1992). Bundles: end-grafted polymers in poor solvent. *Europhys. Lett.*, 19, 79–84.

Rowat, A. C., Foster, L. J., Nielson, M. M., Weiss, M. and Ipsen, J. H. (2004). Characterization of the elastic properties of the nuclear envelope. *J. R. Soc. Interface*, 2004.0022.

Rowat, A. C., Lammerding, J. and Ipsen, J. H. (2006). Mechanical properties of the cell nucleus and the effect of emerin deficiency. *Biophys. J.*, 91, 4649–4664.

Różycki, B., Lipowsky, R. and Weikl, T. R. (2006). Adhesion of membranes with active stickers. *Phys. Rev. Lett.*, 96, 048101.

Russ, C., Heimburg, T. and von Grünberg, H. H. (2003). The effect of lipid demixing on the electrostatic interaction of planar membranes across a salt solution. *Biophys. J.*, 84, 3730–3742.

Sackmann, E. (1990). Molecular and global structure and dynamics of membranes and lipid bilayers. *Can. J. Phys.*, 68, 999–1012.

Saenger, W. (1984). *Principles of Nucleic Acid Structure*. New York, NY: Springer.

Safran, S. A. (1994). *Statistical Thermodynamics of Surfaces, Interfaces, and Membranes*. Reading, MA: Addison-Wesley.

Safran, S. A., Pincus, P. and Andelman, D. (1990). Theory of spontaneous vesicle formation in surfactant mixtures. *Science*, 248, 354–356.

Sandre, O., Moreaux, L. and Brochard-Wyart, F. (1999). Dynamics of transient pores in stretchect vesicles. *Proc. Nat. Acad. Sci. USA*, 96, 10 591–10 596.

Sanford, R. J. (2003). *Principles of Fracture Mechanics*. Upper Saddle River, NJ: Pearson.

Sanger, J. M., Sanger, J. W. and Southwick, F. S. (1992). Host cell actin assembly is necessary and likely to provide the propulsive force for intracellular movement of *Listeria monocytogenes*. *Infect. Immun.*, 60, 3609–3619.

Sarda, S., Pointu, D., Pincet, F. and Henry, N. (2004). Specific recognition of macroscopic objects by the cell surface: evidence for a receptor density threshold revealed by micrometric particle binding characteristics. *Biophys. J.*, 86, 3291–3303.

Satcher, R. L. Jr. and Dewey, C. F. Jr. (1996). Theoretical estimates of mechanical properties of the endothelial cell cytoskeleton. *Biophys. J.*, 71, 109–118.

Sato, M., Schwarz, W. H. and Pollard, T. D. (1987). Dependence of the mechanical properties of actin/α-actinin gels on deformation rate. *Nature*, 325, 828–830.

Sawin, K. E. and Mitchison, T. J. (1991). Poleward microtubule flux in mitotic spindles assembled *in vitro*. *J. Cell Biol.*, 112, 941–954.

Saxton, M. J. and Jacobson, K. (1997). Single-particle tracking: applications to membrane dynamics. *Annu. Rev. Biophys. Biomol. Struct.*, 26, 373–399.

Schaap, I. A. T., Carrasco, C., de Pablo, P. J. and MacKintosh, F. C. (2006). Elastic response, buckling and instability of microtubules under radial indentation. *Biophys. J.*, 91, 1521–1531.

Schär-Zammaretti, P. and Ubbink, J. (2003). The cell wall of lactic acid bacteria: surface constituents and macromolecular conformations. *Biophys. J.*, 85, 4076–4092.

Scheffer, L., Bitler, A., Ben-Jacob, E. and Korenstein, R. (2001). Atomic force pulling: probing the local elasticity of the cell membrane. *Eur. Biophys. J.*, 30, 83–90.

Scheffers, D.-J., Jones, L. J. F. and Errington, J. (2004). Several distinct localization patterns for pennicilin-binding proteins in *Bacillus subtilis*. *Mol. Microbiol.*, 51, 749–764.

Schlaeppi, J.-M., Pooley, H. M. and Karamata, D. (1982). Identification of cell wall subunits in *Bacillus subtilis* and analysis of their segregation during growth. *J. Bacteriol.*, 149, 329–337.

Schmidt, C. F., Bärmann, M., Isenberg, G. and Sackmann, E. (1989). Chain dynamics, mesh size, and diffusive transport of polymerized actin. A quasielastic light scattering and microfluorescence study. *Macromolecules*, 22, 3638–3649.

Schmidt, C. F., Svoboda, K., Lei, N. *et al.* (1993). Existence of a flat phase in red cell membrane skeletons. *Science*, 259, 952–955.

Schneider, M. B., Jenkins, J. T. and Webb, W. W. (1984). Thermal fluctuations of large cylindrical phospholipid vesicles. *Biophys. J.*, 45, 891–899.

Schnurr, B., Gittes, F., MacKintosh, F. C. and Schmidt, C. F. (1997). Determining microscopic viscoelasticity in flexible and semiflexible polymer networks from thermal fluctuations. *Macromolecules*, 30, 7781–7792.

Schopf, J. W. (1968). Microflora of the Bitter Springs Formation, Late Precambrian, central Australia. *J. Paleontology*, 42, 651–688.

Schopf, J. W. (1993). Microfossils of the early archean Apex chert: new evidence of the antiquity of life. *Science*, 260, 640–646.

Schopf, J. W. and Packer, B. M. (1987). Early Archean (3.3-billion to 3.5-billion-year-old) microfossils from Warrawoona Group, Australia. *Science*, 237, 70–73.

Selhuber-Unkel, C., Erdmann, T., López-Garcia, M. *et al.* (2010). Cell adhesion strength is controlled by intermolecular spacing of adhesion receptors. *Biophys. J.*, 98, 543–551.

Seifert, U. (1991). Vesicles of toroidal topology. *Phys. Rev. Lett.*, 66, 2404–2407.

Seifert, U. (1993). Curvature-induced lateral phase segregation in two-component vesicles. *Phys. Rev. Lett.*, 70, 1335–1338.

Seifert, U. (1995). Self-consistent theory of bound vesicles. *Phys. Rev. Lett.*, 74, 5060–5063.

Seifert, U. and Lipowsky, R. (1990). Adhesion of vesicles. *Phys. Rev. A*, 42, 4768–4771.

Seifert, U., Berndl, K. and Lipowsky, R. (1991). Shape transformations of vesicles: phase diagram for spontaneous-curvature and bilayer-couple models. *Phys. Rev. A*, 44, 1182–1202.

Selve, N. and Wegner, A. (1986). Rate of treadmilling of actin filaments *in vitro*. *J. Mol. Biol.*, 187, 627–631.

Serquera, D., Lee, W., Settanni, G. *et al.* (2010). Mechanical unfolding of an ankyrin repeat protein. *Biophys. J.*, 98, 1294–1301.

Servuss, R. M. and Helfrich, W. (1989). Mutual adhesion of lecithin membranes at ultralow tensions. *J. Phys. France.*, 50, 809–827.

Servuss, R. M., Harbich, W. and Helfrich, W. (1976). Measurement of the curvature-elastic modulus of egg lecithin bilayers. *Biochim. Biophys. Acta*, 436, 900–903.

Shao, J.-Y., and Hochmuth, R. M. (1999). Mechanical anchoring strength of *L*-selectin, β_2 integrins and CD45 to neutrophil cytoskeleton and membrane. *Biophys. J.*, 77, 587–596.

Sharpe, M. E., Hauser, P. M., Sharpe, R. G. and Errington, J. (1998). *Bacillus subtilis* cell cycle as studied by fluorescence microscopy: constancy of cell length at initiation of DNA replication and evidence for active nucleoid partitioning. *J. Bacteriol.*, 180, 547–555.

Sheetz, M. P. (1983). Membrane skeletal dynamics role in modulation of red cell deformability, mobility of transmembrane proteins and shape. *Semin. Hematol.*, 20, 175–188.

Sheetz, M. P. (2001). Cell control by membrane–cytoskeleton adhesion. *Nat. Rev. Mol. Cell Biol.*, 2, 392–396.

Sheetz, M. P. and Singer, S. J. (1974). Biological membranes as bilayer couples. A molecular mechanism of drug–erythrocyte interactions. *Proc. Natl. Acad. Sci. USA*, 71, 4457–4461.

Shen, Z. L., Dodge, M. R., Kahn, H., Ballarini, R. and Eppell, S. I. (2008). Stress–strain experiments on individual collagen fibrils. *Biophys. J.*, 95, 3956–3963.

Sheterline, P., Clayton, J. and Sparrow, J. C. (1998). *Actin*, 4th edn. Oxford: Oxford University Press.

Shih, Y.-L., Le, T. and Rothfield, L. (2003). Division site selection in *Escherichia coli* involves dynamic redistribution of Min proteins within coiled structures that extend between the two cell poles. *Proc. Nat. Acad. Sci. USA*, 100, 7865–7870.

Shillcock, J. C. and Boal, D. H. (1996). Entropy-driven instability and rupture of fluid membranes. *Biophys. J.*, 71, 317–326.

Shillcock, J. C. and Seifert, U. (1998a). Thermally induced proliferation of pores in a model fluid membrane. *Biophys. J.*, 74, 1754–1766.

Shillcock, J. C. and Seifert, U. (1998b). Escape from a metastable well under a time-ramped force. *Phys. Rev. E*, 57, 7301–7304.

Shlyakhtenko, L. S., Gall, A. A., Weimer, J. J., Hawn, D. D. and Lyubchenko, Y. L. (1999). Atomic force microscopy imaging of DNA covalently immobilized on a functionalized mica substrate. *Biophys. J.*, 77, 568–576.

Shraiman, B. I. (2005). Mechanical feedback as possible regulator of tissue growth. *Proc. Nat. Acad. Sci USA*., 102, 3318–3323.

Siegel, D. P. and Kozlov, M. M. (2004). The Gaussian curvature elastic modulus of N-monomethylated dioleoylphosphatidylethanolamine: relevance to membrane fusion and lipid phase behavior. *Biophys. J.*, 87, 366–374.

Siegel-Gaskins, D. and Crosson, S. (2008). Tightly regulated and heritable division control in single bacterial cells. *Biophys J.*, 95, 2063–2072.

Simons, K. and Ikonen, E. (1997). Functional rafts in cell membranes. *Nature*, 387, 569–572.

Simson, R., Wallraff, E., Faix, J. *et al.* (1998). Membrane bending modulus and adhesion energy of wild-type and mutant cells of *Dictyostelium* lacking talin or cortexillins. *Biophys. J.*, 74, 514–522.

Sit, P. S., Spector, A. A., Lue, A. J.-C., Popel, A. S. and Brownell, W. E. (1997). Micropipette aspiration on the outer hair cell wall. *Biophys. J.*, 72, 2812–2819.

Skotheim, J. M., Di Talia, S., Siggia, E. D. and Cross, F. R. (2008). Positive feedback of G1 cyclins ensures coherent cell cycle entry. *Nature*, 454, 291–296.

Sleep, J., Wilson, D., Simmons, R. and Gratzer, W. (1999). Elasticity of the red cell membrane and its relation to hemolytic disorders: an optical tweezers study. *Biophys. J.*, 77, 3085–3095.

Small, J. V., Herzog, M. and Anderson, K. (1995). Actin filament organization in the fish keratocyte lamellipodium. *J. Cell Biol.*, 129, 1275–1286.

Snyder, L. and Champness, W. (2007). *Molecular Genetics of Bacteria,* 2nd edn. Washington: ASM Press.

Soga, K. G., Guo, H. and Zuckermann, M. (1995). Polymer brushes in a poor solvent. *Europhys. Lett.*, 29, 531–536.

Southam, G., Firtel, M., Blackford, B. L. *et al.* (1993). Transmission electron microscopy, scanning tunneling microscopy and atomic force microscopy of the cell envelope layers of the archeabacterium *Methanospirillum hungatei* GP1. *J. Bacteriol.*, 175, 1946–1955.

Stauffer, D. and Aharony, A. (1992). *Introduction to Percolation Theory*, 2nd edn. London: Taylor and Francis.

Steck, T. L. (1989). Red cell shape. In *Cell Shape: Determinants, Regulation and Regulatory Role*, eds. W. D. Stein and F. Bronner, pp. 205–246. New York, NY: Academic Press.

Stigter, D. and Bustamante, C. (1998). Theory for the hydrodynamic and electrophoretic stretch of tethered B-DNA. *Biophys. J.*, 75, 1197–1210.

Stokke, T. and Brant, D. A. (1990). The reliability of wormlike polysaccharide chain dimensions estimated from electron micrographs. *Biopolymers*, 30, 1161–1181.

Stokke, B. T., Mikkelsen, A. and Elgsaeter, A. (1985a). Human erythrocyte spectrin dimer intrinsic viscosity: temperature dependence and implications for the molecular basis of the membrane free energy. *Biochim. Biophys. Acta*, 816, 102–110.

Stokke, B. T., Mikkelsen, A. and Elgsaeter, A. (1985b). Some viscoelastic properties of human erythrocyte spectrin networks end-linked *in vitro*. *Biochim. Biophys. Acta*, 816, 111–121.

Storm, C., Pastore, J. J., MacKintosh, F. C., Lubensky, T. C. and Janmey, P. A. (2005). Nonlinear elasticity in biological gels. *Nature*, 435, 191–194.

Strick, T. R., Allemand, J.-F., Bensimon, D., Bensimon, A. and Croquette, V. (1996). The elasticity of a single supercoiled DNA molecule. *Science*, 271, 1835–1837.

Stricker, J., Maddox, P. Salmon, E. D. and Erickson, H. P. (2002). Rapid assembly dynamics of the *Escherichia coli* FtsZ-ring demonstrated by fluorescence recovery after photobleaching. *Proc. Nat. Acad. Sci. USA*, 99, 3171–3175.

Sun, Q. and Margolin, W. (1998). FtsZ dynamics during the division cycle of live *Escherichia coli* cells. *J. Bacteriol.*, 180, 2050–2056.

Sun, Y.-L., Luo, Z.-P., Fertala, A. and An, K.-N. (2002). Direct quantification of the flexibility of type I collagen monomer. *Biochem. Biophys. Res. Comm.*, 295, 382–386.

Suzuki, A., Maeda, T. and Ito, T. (1991). Formation of liquid crystalline phase of actin filament solutions and its dependence on filament length as studied by optical birefringence. *Biophys. J.*, 59, 25–30.

Svetina, S. and Zeks, B. (1985). Bilayer couple as a possible mechanism of biological shape formation. *Biomed. Biochim. Acta*, 44, 979–986.

Svetina, S., Brumen, M. and Zeks, B. (1985). Lipid bilayer elasticity and the bilayer interpretation of red cell shape transformations and lysis. *Stud. Biophys.*, 110, 177.

Svoboda, K. and Block, S. M. (1994). Force and velocity measured for single kinesin molecules. *Cell*, 77, 773–784.

Svoboda, K., Schmidt, C. F., Branton, D. and Block, S. M. (1992). Conformation and elasticity of the isolated red blood cell membrane skeleton. *Biophys. J.*, 63, 784–793.

Svoboda, K., Schmidt, C. F., Schnapp, B. J. and Block, S. M. (1993). Direct observation of kinesin stepping by optical trapping interferometry. *Nature*, 365, 721–727.

Svoboda, K., Mitra, P. P. and Block, S. M. (1994). Fluctuation analysis of motor protein movement and single enzyme kinetics. *Proc. Natl. Acad. Sci. USA*, 91, 11 782–11 786.

Swift, D. G., Posner, R. G. and Hammer, D. A. (1998). Kinetics of adhesion of IgE-sensitized rat basophilic leukemia cells to surface-immobilized antigen in Couette flow. *Biophys. J.*, 75, 2597–2611.

Szule, J. A., Fuller, N. L. and Rand, R. P. (2002). The effects of acyl chain length and saturation of diacylglycerols and phosphatidylcholines on membrane monolayer curvature. *Biophys. J.*, 83, 977–984.

Taketomi, H., Ueda, Y. and Gö, N. (1975). Studies on protein folding, unfolding and fluctuations by computer simulation. *Int. J. Pept. Protein Res.*, 7, 445–459.

Takeuchi, M., Miyamoto, H., Sako, Y., Komizu, H. and Kusimi, A. (1998). Structure of the erythrocyte membrane skeleton as observed by atomic force microscopy. *Biophys. J.*, 74, 2171–2183.

Tanford, C. (1980). *The Hydrophobic Effect*, 2nd edn. New York, NY: Wiley.

Tang, J. X., Janmey, P. A., Stossel, T. P. and Ito, T. (1999). Thiol oxidation of actin produces dimers that enhance the elasticity of the F-actin network. *Biophys. J.*, 76, 2208–2215.

Tang, Q. and Edidin, M. (2003). Lowering the barriers to random walks on the cell surface. *Biophys. J.*, 84, 400–407.

Taupin, C., Dvolaitzky, M. and Sauterey, C. (1975). Osmotic pressure induced pores in phospholipid vesicles. *Biochemistry*, 14, 4771–4775.

Taylor, W. H. and Hagerman, P. J. (1990). Application of the method of phage T4 DNA ligase-catalyzed ring-closure to the study of DNA structure II. NaCl-dependence of DNA flexibility and helical repeat. *J. Mol. Biol.*, 212, 363–376.

Tempel, M., Isenberg, G. and Sackmann, E. (1996). Temperature-induced sol-gel transition and microgel formation in α-actinin cross-linked actin networks: a rheoological study. *Phys. Rev. E*, 54, 1802–1810.

Tessier, F., Boal, D. H. and Discher, D. E. (2003). Networks with fourfold connectivity in two dimensions. *Phys. Rev. E*, 67, 011903.

Theriot, J. A. and Mitchison, T. J. (1991). Actin microfilament dynamics in locomoting cells. *Nature*, 352, 126–131.

Theriot, J. A., Mitchison, T. J., Tilney, L. G. and Portnoy, D. A. (1992). The rate of actin-based motility of intracellular *Listeria monocytogenes* equals the rate of actin polymerization. *Nature*, 357, 257–260.

Thomas, P. D. and Dill, K. A. (1996). An iterative method for extracting energy-like quantities from protein structures. *Proc. Nat. Acad. Sci. USA*, 93, 11628–11 633.

Thorpe, M. F. (1986). Elastic properties of network glasses. *Ann. NY. Acad. Sci.*, 484, 206–213.

Thwaites, J. J. (1993). Growth and control of the cell wall: a mechanical model for *Bacillus subtilis*. In *Bacterial Growth and Lysis*, eds. M. A. de Pedro, V.-J. Höltje and W. Löffelhardt. New York, NY: Plenum.

Thwaites, J. J. and Surana, U. C. (1991). Mechanical properties of *Bacillus subtilis* cell walls: effects of removing residual culture medium. *J. Bacteriol.*, 173, 197–203.

Tidball, J. G. (1986). Energy stored and dissipated in skeletal muscle basement membranes during sinusoidal oscillations. *Biophys. J.*, 50, 1127–1138.

Tolomeo, J. A., Steele, C. R. and Holley, M. C. (1996). Mechanical properties of the lateral cortex of mammalian auditory outer hair cells. *Biophys. J.*, 71, 421–429.

Tran-Son-Tay, R. (1993). In *Physical Forces and the Mammalian Cell*, ed. J. A. Frangos. San Diego, CA: Academic Press.

Treloar, R. L. G. (1975). *The Physics of Rubber Elasticity*. Oxford: Oxford University Press.

Trueba, F. J. and Woldringh, C. L. (1980). Changes in cell diameter during the division cycle of *Escherichia coli*. *J. Bacteriol.*, 142, 869–878.

Tsafrir, I., Guedeau-Boudeville, M.-A., Kandel, D. and Stavans, J. (2001). Coiling instability of multilamellar membrane tubes with anchored polymers. *Phys. Rev. E*, 63, 031603.

Tsuda, Y., Yasutake, H., Ishijima, A. and Yanagida, T. (1996). Torsional rigidity of single actin filaments and actin–actin bond breaking force under torsion measured directly by *in vitro* micromanipulation. *Proc. Nat. Acad. Sci. USA*, 93, 12937–12 942.

Turekian, K. K. (1976). *Oceans*, 2nd edn. Englewood Cliffs, NJ: Prentice-Hall.

Tuszynski, J. A. (2008). *Molecular and Cellular Biophysics*. Boca Raton FL: Chapman and Hall/CRC.

Tzur, A., Kafri, R., LeBleu, V. S., Lahav, G. and Kirschner, M.W. (2009). Cell growth and size homeostasis in proliferating animal cells. *Science*, 325, 167–171.

Umeda, A. and Amako, K. (1983). Growth of the surface of *Corynebacterium diptheriae*. *Microbiol. Immunol.*, 27, 663–671.

Ungewickell, E. and Branton, D. (1981). Assembly units of clathrin coats. *Nature*, 289, 420–422.

Ursitti, J. A. and Wade, J. B. (1993). Ultrastructure and immunocytochemistry of the isolated human erythrocyte membrane skeleton. *Cell Motil. Cytoskeleton*, 25, 30–42.

Vadillo-Rodriguez, V., Schooling, S. R. and Dutcher, J. R. (2009). In situ characterization of differences in the viscoelastic response of individual Gram-negative and Gram-positive bacterial cells. *J. Bacteriol.*, 191, 5518–5525.

Valentine, M. T., Perlman, Z. E., Mitchison, T. J. and Weitz, D. A. (2005). Mechanical properties of *Xenopus* egg cytoplasmic extracts. *Biophys. J.*, 88, 680–689.

van der Ploeg, P. and Berendsen, H. J. C. (1983). Molecular dynamics of a bilayer membrane. *Mol. Phys.*, 49, 233–248.

van der Rijt, J. A. J., van der Werf, K. O., Bennink, M. L., Dijkstra, P. J. and Feijen, J. (2006). Micromechanical testing of individual collagen fibrils. *Macromol. Biosci.*, 6, 697–702.

Van de Ven, T. G. M. (1989). *Colloidal Dynamics*. London: Academic Press.

Vance, D. E. and Vance, J. E. (1996). *Biochemistry of Lipids, Lipoproteins and Membranes*. Amsterdam: Elsevier.

Veatch, S. L. and Keller, S. L. (2003). Separation of liquid phases in giant vesicles of ternary mixtures of phospholipids and cholesterol. *Biophys. J.*, 85, 3074–3083.

Veatch, S. L., Leung, S. S. W., Hancock, R. E. W. and Thewalt, J. L. (2007). Fluorescent probes alter miscibility phase boundaries in ternary vesicles. *J. Phys. Chem. B*, 111, 502–504.

Veatch, S. L., Ciuta, P., Sengupta, P. *et al.* (2008). Critical fluctuations in plasma membrane vesicles. *ACS Chem. Biol.*, 3, 287–293.

Veerman, J. A. C. and Frenkel, D. (1992). Phase behavior of disklike hard-core mesogens. *Phys. Rev. A*, 45, 5632–5648.

Veigel, C., Bartoo, M. L., White, D. C. S., Sparrow, J. C. and Molley, J. E. (1998). The stiffness of rabbit skeletal actomyosin cross-bridges determined with an optical tweezers transducer. *Biophys. J.*, 75, 1424–1438.

Venier, P., Maggs, A. C., Carlier, M.-F. and Pantaloni, D. (1994). Analysis of microtubule rigidity using hydrodynamic flow and thermal fluctuations. *J. Biol. Chem.*, 269, 13 353–13 360.

Verde, F., Dogterom, M., Stelzer, E., Karsenti, E. and Leibler, S. (1992). Control of microtubule dynamics and length by cyclin A- and cyclin B-dependent kinases in *Xenopus* egg extracts. *J. Cell Biol.*, 118, 1097–1108.

Vertogen, G. and de Jeu, W. H. (1988). *Thermotropic Liquid Crystals, Fundamentals*. Berlin: Springer-Verlag.

Verway, E. J. W. and Overbeek, J. Th. G. (1948). *Theory of Stability of Lyophobic Colloids*. Amsterdam: Elsevier.

Verwer, R. W. H., Nanninga, N., Keck, W. and Schwarz, U. (1978). Arrangement of glycan chains in the sacculus of *E. coli*. *J. Bacteriol.*, 136, 723–729.

Vogel, S. (1998). *Cat's Paws and Catapults*. New York, NY: Norton.

Walker, R. A., O'Brien, E. T., Pryer, N. K. *et al.* (1988). Dynamic instability of individual microtubules analysed by video light microscopy: rate constants and transition frequencies. *J. Cell Biol.*, 107, 1437–1448.

Wang, S., Arellano-Santoyo, H., Combs, P. A. and Shaevitz, J. W. (2010a). Actin-like cytoskeleton filaments contribute to cell mechanics in bacteria. *Proc. Nat. Acad. Sci. USA*, 107, 9182–9185.

Wang, H., Robinson, R. C. and Burtnick, L. D. (2010b). The structure of native G-actin. *Cytoskeleton*, 67, 456–465.

Walsh, M. M. and Lowe, D. R. (1985). Filamentous microfossils from the 3,500 Myr-old Onverwacht Group, Barberton Mountain Land, South Africa. *Nature*, 314, 530–532.

Waugh, R. and Evans, E. A. (1976). Viscoelastic properties of erythrocyte membranes of different vertebrate animals. *Microvasc. Res.*, 12, 291–304.

Waugh, R. and Evans, E. A. (1979). Thermoelasticity of red blood cell membrane. *Biophys. J.*, 26, 115–132.

Waugh, R. E. and Agre, P. (1988). Reductions of erythrocyte membrane viscoelastic coefficients reflect spectrin deficiencies in hereditary spherocytosis. *J. Clin. Invest.*, 81, 133–141.

Waugh, R. E. and Bauserman, R. G. (1995). Physical measurements of bilayer-skeletal separation forces. *Ann. Biomed. Eng.*, 23, 308–321.

Waugh, R. E., Song, J., Svetina, S. and Zeks, B. (1992). Local and nonlocal curvature elasticity in bilayer membranes by tether formation from lecithin vesicles. *Biophys. J.*, 61, 974–982.

Weast, R. E., ed. (1970). *Handbook of Chemistry and Physics*, 50th edn. Boca Raton, FL: CRC Press.

Weber, G. (1975). Energetics of ligand binding to proteins. *Adv. Prot. Chem.*, 29, 1–83.

Weber, P. C., Wendolski, J. J., Pantoliano, M. W. and Salemme, F. R. (1992). Crystallographic and thermodynamic comparison of natural and synthetic ligands bound to streptavidin. *J. Am. Chem. Soc.*, 114, 3197–3200.

Wegner, A. (1976). Head to tail polymerization of actin. *J. Mol. Biol.*, 108, 139–150.

Wegner, A. (1982). Treadmilling of actin at physiological salt concentrations. *J. Mol. Biol.*, 161, 607–615.

Weikl, T. R., Netz, R. R. and Lipowsky, R. (2000). Unbinding transitions and phase separation of multicomponent membranes. *Phys. Rev. E*, 62, R45–R48.

Weston, R. E. Jr. and Schwarz, H. A. (1972). *Chemical Kinetics*. Englewood Cliffs, NJ: Prentice-Hall.

Wientjes, F. B., Woldringh, C. L. and Nanninga, N. (1991). Amount of peptidoglycan in cell walls of gram-negative bacteria. *J. Bacteriol.*, 173, 7684–7691.

Wilhelm, C., Winterhalter, M., Zimmermann, U. and Benz, R. (1993). Kinetics of pore size during irreversible electrical breakdown of lipid bilayer membranes. *Biophys. J.*, 64, 121–128.

Wilhelm, J. and Frey, E. (2003). Elasticity of stiff polymer networks. *Phys. Rev. Lett.*, 91, 108103.

Wintz, W., Döbereiner, H.-G. and Seifert, U. (1996). Starfish vesicles. *Europhys. Lett.*, 33, 403–408.

Wintz, W., Everaers, R. and Seifert, U. (1997). Mesh collapse in two-dimensional elastic networks under compression. *J. Phys. I France*, 7, 1097–1111.

Wolfe, J., Dowgert, M. F. and Steponkus, P. L. (1985). Dynamics of membrane exchange of the plasma membrane and the lysis of isolated protoplasts during rapid expansion in area. *J. Membrane Biol.*, 86, 127–138.

Wong, J. Y., Kuhl, T. L., Israelachvili, J. N., Mullah, N. and Zapilsky, S. (1997). Direct measurement of a tethered ligand-receptor interaction potential. *Science*, 275, 820–822.

Wu, P., Song, L., Clendenning, J. B. *et al.* (1988). Interaction of chloroquine with linear and supercoiled DNAs. Effects on the torsional dynamics, rigidity and twist energy parameter. *Biochem.*, 27, 8128–8144.

Xu, J., Wirtz, D. and Pollard, T.D. (1998a). Dynamic cross-linking by α-actinin determines the mechanical properties of actin filament networks. *J. Biol. Chem.*, 273, 9570–9576.

Xu, J., Schwarz, W. H., Käs, J. A. *et al.* (1998b). Mechanical properties of actin filament networks depend on preparation, polymerization conditions and storage of actin monomers. *Biophys. J.*, 74, 2731–2740.

Xu, W., Mulhern, P. J., Blackford, B. L. *et al.* (1996). Modeling and measuring the elastic properties of an archeal surface, the sheath of *Methanospirillum hungatei*, and the implication for methane production. *J. Bacteriol.*, 178, 3106–3112.

Yang, L., van der Werf, K. O., Fitié, C. E. C. *et al.* (2008). Mechanical properties of native and cross-linked Type I collagen fibrils. *Biophys. J.*, 94, 2204–2211.

Yang, Y., Tobias, I. and Olsen, W. K. (1993). Finite element analysis of DNA supercoiling. *J. Chem. Phys.*, 98, 1673–1686.

Yao, X., Jericho, M., Pink, D. and Beveridge, T. (1999). Thickness and elasticity of Gram-negative murein sacculi measured by atomic force microscopy. *J. Bacteriol.*, 181, 6865–6875.

Yasuda, R., Miyata, H. and Kinosita, K. Jr. (1996). Direct measurement of the torsional rigidity of single actin filaments. *J. Mol. Biol.*, 263, 227–236.

Yeung, A. and Evans, E. (1995). Unexpected dynamics in shape fluctuations of bilayer vesicles. *J. Phys. II France*, 5, 1501–1523.

Yeung, C., Balazs, A. C. and Jasnow, D. (1993). Lateral instabilities in a grafted layer in a poor solvent. *Macromolecules*, 26, 1914–1921.

Yi, L., Chang, T. and Ru, C. (2008). Buckling of microtubules under bending and torsion. *J. Ap. Phys.*, 103, 103516.

Ying, J., Ling, Y., Westfield, L. A., Sadler, J. E. and Shao, J.-Y. (2010). Unfolding the A2 domain of Von Willebrand factor with the optical trap. *Biophys. J.*, 98, 1685–1693.

Yoshimura, H., Nishio, T., Mishashi, K., Kinosita, K. Jr. and Ikegami, A. (1984). Torsional motion of eosin-labeled *F*-actin as detected in the time-resolved anisotropy decay of the probe in the sub-millisecond time range. *J. Mol. Biol.*, 179, 453–467.

Young, T. (1805). *Phil. Trans. R. Soc. Lond.*, 95, 65.

Yue, K., Fiebig, K. M., Thomas, P. D. *et al.* (1995). A test of lattice protein folding algorithms. *Proc. Nat. Acad. Sci. USA*, 92, 325–329.

Zhang, Z., Davis, H. T. and Kroll, D. M. (1996). Molecular dynamics simulations of tethered membranes with periodic boundary conditions. *Phys. Rev. E*, 53, 1422–1429.

Zhelev, D. V. (1998). Material property characteristics for lipid bilayers containing lysolipid. *Biophys. J.*, 75, 321–330.

Zhelev, D. V. and Needham, D. (1993). Tension-stabilized pores in giant vesicles: determination of pore size and pore line tension. *Biochim. Biophys. Acta*, 1147, 89–104.

Zhou, Y. and Raphael, R. M. (2005). Effect of salicylate on the elasticity, bending stiffness, and strength of SOPC membranes. *Biophys. J.*, 89, 1789–1801.

Zhou, Z. and Joós, B. (1997). Mechanisms of membrane rupture: from cracks to pores. *Phys. Rev. B*, 56, 2997–3009.

Zhou, Z., Lai, P.-Y. and Joós, B. (2005). Elasticity and stability of a helical filament. *Phys. Rev. E.*, 71, 052801; Erratum (2006), *Phys. Rev. E.*, 73, 039910.

Zhu, Q. and Asaro, R. J. (2008). Spectrin folding versus unfolding reactions and RBC membrane stiffness. *Biophys. J.*, 94, 2529–2545.

Zhu, T. F. and Szostak, J. W. (2009). Coupled growth and division of model protocell membranes. *J. Am. Chem. Soc.*, 131, 5705–5713.

Ziemann, F., Rädler, J. and Sackmann, E. (1994). Local measurements of viscoelastic moduli of entangled actin networks using an oscillating bead microrheometer. *Biophys. J.*, 66, 2210–2216.

Zilker, A., Ziegler, M. and Sackmann, E. (1992). Spectral analysis of erythrocyte flickering in the 0.3–4 μm^{-1} regime by microinterferometry combined with fast image processing. *Phys. Rev. A*, 46, 7998–8001.

Zipperle, G. F. Jr., Ezzell, J. W. Jr. and Doyle, R. J. (1984). Glucosamine substitution and muramidase susceptibility in *Bacillus anthracis*. *Can. J. Microbiol.*, 30, 553–559.

Zuckermann, D. and Bruinsma, R. (1995). Statistical mechanics of membrane adhesion by reversible molecular bonds. *Phys. Rev. Lett.*, 74, 3900–3903.

Index